David William Arthur Sharp was born in Folkestone in 1931 and educated at Ashford Grammar School, Harvey Grammar School, Folkestone and Sidney Sussex College, Cambridge. From 1957 until 1961 he was a lecturer at the Imperial College of Science and Technology, London, and between 1965 and 1967 he was Professor of Chemistry at the University of Strathclyde, Glasgow. From 1968 he was Professor of Chemistry at the University of Glasgow and since 1997 has been Emeritus Professor at that university. He was a contributor to the third, fourth and fifth editions of *Miall's Dictionary* and editor of the fifth edition. He was editor of the first edition of *The Penguin Dictionary of Chemistry*. He has written many papers for learned journals in his field.

THIRD EDITION

The Penguin Dictionary of
CHEMISTRY

Edited by D. W. A. Sharp
M.A., PH.D., C.CHEM., F.R.S.C., F.R.S.E.

PENGUIN BOOKS

PENGUIN BOOKS

Published by the Penguin Group
Penguin Books Ltd, 80 Strand, London WC2R 0RL, England
Penguin Group (USA) Inc., 375 Hudson Street, New York, New York 10014, USA
Penguin Group (Canada), 10 Alcorn Avenue, Toronto, Ontario, Canada M4V 3B2
(a division of Pearson Penguin Canada Inc.)
Penguin Ireland, 25 St Stephen's Green, Dublin 2, Ireland
(a division of Penguin Books Ltd)
Penguin Group (Australia), 250 Camberwell Road, Camberwell, Victoria 3124, Australia
(a division of Pearson Australia Group Pty Ltd)
Penguin Books India Pvt Ltd, 11 Community Centre, Panchsheel Park, New Delhi – 110 017, India
Penguin Group (NZ), cnr Airborne and Rosedale Roads, Albany, Auckland 1310, New Zealand
(a division of Pearson New Zealand Ltd)
Penguin Books (South Africa) (Pty) Ltd, 24 Sturdee Avenue, Rosebank 2196, South Africa

Penguin Books Ltd, Registered Offices: 80 Strand, London WC2R 0RL, England

www.penguin.com

An abridged edition of *Miall's Dictionary of Chemistry 5th Edition*, edited by
Professor D. W. A. Sharp, first published by Longman Group Ltd 1981
First published in Great Britain by Penguin Books 1983
Second edition 1990
Third edition 2003
5

Fifth Edition of *Miall's Dictionary of Chemistry*, copyright © Longman Group Ltd, 1981
This adaptation copyright © Penguin Books Ltd, 1983, 1990, 2003
All rights reserved

This book was produced using Librios® authoring and content management technology.
Printed in England by Clays Ltd, St Ives plc

ISBN-13: 978–0–140–51445–2
ISBN-10: 0–140–51445–7

Contents

The new edition of the *Dictionary of Chemistry*, as before, aims to provide an explanation of many of the terms used in chemistry with brief accounts of important substances and chemical operations. The Dictionary is intended for use in schools, colleges, and universities from the first study of the subject up to about the second year at University. It is also intended for use in the workplace and laboratories and for those in related disciplines who need reference to chemical terms.

Since the last edition of the *Penguin Dictionary* there has been a great explosion in information and use of chemistry in Biochemistry and Solid State Chemistry and Physics, and the *Dictionary of Chemistry* has been updated in these subjects; but in order to save space, some references for further information are now given to the *Penguin Dictionary of Biology*, 10th edn, Editors M. Thain and M. Hickman, and to the *Penguin Dictionary of Physics*, 3rd edn, Editor V. Illingworth. Other general reference books include: G. W. C. Kaye and T. H. Laby, *Tables of Physical and Chemical Constants*; *Handbook of Chemistry* (CRC Press); P. W. Atkins, *Physical Chemistry* (Oxford University Press); Cotton, Wilkinson, Murillo and Bochmann, *Advanced Inorganic Chemistry*; *Organic Chemistry*; and *The Merck Index*. The details of materials mainly developed as drugs and pharmaceuticals have not been increased as specialized volumes (*Pharmacopoeia*) are available. As many users come across pesticides and herbicides, their listing has been updated, although specialized listings are available elsewhere.

As previously, the nomenclature used is that most likely to be met with by the user. For simple compounds systematic nomenclature is generally used, although this may sometimes give difficulty to the non-specialist. The number of structures in the text has been reduced as the relation between systematic nomenclature, structure and properties may only become clear at very advanced levels. Entries on inorganic derivatives immediately follow the relevant headword alphabetically, so, for example, the entry on 'aluminium' is followed by 'aluminium, organic derivatives' and then by 'aluminium acetate'.

The contributors to previous editions of the Dictionary are acknowledged and particular note must be made of the contribution of L. M. Miall and his fellow editors of the *New Dictionary of Chemistry* from which the *Penguin Dictionary* developed. As before, errors and omissions are totally my responsibility and I would appreciate receiving note of them.

David W. A. Sharp
Chemistry Department, University of Glasgow
Glasgow, G12 8QQ

Chemical Safety

I urge all users of the *Dictionary of Chemistry* to exercise the utmost care if they have access to chemicals. In the Dictionary I have emphasized the harmful effects of some key materials, but most materials, even pure nitrogen and excess of water, can prove fatal in certain circumstances, and no material should be treated lightly unless full health, medical and toxicity data are available from a specialized source, and specialized advice is available. In particular, vapours should not be inhaled and direct skin, oral and eye contact should be totally avoided.

Abbreviations and Nomenclature

Abbreviations and symbols commonly used in this book include:

Å	Ångström unit, 10^{-10}m	ml	millilitre
Ar	aromatic group	mol.wt.	molecular weight
at.mass	relative/ atomic mass (atomic weight)	m.p.	melting point
		nm	nanometres
at.no.	atomic number	o	ortho
b.p.	boiling point	oz	ounce (used for platinum metals)
Bu	butyl group		
d	density, ρ kg m^3	p	para
D-, DL-	see optical activity, chirality	p.a.	per annum
		Ph	phenyl group
E	energy	pH	see hydrogen ion concentration
Et	ethyl group		
G	Gibbs function (see free energy)	R	alkyl group
		R	gas constant
g	gram	R	see optical activity, chirality
H	enthalpy		
h	Planck's constant	S	entropy
L-	see optical activity, chirality	s	see optical activity, chirality
l	litre	t	tert
lbs.	pounds	X (in formulae)	any non-metal (usually a halogen
M	mega (10^6, million)		
M	molar	μ	micron, 10^{-6} m
M (in formulae)	any metal	ν	frequency
Me	methyl group	(+)- and (-)-	see optical activity, chirality

Note that small capitals are used for cross-referencing throughout.

AAS ATOMIC ABSORPTION SPECTROSCOPY.

abamectin (averectin) Insecticide, acaricide, from *Streptomyces* sp.

abherents (release agents, parting agents) Liquid or solid films which reduce or prevent adhesion between surfaces; solid–solid, solid–paste, solid–liquid. Waxes, metallic soaps, glycerides (particularly stearates), polyvinyl alcohol, polyethene, silicones, and fluorocarbons are all used as abherents in metal, rubber, food, polymer, paper and glass processing.

abietic acid, $C_{20}H_{30}O_2$. A crystalline diterpenoid carboxylic acid m.p. 172–175°C, obtained from pine rosin by acid treatment. The commercial product may be glassy and of low m.p. Used extensively in the manufacture of plastics, paints, varnishes, paper sizes, detergents, lacquers, soaps, perfumes.

ab-initio calculations Quantum mechanical calculations which do not rely upon experimental data for solution. Involve linear combinations of atomic orbitals.

ablation Erosion and disintegration due to heat. Ablation-resistant materials, particularly Nylon fibres in a phenolic resin, are used to protect space vehicles during re-entry into the atmosphere. Decomposition to gaseous products with a porous refractory residue is desirable.

abrasives Hard materials used to disintegrate other materials. Amongst those most widely used are SiC, Al_2O_3 (often containing Ti), diamond, tungsten carbides, BN and metal abrasives, e.g. steel wool.

abscisic acid An important plant growth substance. See DICTIONARY OF BIOLOGY.

absolute alcohol 100% ethanol but with some chemical impurities.

absolute configuration The actual arrangement of substituents about a chiral atom. See CHIRALITY.

absolute temperature A temperature on the 'absolute' or Kelvin scale usually denoted by T; the zero of this scale is the temperature at which a perfect gas would occupy zero volume if it could be cooled indefinitely without liquefaction or solidification.

absolute zero 0 K is –273.15°C. 0°C is f.pt of water (273.15 K). The interconversion factor between the Kelvin and Centigrade/Celsius scales is T(K) = T°C + 273.15.

absorptiometer An instrument used to measure the absorption of radiation (light). Also the name of the apparatus used to determine the solubility of a gas in a liquid.

absorption See GAS ABSORPTION.

absorption bands See ABSORPTION OF LIGHT.

absorption coefficient of a gas The volume of gas measured at 0°C and 760mm pressure which will dissolve in 1 cm³ of a liquid. The absorption coefficients in water at 0°C for several common gases are as follows: N_2, 0.024; O_2, 0.049; C_2H_4, 0.25; CO_2, 1.71; H_2S, 4.68; SO_2, 79.8; HCl, 506; NH_3, 1300.

absorption coefficient of light See LAMBERT'S LAW and BEER'S LAW.

absorption column (absorption tower) The equipment normally used for absorption of gases consisting of a column or tower, usually circular in cross-section, the absorbing liquid passing down COUNTER-CURRENT to the gas passing upwards.

When the main object of the absorption is to remove impurities these columns are often referred to as SCRUBBERS.

absorption of light When light falls on the surface of a transparent substance part of the light is reflected, the remainder is transmitted unchanged. If, however, the light falls upon the surface of a black substance, e.g. lamp black, all wavelengths are absorbed in a general absorption process. Many substances appear coloured because they have absorbed selectively all the wavelengths of white light except those corresponding to the particular colour which appears. If the spectrum of the light transmitted by a coloured substance is examined it is found that certain wavelengths, called *absorption bands*, are missing. Absorption or emission corresponds to transitions between energy states. For gases in the atomic state the spectrum of the transmitted light shows dark lines rather than bands. These *absorption lines* correspond to the wavelengths of light absorbed by the atoms. General principles apply to all radiation.

Relationships between the intensity of incident light, sample thickness, concentration and intensity of transmitted light are embodied in BEER'S LAW and LAMBERT'S LAW. It is generally more meaningful to use frequency rather than wavelength as frequency corresponds to the difference in energy between the energy levels involved.

absorption spectroscopy Any spectroscopic technique, qualitative or quantitative, depending upon the measurement of an absorption spectrum.

absorption tower See ABSORPTION COLUMN.

ABS plastics A group of plastic materials based on blended copolymers of styrene–acrylonitrile (70:30) and butadiene–acrylonitrile (65:35) and on block copolymers of styrene and acrylonitrile with polybutadiene. Used in pipe (25%), appliances (20%), and automotive parts (15%).

abundance of elements See ELEMENTS, ABUNDANCE OF.

Ac Actinium.
Also used for ethanoate (acetate).

AC 94377 Plant growth regulator, phthalimide type.

AC 263222 Herbicide, imidazolinone type.

acac Abbreviation for acetylacetonato group.

acacia See GUM ARABIC.

accelerator mass spectrometer Technique to measure ratios of isotopes. Originally used in carbon dating, now used generally in chemistry.

accelerators Substances used to accelerate desired cross-linking reactions in polymers. In particular they assist the VULCANIZATION of rubber, in some cases conferring on it increased resistance to wear. Many types of organic compounds can be used, e.g. DIPHENYLGUANIDINE, the thiazoles, e.g. MERCAPTOBENZTHIAZOLE, and the THIURAM DISULPHIDES. The term is also used for those substances which act as catalysts by increasing the rate at which thermosetting resins cure or harden.

acceptor An atom, molecule, or ion that is electron deficient and which can form a CO-ORDINATE BOND with an electron donor. Thus in the complex ion $[Co(NH_3)_6]^{3+}$ the cobalt(III) ion is an acceptor and the ammonia the electron donor. π-acceptors are molecules or atoms that accept electrons into π, p or d orbitals, e.g. metal–alkene (olefin) complexes.

$Co \longleftarrow NH_3$

σ bond   π back bond

Also a substance which, while normally not oxidized by oxygen or reduced by hydrogen, can be oxidized or reduced in the presence of a substance itself undergoing oxidation or reduction.

accumulator See LEAD ACCUMULATOR.

acenaphthene, $C_{12}H_{10}$. Colourless needles, m.p. 95°C, b.p. 278°C. Dyestuff intermediate, used in plastics. Oxidized $(-H_2)$ to acenaphthylene.

acephate Insecticide, arganophosphate.

acetal, $C_6H_{14}O_2$, $CH_3.CH(OEt)_2$. A pleasant-smelling liquid, b.p. 104–105°C. Prepared by mixing ethanal and ethanol in the presence of a catalyst, such as HCl, or by passing C_2H_2 into EtOH in the presence of a catalyst.

acetaldehyde See ETHANAL.

acetal resin See ALDEHYDE POLYMERS.

acetals Compounds of the general formula

$$R^3R^4C(OR^1)(OR^2)$$

Derived from an aldehyde or ketone and an alcohol using an acid catalyst. Some acetals are useful solvents.

acetamide See ETHANAMIDE.

acetanilide, C_8H_9NO, $PhNH \cdot CO \cdot CH_3$. White crystals, m.p. 114°C. Manufactured from aniline and excess ethanoic acid or ethanoic anhydride. Chief uses are in the manufacture of dyestuff intermediates, in the manufacture of rubber, and as a peroxide stabilizer. Hydrolysed to aniline.

acetate fibres See CELLULOSE ACETATE PLASTICS.

acetates Salts or esters of ETHANOIC ACID.

acetic acid See ETHANOIC ACID.

acetic anhydride See ETHANOIC ANHYDRIDE.

acetic ester, $CH_3C(O)OC_2H_5$. See ETHYL ETHANOATE.

acetins Acetates (ethanoates) of glycerol. There are five possible acetates: two mono-, two di-, and one tri-acetate. The commercial acetins are mixtures of the various acetates and form colourless or slightly brown syrupy liquids. Prepared from glycerol, ethanoic acid and sulphuric acid.

Monoacetin contains chiefly the 1-acetate $(CH_2OH \cdot CHOH \cdot CH_2O(O)CCH_3)$. Used as a solvent for the dyes employed in printing paper bags.

Diacetin is chiefly the 1:3-diacetate $(CH_3O(O)CH_2 \cdot CHOH \cdot CH_2O(O)CCH_3)$. Used as plasticizer for cellulose acetate lacquers and as a solvent for basic dyes.

Triacetin is about 90% glycerol triacetate and 10% diacetate. Used as a plasticizer for lacquers and as a solvent for gums and resins and as a fixative for perfumes.

acetoacetic acid (acetonemonocarboxylic acid, 3-oxobutanoic acid), $C_4H_6O_3$, $CH_3CO \cdot CH_2COOH$. A colourless and strongly acid syrup. It is unstable, and decomposes into propanone and carbon dioxide below 100°C. Prepared from ETHYL ACETOACETATE.

acetoacetic ester See ETHYL ACETOACETATE.

acetoin (3-hydroxy-2-butanone), $C_4H_8O_2$, $CH_3CH(OH)COCH_3$. M.p. 15°C, b.p. 148°C. Produced from propene and butene glycols by the action of bacteria, and from ethanal by yeast and by the reduction of diacetyl(2,3-butanedione). When distilled, it forms diacetyl.

acetol See HYDROXYPROPANONE.

acetolysis The process of removing acetyl groups from an organic compound. It is usually carried out by heating the acetyl compound with aqueous or alcoholic alkalis, whereby the acetyl groups are removed as ethanoic acid.

acetone See PROPANONE.

acetone alcohol See HYDROXYPROPANONE.

acetone bodies See KETONE BODIES.

acetone dicarboxylic acid (3-oxo-glutaric acid), $C_5H_6O_5$, $CO \cdot (CH_2C(O)OH)_2$. Colourless needles, m.p. 135°C (decomp.). Prepared by the action of sulphuric acid on citric acid. Readily decomposed by boiling water, acids, or alkalis to acetone (propanone) and carbon dioxide. The acid or its diethyl ester reacts with sodium in a manner similar to acetoacetic ester. Used in organic syntheses.

acetone monocarboxylic acid See ACETOACETIC ACID.

acetonitrile (methyl cyanide, ethanenitrile), CH_3CN. Poisonous liquid, b.p. 82°C. Prepared from ethyne and ammonia or by dehydration of ethanamide. Widely used as solvent for dissolving inorganic and organic compounds, especially when a non-aqueous polar solvent of high dielectric constant is required, e.g. for ionic reactions.

acetonyl The group CH_3COCH_2-.

acetonylacetone See HEXAN-2,5-DIONE.

acetophenone, C_8H_8O, $PhCOCH_3$. Colourless plates, m.p. 20°C. Odour resembling bitter almonds. Prepared by the action of ethanoyl chloride upon benzene in the presence of aluminium chloride. Typical ketone. Used as a solvent for cellulose ethers, in confectionery and as a polymerization catalyst.

acetoxy The group CH_3COO-.

acetylacetonates Metal derivatives of acetylacetone generally containing the grouping shown (Macac) with some delocalization in the ring. The number of acetylacetonate groups bonded to a metal can vary from 1 to 4 (see over).

$$
\begin{array}{c}
\text{R}'\\
\diagdown\\
\text{C}-\text{O}\\
\end{array}
\quad
\text{R}''\text{C}\cdots\text{M}\\
\text{R}'''\\
\text{C}-\text{O}
$$

acetylacetone (pentan-2, 4-dione) acac,
$C_5H_8O_2$, $CH_3COCH_2COCH_3$. A diketone with
enolic properties. The enolate ion acac⁻ (see
ENOLS) forms stable complexes with many metals, e.g. tris(acetylacetonato)iron(III), Feacac₃.
Metal derivatives are generally soluble in organic solvents and often appreciably volatile; they
are used in solvent extraction and mass spectrometry. Acetylacetone is used as a drier in varnishes.

Derivatives of acetylacetone, e.g.
$CF_3COCH_2COCH_3$, $CF_3COCH_2COCF_3$ and
thienyl-$COCH_2COCF_3$ form particularly stable
derivatives.

acetylation (ethanoylation) A process for
introducing acetyl groups into an organic compound containing e.g. –OH, –NH₂ or –SH groups.
It is carried out by heating the compound with
ethanoic anhydride or ethanoyl chloride, usually in presence of an inert solvent such as benzene
or ethanoic acid. In many cases, zinc chloride or
pyridine is used to speed up the reaction.

acetyl chloride See ETHANOYL CHLORIDE.

acetylcholine, $C_7H_{17}NO_3$,
$(CH_3)_3N^+CH_2CH_2O(O)CCH_3OH^-$. A zwitterion. A neurotransmitter in the peripheral
and central nervous systems.

acetyl coenzyme A A reactive thioester of
fundamental importance in metabolism, biosynthesis and as a neurotransmitter. Acetyl
coenzyme A is the principal substrate of the CIT-
RIC ACID CYCLE. It also acts as an acetylating
agent, e.g. in the formation of acetylcholine. It
has a role in three major biosynthetic pathways.
Contains Ni, Cu, and Fe atoms. See DICTIONARY
OF BIOLOGY.

acetylene See ETHYNE.

acetylene black (cuprene) A form of carbon black prepared by pyrolysis of ethyne.

acetylene complexes Co-ordination compounds similar to ALKENE COMPLEXES.

acetylene dicarboxylic acid,
$HO(O)CC\equiv CC(O)OH$.
See DIMETHYLACETYLENEDICARBOXYLATE.

acetylene dichloride See DICHLOROETHENES.

acetylene tetrachloride See sym-TETRA-
CHLOROETHANE.

acetylides CARBIDES containing C_2^{2-} or C_2R^-
species. Formed by more electropositive elements (e.g. K, Ca, Al) and by some transition elements (e.g. Cu, Ag, Au). Hydrolysed to ethyne.
Many transition metal acetylides are explosive.

Also metal derivatives of terminal alkynes, RC_2H. Transition metals form complex
acetylides (e.g. $[M(C\equiv CR)_n]^{x-}$) often containing
the metal in low oxidation states.

acetylsalicylic acid See ASPIRIN.

achiral The molecule of a compound that is
not OPTICALLY ACTIVE is achiral.

achromatic indicators Substances which
give a grey end-point and generally find application in the titration of turbid liquids.

acid An acid on the aqueous system is simply
defined as a substance which is capable of forming hydrogen ions when dissolved in water. Most
inorganic acids may be regarded as a compound
of an acidic oxide and water; where the oxide
concerned is that of a metal. A base is a proton acceptor and an oxide may exhibit amphoteric
character, that is act sometimes as an acid and
sometimes as a base. Aqueous solutions of acids
have a sharp taste, turn litmus red, liberate carbon dioxide from a metallic carbonate and give
reactions characteristic of the anion present.

Since free protons do not exist except in the
gas phase, acidic properties can only be shown
when the solvent can act as a proton acceptor,
i.e. as a base. Thus aqueous solutions of acids
contain the hydroxonium ion, H_3O^+. By
Brönsted–Lowry acid–base theory an acid is a
proton donor and a base an electron acceptor
(see CONJUGATE BASE).

Acids can also exist in non-aqueous solvents.
Since ammonia can also solvate a proton to give
the ammonium ion, NH_4^+, substances which
dissolve in ammonia to give the ammonium
ion, e.g. NH_4Cl, are acids in that system.

Liquid water is ionized

$$2H_2O \rightleftharpoons H_3O^+ + OH^-$$

this ionization being the reverse of the neutralization reaction in water; substances giving hydroxyl ions are bases in water. Liquid ammonia
is ionized

$$2NH_3 \rightleftharpoons NH_4^+ + NH_2^-$$

and ionic amides are bases of this system.

The concept of acids and bases has been extended to solvents which are ionized and yet do

not contain hydrogen: a substance giving the appropriate positive ion is an acid in that system. Thus bromine trifluoride ionizes

$$2BrF_3 \rightleftharpoons BrF_2^+ + BrF_4^-$$

and a substance giving the BrF_2^+ ion in solution, e.g. $BrF_3\ SbF_5$, is an acid in the system.

Typical organic acids contain the –C(O)OH group, but many other acid groupings, e.g. the sulphonic group –S(O)$_2$OH, give acidic properties to organic compounds. Phenols have acidic properties and are classified with enols as pseudo-acids.

The term acid was extended by Lewis to include substances which are electron acceptors. Thus $AlCl_3$ can accept a pair of electrons from a chloride ion forming the $[AlCl_4]^-$ ion and is a Lewis acid.

The 'strength' of an acid is measured by the value of its dissociation constant, 'strong' acids, e.g. HCl, HNO$_3$, being substantially fully ionized in solution and 'weak' acids predominately unionized.

acid–base indicator A substance, a weak acid or weak base, that has a different colour in acid or base solution. The colour change is due to a marked difference in colour between the undissociated and ionic forms. For a good indicator the colour change must occur between narrow limits of pH, e.g. methyl orange is red at pH 3.1 and changes to yellow at pH 4.4.

acid dyes Dyestuffs containing an aromatic chromophoric group and a group conferring solubility in water, generally the SO$_3$H group as its sodium salt. They are relatively simple in application. The types of acid dyestuffs are:

Simple acid dyes contain no polyvalent metals.

Mordant acid dyes combine simultaneously with the mordanting agent (generally Cr(OH)$_3$) and the fibre; the dyestuff generally contains ortho OH–azo or OH–OH groups.

Premetallized acid dyes are similar to the simple acid dyes but are already complexed to a metal.

acid egg A non-mechanical pump for the handling of highly corrosive liquids. Liquid is admitted to a corrosion-resistant vessel and forced into the delivery line by compressed air.

acid exchange resins See ION EXCHANGE.

acidity constant An acid dissociates RC(O)OH + H$_2$O $\rightleftharpoons$ [RCO$_2$]$^-$ + H$_3$O$^+$. K_a the dissociation constant is [RCO$_2^-$][H$_2$O]/[RC(O)OH] where bracketed terms are activities or practically concentrations. [H$_2$O] is constant and is neglected and the equilibrium constant K_a is the acidity constant – a measure of the dissociation of the acid.

acid oil The alkali extract of phenol derivatives formed in the production of gasoline by cracking operations. Also referred to as CRESYLIC ACIDS or 'PHENOLS'.

acid rain Rain containing acidic impurities (e.g. SO$_2$, nitrogen oxides) generally from industrial sources or internal-combustion engines.

acid sludge A complex acid residue produced when kerosines, lubricating oils or other petroleum products are refined with sulphuric acid or oleum. It consists of hydrocarbons, sulphonic acids and free sulphuric acid. Petroleum sulphonates can be recovered by extraction with alkali and used for the metal processing, textile and leather industries.

acifluorfen Herbicide, diphenyl ether type.

aclonifen Herbicide, phenoxyaniline type.

aconitic acid, $C_6H_6O_6$, *trans*-(HO)OCCH = C(C(O)OH)CH$_2$.C(O)OH. Prepared by dehydrating citric acid with 50% sulphuric acid.

acridine, $C_{13}H_9N$. Colourless needles, m.p. 111°C.

Parent compound of a series of dyestuffs.

Acrilan A brand name for a synthetic fibre, based on a copolymer of acrylonitrile with minor proportions of other unspecified vinyl monomers. See also PROPENENITRILE.

acrinathrin Acaricide, insecticide, cyclopropyl type.

acrolein See PROPENAL. Herbicide.

acrolein polymers (propenal polymers) Polymers of CH$_2$ = CHCHO generally formed by free-radical polymerization. Polymerization is generally through the vinyl group and the aldehyde groups are present as acetals. Polymerization with strong base causes polymerization through the carbonyl group. Acrolein polymers can be modified chemically and used as thickening agents and protective colloids. Also used in plastics and lacquers.

acrylamide, $CH_2=CHC(O)NH_2$. Used as polymer and in synthesis, dyestuffs, paper, textiles. Neurotoxin.

acrylamide polymers See POLYACRYLAMIDE.

acrylate resins and plastics Generally used to refer to polymeric methyl acrylate (methyl propenoate) and polymeric methyl methylacrylate (methacrylate) (methyl 2-methylpropenoate). Polymeric methyl acrylate is used, principally in an emulsion form, in textile and leather finishes, lacquers, paints, adhesives and safety glass interlayers and as absorbent polymer. Polymeric methyl methylacrylate gives a clear solid material (Perspex) and is used in injection moulding and extrusion.

acrylic acid See PROPENOIC ACID.

acrylic acid polymers Acrylic acid, $CH_2=CHCO_2H$, and methacrylic acid, $CH_2=CMeCO_2H$ undergo free radical polymerization to give polymers used as absorbent polymers, thickeners, in textile treatment, as drilling-mud additives, as flocculating agents, in paper making, and if copolymerized with e.g. divinylbenzene as ion-exchange resins.

acrylonitrile See PROPENENITRILE.

acrylonitrile polymers See POLYACRYLONITRILE.

ACT Activated Complex Theory. See ACTIVATION ENERGY.

actin Major protein of muscle and important constituent of eukaryotic cells. Many forms.

actinides The elements actinium, thorium, protactinium, uranium, neptunium, plutonium, americium, curium, berkelium, californium, einsteinium, fermium, mendelevium, nobelium, and lawrencium are collectively known as actinides. Higher atomic number elements show resemblance to the corresponding transition metals. Those with atomic numbers 93 and above are artificial and are produced by irradiation of uranium or other artificial elements with neutrons, alpha particles, or carbon, nitrogen or heavier ions. In the actinide elements the 5f shell is being filled and they are thus analogous to the lanthanides or rare-earth elements. There is less shielding of the 5f electrons than of the 4f and oxidation states greater than three are common, particularly among lighter members of the series. Amongst the heavier elements the +2 state becomes stable. Organoactinide complexes are known and many are active hydrogenation catalysts.

All of the actinides are radioactive and present some health hazards; with the more radioactive elements all handling has to be by remote control and no contact with the operator is permissible.

actinium (Ac) Radioactive metallic element, at.no. 89, at.mass ^{227}Ac 227.03, m.p. 1051°C, b.p. 3198°C, density ρ 10 060, electronic configuration [Rn] $6d^1 7s^2$. Occurs naturally as a minor constituent in uranium ores but difficult to separate and best prepared by bombardment of radium with neutrons. Separated from other species by ion exchange or solvent extraction. The most stable isotope ^{227}Ac is very radioactive ($t_{1/2}$ 22 years), the metal glows blue and is oxidized in air. It has been prepared by reduction of AcF_3 with lithium vapour at 1200°C and is silvery white in colour.

actinium compounds Element of Group 3. Because of its intense radioactivity very few actinium compounds have been isolated. In its compounds the element is tripositive, forming very similar compounds to LANTHANUM. Actinium salts are colourless.

activated adsorption Chemisorption often has an activation energy associated with it and is sometimes referred to as activated adsorption.

activated carbon See ACTIVE CARBON.

activated clay See BLEACHING EARTHS.

activated complex See TRANSITION STATE THEORY.

activated molecule A molecule which has acquired more energy than the average amount possessed by other molecules, and is therefore in a more reactive condition, is said to be activated. Molecules may be activated by light or by heat or during formation. Molecules will only react on collision when they possess more than a certain minimum amount of energy.

activation analysis An analytical technique in which an artificial radioactive isotope is formed by irradiation (generally with neutrons) from the stable element to be determined and the amount of the artificial isotope is then estimated from its radioactivity. See NEUTRON ACTIVATION ANALYSIS.

activation energy The minimum energy which reacting species must possess in order to be able to form an 'activated complex' or transition state before proceeding to the products. The activation energy (E_a) may be derived from the temperature dependence of the reaction rate using the ARRHENIUS EQUATION. The Gibbs activation energy of a reaction may be divided into enthalpy and entropy terms.

active carbon Carbon (charcoal) treated at high temperature with steam, air or CO_2. Used as an adsorbent for removal of small traces of impurities from a gas or liquid. Used extensively in water and waste water treatment, air pollution control, as a catalyst, sugar refining, purification of chemicals and gases (gas masks), dry cleaning, rubber reclamation, cigarette filters. See CARBON BLACK.

active earths See BLEACHING EARTHS.

active sites, centres In heterogeneous catalysis adsorption occurs particularly at active centres on the surface of the catalyst. Active centres are the specific sites at which adsorption occurs. The concept of active centres has been extended to enzymes and bacterial action.

active transport The biochemical transport of substances, usually against a concentration gradient, other than by osmosis or diffusion.

activity A thermodynamic quantity that measures the effective concentration or intensity of a particular substance in a given chemical system. The *absolute activity*, $a°$, of a substance is given by $\mu = RT \ln a°$, where μ is the CHEMICAL POTENTIAL of the substance, R is the gas constant, T the absolute temperature. The *relative activity*, a, is given by $\mu = \mu° + RT \ln a$; $\mu°$ being the chemical potential of the substance in its STANDARD STATE. For dilute, ideal solutions the activity is directly proportional to the concentration; for ideal gases, activity is proportional to the partial pressure of the gas.

activity coefficient (f or γ) A dimensionless factor by which the concentration (c) of a substance must be multiplied to give an exact measure of its thermodynamic activity (a) in a chemical system, i.e. $a = fc$. It is a measure of the deviation from ideal behaviour of the solution; f is unity for an ideal mixture and greater or less than unity for a non-ideal system. Activity coefficients of *electrolytes* are considered to be the geometrical mean of the single ion activities, the latter being hypothetical quantities which cannot be determined separately.

activity series The elements arranged in order of their ELECTRODE POTENTIALS. Normally a more active element higher in the series will displace a lower element from its compounds in solution.

actomyosin The most important protein of muscle. Formed from actin and myosin in the presence of Cu^{2+}.

acyclic See ALIPHATIC. Containing chains, possibly with branches but no rings.

acyl The general name for organic acid groups, which are the residues of carboxylic acids after removal of the –OH group, e.g. ethanoyl chloride, $CH_3C(O)\cdot Cl$, is the acyl chloride formed from ethanoic acid, $CH_3C(O)\cdot OH$. The names of the individual acyl groups are formed by replacing the -*ic* of the acid by -*yl*.

acylation A chemical transformation which substitutes the acyl (RCO–) group into a molecule, generally for an active hydrogen of e.g. an –OH group.

acylium ions, $RC\equiv O^+$. Intermediates in many reactions, e.g. Friedel–Crafts reactions.

acyloin condensation The formation of an acyloin, very often cyclic, by condensing two molecules of ester with sodium.

acyloins 1,2-Ketoalcohols of the type $R\cdot C(O)\cdot CHR'OH$. See ACETOIN, BENZOIN.

adamantane, $C_{10}H_{16}$. Colourless hydrocarbon, m.p. 269°C, but subliming readily at room temperature and atmospheric pressure. Occurs (up to 0.0004%) in some petroleum fractions, together with alkylated adamantanes. Has a rigid system composed of three fused chair cyclohexane rings, having the same configuration as the diamond lattice. Synthesized from tetrahydrodicyclopentadiene. 1-Adamantanamine hydrochloride has been found useful against viral infections; also used in treating Parkinson's disease. Adamantyl derivatives are used as lubricants and resins.

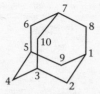

Adams' catalyst (platinum oxide hydrate), $PtO_2 nH_2O$. Produced by fusion of H_2PtCl_6 with sodium nitrate at 500–550°C and leaching of the cooled melt with water. Stable in air, activated by hydrogen. Used as a hydrogenation catalyst for converting alkenes to alkanes at low pressure and temperature. Often used on SiO_2.

$$HOP(O)(OH)\text{-}O\text{-}[P(O)(OH)O]_2\text{-}CH_2$$

adenine

ATP

adatom An adsorbed atom on a surface. May be an important site for catalysis.

addition reactions Reactions in which an unsaturated system (C=C, C≡C, C=O, etc.) is saturated or part saturated by addition of a molecule across the multiple bond. Examples include the reaction of bromine with ethene to give 1,2-dibromoethane, hydrogen cyanide addition to an aldehyde giving a cyanhydrin, the Diels–Alder reaction, and addition polymerization.

Also used in inorganic chemistry, e.g. the reaction of BF_3 with nucleophiles, e.g. ammonia, ether, to form complexes where the coordination number or the oxidation state of the atom becomes more positive.

additives Compounds, usually added in small amounts, which will confer specific properties on the bulk material, e.g. anti-foaming additives for lubricating oils; materials added to polymers to enhance workability.

additive volumes, law of The volume occupied by a mixture of gases is equal to the sum of the volumes which would be occupied by the constituents under the same conditions of temperature and pressure.

adduct A phase (often a compound) formed by direct combination, generally in simple proportions, of two or more different compounds or elements.

adenine (6-aminopurine), $C_5H_5N_5$. Vitamin B_4. A constituent of the nucleic acid portion of nucleoproteins, and, combined, as adenosine pyrophosphate, it plays an important part in many metabolic processes.

adenosine See NUCLEOSIDES.

adenosine diphosphate (ADP) Adenosine 5′-diphosphate (pyrophosphate). A precursor of ATP and is also formed from it during processes in which ATP is involved.

adenosine monophosphate (AMP) Normally adenosine 5′-phosphate (muscle adenylic acid), an important structural component of NUCLEIC ACIDS and of several coenzymes. See also CYCLIC AMP.

adenosine triphosphate, ATP, adenosine nucleotide triphosphate The most important of the so-called 'high-energy compounds', a group of naturally occurring organic phosphates characterized by high free energies of hydrolysis, and playing a fundamental role in biosynthesis, active transport and muscle action. ATP is the primary source of energy in the metabolism of plant, animal and bacterial cells. See DICTIONARY OF BIOLOGY and GLYCOLYSIS.

adenylic acid See ADENOSINE MONOPHOSPHATE.

adhesion agents Any additive which will improve the adhesivity of a material. Surface-active agents used to prevent stripping of an adhesive from a surface by water. They are widely used in road bitumens, particularly CUTBACK BITUMENS, used in surface dressings and in coated macadam surfacings.

Soaps of heavy metals have been used but cationic surface-active agents have proved more suitable, notably organic amines of relatively high molar mass.

adhesive A material which will wet two surfaces which are to be joined and which will subsequently solidify to form a join. Examples include glues (starches, proteins), thermoplastic resins, thermosetting resins, rubber, asphalt, sodium silicates. Used in construction, packaging, textiles, furniture and electrical goods.

adiabatic change A change in which heat is neither allowed to escape from, nor is added to, the system during the change.

adipic acid (hexane-1,6 dioic acid), $C_6H_{10}O_4$, $H(O)OC \cdot [CH_2]_4 \cdot C(O)OH$. M.p. 153°C. Present in beet-juice; manufactured by the air or nitric acid oxidation of cyclohexanol or cyclohexane or from benzene via phenol and cyclohexanone. Distillation of its calcium salt gives cyclopentanone. It forms long chain polymers with diamines and is very largely used in the manufacture of Nylon. It is also used for the manufacture of plasticizers and certain vinyl and urethane plastics.

adipocene Wax-like fatty acids obtained from body fats after death.

adiponitrile (1,4-dicyanobutane), $C_6H_8N_2$. Formed from 1,4-dichlorobutane and sodium cyanide. Used in Nylon production, present in gland and sympathetic nerve endings.

adlayer An adsorbed layer.

ADP See ADENOSINE DIPHOSPHATE.

adrenaline (epinephrine U.S. usage), $C_9H_{13}NO_3$,(R)-4-1[1-hydroxy-2(methylamino)-ethyl]benzene-3,4-diol. M.p. 212°C. It is secreted along with NORADRENALINE by the adrenal medulla. Induces cellular activation of phosphorylase which promotes catabolism of glycogen to glucose. Used in treatment of allergic reactions and circulatory collapse.

adsorbate The substance which is adsorbed on to the surface of the adsorbent.

adsorbent The substance upon whose surface the process of adsorption occurs.

adsorption Because of the bonding of the atoms at the surface of a solid or a liquid compared with atoms in the bulk, the surface of a solid or liquid is potentially an active site. The process of adsorption, in which foreign atoms or molecules become attached to the surface, lowers the free energy of the surface. Thus, e.g. adsorption of detergent molecules at liquid surfaces lowers the surface tension of the liquid.

Adsorption may in principle occur at all surfaces; its magnitude is particularly noticeable when porous solids, which have a high surface area, such as silica gel or charcoal, are contacted with gases or liquids. Adsorption processes may involve either simple unimolecular adsorbate layers or multilayers; the forces which bind the adsorbate to the surface may be physical or chemical (chemisorption) in nature.

Adsorption is of technical importance in processes such as the purification of materials, drying of gases, control of factory effluents, production of high vacua, etc. Adsorption phenomena are the basis of heterogeneous catalysis and colloidal and emulsification behaviour.

adsorption, industrial Adsorption is of considerable industrial importance, being employed, for example, in the refining of lubricating oils, the decolorizing of sugar solutions and the purifying and drying of gases. Industrial adsorption may employ *batch*, *fixed bed*, or *continuous countercurrent* operations.

adsorption indicator An indicator which functions by adsorption on the surface of a precipitate. Thus in precipitating AgCl in the presence of excess Cl^- the surface has a layer of Cl^- and is negatively charged; in the presence of the indicator fluorescein the indicator is in solution. As soon as there is excess of Ag^+ the precipitate takes on a positive charge and the presence of the fluorescein anion as counter ion gives the precipitate a pinkish-red colour.

adsorption isotherms Empirical models for adsorption surfaces.

adsorption (physisorption) Some adsorption processes are not chemically specific and are readily reversible. The forces of attraction between the adsorbate and adsorbent are weak and similar in nature to those responsible for the cohesion of molecules in the liquid state, VAN DER WAALS' FORCES. Physisorption may involve the formation of multiple layers at the adsorbent surface.

aerobic metabolism Processes in which inhaled oxygen takes part.

aerosol (aerogel) A dispersion in which a finely divided solid is suspended in air or a gas and the particles are of colloidal dimensions, e.g. smoke. Aerogels are porous materials. Aerosols can often be formed by the rapid condensation of a vapour such as that of a metal oxide.

Aerosol sprays consist of a material dissolved or suspended in a liquid which when pressure is released volatilizes to produce a fine spray. The spray carries the active material. Used in hair lacquers, paints, etc.; the propellant should be inert and non-inflammable. Chlorofluorocarbons have been used extensively but are now being replaced. U.S. production 1997 3100 million units.

AES Auger electron spectroscopy.

affinity, chemical The term 'affinity of a chemical reaction' is used to denote the change of free energy in the system when the reaction takes place under standard conditions; this is a true measure of the tendency to react.

affinity chromatography A method of purifying natural macromolecules, especially proteins. Depends on covalently attaching a specific species to an insoluble inert support. The adsorbed species must have a special and unique affinity for the macromolecule, so that on passage in solution down a column of the material the macromolecule is preferentially retarded and thus separated from contaminating molecules.

aflatoxins Group of secondary fungal metabolites. Toxic, carcinogens.

Ag Symbol for silver.

agar (agar-agar) A SEAWEED COLLOID. A mixture of two polysaccharides, agarose and agaropectin. Hydrolysed to galactose.

Agar occurs as a cell-wall constituent of the red marine algae *Rhodophyceae*. Used as a solid medium for cultivating micro-organisms, as a thickener, emulsion stabilizer in the food industry and as a laxative.

agate A form of SiO_2 similar to chalcedony with coloured bands of purple or brown. Used as a gemstone, for making pestles and mortars, and as a bearing surface in scientific instruments.

age hardening See PRECIPITATION HARDENING.

Agent Orange Defoliant composed of 2,4-dichlorophenoxy ethanoic acid and 2,4,5-trichlorophenoxy ethanoic acid.

aggregate Pieces of material which are, or can be, united by a cementitious substance or bond, to form a solid mass. In concrete the aggregate consists of crushed stone, slag or clinker and sand, united by Portland or other cement. In bricks and other ceramic materials, particles of non-plastic materials form the aggregate; they are united by plastic clay when in the green or unfired state and later (in the burned state) by a crude glass which fuses and, on cooling, solidifies and forms the bond. The sizes, shapes and mechanical properties of the pieces forming an aggregate have a great influence on the properties of the bonded product.

aglycon(e) See GLYCOSIDES.

agonist A molecule (material) that activates a physiological system. An antagonist blocks a system.

agostic Interaction between an electronically deficient metal atom and an adjacent C–H (or even Si–H).

air The volume composition of normal dry air is N_2 (78.08%), O_2 (20.94%), Ar (0.934%), CO_2 (0.03%), Ne (0.0018%), Kr (0.0001%), Xe (0.000009%), Rn (6×10^{-18} %). The proportion of CO_2 can vary and the content of water vapour varies widely according to atmospheric conditions. Air may also contain many other molecules – often by pollution – e.g. ozone, nitrogen oxides, or SO_2; or from natural processes, e.g. CH_4. The physical constants of dry air are d 1.746 g l^{-1} at 200 K; specific heat capacity C_p 0.2396, C_v 0.1707 (ratio 1.403). Air can be liquefied by cooling (under pressure) and by allowing rapid expansion of gas; the liquid is pale blue because of the presence of O_2. The b.p. varies from –192°C (air) to –185°C (after N_2 evaporation), d approx. 0.9. Liquid air (and oxygen) must not be used in the preparation of low-temperature baths with organic liquids or for cooling organic materials since extremely explosive mixtures are formed.

air-cooled heat exchangers The usual arrangement comprises a bank of externally finned tubes, the fluid to be cooled passing through the tubes with air being forced or induced over the tubes by a fan.

air filters Filters designed to remove atmospheric dust from an air stream (or process dust from air or gas). In general they consist of a space packed with a filtering medium such as glass fibre or slag wool through which air is passed.

air–fuel ratio The normal way of expressing the composition of a mixture of fuel and combustion air, either on a volume or weight basis. The theoretical air–fuel ratio contains just sufficient oxygen to produce complete oxidation of the fuel components.

air hardening A method of hardening steel by allowing it to cool naturally in air or an air blast (air quenching).

air-lift agitator A plant for the leaching of fine solids.

air-lift pump A device for pumping corrosive liquids in which a U-shaped tube is used in conjunction with injected air.

air quenching See AIR HARDENING.

AKH-7088 Herbicide, trifluoromethyl derivative.

Al Aluminium.

alachlor Herbicide, acetanilide type.

α-alanine (L-2-aminopropanoic acid), $C_3H_7NO_2$, $CH_3.CH(NH_2)C(O)OH$. M.p. 297°C. One of the naturally occurring amino-acids obtained by the hydrolysis of proteins.

β-alanine (3-aminopropanoic acid), $C_3H_7NO_2$, $H_2N.CH_2.CH_2.C(O)OH$. M.p. 200°C.

Prepared by heating propenoic acid under pressure with ammonia. A step in the manufacture of PANTOTHENIC ACID both in the laboratory and the living organism. Not present in proteins.

alanylcarb Insecticide, carbamate type.

albumins See PROTEINS.

alcohol See ETHANOL AND ALCOHOLS.

alcoholometry The process of determining the proportions of ethanol in a liquid. Measurements of specific gravity are often used, but accurate measurement in breath, blood samples etc. are determined by chromatography. Often expressed in degrees (percentage) proof.

alcohols Organic compounds containing one or more hydroxyl groups attached directly to carbon atoms. Aromatic derivatives having hydroxyl groups joined directly to the carbon atoms of the ring are called phenols and have distinctive properties. Compounds with one -OH group are called monohydric alcohols; those with more than one are di-, tri-, or polyhydric alcohols. Aliphatic dihydric alcohols are known as glycols; the most important trihydric alcohol is glycerol; tetra-, penta- and hexahydric alcohols are generally derived from sugars. Monohydric alcohols are named from the names of the corresponding paraffin hydrocarbons, e.g. ethanol for ethyl alcohol, $CH_2 \cdot CH_2OH$. Primary alcohols contain $-CH_2OH$; secondary $>CH(OH)$; tertiary $>C(OH)$ groups. The monohydric alcohols form a series whose lower members are odorous liquids and higher members white, odourless solids. They react with metals such as sodium, calcium, and aluminium to give ALKOXIDES, and with acids to give esters. Sulphuric acid and other agents remove water to form alkenes. Primary alcohols on oxidation give aldehydes; secondary alcohols give ketones (see OPPENAUER and DESS–MARTIN OXIDATIONS), while tertiary alcohols break down to give compounds containing fewer carbon atoms than the original alcohol.

alcoholysis A reaction in which an alcohol plays a part similar to that of water in hydrolysis. It is most frequently applied to the reaction between an alcohol and an ester whereby the alcohol replaces the original alcohol of the ester. Such reactions usually proceed more rapidly in the presence of small amounts of a sodium alkoxide. In many cases the reaction is reversible.

aldehyde polymers Polymers with (—CHROX—) backbones formed by polymerization of the aldehydes RCHO. The commercially important aldehyde polymers are polyformaldehyde (polymethanal, Delrin) (R = H) formed by polymerization of CH_2O in polar solvents; polyacetaldehyde (polyethanal) (R = CH_3); polychloral (R = CCl_3). These materials are light but hard and rigid and fairly chemically resistant. They are used in machine parts, auto parts, etc. Aldehydes telomerize to trimers, etc. (see under individual headings).

aldehydes Organic compounds containing the group C(O)H joined directly onto another carbon atom: methanal, HCHO, is included in the class, but some of its properties are not those of a typical aldehyde. Aldehydes are usually colourless liquids (aliphatic) or solids (higher aromatic) with characteristic odours, and are oxidized to acids and reduced to primary alcohols. A number of preparations are available: oxidation of a primary alcohol; the selective reduction of carboxylic acid derivatives (e.g. amides, nitriles, acid chlorides) by complex metal hydrides. Aromatic aldehydes may be prepared by heating aromatic hydrocarbons with carbon monoxide, hydrogen chloride, anhydrous aluminium chloride and copper(I) chloride (Gattermann–Koch synthesis). Alkaline solutions of phenols react with chloroform (Reimer–Tiemann reaction) to give phenolic aldehydes via carbenes.

Most aliphatic and aromatic aldehydes give adducts with alkali hydrogen sulphites and form cyanhydrins, ALDOXIMES, semicarbazones and phenylhydrazones. Aldehydes, unlike ketones, restore the pink colour to Schiff's reagent, and reduce ammoniacal silver nitrate and Fehling's solution. They also form acetals with alcohols. Aliphatic aldehydes form aldehyde ammonias, and some undergo the aldol condensation. Aromatic aldehydes form unsaturated acids when heated with the sodium salt of a fatty acid and its anhydride (Perkin reaction). They form Schiff's bases with primary amines and react with phenols and dimethylaniline to give triphenylmethane derivatives. All aromatic aldehydes and some aliphatic aldehydes undergo the CANNIZZARO REACTION.

aldicarb Insecticide, acaricide, nematocide, carbamate type.

aldol (acetaldol, 3-hydroxybutanal, β-hydroxybutyraldehyde), $C_4H_8O_2$, $CH_3CH(OH)CH_2CHO$. A colourless oily liquid, b.p. 83°C/20 mm. Manufactured by treating ethanal with alkali carbonates, lime, borates and other alkaline condensing agents. Decomposes on heating to crotonaldehyde. Reduction with hydrogen under pressure gives 1,3-butylene glycol. Used in rubber processing, perfumery, oil flotation.

aldol condensation A reaction between two molecules of an aliphatic aldehyde to give a β-hydroxyaldehyde. The simplest case is that of ethanal which gives aldol, $CH_3CH(OH)CH_2CHO$. The reaction is carried out at a low temperature in presence of a catalyst, usually potassium cyanide, sodium ethanoate or dilute sodium hydroxide. Only those aldehydes containing the $-CH_2.CHO$ grouping can react in this way: other aldehydes undergo the Cannizzaro reaction. The β-hydroxyaldehydes readily lose water to give unsaturated aldehydes. Aliphatic ketones react in a similar manner with aliphatic ketones or aldehydes and with some aromatic aldehydes to give β-hydroxyketones, which also readily lose water, giving unsaturated ketones.

aldonic acid An acid derived from an aldose by oxidation of the aldehyde group to a carboxyl group, e.g. gluconic acid.

aldose A sugar containing a potential aldehyde (C(O)H) group. The group may be obscured by its inclusion in a ring system. Aldoses are called aldopentoses, aldohexoses, etc·, according to the number of carbon atoms they contain.

aldosterone, $C_{21}H_{28}O_5$. Steroid hormone secreted by the adrenal cortex.

aldoximes Compounds containing the group $-CH=NOH$ formed by the action of hydroxylamine on aldehydes. They are also formed by the oxidation of primary amines by permonosulphuric acid. The aliphatic aldoximes are colourless liquids or low-melting solids, some of which are soluble in water. Aromatic aldoximes are crystalline solids, e.g. benzaldoximes. All aldoximes are capable of existing in two stereoisomeric forms known as the *syn-* and the *anti-* modifications, but in practice only the *anti-* form is isolated from most aliphatic aldehydes. Aromatic aldehydes yield *syn*-aldoximes which may frequently be isomerized by HCl to the *anti-* aldoximes. The *syn*-aldoximes yield acetyl derivatives with ethanoyl chloride but the *anti-* forms are dehydrated to the corresponding nitriles RCN. Dilute mineral acids may regenerate the aldehyde and hydroxylamine hydrochloride.

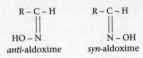

anti-aldoxime   *syn*-aldoxime

aldrin Toxic chlorocarbon insecticide no longer in use.

algin A SEAWEED COLLOID, principally alginic acid which is D-mannuronic acid. Used mainly as polymannuronic acid, the Na salt giving a very viscous solution with water. Used as a stabilizer for food products (ice cream, etc.), pharmaceuticals and as a dressing in the textile industry.

alginic acid See ALGIN.

alicyclic Cyclic carbon compounds that do not possess an aromatic ring. They are cyclic compounds with aliphatic characteristics. Examples include cyclopropane, cyclohexane.

aliphatic A compound having its carbon atoms in chains and not in closed rings. Originally used to describe the fats and fatty acids, which are typical of this structure.

alizarin (1,2-dihydroxyanthraquinone), $C_{14}H_8O_4$. Orange-red dyestuff. Dissolves in alkaline solutions to give purple-red solutions which are precipitated as lakes by heavy metal salts (spot tests for metals). Occurs naturally as a glycoside in madder but produced synthetically by fusing anthraquinone-2-sulphonic acid with NaOH and some $KClO_3$. Dyeing is carried out in boiling aqueous solution.

alkali A hydroxide of one of the ALKALI METALS or ALKALINE EARTH. Also used for such substances as CaO, $Ca(OH)_2$, Na_2CO_3, which give an alkaline solution (pH > 7) in water and react with acids. In aqueous solution the term alkali is virtually synonymous with the term base.

alkalides Compounds of the alkali metals, Na–Cs, containing M^- species.

alkali metals The metals Li, Na, K, Rb, Cs and Fr. All are very electropositive with an s^1 outer electron shell and readily form M^+ ions.

alkaline An aqueous solution with a pH greater than 7.

alkaline earth metals The elements Mg, Ca, Sr, Ba and Ra, which are very electropositive and form M^{2+} ions.

Historically an 'earth' was a non-metallic substance, nearly insoluble in water and unchanged on heating. The alkaline earth oxides, e.g. CaO, have an alkaline reaction in addition to being clearly 'earths'.

alkaloids Organic substances existing in combination with organic acids in great variety in many plants and to a limited extent in animals, and to which many drugs owe their medicinal properties. Many are very poisonous, and even in minute doses produce characteristic pharmacological activity. They vary considerably in their chemical properties and constitutions. All alkaloids contain nitrogen, are basic and combine with acids to form crystalline salts, which are usually water-soluble. The majority of alkaloids are only very sparingly soluble in water, but are usually readily soluble in organic solvents such as alcohol, chloroform and ether.

In the commercial extraction of alkaloids from the drugs in which they exist, the powdered drug, or an alcoholic extract of it, is treated with an alkali such as ammonia or lime to liberate the alkaloid.

Alkaloids are formed from amino-acids combining with methanoate or methionine, or ethanoate or mevalonate, or various higher transformation products of these including shikimic acid and terpenes. Many alkaloids are derivatives of pyridine, quinoline, isoquinoline, or pyrimidine. They are chiral. They include a number of important drugs, e.g. morphine, caffeine, nicotine, atropine, cocaine, hyoscyamine, quinine, strychnine and pilocarpine. Their part in the physiology of plants is not known.

A minor chemical use for many of the commoner alkaloids is the resolution of racemic compounds (often acids) into their optically active enantiomers.

alkanals Aldehydes. Compounds containing RC(O)H groups.

alkanes (paraffins) Aliphatic hydrocarbons of the general formula C_nH_{2n+2}. The first four members are gases, the higher members are liquids and those above $C_{16}H_{34}$ are waxy solids. They are insoluble in water but soluble in chloroform and benzene. They form the chief constituents of petroleum. They are remarkably resistant to chemical action and only chlorine and bromine will react with any ease. These give chloro- and bromo-substituted paraffins respectively. Alkane dehydrogenation produces valuable alkenes. They are formed by the reduction of alkenes or by treating alkyl iodides with sodium in ethereal solution.

alkanolamines (alkylolamines) Hydroxyamines produced by the reaction of an alkene oxide with aqueous ammonia at 50–60°C. They are hygroscopic solids of low m.p., and usually occur as rather viscous, colourless liquids of ammoniacal odour. They are used as accelerators in rubber manufacture, as catalysts in certain polymerization reactions, and as absorbents for acidic gases such as carbon dioxide and hydrogen sulphide. They form soaps with fatty acids; these are almost neutral in reaction and are used as detergents and emulsifying agents. See ETHANOLAMINES, ISOPROPANOLAMINES.

alkanolamine soaps See ALKANOLAMINES.

alkanols Alcohols. Compounds ROH where R is an alkyl group.

alkenes (olefins) Aliphatic hydrocarbons of the general formula C_nH_{2n} containing one double bond; they are isomeric with the cycloalkanes. In physical properties they closely resemble the alkanes; the lower members are gases, the intermediate liquids, while the higher members are waxy solids. They are insoluble in water but soluble in chloroform and benzene. They burn with a smoky, luminous flame. They occur in crude petroleum and in the gases from the cracking of petroleum. Alkenes are more reactive than alkanes, adding groups across the double bond. Reduced to alkanes by hydrogen in presence of catalysts. React with halogens to give dihalides and with HX to give haloalkanes. Dilute chlorine or bromine solutions in water react to give chloro- or bromohydrins. They dissolve in concentrated sulphuric acid to give alkylsulphonic acids and dialkyl sulphates, and are oxidized by potassium permanganate or lead tetra-acetate to glycols. They polymerize with various catalysts or under pressure. May be prepared by heating aliphatic alcohols with sulphuric acid or by passing the alcohol vapour over heated alumina. They are used as fuels and are the starting materials for the preparation of alcohols, alkene oxides containing the three-membered C_2O ring, glycols and other substances. React with water in the presence of mercury compounds to give adducts which can be reacted on to further alcohols. Alkene polymerization can occur with $TiCl_3$-based catalysts (Ziegler–Natta), heterogeneous catalysts ($Cr^{III}{}_nSiO_2$), (Phillips catalysts and homogeneous Group 4 metal derivatives. Widely used polymers.

alkoxides (alcoholates) Metal derivatives of alcohols (e.g. $KOCH_3$, $Al(OC(CH_3)_3)_3$). Alkali metal derivatives can be prepared directly but many derivatives (e.g. transition metals) must be prepared indirectly (e.g. $Ti(OR)_4$ from $TiCl_4$, ROH and a tertiary amine). Although many alkoxides are polymeric, many are volatile. Used as reagents (see ALUMINIUM ORGANIC DERIVATIVES) and as catalysts. Commercially $Al(OPr^i)_3$ is used for reductions, as a paint additive; NaOMe is used as a catalyst and reagent; $Ti(OPr^i)_4$ and other alkyl titanates are used in polymer technology and paints.

alkyd resins Products of the condensation of polyhydric alcohols and polybasic (or sometimes monobasic) acids, e.g. phthalic acid. Used widely in coatings, particularly in paints.

alkyl The residue left when a hydrogen atom is removed from an aliphatic hydrocarbon, e.g. $CH_3 \cdot$, $C_2H_5 \cdot$, etc. The free radicals are known and are intermediates in many reactions, e.g. the thermal degradation of alkanes, as in petrochemical processes.

alkylamides Metal and non-metal derivatives containing ENR_2 groups (e.g. $W(NMe_2)_6$). Prepared from the halide plus amine, silylamide, or lithium alkylamide, $LiNR_2$. Often volatile and less associated than alkoxides. Molecular species with low co-ordination numbers are known.

alkylation Formally the introduction of alkyl groups into hydrocarbon chains or aromatic rings.

The production of branched-chain hydrocarbons by reacting alkenes with iso-alkanes or aromatics, e.g. the production of iso-octane as a high anti-knock gasoline by reacting iso-butane and 1-butene. While alkylation can be effected thermally, milder reaction conditions are possible if sulphuric acid or hydrogen fluoride catalysts are used. For alkylation of aromatic substances see FRIEDEL–CRAFTS REACTION.

alkylidene complexes Transition metal species $M=CR'R$. Formed from metal derivatives and alkanes and of importance in alkene metathesis and related reactions and carbene complexes.

alkylolamines See ALKANOLAMINES.

alkylphenols Made by alkylation of phenol or its homologues. Used to form thermosetting resins (with methanal) and, particularly for long-chain derivatives, polyethers used as surfactants and detergents, with ethylene oxide. Hindered di-t-butyl derivatives are valuable anti-oxidants.

alkynes (acetylenes) Compounds containing $C\equiv C$ groups. Thus methylacetylene, $CH_3C\equiv CH$ is propyne; dimethylacetylene, $MeC\equiv CMe$ is but-2-yne, etc.

allantoin (glyoxyldiureide-5-ureidohydantoin), $C_4H_6N_4O_3$. M.p. 235–236°C. The end product of purine metabolism.

allene See 1,2-PROPADIENE, ALLENES.

allenes 1,2-Dienes having the

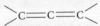

skeleton and general formula C_nH_{2n-2}. They are derivatives of 1,2-propadiene (allene) and are isomeric with alkynes. They undergo typical addition reactions of alkenes. They are not as stable as alkenes with conjugated, or isolated, double bonds. The action of base is usually sufficient to cause isomerization to an alkyne. Allenes are usually colourless liquids with garlic-like odours but the higher members are solids. They are prepared by removal of bromine and hydrogen bromide from 1,2,3-tribromopropane derivatives, usually by KOH followed by zinc and ethanol.

allethrin Insecticide, pyrethroid type.

allo- The prefix allo- before the name of an organic compound indicates that two compounds have the same empirical formula. With sterols and related substances the prefix allo- should indicate that rings A and B are in the *trans* position to each other, but allocholesterol is an exception.

allobarbitone (diallylbarbituric acid) See ALLOXAN.

allomone A compound or mixture acting between different species in the environment which adaptively favours the emitting species, e.g. a defensive secretion, a floral scent attracting pollinating insects. See also KAIROMONE AND PHEROMONE.

allophanic acid (carbamylcarbamic acid), $C_2H_4N_2O_3$, $NH_2CONHC(O)OH$. Unknown in the free state as it breaks down immediately to urea and CO_2. Salts are known and are prepared by treating ethyl allophanate with the appropriate hydroxide. The esters with alcohols and phenols are crystalline solids, sparingly soluble in water and alcohol. Formed by passing cyanic acid into alcohols or a solution of an alcohol or phenol in benzene. The amide of allophanic acid is biuret. Alcohols were sometimes isolated and identified by means of their allophanates.

allose See HEXOSE.

allosteric effects The control of molecular functions of molecules by interaction with other molecules. Allosteric effects control the output from biosynthetic pathways.

allotrope See ALLOTROPY.

allotropy The existence of more than one physical form of an element. The difference may be in bonding, e.g. graphite and diamond, ozone and dioxygen, or in crystal form, e.g. most metallic elements. The various forms are termed allotropes (see POLYMORPHISM).

alloxan (2,4,5,6-tetraoxohydropyrimidine) Prepared by oxidation of uric acid. Used in nutrition studies. Derivatives – allobarbitone and alloxantine – are well known but now little used.

alloxantin, $C_8H_6N_4O_8$. See ALLOXAN.

alloy An intimate association or compound of two or more metals or of a metal and one or more non-metallic elements, the alloy possessing in a marked degree all or most of those characteristics commonly described as metallic. Pure metals are too soft and possess insufficient strength for most engineering purposes. Alloying is one of the common methods of increasing useful properties in metals, e.g. strength and hardness.

alloy elements Elements added to give special properties to alloys, such as resistance to corrosion, heat, creep, etc.

allyl The trivial name given to the propenyl group $CH_2:CHCH_2-$.

allyl alcohol See PROPENOL.

allyl derivatives of metals Two series are known. The h^1 derivatives (with one C bonded to Mn), e.g. $CH_2=CHCH_2Mn(CO)_5$, and the h^3 (with three C bonded to Mn), e.g. $\ll-Mn(CO)_4$. The former are prepared using e.g., Grignard reagents; decarbonylation or frequently direct reaction gives the more important h^3 derivatives (also h^4-butadiene complex plus H^+). Allyl complexes are important catalysts and important as intermediates in many reactions involving metals. Allyl derivatives are the simplest members of a series of complexes of higher homologues.

allylene See PROPYNE.

allylic rearrangement The migration of a double bond in a three carbon system from carbon atoms one and two to carbon atoms two and three. The substituent simultaneously migrates from carbon atom three to carbon atom one.

$$\overset{1}{C} = \overset{2}{C} - \overset{3}{C} - X \rightleftharpoons X - \overset{1}{C} - \overset{2}{C} = \overset{3}{C}$$

The simplest example involves the hydrolysis of crotyl chloride (2-butenyl chloride) to give crotyl alcohol (2-buten-1-ol) and α-methylallyl alcohol (1-buten-3-ol).

allyl isothiocyanate See PROPENYL ISOTHIOCYANATE.

allyl polymers Polymers derived from monomers containing the $CH_2=CHCH_2-$, allyl, propenyl group. An important group of thermosetting polymers, polymerization is initiated thermally, photochemically, by radiation, or by catalysis. Triallyl cyanurate and isocyanurate are copolymerized with polyesters for use in reinforced laminates. Allyl ethers of carbohydrates and other polyols together with styrene and other vinyl monomers are used in automative parts and adhesives and as furniture finishes. Allyl esters, e.g. $RCOOCH_2CH=CH_2$, particularly diallyl phthalate and isophthalate, readily polymerize to cross-linked thermoset materials with good abrasion and chemical resistance. Allyl phosphates form polymers with good flame-resistant properties.

allylthiourea See PROPENYLTHIOUREA.

Alnico alloys Alloys containing Al, Ni, Co and Fe. Used for permanent magnets.

alpha decay Radioactive decay by loss of an alpha particle.

alpha helix The helical structure adopted by most proteins, etc.

alpha particle (α-ray) A He^{2+} ion, i.e. a helium nucleus with no outer electrons. α-Particles are emitted with high velocity ($c. 2 \times 10^9$ cm s^{-1}) by Ra and other radioactive substances. For the detection and estimation of α-particles, see COUNTERS.

α-Particles may be used, either with or without acceleration, as bombarding agents in nuclear disintegration reactions.

The energy, and thus the range, of α-particles, is characteristic of the source of emission.

alternating axis of symmetry An axis of symmetry which combines rotation with reflection through a plane at right angles to the axis.

altrose See HEXOSE.

alum, potash alum (potassium alum), $KAl(SO_4)_2 \cdot 12H_2O$. Crystallizes in large colourless octahedra. Crystals become white on exposure to the air owing to the production of a basic salt. Dehydrated at a dull red heat to a porous friable mass known as *burnt alum*. A mineral approximating in composition to potash alum occurs in nature. An allied mineral, ALUNITE, may be used as a source of alum. Other common starting materials for the manufacture of alum are aluminium schists and shales, which contain aluminium silicate and iron pyrites. The latter furnishes the sulphuric acid required, potassium sulphate or chloride being derived from an independent source. Alum is used in the dyeing industries for the production of mordants and pigments, in dressing leather, sizing paper, in waterproofing fabrics, in fireproofing, in gelatin hardening, porcelain, cements, baking powder and medicinally as a styptic and astringent, and as a coagulant for water purification. See ALUMS.

alumina See ALUMINIUM OXIDE. See also ALUMINATES.

alumina gel See ALUMINIUM HYDROXIDE.

aluminates A group of compounds formally containing the Al^{3+} ion in anionic hydroxy- or oxo-complexes. Species containing $[Al(OH)_4]^-$ and $[(HO)_3Al\text{-}\mu\text{-}O\text{-}Al(OH)_3]^{2-}$ and higher polymers (cationic hydroxy species are also known – see ALUMINIUM HYDROXIDE) are formed from Al^{3+}aq solutions at high pH. Solid salts are formed by crystallization or precipitation.

Aluminates formed by fusing together a metal oxide and Al_2O_3 are MIXED METAL OXIDES and contain Al^{3+} in tetrahedral or octahedral coordination. E.g. spinel, $MgAl_2O_4$. $NaAlO_2$ (Al_2O_3 plus Na oxalate, prepared commercially by dissolving bauxite in NaOH solution) is used as a mordant, in brick manufacture, for paper sizing and in water softening. Barium aluminate is an efficient but costly water-softening agent. Calcium aluminates are constituents of Portland cements. So-called β-alumina is an aluminate of ideal composition $Na_2O \cdot 11Al_2O_3$. It is an ionic conductor. There are variable amounts of Na^+ which can be replaced by other cations.

aluminium (Al) Metallic element, at. no.13, at.mass. 26.982, m.p. 660.32°C, b.p. 2519°C, density ρ 2698, crustal abundance 82 000

p.p.m., electronic configuration [Ne] $3s^23p^1$. Occurs naturally in many silicates (see ALUMINOSILICATES) including clays, and as the hydrated oxide (see BAUXITE). Normally present in ground water to only a limited extent but acid rain causes an increase in concentration. Manufactured by the electrolysis of Al_2O_3 dissolved in Na_3AlF_6 (cryolite) with added Li_2CO_3 using carbon electrodes; the product is 99% pure but contains some Fe and Si. Much aluminium is recovered and reused. Very pure Al is obtained by electrolysis of compounds such as $NaAlF(C_2H_5)_3$. Al reduces many oxides (GOLDSCHMIDT, THERMITE reactions). The pure metal is coated with an oxide film and almost unattacked by water but traces of impurity induce corrosion: corroded by salt or sea water; dissolved by acids. The metal, or more generally its alloys are used in applications requiring low density and high strength, e.g. in engineering, aircraft, kitchen utensils, overhead cables. Al film is used as a mirror coating and as a wrapping material, plate-like particles coated with iron oxide are used in metallic paints. Aluminium salts are important in water purification (see ALUM), as catalysts (see ALUMINIUM ORGANIC DERIVATIVES and OXIDE); the oxide is important in refractories, abrasives, cement, ceramics, antacids, anti-perspirants and gemstones (ruby – also lasers).

aluminium, organic derivatives, aluminium alkyls Formed Al plus R_2Hg, $AlCl_3$ plus RMgX, or on a commercial scale =Al–H plus C_nH_{2n} (alkene) to give =AlC_nH_{2n+1}. The latter reaction can be carried out by the direct reaction Al plus H_2 plus AlR_3 to give AlR_2H which can react with alkenes. All aluminium organic derivatives are easily oxidized; the lower alkyls are spontaneously inflammable. The lower alkyls are polymeric liquids with bridging alkyl groups with multicentre bonds. Complexes, AlR_3L, are readily formed. Aluminium alkyls are used for introduction of alkyl groups and industrially for polymerization of alkenes (Ziegler–Natta polymerization) and telomerization of alkenes to medium-chain derivatives for detergents and fats. These processes operate by insertion of an alkene into AlR bonds.

aluminium acetate See ALUMINIUM ETHANOATE.

aluminium alkoxides, $Al(OR)_3$. Formed by reaction between Al and the alcohol in the presence of $HgCl_2$ as catalyst or from $AlCl_3$ and RONa. The alkoxides are polymeric $[Al(OBu^t)_3]_2$, $[Al(OPr^i)_3]_4$, fairly low-melting solids readily hydrolysed by water. Used as specific reducing agents for C=O groups, for formation of esters from aldehydes by the Cannizzaro reaction (Al(OPr^i)$_3$), oxidation of secondary alcohols to ketones in the presence of excess Me$_2$CO (Oppenauer oxidation – Al(OBu^t)$_3$), and with co-catalysts, e.g. TiCl$_4$, as alkene polymerization catalysts. The most important aluminium alkoxides are Al(OBu^t)$_3$ and Al(OPr^i)$_3$.

aluminium alkyls See ALUMINIUM ORGANIC DERIVATIVES.

aluminium alloys Important materials for their combination of light weight, strength and generally good corrosion resistance.

aluminium borate, AlBO$_3$. Used as polymerization catalyst and in glass.

aluminium bromide, AlBr$_3$. Very similar to AlCl$_3$, formed from Al plus Br$_2$ or Br$_2$ over Al$_2$O$_3$ plus carbon. Forms hydrates.

aluminium t-butoxide (aluminium t-butylate), Al(OC(CH$_3$)$_3$)$_3$, Al(OBu^t)$_3$ See ALUMINIUM ALKOXIDES.

aluminium chemistry Al is an element of group 13. The chemistry at normal temperature is that of Al(III) ($E°$ Al^{3+}→Al in acid solution -1.66 volts) although gaseous Al(I) compounds (e.g. AlCl$_3$ plus Al gives AlCl, at high temperatures) are known. The normal co-ordination arrangements for solid Al(III) are octahedral (AlCl$_3$) and tetrahedral [AlCl$_4$]$^-$. ALUMINIUM ORGANIC DERIVATIVES are common and of industrial importance. ALUMINIUM OXIDE and ALUMINOSILICATES (in which Al may be 6-coordinate or 4-co-ordinate, replacing Si in [SiO$_4$]$^{4-}$ tetrahedra) are of great importance.

aluminium chloride *Aluminium trichloride* (AlCl$_3$) Is the stable chloride at room temperature. Colourless or yellowish solid ρ 2440, sublimes 180°C. Soluble in water and many organic solvents. Prepared from Al and Cl$_2$ or HCl or Cl$_2$ over Al$_2$O$_3$ plus carbon. Forms complexes AlCl$_3$L with e.g., H$_2$O, H$_2$S and amines, and [AlCl$_4$]$^-$ [Al$_2$Cl$_7$]$^-$ with chloride ions. Solid AlCl$_3$ contains octahedrally co-ordinated Al, dimeric, [Cl$_2$Al(μ-Cl)$_2$AlCl$_2$] in the gas phase. Forms volatile, chloride-bridged species with many halides. Used as a source of aluminium salts, and (also in the laboratory) in, e.g. Friedel–Crafts syntheses, as a disinfectant and wood preservative.

Aluminium monochloride (AlCl) Is formed at very high temperatures from AlCl$_3$ and Al. The reaction is reversed on cooling.

aluminium ethanoate (aluminium acetate) Al(O$_2$CCH$_3$)$_3$ Prepared from Al(OH)$_3$ and ethanoic acid or Pb(O$_2$CCH$_3$)$_2$ or Ba(O$_2$CCH$_3$)$_2$ and Al$_2$(SO$_4$)$_3$. Normally occurs in the solid state as a partially hydrolysed material. The ethanoate is soluble in water to give extensively hydrolysed solutions. Used in dyeing as a mordant, often with Al$_2$(SO$_4$)$_3$ ('red liquor') or Al(NO$_3$)$_3$. Also used in sizing paper, hardening cardboard, fireproofing paper and fabrics, and in pharmacy as an antiseptic and astringent.

aluminium ethoxide (aluminium ethylate), (CH$_3$CH$_2$O)$_3$Al. See ALUMINIUM ALKOXIDES.

aluminium fluoride, AlF$_3$. A colourless substance, ρ 3070, sublimation pt. 1257°C, prepared by the action of HF on Al(OH)$_3$ or by heating (NH$_4$)$_3$AlF$_6$ at 600°C. An inert material, relatively insoluble in most solvents, unattacked by acid and alkali. Hydrates (3 or 9 H$_2$O) formed from aqueous solution. The AlF$_6$$^{3-}$ ion is formed in conc. HF solution and salts, e.g. Na$_3$AlF$_6$ (CRYOLITE), are known which contain 6-co-ordinate Al. Condensed octahedrally co-ordinated AlF$_6$ units are found in, e.g. Tl$_2$AlF$_5$, NH$_4$AlF$_4$ and AlF$_3$. Used in ceramics and as a flux.

aluminium hydride, AlH$_3$. A white polymeric substance (H$_2$SO$_4$ plus LiAlH$_4$ in THF). A useful reducing agent (e.g. RCN gives RCH$_2$NH$_2$); reacts with alcohols to give alkoxy hydrides, (RO)$_n$AlH$_{3-n}$; forms complex hydrides including the important tetrahedral tetrahydroaluminate ion, [AlH$_4$]$^-$. LiAlH$_4$ (LiH plus AlCl$_3$ in ether) is a very important reducing agent (e.g. RCOOH gives RCH$_2$OH) and starting point for the formation of Al derivatives, e.g. RNH$_2$ gives Li[Al(NHR)$_4$]. Alkoxy derivatives, e.g. Li[AlH(OR)$_3$] and Na[AlH(OMe)$_2$(OEt)] are good and often specific reducing agents.

aluminium hydroxide A white or yellowish gelatinous mass precipitated from solutions of aluminium salts by aqueous ammonia and also existing as mineral forms. Formally hydrates of aluminium oxide (alumina). Al(OH)$_3$ has two forms, α, bayerite, and γ, gibbsite, hydrargillite; AlO(OH) also exists in two forms, α, diaspore, and γ, boehmite. On heating (see ALUMINIUM OXIDE), diaspore gives α-Al$_2$O$_3$ but other hydroxides and most Al salts initially give γ-Al$_2$O$_3$ and only give α-Al$_2$O$_3$ on strong heating. Gels of aluminium hydroxide partially dried and converted to γ-Al$_2$O$_3$ are valuable drying agents, catalysts and absorbents. Alumina gel is used medicinally as an antacid. Sols of aluminium hydroxide are formed by dialysis of aqueous solutions of aluminium salts. Aluminium hydroxide is amphoteric, reacting with acids and in alkalis. Used in glass, fireclay, paper.

aluminium isopropoxide (aluminium isopropylate), $Al(OPr^i)_3$. M.p. 125°C. See ALUMINIUM ALKOXIDES.

aluminium nitrate, $Al(NO_3)_3$, nH_2O (n commonly 9). See ALUMINIUM OXY-ACID SALTS.

aluminium nitride, AlN. Formed from bauxite, carbon and N_2. Used in steel and semiconductors.

aluminium oxide (alumina), Al_2O_3. The oxide of aluminium formed by heating aluminium hydroxides or most aluminium salts of oxyacids. Occurs naturally as the α-form, corundum, obtained by heating α-AlO(OH), m.p. 2045°C, b.p. 2980°C, almost insoluble in water. After strong ignition it becomes very inert. For structure see CORUNDUM. Moderate heat on other aluminium hydroxides produces γ-Al_2O_3 with a defect spinel structure (Al in both tetrahedral and octahedral co-ordination); γ-Al_2O_3 is a very important catalyst, because of its adsorptive power (activated alumina), and is used in the manufacture of artificial gemstones which are crystalline and contain traces of other metals (ruby and sapphire). α-Al_2O_3 is used as an abrasive (EMERY) and as an inert material. Used in ceramics and chromatography. β-alumina is an ALUMINATE. For hydrated aluminium oxide see ALUMINIUM HYDROXIDE. Can form useful fibres.

aluminium oxy-acid salts Aluminium oxyacid salts are soluble in water (except the phosphate) and often form hydrolysed solutions. They are generally heavily hydrated. Used as cements and fluxes. The most important salts are the ETHANOATE (acetate), the nitrate (used as a mordant), and the SULPHATE (used in paper sizing and water treatment. See ALUM). Carboxylates, e.g. oleate, palmitate, stearate are used in dyeing, tanning, waterproofing and in paint and varnish dryers. U.S. use sulphate 1999 1195 ktonnes. Carboxylates, e.g. oleate, palmitate, stearate are used in dyeing, tanning, waterproofing and in paint and varnish dryers.

aluminium silicates See ALUMINO-SILICATES.

aluminium sulphate, $Al_2(SO_4)_3$, nH_2O (n = 0, 6, 10, 16, 18, 27). See ALUMINIUM OXY-ACID SALTS. Occurs naturally as ALUNITE. See ALUMS.

Used in tanning, sizing, water purification, as mordant. U.S. production 1997 540 ktonnes.

aluminium *t*-butoxide (aluminium *t*-butylate), $Al(OC(CH_3)_3)_3$, $Al(OBu^t)_3$. See ALUMINIUM ALKOXIDES.

aluminoferric Impure $Al_2(SO_4)_3$ containing some Fe^{II} and Fe^{III}. Used in water and sewage purification.

aluminon (ammonium aurinetricarboxylate) Forms coloured lakes with many metals. Used for estimation of aluminium.

aluminosilicates Silicates containing aluminium. The most important aluminosilicates have Al in tetrahedral positions replacing Si. An M^+ cation (or equivalent, e.g. ½(M^{2+})) must be present to balance the charge for every Si replaced by Al. Silicates with framework structures based on SiO_2 must contain AlO_4 groups. Important aluminosilicates of this type are feldspars, micas, zeolites and some clay minerals. Al can also be present in octahedral co-ordination in silicates when the Al functions as a normal cation. Used in cements, glasses, paints.

aluminous cement A hydraulic cement closely resembling Portland cement, but composed chiefly of calcium aluminate.

aluminoxane See METHYLALUMINOXANE.

aluminum ALUMINIUM, U.S. usage.

alums A group of crystalline double sulphates with the general formula $M^I M^{III}(SO_4)_2,12H_2O$. The sulphate group may be replaced by SeO_4^{2-}, BeF_4^{2-}, or $ZnCl_4^{2-}$. The crystals are made up of $[M^I(H_2O)_6]^+$ and $[M^{III}(H_2O)_6]^{3+}$ cations and SO_4^{2-} anions and in solution the alums behave chemically as mixtures of the two constituent sulphates. $KAl(SO_4)_2,12H_2O$ is normally known as potash alum and the sodium and ammonium derivatives of aluminium sulphates are also commercial materials. Used in dyeing, tanning, electroplating.

alumstone See ALUNITE.

alundum An artificial form of corundum (α-Al_2O_3), made by fusing calcined bauxite in an electric furnace, and allowing the molten product to cool rather rapidly.

Alundum is used for highly refractory bricks (m.p. 2000–2100°C), crucibles, refractory cement and muffles; also for small laboratory apparatus used at high temperatures (combustion tubes, pyrometer tubes, etc.).

alunite (alumstone) Mineral $KAl_3(SO_4)_2$ $(OH)_6$; the potassium may be partially replaced by sodium. Used commercially as a source of potash alum and of potassium sulphate.

Am Americium.

amalgam Mercury–metal compounds which may be liquid or solid. See MERCURY AMALGAMS.

amalgamation The process of forming an AMALGAM.

amanitins A group of very toxic cyclic peptides found in the mushroom *Amanita phalloides*.

amantadine hydrochloride (amino-adamantane) The first anti-viral drug.

amaranth An important red azo dye used for wool, silk and colour photography.

amatol A mixture of NH_4NO_3 and TNT used as an explosive.

amberlite Group of ion-exchange resins.

ambident anions Anions which are capable of reaction at two or more different sites, e.g. NO_2 (from $AgNO_2$) can react with RX to give nitrites RONO and nitroalkanes, RNO_2.

ambident ligands Ligands which can use more than one co-ordinating site. E.g. NH_2CH_2COOH can co-ordinate through N or O.

amblygonite, $LiAl(F,OH)PO_4$. A lithium ore.

americium, Am. Actinide element, at.no. 95, at.mass ^{241}Am 241.06, m.p. 1176°C, b.p. 2011°C, density ρ 13 676, electronic configuration [Rn] $5f^7 7s^2$. The most stable isotopes are ^{243}Am (7650 years) and ^{241}Am (433 years) formed by multiple neutron irradiation of ^{239}Pu. Purified by ion exchange. Metallic Am is obtained by reduction of AmF_3 with barium at about 1200°C. The metal is electropositive, silvery in appearance and reacts with acids. Hazardous to health because of its radioactivity. ^{243}Am is a target for use in producing higher actinides. ^{241}Am is used as a radiation source for gauging thickness.

americium compounds Americium is a typical actinide showing similar chemistry to uranium. Aqueous Am(III) is oxidized only with difficulty, Am^{4+} disproportionates to Am^{3+} and AmO_2^+. Americium(II) halides (Cl, Br, I) are prepared from the metal and HgX_2 at 300–400°C, all of the trihalides are known and tan AmF_4 is formed from fluorine and AmO_2. AmO_2 is the stable oxide in air, Am_2O_3 is also known and mixed oxides, e.g. $Li_6Am^{VI}O_6$ and $Li_3Am^VO_4$ are made by solid-state reactions using oxides, peroxides and oxygen. Oxidation to aqueous AmO_2^{2+} is with peroxydisulphate. Am^{VI} highest state.

ametryne Herbicide, triazine type.

amides Organic compounds derived from ammonia by substitution of one or more of its hydrogen atoms by organic acid groups: $RCONH_2$ (primary); $(RCO)_2NH$ (secondary); $(RCO)_3N$ (tertiary). The amides are crystalline solids soluble in alcohol and ether; some are also soluble in water. Primary amides are prepared by the action of ammonia or amines on acid chlorides (see SCHOTTEN–BAUMANN REACTION), anhydrides, or esters. Some amides are prepared by distillation of the ammonium salt of the appropriate acid. Secondary and tertiary amides are formed by treating nitriles or primary amides with organic acids or their anhydrides. Primary amides react with nitrous acid to give carboxylic acids: in many cases, heating with mineral acids or alkalis has the same effect. Amides are weakly basic. Dehydrated by P_4O_{10} to nitriles. Compounds are formed with metals such as sodium, potassium, and mercury. HClO and HBrO react to give N-chloro- or bromoamides; these give amines when treated with alkali. Alkylated amides, particularly dimethylacetamide (DMA) and dimethylformamide (DMF) are good solvents for conducting replacement reactions involving ionic reagents (e.g. replacement of a chlorine atom by fluorine using KF).

Inorganic amides contain NH_2^-, which may be covalently bonded, and bridges. They are formed by the action of ammonia on metals or by the ammonolysis of nitrides. The heavy metal amides are prepared by metathetical reactions in liquid ammonia, e.g.

$$Cd(SCN)_2 + 2KNH_2 \rightarrow Cd(NH_2)_2 + 2KSCN.$$

The alkali amides are stable, crystalline salts; the heavy metal amides are often explosive.

The amides are the bases of the ammonia system.

amidines, $RC(=NH)NH_2$. Strong monoacid bases forming salts with strong acids. Imido derivatives of amides to which they are converted by acid hydrolysis. Formed by addition of R_2NH to ketenimines.

amido See AMINO.

amidol See AMINOPHENOLS.

amidone See METHADONE.

amido sulfuron Herbicide, sulphonylurea type.

amine oxides, R_3NO. Prepared by alkylation of hydroxylamines or oxidation of tertiary amines with ozone or hydrogen peroxide. Long-chain aliphatic amine oxides are used in detergents and shampoos (alkyl-*bis*(2-hydroxyethyl)amine oxides and alkyldimethylamine oxides). Some amine oxides are physiologically active.

amines Organic compounds derived from ammonia by replacement of one or more of its hydrogen atoms by hydrocarbon groups. Replacement of one, two and three hydrogen atoms results in primary, secondary and tertiary amines respectively. Aliphatic, aromatic and heterocyclic and mixed aliphatic–aromatic amines are known. Primary amines are obtained by the action of ammonia on alcohols or halogeno derivatives of hydrocarbons. The ammonia may be replaced by primary and secondary amines to give secondary and tertiary amines. Some aldehydes react with ammonia and hydrogen under pressure in presence of nickel catalysts to give amines. Primary amines, RCH_2NH_2, are also obtained by reduction of nitro-compounds, nitriles, ketoximes and amides, or by treating amides with NaOBr. Secondary amines, $(RCH_2)_2NH$, are obtained by reducing Schiff's bases. A general preparation of amines is from potassium phthalimide and a halogen compound, with subsequent hydrolysis of the product. The aliphatic amines are strong bases; the lower members are soluble in water and are stronger bases than ammonia; they have ammoniacal or fishy odours. The higher members are odourless solids. Aromatic amines generally prepared by reduction of nitro-compounds are not such strong bases. They are not generally soluble in water, and have characteristic odours. All types form crystalline salts with acids; complexes with, e.g. $SnCl_4$ and $HgCl_2$ and weak complexes with other metal halides. With nitrous acid, primary aliphatic amines give alcohols (Hinsberg test). Aromatic amines give diazonium salts in the cold, but these decompose to phenols if the reaction is carried out at higher temperatures. Secondary amines of all types give nitrosamines. Schiff's bases, or azomethines, are formed by the action of aldehydes on primary or secondary amines. All types except tertiary amines, $(RCH_2)_3N$, give ethanoyl (acetyl) compounds with ethanoyl chloride or anhydride. These are usually crystalline, sparingly soluble solids. These amines also yield sulphonamides with aromatic sulphonyl chlorides. Primary amines react with chloroform and potassium hydroxide to give isocyanides or carbylamines. Primary and secondary amines yield derivatives of thiocarbamic acid with alcoholic carbon disulphide, while aromatic amines give substituted thioureas. The aromatic amines are of great importance as dyestuff intermediates. Amines are used in synthesis, tertiary amines are used to prepare amine oxide surfactants and quaternary biocides.

amino- A compound containing an amino group ($-NH_2$) joined directly to a carbon atom. Formerly the prefix amido was also used but this is now usually restricted to compounds containing the amide group ($-CO \cdot NH_2$).

aminoacetal, $C_6H_{15}NO_2$, $H_2N \cdot CH_2 \cdot CH(OC_2H_5)_2$. Colourless oily liquid with an ammoniacal odour, b.p. 172–174°C. Prepared by the action of ammonia on chloroacetal, $ClCH_2 \cdot CH(OEt)_2$. Hydrochloric acid converts it to aminoacetaldehyde. Condensation with aromatic aldehydes gives isoquinoline derivatives.

aminoacetic acid See GLYCINE.

amino-acids A large class of organic compounds containing both the carboxyl, C(O)OH, and the amino, NH_2, group, e.g. glycine, $H_2N \cdot CH_2 \cdot COOH$. Their chief importance lies in the fact that proteins are built up entirely of amino-acid groupings by condensation between the NH_2 and COOH groups, the amino-acids in these proteins being generally α-amino-acids, with the amino group attached to the α-carbon atom as the carboxyl group and with the same L-configuration of asymmetric groups about the α-carbon atom. The following amino-acids are found in very variable proportions as constituents of most proteins (polymeric amino-acids):

Ala	alanine
Arg	arginine
Asp(NH₂) or Asn	asparagine
Asp	aspartic acid
CySH	cysteine
CyS	cystine (half)
Glu	glutamic acid
Glu(NH₂) or Gln	glutamine
Gly	glycine
His	histidine
HyLys	hydroxylysine
HyPro	hydroxyproline

Ileu	isoleucine
Leu	leucine
Lys	lysine
Met	methionine
Phe	phenylalanine
Pro	proline
Ser	serine
Thr	threonine
Try	tryptophan
Tyr	tyrosine
Val	valine

The listed amino-acids have the basic structure $H_2NCHRC(O)OH$. L-pyrrolysin and seleno-cysteine derivatives are also known. Other amino-acids, e.g. $H_2NCHRCHR'C(O)OH$ form different polymers.

In representations of peptides the shortened form is used and the amino-acid first listed has the free amino group.

Certain other amino-acids including β-amino-acids occur in a few proteins, and others, not necessarily α- or L-amino-acids, are found naturally in the free state or as constituents of peptides.

The amino-acids are colourless, crystalline substances which melt with decomposition. They are mostly soluble in water and insoluble in alcohol. Have been found in carbon-containing meteorites.

As constituents of proteins the amino-acids are important constituents of the food of animals. Certain amino-acids can be made in the body from ammonia and non-nitrogenous sources; others can be made from other amino-acids, e.g. tyrosine from phenylalanine and cystine from methionine, but many are essential ingredients of the diet. The list of essential amino-acids depends partly on the species. See also PEPTIDES and PROTEINS.

amino-acid analysis Peptides are hydrolysed by acids (e.g. 6MHCl) which destroy some amino-acids (e.g. tryptophan). Alkaline solution also hydrolyses peptides and destroys other amino-acids (e.g. arginine) and causes racemization. Enzymic hydrolysis is slow and incomplete. Use of specific methods allows degradation and hence determination of amino-acid sequence. Hydrolysed amino-acids are analysed by, e.g. ion-exchange chromatography, paper chromatography, paper electrophoresis, n.m.r. The whole process of sequencing and the reverse process of synthesis can be automated. Sequences of amino-acids can also be determined by specific reactions, e.g. reaction of the C-terminal with hydrazine; reduction of the terminal-COOH to CH_2OH with $LiBH_4$ or $LiAlH_4$;

attack of carboxypeptidase on a free α-COOH. See, e.g. 1-FLUORO-2,4-DINITROBENZENE.

1-aminoanthraquinone (α-aminoan-thraquinone), $C_{14}H_9NO_2$. Red prisms, m.p. 252°C. A typical aromatic amine. Prepared by the prolonged action of concentrated ammonia solution at a high temperature upon anthraquinone-1-sulphonic acid in the presence of $BaCl_2$ and by reduction of the corresponding nitro compound or by amination of chloroanthraquinone.

Used in the dispersed form as a dye for acetate silk, though it has no affinity for other fibres. It is also used as a starting point for alkyl- or acyl-aminoanthraquinones which are used either as vat dyes or, after sulphonation, as acid wool dyes.

2-aminoanthraquinone (β-aminoan-thraquinone), $C_{14}H_9NO_2$. Orange-red, m.p. 302°C. Typical aromatic amine. Prepared by the prolonged action of ammonia 200°C on anthraquinone-2-sulphonic acid in the presence of MnO_2 or a barium salt. It can also be prepared by amination of 2-chloroanthraquinone.

Used as an intermediate in the preparation of flavanthrene and indanthrene.

aminoazobenzene, $Ph \cdot N{:}N \cdot (4\text{-}H_2NC_6H_4)$, $C_{12}H_{11}N_3$. Brownish-yellow needles, m.p. 127°C. The hydrochloride forms steel-blue needles.

Prepared by an intramolecular transformation of diazoaminobenzene in the presence of aniline hydrochloride, or in one stage by diazotizing a solution of aniline and aniline hydrochloride with a deficiency of nitrous acid.

It is used as a first component in the preparation of azo-dyes.

aminoazo-dyes See AZO-DYES.

2-aminobenzoic acid See ANTHRANILIC ACID.

4-aminobenzoic acid (PAB), $C_7H_7NO_2$. Yellowish-red crystals, m.p. 186–187°C. 4-Aminobenzoic acid is an essential metabolite for certain bacterial cells and is regarded as a member of the vitamin B group. The ethyl ester is used as a local anaesthetic.

aminocaproic acid (6-aminohexanoic acid), $H_2N(CH_2)_5C(O)OH$, $C_6H_{13}NO_2$. Prepared from ε-benzoylaminocapronitrile or from 1-hydroxycyclohexylhydroperoxide. M.p. 205°C.

1-(2-aminoethyl) piperazine, $C_6H_{15}N_3$. Curing agent for epoxy resins, dyestuff intermediate, corrosion inhibitor. Hemostatic agent.

aminoethyl alcohol See ETHANOLAMINES.

6-amino hexanoic acid See AMINO CAPROIC ACID.

5-aminolaevulinic acid (5-amino-4-oxopentanoic acid), $C_5H_9NO_3$, $H_2N \cdot CH_2CO \cdot CH_2CH_2C(O)OH$. The base unit in the biosynthesis of PORPHYRINS and so furnishing all the carbon and nitrogen atoms in the heme of hemoglobin, myoglobin, cytochromes, catalase and peroxidase as well as the dihydroporphyrin ring of chlorophyll. It is also incorporated into the corrin ring of vitamin B_{12}. 5-Aminolaevulinic acid is formed from succinyl coenzyme A and glycine. Condensation of two molecules yields PORPHOBILINOGEN (a pyrrole derivative) and four molecules of this condense to give the porphyrin skeleton.

2-amino-2-methyl-1-propanol, $C_4H_{11}NO$. Prepared by reduction of nitro compound and used in surface agents, coating polishes, hair sprays.

aminomethylation See MANNICH REACTION.

aminonaphthols, $C_{10}H_9NO$. Usually prepared by reduction of the nitronaphthols. The sulphonated aminonaphthols are valuable dyestuffs intermediates.

aminophenols Prepared by reduction of nitrophenols. Important compounds include 2-aminophenol, 4-aminophenol (rodinol), 4-methylamino phenol hemisulphate (metol), N-4-hydroxyphenylglycine (glycin) and 2,4-diaminophenol dihydrochloride (amidol). Used extensively in dyestuff synthesis and as photographic developers.

amino resins and plastics An important group of organic nitrogen-rich polymers formed by condensation of amino derivatives (urea, melamine, thiourea, dicyanodiamide, acrylamide, aniline) with methanal.

4-aminosalicylic acid (PAS), $C_7H_7NO_3$. M.p. 150°C. Prepared by heating 3-aminophenol with ammonium carbonate or potassium bicarbonate in solution under pressure. Has been used to treat tuberculosis infections.

aminotoluene See TOLUIDINES.

amitraz Insecticide, acaricide, amidine derivative.

amitrole Herbicide, triazole derivative.

ammines Complexes containing ammonia co-ordinated to the acceptor atom, e.g. $[Co(NH_3)_6]Cl_3$, hexamminecobalt(III) chloride.

ammonia, NH_3. The most important of the nitrogen hydrides. Colourless gas with a characteristic smell. Readily liquefied by cooling or compression, b.p. −33.5°C, m.p. −77.7°C. NH_3 is very soluble in water, the saturated solution at room temperature has ρ 0.88 ('880 ammonia'); hydrates $NH_3, 2H_2O$, NH_3, H_2O and $2NH_3, H_2O$ are known. NH_3 occurs naturally in some gases from hot springs. It is manufactured (U.S. production 1999 1.91 megatonnes) by the Haber process (N_2 and H_2 at 500°C and 300 atm in presence of a catalyst (increasingly Rh based), the H_2 is generally from the steam-reforming process) and used to produce HNO_3 (25%), urea (15%), NH_4NO_3 (15%), Nylon (10%) the ultimate products being fertilizers (80%), explosives (5%) and plastics, foams and films (10%). Prepared in the laboratory from an ammonium salt and a strong alkali or by hydrolysis of a nitride. Just burns in oxygen; $NH_3:O_2$ mixture gives NO over catalysts, oxidized to N_2 over heated metal oxides. NH_3 co-ordinates to many elements and forms AMMINES; reacts with Na or K to give amides MNH_2. Liquid ammonia is a good ionizing solvent (self-dissociation $2NH_3 \rightleftharpoons NH_4^+ + NH_2^-$ pK = 30; NH_4^+ salts acids; NH_2^- salts bases). It is generally a better solvent for organic compounds than water but a poorer solvent for inorganic compounds (unless NH_3 complexes formed). Many metals dissolve to give blue reducing solutions containing solvated electrons. Detected by smell, blackening of $Hg_2(NO_3)_2$ paper and Nessler's reagent.

ammonia–soda process See SODIUM CARBONATE.

ammonium The unipositive cation $[NH_4]^+$ which forms a group of, generally, soluble salts similar to the alkali metal salts (formed NH_3 and acid). Most of the salts have strong hydrogen bonding between cation and anion. Substituted ammonium cations, e.g. $[Me_4N]^+$, $[Me_3NH]^+$ are formed from organic amines and acids.

ammonium bicarbonate See AMMONIUM HYDROGEN CARBONATE.

ammonium bromide, NH_4Br. A colourless crystalline solid. Turns yellow in air. Readily soluble in water, sublimes on heating. Prepared HBr and NH_3 in the gaseous state or solution.

ammonium carbonate, $(NH_4)_2CO_3$. Pure $(NH_4)_2CO_3$ is formed from the commercial salt (see below) and NH_3. It is very soluble in water and decomposes to NH_3, CO_2 and H_2O on heating. Commercial ammonium carbonate is NH_4HCO_3. NH_2COONH_4, obtained from

$CaCO_3$ and NH_4Cl or $(NH_4)_2SO_4$ by heating and sublimation. It smells of NH_3, decomposes in moist air to NH_4HCO_3. Used in baking powders, fire extinguishers, smelling salts (sal volatile), dyeing, in wool treatment and as an expectorant.

ammonium chloride (sal ammoniac), NH_4Cl. A white crystalline solid. Readily soluble in water, sublimes on heating. Obtained from NH_3 and HCl or commercially from $CaCl_2$ solution (from Solvay process) with NH_3 and CO_2 or by recrystallization from solutions from the process, or by crystallization from $(NH_4)_2SO_4$ and NaCl (Na_2SO_4 crystallizes preferentially). Used in dry cells, as a mordant, and as a soldering and galvanizing flux. Not, generally, suitable as a fertilizer.

ammonium chromate, $(NH_4)_2CrO_4$. A golden yellow solid (NH_3 plus CrO_3 plus H_2O). Soluble in water, decomposes on heating to the dichromate, NH_3, and H_2O. Used in textile production.

ammonium dichromate, $(NH_4)_2Cr_2O_7$. A red crystalline solid (NH_3 plus excess CrO_3 plus H_2O). Very soluble in water. Decomposes spectacularly on heating to N_2, H_2O and Cr_2O_3. Used in lithography and as a catalyst.

ammonium fluoride, NH_4F. A white deliquescent solid (NH_3 and gaseous HF). The solutions are extensively hydrolysed and the solid is hydrolysed on standing in moist air. Strongly hydrogen bonded in the solid, sublimes on heating. Used as a disinfectant and wood preservative, and for glass etching, soluble in ethanol. Adds excess HF to give the hydrogen difluoride NH_4HF_2.

ammonium fluoroborate, NH_4BF_4. Colourless crystals (HBF_4 plus aqueous NH_3) used as an electrolyte.

ammonium hexachlorostannate (IV) (chlorostannate, pink salt), $(NH_4)_2SnCl_6$. Precipitated from concentrated H_2SnCl_6 solution with NH_4Cl; used as a mordant.

ammonium hydrogen carbonate (ammonium bicarbonate), NH_4HCO_3. A white crystalline solid (NH_3 plus CO_2 plus water vapour; CO_2 and steam on ammonium carbonate solution; decomposition of commercial ammonium carbonate). A constituent of commercial ammonium carbonate and formed by decay of nitrogenous materials and present in guano. Used in baking powders. More stable as a solid than the carbonate and hence frequently used medicinally in place of the carbonate.

ammonium hydrogen fluoride (spirit of hartshorn), NH_4HF_2. See AMMONIUM FLUORIDE.

ammonium hydroxide An aqueous solution of NH_3 (see AMMONIA), postulated to contain the weak base NH_4OH but mainly contains ammonia hydrates.

ammonium iodide, NH_4I. Colourless salt (aqueous KI plus $(NH_4)_2SO_4$ plus ethanol, K_2SO_4 precipitated). Very soluble in water, sublimes, with partial oxidation in air. Readily forms polyiodides. Used in the photographic industry.

ammonium iron sulphate See IRON AMMONIUM SULPHATE and IRON ALUMS.

ammonium molybdates A range of salts formed from NH_3 plus molybdic acid. The commercial salt is $(NH_4)_6Mo_7O_{24},4H_2O$. Used in phosphate determinations.

ammonium nitrate, NH_4NO_3. Colourless crystals, sublimes (NH_3 plus HNO_3). On gentle heating decomposes to N_2O. U.S. production 1999 8.2 megatonnes. Used mainly as a fertilizer but also, with a detonator, as an explosive or constituent of explosives (e.g. amatol, slurry explosives). Has been known to detonate spontaneously.

ammonium nitrite, NH_4NO_2. Yellowish solid formed by part oxidation of NH_3 or from $Ba(NO_2)_2$ and $(NH_4)_2SO_4$. Decomposes on heating to N_2 and H_2O.

ammonium perchlorate, NH_4ClO_4. White crystalline solid (NH_3 plus $HClO_4$). Used as an oxidizer in solid propellants.

ammonium persulphate (ammonium peroxodisulphate), $(NH_4)_2S_2O_8$. Colourless solid, soluble in water, prepared by electrolysis of cooled saturated $(NH_4)_2SO_4$ in dil. H_2SO_4. A powerful oxidizing agent. Used as a bleach for flour. An ammoniacal solution of ammonium persulphate is used to strip brass plating from iron.

ammonium phosphates, $NH_4H_2PO_4$, $(NH_4)_2HPO_4$ and $(NH_4)_3PO_4$. Formed from NH_3 and phosphoric acid and used as important fertilizers also in fireproofing and as fluxes. The two hydrogen phosphates are used medicinally as diuretics. All of the ammonium phosphates lose NH_3 on heating. Other ammonium phosphates include $NH_4NaHPO_4\cdot4H_2O$, microcosmic salt.

ammonium phosphomolybdate, $(NH_4)_3 Mo_{12}PO_{40}.xH_2O$. The bright yellow precipitate formed from a phosphate, ammonium molybdate and HNO_3 in solution. Used as a test for phosphates.

ammonium sulphamate Herbicide. Fire retardant, used in production of N_2O and in cements.

ammonium sulphate, $(NH_4)_2SO_4$. Colourless crystalline solid produced from NH_3 and H_2SO_4 or NH_3 and CO_2 and a solution of $CaSO_4$ (gypsum or anhydrite) in water ($CaCO_3$ precipitates). Used as a fertilizer, although tending to be replaced by fertilizers with higher nitrogen contents. Decomposes on heating, firstly to NH_4HSO_4 and then to N_2, NH_3, SO_2 and H_2O. U.S. production 1999 2.9 megatonnes.

ammonium sulphides, $(NH_4)_2S$ (NH_3 and H_2S at $-18°C$) and NH_4HS (NH_3 and H_2S in CH_3OOCCH_3 solution) are relatively unstable. Yellow polysulphides are formed by oxidizing the sulphides, by passing H_2S through a strong aqueous solution of ammonia, or from solutions of $[NH_4]_2S$ and sulphur. Used to dissolve metal sulphides, particularly in qualitative analysis ('yellow ammonium sulphide').

ammonium thiocyanate, NH_4NCS. Colourless solid, soluble in water, prepared from CS_2 and NH_3 in EtOH or HCN and yellow ammonium sulphide. Forms isomeric thiourea, $SC(NH_2)_2$, on heating. Used as a source of NCS⁻ ions, in explosives and matches, and in photography.

amorphous Substances which do not have an extended ordered arrangement of the constituent molecules or ions and are thus without crystalline form. Amorphous substances show no cleavage planes and give an irregular or conchoidal fracture. Many apparently amorphous substances have a microcrystalline structure. Glasses and supercooled liquids are amorphous.

AMP See ADENOSINE MONOPHOSPHATE.

ampere (A) The measure of electric current, one coulomb per second. For electrons 1 A is 6×10^{18} electrons per second.

amperometric titration A method of analysis in which current flowing through a cell is plotted against added titrant. There are sharp breaks in the curves at the end-points.

amphetamine (β-aminopropylbenzene), $C_9H_{13}N$, $PhCH_2CH(CH_3)NH_2$. Colourless liquid, b.p. 200°C (decomp.). Prepared by the reduction of phenylacetone oxime. Has addictive properties.

amphiboles Silicates containing double chains of stoichiometry $[Si_4O_{11}]_n^{6n-}$, e.g. tremolite $(OH)_2Ca_2Mg_5(Si_4O_{11})_2$. See ASBESTOS.

amphipathic Molecules with both polar and non-polar ends which dissolve in polar and non-polar solvents.

amphiphiles Hollow self-assembled aggregates formed by, e.g. surfactants.

amphiprotic Solvents (e.g. H_2O, MeOH) which both produce protons (protogenic, acidic) and react with protons (protophilic, basic).

ampholyte An amphoteric electrolyte, a substance which can behave as both an acid and a base.

amphoteric electrolyte See AMPHOLYTE.

amphoteric oxide or hydroxide An oxide or hydroxide which can have acid or basic functions, i.e. can combine with either an acid or a base to form a salt. ZnO reacts with acids to form zinc salts (basic function) and with alkalis to form zincates (e.g. $Na_2Zn(OH)_4$) (acidic function). $Al(OH)_3$ is an amphoteric hydroxide.

amu, atomic mass units Units of relative atomic mass in terms of standard ^{12}C. Quoted in daltons.

amygdalin, $C_{20}H_{27}NO_{11}$. Cyanophoric glycoside m.p. 215°C found in the kernels of most fruits belonging to the *Rosaceae*, particularly in bitter almonds. A gentiobiose mandelonitrile, $C_6H_5CH(CN)\cdot O\cdot C_6H_{10}O_4\cdot O\cdot C_6H_{11}O_5$. It is widely used as a flavouring material in the form of essence of bitter almonds.

amyl (iso 3-methylbutyl) The group $(CH_3)_2CHCH_2CH_2-$.

amyl (normal, *n*-amyl, pentyl) The group $CH_3\cdot CH_2\cdot CH_2\cdot CH_2\cdot CH_2-$.

amyl (secondary, 1-methylbutyl) The group $CH_3CH_2CH_2CH(CH_3)-$.

amyl (tertiary, 1,1-dimethylpropyl) The group $CH_3CH_2C(CH_3)_2-$.

amyl acetate, amyl ethanoate Colourless, volatile liquid with a strong pear-like odour, b.p. 138.5°C. Manufactured by heating 1-pentanol with potassium ethanoate and sulphuric acid or with ethyl ethanoate in the presence of a little sulphuric acid. Commercial amyl acetate is usually a mixture of isoamyl acetate, $(CH_3)_2CHCH_2CH_2O(O)CCH_3$, with varying amounts of secondary amyl acetate, $CH_3CH_2CH_2CHMeO(O)CH_3$. It is used as a solvent for cellulose acetate lacquers and paints: an alcoholic solution is used for flavouring purposes as 'essence of Jargonelle pears'.

amyl alcohols See PENTANOLS.

amylases (diastase) Enzymes that hydrolyse polysaccharides. They are widely distributed in animal and plant tissues. Used in starch processing, baking, animal feed, sewage treatment.

amylene hydrate, $CH_3CH_2C(CH_3)_2OH$. See PENTANOLS.

amyl ether (11´-oxy*bis*pentane, pentyl ether) Colourless liquid with a pleasant odour, b.p. 172.5–173°C. Manufactured by heating amyl alcohol with sulphuric acid. Commercial amyl ether contains isoamyl ether, $[(CH_3)_2CHCH_2CH_2]_2O$, with varying amounts of secondary amyl ether, $(CH_3CH_2CH_2CHMe)_2O$. Used as a solvent.

t-amylmethyl ether, $MeCH_2C(Me)_2OMe$. Gasoline fuel additive.

amylobarbitone (5-ethyl-5-iso-amyl-barbituric acid), $C_{11}H_{18}N_2O_3$. A white crystalline powder, m.p. 155–158°C. An intermediate-acting barbiturate, used as a hypnotic and sedative.

amylocaine hydrochloride Local anaesthetic.

amylopectin See STARCH.

amylose See STARCH.

4-t-amylphenol, $C_{11}H_{16}O$, $CH_3CH_2CMe_2C_6H_4OH$. M.p. 95°C, b.p. 265°C. Made by alkylation of phenol. Used to form thermosetting resins (with methanal) and surfactants (with ethylene oxide).

anabolic agents Drugs which promote storage of protein and stimulate tissue metabolism, often used in convalescence and illegally to promote athletic performance. Androgenic steroid hormones, e.g. TESTOSTERONE, have marked anabolic properties, and a number of synthetic analogues are in therapeutic use.

anabolism See METABOLISM.

anaerobic metabolism Natural process without participation of inhaled oxygen. The energy source is glycolysis – conversion of glucose to lactic acid. Can occur at volcanic vents and in the gut with formation of methane.

anaesthetics Compounds used to abolish the perception of pain and other stimuli. General anaesthetics produce total anaesthesia, e.g. N_2O, ethylvinyl ether, halothane, cyclopropane. Local anaesthetics act only at the site of application by freezing, e.g. ethyl chloride or by affecting the nerves, e.g. cocaine.

analgesics Compounds which reduce pain perception, e.g. morphine.

analysis The identification of a substance or of the components in a mixture of substances is termed qualitative analysis. Analysis of inorganic substances is carried out by various physical methods; e.g. metals are analysed by spark spectra, mixtures may be analysed by ultraviolet or infra-red spectroscopy, or by identification by X-ray analysis. Chemical analysis is carried out using various organic reagents which are more or less specific for a given metal (see, e.g. ALUMINON, CUPFERRON, MAGNESON).

Organic qualitative analysis is carried out by identification of the type of compound by physical methods or by specific reactions followed by complete identification by preparation of derivatives or by physical methods. The physical methods most used are ultra-violet, infra-red, or nuclear magnetic resonance spectroscopy, mass spectrometry, or molecular weight or relative molecular mass determination. Chromatography is much used.

Quantitative analysis is the estimation of the amount of element or group present in a mixture or compound. This is done by various methods: in volumetric analysis a titration, in gravimetric analysis a precipitation followed by a weighing, in colorimetric analysis the estimation of a coloured species. Other quantitative methods include infra-red spectroscopy, estimation of the opalescence of a precipitate (turbidimetry, nephelometry and fluorimetry), estimation of optical rotation, electrolytic decomposition, potentiometric, conductometric and amperometric titrations, and polarography. Organic quantitative analysis is generally carried out by physical methods or by conversion to known derivatives which can be estimated by weighing or by titration.

anaplerotic sequences Ancillary routes to the catabolic cycles in organisms, which operate to maintain levels of intermediates in these cycles despite tapping off for biosynthesis.

anatase, octahedrite, TiO_2. A steel blue or yellow form of TiO_2.

anation Replacement of an uncharged ligand, e.g. H_2O, in a complex by an anion, e.g. Cl^-.

anchimeric assistance Assistance in expulsion of a leaving group by a migratory group in the reacting molecule.

ancymidol Plant growth regulator, pyrimidinyl derivative.

andalusite, Al_2SiO_5. A form of aluminium silicate. Used as a refractory material in the manufacture of electrical porcelain (spark plugs), firebricks and for metallurgical furnace linings.

Andrews titration An important titration for the estimation of reducing agents. The reducing agent is dissolved in concentrated hydrochloric acid and titrated with aqueous potassium iodate(V) solution. A drop of tetrachloromethane is added to the solution and the end point is indicated by the disappearance of the iodine colour from this layer. The reducing agent is oxidized and the iodate reduced to ICl, i.e. a 4-electron change.

androsterone (androstenedione, 3α-hydroxy-5α-androstan-17-one), $C_{19}H_{30}O_2$. M.p. 183°C. Transformed to testosterone in the body by enzymes.

anemometer An instrument for measurement of gas velocity.

anethole (trans-1-methoxy-4-prop-1-enylbenzene), $C_{10}H_{12}O$. White leaflets, with a strong smell and sweet taste, m.p. 22°C, b.p. 235°C. The chief constituent of anise and fennel oils. It can also be prepared from anisole (methoxybenzene). It is widely used for flavouring pharmaceuticals and dentifrices, and in perfumery.

aneurine See THIAMINE.

angelic acid (Z-2-methyl-2-butenoic acid), $C_5H_8O_2$. M.p. 45–46°C, b.p. 185°C. It occurs free in the roots of *Angelica archangelica*. When heated alone or with acids or alkalis it is converted to tiglic acid, (E)-isomer.

angiotensins Peptides influencing blood pressure. See DICTIONARY OF BIOLOGY.

angle strain Reactivity (often in cyclic compounds) arising from deviations from the ideal angles for orbital (electronic) factors, e.g. the CCC angle of cyclopropane is 60° compared with the normal tetrahedral angle of 109° for carbon.

Ångstrom unit (Å) A non-S.I. unit of length equivalent to 10^{-8} cm, (10^{-10} m). Used to describe molecular and lattice dimensions and wavelengths of visible light. 1 nm=10 Å.

anharmonicity Harmonic vibrations are characterized by a restoring force which is proportional to the displacement with a parabolic potential energy curve. While at very small amplitudes the vibrations of the atomic nuclei in molecules are harmonic, at larger amplitudes, especially those approaching the dissociation point of the bond, the vibrations are markedly anharmonic as shown by band spectra. The deviations from harmonic behaviour are expressed by the anharmonicity constant.

anhydride A substance formed by elimination of one or more molecules of water from one or more molecules of an acid, or (less frequently) a base. Thus phthalic acid on heating gives phthalic anhydride. Similarly, sulphur trioxide, SO_3, is the anhydride of sulphuric acid, H_2SO_4.

anhydrite Mineral $CaSO_4$ generally associated with some gypsum, $CaSO_4 . 2H_2O$, and rock-salt, NaCl. Converted by water to gypsum with a 60 % increase in volume. Used in the manufacture of sulphuric acid and ammonium sulphate.

anhydro A prefix for an organic compound indicating that one or more molecules of hydrogen have been removed.

anhydrone See MAGNESIUM PERCHLORATE.

anilides A name applied to the acyl derivatives of aniline, of which acetanilide is the commonest example.

aniline See PHENYLAMINE.

anilofos Herbicide, organophosphate.

anils (N-phenylimides) See SCHIFF'S BASES.

anion An atom which has gained one or more electrons, or a negatively charged group of atoms. Anions are present in the solid state, e.g. Cl^- in NaCl, and in solutions and melts. In electrolysis anions travel to the anode where they are discharged.

anionic polymerization Polymerization induced by anions (bases).

anisaldehyde (4-methoxybenzaldehyde), $C_8H_8O_2$. A colourless liquid, b.p. 248°C. It occurs in aniseed and is used in synthetic perfumery under the name 'aubepine' or artificial hawthorn.

o-anisidine (2-methoxyaniline),

C_7H_9NO. A colourless oil, m.p. 2.5°C, b.p. 225°C. Prepared by the reduction of o-nitroanisole with iron and hydrochloric acid. It is used in making azo dyestuffs. Probably carcinogenic.

anisole, methoxybenzene,

C_7H_8O, $PhOCH_3$. Colourless liquid, b.p. 155°C. Prepared by the action of dimethyl sulphate on a solution of phenol in excess of alkali. Used in perfumery.

anisotropic A substance is anisotropic when any of its physical properties (e.g. thermal or electrical conductivity, refractive index) are different in different principal directions. Crystals, other than those belonging to the cubic system, are anisotropic. Certain substances (e.g. 4-azoxyanisole) melt to form a cloudy anisotropic liquid ('liquid crystals'), which again 'melts' at a higher temperature to a normal (isotropic) liquid.

anisyl The 4-methoxybenzyl group.

annealing The heating and controlled cooling of a substance (generally a metal or glass) to relieve stresses. In general ordered structures are produced by annealing.

annulation The formation of ring compounds in organic synthesis.

annulenes Simple conjugated cyclic polyalkenes: a prefix [n] is added to indicate the number of carbon atoms in the cycle. Thus [18]-annulene is a reasonably stable brown-red crystalline solid which has cis and trans double bonds. Some annulenes are aromatic in character.

anode The electrode which carries the positive charge in an electrochemical cell. In electrolysis the anode is the electrode at which the negative ions are discharged.

anodic oxidation Process of oxidation in solution by means of an electric current.

anodizing Process of anodic oxidation of metals with the object of giving protection against corrosion. Commonly used for aluminium sheet which is made the anode of a sulphuric-chromate(VI) bath, and becomes coated with a layer of porous oxide. The pores are sealed by dipping in hot water; the layer may be coloured by adsorbed dyes.

anomers The specific term used to describe carbohydrate stereoisomers differing only in configuration at the hemiacetal carbon atom(*). These stereoisomers are referred to as the α and β anomers and for the pyranose forms of D-glucose, for example, both forms are known.

α-D-Glucopyranose   β-D-Glucopyranose

anorthic system The TRICLINIC crystal system.

anoxic Without oxygen.

anserine (β-alanylmethylhistidine),

$C_{10}H_{16}N_4O_3$. M.p. 238–239°C. A dipeptide constituent of the muscles of various species.

antabuse A drug used in the treatment of alcoholism. See DISULPHIRAM.

antacids Materials used to reduce the amount of acid in the stomach. There can be direct neutralization ($NaHCO_3$, MgO); buffering (sodium citrate, magnesium silicate); absorption of $H^+(Al(OH)_3)$; ion exchange (zeolites).

antagonists A material that reduces tendency to reaction. See DICTIONARY OF BIOLOGY.

anthelmintic Drugs for treatment of parasitic worms.

anthocyanidines The coloured aglycones (sugar-free compounds) obtained by hydrolysis of the glycoside-containing anthocyanines. See ANTHOCYANINES.

anthocyanines Dyestuffs causing colour (red, blue) in plants. Glycosides which yield anthocyanidines on hydrolysis. Involved naturally in photosynthesis. Pelargonidin, cyanidin and delphinidin are important examples.

anthracene, $C_{14}H_{10}$. Colourless plates with slight blue fluorescence. M.p. 217°C, b.p. 354–355°C. Polyaromatic hydrocarbon.

Present in coal tar (½ %), from which it is separated by fractional distillation. Can be chlorinated, sulphonated and nitrated by a mixture of nitric acid, ethanoic acid and ethanoic anhydride. Treatment with sulphuric and nitric acids gives anthraquinone derivatives.

Of little use commercially except as a route to anthraquinone. For this purpose it is oxidized with acid potassium dichromate solution, or better, by a catalytic air oxidation at 180–280°C, using vanadates or other metal oxide catalysts.

anthracite A bright, hard coal with a carbon content greater than 93 %.

anthrahydroquinone See OXANTHROL.

anthralin See DITHRANOL.

anthranilic acid (2-aminobenzoic acid), $C_7H_7NO_2$. Colourless leaflets, m.p. 145°C. It is both acidic and basic.

It is prepared by the action of sodium hydroxide and sodium hypochlorite (NaOCl) on phthalimide (Hofmann reaction). When heated with soda lime it gives aniline.

It is an important dyestuffs intermediate. It condenses with chloroethanoic acid to give phenylglycine-*o*-carboxylic acid for the synthesis of indigo. It can be diazotized and used as a first component in azo-dyes; it condenses also with chloroanthraquinones to give intermediates for anthraquinone dyes.

Decomposition of its internal salt (a zwitterion) provides a convenient source of benzyne for organic synthesis.

anthranol (9-hydroxyanthracene), $C_{14}H_{10}O$. Golden brown needles which sinter at 120°C.

Isomeric with anthrone, and behaves in its reactions as a typical hydroxylic compound. The equilibrium mixture between the two compounds consists mainly of the keto form; anthranol is largely converted into anthrone on heating.

It is used in the preparation of benzanthrone by heating with glycerol and sulphuric acid (Skraup's reaction). Dyestuff intermediate.

anthraquinone (9,10-dioxo-anthracene), $C_{14}H_8O_2$. Colourless needles or prisms which can be purified by sublimation or steam distillation. M.p. 285°C. It is manufactured by condensing benzene with phthalic anhydride in the presence of aluminium chloride to give 2-benzoyl benzoic acid; this is then converted to anthraquinone by heating with concentrated sulphuric acid at 120–150°C. Another method is the condensation of 1,4-naphthoquinone with butadiene, the former being obtained by the vapour-phase oxidation of naphthalene.

Anthraquinone is a very stable compound which more closely resembles a diketone than a quinone: it yields a monoxide with difficulty.

Reduction of anthraquinone gives oxanthrol, anthrahydroquinone, dianthryl, ANTHRONE, and finally anthracene.

It is the parent compound of the anthraquinone vat dyes, in which the dyeing is carried out by immersion in the reduced vat solution followed by air oxidation to the original insoluble compound.

Anthraquinone can be directly brominated and chlorinated to the tetrahalo(1,4,5,8) stage. Bird repellant.

anthraquinone dyes A large group of dyestuffs based on anthraquinone. There are two main types; the alizarin and indanthrene dyestuffs. They are prepared via the sulphonic acids and/or amino derivatives.

anthraquinone sulphonic acids Prepared by sulphonation of anthraquinone. The anthraquinone sulphonic acids are very important, as nearly all anthraquinone derivatives are obtained from them.

Anthraquinone-1-sulphonic acid. Colourless leaflets, m.p. 214°C. Used in the preparation of 1-aminoanthraquinone.

Anthraquinone-2-sulphonic acid. The sodium salt, commonly called 'silver salt', is used for the preparation of alizarin and 2-aminoanthraquinone and to prepare FIESER'S SOLUTION. Aminoanthraquinone derivatives are the basis of many dyestuffs.

anthrone (9(10-H) anthracenone), $C_{14}H_{10}O$. Colourless needles, m.p. 154°C. Prepared by reduction of anthraquinone with tin and hydrochloric acid in glacial ethanoic acid or by prolonged treatment with sodium dithionite. On heating it is partly converted into the enol anthranol.

anti-aromatic Conjugate, cyclic systems which are thermodynamically less stable than the corresponding acyclic analogues; CYCLO-BUTADIENE and CYCLO-OCTATETRAENE are anti-aromatic, since conjugation of the respective acyclic butadiene and octatetraene results in diminished stability.

antibacterials Used to reduce bacterial concentrations. Often polymeric surfactants.

antibiotic An organic substance which is produced by micro-organisms or has a molecular structure similar to naturally occurring substances and is capable at low concentration of inhibiting the growth of, or destroying, another micro-organism. Antibiotics have been isolated from numerous sources, but principally from bacteria (e.g. BACITRACIN, POLYMYXIN, GRAM-ICIDIN), actinomycetes (e.g. TETRACYCLINE, STREPTOMYCIN, chloramphenicol) and fungi (e.g. PENICILLINS, CEPHALOSPORINS). Bacterial antibiotics are mostly polypeptides.

Many antibiotics are unsuitable for therapeutic use, frequently because of their general toxicity or as a result of other drawbacks such as instability, inadequate solubility or malabsorption.

antibodies Protein molecules present in the serum which are formed in the body in response to the presence of foreign substances called ANTIGENS. For each antigen there is a specific antibody. The effect is to agglutinate or precipitate the antigens and, if cellular, prepare them for the action of COMPLEMENT. The antibody–antigen mechanism is the basis of the immune response. Monoclonal antibodies are engineered versions of antibodies. See DICTIONARY OF BIOLOGY.

antibonding orbitals See MOLECULAR ORBITALS.

anticodon The specific sequence of bases that allows t-RNA to bond to a specific site of RNA. See NUCLEIC ACIDS and DICTIONARY OF BIOLOGY.

anticonformation The form of a molecule in which particular substituents are opposed. Compare ANTI-ISOMERS.

$anti$-$CH_2I.CH_2I$

antiferromagnetism Positive magnetic behaviour in which the magnetic susceptibility drops with decreasing temperature. There is generally a characteristic temperature, the Néel point, above which magnetic behaviour is normal.

anti-fluorite structure Adopted by compounds M_2X with the cations occupying the F^- positions of the FLUORITE structure and the anions occupying the Ca^{2+} positions. Adopted by, e.g. K_2S and K_2O.

anti-foaming agents Substances which prevent the formation of foams. They are strongly adsorbed by the liquid medium but they do not have the electrical or mechanical properties required to form a foam. Examples of anti-foaming agents are polyamides which are used in boiler-feed water and octanol used in electroplating baths and in paper making. Low concentrations of silicones also find quite general use. For lubricating oils polysiloxanes are widely used.

antifreeze additives This can be any additive which will lower the freezing point of the liquid concerned, but it is generally applied to materials used to prevent freezing in the coolant system of internal combustion engines. ETHYLENE GLYCOL 1,2-dihydroxyethane is the principal component of anti-freeze liquids although alcohols may also be employed.

antigens Macromolecular proteins which, when injected into the blood of an animal, stimulate the production of ANTIBODIES. Antigens may be bacteria, viruses (living or dead), foodstuffs, pollens, proteins, some polysaccharides, or nucleic acids.

antihistamines Substances which antagonize the physiological and pharmacological effects of HISTAMINE.

anti-isomer An isomer with particular, named, substituents on opposite sides of a molecule, e.g. for planar $[(MeSCH_2CH_2SMe)PtCl_2]$.

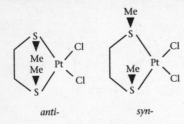

anti-   *syn-*

See ISOMERISM.

anti-isomorphism Having the same crystal lattice but with the relative positions of anions and cations interchanged, e.g. ThO_2 and Li_2O (with CaF_2 structure). See ANTI-FLUORITE.

anti-knock additives Some substances, e.g. ethanol added to gasoline, are able to inhibit the precombustion oxidation chain known to produce KNOCKING. A mixture of lead alkyls with dibromoethane and dichloroethane (known as ethyl fluid) has been added to gasoline, the lead being removed in the exhaust as volatile lead compounds but its use is being phased out. Other means, e.g. changing the fuel mix, are increasingly being used to reduce knocking.

anti-knock value See OCTANE NUMBER.

antimatter See ANTIPARTICLES, HYDROGEN.

antimonates Antimonates(III), M^ISbO_2, are formed from alkalis and Sb_2O_3. The free acid is not known. Antimonates(V) are formed from hydrated Sb_2O_5 and contain the $[Sb(OH)_6]^-$ ion. Mixed antimony oxides M^ISbO_3, $M^{III}SbO_4$ and $M^{II}_2Sb_2O_7$ are known and contain SbO_6 octahedra.

antimony, Sb. Non-metal, metalloid element, at.no. 51, at.mass 121.76, m.p. 630.63°C, b.p. 1587°C, relative density ρ 6691, crustal abundance 0.2 p.p.m., electronic configuration $[Kr]4d^{10}5s^25p^3$. Main source is stibnite, Sb_2S_3. Extracted by reduction with Fe or C. The most stable form of the element has a metallic appearance, white or bluish white in colour. This form has a layer structure with three near neighbours and three at a greater distance. Sb will burn in air but is unattacked by water or dilute acids; it is attacked by oxidizing acids and by halogens. Sb is widely used in alloys (e.g. typemetal); its use in semi-conductors is important. Antimony compounds are used in flameproofing, paints, ceramics, enamels, glass, pottery, rubber technology and dyestuffs. Antimony derivatives have considerable toxicity.

antimony bromides Antimony tribromide $SbBr_3$ (Br_2 plus Sb) m.p. 97°C, b.p. 280°C, similar to $SbCl_3$. Higher bromides are not known although complexes, M_2SbBr_6 (contains Sb(III) and Sb(V)) and $MSbBr_6$ are known.

antimony chemistry Antimony is a metalloid of Group 15. Its chemistry is dominated by the +5 and +3 oxidation states which are generally covalent although some pseudocationic chemistry of the $(SbO)^+$ and Sb^{3+} species is known (cations are co-ordinated by oxygen). Antimonides are formed with metals and polymeric cations and anions, e.g. Sb_4^{2-}, Sb_3^{3-}, are known. Alleged Sb(IV) compounds contain Sb(III) and Sb(V).

antimony chlorides
Antimony trichloride ($SbCl_3$) Butter of antimony. Colourless solid m.p. 73°C, b.p. 283°C (Sb compound in conc. HCl followed by distillation). Hydrolysed by water to basic chlorides. Forms many complexes including $[SbCl_4]^-$ (polymeric), $[SbCl_5]^{2-}$ (square pyramidal), and $[SbCl_6]^{3-}$, and complexes with, e.g. N-bonded ligands.
Antimony pentachloride ($SbCl_5$). M.p. 7°C, b.p. 79°C (Sb or $SbCl_3$ plus Cl_2). Readily hydrolysed by water, forms complexes, e.g. $[SbCl_6]^-$. Mixed antimony(III)/(V) complexes occur as salts M_2SbCl_6. $SbCl_5$ is used extensively as a chlorinating agent.

antimony fluorides
Antimony trifluoride (SbF_3) M.p. 292°C, sublimes 319°C (Sb_2O_3 in HF solution). Forms complexes $[SbF_4]^-$ containing polymeric species. Widely used as a fairly mild fluorinating agent to produce CF bonds.

Antimony pentafluoride (SbF_5) M.p. 7°C, b.p. 150°C is an associated liquid (Sb plus F_2 or $SbCl_5$ plus HF). Forms many complexes and complex ions including $[SbF_6]^-$, $[Sb_2F_{11}]^-$ and is a very powerful fluoride ion acceptor. Greatly enhances the dissociation of, e.g. HF and HSO_3F by forming anionic species (magic acid, super acid). Used as a fluorinating agent (sometimes in the form of its graphite intercalation compound).

antimony hydride (stibine), SbH_3. An unstable colourless gas, m.p. –88°C, b.p. –17°C (Sb compound plus Zn plus HCl or HCl on Mg–Sb alloy), decomposes easily to Sb.

antimony iodide, SbI_3. Red or yellow solid m.p. 170°C, similar in properties to $SbCl_3$.

antimony organic derivatives (stibines), SbR_3. A large range of such derivatives are prepared from, e.g. Grignard reagents and $SbCl_3$. Good donors, form tetra-organostibonium ions $[R_4Sb]^+$. SbR_5 derivatives are also known.

antimony oxides
Antimony trioxide (Sb_2O_3) The white pure oxide (Sb plus steam; Sb_2S_3 plus O_2) turns yellow on heating, m.p. 656°C, subl. 1550°C. Solid Sb_2O_3 contains Sb_4O_6 molecules, it forms SbO_2 on heating in air. Sb_2O_3 is reduced to Sb by H_2 at red heat, is insoluble in water, dissolves in acids (to give solutions. of, e.g. $SbCl_3$, $Sb_2(SO_4)_3$) and in alkalis (to give antimonates(III)). Used as an opacifier in paints and as a flame retardant in plastics.
Antimony dioxide (SbO_2) (Heat on Sb_2O_3 in air) is a yellowish compound containing Sb(III) and Sb(V).
Antimony pentoxide (Sb_2O_5) Is yellow (Sb plus conc. HNO_3). Decomposes to SbO_2 on heating. Gives antimonates(V) with alkalis. Used as a flame retardant.

antimony potassium tartrate (tartar emetic), $KSbO(C_4H_4O_6)$,$1.5H_2O$. Used medicinally as an emetic and against parasites.

antimony sulphates The normal salt, $Sb_2(SO_4)_3$ (Sb_2O_3 plus conc. H_2SO_4) is readily available. It is hydrolysed by water. Complex species are present in H_2SO_4 solutions (e.g. $[SbO]^+$, $[Sb(OH)_2]^+$, $[Sb_3O_9]^{3-}$) of antimony (III) and (V).

antimony sulphides The trisulphide, Sb_2S_3, occurs in a black (stibnite) and a red (H_2S plus an Sb(III) compound in HCl) form. It is transparent to i.r. radiation. Forms thioantimonates(III), e.g. $[SbS_3]^{3-}$ with excess of sulphide ion. Antimony(V) sulphide is very unstable, readily loses sulphur on heating. Thioantimonates(V), M_3SbS_4 (Sb_2S_3 plus S plus alkali) give Sb_2S_5 with acid.

antimonyl derivatives Compounds apparently containing the $[SbO]^+$ grouping although the actual species present are much more complex (see ANTIMONY SULPHATES).

antioxidants Substances which slow the oxidation rate in autoxidizable substances. Present in fruits and vegetables; protectors against degenerative disease. They are added for the protection of foods, particularly fats; for stopping the ageing or deterioration of rubber and many plastics; for inhibiting gum formation in cracked petroleum; and for preserving many other products. Many antioxidants are highly substituted phenols, aromatic amines, or sulphur compounds. Certain sequestering agents act as antioxidants by inactivating metals that may catalyse oxidation. Many organic raw materials contain natural antioxidants. The u.v. rays of sunlight are responsible for activating rubbers etc. to oxidation by air. Hence u.v. absorbers are added if exposed to sunlight.

antiparticle An elementary particle with charge properties opposite to those normally found, e.g. a positive electron.

antiperspirants Materials that inhibit sweating or the effects of sweating. Generally aluminium compounds.

antipyretics Drugs used to reduce the temperature in the case of fever, e.g. salicylates.

antiseptics Materials which stop the growth of micro-organisms. Generally applied to living tissue. I_2 is an example.

antistatic agents Compounds added to materials (particularly fibres, plastics) to reduce charge accumulation with resultant dust and dirt build up. Examples are metals, water and hygroscopic salts.

antitoxins ANTIBODIES formed (usually over a period of months) in blood to which bacterial toxins have been introduced.

apatite, $Ca_5(PO_4)_3F$. A phosphate mineral containing fluoride and phosphate anions. Used for the production of phosphorus and its compounds including superphosphates. Hydroxyapatite, $Ca_5(OH)(PO_4)_3$ is isostructural with apatite. These compounds are of importance as components of teeth and bone. The mineral frequently contains useful quantities of uranium. Synthetic apatites have importance as ceramics and in dentistry.

apo Prefix indicating a derived compound; thus apomorphine is derived from morphine.

apoenzyme The peptide portion of an enzyme without the cofactor. See DICTIONARY OF BIOLOGY.

apomorphine, $C_{17}H_4NO_2$. White crystals. Its hydrochloride is prepared by heating morphine with hydrochloric acid under pressure. It is a potent emetic.

aprotic solvent A solvent which does not contain protons which could be released as H^+.

aqua regia A mixture of conc. HNO_3 with three to four times its volume of conc. HCl, so called because it will dissolve 'noble' metals, such as Pt and Au. Its strong oxidizing action is due to nitrosyl chloride (NOCl) and chlorine produced by interaction of the two acids. Tends to form chloro complexes.

aquation The process of replacement of other ligands by water, e.g.

$$[Co(NH_3)_4(H_2O)(NO_3)]^{2+} \rightarrow$$
$$[(Co(NH_3)_4(H_2O)_2]^{3+}.$$

The process of complexing by water molecules.

aqueous ammonia A solution of ammonia in water. See AMMONIUM HYDROXIDE.

aquo ions (aqua ions) Hydrated metal ions in aqueous solution; also applied to hydrated ions in complexes, e.g. $[Co(H_2O)_6]^{2+}$. Aquo ions can lose protons and act as acids, e.g.

$$[Fe(H_2O)_6]^{3+} + H_2O \rightarrow [Fe(H_2O)_5(OH)]^{2+} + H_3O^+$$

Ar Argon. Also used as an abbreviation for an aromatic group.

L-arabinose, $C_5H_{10}O_5$. The pentose sugar of the hemicelluloses, gums and mucilages, and of some glycosides, in particular the vicianosides. It crystallizes in the β-pyranose form, m.p. 160°C. Used as a culture medium.

D-Arabinose is found in the glycoside barbaloin and in the polysaccharides of the tubercle bacillus.

arachidic acid (n-eicosanoic acid), $C_{20}H_{40}O_2$, $CH_3 \cdot [CH_2]_{18} \cdot C(O)OH$. M.p. 75°C. A fatty acid occurring as glycerides in peanut and other vegetable oils.

arachidonic acid, $C_{20}H_{32}O_2$. *Cis,cis,cis,cis-*5,8,11,14-eicosatetraenoic acid. $CH_3[CH_2]_4[CH=CHCH_2]_4CH_2CH_2C(O)OH$. Arachidonic acid is an 'ESSENTIAL FATTY ACID' in animal metabolism and is a biological precursor of PROSTAGLANDINS E_2 and F_{2a}.

aragonite, $CaCO_3$. A form of calcium carbonate.

aramides Polymeric species containing [NH-C(O)-] groups for heavy duty uses. Examples – Kevlar and Nomex.

arbutin M.p. 195°C. A typical phenolic glycoside, hydroquinone-β-D-glucopyranoside, $C_6H_{11}O_5 \cdot O \cdot C_6H_4OH$. A colourless, crystalline,

bitter substance obtained from the leaves of most *Ericaceae*. Antiseptic and diuretic.

arc spectrum The emission spectrum obtained when a substance has been excited by application of an electric arc, rather than an electric spark which gives a *spark spectrum*.

Arctons See FREONS.

π-arene complexes Complexes in which an aromatic system is bonded to a metal through its π-electrons. Generally only applied to complexes of uncharged aromatic systems, e.g. $[(C_6H_6)_2Cr]$ but formally applied to any complex of an aromatic system, e.g. $[(C_5H_5)_2Fe]$ as a complex of $(C_5H_5)^-$.

argentates See SILVER OXIDES.

argentite, Ag_2S. An important silver ore.

argentous and argentic Compounds of silver(I) and silver(II).

arginine (D-2-amino-5-guanidino pentanoic acid), $C_6H_{14}N_4O_2$, $H_2NC(=NH)NH[CH_2]_3CHNH_2C(O)OH$. M.p. 207°C. Arginine is one of the essential amino-acids. It is obtained in particularly high concentration from proteins belonging to the protamine and histone classes. It plays an important role in the production of urea as an excretory product and in nitric oxide cell signalling.

argol Crude potassium hydrogen tartrate obtained from wine.

argon, Ar. Noble gas. Gaseous element, at. no. 18, at.mass 39.948, m.p. –189.35°C, b.p. –185.85°C, density ρ 1.633 g 1^{-1}, atmospheric abundance 9300 p.p.m., crustal abundance 1.2 p.p.m., electronic configuration [Ne] $3s^23p^6$. The most abundant of the noble gases (0.93 % of dry air). Separated by fractionation from liquid air and used in light bulbs, as an inert shield in arc welding and for metal and semi-conductor production (Ti, Si). Argon has outer electronic configuration $3s^23p^6$ and although it is soluble in water and forms some clathrates forms no true compounds except excited-state derivatives. Forms $Cr(CO)_5Ar$ in a matrix.

Arndt–Eistert synthesis A procedure for converting a carboxylic acid to its next higher homologue, or to a derivative of a homologous acid, e.g. ester or amide.

$$RC(O)OH \longrightarrow RC(O)Cl$$

$$\xrightarrow[\text{Et}_2\text{O}]{\text{CH}_2\text{N}_2} RC(O)CH_2N_2$$

$$\xrightarrow{\text{Ag}^+} RCH_2C(O)OH$$

aromatic The property (aromaticity) which allows benzene to undergo electrophilic substitution reactions, to account for its stability over open-chain conjugated derivatives. Compounds containing a benzene skeleton are said to be aromatic, whilst non-benzenoid compounds such as many heterocyclic compounds which also undergo electrophilic substitution are termed aromatic using this criterion only. Other definitions of aromaticity include a molecular orbital description which connects aromatic stability with the number of π electrons $(4n + 2)$ associated with a planar cyclic conjugated system of double and single bonds (Hückel's Rule), e.g. azulene 10 π electrons, $n =$ 2: tropylium cation 6 π electrons, $n =$ I. Aromaticity has also been defined in terms of effect on the n.m.r. spectrum of the compound under consideration. Metallo-clusters, e.g. Al_4^{2-}, Hg_4^{6-}, $GeAl_3^{-}$, have aromatic character.

aromatic hydrocarbons Compounds containing a cyclic delocalized π-electron system.

Arosolvan process A process for the extraction of benzene and toluene from a mixture of aromatic and saturated hydrocarbons using a mixture of water and *N*-methylpyrrolidone. The process is used when naphtha is cracked to produce alkenes. To prevent extraction of alkenes these are saturated by hydrogenation prior to extraction.

Arrhenius equation The variation in the rate of a chemical reaction with temperature can be represented quantitatively by the Arrhenius equation

$$k = A \exp(-E_a/RT)$$

where k is the rate constant, A and E_a are constants, termed the pre-exponential factor and the ACTIVATION ENERGY, of the particular reaction being considered (R is the gas constant). The pre-exponential factor A is a collision frequency factor, which is comparatively insensitive to temperature. $A = pZ$, where Z is the collision frequency and p is a 'steric factor', which is a measure of the efficiency of the collision for effective reaction. The Arrhenius equation applies to processes such as diffusion, electrolytic conduction, viscous flow, etc., as well as chemical reactions.

arsenates Arsenates(III) are formed from As_2O_3 and base but the free acid is unknown. Arsenates(III) containing the pyramidal $(AsO_3)^{3-}$ and more complex species are known. Sodium and copper arsenates(III) (emerald green, Scheele's green) are used as insecticides. Sodium arsenate(III) is used as a standard reducing solution. Arsenates(V) are formed as the free acid $H_3AsO_4,\frac{1}{2}H_2O$ (As_2O_3 plus HNO_3), condensed arsenates are less stable to hydrolysis than condensed phosphates. $Na_2HAsO_4,7H_2O$ is used commercially in calico-printing.

arsenic, As. Metalloid element, at.no. 33, at.mass 74.922, m.p. 817°C (36 atm), subl. pt. 603°C, density ρ 5750, crustal abundance 1.5 p.p.m., electronic configuration [Ar] $3d^{10}4s^24p^3$. The principal sources of As are as impurities in sulphide ores, orpiment As_2S_3, realgar As_4S_4, arsenolite As_2O_3, mispickel FeAsS, nickel glance NiAsS, cobalt bloom $Co_3(AsO_4)_2,8H_2O$. The ores are converted to As_2O_3 (white arsenic) by roasting in air and As_2O_3 is used as a source of As compounds, As_2O_3 is reduced to As with hydrogen or carbon. Normally bright and metallic in appearance with a double-layer structure (3 contacts in plane, 3 next near neighbours). Yellow As (As_4 units) is formed by rapid condensation of the vapour; the vapour contains As_4 and As_2 units. Arsenic burns in air to As_2O_3 and reacts with halogens, conc. oxidizing acids and fused alkalis. Arsenic and its compounds are used in insecticides (see ARSENATES), as a dopant in solid-state devices (transistors, lasers), and in alloys where it has a hardening effect. Arsenic compounds in quantity are poisonous. Ground water may contain sufficient arsenic to give health problems.

arsenic bromide, $AsBr_3$. See ARSENIC HALIDES.

arsenic chemistry Arsenic is an element of Group 15. It behaves as a metalloid and there is little evidence of cation chemistry although complex anions, e.g. $[AsO_4]^{3-}$, $[AsF_6]^{-}$ are well known. The stable oxidation states are +5 (co-ordination numbers 4, 5 and 6) and +3 (co-ordination numbers 3, 4). Arsenides are formed with many metals.

arsenic chlorides, $AsCl_3$ and $AsCl_5$. See ARSENIC HALIDES.

arsenic fluorides, AsF_3 and AsF_5. See ARSENIC HALIDES.

arsenic halides Arsenic forms AsF_3 (As_2O_3 plus HF or CaF_2 and H_2SO_4) used as a mild fluorinating agent (b.p. 57°C), AsF_5 (As plus F_2) (b.p. -53°C), both fluorides are readily hydrolysed by water and can act as acceptors with, e.g. F^- to give $[AsF_6]^-$ (hexafluoroarsenates). $AsCl_3$ is the only stable chloride (As or As_2O_3 plus Cl_2), b.p. 130°C. It forms many complexes in which it can act as acceptor and donor and is used as a starting point for organoarsenic compounds. A very unstable $AsCl_5$ ($AsCl_3$ plus Cl_2) is formed at low temperatures. $AsBr_3$ and AsI_3 are the only bromides and iodides and are very similar to $AsCl_3$.

arsenic hydride (arsine), AsH_3. M.p. -116.3°C, b.p. -55°C. An unstable poisonous gas (metal arsenide plus acid or arsenic compounds plus Zn with dil. acid; $AsCl_3$ plus $LiAlH_4$) which decomposes to As on heating (Marsh test). Very poisonous, a strong reducing agent; gives arsenides with metal solutions. Other hydrides are known.

arsenic iodide, AsI_3. See ARSENIC HALIDES.

arsenic, organic derivatives Many organoarsenic compounds, e.g. R_3As, arsines, R_5As are known and are formed from, e.g. Grignard reagents. R_3As are good donors and extensive ranges of complexes are known. Compounds containing As–As bonds are known.

arsenic oxides
Arsenic(III) oxide (As_2O_3) (White arsenic). Formed by burning As in air, contains As_4O_6 molecules or AsO_3 units joined by oxygen bridges. As_2O_3 is used in glass manufacture and as arsenates(III) in insecticides, weedkillers and defoliants. It is used as a standard reducing agent (with I_2 solution). It gives arsenates(III) with alkalis.
Arsenic(V) oxide (As_2O_5) Is a rather indefinite compound (As_2O_3 plus conc. HNO_3) which loses O_2 and forms arsenates(V) with alkalis.

arsenic sulphides The sulphides As_4S_3, As_4S_4 (REALGAR), As_2S_3 and As_2S_5 can be prepared by direct reaction of the elements or (As_2S_3 and As_2S_5) precipitated from acidified As(III) or As(V) solutions with H_2S. As_4S_3, As_4S_4 and As_2S_3 all have cage structures with As–As and As–S–As linkages. As_4S_4 is used in tanning and in pyrotechny, As_2S_3 is transparent to i.r. and has optical uses. Thioarsenates(III), e.g. K_3AsS_3, and (V), e.g. K_3AsS_4, are formed from the sulphides and alkalis or alkali sulphides.

arsenides Compounds between metals and arsenic.

arsenopyrite See MISPICKEL.

arsine, AsH_3. See ARSENIC HYDRIDE.

arsines See ARSENIC (ORGANIC DERIVATIVES).

aryl When a hydrogen atom is removed from a hydrocarbon of the benzene series, the residue is called an aryl group.

arynes The transient intermediates, having two hydrogen atoms removed from adjacent non-ring junction carbon atoms of the corresponding aromatic compounds. Examples known include benzene, naphthalene, pyridine and coumarone-type compounds. The existence of arynes is inferred from the identity of products resulting from the trapping of such species in Diels–Alder reactions, and from the isomer orientation of elimination–addition reaction products of haloaromatic compounds and from spectroscopy. See BENZYNE.

As Arsenic.

asbestos The fibrous varieties of a number of silicate minerals containing SiO_4 groups linked into chains. The most important are the fibrous amphiboles (tremolites, crocidolites, blue) and fibrous talcs and clay minerals (chrysotile serpentine, cummingtonite, grunerite (amosite, brown) anthophyllite). The fibres can be spun and woven into textiles and used in fabrics for heat and electrical insulation and fireproof fabrics. Formerly used as a binder in asbestos-cement building materials and in filtration. Asbestos is now replaced by other fibrous materials. Inhalation of asbestos fibres can cause asbestosis, a very serious lung disease.

ascorbic acid (vitamin C), $C_6H_8O_6$.

L-ascorbic acid

M.p. 190–192°C. The enolic form of 3-oxo-L-gulofuranolactone. It can be prepared by synthesis from glucose, or extracted from plant sources such as rose hips, blackcurrants or citrus fruits. Easily oxidized, antioxidant and antimicrobial. It is essential for the formation of collagen and intercellular material, bone and teeth, and for the healing of wounds. Used as a photographic developing agent in alkaline solution.

asparagine (2-aminosuccinamic acid), $C_4H_8N_2O$, $H_2N \cdot C(O) \cdot CH_2 \cdot CH(NH_2) \cdot C(O)OH$. M.p. 234–235°C. Amino acid. Hydrolyses to aspartic acid. L-asparagine can be prepared from lupin seedlings, and DL-asparagine is synthesized from ammonia and maleic anhydride. L-asparagine is very widely distributed in plants, being found in all the *Leguminosae* and *Gramineae*, and in many other seeds, roots and buds.

aspartame The ester of a dipeptide, $H_2N \cdot CH \cdot (CH_2CO_2H) \cdot CONH \cdot CH(CH_2Ph) \cdot C(O)OMe$, used as an artificial sweetener.

aspartic acid (aminosuccinic acid), $C_4H_7NO_4HO_2C \cdot CH_2 . CH(NH_2) \cdot CO_2H$. M.p. 271°C. The naturally occurring substance is L-aspartic acid. One of the amino acids obtained by the hydrolysis of proteins. Used as dispersing agent and in forming aspartame.

asphalt In the U.K. the term asphalt generally refers to a natural mixture of BITUMEN with a proportion of mineral matter, e.g. rock asphalt, or to a mechanical mixture of bitumen, aggregate, sand and filler, i.e. hot-rolled asphalt.

In the U.S. and elsewhere the term asphalt is commonly applied to the material known in Britain as bitumen or asphaltic bitumen.

asphalt emulsions See ASPHALTIC BITUMEN.

asphaltenes Compounds of high molecular mass consisting largely of highly condensed aromatic ring structures. Precipitated from AS-PHALTIC BITUMEN by saturated hydrocarbon solvents or aromatic-free petroleum spirit. Maltenes are similar but have lower mass and are more soluble.

asphaltic bitumen A black or brown viscous liquid or solid, consisting of hydrocarbons and their derivatives, obtained from crude oil as a distillation residue or from natural sources or combined with mineral matter in natural asphalt. It possesses good waterproofing and adhesion properties and is used largely for road construction, waterproof or protective coating and electrical insulation. U.S. use 28 billion kg p.a.

asphaltites Naturally occurring, hard bitumens of high purity containing only 1–2 % of mineral matter.

aspirin (2-O-acetylsalicylic acid, 2-acetylbenzoic acid), $C_9H_8O_4$. Colourless crystals, m.p. 135–138°C. Manufactured by the action of ethanoic anhydride on salicylic acid. Widely used in tablet form as the acid or a salt as an analgesic and antipyretic; has many useful medical effects.

associated liquids Liquids which are composed wholly or partly of loosely combined aggregates of two or more molecules, rather than, as in the case of normal unassociated liquids, single molecules. Water and ethanol are examples of associated liquids, which show anomalous properties such as relatively high boiling points. Association in liquids arises from bonding between adjacent molecules, particularly hydrogen bonds in water and other hydroxylic compounds.

association A term applied to the combination of molecules of one substance with another to form more complex molecules. Substances showing this effect are said to be associated. The phenomenon is encountered in pure liquids, in vapours and in solutions and may be determined by ordinary methods of molecular mass determination, e.g. depression of freezing point.

astatine, At. Radioactive halogen, at.no. 85, at.mass ^{210}At 209.99, m.p. 302°C, crustal abundance, trace, electronic configuration [Xe] $5d^{10}6s^26p^5$. Formed in natural radioactive decay series but studied on ^{211}At formed by α-bombardment of ^{209}Bi. Element of Group 17. Oxidation states established include −1 (At^- formed by reduction with SO_2 or Zn, AgAt insoluble), +1 ($[AtO]^-$ formed from At^-, water and Br_2 or Fe^{3+} also $[Atpy_2]^+$), and +5 ($[AtO_3]^-$ from $[AtO]^-$ and hypochlorite or $[S_2O_8]^{2-}$).

asulam Herbicide, carbamate type.

asymmetric induction The selective synthesis of one diastereomeric form of a compound, resulting from the influence of an existing chiral centre adjacent to the developing asymmetric carbon atom. This usually arises because, for steric reasons, the incoming atom or group does not have equal access to both sides of the molecule.

asymmetry See CHIRALITY.

At Astatine.

atactic polymer If in crystalline polymers it is considered that the carbon atoms in any chain lie in the same plane with a regular arrangement of the substituents, the structure may be either ISOTACTIC or SYNTACTIC. If there is no stereospecificity of the substituents with respect to the carbon atoms, then the polymer is said to be atactic.

atmosphere See AIR. Also the unit of pressure of a gas 1 atm = 101325 Nm^{-2}.

atom The smallest part of an element which can exist as a stable entity. Often behaves as a hard sphere.

atomic absorption spectroscopy (AAS) Analytical technique in which characteristic radiation is absorbed by non-excited atoms in the vapour state. The sample is volatilized in a flame or by other means – e.g. laser. Used particularly to determine metals.

atomic emission spectroscopy An analytical technique used for the determination of trace metals. The sample is vaporized and excited in an arc or flame and the emission spectra observed.

atomic energy Mass and energy are equivalent, $E = mc^2$ (E, energy in ergs; m, mass in grams; c, velocity of light $= 3 \times 10^{10}$ cm/s^{-1}). In the building up of nuclei, part of the mass of the constituent particles appears as binding energy; the ratio (mass defect) : (atomic number) is known as the packing fraction. The elements of low atomic number and those of very high atomic number have a negative packing fraction; those of intermediate atomic number have a positive packing fraction and are more stable. If very heavy nuclei are split to form nuclei of the more stable elements there is a loss of mass; there is a similar loss of mass if light nuclei are fused to give intermediate elements. This loss in mass is evolved as energy.

The effect of a slow or 'thermal' neutron on a nucleus of ^{235}U is to split it into one or more neutrons and into large fragments of approximately equal mass. There is a liberation of energy equal to the loss in total mass. If the neutrons produced effect further fissions, a 'chain-reaction' of successive fissions may be set up. ^{232}U, ^{233}U, ^{235}U, ^{239}Pu, ^{241}Am and ^{242}Am undergo fission with thermal neutrons; of these isotopes ^{235}U and ^{239}Pu are the most important as they are most readily obtainable. Other heavy nuclei require fast neutrons to induce fission; such neutrons are much more difficult to control into a self-sustaining chain reaction.

The rapid fission of a mass of ^{235}U or another heavy nucleus is the principle of the atomic bomb, the energy liberated being the destructive power. For useful energy the reaction has to be moderated; this is done in a reactor where moderators such as water, heavy water, graphite, beryllium, etc., reduce the number of neutrons and slow those present to the most useful energies. The heat produced in a reactor is removed by normal heat-exchange methods. The neutrons in a reactor may be used for the formation of new isotopes, e.g. the transuranic elements, further fissile materials (^{239}Pu from ^{238}U), or of the radioactive isotopes used as tracers. The engineering difficulties in building a reactor are high, owing to the destructive effect of neutrons on normal structural materials and to the necessity to use materials which will absorb neutrons selectively. Further difficulties arise from the separation and disposal of the often intensely radioactive fission products.

If light nuclei can be induced to fuse together to produce heavier nuclei, energy will again be liberated. Such fusion can take place only when the nuclei have high thermal energies (temperatures 10^8 °C). Temperatures of this order occur in the centre of stars, where atomic fusions are the major source of energy, and in an atomic bomb. The high temperatures produced in an atomic explosion are used to effect the fusion of very light nuclei (^{1}H, ^{2}H, ^{3}H) in a hydrogen bomb. The practical difficulties in creating a fusion process in a controlled manner are even greater than those of deriving energy from an atomic fission and it is not yet possible to get useful energy from a fusion process.

atomic heat See DULONG AND PETIT'S LAW.

atomic mass (z) See ATOMIC WEIGHT.

atomic number The atomic number of an element is the number of unit positive charges (protons) contained in the nucleus of an atom of the element. It is the property of an element which determines the position of the element in the PERIODIC TABLE. The atomic number can be determined directly from the X-ray spectrum of the element.

atomic orbital The energy levels of electrons in an atom which may be described by the four *quantum numbers*. In wave mechanics, the energy of a particular system may be described by the SCHRÖDINGER EQUATION and the wave function Ψ_{nlm} may be used to represent a solution of the wave equation in terms of these quantum numbers. Wave functions may be used to describe electron distribution and are thus sometimes referred to as atomic orbitals.

The wave function Ψ_{100} ($n = 1, l = 0, m = 0$) corresponds to a spherical electronic distribution around the nucleus and is an example of an s orbital. Solutions of other wave functions may be described in terms of p, d and f orbitals.

atomic radii Half the closest distance of approach of atoms in the structure of the elements. This is easily defined for regular structures, e.g. close-packed metals, but is less easy to define in elements with irregular structures, e.g. As. The values may differ between allotropes (e.g. C–C 1.54 Å in diamond and 1.42 Å in planes of graphite). Atomic radii are very different from IONIC and COVALENT radii.

atomic spectrum Absorption of energy can occur only between allowed energy levels. Absorption of energy by an atom can lead only to specific changes to excited states (e.g. with transfer of electrons from one energy level to others). The reverse of this process is generally accompanied by emission of energy of a single definite frequency corresponding to the energy involved in the process. The spectrum of an atom thus consists of a series of discrete lines, each line at a frequency corresponding to the appropriate electronic transition. The line spectrum is characteristic of an element in its atomic state.

atomic weights (at.wt., relative atomic masses) Defined relative to ^{12}C as 12.0000, atomic weights are ratios of masses. The atomic mass unit $(10^{-3}$ kg mol$^{-1})$/L is 1.660×10^{-27} kg. Unit of at.wt. the dalton. Values for each element are given under the heading of the element.

The relative atomic mass of a particular sample depends on the origin of the sample.

ATP See ADENOSINE TRIPHOSPHATE.

ATR Attenuated total reflectance – a system for studying the i.r. spectra of solids, particularly of surfaces.

atrazine (2-chloro-4-ethylamino-6-isopropylamino-1,3,5-triazine), $C_8H_{14}ClN_5$. A soil-acting herbicide. Now being banned.

atrolactic acid (2-hydroxy-2-phenylpropanoic acid), $C_9H_{10}O_3$, PhC(OH)(Me)C(O)OH. Colourless racemate, m.p. 94.5°C, soluble in water and polar organic solvents. Prepared from acetophenone cyanhydrin.

atropine ((±)-hyoscyamine), $C_{17}H_{23}NO_3$. M.p. 114–116°C. Alkaloid. Prepared by racemization of hyoscyamine. It and its salts are used to dilate the pupil of the eye and for other medical uses.

atropoisomerism Chirality arising from restricted rotation about single bonds.

atropoisomers Isomers arising from conformational effects. Often only arise at low temperatures.

attapulgite A detergent clay. See FULLER'S EARTH.

attrition mill See SIZE REDUCTION EQUIPMENT.

Au Gold.

Aufbau principle In building up the electronic configuration of an atom or a molecule in its ground state, the electrons are placed in the orbitals in order of increasing energy.

Auger spectroscopy (Auger electron spectroscopy) The photoinduced ejection of an electron from an atom causes the creation of an electron hole or vacancy (see PHOTO-ELECTRON SPECTROSCOPY). This process is followed by an internal electronic reorganization in which an electron from a higher energy level falls into the vacancy. The excess energy associated with this process is released either as X-ray fluorescence, or by the ejection of a second electron (the Auger electron). The kinetic energy of the auger electron (E_A) is characteristic of the element and its environment. Auger spectroscopy is particularly useful with the lighter elements and may be used to identify elements in the surface layers of solid samples. Scanning Auger electron microscopy, SAM – the surface is scanned.

auramine, Basic Yellow 2 Important yellow triarylmethane basic dyestuff used for wool or cotton and as a biological stain $[H_2NC(C_6H_4NMe_2-p)_2]^+Cl^-$

aurates Gold-containing oxo compounds.

auric Gold(III) compounds.

aurine (rosolic acid), $(4-C_6H_4OH)_2-(4-C_6H_4O)C$. Yellow dyestuff. Its lakes are used for staining paper, and it is also used as an indicator in alkalimetry.

aurous Gold(I) compounds.

austenite See IRON, STEEL.

auto-catalysis The process whereby the reaction products catalyse the further reaction of the reactants.

autoclave Metal equipment, sometimes lined with glass or inert plastic, used for reaction between gases under pressure and liquid or solid reactants.

autofining process A process for the catalytic DESULPHURIZATION of oil fractions, the hydrogen for the reaction being obtained from the dehydrogenation of naphthenes in the oil being treated.

autolysis The self-destruction which occurs in living systems after death. It is mainly due to the release of enzymes released from the LYSOSOMES.

autoxidation Slow oxidation by atmospheric oxygen, often catalysed by light and other components in the system. Occurs particularly in foodstuffs, oils, living organisms. Propagated by peroxy radicals.

autunite, $Ca(UO_2)_2(PO_4)_2, nH_2O$. A uranium mineral.

auxins Hormones which regulate growth. See DICTIONARY OF BIOLOGY.

auxochrome A group containing lone pairs of electrons and thus modifying the colour of a chromophore.

aviation turbo-fuels Fuels used for aircraft jet engines, consisting either of KEROSINE or 'wide-cut' mixtures of NAPHTHA and kerosine. There are rigorous specifications for various physical and chemical properties.

Avogadro's constant (N_A, L) The number of particles (atoms or molecules) in one mole of any pure substance. $L = 6.022 \times 10^{23}$ mol^{-1}. It has been determined by many methods including measurements of Brownian movement, electronic charge and the counting of α-particles. Called Loschmidt's constant in German literature.

axes of symmetry See SYMMETRY ELEMENTS.

axial See CONFORMATION.

axial ratios The ratio of the cell dimensions (a, b and c) in a crystal; b being taken as unity. In triclinic, monoclinic and orthorhombic systems

$$a \neq b \neq c$$

the axial ratios are, therefore, quoted as

$$a/b : 1 : c/b$$

but the angles between the cell sides must also be quoted. In hexagonal and tetragonal systems

$$a = b \neq c$$

only $c:a$ is required. In rhombohedral and cubic systems $a = b = c$. See CRYSTAL STRUCTURE AND CRYSTAL SYSTEMS.

axis of symmetry See CRYSTAL SYMMETRY AND SYMMETRY ELEMENTS.

azaconazole Fungicide, azole derivative.

aza crown ethers Crown ethers containing ring nitrogen.

azadirachtin Insecticide, natural product.

azafemdin Herbicide, azole derivative.

azamethiphos Insecticide, organophosphate type.

azelaic acid (lepargylic acid), $C_9H_{16}O_4$, $HO(O)C \cdot (CH_2)_7C(O)OH$. Colourless plates, m.p. 106°C. Made by the oxidation of oleic acid with ozones.

azeotrope The mixed product of distillation of many mixtures.

azeotropic distillation If two substances have similar boiling points or form an azeotrope, separation by fractional distillation is either very difficult or impossible. The method of azeotropic distillation consists of adding a third substance which forms AZEOTROPIC MIXTURES with the other two; these azeotropes are then further treated to obtain pure compounds. Used in separating ethanol and water using benzene.

azeotropic mixtures Mixtures of liquids when distilled may reach a stage at which the composition of the liquid is the same as that of the vapour.

Many liquid mixtures exhibit a minimum boiling point (e.g. methanol and chloroform; n-propanol and water) whilst others show a maximum b.p. (e.g. hydrochloric acid and water) when vapour pressures are plotted as a function of change in composition of the mixture. An example of a system ($A + B$) where a minimum is seen is shown in the figure. The upper curve represents the composition of the vapour phase whilst the lower that of the liquid. When the liquid of composition x boils at temperature t it is seen that the composition of the vapour (v) differs in that it is richer in constituent A than the liquid, and an increase in the b.p. would eventually give pure B.

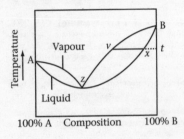

100% A   Composition   100% B

At z in the curve, however (the minimum of vapour pressure), the solution and vapour are in equilibrium and the liquid at this point will distil without any change in composition. The mixture at z is said to be azeotropic or a constant boiling mixture. The composition of the azeotropic mixture does vary with pressure.

azetidine, $CH_2CH_2CH_2NH$. Colourless liquid, b.p. 61°C, smelling of ammonia and soluble in polar solvents. Azetidine-2-carboxylic acid has been isolated from leaves of lily-of-the-valley, and arises biosynthetically from methionine. The fused azetidine ring is present in the penicillin structure.

azides Derivatives of hydrazoic acid, HN_3. Heavy metal azides (often insoluble) are extremely explosive and are used in percussion caps. Alkali azides, e.g. NaN_3 ($NaNH_2$ plus N_2O) are stable but decompose to, e.g., Na and N_2 on heating. The azide group is a good complexing agent. The azide ion is linear. Organic azides are generally relatively stable.

azidodithiocarbonates Compounds containing the $-SCSN_3$ group. A pseudohalide.

azidothymidine (AZT) Anti-viral drug used against HIV.

azimsulfuron Herbicide, sulfonyl urea type.

azimuthal quantum number (l) Also called the angular momentum quantum number. See ELECTRONIC CONFIGURATION.

azines Compounds containing $C = N - N = C$ groups prepared from $C = O$ compounds and hydrazine. Also used for compounds containing multiple N_2 groups and azine dyes.

azinphos-methyl (ethyl) Insecticides, acaricides, organophosphate type.

aziridine See ETHYLENE IMINE.

azobenzene, $C_{12}H_{10}N_2$, PhN = NPh. Orange-red crystals, m.p. 68°C. Reduces to aniline and also hydrazobenzene which can revert to benzidine under the reaction conditions. Prepared by the partial reduction of nitrobenzene.

It is the simplest compound containing the $-N=N-$ chromophore, but is of no practical importance since it lacks affinity for fibres. Normally it exists in the most stable *trans* form, but may be converted to a less stable, bright red *cis* form, m.p. 71.5°C by ultra-violet irradiation. *Cis* to *trans* isomerization occurs rapidly on heating.

azobisisobutyronitrile (AIBN) $[Me_2 C(CN)N]_2$. White solid, m.p. 102°C, decomposes slowly into cyanopropyl radicals; and used for initiation of radical reactions.

azo-compounds A large group of compounds which contain the $-N=N-$ chromophore attached to two aromatic nuclei and which absorb light in the visible range of the spectrum. If the aromatic nuclei contain amino, hydroxyl, sulphonic acid or other salt-forming groups the compound will have some affinity for animal, vegetable or artificial fibres and will thus be a dyestuff.

Azo-compounds can be obtained by reduction of nitro-compounds, or by oxidation of hydrazo-compounds. They are usually prepared, however, by reacting a phenol or amine with a diazonium salt. The coupling usually takes place in the position para to the hydroxyl or amino group, but if this position is occupied it goes to the ortho position, e.g.

$$PhN_2Cl + PhOH \rightarrow C_6H_5 - N = N - C_6H_4(OH) - 4$$

benzeneazophenol

Azo-compounds are usually very stable, and can be directly chlorinated, nitrated and sulphonated. On vigorous reduction the molecule splits at the azo group to give two molecules of primary amines, e.g. benzene-azophenol gives $PhNH_2$ and $4-HOC_6H_4NH_2$. Aliphatic azo-compounds are used as polymerization catalysts.

azo dyes Important dyestuffs. Contain $-N=N-$ chromophores. Prepared by coupling of a diazonium salt with a phenol or aromatic amine or by coupling heterocyclic hydrazines with aromatic amines. Many azo dyestuffs are sulphonic acid derivatives and are soluble in water. They are used as acid dyes for wool and affinity for cotton is increased by the number of auxochromes or azo groups. Alkylation of phenolic groups increases fastness. Azo dyes are used as pigments, in colour photography and are used as dyestuffs for most materials. They can be used as mordant or pre-metallized dyestuffs.

azoic dyes Water-insoluble azo dyes, also known as ice colours and ingrain colours. In use they are formed within the fibre.

azomethines Schiff's bases.

azophenols Compounds obtained by coupling a diazonium salt with a phenol.

azoxybenzene, $C_{12}H_{10}N_2O$. The mono N-oxide of azobenzene.

azoxystropin Fungicide, pyrimidine type.

AZT See AZIDO-THYMIDINE.

azulene, $C_{10}H_8$. A blue-violet crystalline solid with a naphthalene-like smell, m.p. 99°C.

The generic term 'azulene' was first applied to the blue oils obtained by distillation, oxidation, or acid-treatment of many essential oils. These blue colours are usually due to the presence of either guaiazulene or vetivazulene. The parent hydrocarbon is synthesized by dehydrogenation of a cyclopentanocycloheptanol or the condensation of cyclopentadiene with glutacondialdehyde anil.

Azulene is an aromatic compound and undergoes substitution reactions in the I-position. At 270°C it is transformed into naphthalene.

azurite, $Cu_3(OH)_2(CO_3)_2$. Blue basic carbonate of copper. An artificial form is known as Blue Verditer or Blue Ashes and is used as an artists' colour.

B

B Boron.

Ba Barium.

bacitracin Polypeptide antibiotic.

back bonding Overlap of a filled orbital, generally a d orbital, of an acceptor with a vacant orbital d, p, or π (bonding *or* antibonding) of a ligand. Formally relieves the build-up of negative charge on the central atom. Particularly important in complexes in low oxidation states, e.g. metal carbonyls. See ACCEPTOR.

back titration A method of volumetric analysis where excess of one reagent which reacts with the substance to be determined is added to the solution to be estimated and the excess of reagent is estimated.

baddeleyite One form of ZrO_2 and a principal mineral source of Zr.

Baeyer test Reaction of alkenes with permanganate to give diols. Best carried out using a phase transfer catalyst.

bag filter A type of AIR FILTER.

Bakelite Phenol-formaldehyde (phenol-methanal) plastics and resins; a very widely used range of plastic materials. The first synthetic plastic.

baking powders Substitutes for yeast used in baking depending for their action on the slow evolution of CO_2. They consist of various acid substances (potassium hydrogen tartrate, calcium hydrogen phosphate, sodium hydrogen phosphate) with $NaHCO_3$ or NH_4HCO_3 together with added inert materials such as flour or starch.

BAL (British Anti-Lewisite) See DITHIO-GLYCEROL.

balance An instrument used for weighing. Sensitive balances formally operate on a lever principle, less sensitive balances can use springs. The most sensitive balances use electronic methods and can weigh to a sensitivity of 0.1 μg.

ball mill See SIZE REDUCTION EQUIPMENT.

Balmer series Frequencies of lines in the line spectrum of hydrogen are related simply to each other, and can be expressed by a general formula. One group of lines which occurs in the visible spectrum is termed the Balmer series. Other series are named after Lyman, Paschen, Brackett and Pfund.

All these series may be expressed by the equation

$$v = 109\,678.8/n^2 - 109\,678.8/m^2,$$

where v is the frequency, and n and m are integers. Thus, in the Lyman series, $n = 1$, $m = 2$, 3, 4, etc. In the Balmer series, $n = 2$; in the Paschen series, $n = 3$; in the Brackett series, $n = 4$; and in the Pfund series, $n = 5$. 109 678 is the Rydberg constant.

band gap See BAND THEORY OF SOLIDS.

band spectrum The absorption of energy by a molecule may lead to the transition of an electron from an orbital of lower to one of higher energy (compare line spectrum), but in addition there may be an increase in the energy of rotation of the molecule as a whole, and in the energy of vibration of the constituent atoms relative to each other. The simultaneous occurrence of these changes gives rise to many possible energy changes. The resulting spectrum is correspondingly complicated, and consists of numerous very closely spaced lines. The frequency of a line is determined by the algebraic sum of the energy changes involved. An electronic transition is associated with so much more energy than vibrational and rotational changes, that it determines the general frequency of the line in the spectrum. The spectrum thus consists of well-defined groups or *bands*, of closely packed lines. Such a spectrum, characteristic of a molecule, is termed a *band spectrum*.

band theory of solids In molecules, molecular orbitals arise by combination of atomic orbitals. The ideas of delocalized electrons can also be applied to large assemblies of atoms in solids. The quantum mechanical treatment of these electrons gives rise to the band theory of solids, e.g. metals, in which there are bands of electron levels, sometimes overlapping, in metals, sometimes with a forbidden energy gap (band gap), e.g. insulator oxides, between them.

bar A unit of pressure used particularly at higher pressures 1 bar = 10^5 N m^{-2} = 10^5 Pascal. 1 bar ≈ 1 atm.

Barbier–Wieland degradation A method for the stepwise degradation of aliphatic acids

$$RCH_2CO_2R' \xrightarrow{PhMgBr} RCH_2C(OH)Ph_2 \xrightarrow{R'OH-H_2O}$$

$$RCH = CPh_2 \xrightarrow{CO_2} RC(O)OH + O = CPh_2$$

when one carbon atom is eliminated in each step. Opposite in objective to the Arndt–Eistert synthesis; especially useful for degrading side chains of steroid and terpenoid molecules.

barbitone (diethylmalonylurea), $C_8H_{12}N_2O_3$. White crystals, m.p. 191°C. A barbituric acid derivative. The sodium salt is administered orally as a sedative.

barbiturates (barbituric acids) Group of central nervous system depressant drugs derived from barbituric acid by replacement of the hydrogen atoms on C_5 by alkyl or aryl groups or by substitution of S for O on C_2. Administered orally as sedatives or intravenously as anaesthetics, but now little used.

barbituric acid (malonylurea), $C_4H_4N_2O_3$.

Crystallizes as the dihydrate in colourless prisms. m.p. 253°C. Prepared by heating malonic ester with urea and sodium ethoxide, or by heating alloxantin with sulphuric acid. It is a dibasic acid. Used in plastics and pharmaceuticals.

barium, Ba. Metallic element, at.no. 56, at. mass 137.33, m.p. 727°C, b.p. 1897°C, density ρ 3594, crustal abundance 500 p.p.m., electronic configuration [Xe] $6s^2$. An alkaline earth occurring principally as $BaSO_4$ (barytes) and as $BaCO_3$ (witherite, alstonite). The metal (bcc) is prepared by electrolysis of fused $BaCl_2$ or $BaCl_2$ plus Na or BaO plus aluminium. It is silver white in colour, oxidizes easily in air. Used as a getter for oxygen removal. Ba salts find extensive use: $BaSO_4$ used in paints (lithopone), X-ray diagnosis, glass, oil well drilling fluids; $BaCO_3$ as a rat poison. World production Ba minerals 1998 6 million tonnes.

barium bromide, $BaBr_2$. See BARIUM HALIDES.

barium carbonate, $BaCO_3$. A white crystalline solid occurring native (witherite) or prepared from a soluble Ba salt and Na_2CO_3. Used in ceramics, enamels, rubber, paper.

barium chemistry Barium is an element of Group 2. It forms a single series of colourless M^{2+} salts

E^o for $Ba^{2+} \rightarrow Ba$ – 2.90 (volts in acid solution)

The co-ordination number of Ba^{2+} is generally high.

barium chloride, $BaCl_2$. See BARIUM HALIDES.

barium chromate, $BaCrO_4$. A yellow pigment (Ba salt plus Na_2CrO_4) rather insoluble in water.

barium diphenylamine-4-sulphonate A redox indicator, used particularly in CrO_4^{2-}/Fe^{2+} titrations. In the oxidized form the indicator is deep reddish-violet.

barium fluoride, BaF_2. See BARIUM HALIDES. Used as flux.

barium halides, BaX_2. Barium forms a single series of halides. BaF_2 (Ba salt plus fluoride) is only sparingly soluble in water, m.p. 1285°C, b.p. 2137°C. It has some optical uses. $BaCl_2$ forms $BaCl_2.2H_2O$ and is used for the electrolytic production of Ba. $BaBr_2$ and BaI_2 are similar to $BaCl_2$.

barium hydroxide (baryta), $Ba(OH)_2$. A white true hydroxide (BaO plus H_2O) forming hydrates, also pentahydrate. Used in volumetric analysis and in the manufacture of lubricating oils and greases, as a plastic stabilizer, in pigments, as a vulcanization accelerator, gasoline additive, in glasses, refractories.

barium nitrate, $Ba(NO_3)_2$. Colourless barium salt (BaO plus HNO_3) soluble in water.

barium oxides Barium monoxide, BaO, (ignition of $BaCO_3$), white powder. On heating BaO in dry oxygen barium peroxide, BaO_2, is formed; this decomposes to BaO and O_2 at 800°C. BaO_2 is used as a bleaching agent and in H_2O_2 production. Constituent of some superconducting oxides.

barium peroxide, BaO_2. See BARIUM OXIDES.

barium sulphate, $BaSO_4$. Occurs native (heavy spar, barytes) and is a principal source of Ba compounds after reduction to BaS with carbon. Also used as a pigment, etc. (see BARIUM, LITHOPONE). $BaSO_4$ is insoluble in water and is used for the gravimetric estimation of Ba^{2+} and SO_4^{2-}. Administered orally as an X-ray contrast medium.

barium sulphide (black ash), BaS. Important intermediate in Ba chemistry ($BaSO_4$ plus C). Strongly phosphorescent as normally prepared because of impurities.

barium titanate, $BaTiO_3$. An important solid state compound. Piezoelectric, used in transducers, ultrasonics, gramophone pick-ups and for polishing. Prepared from $BaCO_3$ and TiO_2 at high temperatures.

barn In a nuclear reaction, the probability of reaction may be expressed in terms of the nuclear cross section, σ. If N target atoms per cm^3 are bombarded with I_0 particles, the number of particles transmitted is given by the equation

$$I = I_0 e^{-N\alpha x}$$

where x is the thickness of the target sample. The cross section of the target atoms σ is expressed in barns; 1 barn = 10^{-28} m^2.

barometric condenser A plant for condensing steam at pressures below atmospheric.

barrel A measure of volume widely used in the petroleum industry. A barrel is approximately 159 litres or 0.159 m^3 (5.61 ft^3) and contains 35 U.K. gallons.

baryta, $Ba(OH)_2$. BARIUM HYDROXIDE.

barytes (heavy spar), $BaSO_4$. A common mineral, it often forms well-developed orthorhombic crystals. Pure specimens are colourless or white, but brownish or bluish tints may be introduced by impurities. The mineral is used as a source of barium compounds; when finely ground it is used as a pigment and in oil well drilling muds.

base On the simplest view a base is a substance which in aqueous solution reacts with an acid to form a salt and water only, i.e. is a substance which furnishes hydroxyl ions. A more general definition (Lowry, Brønsted), which also applies to non-aqueous solutions, states that a base is a substance with a tendency to gain protons. By this definition, OH^- and the anions of weak acids, e.g. CH_3COO^-, are bases in aqueous solution.

For ionized solvents which do not contain protons a base is a substance which reacts with the acid of that system to give a salt and the solvent. Thus the base $KBrF_4$ reacts with the acid BrF_2SbF_6 to give the salt $KSbF_6$ and BrF_3 in bromine trifluoride.

A Lewis base is a molecule with available electrons, e.g. ammonia.

As an adjective applied to metals base represents the opposite of noble, i.e. a base metal would be attacked by mineral acids.

base exchange An old term used to describe the capacity of soils, zeolites, clays, etc. to exchange their cations (Na^+, K^+, Ca^{2+}) for an equivalent of other cations without undergoing structural change. An example of the general process of ION EXCHANGE.

base-pairing A term used with reference to a nucleic acid double helical structure indicating the specific hydrogen-bonded association of pairs of complementary purine and pyrimidine bases, viz. adenine with thymine or uracil, and guanine with cytosine.

base peak The most intense peak in a mass spectrum.

base strength The tendency of a base to coordinate a proton. Inversely related to the strength of the conjugate acid.

basic or cationic dyes Dyestuffs containing ionic species. Chlorides are generally soluble in water and organic salts, stearates or oleates, are soluble in organic solvents and are used in printing inks. Form insoluble salts with heteropolyanions which have high fastness and brilliant shades. Used in paints and wallpaper pigments. Basic dyestuffs need a mordant with most natural fibres but are widely used for acrylates and polyester fibres.

basic salts Compounds containing O^{2-} and OH^- as well as other anions. Formed by hydrolysis of metal salts and often precipitated by sodium carbonate solution.

basic slag See SLAG.

bastnaesite A mineral, $CeFCO_3$, used as a source of the lighter lanthanides.

bathochromic Shifts absorption to longer wavelength, lower frequency, lower energy. Compare HYPSOCHROMIC.

batteries (electric cells) Systems in which chemical change is used to produce electricity – the systems often being portable. Primary batteries are generally non-rechargeable. They include $MnO_2 \rightarrow Mn^{3+} \rightarrow Mn^{2+} \| Zn \rightarrow Zn^{2+}$ in a NH_4Cl–MnO_2–Zn system (also using ferrate (VI) in place of Mn); $HgO \rightarrow Hg \| Zn \rightarrow Zn^{2+}$ – the alkaline HgO–zinc system; $CuO \rightarrow Cu \| Zn \rightarrow Zn(OH)_4^{2-}$ Lalande cell; Ag_2O/Zn cell, Li/F_2/graphite. Ni–nickel hydrides, lithium–carbon batteries, lithium–cobalt oxides–graphite.

Secondary cells are rechargeable. The most important systems are lead–acid $PbO_2 \rightarrow PbSO_4$; nickel–cadmium $NiO(OH) + Cd \rightarrow Ni(OH)_2 + Cd(OH)_2$; Ag–Zn; Ni–Fe; Ag–Cd; MnO_2–Zn; $LiCoO \rightarrow Li$ intercalation compounds. See LEAD ACCUMULATOR.

Fuel cells involve use of gaseous reactants to produce electricity – most often H_2–O_2 within a porous electrode.

bauxite The principal ore of aluminium, a mixture of minerals formed by the weathering of aluminium-containing rocks and containing various forms of ALUMINIUM HYDROXIDES. Readily attacked by acids but forms Al_2O_3 after heating to 900°C and then becomes almost insoluble; used in the manufacture of aluminous refractories, as a source of Al and Al compounds and for road surfacing.

bayerite, α-$Al(OH)_3$. See ALUMINIUM HYDROXIDE.

bcc Body-centred cubic. See CLOSE-PACKED STRUCTURES.

B.C.F. (bromochlorodifluoromethane), $CBrClF_2$. A gas used in fire extinguishers but being phased out.

Be Beryllium.

Beattie–Bridgeman equation See VAN DER WAALS' EQUATION.

Beckmann rearrangement The conversion of ketoximes to amides by intramolecular rearrangement with e.g., PCl_5, ethanoyl chloride, $SbCl_3$, sulphuric acid, or hydrochloric acid in ethanoic acid.

$$PhC(N \cdot OH)Ph \rightarrow PhC(O) \cdot NH \cdot Ph$$

With the oximes of cyclic ketones, the rearrangement involves an enlargement of the ring by the inclusion of the nitrogen atom. This process is employed in the conversion of cyclohexanone oxime into CAPROLACTAM, the precursor of Nylon-6.

Beckmann thermometer A very sensitive mercury thermometer used for accurate measurements of temperature difference in the determination of molecular weights by freezing-point depression or boiling-point elevation.

becquerel (Bq) Unit of radioactivity. Disintegrations per unit time.

Beer's law The proportion of radiation absorbed depends on the thickness (d) of the absorbing layer, and on the *molecular concentration* (c) of the absorbing substance in the layer. It is an extension of LAMBERT'S LAW, and may be written in the form

$$I = I_0 e^{-Kcd},$$

where I_0 is the intensity of the incident radiation, and I that of the transmitted radiation; and K is a constant, the *molecular absorption coefficient*, which is characteristic of the absorbing substance for radiation of a given frequency. The *molecular extinction coefficient*, α_m, is the thickness in cm of a layer of a molar solution which reduces the intensity of light passing through it to one-tenth of its original value. $\alpha_m = 0.4343 K$. The law does not hold for substances whose absorption spectra and the species present are affected by concentration. See also EXTINCTION COEFFICIENT.

beeswax A mixture of straight chain acids, esters, alcohols and paraffins produced by bees. Used in floor waxes, shoe polishes, leather treatment.

behenic acid (n-docosanic acid), $C_{22}H_{44}O_2$, $CH_3 \cdot (CH_2)_{20} \cdot C(O)OH$. M.p. 80°C. A fatty acid occurring widely in animal and marine oils. Used in lubricating oils, detergents, polymers.

Beilstein (Handbuch der organischen Chemie) The definitive reference series on organic chemistry, no longer being revised.

Beilstein's test Qualitative method for detecting the presence of halogen in an organic compound. A piece of copper gauze is heated in an oxidizing flame until the flame is no longer tinged green; the compound is placed on the gauze, which is reheated. If Cl, Br, or I is present the flame is tinged bright green.

benalaxyl Fungicide, phenylamine derivative.

benazolin Herbicide, thiazole derivative.

bendiocarb Insecticide, carbamate derivative.

Benedict solution Aqueous solution of Na_2CO_3, $CuSO_4$, and sodium citrate used for testing for reducing sugars, particularly sugars, which give red-yellow colours or precipitates.

Benfield process Removal of carbon dioxide from fuel gases, such as those obtained by gasifying coal in the LURGI PROCESS, by countercurrent scrubbing of the gases by hot potassium carbonate solution. Compare catacarb process.

benfluralin Herbicide, dinitroaniline derivative.

benfuracarb Insecticide, carbamate derivative.

benfuresate Herbicide, sulphamate derivative.

benomyl Fungicide, benzamidazole derivative.

benoxacor Herbicide softener, benzoxazine derivative.

bensulfuron-methyl Herbicide, sulphonyl urea derivative.

bensulide Herbicide, organophosphorus derivative.

bensultep Insecticide, thiosulphonate derivative.

bentazone Herbicide, benzothiadiazine derivative.

bentonite A clay-like mineral consisting largely of montmorillonite. Forms gelatinous suspensions at low concentrations and used as a binder for foundry sand and in oil drilling. One form is used for its absorptive properties.

benzalacetophenone See CHALCONES.

benzal chloride (benzylidene chloride, α,α-dichlorotoluene), $C_7H_6Cl_2$, $PhCHCl_2$. Colourless strongly refracting lachrymatory liquid, b.p. 205°C. Prepared by the direct chlorination of toluene in the absence of metallic catalysts.

It is used for the preparation of benzaldehyde and cinnamic acid.

benzaldehyde, C_7H_6O, PhCHO. Colourless refractive liquid with almond-like odour. B.p. 180°C.

It occurs in nature as part of the glucoside, amygdalin. Manufactured from toluene, either by vapour-phase air oxidation over a catalyst, or by chlorination to benzal chloride followed by hydrolysis with boiling water.

It is readily oxidized by air to benzoic acid. With aqueous KOH gives benzyl alcohol and benzoic acid. Gives addition products with hydrogen cyanide and sodium hydrogen sulphite.

Its chief importance is as a source of cinnamic acid by condensation with sodium ethanoate and ethanoic anhydride and as a source of triphenylmethane dyestuffs by condensation with pyrogallol, dimethylaniline, etc. It is also used in the manufacture of perfumes.

benzaldoxime, C_7H_7NO. There are two

$$\begin{array}{cc} C_6H_5.CH & C_6H_5.CH \\ \| & \| \\ \alpha\text{-}syn\text{ N.OH} & \beta\text{-}anti\text{ HO.N} \end{array}$$

stereoisomers. The β-isomer cyclizes to a benz-isooxazole and has the anti-configuration. The α-isomer is stable and is prepared by the action of hydroxylamine on benzaldehyde, m.p. 34°C. The β-isomer is obtained by irradiating a benzene solution of α-isomer, or by saturating an ethereal solution with hydrogen chloride, m.p. 127°C. Prepared from benzaldehyde.

benzalkonium chloride Mixture of alkyl-benzyldimethylammonium chlorides, generally in aqueous solution. A clear, colourless or pale yellow syrupy liquid. Used as an antiseptic detergent and preservative in some pharmaceutical preparations, e.g. eyedrops. The corresponding hydroxide in aqueous solution is a strong base used in phase-transfer chemistry.

benzamide, C_7H_7NO, PhC(O)NH_2. Colourless, m.p. 130°C, b.p. 288°C. Dehydrated by phosphorus pentoxide to benzonitrile, hydrolysed to benzoic acid by dilute acids or alkalis. Forms metallic salts, e.g. silver benzamide, $C_6H_5CONHAg$.

Prepared by the action of ammonia on benzoyl chloride or benzoic esters, or by partial hydrolysis of benzonitrile.

benzanthrone (7-H-benz[de]anthracene-7-one) M.p. 170°C. Prepared by heating anthrone or anthranol with glycerol and sulphuric acid. Dyestuffs intermediate (see over).

O

benzene, C_6H_6. The benzene molecule contains six C atoms arranged in the form of a regular hexagon; the C–C bond length is 1.39 Å. The six π-electrons, or aromatic sextet, represented by the double bonds, are not localized but are distributed uniformly around the ring.

A thin, colourless, highly refractive liquid. Characteristic smell. Highly inflammable, burning with a smoky yellow flame. A good solvent for fats and lower m.w. aromatic compounds. Miscible with alcohol, ether, propanone and ethanoic acid. M.p. 5.49°C, b.p. 80.2°C, d15 0.885. Shows strong absorption bands in the u.v. The vapour is toxic when inhaled over long periods, carcinogenic.

It is the lightest fraction obtained from the distillation of the coal-tar hydrocarbons, but is now manufactured from suitable petroleum fractions by dehydrogenation (54%) and dealkylation processes. Its principal industrial use is as a starting point for other chemicals, particularly ethylbenzene, cumene, cyclohexane, styrene (45%), phenol (20%), and Nylon (17%) precursors. U.S. synthetic production 1999 1.06 million tonnes.

Structurally benzene is the simplest stable compound having aromatic character. The Hückel description of aromaticity was based in part on benzene, a cyclic fully conjugated hydrocarbon having $(4n + 2)$ π-electrons ($n = 1$) in the closed shell (ring).

Benzene can undergo addition reactions which successively saturate the three formal double bonds, e.g. up to 6 chlorine atoms can be added under radical reaction conditions whilst catalytic hydrogenation gives cyclohexane.

The most widely used reactions are those of electrophilic substitution, and under controlled conditions a maximum of three substituting groups, e.g. $-NO_2$ (in the 1,3,5 positions) can be introduced by a nitric acid/sulphuric acid mixture. Hot conc. sulphuric acid gives sulphonation whilst halogens and a Lewis acid catalyst allow, e.g. chlorination or bromination. Other methods are required for introducing fluorine and iodine atoms. Benzene undergoes the Friedel–Crafts reaction.

Nucleophilic substitution of benzene itself is not possible but the halogeno derivatives undergo nucleophilic displacement or elimination reactions (see arynes). Substituents located in the 1,2 positions are called ortho-; 1,3 meta- and 1,4 para-.

Benzene forms a wide range of organometallic complexes with the transition metals, using π-electrons for bonding to the metal, e.g. dibenzene chromium. Metallabenzenes containing MC_5 rings are known and some possess aromatic character.

benzene diazonium salts Containing the $[PhNN]^+$ group.

benzene-1,3-disulphonic acid, 3-C_6H_4 $(SO_3H)_2$, $C_6H_6O_6S_2$. Very deliquescent crystals, $+ 2\frac{1}{2}H_2O$. Prepared by sulphonation of benzene at 225°C. It gives resorcinol on fusion with KOH.

benzene hexachloride (BHC, hexachlorobenzene), C_6Cl_6 and $C_6H_6Cl_6$. The latter is carcinogenic, but was formerly used as a herbicide.

1,4-benzenedicarboxylic acid See TEREPHTHALIC ACID.

benzenesulphonic acid, $C_6H_6O_3S$, $PhSO_3H$. Colourless deliquescent plates with $1\frac{1}{2}H_2O$ (from water), m.p. 43–44°C. Anhydrous acid, m.p. 65–66°C.

Prepared by sulphonation of benzene in the liquid state or by passing benzene vapour into concentrated sulphuric acid at 150–180°C.

Forms water-soluble alkali and alkaline earth metal salts. Heating with KCN gives benzonitrile and phenol is formed by fusion with NaOH or KOH. Further sulphonation at 250°C gives benzene-1,3-disulphonic acid.

benzenetricarboxylic acids See TRIMELLITIC and TRIMESIC ACIDS.

benzfuran (coumarone) ring system

The system numbered as shown.

benzidine (4,4´-diaminobiphenyl), $C_{12}H_{12}N_2$. M.p. 127.5°C. Dibasic, readily diazotizes. Prepared by treating hydrazobenzene with hydrochloric acid, intramolecular rearrangement taking place. Carcinogen. Used for blood detection.

benzidine conversion The intramolecular rearrangement which occurs when the hydrazobenzenes are heated in acid solution. If both *para* positions in the hydrazobenzene are free a 4,4´ benzidine is produced; if one or both para positions are occupied by other groups then either a 2-benzidine or 2- or 4-semidines (ArNHAr´NH₂) are formed.

benzil, $C_{14}H_{10}O_2$, PhC(O)C(O)Ph. M.p. 95°C, b.p. 346–348°C (decomp.). Prepared by the oxidation of benzoin with nitric acid. Converted to $Ph_2C(OH)C(O)(OPh)$ by NaOH in the benzidine acid rearrangement. Used in synthesis.

benzine A term still used generically to describe SPECIAL BOILING POINT SPIRIT (SBP).

benzoates Salts or esters of benzoic acid.

benzocaine (ethyl 4-aminobenzoate), $C_9H_{11}NO_2$. White crystals, m.p. 90–91°C. Prepared from 4-nitrotoluene by way of 4-aminobenzoic acid. Used as a local anaesthetic.

benzodiazine The ring system.

benzodiazipine (diazipine, valium, 7-chloro-2,3-dihydro-1-methyl-5´-phenyl-2H-1,4-benzo-diazepin-2-one), $C_{16}H_{13}ClN_2$. White plates, m.p. 125°C. Diazapam is one of several benzodiazipines used as sedatives but can also be abused.

benzofenap Herbicide, pyrazole derivative.

benzoic acid, $C_7H_6O_2$, PhC(O)OH. Colourless lustrous leaflets, m.p. 122°C, b.p. 249°C.

First described in 1608 when it was sublimed out of gum benzoin; also occurs in many other natural resins. Benzoic acid is manufactured by the air oxidation of toluene in the liquid phase at 150°C and 4–6 atm. in the presence of a cobalt catalyst; by the partial decarboxylation of phthalic anhydride in either the liquid or vapour phase in the presence of water; by the hydrolysis of benzotrichloride (from the chlorination of toluene).

It gives benzene when heated with soda lime.

It is very stable towards oxidizing agents.

Much of the benzoic acid produced is converted to sodium benzoate, which is used as a food preservative (as is the acid) and corrosion inhibitor. Other important uses of the acid are in the manufacture of alkyd resins, plasticizers, caprolactam, dyestuffs and pharmaceuticals.

benzoin, $C_{14}H_{12}O_2$, PhC(O)CH(OH)Ph. The (±)-compound has m.p. 137°C.

Usually prepared by the action of NaCN on benzaldehyde in dilute alcohol. It is oxidized by nitric acid to benzil, and reduced by sodium amalgam to hydrobenzoin PhCHOHCHOHPh, by tin amalgam and hydrochloric acid to desoxybenzoin, $PhCH_2COPh$; and by zinc amalgam to stilbene PhCH=CHPh. It gives an oxime, phenylhydrazone and ethanoyl derivative. The α-oxime is used under the name 'cupron' for the estimation of copper and molybdenum.

The name is also given to a balsamic resin obtained from *Styrax benzoin*, which is carminative and mildly expectorant.

benzole A mixture of predominantly aromatic hydrocarbons produced by the carbonization of coal, obtained from coal gas by adsorption or from coal tar by distillation.

benzonitrile, C_7H_5N, PhCN. Colourless refractive liquid, b.p. 191°C. Slightly soluble in water, miscible with alcohol and ether in all proportions.

Prepared by the dehydration of benzamide. Hydrolysed by dilute acids and alkalis to benzoic acid. Good solvent.

benzophenone, $C_{13}H_{10}O$, PhC(O)Ph. Colourless, m.p. 49°C, b.p. 306°C. Characteristic smell. It is prepared by the action of benzoyl chloride upon benzene in the presence of aluminium chloride (Friedel–Crafts reaction) or by the oxidation of diphenylmethane. It is much used in perfumery. Forms KETYLS with sodium or potassium.

benzo(a)pyrene (1,2-benzpyrene), $C_{20}H_{12}$.

Pale yellow crystals, m.p. 179°C. A constituent of coal tar with strong carcinogenic properties.

benzoquinone, $C_6H_4O_2$. Yellow, m.p. 115.7°C, readily sublimes, volatile in steam, penetrating

odour. Electrolytic reduction gives quinhydrone, reduction with hydrogen sulphide gives hydroquinone. Gives mono- and di-oximes. It is prepared by the oxidation of phenylamine with chromic acid. Used as a source of hydroquinone and of some sulphur dyes.

benzotriazolyl-oxy-tris(dimethylamino)-phosphine hexafluorophosphate (BOP) Activator for coupling in peptide synthesis.

benzotrichloride (*α,α,α*-trichlorotoluene), $C_7H_5Cl_3$, $PhCCl_3$. Colourless liquid, b.p. 213–214°C. Insoluble in water, miscible with organic solvents. It is prepared by chlorination of toluene. When heated with water at 100°C, or with lime, benzoic acid is obtained.

benzoyl The group PhC(O)–.

benzoyl chloride, C_7H_5ClO, PhC(O)Cl. Colourless lachrymatory liquid, b.p. 198°C. It has a pungent smell, and is slowly hydrolysed by cold water to give benzoic acid. With alcohols benzoic esters are obtained. It is prepared by heating benzoic acid with PCl_5 or thionyl chloride. Used as a benzoylating agent (i.e. to add the benzoyl group in place of H).

benzoylglycine See HIPPURIC ACID.

benzoyl peroxide See DIBENZOYL PEROXIDE.

benzyl The group $PhCH_2$–.

benzyl alcohol (*α*-hydroxytoluene), C_7H_8O, $PhCH_2OH$. Colourless liquid, b.p. 205°C. Oxidizes to benzaldehyde and benzoic acid. Prepared by the hydrolysis of benzyl chloride. It is used in perfumery in the form of its esters.

benzylamine (phenylmethylamine, *α*-aminotoluene), C_7H_9N, $PhCH_2NH_2$. A colourless liquid, b.p. 185°C, partly oxidized by air. It behaves as a typical primary amine.

benzylaminopurine Plant growth regulator.

benzyl benzoate, $C_{14}H_{12}O_2$, $PhCH_2OC(O)Ph$. White crystals, m.p. 20°C. It is used as a solvent, in flavouring and for some pharmaceutical uses.

benzyl chloride, C_7H_7Cl, $PhCH_2Cl$. Colourless liquid with a characteristic odour, b.p. 179°C. It is slowly hydrolysed by boiling water, yielding benzyl alcohol.

It is prepared by the direct chlorination of toluene in the presence of PCl_5. It is purified by fractionation from the unchanged toluene and the higher chlorinated products. It is used for benzylating amines and for preparing benzyl alcohol.

benzyl chlorocarbonate (carbobenzoxy chloride), $PhCH_2OC(O)Cl$. Formed from benzyl alcohol and phosgene. Agent for protection of amines, e.g. in peptide synthesis.

benzyl ether, $(PhCH_2)_2O$. Colourless liquid, b.p. 298°C. Prepared by dehydration of benzyl alcohol. Used as a plasticizer for cellulose acetate and a solubilizer for gums, resins and rubbers.

benzylidene chloride See BENZAL CHLORIDE.

benzyne, C_6H_4. The simplest aryne, or dehydroaromatic compound. It is a highly reactive intermediate with an estimated lifetime of 10^{-5}–10^{-4} seconds in the vapour state, and cannot be isolated. A variety of methods are available for its production, including the decomposition of benzenediazonium-2-carboxylate and the base-induced elimination of hydrogen halide from halobenzenes, e.g. phenyllithium and fluorobenzene. The reactivity of benzyne results from the formal incorporation of a triple bond in a six-membered ring: other evidence suggests that benzyne is better represented by a dipolar structure. Complexes to metals. See ARYNES.

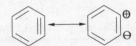

berkelium, Bk. Radioactive element, at.no. 97, at.mass ^{249}Bk 249.08, m.p. 986°C (*β* form), density ρ 14 780 (*α* form), electronic configuration [Rn] $5f^9 7s^2$. ^{249}Bk (314 days) is formed by the action of neutrons on ^{243}Am; ^{247}Bk (10^4 years) is much more stable but can only be produced in an accelerator. Bk is separated by ion exchange. The metal has been prepared by Li reduction of BkF_3, it has a double hexagonal close-packed structure and is a typical electropositive actinide.

berkelium chemistry Generally tripositive. BkF_4 (F_2 on BkF_3) and Bk(IV) formed by BrO_3^- on Bk^{3+} ($Bk^{4+} \rightarrow Bk^{3+} + 1.6V$). Cs_2BkCl_6 precipitated from HCl. All trihalides, BkOCl, Bk_2O_3 and BkO_2 are known.

Berlin green, $FeFe(CN)_6$. See CYANOFERRATES.

Berry mechanism The mechanism postulated for the interchange of substituents in trigonalbipyramidal 5-co-ordinate complexes, e.g. PF_5 and its substituted derivatives.

Berthelot equation See VAN DER WAALS' EQUATION.

berthollide compound Solid phases showing a range of composition.

beryl, $Be_3Al_2Si_6O_{18}$. A beryllium mineral containing hexagonal rings of six linked SiO_4 tetrahedra. Transparent forms of beryl are rare; when coloured green by Cr^{3+} they are known as emerald and when blue-green as aquamarine. Used for the extraction of Be and the manufacture of Be refractories.

beryllates Anionic species formed by addition of hydroxyl ions to Be^{2+}.

beryllia, BeO. BERYLLIUM OXIDE.

beryllium, Be. Metallic element, at.no. 4, at. mass 9.0122, m.p. 1287°C, b.p. 2471°C, density ρ 1848, crustal abundance 2.6 p.p.m., electronic configuration [He] $2s^2$. The lightest alkaline earth. The main source is BERYL and bertandite although Be is present in many other minerals. The metal is obtained by electrolysis, generally of $BeCl_2$ with added NaCl. The metal is grey in colour, quite hard and brittle with an hcp structure. It is fairly resistant to acids because of a layer of oxide. Be is used in alloys, as a light structural material and in nuclear reactors as a reflector or moderator. BeO is used in ceramics and also in reactors. Be compounds are toxic and can cause serious respiratory conditions and dermatitis.

beryllium bromide, $BeBr_2$. See BERYLLIUM HALIDES.

beryllium carbonate Various oxide and hydroxide carbonates are precipitated by addition of carbonate to a Be^{2+} solution.

beryllium chemistry Beryllium is an element of Group 2, electronic configuration $1s^2 2s^2$. The only stable oxidation state is +2

$E°$ for $Be^{2+} \rightarrow Be$ (acid solution – 1.85 volts)

The chemistry of Be is largely that of covalent species although ionic oxygen-co-ordinated species, e.g. $[Be(H_2O)_4]^{2+}$, $[Be_3(OH)_3]^{3+}$ are known. Be compounds with 2-, 3- and 4-co-ordination are known; the latter is the maximum. Many Be compounds achieve co-ordination number 4 by polymerization, e.g. $[BeCl_2]_\infty$.

beryllium chloride, $BeCl_2$. See BERYLLIUM HALIDES.

beryllium ethanoate (beryllium acetate), $Be_4O(O_2CCH_3)_6$. The acetate is typical of the basic beryllium carboxylates ($Be(OH)_2$ plus ethanoic acid). The structures have O at the centre of a tetrahedron of Be with carboxylate spanning each edge of the tetrahedron. $Be(O_2CCH_3)_2$ is formed from $BeCl_2$ and glacial ethanoic acid.

beryllium fluoride, BeF_2. See BERYLLIUM HALIDES.

beryllium halides Beryllium forms a single series of halides. BeF_2 (heat on $(NH_4)_2BeF_4$) is very soluble in water, sublimes 800°C. It forms a glassy mass and gives fluoroberyllates containing $(BeF_4)^{2-}$ (used in glasses) with excess of fluoride ion. $BeCl_2$ (BeO plus carbon with chlorine) forms $BeCl_2.4H_2O$ from water. $(BeCl_4)^{2-}$ is formed from melts, $BeCl_2$ has m.p. 405°C and has an ordered polymeric structure. Catalyst. $BeBr_2$ and BeI_2 are very similar to the chloride.

beryllium hydroxide, $Be(OH)_2$. Precipitated from Be solutions by OH^-. Dissolves in excess of OH^- to give beryllates containing $[Be(OH)_4]^{2-}$. $[Be_2(OH)]^{3+}$, $[Be_3(OH)_3]^{3+}$, further co-ordinated by water, are also formed.

beryllium iodide, BeI_2. See BERYLLIUM HALIDES.

beryllium nitrate, $Be(NO_3)_2.3H_2O$. A readily available beryllium salt (beryllium carbonate plus nitric acid). Extensively hydrolysed in solution, the salt loses HNO_3 on exposure to air. Added in small quantities to solutions used in impregnating incandescent gas mantles.

beryllium oxide (beryllia), BeO. Obtained on ignition of $Be(OH)_2$, or the carbonate, etc. It is a very hard material used in ceramics and reactors and used as an additive in gas mantles and as a catalyst.

beryllium sulphate, $BeSO_4.4H_2O$. A very soluble Be salt (BeO plus H_2SO_4).

Bessemer process Obsolete process for converting pig iron to steel.

beta decay (electron decay) Emission of a nuclear electron as a neutron decays to a proton. Atomic number increases by one.

betaine (trimethylglycine), $C_5H_{11}NO_2$, $Me_3N^+CH_2CO_2^-$. Crystallizes 1 H_2O, which it loses when heated to 100°C, m.p. 293°C. Soluble in water and alcohol. It is a very feeble base.

It occurs in beets and mangolds and many other plants, and can conveniently be prepared from beet molasses. Used in soldering and synthesis.

betaines A group of feebly basic substances, resembling betaine, which occur chiefly in plants. They are intramolecular salts (ZWITTER-IONS) of e.g. quaternary ammonium compounds and include stachydrine, trigonelline and carnitine. Also the name given to the dipolar intermediates, $X^+-C-C-O^-$ assumed to be formed between ylides and a ketone or an aldehyde ($X = R_2S, R_3P, R_3N$).

betamethasone (9α-fluoro-16β-methyl-prednisolone) Steroid derivative used in treating inflammation.

beta particle (beta ray) An ELECTRON emitted during radioactive disintegrations.

BET isotherm See ADSORPTION ISOTHERM.

BHC (benzene hexachloride, HCH, hexachlorocyclohexane), $C_6H_6Cl_6$. Chlorinated insecticide.

BHT 2,6-Di-*tert*-butyl-4-methylphenol.

Bi Bismuth.

biacetyl See DIACETYL.

bicarbonates (hydrogen carbonates) See CARBONIC ACID.

bifenthrin Insecticide, acaricide, pyrethroid.

biferox See DIPHENYL ETHER.

Bi-Gas process A high-pressure operation for the conversion of solid fuel into SUBSTITUTE NATURAL GAS (SNG) using two stages of gasification.

bilanofos Herbicide, alanine derivative.

bilayers Two-sided layers with polar groups formed by, e.g. fatty acid chains interacting with water. Control the passage of materials.

bile acids See CHOLIC ACID.

bile pigments Breakdown products of hemoglobin which are formed during the disintegration of erythrocytes.

bile salts The sodium salts of glycocholic and taurocholic acids, which are present in the bile of animals. They have a very low surface tension, and so act as stabilizers for the emulsion formed by fat particles in the intestine, thus easing further chemical action on the fats. They also serve to hold in solution the otherwise insoluble fatty acids.

bimolecular reaction A chemical reaction in which in one step two molecules or other species react together, e.g. $A + B \rightarrow AB$. The majority of reactions are bimolecular or proceed through a number of bimolecular steps.

binary compound A compound containing two elements, e.g. NiAs.

binder The material in, e.g. paints that protects the substrate and keeps the pigment in place and evenly distributed.

bio-allethrin Insecticide, pyrethroid.

bioassay Quantitative and sometimes qualitative analysis by determining effect on a test organism and comparing response with that of a standard. Used particularly for substances having a large physiological effect.

biocatalysis The use of biological methods, e.g. fermentation, to produce generally pharmaceutical products and intermediates, generally in a chiral form.

biocides Anti-microbials in industrial (including foodstuffs), cosmetics and domestic application. Isothiazolines and betaines are important biocides.

biodegradable Materials, e.g. detergents, polymers that are readily broken down by, e.g. bacteria, atmospheric oxidation, u.v. Alkyl sulphides, ethoxylates are readily biodegradable. Polypropene is very stable.

biogas Combustible gases derived from biomass containing, e.g. CO, H_2.

biogenesis See BIOSYNTHESIS.

biogenic amines An important group of naturally occurring amines derived by enzymic decarboxylation of the natural amino-acids.

bioluminescence Light emission by plant or animal organisms, often due to the reaction of oxygen with an oxidizable substrate (LUCIFER-IN) catalysed by an enzyme (a luciferinase).

biomarkers Fingerprint molecules that are characteristic of particular species. Sometimes found in rocks. Often characterized by GCMS.

biomass Materials derived from plants used as energy or material source.

bioresmethrin Insecticide, pyrethroid.

biose A carbohydrate with two carbon atoms. The only biose is glycollic aldehyde, $CHO \cdot CH_2OH$.

biosensors Devices for determination of a particular substrate using biological or biologically derived sensors.

biosynthesis The processes by which living organisms build up the compounds and structures required for their further growth and reproduction. Ultimately the build-up is from CO_2, H_2O, light or thermal energy and inorganic compounds. The term biogenesis is not quite synonymous, as it may also be used to refer to non-synthetic biological transformations.

biotin See VITAMIN H. Present in yeast, egg yolk, liver and other tissues.

biotransformation The process by which animals metabolize foreign components to allow renal elimination.

biphasic catalysis Use of a two-phase system which on warming gives a single phase in which a catalysed reaction can occur and which on cooling reverts to two phases with products in one phase.

biphenyl See DIPHENYL.

bipy Generally 2,2′-dipyridyl.

bipyridyl See DIPYRIDYL.

Birge–Sponer plot Method of determining dissociation energies by summing the intervals of successive frequencies and extrapolating to dissociation.

bisabolol Terpene alcohol obtained from natural oils. Used in cosmetics.

bisallothrin Insecticide, pyrethroid.

bis-2-chloroethyl formal, $CH_2(OCH_2CH_2Cl)_2$, $C_5H_{10}Cl_2O_2$. Formed from 2-CHLOROETHYL ALCOHOL and methanal. B.p. 105°C/14mm. Used extensively in the formation of polysulphide polymers.

1,8-bis-(dimethylamino)-naphthalene,-1 (N,N,N′,N′-tetramethyl-1,8,-naphthalene-diamine) M.p. 51°C. Strong monoacidic base (pK_a 12.3) which is almost completely non-nucleophilic and valuable for promoting organic elimination reactions (e.g. of alkyl halides to alkenes) without substitution.

2,4-bis(4-aminobenzyl)aniline Used as an isocyanate curing agent and in adhesives.

4,4′-bis-(isocyanatophenyl)methane (methylene-bis(4-phenylisocyanate)) A bifunctional isocyanate used for cross-linking polyurethanes.

Bismarck brown (Basic Brown 1) Basic azo dyestuff, dyes wool (reddish-brown). Used as hair dye, for textiles, leather and as biological stain.

bismuth, Bi. Metallic element, at.no. 83, at. mass 208.98, m.p. 271.4°C, b.p. 1564°C, density ρ 9790, crustal abundance 0.048 p.p.m., electronic configuration [Xe] $4f^{14}5d^{10}6s^26p^3$. Occurs native in sulphide ores, and as Bi_2S_3 (bismuth glance) and in Cu, Sn and Pb ores. The ores are roasted to Bi_2O_3 and reduced to the metal with H_2 or C. The metal is brittle, reddish-white in colour with the As and Sb structure (double layers with 3 near neighbours and 3 far). Bi burns in air to Bi_2O_3 and is only slowly attacked by acids. It is used extensively in low melting alloys and for making castings. BiOCl is used in cosmetics and some Bi compounds are used as catalysts. Bi is used in alloys and in electronics.

bismuth alloys Alloys of bismuth with Pb, Sn and Cd of low m.p. Used for fire protective devices, fuses, solders, etc. The alloys of bismuth with lead and antimony are used for stereotype plates, also in electronic and thermoelectric materials.

bismuthates(V) Formed from, e.g. Na_2O_2 and Bi_2O_3 (impure $NaBiO_3$). Never formed pure, they are very strong oxidizing agents (e.g. Mn^{2+} to MnO_4^- in acid solution).

bismuth bromide, $BiBr_3$. See BISMUTH HALIDES.

bismuth carbonates (basic carbonates), $Bi_2O_2CO_3.\frac{1}{2}H_2O$. ($Bi^{3+}$ plus soluble carbonate) is precipitated from solution. Loses H_2O at 100°C. Used for the relief of indigestion.

bismuth chemistry Bismuth is an element of Group 15. There are two stable oxidation states, +5 (strongly oxidizing in, e.g., $NaBiO_3$, BiF_5) and +3 and some evidence for Bi^+. There are also cluster ions found in e.g. BiCl. There is a considerable chemistry of complex cationic species, e.g. $[Bi_6(OH)_{12}]^{6+}$, $[Bi_6O_6(OH)_3]^{3+}$; the free Bi^{3+} ion does not seem to exist. There is a limited organic chemistry of Bi.

bismuth chlorides, $BiCl_3$, BiCl. See BISMUTH HALIDES.

bismuth fluorides, BiF_5, BiF_3. See BISMUTH HALIDES.

bismuth halides BiF_5 (Bi plus F_2) is a very powerful fluorinating agent, forms $[BiF_6]^-$ (MF plus Bi_2O_3 plus BiF_3) and BiF_3, a white rather insoluble solid (Bi^{3+} plus a fluoride). Bismuth trichloride, $BiCl_3$ (Bi^{3+} and conc. HCl or Bi plus Cl_2). Forms $[BiCl_4]^-$ and $[Bi_2Cl_7]^-$ complex ions and is hydrolysed to BiOCl. $BiCl_2$ ($BiCl_3$ melt plus Bi) consists of $[BiCl_5]^{2-}$, $[Bi_2Cl_8]^{2-}$, $[Bi_9]^{5+}$. Many of these species contain Bi_n clusters. $BiBr_3$ and BiI_3 are similar to $BiCl_3$.

bismuthides Compounds of Bi with other elements. Group 13 bismuthides have electronic applications. Form Zintl phases.

bismuth iodide, BiI_3. See BISMUTH HALIDES.

bismuth nitrates The most important bismuth compound. $Bi(NO_3)_3.5H_2O$ (Bi plus HNO_3) forms other hydrates and an oxide nitrate, $BiONO_3.H_2O$. The nitrate is probably coordinated to Bi^{3+}.

bismuth oxides Bi_2O_3 (Bi^{3+} plus alkali to give $Bi(OH)_3$ then dehydration) is the only definite oxide. It is yellow and dissolves in acids to give salts (e.g. nitrate, sulphate). Used in porcelain glazes and stained glass. Bi_2O_5 (oxidizing agents on Bi_2O_3) is unstable and has never been obtained pure.

bismuth selenides Series of compounds including $[Bi_4Se_7]^{2-}$.

bismuth sulphates $Bi_2(SO_4)_3$ (white), $Bi_2(OH)_2SO_4$ (white), $Bi_2O_2SO_4$ (yellow), are formed from sulphuric acid solution.

bismuth sulphides Bi_2S_3 (Bi^{3+} solution plus H_2S – dark brown) and BiS_2 (Bi plus S) are known.

bisphenol A (2,2-*bis*(4-hydroxyphenyl) propene, 4,4´-isopropylidenediphenol), $C_{15}H_{16}O_2$. White flaky solid, m.p. 152–153°C, b.p. 220°C/4mm. Prepared by condensation of phenol and propanone under acid conditions. Important material for production of epoxy resins (e.g. with epichlorhydrin) and modified phenolic resins.

bisphosphonates Analogues of pyrophosphates with bridging carbon. Used in water softening and as pharmaceuticals.

bitumen See ASPHALTIC BITUMEN.

bitumen emulsions Oil-in-water type emulsions prepared by passing asphaltic bitumen of suitable viscosity, along with an aqueous phase containing an emulsifier, through a colloid mill or other homogenizer. Bitumen emulsions flow easily at ambient temperature and are used without heating. They are used extensively for surface dressing and screed coating of roads and for soil stabilization.

bituminous coals The term applied to a wide variety of coals containing about 75–91 % carbon on a dry, ash-free basis.

bituminous mastic A mixture of asphaltic bitumen with inert fillers.

bituminous plastics Compositions based on asphaltic bitumen, coal tar pitches and petroleum still residues which are used as thermoplastic moulding materials and as flooring compositions.

biuret (ureidoformamide), $C_2H_5N_3O_2$, $NH_2 \cdot CO \cdot NH \cdot CO \cdot NH_2$. Crystallizes with $1\ H_2O$, m.p. 193°C (decomp.). Formed by the action of heat on urea.

biuret reaction Substances containing two C(O)NH– groups attached to one another, or to the same N or C atom, give a violet or pink colour when treated with sodium hydroxide and copper sulphate. The reaction therefore serves as a test for biuret, oxamide, peptides and proteins.

bixin, $C_{25}H_{30}O_4$. Carotenoid carboxylic acid derivative. Violet-red needles, m.p. 198°C. Used for colouring foodstuffs.

Bk Berkelium.

blackash See BARIUM SULPHIDE.

black lead POWDERED GRAPHITE used for blackening and polishing.

blanc fixe Precipitated $BaSO_4$ used as a fine white pigment in coating paper. It does not absorb ink.

Blanc's rule Empirical rule. All dicarboxylic acids, with the exception of oxalic and malonic acids, up to and including 1:5 acids, give anhydrides when treated with ethanoic anhydride and subsequently distilled; 1:6 and higher acids lose CO_2 to give ketones. Thus glutaric acid gives glutaric anhydride, adipic acid gives cyclopentanone. There are exceptions to this rule, presumably due to steric hindrance, and to ring strain.

bleach bath A bath used in photography to convert developed silver back to silver halide prior to removal (colour) or sulphide toning. $K_3Fe(CN)_6$ and KBr is a commonly used bleach bath.

bleaching agents Chemicals, generally oxidants, which remove colours, e.g. bleaching powder, chlorine, chlorates(I), perborates, hydrogen peroxide. Brightening agents, optical bleaches, act differently in that they convert u.v. radiation to blue light and thus mask colours. Typical brighteners are diaminostilbenedisulphonic acid derivatives. See FLUORESCENT BRIGHTENING AGENTS.

bleaching earths There are two types: **1.** FULLER'S EARTH possessing absorptive properties in the natural state, **2.** *activated clay* in which the absorptive powers which are low in the raw state are enhanced by heat or chemical treatment (usually a mild acid leach). Used for purification and bleaching.

bleaching powder (chloride of lime) Manufactured from Cl_2 and $Ca(OH)_2$ (slaked lime). The overall composition approximates to $CaCl(OCl)$. Yields Cl_2 with dilute acid and sold on the basis of 'available chlorine' (about 35%). Used in bleaching.

blende (sphalerite), ZnS. Generally contains some Fe which gives a dark colour. See ZINC BLENDE.

block copolymerization Copolymerization in which individual units are joined together in relatively long sequences. Branched copolymers of this type are graft copolymers. Copolymers generally have different physical properties from the formal constituents.

blooming The white cloudy appearance sometimes seen on the surface of vulcanized rubber and caused by the separation of sulphur crystals.

blowing agents Gas-forming compounds used for making foam and sponge rubber and plastics. Agents used include carbonates (e.g. Na_2CO_3), azo and nitroso compounds, and compounds also used as refrigerants.

blown bitumen, oxidized bitumen A bitumen produced by air blowing a selected soft bitumen at about 300°C, often in the presence of catalysts. The product is more rubber-like than petroleum bitumen.

blow-off See LIFT-OFF.

blue-john Blue or violet FLUORITE.

blue print paper Light-sensitive paper (turns blue) impregnated with $[Fe(CN)_6]^{3-}$ and iron(III) citrate or oxalate. The paper is fixed by washing with water. Brown print paper contains a metal (Pt, Pd, Au, Hg, Cu) salt and Fe(III). Diazo (ozalid) papers are impregnated with azo-

compounds which couple under the influence of light.

blue vitriol, $CuSO_4.5H_2O$. See COPPER SULPHATE.

blue water gas See WATER GAS.

BMC A very powerful perchlorinating agent (to replace all H by Cl) prepared from S_2Cl_2, SO_2Cl_2 and $AlCl_3$.

boat form See CONFORMATION.

BOC The *t*-butoxycarbonyl group used to protect amino groups, particularly in peptides. Generally introduced with di-*t*-butyl dicarbonate amino groups.

body-centred lattice A lattice having a point (practically atom or molecule) at the corners of the cell and at the body centre. Body-centred cubic lattices (bcc) are adopted by many METALS.

boehmite, γ-AlO(OH). See ALUMINIUM HYDROXIDE.

Bohr frequency condition When an atom changes energy by ΔE, energy is lost as a photon of frequency v where $\Delta E = hv$.

bohrium, Bh. Actinide element, at.no. 107. Prepared in linear accelerator. ^{267}Bh (half-life 17s) longest lived isotope. BhO_3Cl has been partially characterized.

Bohr magneton (μ) The unit of magnetic moment. The theoretical values of magnetic moments for unpaired electrons, taking spin only into account are: $1 = 1.73$ B.M., $2 = 2.83$, $3 = 3.87$, $4 = 4.90$, $5 = 5.92$, $6 = 6.93$, $7 = 7.94$. The observed values may be altered by metal–metal bonding, FERROMAGNETISM, ANTIFERROMAGNETISM, spin-orbit coupling and other effects. $\mu B = eh/2m$. $e = 9.27 \times 10^{-24}$ JT^{-1}.

boiling point The temperature at which the vapour pressure of a liquid is equal to that of the atmosphere. The temperature at which a liquid boils.

boiling point, elevation of The increase in boiling point of a solution, compared with that of the pure solvent, due to a dissolved substance. The elevation of boiling point is proportional to the amount of a particular solute. Molecular proportions of different solutes produce the same elevation. The following values (ebullioscopic constants) represent the elevation of boiling point produced in different solvents by dissolving 1 mol of any solute in 100g of solvent. This is known as the molecular elevation of boiling point. See RAOULT'S LAW.

Solvent	Molecular elevation
water	5.2°
chloroform	38.8°
ether	21.1°
acetone (propanone)	17.2°
benzene	25.7°
ethanol	11.5°

boiling-point diagram A graph showing equilibrium compositions of liquid and vapour plotted against temperature at any given pressure. See AZEOTROPIC MIXTURES.

Boltzmann constant (k) A fundamental constant. It is the gas constant per molecule, equal to R divided by Avogadro's number L. 1.381×10^{-23} J k^{-1}.

bomb calorimeter (adiabatic bomb calorimeter) An instrument for measuring the heat of combustion of a substance in which the heat liberated, when the sample is ignited in an atmosphere, e.g. of oxygen under pressure, is measured. Used to obtain precise thermochemical data under conditions in which there is no net loss of heat to the surroundings.

bond The linkage between atoms in molecules and between molecules and ions in crystals. Also the binding substance in aggregates.

bond angle The angle between the bonds (lines of interaction) with adjacent atoms.

bond energy The energy required to break a particular bond and, precisely, to produce the products in particular electronic states. Bond energies of multiple bonds are generally greater than those of single bonds.

For values of bond energies see, e.g. KAYE AND LABY and CRC HANDBOOK.

bonding orbitals See MOLECULAR ORBITALS.

bond order The net bonding in terms of electron pairs between atoms. The higher the bond order the lower the bond length and the greater the bond strength.

Bone and Wheeler apparatus An apparatus for the analysis of fuel and flue gases by absorption.

bone ash A white or creamy powder obtained by calcining bones. Essentially tricalcium phosphate with some calcium carbonate. It is a characteristic constituent of English china ware, mostly as BONE CHINA.

bone black (animal charcoal) The residue remaining after the destructive distillation of degreased bones. The black residue, bone black, contains about 10% amorphous carbon, disseminated through a very porous substrate of calcium phosphate (80 %), and carbonate, etc. Treatment with mineral acid dissolves away the salts and leaves a charcoal known as Ivory Black, used in sugar refining as a decolorizing agent.

bone china The chief English chinaware made of a mixture of china clay, ball clay, flint, Cornish stone and bone ash in various proportions. It resembles true porcelain, but is less refractory and not quite so hard. It permits special styles of decoration which are inapplicable to hard porcelain.

boost fluids Fluids (e.g. water, methanol) added to both aviation gasoline and aviation turbo-fuel to obtain extra power or thrust for short periods, normally during aircraft take-off.

borane Formally BH_3, e.g. borane pyridine is $H_3BNC_5H_5$. BH_3 is stable only when complexed.

borane anions The anions $B_nH_{nx}^-$ formally formed from boron hydrides and having structures containing bonding similar to that of the boron hydrides. Prepared by the action of bases on boron hydrides or by hydride on, e.g. a borate ester or diborane. Specific salts are obtained by metathesis. Mixed anions, e.g. $[BH_3CN]^-$ are also known. The simplest borane anion is tetrahedral BH_4^-, the borohydride ion (NaH plus $B(OMe)_3$ gives $NaBH_4$; LiH plus B_2H_6 gives $LiBH_4$; $AlCl_3$ plus $NaBH_4$ gives $Al(BH_4)_3$). $NaBH_4$ is white crystalline, stable in air, soluble in water and is widely used as a reducing agent, e.g. $-COOH$ gives $-CH_2OH$. Covalent borohydrides are generally inflammable. Complex borohydrides are known for, e.g. Zr and Cu. The $[B_3H_8]^-$ ion ($NaBH_4$ plus B_2H_6 [diglyme 100°C] gives NaB_3H_8) is a further simple species. Borane anions with polyhedral (closo) structures with triangular faces include $[B_9H_9]^{2-}$, $[B_{10}H_{10}]^{2-}$, $[B_{12}H_{12}]^{2-}$. $B_{10}H_{14}$ with Et_3N in boiling xylene.

Species with a vacant site – nido – and an open edge – arachno – are also known.

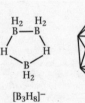

$$[B_3H_8]^-$$

$$[B_{10}H_{10}]^{2-}$$

boranes BORON HYDRIDES. The simplest borane is BH_3 but this has only transitory existence and is normally dimerized.

borates Salts of boric acid H_3BO_3. Contain planar trigonal BO_3 and tetrahedral BO_4 groups either of which may be discrete or linked. Hydroxyl groups may be co-ordinated to boron. Pyroborates, e.g. $Co_2B_2O_5$, contain two BO_3 linked through one oxygen. Polyborates contain infinite chains or rings, e.g. B_3O_3 rings in metaborates $(B_3O_6)^{3-}$ or linear $[BO_2]_n^{n-}$ (in $LiBO_2$), two rings linked through a tetrahedral boron in $K(B_5O_6(OH)_4.2H_2O$. See also BORAX, BORIC ACID. Borates readily form glasses, e.g. Pyrex. H_2O_2 or Na_2O_2 gives peroxyborates used in washing powders. Borate esters ($B(OH)_3$ plus ROH plus H_2SO_4) are known for most alcohols; anionic derivatives, e.g. $Na(BH(OR)_3)$ are formed and acyl borates $B(O_2CR)_3$ are also known. $(HO)_2B\cdot B(OH)_2$ forms esters.

borax, $Na_2[B_4O_5(OH)_4].8H_2O$. Occurs naturally as kernite ($Na_2B_4O_5(OH)_4.2H_2O$) and tincal ($Na_2B_4O_5(OH)_4.8H_2O$). The anion is strongly hydrogen bonded to water. Used as a source of boron compounds in the manufacture of glasses and fire retardants.

$$[B_4O_5(OH)_4]^{2-}$$

Bordeaux mixture An aqueous mixture of $CuSO_4$ and $Ca(OH)_2$ made up in water and used as a fungicide.

boric acid (orthoboric acid), $B(OH)_3$. A white crystalline solid with a characteristic greasy feel. The crystal contains planar $B(OH)_3$ units linked in layers by hydrogen bonding. Obtained by treating borax, colemanite, or other natural borate with a mineral acid or by hydrolysis of boron halides, hydrides, etc. On heating, H_3BO_3 gives polymeric species, e.g. $[B_4O_5(OH)_4]^{2-}$, $[B_5O_6(OH)_4]^-$ predominantly as ring polymers. It acts as an acid to OH^- forming

$[B(OH)_4]^-$ with water. The $B(OH)_3$ is volatile in steam and the solubility in water increases markedly with temperature. The salts of $B(OH)_3$ and its polymers are BORATES. It is a very weak acid but forms complexes with polyols (e.g. glycerol) which can be titrated using normal indicators. Boric acid is mildly bacteriostatic as a powder and in solution forms adducts with hydrogen peroxide – perborates.

borides Borides are formed by most elements, by direct interaction of the elements, by reduction of metal oxides with C and boron carbide, and by electrolysis of melts. They are hard refractory materials, chemically inert, which are used industrially where inertness is of importance. The bonding is complex with B_6, B_{12} clusters and networks. Some borides have high conductivities.

borinates, R_2BX. C–C coupling can give R_2CO and also alcohols (cf. boronates).

borine derivatives Derivatives of borane BH_3.

bornane, $C_{16}H_{18}$. The fundamental terpene.

D-Borneol (2-hydroxybornane), $C_{10}H_{18}O$. The secondary alcohol related to the ketone CAMPHOR.

Borneol and isoborneol are respectively the *endo* and *exo* forms of the alcohol. Used in perfumery.

Born equation Work in moving an ion from vacuum into a solvent – Gibbs' energy of solvation.

$$\Delta_{solv}G^\theta = -\frac{Z_i^2 e^2 N_A}{8\pi\varepsilon_0 r_i}\left(1 - \frac{1}{\varepsilon_r}\right)$$

Born–Haber cycle A thermodynamic cycle derived by application of HESS'S LAW. Commonly used to calculate lattice energies of ionic solids and average bond energies of covalent compounds. E.g. for NaCl:

$$\text{NaCl(s)} \xrightarrow{\quad -U_0 \quad} \text{Na}^+(g) + \text{Cl}^-(g)$$

$$\text{Na(s)} + \tfrac{1}{2}\text{Cl}_2(g) \xrightarrow{\quad S + \frac{1}{2}D \quad} \text{Na(g)} + \text{Cl(g)}$$

(with ΔH_f, I, and $-E$ labelled)

S = Heat of sublimation of sodium
D = Dissociation energy of chlorine
I = Ionization energy of sodium
E = Electron affinity of chlorine
U_0 = Lattice energy of sodium chloride
ΔH_f° = Heat of formation of sodium chloride.

From the Born–Haber cycle it follows that

$$\Delta H_f^{\circ} = S + \tfrac{1}{2}D + I - E - U_0$$

All terms in the equation can be determined experimentally except U_0, which can thus be calculated.

Similar cycles may be drawn for covalent compounds. E.g. PCl_5:

$$\text{P(s)} + \tfrac{5}{2}\text{Cl}_2(g) \longrightarrow \text{PCl}_5(g)$$

$$\text{P(g)} + 5\text{Cl(g)}$$

(with S, $\tfrac{5}{2}D$, and $-5B$ labelled)

B = Average bond energy of P–Cl bond

From the cycle it follows that:

$$\Delta H_f^{\circ} = S + \tfrac{5}{2}D - 5B$$

bornite, Cu_5FeS_4. An important Cu ore, dark bronze in colour. Often occurs mixed with chalcopyrite, $CuFeS_2$.

Born–Landé equation A derived equation for lattice energies of ionic solids as a balance between attractive and repulsive forces and including a term related to the lattice type – Madelung constant.

bornyl and isobornyl chlorides (2-chlorobornanes), $C_{10}H_{17}Cl$. Monochloro derivatives related to CAMPHOR. Bornyl chloride is the exo-chloride (Cl facing in cage).

borohydrides, $M(BH_4)_n$. See BORANE ANIONS.

boron, B. Non-metallic element, at.no. 5, at. mass 10.811, m.p. 2075°C, b.p. 4000°C, density ρ 2340, crustal abundance 10 p.p.m., electronic configuration [He] $2s^2 2p^1$. Occurs in available forms as borates (rasorite, borax and colemanite). The element is obtained in a rather impure

form by reduction of B_2O_3 with Mg. Purer forms result from reduction or pyrolysis of the halides. The various forms all contain icosahedral B_{12} units. The chief use of boron is as borosilicates in enamels and glasses; ^{10}B is used in nuclear reactors. Boron filaments (in epoxy resins and Al matrices) and boron-containing materials find use as lightweight components. Boron itself is very inert and is only slowly attacked by oxidizing agents.

boron, organic derivatives See ORGANOBORANES, BORINATES, BORONATES.

boronates, RBX_2. Aryl derivatives, $ArB(OH)_2$, are used in the synthesis of diaryls, Ar–Ar, by coupling under the influence of a Pd or Pt catalyst (Suzuki reaction) (cf. borinates).

boron bromides *Boron tribromide*, BBr_3. Prepared by passing bromine over boron. M.p. –46°C, b.p. 91°C, and has very similar properties to boron chloride but is a stronger Lewis acid. *Diboron tetrabromide*, B_2Br_4, also known.

boron chemistry Element of Group 13. The chemistry of boron is entirely that of covalent B–X bonds except for some complexed cationic and anionic species. Boron is commonly in the +3 oxidation state, 3- or 4-co-ordinate. B–B bonds are readily formed and boron clusters (e.g. B_{12} in the element, boron hydrides) are well established. BX_3 derivatives are strong Lewis acids. Organoboranes and boron hydrides are important synthetic reagents.

boron chlorides
Boron trichloride, BCl_3. Colourless mobile liquid, m.p. –107°C, b.p. 12.5°C. Obtained directly from the elements or by heating B_2O_3 with PCl_5 in a sealed tube. Readily hydrolysed by water to boric acid. Tetrachloroborates containing the BCl_4^- ion are prepared by addition of BCl_3 to metal chlorides.

Diboron tetrachloride, B_2Cl_4. M.p. –93°C, b.p. 55°C, is obtained by passing BCl_3 vapour through a glow discharge or by interaction of boron monoxide with BCl_3. Decomposes above 0° to *tetraboron tetrachloride*, B_4Cl_4 and other involatile chlorides including B_9Cl_9 and B_8Cl_8.

boron fluorides The normal fluoride is BF_3 (CaF_2 plus B_2O_3 plus H_2SO_4; heat on a diazonium fluoroborate) a colourless gas b.p. –100°C which fumes in moist air and gives boric acid and fluoroboric acid, HBF_4, with water. BF_3 is a strong Lewis acid and forms adducts with many donors (e.g. $F_3B \cdot NMe_3$). It is a very powerful catalyst in Friedel–Crafts and other reactions and is used as a starting point for the preparation of organoboron derivatives. Reacts with B_2O_3 to (FBO)$_3$. Lower fluorides are also known. B_2F_4 (BO plus SF$_4$) contains a single B–B bond. B plus BF$_3$ at high temperatures gives B_3F_5, B_8F_{12}, etc.

boron halides Boron forms BX$_3$, B_2X_4 and BX$_4^-$ with each halogen and also some lower halides. The order of Lewis acidity is BBr$_3$ > BCl$_3$ > BF$_3$. Each of the halides is rapidly hydrolysed by water. See under individual entries. Lower halides with B–B bonds are known.

boron hydrides A group of compounds containing B–H, B–B and B–H–B linkages. The

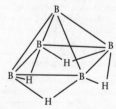

$$B_2H_6$$

simplest members are B_2H_6 (diborane(6)) m.p. –164.8°C, b.p. –92.6°C; B_4H_{10} (tetraborane (10)) m.p. –120°C, b.p. 18°C; B_5H_9 (pentaborane (9)) m.p. –46.6°C, b.p. 48°C; B_5H_{11} (pentaborane(11)) m.p. –123°C, b.p. 63°C; B_6H_{10} (hexaborane(10)) m.p. –62.3°C, b.p. 108°C; B_6H_{12} (hexaborane (12)) m.p. –82.3°C. Higher boron hydrides up to $B_{20}H_{26}$ are known. Diborane is prepared from MH, LiAlH$_4$, or NaBH$_4$ and BF$_3$. The higher hydrides are obtained from borohydrides. All of the boranes are electron deficient and the structures can be rationalized in terms of multicentre bonds involving B and H atoms. Diborane reacts with NaH to give sodium borohydride, NaBH$_4$, with anions and ligands to give complexes, (BH$_3$CN)$^-$, (BH$_3$CO). Olefins give insertion, e.g. B_2H_6 plus C_2H_4 gives B(C$_2$H$_5$)$_3$, in HYDROBORATION REACTIONS. Boron trialkyls react with B_2H_6 to give alkyl boranes. Acetylenes give carboranes (carbaboranes).

boron iodides
Boron tri-iodide, BI$_3$. (BCl$_3$ plus HI at red heat or I$_2$ plus NaBH$_4$), m.p. 43°C, b.p. 210°C. It has very similar properties to boron trichloride.

Boron sub-iodide, B_2I_4. Prepared by the action of an electrical discharge on BI$_3$ vapour.

boron neutron capture ^{10}B has a high cross-section for neutrons. Used in reactors to moderate the neutron flux. ^{10}B and thermal neutrons generate ^{7}Li and α-particles which can destroy nearby cells. Used in neutron capture therapy in conjunction with boron derivatives bonded to tumour-specific antigens.

boron nitride, BN. Prepared by the action of nitrogen or ammonia on boron at high temperatures (thus part of the amide series B(NH$_2$)$_3$, B_2(NH)$_3$, BN). Has graphite-like and diamond-like modifications. It is almost as hard as diamond and can form nanotubes.

boron–nitrogen compounds Aminoborane derivatives (e.g. B(NMe$_2$)$_3$) are formed by reaction between boron halides and amines. Amino derivatives of borane clusters and also cyclic (RBNR^i)$_n$ polymers (borazines) are also known.

boron oxides
Boron(III) oxide, B_2O_3. Obtained by ignition of boric acid. Combines with water to reform B(OH)$_3$. The fused oxide dissolves metal oxides to give borates with trigonal planar and tetrahedral boron.

Some lower oxide derivatives containing B–B bonds are known.

borosilicates Boron-containing silicates found naturally, e.g. danburite, CaB$_2$Si$_2$O$_8$, and prepared as glasses by fusing together B_2O_3, SiO$_2$ and a metal oxide (the latter is not a necessary constituent). Pyrex glasses are borosilicate glasses.

bort See DIAMOND.

boson Particle with integral spin (e.g. photon). Compare FERMION.

boundary layer When a fluid flows over a surface, due to the frictional resistance there is a layer of fluid in the immediate vicinity of the surface which has a velocity smaller than that in the bulk of the fluid. This is known as the boundary layer. In the part of the layer nearest the surface flow is laminar, that is, the layers of fluid move parallel to each other and there is no mixing between them. Transfer of heat or material across the boundary layer can occur only by the relatively slow process of molecular diffusion, and it is therefore of the greatest technical importance in that it determines the heat and mass transfer properties of the system.

Boyle's law At constant temperature the volume of a given mass of gas is inversely proportional to the pressure. The law is not accurately obeyed at high pressures because of the finite size of molecules and the existence of intermolecular forces. See VAN DER WAALS' EQUATION.

Br Bromine.

bradykinin A nonapeptide occurring in normal blood plasma as the precursor bradykininogen. Obtained from bovine plasma.

Brady's reagent See 2,4-DINITROPHENYL-HYDRAZINE.

Bragg equation When a beam of monochromatic X-rays of wavelength λ impinges on a crystal, strong scattering occurs in certain directions only: this is the phenomenon of X-RAY DIFFRACTION. For diffraction the path difference between waves scattered from successive planes of atoms in the crystal must equal an integral number of wavelengths, n ($n = 1,2,3,...$). This is expressed by the equation $n\lambda = 2d \sin \theta$ where d is the distance separating successive planes in the crystal and θ is the angle which the incident beam of X-rays makes with the same planes. The basis of X-ray diffraction.

Bragg scattering Coherent elastic scattering of monochromatic neutrons by a set of crystal planes.

branched copolymers See BLOCK CO-POLYMERIZATION.

brass Cu–Zn alloy containing up to 40% Zn. There are two main groups of alloys: **1**. the α-brasses contain up to 30% Zn; the structure is single-phase α-solid solution which is very ductile and so is suitable for severe cold forming, e.g. thin sheet, wire, tubes, cartridge cases, etc., **2**. the $\alpha\beta$-brasses contain about 40% Zn; the structure is two-phased ($\alpha + \beta$) since the solubility of Zn in Cu has been exceeded. The β-phase is an ordered or disordered bcc structure and is stronger than α but is lacking in ductility, making this alloy unsuitable for cold shaping. It is widely used for castings of all sizes and also for hot forming, particularly extrusion. Additions of small quantities of other metals give improved properties.

The γ-brass structure is adopted by a series of alloys with an electron:atom ratio of 21:13, e.g. Cu_5Zn_8, Cu_9Al_4, Cu_3Sn_8.

brassidic acid See ERUCIC ACID.

braunite Brown Mn_2O_3 containing some SiO_2. Used as a source of manganese and as a colourizing agent for bricks and pottery.

Bravais lattices The 14 different crystal lattices that can exist in solids: 3 cubic, 2

tetragonal, 4 orthorhombic, 2 monoclinic, 1 triclinic, 1 hexagonal, 1 trigonal.

brazing metal See SOLDER.

Bredt's rule A double bond cannot be introduced at the bridgehead carbon of a bridged bicyclic or polycyclic system with small- or medium-sized rings.

breunnerite Impure magnesite ($MgCO_3$) containing 5–30% iron carbonate. Also known as giobertite and mesitite. Used as a source of MgO and for forming refractories when the Fe produces stronger (although less refractory) bricks.

brighteners, optical See BLEACHING AGENTS.

bright stock Viscous lubricating oil obtained from crude oil or the residue from steam distillation (steam refined stock).

brilliant green A basic dyestuff, bis-(4-diethylaminophenyl)phenylmethyl hydrogen sulphate, also used as a bactericide.

Brillouin zones Electronic theory of metals divides the electronic states of a metal into a series of broad energy levels known as Brillouin zones.

brine Concentrated NaCl in water. Natural NaCl is generally extracted as brine.

British Standards National standards involving test methods, specifications for quality, safety, performance, etc., and codes of practice. Standards are prepared by the British Standards Institution. Ref.: BRITISH STANDARDS YEAR BOOK.

British thermal unit (Btu) A non-S.I. unit of energy. Widely used industrial heat unit; the amount of heat required to raise 1 lb of water through 1°F under specified conditions. The value corresponding to the international table calorie is 1055.06 joules, and this is the value used by the British Standards Institution.

bromacil Uracil derivative.

bromal See TRIBROMOETHANAL.

bromates Salts containing bromine oxyanions; hypobromates, $[BrO]^-$, bromates(I); bromates (the usual meaning), $[BrO_3]^-$, bromates(V); perbromates, $[BrO_4]^-$, bromates(VII). Bromates(I) are formed from Br_2 and base, but BrO^- disproportionates to Br^- and $[BrO_3]^-$ above 0°C. Bromates(V) are stable at elevated temperatures ($Ba(BrO_3)_2$ plus H_2SO_4 gives $HBrO_3$, bromic acid). Bromates(VII) are formed by electrolytic, XeF_2, or F_2 oxidation of $[BrO_3]^-$; the free acid $HBrO_4$, $2H_2O$ can be crystallized. $[BrO_4]^-$ is a strong but sluggish oxidizing agent.

$E°$ (acid solution) $[BrO_4]^- \rightarrow [BrO_3]^- + 1.76$ volts

$E°$ (acid solution) $[BrO_3]^- \rightarrow [BrO]^- + 0.56$ volts

$[BrO]^- \rightarrow Br_2 + 0.56$ volts

bromelin (bromelain) A potent proteolytic enzyme (glycoprotein) obtained from the stems of *Ananas comosus* (pineapple). Used to tenderize food.

bromic acid, $HBrO_3$. See BROMATES.

bromides Derivatives of HBr prepared by solution of the metals in aq. HBr or metal plus Br_2 (direct or in methanol) or HBr plus alkali carbonate (Na, K, etc.). Some bromides (NH_4, K, Na) are used medicinally as sedatives.

brominated biphenyls (diphenyls) See DIPHENYL.

bromine, Br. Non-metallic element, at.no. 35, at.mass 79.904, m.p. $-7.2°C$, b.p. $58.8°C$, density ρ 3100, crustal abundance 0.37 p.p.m., electronic configuration [Ar] $4s^2 4p^5$. Bromine is a dark red liquid; the vapour is red and poisonous, containing Br_2 molecules. Bromine occurs naturally as bromides in sea-water and in some natural brines and salt deposits. It is obtained by conversion to Br_2 with Cl_2 followed by sweeping out with air. Br_2 is very reactive, acting as an oxidizing agent and forming bromides with many elements. Its chief use is for making ethylene dibromide but is also used in the manufacture of fumigants, flameproofing agents, disinfectants and water-purification materials, photographic materials. U.S. use 1999 257 kilotonnes.

bromine chemistry Bromine is a typical halogen of Group 17, a non-metal. The most stable oxidation state is -1 occurring as the Br^- ion

$E°$ $\frac{1}{2}Br_2 \rightarrow Br^- + 1.07$ volts in acid solution

and in many covalent bromides. The covalent bromine group is generally bonded to only one other atom but can bridge two or three metals. There is no simple cationic chemistry but unstable species, e.g. $[Brpy_2]^+$ ($AgClO_4$ plus Br_2 plus pyridine) and $[BrF_2]^+$ (in some BrF_3 adducts) are known. Bromine forms covalent compounds in oxidation states +1 (BrF), +3 (BrF_3, $[BrF_4]^-$), +5 (BrF_5, $[BrO_3]^-$) and +7 ($[BrO_4]^-$) and as indicated most of these also form anionic species. See BROMATES.

bromine chlorides See BROMINE HALIDES.

bromine fluorides See BROMINE HALIDES.

bromine halides Fluorides, BrF, m.p. $-33°C$, b.p. $20°C$, readily decomposes to Br_2 and BrF_3; BrF_3, m.p. $9°C$, b.p. $126°C$, T structure, a yellow liquid which has found great use as a combined fluorinating agent and non-aqueous solvent. BrF_5, m.p. $-60°C$, b.p. $41°C$ square-pyramidal; chloride BrCl red brown; iodide BrI. Complex haloanions and oxide halides are also known.

bromine oxides Bromine oxides are relatively unstable. Br_2O, m.p. $-17°C$ (Br_2 on HgO); BrO_2, Br_3O_5 are even less stable (Br_2 plus O_2 in discharge tube). The oxides are formally acidic and a range of oxyanions is known – see BROMATES.

bromoacetic acid See BROMOETHANOIC ACID.

bromoacetone (bromopropanone), $CH_3C(O)CH_2Br$. Colourless liquid which rapidly becomes violet in colour; lachrymator; b.p. $136°/725$ mm. Manufactured by treating propanone with bromine, best in the presence of base. It is not very stable and decomposes on standing.

bromoacids, complex Many elements, e.g. Pt, form complex anions, e.g. $[PtBr_6]^{2-}$, in concentrated HBr solution. The free acids are generally not stable.

bromobenzenes

Bromobenzene, C_6H_5Br. B.p. $155°C$. Prepared from benzene by direct bromination in the presence of a carrier (I_2, Fe, $AlCl_3$) or by treatment of the diazonium salt with CuBr. Used for making diphenyl and diphenyl ether and derivatives, and for introduction of a phenyl group into a molecule via a Grignard reagent; the halogen atom can be removed by treatment with magnesium, sodium or copper.

The disubstituted derivatives $C_6H_4Br_2$ 2-*dibromobenzene*, m.p. $7.8°C$, b.p. $224°C$, and 4-*dibromobenzene*, m.p. $89°C$, b.p. $219°C$, also yield substituted diphenyl derivatives by coupling.

bromoethanoic acid (bromoacetic acid), $CH_2Br \cdot COOH$. White crystalline solid, m.p. $50°C$, b.p. $208°C$. Soluble in water and ethanol. Prepared by the action of dry bromine on dry ethanoic acid in presence of small amounts of red phosphorus. Produces sores upon the skin; used in chemical syntheses. See REFORMATSKI REACTION.

bromoform See TRIBROMOMETHANE.

bromomethane (methyl bromide), CH_3Br. B.p. $5°C$. Generated from sodium bromide, methanol and conc. sulphuric acid and naturally by plants. Used for methylation and commercially, very widely, as a fumigant, but use is being reduced.

bromonaphthalenes, $C_{10}H_7Br$.

1-Bromonaphthalene, m.p. 5°C, b.p. 279°C.

2-Bromonaphthalene, m.p. 59°C, b.p. 282°C.

The 1-isomer can be prepared by the direct action of bromine on naphthalene. The 2-isomer is obtained by treating diazotized 2-naphthylamine with CuBr. 1-Bromonaphthalene is used as a standard for refractive index measurements.

N-bromosuccinimide (NBS), $C_4H_4BrNO_2$.

A white solid, m.p. 178°C. Prepared from bromine on succinimide. Brominating agent which will replace activated hydrogen atoms in benzylic or allylic positions, and also those on a carbon atom α to a carbonyl group. Activating influences can produce nuclear substitution in a benzene ring and certain heterocyclic compounds; also used in the oxidation of secondary alcohols to ketones.

bromothymol blue Indicator pH range 6.0 (yellow) to 7.6 (blue).

bromotrifluoromethane See BTM.

Brönsted–Lowry classification See ACIDS.

bronze An alloy of copper and tin. The Sn hardens and strengthens the Cu. *Gun metal, phosphor bronze, aluminium bronze,* and *beryllium bronze* all have particular uses.

bronzes Compounds with metallic appearance and properties formed by reduction of heavy transition metal oxides with non-stoichiometric amounts of, e.g. alkali metals. Formed particularly by Nb, Ta and W. E.g. $Sr_{0.8}NbO_3$ and Na_xWO_3.

brosyl, Bs The trivial abbreviation for the 4-bromobenzenesulphonyl group: hence brosylate, etc.

$$4 - BrC_6H_4SO_2-$$

Brownian movement The rapid and random movement of particles of a colloidal sol, observed brightly lit against a dark ground. First observed with a pollen suspension. The Brownian movement is due to the impact on the dispersed particles of the molecules of the dispersion medium. As the particles increase in size, the probability of unequal bombardment from different sides decreases, and eventually collisions from all sides cancel out and the Brownian movement becomes imperceptible at a particle size of about 3–4μ. From the characteristics of the movement, Perrin calculated Avogadro's number L.

brown print paper See BLUE PRINT PAPER.

brown ring test A test for NITRATES.

brucine, $C_{23}H_{26}N_2O_4$. Alkaloid. The dimethoxy derivative of strychnine. Used to resolve racemic mixtures.

Brunauer–Emmett–Teller (BET) Method of surface measurement by adsorption of a monolayer of gas.

brunswick black A black paint.

brunswick green A green pigment obtained from lead ethanoate, $FeSO_4$, $K_4Fe(CN)_6$ and $Na_2Cr_2O_7$ solutions together with $BaSO_4$ as diluent.

Bs See BROSYL.

BTM (bromotrifluoromethane), CF_3Br. A colourless gas, b.p. −59°C/740 mm. Has been used as a relatively non-toxic propellant gas in firefighting apparatus, e.g. dry-powder extinguishers. Made by the bromination of fluoroform, CHF_3. Use being phased out.

Btu See BRITISH THERMAL UNIT.

BTX A mixture of low boiling point aromatics, i.e. benzene, toluene and xylenes.

bubble Region in which air and vapour are trapped by a thin film. Stability is conferred by surface tension. Of importance in the action of surfactants.

bubble-cap plate (bubble-cap tray) A vapour–liquid contacting device used in distillation and absorption columns.

Bucherer reaction The interconversion of 2-naphthol and 2-naphthylamine by the action of alkali and ammonia facilitated in the presence of $[HSO_3]^-$. This reaction is exceptional for the ease with which an aromatic C–OH bond is broken. It is not of general application.

buckminsterfullerene (bucky balls), C_{60} and others. A form of carbon with the geometry of a soccer ball. Made by evaporation of pure graphite electrodes or by use of lasers. Compare NANOTUBES.

buffer solutions It is often desirable to prepare a solution of definite pH, made up in such a way that this pH alters only gradually with the addition of alkali or acid. Such a solution is called a buffer solution, and generally consists of a solution of a salt of a weak acid in the presence of the free acid itself, e.g. sodium ethanoate and ethanoic acid. The pH of the solution is determined by the dissociation equilibrium of the free acid:

$$\text{e.g.} \frac{[H^+][CH_3COO^-]}{[CH_3COOH]} = k$$

The sodium ethanoate which is largely dissociated, serves as a source of ethanoate ions, which combine with any hydrogen ions which may be added to the solution to yield more of the acid. The addition of hydrogen ions has therefore much less effect on such a solution than it would have on water. In a similar manner, the solution of the salt of a strong acid and a weak base, in the presence of a weak base, has a pH that is insensitive to additions of alkali.

bufotenin (5-Hydroxy-3-dimethylamino-ethylindole), $C_{12}H_{16}N_2O$. M.p. 146–147°C. Alkaloid, typical toad venom.

bufotoxin The suberylarginine ester of a steroid isolated from toad venom.

buna rubbers Synthetic elastomers based on butadiene copolymers. Buna-N, nitrile rubber, NBR, is a butadiene (70)–acrylonitrile (30)–copolymer. Buna-S, SAR, is a butadiene (70)–styrene (30)–copolymer. These synthetics have very similar properties to natural rubber.

Bunte salts S-thiosulphonates $[RSSO_3]^-$ formed from alkyl halides and thiosulphates. Hydrolyse with acids to thiols, disulphides, etc.

burette Apparatus used for measuring the quantity of a liquid or gas in a chemical operation. In volumetric analysis the burette generally consists of a graduated vertical tube, provided with a tap at the lower end, by means of which the amount of liquid which is allowed to flow from the graduated tube may be controlled. In more precise work, a weight burette is employed. It consists of a flat-bottomed flask provided with a ground stopper, and a narrow side limb provided with a ground glass cap. The burette is weighed before and after releasing the requisite amount of liquid from the narrow side limb.

In gas analysis the burettes are generally vertical graduated tubes provided with a tap at the upper end. The lower end is connected by means of tubing to a reservoir containing mercury or water, by means of which the pressure on the gas enclosed between the tap and the liquid surface may be adjusted and ascertained.

Burettes can be automated for continuous use.

burners For the combustion of solid, gaseous or liquid fuels on an industrial scale special burners are usually necessary.

Solid fuels. Large coal-fired equipment normally uses pulverized fuel blown into the combustion chamber by a blast of air, similar to oil droplets.

Liquid fuels. Industrial burners for liquid fuels usually atomize the fuels in hot air so that droplets will evaporate during combustion. For more volatile fuels such as kerosine, *vaporizing burners* of various types are employed, usually for domestic purposes.

Gaseous fuels. Gas burners can be *diffusion flame burners* or *pre-aerated burners*. Diffusion flame burners may be relatively simple, with fuel gas burning at an orifice in the presence of surrounding air which diffuses into the gas. These flames are normally luminous because of the incandescence of carbon particles produced by cracking of the fuel. Pre-aerated burners are designed so that gas and air are premixed in the burner and emerge as a flammable mixture. The laboratory bunsen burner is a low temperature aerated burner.

burning oils Kerosines for domestic use. See FUEL OILS.

burning velocity The speed at which the flame front in a burning gas mixture enters the mixture of unburned gases, i.e. the flame speed. A *stationary flame* in a burner is maintained due to the balance of a steady flow of reactants by an equal and opposite burning velocity. If burning velocity of the fuel/air mixture is increased for a particular appliance there is FLASH-BACK, while if the gas flow rate is increased there is LIFT-OFF of the flame.

butadiene (1,3-butadiene),
$CH_2:CH\cdot CH:CH_2$. Colourless gas; b.p. –5°C. Prepared by passing the vapours of 1,3-butylene glycol, butylene oxide or cyclobutanol over heated catalysts. Manufactured by the catalytic cracking of *n*-butenes from oil or the catalytic dehydrogenation of butane. Also made by the dehydrogenation of ethanol to ethanal, followed by the condensation of ethanol and ethanal. When heated, e.g. with sodium, it polymerizes

to a rubber-like material. Used for the preparation of artificial rubbers (styrene/butadiene 50%, polybutadiene 20%, nitrile rubbers 10%). Also oligomerized to 4-vinylcyclohexene-1 by heat and to cyclo-octa-1,5-diene and isomeric forms of 1,5,9-cyclododecatriene and higher oligomers by transition metal catalysts. Forms complexes, e.g. butadiene–Fe(CO)$_3$, with metal compounds. U.S. production 1999 1.9 megatonnes. Carcinogenic.

butadiene polymers Polymeric elastomers derived from butadiene and its derivatives. Polybutadiene and particularly copolymers between butadiene and acrylonitrile or styrene are used as rubbers. See BUNA RUBBERS.

butadienes Unsaturated hydrocarbons containing the grouping, $R^1R^2C=CH – HC=CR^3R^4$. The simplest member, C_4H_6, is a gas; the substituted butadienes are liquids or solids which with alkyl substituents readily polymerize to rubber-like solids. They are formed by removal of water or halogen from suitable glycols or chloroparaffins, or in certain cases from alkynes. They are very reactive and combine with either one or two molecules of halogens, halogen acids, or HOCl and also with N$_2$O$_4$ and SO$_2$. Give Diels–Alder adducts with maleic anhydride (derivatives of tetrahydrophthalic anhydride). See BUTADIENE, CHLOROPRENE, ISOPRENE.

1,2 Dienes are unconjugated. See ALLENES.

butaldehydes See BUTANALS.

butanals (butaldehydes, butyraldehydes)
n-Butanal, CH$_3$CH$_2$CH$_2$CHO. Colourless liquid, ethylene oxide and Et$_3$Al, with a pungent odour, b.p. 75°C. Manufactured by reduction of crotonaldehyde with hydrogen and a metallic catalyst, by passing the vapour of *n*-butanol over heated copper oxide or by oxo-process from propene. Used in the preparation of rubber accelerators and as solvent.

Isobutanal, Me$_2$CHCHO. B.p. 64°C.

butane, C$_4$H$_{10}$. The lowest member of the paraffin series to exhibit isomerism, there being two possible isomers.

n-Butane, CH$_3$CH$_2$CH$_2$CH$_3$. Colourless gas with a faint odour, b.p. –0.3°C. It occurs in natural gas, and is obtained in large amounts during the cracking of petroleum. For chemical properties see PARAFFINS. Used in refrigeration plant; it is also compressed in cylinders for use as an illuminant, or for heating purposes, e.g. Calor gas. It has a high calorific value.

Iso-butane (2-methylpropane), CH$_3$CH(CH$_3$)CH$_3$.

Colourless gas, b.p. –10.3°C. It also occurs in natural gas and petroleum gas and may be separated from *n*-butane by fractional distillation under pressure.

butanediols (butylene glycols, di-hydroxybutanes), C$_4$H$_{10}$O$_2$. There are five glycols of this formula, three chiral. They are colourless, rather viscous liquids. The important isomers are:

1,3-Dihydroxybutane (*β-butylene glycol*), CH$_3$CH (OH)CH$_2$CH$_2$OH. B.p. 204°C. Manufactured by catalytic or enzymatic reduction of aldol. Used to prepare butadiene. Used in brake fluids, in gelling agents and as an intermediate in plasticizers.

2,3-Dihydroxybutane (*ψ-butylene glycol*), CH$_3$CH (OH)CH(OH)·CH$_3$. The glycol produced by fermentation consists largely of the optically inactive meso-form, while the synthetic glycol is mainly the optically inactive (±)-form; b.p. 177–180°C. Manufactured by the fermentation of potato mash or molasses or from epoxybutanes.

1,4-Dihydroxybutane, HOCH$_2$CH$_2$CH$_2$CH$_2$OH. B.p. 228°C. Prepared ethyne plus methanal, hydrogenated to butanediol. Used in production of γ-butyrolactone and 2-pyrrolidone. Widely used in polyurethane products.

butanoic acids (butyric acids), C$_4$H$_8$O$_2$. Two acids are known.

Butanoic acid, CH$_3$CH$_2$CH$_2$COOH. Colourless syrupy liquid with a strong odour of rancid butter; b.p. 162°C. Occurs in butter as the glyceryl ester. Prepared by oxidation of 1-butanol or by the fermentation of sugary or starchy materials by *B. subtilis*, etc. Oxidized by nitric acid to succinic acid. Cellulose derivatives are used in lacquers and as moulding plastics. Butanoates are used in flavouring and as plasticizers.

Isobutyric acid (*dimethylacetic acid, 2-methylpropanoic acid*), (CH$_3$)$_2$CH·COOH. Colourless syrupy liquid with an unpleasant odour; b.p. 154°C. Prepared by oxidation of 2-methylpropanol with K$_2$Cr$_2$O$_7$ and H$_2$SO$_4$. Salts soluble in water. Used in alkaline solution for SWEETENING gasoline.

butanols (butyl alcohols), C$_4$H$_{10}$O. There are four butanols. The liquid alcohols are important solvents for resins and lacquers; they are used to prepare butyl ethanoates, also important lacquer solvents: other butyl esters are used in artificial flavouring essences and perfumes.

Normal butyl alcohol (*propyl carbinol, n-butanol, 1-butanol*), CH$_3$CH$_2$CH$_2$CH$_2$OH. B.p. 117°C.

Manufactured by reduction of crotonaldehyde (2-butenal) with H_2 and a metallic catalyst. Forms esters with acids and is oxidized first to butanal and then to butanoic acid. Used in resins, plasticizers and as solvent and in food industry.

Isobutyl alcohol (*isobutanol, 2-methyl-propanol, isopropyl carbinol*), Me_2CHCH_2OH. B.p. 108°C. Occurs in fusel-oil. Oxidized by potassium permanganate to 2-methylpropanoic acid; dehydrated by strong sulphuric acid to 2-methylpropene.

Secondary butyl alcohol (*methylethyl carbinol, 2-butanol*), $CH_3CH_2CH(Me)OH$. B.p. 100°C. Manufactured from the butane–butene fraction of the gas from the cracking of petroleum. Used to prepare butanone.

Tertiary butyl alcohol (*trimethyl carbinol, tertiary butanol, 2-methyl-2-propanol*), Me_3COH. Colourless prisms, m.p. 25°C, b.p. 83°C. Prepared by absorbing isobutene (2-methylpropene) in sulphuric acid, neutralizing and steam distilling the liquor. Stabilizer for MeOH in petrol. Converted to isobutene by heating with oxalic acid. Potassium-*t*-butoxide is a very strong base. Used as a denaturant for ethanol.

butanone (methyl ethyl ketone, M.E.K.), C_4H_8O, $CH_3COCH_2CH_3$. Colourless liquid with a pleasant odour, b.p. 80°C. It occurs with propanone in the products of the destructive distillation of wood. Manufactured by the liquid or vapour-phase dehydrogenation of 2-butanol over a catalyst. Used as a solvent, particularly for vinyl and acrylic resins, and for nitrocellulose and cellulose acetate, also for the dewaxing of lubricating oils.

butanoyl Groups formed from the butanoic acids by removal of the OH. Thus 1-butanoyl, *n*-butyryl, $CH_3CH_2CH_2C(O)-$, 2-methyl-propanoyl, isobutyryl, $(CH_3)_2CHC(O)-$.

2-butenal (crotonaldehyde), C_4H_6O, $CH_3 \cdot CH:CH \cdot CHO$. Colourless lachrymatory liquid with a pungent odour. B.p. 104°C. Manufactured by the thermal dehydration of aldol. May be oxidized to 2-BUTENOIC ACID and reduced to crotonyl alcohol and 1-butanol; oxidized by oxygen in the presence of V_2O_5 to maleic anhydride. It is an intermediate in the production of 1-butanol from ethanol.

2-butene-1,4-diol (butenediol), $HOCH_2CH= CHCH_2OH$. Colourless stable liquid used in the manufacture of insecticides, resins and pharmaceuticals. Both *cis*- and *trans*-isomers are known. Prepared by hydrogenation of butynediol over a catalyst.

butenedioic acids See *CIS*-MALEIC, *TRANS*-FUMARIC.

butene polymers Isobutene $Me_2C=CH_2$ readily undergoes ionic polymerization and incorporation of some isoprene or butadiene gives cross-linking and elastomers – BUTYL RUBBERS. Halogenation of butyl rubbers extends the usefulness of these elastomers. 1-butene, $CH_3CH_2CH=CH_2$ can be polymerized with Ziegler–Natta catalysts to a useful material.

butenes, C_4H_8. Colourless gases with unpleasant odours. There are three isomers obtained from the appropriate butanols. All three are present in the gas from the cracking of petroleum.

1-Butene (*α-butylene*), $CH_3CH_2CH:CH_2$. Prepared by passing the vapour of 1-butanol over heated alumina.

2-Butene (*β-butylene*), $CH_3CH:CHCH_3$. Prepared by heating 2-butanol with sulphuric acid. It occurs as *cis*- and *trans*-isomers.

2-Methylpropene (*isobutylene, isobutene*), $Me_2C:CH_2$. Prepared by heating *t*-butanol with oxalic acid.

For general reactions see OLEFINS. The butenes are used to prepare 2-butanol. 1-Butene and isobutene are formed into widely used polymers.

2-butenoic acids (crotonic acids, 2-methylacrylic acids), $C_4H_6O_2$.

α-crotonic acid ((*E*)-*trans-crotonic acid, trans-Me-C(O)OH*) M.p. 72°C, b.p. 180°C. Prepared by the oxidation of crotonaldehyde or by heating the substance formed by the action of ethanol on malonic acid. Used in resins, surface coatings, plastic resins and pharmaceuticals.

Z-butenoic acid (*isocrotonic acid, β-crotonic acid, cis-crotonic acid*) Colourless needles; m.p. 14°C, b.p. 169°C. Prepared by distilling β-hydroxyglutaric acid under reduced pressure or plus methanol. Converted to E-2-butenoic acid by heating at 180°C, or by the action of bromine and sunlight on an aqueous solution. Used in dyestuffs and paper.

butoxy carbonyl group See BOC.

butter of antimony, $SbCl_3$. See ANTIMONY CHLORIDES.

butyl, iso The 2-methylpropyl group Me_2CHCH_2-. See ISOBUTYL.

butyl, normal (*n*-butyl) The group $CH_3CH_2CH_2CH_2-$.

butyl, secondary (*sec*.-butyl) The 1-methylpropyl group, CH_3CH_2CHMe-.

butyl, tertiary (*tert.*-butyl, *t*-butyl) The group Me_3C-.

butyl acetates See BUTYL ETHANOATES.

butyl alcohols See BUTANOLS.

butyl Cellosolve A trade name for ethylene glycol monobutyl ether.

butylene glycols See BUTANEDIOLS.

butylenes See BUTENES.

butyl ethanoates (butyl acetates) There are four esters corresponding to the four butanols. They are colourless liquids with fruity odours: the normal, iso- and secondary butyl ethanoates are important solvents for cellulose lacquers.

Normal butyl ethanoate,
$CH_3CH_2CH_2CH_2O\cdot(O)CCH_3$. B.p. 126°C.
Isobutyl ethanoate,
$(CH_3)_2CHCH_2O\cdot(O)CCH_3$. B.p. 118°C.
Secondary butyl ethanoate,
$CH_3CH_2CH(Me)O\cdot(O)CCH_3$. B.p. 112–113°C.
Manufactured by heating 2-butene with ethanoic and sulphuric acids in water.
t-butyl ethanoate,
$(CH_3)_3CO\cdot(O)CCH_3$.
B.p. 98°C. Used as a gasoline additive and in synthesis and as a protecting group.

***t*-butyl hydroperoxide**, $(CH_3)_3C(O)OH$. Prepared from butanol and H_2O_2. Polymerization catalyst.

***t*-butyl hypochlorite**, $(CH_3)_3COCl$. A yellow liquid with a lachrymatory action, which is prepared by passing chlorine into an aqueous alkaline solution of *t*-butanol. It may be distilled, b.p. 77–78°C, but is liable to decompose violently. It is used for *N*- and *C*-chlorinations (reaction similar to *N*-bromosuccinimide) and for dehydrating alcohols.

butyl lithium See LITHIUM ORGANIC DERIVATIVES. Used in organic syntheses, as polymerization catalyst (ethanal, isoprene, butadiene) and for the preparation of LiH and lithio derivatives.

4-*t*-butylphenol, $4\text{-}Me_3CC_6H_4OH$, $C_{10}H_{14}O$. M.p. 98°C, b.p. 239°C. Insoluble in water, soluble in alkaline solutions. Made with phenol plus 2-methylpropene. Used in phenolic coating materials (condensation with methanal).

butyl rubber A copolymer of $Me_2C:CH_2$ (isobutene) and $H_2C:C(Me)CH:CH_2$ (isoprene) (1–3%) that can be vulcanized in the normal way with sulphur. Less permeable than natural rubber and used in tyre inner tubes. It is resistant to atmospheric ozone and so can be used out-of-doors in black sheet form for roofing and as a water-impermeable membrane to prevent seepage from reservoirs.

2-butyne-1,4-diol (butynediol), $HOCH_2C\equiv CCH_2OH$. White solid, m.p. 58°C, b.p. 238°C prepared by the high-pressure reaction between ethyne and methanol and also from $BrMgCCMgBr$ and methanal. Used in electroplating (Ni), as a corrosion inhibitor, and in paint and varnish removal.

butyric acids See BUTANOIC ACIDS.

γ-butyrolactone (4-hydroxybutanoic acid lactone), $C_4H_6O_2$. Colourless liquid with a pleasant odour; b.p. 206°C. Prepared from ethyne

plus methanol. It is used as a solvent for various polymers and as an intermediate in the preparation of, e.g. polyvinylpyrrolidone and piperidine.

butyrone See HEPTANONES.

butyryl See BUTANOYL.

C

C Carbon.

c The speed of light in vacuum $2.997 \times 10^8 \, \mathrm{ms^{-1}}$.

CA Chemical Abstracts (Registry Number, etc.).

Ca Calcium.

cacodyl derivatives Organoarsenic compounds containing the $AsMe_2$ grouping. Cacodyl is $Me_2AsAsMe_2$.

cadaverine (pentamethylenediamine), $H_2N\cdot[CH_2]_5\cdot NH_2$. A syrupy fuming liquid, b.p. 178–180°C. Cadaverine is found, associated with putrescine, in putrefying tissues, being formed by bacterial action from the amino-acid lysine. The free base is poisonous.

cadmium, Cd. Metallic element, at.no. 48, at.mass 112.41, m.p. 321.07°C, b.p. 767°C, density ρ 8680, crustal abundance 0.11 p.p.m., electronic configuration [Kr] $4d^{10}5s^2$. Cadmium occurs in Zn and also Pb and Cu ores but also as greenockite, CdS. Cd is more volatile than Zn and is collected in the dust in flues of furnaces for producing zinc. CdO is reduced with C or Cd^{2+} is reduced in solution with zinc dust. Cd is a white lustrous metal with a deformed hcp structure with 6 near (in plane) and 6 far neighbours. The metal reacts with oxygen on heating and with acids. Cd is used extensively for electroplating (35%) and in alloys, Ni–Cd batteries, atomic reactors; cadmium compounds are used as pigments, plasticizers and as phosphors in colour TV tubes. Cadmium telluride has many electronic applications. Cadmium compounds are toxic but seem necessary in carbonic anhydrase.

cadmium, organic derivatives Derivatives R_2Cd and RCdX are prepared from RLi and CdX_2 or Cd plus RI in a polar solvent. They are fairly unstable thermally. The reactions are similar to those of Grignard reagents but are used specifically to prepare ketones RCOR' from acyl chlorides RCOCl.

cadmium chemistry Cadmium is an electropositive element of Group 12. The only stable oxidation state is +2 frequently in a mainly ionic state

$E°$ for $Cd^{2+} \to Cd$ (in acid solution) –0.402 volts

An unstable lower oxidation state is found in $Cd–CdCl_2$ and $Cd–CdCl_2–AlCl_3$ melts and may contain Cd–Cd bonds. The common co-ordination number for Cd^{2+} is 6 (octahedral) although 4 and 5 co-ordination are known.

cadmium chloride, $CdCl_2$. See CADMIUM HALIDES. The $CdCl_2$ structure is an important lattice type, being a layer structure based on ccp of Cl. Used in photography, dyeing, pigments, lubricants.

cadmium halides All dihalides, CdX_2, known. CdF_2 is ionic with the rutile structure. $CdCl_2$, m.p. 868°C (structure see CADMIUM CHLORIDE) forms hydrates and chlorocomplexes. The bromide and iodide are similar to the chloride. CdF_2 is used in phosphors and glasses.

cadmium hydroxide See CADMIUM OXIDE.

cadmium iodide, CdI_2. See CADMIUM HALIDES. The CdI_2 is an important lattice type being a layer structure based on hcp of I^-.

cadmium oxide, CdO, **cadmium hydroxide**, $Cd(OH)_2$. The hydroxide is precipitated from aqueous solution by OH^-, it does not dissolve in excess OH^-. Ignition of $Cd(OH)_2$ or $CdCO_3$ gives CdO which varies in colour from red-brown to black because of lattice defects. CdO is used in phosphors, semi-conductors, electronics applications.

cadmium oxy-acid salts Colourless salts prepared by solution of the oxide or carbonate in the appropriate acid $[Cd(O_2CCH_3)_2.2H_2O, Cd(NO_3)_2.4H_2O, Cd(ClO_4)_2.6H_2O, 3CdSO_4.8H_2O]$. The carbonate, $CdCO_3$, is precipitated from an aqueous solution of Cd^{2-} by CO_3^{2-}; the other salts are soluble in water although the solutions are hydrolysed.

cadmium red, orange, scarlet Pigments consisting of CdS(Se) prepared from a solution of $CdSO_4$ with added BaS and Se.

cadmium sulphate, $3CdSO_4.8H_2O$. The most important Cd salt. See CADMIUM OXY-ACID SALTS. Used in glasses.

cadmium sulphide, CdS. Occurs naturally as greenockite and obtained as a yellow precipitate from Cd^{2+}(aq) and H_2S. Used as a yellow pigment.

cadmium yellow Natural cadmium sulphide. Used as a pigment.

cadusafos Nematocide, insecticide, organophosphate derivative.

caesium (**cesium** (U.S. usage)), Cs. Metallic element, at.no. 55, at.mass 132.91, m.p. 28.5°C, b.p. 671°C, density ρ 1813, crustal abundance 3 p.p.m., electronic configuration [Xe] $6s^1$. An alkali metal occurring in traces in salt deposits and as pollucite $Cs(AlSi_2O_6)$,xH_2O and as a by-product of lithium production. Separated from the other alkali metals by ion exchange and fractional crystallization, the metal (bcc) is obtained by electrolysis or better by $CsAlO_2$ (from caesium alum) and Mg. The metal is extremely reactive and reacts violently with water, oxygen, halogens, etc. The metal is used as an oxygen-getter, in photoelectric cells, as a hydrogenation polymerization catalyst and in SO_2 oxidation. CsOH is a battery electrolyte. ^{137}Cs is an important fission product used in deep-ray therapy.

caesium chemistry Caesium is an element of Group I. It forms a single series of components in the +1 oxidation state although Cs^- compounds are known; most compounds are predominantly ionic. It also forms a series of lower oxides, e.g. Cs_7O, $Cs_{11}O_3$ containing cluster cations.

caesium chloride, CsCl. A typical alkali halide formed from Cs_2CO_3 and HCl. The structure is that adopted by many compounds AX with a large radius ratio of A : X. Each ion has a co-ordination number eight. Adopted by CsCl, CsI, AgLi, HgTl, etc.

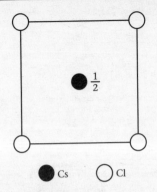

$\bullet$ Cs $\quad$ $\bigcirc$ Cl

caesium fluoride, CsF. Important in organic fluorination reactions and with other halides in coupling and addition reactions.

caffeine (1,3,7-trimethylxanthine, theine), $C_8H_{10}N_4O_2$. An alkaloid occurring in tea, coffee and guarana. It is also manufactured by the methylation of theobromine and by the condensation of ethanoic acid with urea. M.p. (anhydrous) 235°C, sublimes at 176°C. Odourless, and with a very bitter taste. Caffeine acts as a stimulant and diuretic, and is a constituent of cola drinks, tea and coffee.

cafsentrole Herbicide, triazole derivative.

cage compounds CLATHRATE COMPOUNDS.

calamine, $ZnCO_3$ (British), $ZnSiO_4.H_2O$ (U.S.). Medicinal calamine is a basic zinc carbonate coloured with Fe_2O_3. Formerly obtained from the carbonate, now obtained by precipitation. Used either in lotions or as a powder for sunburn, sore skin and dermatitis.

calciferol VITAMIN D constituent.

calcined bauxite A very hard, almost insoluble form of alumina produced by calcining BAUXITE at temperatures above 1175K. It has a high resistance to abrasion and polishing and is used, often in conjunction with epoxy resin/bitumen mixes, as aggregate for highly skid-resistant road surfacings in specially dangerous locations.

calcite A form of $CaCO_3$. In the massive condition it forms limestone, marble, chalk. The crystals have many local names, including calc-spar, Iceland spar, dogtooth spar. Calcite is the most stable form of calcium carbonate. Compare ARAGONITE.

calcium, Ca. Metallic element, at.no. 20, at. mass 40.078, m.p. 842°C, b.p. 1484°C, density ρ 1550, crustal abundance 41 000 p.p.m., electronic configuration [Ar] $4s^2$. Calcium compounds are widely distributed in nature occurring as $CaCO_3$ (marble, limestone, chalk, calcite, aragonite, dolomite), $CaSO_4$ (gypsum), $Ca_3(PO_4)_2$, halides and silicates (various feldspars, anorthite). The metal is obtained by electrolysis of the fused chloride, or by $CaCO_3$ plus Al at high temperature and pressure. It is a soft silvery white metal, ccp at room temperature, hcp above 450°C. It tarnishes rapidly in air, reacts violently with water and combines with oxygen and the halogens. Ca is used as a reducing agent in the preparation of metals such as Th, V, Zr, as a deoxidizer and as an alloying agent. Calcium oxide is an important base in the chemical industry. Ca^{2+} ions are important in biological systems and are a major component in some enzymes, bone and teeth.

calcium acetylide, CaC_2. See CALCIUM CARBIDE.

calcium aluminates Present in cements, particularly in ALUMINOUS CEMENTS.

calcium bromide, $CaBr_2$. Very similar to the chloride.

calcium carbide (calcium acetylide), CaC_2. Prepared commercially by the action of an electric arc on coke and CaO in an electric furnace. Gives ethyne, H_2C_2, with water. The structure is ionic containing C_2^{2-} ions. Used for producing ethyne and calcium cyanamide.

calcium carbonate, $CaCO_3$. Occurs naturally (see CALCIUM). Precipitated from $Ca(OH)_2$ solution by CO_2 but dissolves in excess CO_2 to form the hydrogen carbonate $Ca(HCO_3)_2$. Precipitated calcium carbonate is widely used in paper (64%) and also as a filler in plastics, pharmaceuticals, dentifrices. World capacity 2000 6.5 megatonnes p.a.

calcium chemistry Calcium is a Group 2 element, electronic configuration [Ar]$4s^2$. It shows a single oxidation state of +2 ($E°$ for Ca^{2+} to Ca in acid solution –2.76 volts). The Ca^{2+} ion shows normal co-ordination arrangements of octahedral and cubic. An extensive range of salts is known and are used. Complexes are formed with oxygen-ligands particularly chelating ligands such as EDTA and polyphosphates (remove hardness of water).

calcium chloride, $CaCl_2$. Occurs naturally as tachydrite, $CaCl_2.2MgCl_2.12H_2O$ and in sea and mineral waters. Prepared pure from $CaCO_3$ and HCl and gives $CaCl_2.6H_2O$, m.p. 30°C, $CaCl_2.4H_2O$, $CaCl_2.2H_2O$ and the anhydrous salt. Gives adducts with alcohols and ammonia. Obtained commercially from natural brines and from the ammonia–soda process for Na_2CO_3. Used for de-icing roads, dust control, as an antifreeze in concrete mixes and in solution in refrigeration plants. U.S. demand 1999 900 000 tonnes.

calcium cyanamide, CaNCN. Prepared from N_2 and CaC_2 at 1100°C. Used as a fertilizer; on hydrolysis gives NH_3 and urea in the soil. Gives cyanides on fusion with C and Na_2CO_3 at 1200°C.

calcium ethanoate, $Ca(O_2CCH_3)_2.2$ or $1H_2O$. $CaCO_3$ plus ethanoic acid. Used in the production of other ethanoates and in calico printing, and tanning.

calcium fluoride (fluorite, fluorspar), CaF_2. Occurs naturally as fluorspar. Relatively insoluble in water; m.p. 2533°C. Used in optical systems for u.v. and near i.r. radiation and also to lower the melting point of molten salt baths used in Ca and Al production. The structure is ionic (see FLUORITE).

calcium glycerophosphate, $CaC_3H_5(OH)_2 PO_4.H_2O$. A constituent of many proprietary nerve tonics and dentifrices.

calcium hydride, CaH_2. White solid (Ca (solid in NH_3) and hydrogen). Releases hydrogen with water (hydrolith), used as a drying agent and a reducing agent in metallurgy.

calcium hydrogen carbonate, $Ca(HCO_3)_2$. See CALCIUM CARBONATE.

calcium hydrogen sulphite, $Ca(HSO_3)_2$. Known only in solution. Germicide, preservative, disinfectant.

calcium hydroxide (slaked lime), $Ca(OH)_2$. The powder from the action of an equivalent quantity of water on CaO is slaked lime; a suspension with excess water is milk of lime, a clear solution of $Ca(OH)_2$ in water is limewater. Used as an industrial alkali and in the preparation of mortar (slaked lime plus sand) which sets to a solid by reconversion of the $Ca(OH)_2$ to $CaCO_3$, fireproofing, water treatment.

calcium iodide, CaI_2. Forms a hexahydrate. Similar to CALCIUM CHLORIDE.

calcium nitrate, $Ca(NO_3)_2.4H_2O$. A deliquescent solid.

calcium oxalate, $CaC_2O_4.H_2O$. One of the least soluble calcium salts and used (by estimation of the oxalate) in the estimation of Ca^{2+}. Loses H_2O at 180°C, gives $CaCO_3$ and eventually CaO on heating in air. Occurs in rhubarb and other plants.

calcium oxide (lime, quicklime), CaO. A white solid with the NaCl structure produced by heating $CaCO_3$ to high temperatures. Commercially, lime is produced by heating $CaCO_3$ to a temperature between 900°C and 1200°C.

Although much lime is used in building and agriculture, the main use is in the chemical process industries. Thus large quantities are used in the manufacture of Na_2CO_3 and NaOH, SO_2 absorption, steel (45%), refractories, CaC_2, glass, ceramics, pulp and paper, and sugar; it is also employed in the treatment of water and sewage and in ore concentration and refining and soil stabilization. U.S. production CaO plus $Ca(OH)_2$ 1999 22.8 megatonnes.

calcium perchlorate, $Ca(ClO_4)_2.6H_2O$. A soluble calcium salt.

calcium peroxide, $CaO_2.8H_2O$. Formed anhydrous at high concentrations ($Ca(OH)_2$ plus H_2O_2). Manufactured for use as an antiseptic from $Ca(OH)_2$ and Na_2O_2 washed with icewater.

calcium phosphates The orthophosphate, $Ca_3(PO_4)_2$ occurs naturally and is precipitated from a Ca^{2+} solution by phosphate (used as a polymerization aid for styrene and nutritional supplement). $CaHPO_4.2H_2O$ ($CaCl_2$ solution plus Na_2HPO_4) and $Ca(H_2PO_4)_2.H_2O$ (other phosphates plus H_3PO_4) are also known. The latter compound is the principal active ingredient of SUPERPHOSPHATES (fertilizers). Calcium phosphates are used in foodstuffs. APATITE is a calcium fluoride phosphate constituent of teeth.

calcium silicates Important minerals also present in slag and cements. Used in glass filters.

calcium sulphate, $CaSO_4$. Occurs naturally as anhydrite, and as $CaSO_4.2H_2O$ as gypsum, selenite, satin-spar, alabaster. Natural anhydrite and $CaSO_4$ prepared by dehydrating the dihydrate above 650°C is only slowly soluble in water and is used as a filler for paper and other materials (see also SATIN WHITE). $CaSO_4$ prepared by dehydration of the dihydrate at 60–90°C is used as a drying agent. $CaSO_4.0.5H_2O$, prepared by heating gypsum at 130°C, is used as PLASTER OF PARIS. Gypsum and anhydrite are used for H_2SO_4 production and gypsum is used

as a soil additive and as an inert additive to pharmaceuticals and insecticides.

calcium sulphide, CaS. Formed from H_2S and limewater or $CaSO_4$ and charcoal at 1000°C. The impure material is luminescent. Used as a depilatory. Calcium polysulphides are used as fungicides, insecticides, acaricides.

calcium sulphite, $CaSO_3$. Formed from SO_2 and limewater or $CaCl_2$ plus a sulphite as a precipitate, dissolves in excess SO_2. Used for sterilization and for dissolving the lignin in wood pulp in paper manufacture.

calcium superphosphate See SUPERPHOSPHATES.

calcium titanate (perovskite), $CaTiO_3$. A mixed metal oxide which has given its name to the PEROVSKITE structure.

calgon Sodium metaphosphate used as a washing powder and cleansing agent. See PHOSPHORUS OXYACIDS.

caliche Impure SODIUM NITRATE.

californium, Cf. Radioactive element, at.no. 98, at.mass ^{252}Cf 252.08, m.p. 900°C, density ρ 15 100, electronic configuration [Rn] $5f^{10}7s^2$. ^{252}Cf (961 days) is formed by neutron bombardment of ^{243}Am and ^{244}Cm and ^{249}Cf (360 years) by decay of ^{249}Bk. Cf is purified by ion-exchange chromatography. The silver-grey metal has been obtained by reduction of Cf_2O_3 with lanthanum metal, double-hexagonal close-packed and ccp structures seem established. ^{242}Cf is a potential source of ^{248}Cm. ^{252}Cf is a neutron source used in reactor fuel rods.

californium compounds Californium is the first actinide to show the +2 state, found in $CfBr_2$ formed by reduction of $CfBr_3$ with hydrogen. The normal oxidation state is +3 found in the halides ($CfBr_3$ light green), and the oxide Cf_2O_3. CfO_2 and CfF_4 are formed by strong oxidation.

calixarenes Cone-shaped molecules used in synthesis. A metal atom can be added across the mouth of the cone.

calomel, Hg_2Cl_2. Mercury(I) chloride.

calomel electrode The hydrogen electrode is not easy to use as a general standard for laboratory use, and it is often replaced by a calomel electrode which has a potential known in terms of that of the hydrogen electrode. It consists of a pool of mercury covered with a paste of Hg_2Cl_2 (calomel) and Hg, in contact with a standard solution of KCl saturated with Hg_2Cl_2. The potential of the cell is developed on the Hg and a contact is made with it by means of a platinum wire. Three types of calomel electrode have been used, viz. with 0.1 M, 1.0 M and saturated KCl. Their respective potentials in volts on the hydrogen scale at 298 K are: –0.3338, –0.2800, –0.2415.

Calor gas (Calor propane) (British nomenclature) The trade name for a LIQUEFIED PETROLEUM GAS (LPG) sold in cylinders for domestic and industrial heating. Calor gas is the name used for commercial butane while commercial propane is sold as Calor propane.

calorie A non-S.I. heat unit. Defined in terms of the joule. The thermochemical calorie is 4.184 J.

For fuels it is quite usual to employ kilocalories per kilogram as a measure of calorific value while for foods the unit used is kilocalories per gram. This, however, is often abbreviated to 'Calories', so that a value for carbohydrates of 4.1 'Calories' per gram is 4100 calories per gram.

calorific value The amount of heat given out by complete combustion of unit weight of a solid or liquid or unit volume of a gas in a suitable CALORIMETER.

calorimeter A device for the measurement of calorific values of fuels and foods or for the measurement of heats of chemical reactions. For solid and liquid fuels the BOMB CALORIMETER is used, while for gases a suitable GAS CALORIMETER is used. See CALORIE.

campesterol ((24R)-24-methyl-5-cholesten-3β-ol), $C_{28}H_{48}O$. M.p. 158°C. Occurs with SITOSTEROL in many plants, e.g. rape seed.

camphane See BORNANE.

camphene, $C_{10}H_{16}$. M.p. 51°C, b.p. 159°C. See CAMPHOR, 2,2-DIMETHYL-3-METHYLENE-NORBORNANE. Present in oils, produced artificially by the elimination of hydrogen chloride from bornyl chloride by the dehydrogenation of borneol and isoborneol and by the action of ethanoic anhydride on bornylamine.

camphor (2-oxo-bornane), $C_{10}H_{16}O$. M.p. 179°C, b.p. 209°C. Ordinary commercial camphor is (+)-camphor, from the wood of the camphor tree, *Cinnamomum camphora*. Camphor is

used in the manufacture of celluloid and explosives, and as a plasticizer. It is manufactured from pinene through bornyl chloride to camphene, which is either directly oxidized to camphor or is hydrated to isoborneol, which is then oxidized to camphor.

Related to a series of compounds, e.g. camphane (parent hydrocarbon), camphene (methylene derivative), borneol and isoborneol (hydroxy derivatives), bornyl and isobornyl chlorides. It is a popular remedy for colds and is a constituent of many liniments.

canavanine, $C_5H_{12}N_4O_3$. An amino-acid,

$H_2N(HN)C \cdot NH \cdot O \cdot CH_2 \cdot CH_2 \cdot CH(NH_2) \cdot C(O)OH$

m.p. 184°C, isolated from jackbean meal.

cane sugar SUCROSE.

cannabis (Indian hemp) Recreational drug the constituents of which have been used as sedatives and for other medical applications.

Cannizzaro reaction Two molecules of many aldehydes, under the influence of dilute alkalis, will interact, so that one is reduced to the corresponding alcohol, while the other is oxidized to the acid. Benzaldehyde gives benzyl alcohol and benzoic acid. Compare the aldol condensation.

canonical form A single structure based on

classical valency theory cannot be written for, e.g. the benzene molecule, but instead a number of structures of very similar energy, e.g. the Kekulé structures (canonical forms) can be envisaged.

capillary condensation When a substance wets the wall of a capillary tube in which it is contained, and possesses a concave meniscus, the vapour pressure above the meniscus is less than that in contact with a plane surface of the liquid at the same temperature. This is expressed by the equation

$$\log_\varepsilon \frac{P}{P_s} = 2\frac{\sigma v}{rRT}$$

(p is the pressure at the concave surface, p_s the normal vapour pressure at the same absolute temperature T, σ the surface tension, r the radius of the capillary tube, R the gas constant and v the mol. volume of the liquid). There is thus a considerable tendency for a vapour to condense in the fine capillaries of a porous material such as charcoal. Capillary condensation is the mechanism of the sorption of water by silica gel, and probably also at high humidities by charcoal, but it is unsatisfactory as a general theory of physical adsorption.

capillary electrophoresis Analytical technique for separation of inorganic ions by applying a voltage across buffered filled capillaries.

capric acid (n-decanoic acid), $C_{10}H_{20}O_2$, $CH_3\cdot[CH_2]_8\cdot C(O)OH$. M.p. 31.5°C, b.p. 268–270°C. A fatty acid, occurring in wool as the potassium salt, as esters in fusel oil, and as glycerides in cows' and goats' milk and coconut and palm oils. Esters are used in fruit flavours and in perfumery.

caproic acid (n-hexanoic acid), $C_6H_{12}O_2$, $CH_3\cdot[CH_2]_4\cdot CO_2H$. An oil, m.p. –3.4°C, b.p. 205°C. It occurs as glycerides in the milk of cows and goats, in coconut oil and palm oil. Used in synthesis, flavouring, perfumery.

caprolactam, $C_6H_{11}NO$. Prepared by Beckmann rearrangement of cyclohexanone oxime from adiponitrile from butadiene. M.p. 68–70°C, b.p. 139°C/12 mm. On heating it gives polyamides. Used in the manufacture of Nylon[6]. Cyclohexanone oxime is formed from cyclohexane and nitrosyl chloride.

caprolactone (2-oxepanone) (6-hexanolactone) M.p. –10°C. Forms thermoplastics used in orthopaedic surgery.

capryl alcohol See 2-OCTANOL.

caprylic acid See OCTANOIC ACID.

capsicum Dried ripe fruit of various species of *Capsicum*. Contains capsaicin. Used as a liniment or ointment for lumbago and rheumatism.

captafol Fungicide, trihalomethylthio derivative.

captan Fungicide, biocide, thiotrichloromethyl derivative.

carageenin Hydrocolloid from seaweed used as a thickener.

caramel The brown substance obtained by heating cane sugar or other carbohydrate materials. Its chemical nature is unknown, and its reactions vary with its method of preparation. It is soluble in water and is used as a colouring agent for foodstuffs and drinks.

carbaboranes See CARBORANES.

carbachol (carbamylcholine chloride), $[H_2N\cdot COCH_2CH_2NMe_3]^+Cl^-$, $C_6H_{15}ClN_2O$. Colourless, hygroscopic, m.p. 210–212°C (decomp.). Has physiological action similar to acetylcholine. Used in veterinary medicine.

carbamates Salts and esters of carbamic acid.

carbamic acid, $H_2N\cdot C(O)OH$. The acid is not known in the free state, but salts and esters exist. Ammonium carbamate is formed by the action of dry carbon dioxide on dry ammonia. Commercial AMMONIUM CARBONATE contains appreciable quantities of ammonium carbamate. Urethanes, widely used in plastics technology, are esters of carbamic acid.

carbamide See UREA.

carbamido- The group –NHC(O)NH_2.

carbamyl- The group –C(O)NH_2.

carbanions, R_3C^-. Produced by cleavage of certain C–H; C–halogen, C–metal and C–C bonds. They have high reactivity with air, water, etc. In many cases the existence of these negatively charged species can only be inferred from mechanistic studies. The delocalization of charge (unshared electron pair) over multiple bond systems increases their stability: e.g. triphenylmethyl sodium and cyclopentadienyl sodium are relatively stable.

carbaryl Insecticide and plant growth regulator, carbamate derivative.

carbazole (9-azafluorene), $C_{12}H_9N$. Occurs with anthracene in the solid which separates from anthracene oil. Manufactured from

2-biphenylamine. Used in photography and as dyestuff intermediate. M.p. 238°C, b.p. 335°C.

carbendazim Fungicide, benzamidazole derivative.

carbene, $R_2C:$. Reactive species, containing a formally divalent carbon atom. Produced, e.g., by photolysis of diazoalkanes or by an α-elimination of HX from $H(X)CR_2$. Dihalo-carbenes have great synthetic value, e.g. $Cl_2C:$ may be generated by the thermal decomposition of $Cl_3C \cdot CO_2Na$ and used *in situ*. The reaction of base with haloforms under Riemer–Tiemann conditions, and the Arndt–Eistert synthesis, are reactions which proceed via carbene intermediates. Can act as ligands to transition metals when stable complexes are often formed. $R_2C:$ may be spin paired (singlet) or spin free (triplet).

carbenium ions (carbonium ions) Positively-charged species containing a trivalent carbon atom R_3C^+. They exhibit a broad spectrum of reactivity and stability, the latter decreasing in the order tertiary > secondary > primary. They are present as intermediates in many reactions, e.g. the dissolution of alkenes in strong acids, the dehydrohalogenation of alkyl halides and certain solvolytic reactions. Carbenium ions may be detected by spectroscopic (n.m.r.) methods; tropylium and triphenylmethyl carbenium ions are stable and may easily be isolated; alkyl carbenium ions (as SbF_6^- salts) can be prepared *in situ* from the corresponding alkyl fluoride and SbF_5 at low temperature in SO_2 or $ClSO_2F$, or from alcohols or alkenes in HF/SbF_5 or FSO_3H/SbF_5 solutions and are powerful cationic-type alkylating agents. CCl_3^+ is a fairly stable species. All carbenium ions have a strong affinity for nucleophiles, e.g. water.

carbetamide Herbicide, carbamate derivative.

carbethoxy- The group $-C(O) \cdot OCH_2CH_3$.

carbide formers Alloying elements added to steel (e.g. Mn, Cr, Mo, W, Ti, V, Nb) which stabilize carbides and confer wear resistance when present in a coarse form or creep resistance when present in a fine coherent form.

carbides Derivatives of carbon with elements of lower electronegativity. Carbides are formed by direct interaction of the elements or metal plus suitable hydrocarbon. The most electropositive elements form salt-like carbides which are essentially ionic. Be_2C and Al_4C_3 contain C^{4-} and give CH_4 on hydrolysis. Group I, II, Cu group, Zn, Cd, and Al, La, Pr, Th, form ACETYLIDES containing $[C_2]^{2-}$ ions (HC_2H with water). Other lanthanide and actinide carbides contain C_2 units but give hydrocarbons in addition to ethyne on hydrolysis. Interstitial carbides, e.g. Fe_3C, W_2C, have carbon in holes in approximately cp metal lattices: there are related examples of carbon atoms in metal clusters, e.g. $[Ru_6C(CO)_{14}(mesitylene)]$. Carbides formed with elements close to carbon in the periodic table, e.g. SiC, $B_{13}C_2$, are essentially covalent and are generally extremely hard and infusible. WC, TiC, TaC and NbC are of major importance for use in cutting tools and wear-resistant materials.

carbinol Obsolete name for methanol.

carbitols The trivial name given to the monoalkyl ethers of diethylene glycol, e.g. butylcarbitol, $C_4H_9OCH_2CH_2OCH_2CH_2OH$. See also CELLOSOLVE and DIETHYLENEGLYCOL MONOETHYLETHER. 2-(2-ethoxy)ethanol formed from ethylene oxide plus ethanol. Used as solvents.

carbobenzoxy chloride See BENZYL CHLOROCARBONATE.

carbocyclic A generic term relating to organic compounds which are derived from one or more rings of carbon atoms, e.g. benzene, naphthalene, diphenyl.

carbodiimides Derivatives $RN=C=NR'$ (see DICYCLOHEXYLCARBODIIMIDE). MeNCNMe is formed from MeNHC(S)NHMe and HgO; it is stable at low temperatures. Carbodiimides are formed by heating isocyanates and can be polymerized in the presence of catalysts. Used in synthesis of amides.

carbofuran Insecticide, nematocide, carbamide derivative.

carbohydrases Enzymes which act on carbohydrates.

carbohydrates One of the principal classes of naturally occurring organic compounds. An approximate formula is $(CH_2O)_x$ and for various values of x includes sugars, starches and cellulose, each essential to plant and animal life. They are produced in plants as the result of photosynthesis, the chlorophyll-catalysed combination of carbon dioxide and water under the influence of light. The simple carbohydrates are mono-, di- or polysaccharides, having heterocyclic repeating units usually containing 5 or 6 carbon atoms joined through oxygen linkages.

For example, sucrose and lactose are disaccharides which are simple carbohydrates; cellulose, a polysaccharide, is a polymer containing approximately 2000–3000 glucose units per molecule. As many of the carbon atoms are asymmetric the carbohydrates can exist in many stereochemical and structural modifications.

carbolan dyes A range of azo or anthraquinone dyes.

carbolic acid An old name for PHENOL.

carbomethoxy- The group $-C(O)\cdot OCH_3$.

carbon, C. Non-metallic element, at.no. 6, at.mass 12.011, m.p. 4492°C (graphite) 4440°C (diamond), density ρ 220 (graphite), crustal abundance 480 p.p.m., electronic configuration [He] $2s^2 2p^2$. Occurs in both free and combined (CO_2, $(CO_3)^{2-}$, etc.) states. The crystalline forms are diamond (metastable) and graphite, fullerenes – for details of structure see under individual headings. C_2 occurs in space. Various forms of amorphous carbon are described: most contain microcrystalline material. ACTIVE CARBON and CARBON BLACK are widely used in industry. Steel is an iron-carbon system. Carbon fibres, produced by pyrolysis, confer great strength on plastic materials. ^{12}C is the standard of relative atomic mass, ^{13}C has nuclear spin of ½ and can be used for n.m.r. measurements, ^{14}C is radioactive and is used for CARBON DATING since it is formed in the atmosphere and absorbed only by living materials. Carbon burns in oxygen and reacts with halogens.

carbonates Salts of the weak acid H_2CO_3. The normal carbonates include many minerals and compounds of great technical importance (e.g. $CaCO_3$, $MgCO_3$, Na_2CO_3, $ZnCO_3$). The hydrogen carbonates contain the $[HCO_3]^-$ ion and important salts include $NaHCO_3$ and $Ca(HCO_3)_2$. The carbonates of the alkali and alkaline earth metals are stable (the latter insoluble). Most other metal carbonates are precipitated as basic carbonates, e.g. $2PbCO_3 \cdot Pb(OH)_2$. All carbonates except the alkali metal carbonates lose CO_2 on strong heating. Carbonate esters are important solvents particularly in batteries and reactants. Di (and poly) carbonates are known and are used in synthesis and form resistant polymeric materials used in, e.g. lenses.

carbonation Addition (frequently insertion) of CO_2 into derivatives such as RMgX to give $RC(O)OMgX$.

carbon black A range of technologically useful materials obtained by the incomplete combustion of natural gas or liquid hydrocarbons. The different varieties vary in particle size and colour value. Probably carcinogenic. Used in rubber manufacture, as a filler (tyres 65%, other rubber 25%), as a pigment for paint, ink, polish, rubber, leather, etc. and for decolorizing. Global capacity 1999 8.5 megatonnes.

carbon chalcogenides Carbon forms CS_2 (see CARBON DISULPHIDE), CSSe, CSe_2, CSTe. These are all fairly similar to CS_2.

carbon chemistry Carbon is an element of Group 14. It shows a formal oxidation state of +2 in CO and in carbenes but this state is relatively reactive. The majority of carbon compounds are in the +4 oxidation state and the chemistry is dominated by the tendency to catenation (C–C bond formation) and to the formation of multiple bonds (C=C in alkenes, C≡C in alkynes, conjugated bonds and aromatic systems) and the formation of stable bonds to most other elements (see element organic derivatives) but particularly to O, N, S, halogens and H. Anionic derivatives of carbon are found in carbanions, carbides and carbenes; cations in carbenium ions. Basis of life as on Earth.

carbon dating ^{14}C is continuously formed in the atmosphere by the reaction

$$^{14}N + {}^1n \rightarrow {}^{14}C + {}^1H$$

The ^{14}C exchanges with ^{12}C in living organisms, but exchange ceases on death. The radioactive ^{14}C content decays with a half-life of 5730 years. Hence the age of a once-living material may be established by determining the content of ^{14}C.

carbon dioxide, CO_2. See CARBON OXIDES.

carbon disulphide, CS_2. M.p. –112°C, b.p. 46°C. An important solvent and reagent (C plus S). Pure CS_2 is pleasant smelling. It is toxic and exceedingly inflammable. Used in the manufacture of Rayon and CCl_4 and as solvent. Adds SH^- to give TRITHIOCARBONATES, $(CS_3)^{2-}$, OR^- to give XANTHATES, $(ROCS_2)^-$, amines to give DITHIOCARBAMATES, $(RHNCS_2)^-$.

carbon electrodes Rods used in arcs and in electrolysis (e.g. F_2 production).

carbon fibres Graphite fibres used in the strengthening of composites. Precursor fibres, e.g. polyacrylonitrile, are carbonized under extreme heat.

carbon fluorides See FLUOROCARBONS.

carbonic acid, H_2CO_3. A very weak dibasic acid formed from CO_2 and H_2O, the anhydrous acid can be prepared in ether at –30°C. The acid forms CARBONATES and hydrogen carbonates as salts. Stable compound with half-life for decomposition 180 000 years.

carbonic anhydrase Important enzyme which reversibly catalyses the reaction

$$H_2O + CO_2 \rightleftharpoons H_2CO_3$$

in an organism. In respiration, it catalyses the formation of carbon dioxide from hydrogen carbonate in the blood vessels of the lung, whereas in kidney function it promotes the formation of carbonic acid. The enzyme contains Zn as a prosthetic group. See DICTIONARY OF BIOLOGY.

carbonium ions See CARBENIUM IONS.

carbonization The process of destructive distillation of coal, decomposition by heat in the absence of air. This produces a residue of coke along with coal gas, coal tar and other liquors. In high temperature carbonization (HTC) coal is heated in refractory retorts to about 1100°C and almost all the volatile matter is driven off leaving a non-reactive coke. HTC was used to produce TOWN GAS in continuous vertical retorts and for hard, metallurgical coke in coke ovens. Low-temperature carbonization (LTC) is used to produce 'semi-coke' or 'smokeless fuel' by heating coal to about 900 K.

carbon monoxide, CO. See CARBON OXIDES.

carbon oxides
Carbon monoxide, CO. M.p. –205°C, b.p. –191°C. Colourless, odourless, toxic gas formed by incomplete combustion of C or carbon compounds, C plus CO_2 at 900–1000°C, C plus H_2O (with H_2 WATER GAS), methanoic acid plus conc. H_2SO_4. Low levels are produced in the body. Burns in air with pale blue flame and can react with oxygen explosively. Acts as a reducing agent (e.g. PbO or Fe_2O_3 to Pb or Fe at high temperatures). $[IO_3]^-$ to I_2 at 90°C (used in analysis) gives carbonyls with many transition metals. Cl_2 gives phosgene. Used industrially in synthesis (e.g. carbonylation of alkynes to carboxylic acids – synthesis gas), and in metallurgical reduction processes and the purification of nickel.

Carbon dioxide, CO_2. Sublimes –78.5°C. A colourless gas at room temperature, occurs naturally in the atmosphere and plays an important part in animal and plant respiration. Produced by the complete combustion of carbon-based fuels (industrially from flue gases and from synthesis gas used in ammonia production) and by heating metal carbonates or by the action of acid on carbonates. Soluble in water to give mainly a hydrate and the weak acid, H_2CO_3. Detected by the formation of a white precipitate with a solution of $Ca(OH)_2$ (limewater). Used in mineral waters under pressure, as a refrigerant (the solid 'dry ice' or as the liquid), as a chemical reagent (RMgX gives RCO_2MgX), as an inert gas, as a fire extinguisher. Has a linear molecule, O=C=O. Atmospheric CO_2 absorbs i.r. radiation and is a major cause of the greenhouse effect. Supercritical CO_2 of use as a solvent (e.g. in dry-cleaning, to decaffeinate coffee and in semiconductor manufacture). Emissions 1998 7.4 billion tonnes carbon equivalent.

Carbon suboxide, C_3O_2, OCCCO. M.p. –111°C, b.p. 6.8°C. A toxic gas (malonic acid plus P_2O_5) which polymerizes at room temperature. Reforms malonic acid with water.

$C_{12}O_9$ mellitic anhydride.

carbon residue The tendency for diesel fuel, fuel oil, or lubricating oil to form carbon on heating or burning in the absence of excess air can be measured by evaluating the carbon residue.

carbon sulphides See CARBON CHALCOGENIDES.

carbon tetrabromide, CBr_4. M.p. 94°C, b.p. 190°C. White crystalline solid (Br_2 plus CS_2 plus I_2).

carbon tetrachloride (tetrachloromethane, perchloromethane), CCl_4. M.p. –23°C, b.p. 76.5°C. Manufactured from CS_2 plus Cl_2 or by chlorination of CH_4 and other hydrocarbons. In presence of H_2O gives HCl and phosgene Cl_2CO. Relatively toxic to the liver. The simplest member of the series of perchlorocarbons. Used as a reagent and solvent (use tends to be replaced by higher homologues and chloroethane derivatives), as a fumigant and in fire extinguishers.

carbon tetrafluoride, CF_4. M.p. –150°C, b.p. –128°C. An inert gas (C plus F_2). Used in refining and as insulator.

carbon value Measure of the tendency of an oil to form carbon when used as lubricant.

carbonylate ions Ions containing metal carbonyls. Anions are prepared by the action of a base or an alkali metal on the carbonyl (e.g. $Mn_2(CO)_{10}$ plus pyridine or Na gives $[Mn(CO)_5]^-$). Cations are prepared from a Lewis acid and a co-ordinating group on the carbonyl halide (e.g. $Mn(CO)_5Cl$ plus $AlCl_3$ plus CO gives

$[Mn(CO)_6]^+$). These ions generally have a noble gas electronic structure (both $[Mn(CO)_5]^-$ and $[Mn(CO)_6]^+$ have the Kr electronic structure for Mn).

carbonylation The reaction of an organic or intermediate organometallic compound, with CO. E.g. methanol will react to ethanoic acid and methyl esters over Co or Rh catalysts. Alkynes will react with carbon monoxide in the presence of a metal carbonyl (e.g. $Ni(CO)_4$) and water to give propenoic acids, ($R \cdot CH = CH \cdot C(O)OH$); with alcohols ($R'OH$) to give propenoic esters, $RCH:CHC(O)OR'$; and with amines ($R'NH_2$) to give propenoic amides $RCH:CHC(O)NHR'$. Using alternative catalysts, e.g. $Fe(CO)_5$, alkynes and carbon monoxide will produce cyclopentadienones or hydroquinols. A commercially important variation of this reaction is hydroformylation (the OXO REACTION).

carbonyl chloride, Cl_2CO. See CARBONYL DERIVATIVES.

carbonyl derivatives Derivatives containing the C=O group. The category includes ALDEHYDES, RCHO; KETONES, RR'CO; CARBOXYLIC ACIDS; METAL CARBONYLS; and CARBONYL HALIDES. The halides, all slowly hydrolysable, X_2CO; carbonyl fluoride, F_2CO, m.p. $-114°C$, b.p. $-83°C$ (Cl_2CO plus KF); carbonyl chloride, Cl_2CO, phosgene, m.p. $-127.9°C$, b.p. $8°C$, a colourless poisonous gas (CO plus Cl_2 plus light or over a catalyst – often charcoal). Gives urea, $OC(NH_2)_2$, and NH_4Cl with ammonia, used to form toluene diisocyanate used in polyurethane plastics, and polycarbonates and metal recovery. Carbonyl chalcogenides include SCO (CO plus S). An explosive, inflammable gas.

carbonyl group The >C=O group in such compounds as aldehydes, ketones, metallic carbonyls and phosgene. For the general reactions of this group see, e.g. ALDEHYDES and KETONES.

carbonyl halides Applied to either the derivatives X_2CO (see CARBONYL DERIVATIVES) or to metal carbonyl derivatives, e.g. $Mn(CO)_5Cl$.

carbonyl hydrides The metal carbonyl hydrides, e.g. $HMn(CO)_5$. These compounds have a weak acidic character and form carbonylate anions, e.g. $[Mn(CO)_5]^-$. The H–M bond is very short.

carbonyloxime Misnomer for FULMINIC ACID.

carbonyls See METAL CARBONYLS.

carboplatin Platinum(II) complex. Antitumour drug.

carboranes (carbaboranes) Derivatives of the borane anions in which a CH group has formally replaced a BH^-. Prepared by the action of alkynes and borane derivatives. The series contain closo (polyhedral) carboranes, e.g. 1,2- and 1,7-$B_{10}C_2H_{12}$ (based on an icosahedron), $B_7C_2H_9$ and nido (open) carboranes, e.g. $B_7C_2H_{13}$. Nido carborane anions, e.g. $[B_9C_2H_{11}]^{2-}$ form metal complexes, e.g. $[(B_9C_2H_{11})_2Fe]^{2-}$ through the open pentagonal faces of the ions (dicarbollide derivatives). Used in compounds for boron neutron capture therapy.

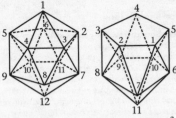

closo -$[B_{10}C_2H_{12}]$
(C at 1,2- or 1,7-)

nido -$[B_9C_2H_{11}]^{2-}$
(C at 1,2- or 3,5-)

carborundum See SILICON CARBIDE.

carbosulfan Insecticide, carbamate derivative.

Carbowaxes Trade name for a wide range of waxy polyethylene glycols of the general formula $HOCH_2(CH_2OCH_2)_xCH_2OH$. Carbowaxes of high molecular weight are used in rubber mixes, as dispersing agents in the textile industry and for pharmaceutical purposes. Also used as the stationary phase in GLC columns.

carboxin Fungicide, carboxamide derivative.

carboxy- A prefix denoting that the substance contains a carboxyl group, e.g. –C(O)OH.

carboxyhemoglobin The combination of hemoglobin with CO instead of O_2, by coordination, gives carboxyhemoglobin. The affinity of hemoglobin for carbon monoxide is, in humans, 500 times as great as for oxygen; this accounts for the high toxicity of carbon monoxide as it prevents hemoglobin from fulfilling its natural function. (Does not contain a –C(O)OH group, see CARBOXY-.)

carboxylase An enzyme present in yeasts, bacteria and plants which catalyses the decarboxylation of α-ketoacids, in particular pyruvic acid, which is converted to ethanal.

Decarboxylases, which are highly specific for individual amino-acids, decarboxylate these to amines.

carboxylation Addition of CO_2.

carboxyl group The $-C(O)OH$ group.

carboxylic acids Organic compounds containing one or more carboxyl ($-C(O)OH$) groups; the number of these groups is indicated by the prefixes mono-, di-, tri-, etc. These acids are much weaker than the mineral acids such as hydrochloric, but they are partially ionized in aqueous solution and all liberate carbon dioxide from metallic carbonates and hydrogen carbonates. The strength of the acid is markedly affected by the nature of the remainder of the molecule, e.g. $Cl_3CC(O)OH$, trichloroethanoic acid is a very strong acid. They form salts with metals and with organic bases, and esters with alcohols. Anhydrides containing $-C(O) \cdot O \cdot C(O)-$ groups are formed from most acids by elimination of a molecule of water between two carboxyl groups. This elimination may be either inter- or intramolecular. The hydroxyl ($-OH$) group is replaceable by halogens (with, e.g. PCl_5) to give acyl halides. The acids are formed by oxidation of the corresponding alcohol or aldehyde; by hydrolysis of nitriles with alkalis; or by treating Grignard reagents with carbon dioxide and decomposing the product with sulphuric acid. Many of these acids occur naturally in plants and animals, either free or as salts or esters.

carbylamine reaction A qualitative test for primary amines carried out by warming the suspected amine with chloroform and an alcoholic solution of potassium hydroxide. Under these conditions primary amines produce the intensely nauseating odour characteristic of the carbylamines (ISONITRILES).

carbylamines See ISONITRILES.

carbyne derivatives Metal derivatives containing a $M{\equiv}CR$ grouping.

carceplexes Two cup-shaped molecules linked together to form a molecular trap (a carcerand). Used to examine unstable species.

carcinogens Substances capable of causing the development of malignant cells in animals or plants. Many amino aromatic, nitro and other derivatives, including methyl (chloromethyl) ether, ethyleneimine and propiolactone, are carcinogenic and present danger. See DICTIONARY OF BIOLOGY.

carenes Dicyclic terpenes. Constituents of turpentine.

carfentriazine-ethyl Herbicide, triazole derivative.

Carius method The quantitative determination of S and halogens in covalent (organic) compounds by complete oxidation of the compound with conc. nitric acid and subsequent estimation of precipitated AgX or $BaSO_4$.

Carius tube A thick-walled glass tube originally used in the CARIUS METHOD but used in any reaction involving volatile materials.

carmine See CARMINIC ACID.

carminic acid, $C_{22}H_{20}O_{13}$. The red colouring matter of cochineal obtained from the *Coccus* insects. Used as a stain and dyestuff including foodstuffs.

carnallite, $KCl \cdot MgCl_2.6H_2O$. A principal component of many salt deposits. Used as a source of potassium salts.

carnauba wax The dried exudation from the leaves of the South American copaiba palm. The principal constituent is melissyl cerotate. It is used in many polishes and varnishes, as when mixed with other waxes it makes them harder and gives more lustre. Widely used in cosmetics and for coating pharmaceuticals.

carnitine, $C_7H_{15}NO_3$, $(CH_3)_3N^+CH_2CHOHCH_2C(O)O^-$. Isolated from skeletal muscle. It acts as a carrier for ethanoyl groups and fatty acyl groups across the mitochondrial membrane during the biosynthesis or oxidation of fatty acids. Used in pharmaceuticals.

carnosine (N-β-alanylhistidine), $C_9H_{14}N_4O_3$. M.p. 246–250°C (decomp.). A dipeptide present in mammalian muscle. Contains the amino-acid β-alanine which is not found in proteins (cf. conserine).

Carnot cycle A hypothetical scheme for an ideal heat machine. Shows that the maximum efficiency for the conversion of heat into work depends only on the two temperatures between which the heat engine works, and not at all on the nature of the substance employed. Efficiency of the heat engine is work done divided by the heat supplied from the hot source.

carnotite, $KUO_2VO_4.1.5H_2O$. An important uranium ore.

Caro's acid, H_2SO_5. See PERMONOSULPHURIC ACID.

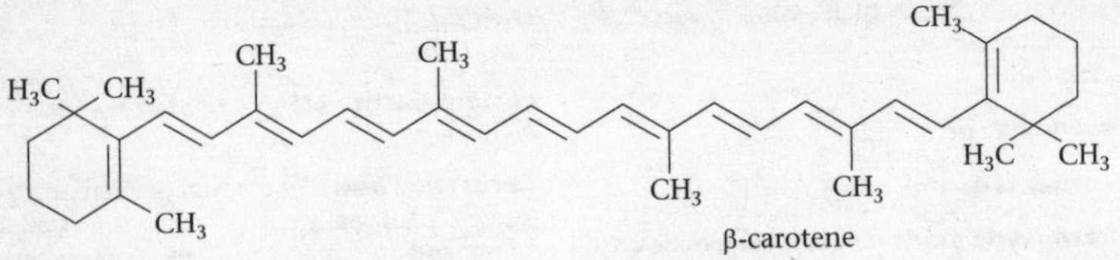

β-carotene

carotene, $C_{40}H_{56}$. M.p. 181–182°C. Carotenes are colouring matters of green plants and are found in many animal and bacterial tissues. It is the chief yellow colouring matter of carrots, of butter and of yolk of egg. It exists in several forms, the most common being β-carotene, the primary source of vitamin A for humans and animals. Carotene is found in plant chloroplasts. In animals it serves as a precursor of vitamin A. Used in foodstuffs.

carotenoids Pigments related to carotene which occur in plants, particularly in the leaves, and in certain animal tissues. Important in photosynthesis, protective antioxidants; play a part in visual systems. Used in cosmetics and foodstuffs.

carrageenan See CARAGEENIN.

cartap Insecticide, carbamate type.

carvacrol (2-hydroxycymene), $C_{10}H_{14}O$. A colourless liquid, b.p. 237–238°C; a constituent of many essential oils. Has been used against infection.

carveol, $C_{10}H_{16}O$. A monocyclic terpene alcohol.

carvestrene The racemic form of SYLVESTRENE.

carvone, $C_{10}H_{14}O$. A ketone derived from the terpene limonene. B.p. 230°C. Found in caraway and dill seeds.

caryophyllene, $C_{15}H_{24}$. Sesquiterpene hydrocarbon which is the main hydrocarbon constituent of oil of cloves. Colourless oil, b.p. 123°–125°C/10 mm. Used in perfumery.

casein A phosphoprotein which occurs in the milk of mammals as a suspension of calcium caseinate. It can be obtained from milk by treatment with mineral acid or rennet. See also RENNIN.

casing head gas Natural gas evolved at an oil well head.

cassel yellow See LEAD CHLORIDES.

cassiterite (tinstone), SnO_2. The principal ore of tin. See TIN OXIDES.

castile soap Made from olive oil.

cast iron (pig iron) See IRON. Iron containing 2.5–4% C either in the form of cementite, Fe_3C (white cast iron, brittle but wear resistant) or graphite (grey cast iron, low strength). Alloying metals are added to improve properties when materials may replace steel.

Castner–Kellner cell An electrolytic cell for the production of SODIUM HYDROXIDE.

castor oil Oil expressed from the seeds of *Ricinus communis*. It consists almost entirely of the glyceryl esters of ricinoleic acids. Used industrially for the manufacture of hydrated castor oil and also in the production of polyurethanes and coatings.

catabolism See METABOLISM.

Catacarb process An extraction process used to remove carbon dioxide from process gases by scrubbing the hot gases with potassium carbonate solution containing additives which increase the hydration rate of the gas in the solution. The Vetrocoke process is similar. See BENFIELD PROCESS.

catalase Enzymes which destroy hydrogen peroxide. They contain hematin or related compounds as the prosthetic group. Used commercially to convert latex to rubber and to remove H_2O_2 from food. See also PEROXIDASES.

catalysis The use of a CATALYST to affect the rate of a reaction.

catalyst A substance which when added to a reaction mixture changes the rate of attainment of equilibrium in the system, formally without itself undergoing a permanent chemical change. Catalysts may be homogeneous or heterogeneous, but all comply with the following criteria:

   1 Although the catalyst affects the rate of reaction, it cannot affect the position of equilibrium in a reversible reaction.

   2 In theory the catalyst can be recovered

chemically unchanged at the end of the reaction, although it may be changed physically.

The efficiency of a catalyst is expressed in terms of the turnover number. Catalysts are often supported on silica and other oxides.

catalytic converters Generally applied to catalyst systems in automobiles designed to convert hydrocarbons, CO, NO, to harmless materials. Generally contain Pd or other metal species on a ceramic substrate.

catalytic cracking The most widely used method of producing gasoline from heavy petroleum distillates is to subject the heavy fractions to catalytic cracking. Nowadays fluidized bed catalytic cracking is almost universally used, where the powdered catalyst is moved through the system in a fluidized state.

Many catalysts have been used but the standard catalysts are generally mixtures of silica and alumina or natural or synthetic aluminium silicate zeolites.

The major product of cracking is gasoline but gases and light and heavy gas oils are produced in proportions depending on feedstocks and operating conditions. The gasoline obtained is usually of a fairly high OCTANE NUMBER due to the high proportion of unsaturated hydrocarbons. See CRACKED GASOLINE.

catalytic reforming Catalytic reforming processes are employed to improve the OCTANE NUMBER of straight-run gasoline by increasing the proportion of aromatics and branched-chain alkanes. Feedstocks consist mainly of alkanes and NAPHTHENES and the main reactions are dehydrogenation of naphthenes to produce aromatics, isomerization of alkanes and naphthenes, dehydrocyclization of *n*-alkanes yielding aromatics and hydrocracking.

Earlier processes (e.g. hydroforming) used MoO_3–Al_2O_3 catalysts but platinum-based catalysts are now extensively used, enabling longer on-stream times before catalyst replacement.

catechol See 1,2-DIHYDROXYBENZENE.

catecholamines Amines found in urine and plasma; dopamine, adrenaline and their metabolites are examples.

catenanes Compounds containing interlocking rings. Proteins of viral coatings form catenanes that form coatings known as capsids.

catenation The formation of element–element bonds as in hydrocarbons and higher silanes. Carbon is the element showing the greatest tendency to catenation.

catforming Catalytic reforming.

cathode The electrode which carries the negative charge in an electrochemical cell. In electrolysis the positively charged ions are discharged at the cathode.

cathodic protection A method of metal protection by connection to a more active metal which is itself oxidized. Used to prevent corrosion of submerged or underground metal work. For example, a moored ship may be made the cathode of a cell by suspending magnesium anodes some distance below the ship and short-circuiting them to the hull. Alternatively in fresh water, which is a poor conductor, direct current supply may be used instead of electrodes. Thus a negative lead may be connected to an underground pipeline while a positive lead is connected to a graphite or iron anode some distance away, moist soil acting as the electrolyte.

cation An atom which has lost one or more electrons or a positively charged group of atoms, e.g. $[NH_4]^+$. Cations are present in the solid state, e.g. Na^+ in NaCl, and in solutions and melts. In electrolysis cations travel to the cathode.

cationic detergents Detergents which, in aqueous solution, form positively charged surface-active ions. They are often based on quaternary ammonium compounds, derivatives of pyridine or fatty acid amides. Although not used extensively as detergents they are used in the textile industry, while those with bactericidal properties are used in disinfecting detergents.

cationic polymerization Polymerization initiated by acids – positive ions.

caustic potash, KOH. POTASSIUM HYDROXIDE.

caustic soda, NaOH. SODIUM HYDROXIDE.

ccp Cubic close-packed. See CLOSE-PACKED STRUCTURES.

Cd Cadmium.

C.D. Circular dichroism.

Ce Cerium.

celestine, $SrSO_4$. Native strontium sulphate, a major source of Sr compounds.

cell A device used as a source of, or for the use of, electrical energy. Electrochemical cells consist of two electrodes dipping into one or more electrolyte solutions. If the electrodes are in separate solutions, electrical contact between

← β-Glucose → unit

← Cellobiose → unit

solutions is achieved by means of a salt bridge, porous disc or ionic conductor. The electrodes may be of the same metal as the electrolyte or of another inert material, e.g. platinum.

cell dimensions The dimensions of the unit cell in a crystal.

cellobiose, $C_{12}H_{22}O_{11}$.

4-[β-D-Glucopyranosido]-D-glucopyranose. It is the disaccharide that forms the repeating unit of cellulose through the terminal –OH groups. Hydrolysed to β-D-glucose.

cellophane A transparent sheet cellulose prepared by extruding cellulose xanthate solution into a bath of acid. Sensitive to humidity although can be rendered waterproof by varnishing. Softened by glycerol. Highly inflammable. Used as a wrapping material for foodstuffs and cigarettes but being replaced by polypropene.

Cellosolve Ethylene glycol monoethyl ether. Oxitol.

Cellosolve acetate A trade name for ethylene glycol monoethyl ether acetate. Important solvent.

cell potential The potential difference in volts of two connected electrodes which have not reached chemical equilibrium. ˙v.FE = $\Delta_T G$.

cellular plastics The various forms of foamed and expanded plastics which have a form containing cells formed by gases (blowing agents). Have light weight and frequently favourable thermal properties. Used in packaging, thermal insulation, lightweight partitions and roofing. Plastics used include polystyrene, polyurethanes, PVC, polyethylene, urea and phenolic resins. Foaming gases were chlorofluorocarbons but now non-chlorinated species used.

celluloid A little-used inflammable thermoplastic based on cellulose nitrate plasticized with camphor.

cellulose, $(C_6H_{10}O_5)_x$. The chief constituent of the cell wall of all plants, and the most abundant organic substance found in nature. It is a polymer of glucose, with over 3500 repeat units in a chain. The glucoside linkage is β, whereas in starch it is α. Contains cellobiose units.

Strong acids completely hydrolyse cellulose to glucose; very mild hydrolysis gives hydrocelluloses with shorter chains and lower viscosity and tensile strength. Under special conditions a large yield of cellobiose is obtained.

Cellulose dissolves in strong mineral acids, in NaOH and in cuprammonium solution. It forms a triacetate (tri-ethanoate), a trinitrate and ethers with methyl, ethyl and benzyl alcohols; all of these are of great industrial importance. Its largest use is in the RAYON industry. Wood pulp and cotton linters are the most important sources from which it is prepared commercially.

Carboxymethyl cellulose sodium from alkali cellulose and sodium chloroethanoate is used in drilling fluids, detergents, resins, adhesives and as a food stabilizer.

cellulose acetate (ethanoate) plastics Formed by treatment of cellulose (cotton, purified wood pulp) with ethanoic anhydride, ethanoic acid and sulphuric acid. Heavy acetylation gives the triacetate used as a plastic material. Part acetylation also gives very useful materials. Processed by injection moulding. Used to form fibres and for use in textiles.

Has relatively low strength and is moisture sensitive.

cellulose ethers Formed by alkyl- and aryl-halides on cellulose in alkali solution. Used as plastic materials. Methyl cellulose is water soluble and is used as an emulsifying, sizing and priming material.

cellulose nitrate, nitrocellulose Prepared by treating cellulose (cotton waste or wood pulp) with HNO_3 or HNO_3 plus sulphuric or phosphoric acid. If the N content is greater than 13% it is known as gun cotton or smokeless powder (cellulose trinitrate); collodion cotton pyroxylin is a rather less nitrated species. Very inflammable, shipped as the moist material. Used as a propellant and in blasting gelatine. Mainly used as a lacquer and for making some solid objects, although these are very inflammable. Formerly used as films.

cellulose phosphate (sodium) Used as an ion exchanger.

Celsius scale Temperature scale commonly used in Europe. Degrees Celsius are degrees Kelvin minus 273.15.

cement A substance used to make other substances cohere. There are many kinds of cement, but they may be divided into two main groups: **1** cements used to bind together numerous particles so as to form a coherent mass of considerable strength and **2** used to unite two or more separate masses (e.g. portions of a broken article); this second class includes glues and various organic cements. Cements in the first group include Portland cement, high alumina cement and many others. Cements which form a plastic paste with water and set hard on standing are known as hydraulic cements. Cements are also known by their chief constituent, e.g. calcareous cements, aluminous cements, siliceous cements; by their characteristic property, e.g. acid-resisting cements, quick-setting cements, rapid-hardening cements; by their reputed origin, e.g. Roman cement, and by their fancied resemblance to some other material, e.g. Portland cement which was, at one time, supposed to resemble Portland stone. Cements which can be used at high temperatures, e.g. for repairing furnaces, are known as refractory cements. The term 'cement' is also used for any argillaceous limestone capable of producing cement when sintered. Composition varies for different uses.

Cements are commonly made by heating a mixture of limestone and clay to about 1700°C.

The product is ground with gypsum. Chemically cements consist of a mixture of calcium silicates and aluminates with some sulphate present.

cement fondu, ciment fondu A hydraulic cement composed chiefly of calcium aluminate.

cement, high alumina A rapid-hardening, resistant, dark-coloured cement produced by heating a mixture of bauxite and limestone to about 1850 K.

cement, quick-setting PORTLAND or other which has a shorter setting time than ordinary cement.

cement, rapid-hardening PORTLAND or other cements which harden rapidly.

cement, supersulphated A cement with high sulphate resistance used largely for work underground and for work in sea water.

cementite, Fe_3C. See IRON.

centre of symmetry See SYMMETRY ELEMENTS.

centrifugal pump The most widely used pump in the chemical and process industries. It consists of a rotor with a number of curved vanes rotating inside a flat, cylindrical casing. Liquid enters the centre of the casing along the rotor axis and is swept along the vanes into the volute which is a ring-shaped chamber.

centrifugal separators See CENTRIFUGES.

centrifugation Separation of substances by means of a CENTRIFUGE.

centrifuges Machines which employ centrifugal force to obtain high rates of sedimentation or filtration in order to separate a solid and a liquid or two immiscible liquids. See also ULTRACENTRIFUGE.

cephalins (kephalins) Phospholipids similar to lecithins except that ethanolamine or serine replaces choline in the molecule. They are found in all animal and vegetable tissues. They are easily hydrolysed to fatty acid, etc.

cephalosporin C ANTIBIOTIC produced by *Cephalosporium acremonium*. Converted chemically to other semi-synthetic antibiotics.

ceramics Engineering materials and products made from inorganic chemicals (excluding metals and alloys) by high-temperature processing. They are resistant to chemical attack, can range from semi-conductors to insulators and include refractories, glasses, cement and cement products, vitreous enamels, abrasives, pottery, china, porcelain, clay wares, alumina, etc. Withstand high temperatures but brittle.

cerebrosides A group of lipids which occur most abundantly in the membranes of nerves and brain cells.

ceresin The name formerly given to a hard, white, odourless wax obtained from fully refined OZOKERITE. The term now often refers to certain forms of hard, brittle waxes obtained from petroleum.

cerium, Ce. Metallic element, at.no. 58, at. mass 140.12, m.p. 798°C, b.p. 3443°C, density ρ 6770, crustal abundance 68 p.p.m., electronic configuration [Xe] $4f^15d^16s^2$. The commonest of the LANTHANIDES, obtained in large quantities after extraction of thorium from monazite. The metal is used in alloys for improving the malleability of cast iron and the mechanical properties of Mg alloys; Ce alloys are pyrophoric; CeO_2 is used as a glass polish, ceramic coatings, in incandescent gas mantles and catalysts. Cerium sulphate is a catalyst in 'self-cleaning ovens'.

cerium compounds Cerium shows oxidation states +4 and +3. Ce(III) compounds are typical LANTHANIDE COMPOUNDS.

Ce^{3+} (f^1 colourless) $\rightarrow$ Ce (-2.48 volts in acid) $CeF_4(CeO_2 + F_2)$ is the only cerium(IV) halide although hexachlorocerates(IV) containing the $CeCl_6^{2-}$ ion are stable. Most Ce compounds give CeO_2 on heating in oxygen, intermediate phases are present in the Ce_2O_3–CeO_2 system. Various cerium(IV) sulphates, e.g. $Ce(SO_4)_2 \cdot 2H_2SO_4$ are obtained from CeO_2 and H_2SO_4 and are used in analysis; complex nitrates, $M_2Ce(NO_3)_6$, are also stable. The Ce(IV) state is the most stable +4 state of the lanthanides and is the only +4 state stable in aqueous solution. Ce^{4+} (yellow-orange) $\rightarrow Ce^{3+}$ (+ 1.28 volts in HCl). CeI_2 (CeI_3 + Ce) is Ce^{3+}, $2I^-$, e^-. Organocerium compounds used in synthesis.

cermets Pressed or sintered ceramic–metal mixtures, e.g. $Cr–Al_2O_3$, used in high-temperature applications, e.g. jet engines, as anodes in fuel cells and in oxidation of hydrogen or hydrocarbon fuels.

cerotic acid (hexacosanoic acid), $CH_3 \cdot [CH_2]_{24} \cdot C(O)OH$. M.p. 88°C. Occurs free in many waxes.

cerussite, $PbCO_3$. A native lead carbonate used as a Pb ore.

ceryl alcohol (1-hexacosanol), $C_{26}H_{54}O$, $CH_3 \cdot [CH_2]_{24} \cdot CH_2OH$. Colourless crystals, m.p. 79°C. Occurs as esters in various waxes, ceryl palmitate being the chief component of opium wax and ceryl cerotate of Chinese wax.

cesium The North American spelling for CAESIUM.

cetane See HEXADECANE, $C_{16}H_{34}$.

cetane number See KNOCK RATING.

cetrimide (cetyltrimethylammonium bromide, CTAB), $[C_{16}H_{33}N(CH_3)_3]^+Br^-$. The commercial product is made by the condensation of cetyl bromide with trimethylamine. Contains other alkyl ammonium bromides. A creamy-white powder, soluble in water to give a readily foaming solution. A cationic detergent and wetting agent.

cetyl alcohol See HEXADECANOL.

cetyl trimethylammonium bromide See CETRIMIDE.

Cf Californium.

CFC Chlorofluorocarbons. See FREONS.

CGA 245704 Plant activator, thiadiazole derivative.

CGA 50439 Acaricide, thiazole derivative.

chain reactions Reactions which proceed by means of a chain reaction mechanism, where initial products react further with the reactants or other products all producing active intermediates. Example, the hydrogen–bromine reaction.

(a) $Br_2 \rightarrow 2Br\cdot$
(b) $Br\cdot + H_2 \rightarrow HBr + H\cdot$
(c) $H\cdot + Br_2 \rightarrow HBr + Br\cdot$
(d) $H\cdot + HBr \rightarrow H_2 + Br\cdot$
(e) $2Br\cdot \rightarrow Br_2$

The first stage (a) of the reaction represents the dissociation of bromine into bromine atoms. Both steps (b) and (c) lead to production of HBr, and since bromine atoms are regenerated the process can be repeated. Reaction (d) leads to an inhibition in the chain reaction and step (e) represents chain termination.

Free radical reactions (e.g. the decomposition of ethanal) proceed via chain mechanisms as does polymerization.

In nuclear chemistry, a fission reaction (see ATOMIC ENERGY) may be initiated by a neutron and may also result in the production of one or more neutrons, which if they react in like manner could start a chain reaction. Normally, moderators such as cadmium rods which absorb neutrons are placed in the reactor to control the rate of fission.

chair form See CONFORMATION.

chalcogens The elements oxygen, sulphur, selenium, tellurium, polonium of Group 16. Chalcogenides contain the X^{2-} species.

chalcones The trivial name given to the α,β-unsaturated ketones ArCH=CHCOAr′ obtained by condensing an aromatic aldehyde with an aryl methyl ketone in the presence of base. BENZALACETOPHENONE Ph·CH=CHC(O)Ph is the simplest example.

chalconides Chalcogenides. See CHALCOGENS.

chalcopyrite (copper pyrites), $CuFeS_2$. The most important ore of copper, brassy yellow in colour with a metallic lustre.

chalk A naturally occurring form of $CaCO_3$ of marine origin. Also reprecipitated calcium carbonate of wide use.

chaperone protein A protein or sub-unit of a protein that transports an essential element in a process. Often transports a metal–metallachaperone.

charcoal A form of carbon produced by slow combustion of wood in a deficiency of oxygen. Formerly used as a fuel and reducing agent. Used to absorb gases. Has a low-order graphitic structure.

charge transfer band or spectrum An absorption band for which the ground and excited states correspond to transfer of an electron from one atom or group to another.

charge transfer complexes Complexes in which there is weak interaction, charge transfer, between the donor and the acceptor properties which are generally induced in the acceptor. Examples of charge transfer complexes include the complexes formed between aromatic derivatives and halogens. The class is practically indistinguishable from true complexes. Charge (electron) transfer takes place in all complexes.

Charles's law At constant pressure the volume of a given mass of gas is directly proportional to the absolute temperature.

chelate compound A compound in which co-ordinate links complete a closed ring, e.g. in beryllium acetylacetonate.

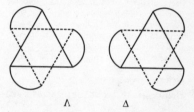

In complexes of chelates there are a number of types of isomerism which may occur. In a tris(ethylenediamine) octahedral complex two optically active isomers occur (often denoted Δ and Λ).

Λ   Δ

A ligand such as ethylenediamine is not planar and has a spiral form (gauche) giving rise to further forms. When the direction of one C–C bond in one ethylenediamine is parallel to the 3-fold axis the isomer is termed the *lel* form, when it is inclined to the axis it is termed the *ob* form.

chelate effect The enhanced stability of a complex containing chelate rings as compared with the complex containing monodentate ligands. E.g. ethylenediamine complexes are more stable than ethylamine complexes.

chelation See CHELATE COMPOUND.

Chemical Abstracts (C.A.) The abstracting service running from 1907. For abstracts of earlier research use British Abstracts. Other services provided include a Registry Numbering System which included 25 million compounds by 2000.

chemical affinity See AFFINITY, CHEMICAL.

chemical development The photographic process of reducing the exposed Ag^+ in the halide grains of a photographic emulsion to Ag.

chemical equivalent The weight of substance which will combine with or displace 8 parts by weight of oxygen. May be related to weights of other elements (e.g. 35.5 g Cl, 1 g H). In redox reactions, expressed as the number of gram equivalent to a Faraday (1 mole electrons).

chemical kinetics The study of reaction rates and their interpretation in terms of mechanism of reaction.

chemical potential (μ) If the total Gibbs free energy G of a system increases by an increment dG when an infinitesimal quantity, dn_A mole, of component A is added to the system under conditions of constant temperature and pressure, the chemical potential of A in the system denoted μ_A, is given by

$$\mu A = (dG/dn_A)_{P,T}$$

μ is therefore sometimes called the partial molar free energy. Chemical potential is a valuable function in the theory of heterogeneous equilibria, since it can readily be proved that, if such a system is not initially at equilibrium, substances will tend to pass from one phase to another (e.g. solid dissolving or crystallizing) until a uniform chemical potential is attained for each substance in every phase.

chemical shift See NUCLEAR MAGNETIC RESONANCE.

chemical vapour deposition, CVD Deposition of coatings from gas phase, e.g. SnS from $SnCl_4$ and H_2S.

chemical warfare agents Materials used, or usable in, chemical warfare include toxic chemicals, flame, incendiaries, smokes and defoliating agents.

chemiluminescence The emission of light during a chemical reaction. e.g. the oxidation of yellow phosphorus. The light emitted by the fire-fly or glow-worm, and luminous combustion, are examples of this very common phenomenon. Used analytically and in 'light studies' where a reaction gives an active intermediate which transfers energy to a dyestuff which fluoresces under the influence of the absorbed energy.

chemioinformatics The process of building up a database of chemical structures or other chemical information.

chemiosmotic hypothesis Chloroplasts and mitochondria have a proton gradient generated across a membrane coupled to ATP synthesis.

chemisorption See ADSORPTION.

chemometrics The analysis of large amounts of data – often leading to experimental design.

chemosynthesis The process of conversion of H_2S and CO_2 into carbon compounds adopted by bacteria in tube-worms in hydrothermal vents.

chemotaxonomy The study of the chemical constituents of plants, animals and other organisms as a guide to their phylogeny and classification.

chemotherapy The selective destruction of pathogenic organisms or control of, e.g. cancer within a host by the use of chemicals.

chevral phases Ternary molybdenum chalcogenides $M_xMo_6X_8$. Some show superconductivity.

chichibabin reaction Amination of pyridine with sodium amide to give 2-aminopyridine.

chicle A resin obtained from the Mexican tree *Achras sapita*; softens at about 50°C, used in chewing gum manufacture.

Chile saltpetre, $NaNO_3$. See SODIUM NITRATE.

china clay (kaolin) A white powdery material arising from the decomposition of feldspars in granites. Used in paper and pottery manufacture and as a filler in textiles and paint. World production 1998 28 megatonnes.

Chinese blue See CYANOFERRATES, PRUSSIAN BLUE.

Chinese white, ZnO. See ZINC OXIDE.

chinomethionat Fungicide, acaricide, quinoxaline derivative.

chiral See CHIRALITY.

chiral catalysis The use of a chiral catalyst to form one enantiomer preferentially. Used in pharmaceutical synthesis.

chiral chromatography Separation of chiral isomers by use of a chiral substrate.

chiral drugs Drugs containing a single enantiomer of the active material. Physiologically only one enantiomer is generally active.

chirality A term which may be applied to any asymmetric object or molecule. The property of non-identity of an object with its mirror image is termed chirality and implies 'handedness', i.e. the object and mirror-image relationship of a right hand to a left hand.

A molecule is chiral if it cannot be superimposed on its mirror image (or if it does not possess an alternating axis of symmetry) and would exhibit OPTICAL ACTIVITY, i.e. lead to the rotation of the plane of polarization of polarized light, circular dichroism. Lactic acid, which has the structure (2 mirror images) shown, exhibits molecular chirality. In this the central carbon

atom is said to be chiral but it is the molecule which is chiral. For designation see CONFIGURATION. Using X-ray crystallography it is possible to determine the absolute configuration which can give rise to particular physical, chemical or physiological effects.

In certain crystals, e.g. in quartz, there is chirality in the crystal structure. Molecular chirality is possible in compounds which have no chiral carbon atoms and yet possess non-superimposable mirror image structures. Restricted rotation about the C=C=C bonds in an allene $abC=C=Cba$ causes chirality and the existence of two optically active forms **1** and **2**.

Restricted rotation about single bonds, atropisomerism, caused by bulky substituents X and Y allows the existence of ENANTIOMERS of the chiral diphenyl derivatives **3** and **4**. E.g. X=NO_2, Y=CO_2H.

chirooptical spectroscopy The differential interaction of a chiral, optically active molecule with right or left circularly polarized light. Observed as differential speed, absorption, fluorescence, or scattering. Can be observed and measured in u.v., visible, or vibrational i.r. See CIRCULAR DICHROISM.

chitin A structural polysaccharide in which

N-acetyl-D-glucosamine units are linked in the β-(1→4)-positions. Forms the integument of crustaceans and insects and is also found in some fungi. Used as a coagulant for water purification and food processing. Hydrolysed by the enzyme chitinase.

chitinase An enzyme of the digestive juices of snail and mould fungi which can hydrolyse chitin and related compounds of crustaceans and fungi. N-acetyl glucosamine.

chitosan A deacetylated form of chitin.

chloral (trichloroacetaldehyde, trichloroethanal), $CCl_3.CHO$. A colourless oily liquid with a pungent odour; b.p. 98°C. Manufactured by the action of chlorine on ethanol; it is also made by the chlorination of ethanal. When allowed to stand, it changes slowly to a white solid. Addition compounds are formed with water (see CHLORAL HYDRATE), ammonia, sodium hydrogen sulphite, alcohols, and some amines and amides. Oxidized by nitric acid to trichloroethanoic acid. Decomposed by alkalis to chloroform and a methanoate; a convenient method of obtaining pure $CHCl_3$. Used as a hypnotic and anaesthetic. Controlled substance.

chloral hydrate (trichloroethylidene glycol), $CCl_3CH(OH)_2$, $C_2H_3Cl_3O_2$. Crystallizes in large, colourless prisms having a peculiar odour; m.p. 57°C, b.p. 97.5°C. Manufactured by adding the calculated amount of water to chloral. For other properties see CHLORAL.

chloralose Rodenticide, glucochloral derivative.

chloramben Herbicide, benzoic acid derivative.

chloramine, $ClNH_2$. See HALOAMINES. Also used to denote organic compounds containing N–Cl bonds.

chloramine T (sodium p-toluenesulphonylchloroamide), $[4-CH_3C_6H_4S(O)_2NCl]$ Na·$3H_2O$. Prepared from 4-toluenesulphonamide and NaOCl. Powerful antiseptic used in dilute aqueous solution for washing wounds.

chloranil (tetrachloro-1,4-benzoquinone), $C_6Cl_4O_2$. Yellow crystals, m.p.

290°C. Prepared by oxidizing phenol with an acid and potassium chlorate(V). Used as a de-hydrogenating agent and in dyestuff synthesis.

chlorates Salts of the various chlorine oxyacids. Chlorates(I), hypochlorites, contain the $[OCl]^-$ ion and are formed from Cl_2 and H_2O, or HgO and Cl_2 and cold NaOH solution. The free acid is formed from H_2O and Cl_2 in the gas phase. Chlorates(III), chlorites, contain the bent $[ClO_2]^-$ ion (ClO_2 plus base). The free acid is a weak acid and cannot be isolated. Both chlorates(I) and (III) are used as bleaching agents. Chlorates(V), chlorates, contain the pyramidal $[ClO_3]^-$ ion and are obtained from Cl_2 plus OH^- in hot solution. The free acid is a strong acid. Chlorates(VII), perchlorates, contain the tetrahedral $[ClO_4]^-$ ion and are obtained by electrolytic oxidation of $[ClO_3]^-$. The free acid is a very strong acid and is widely used as an oxidizing agent; it may explode on contact with organic materials. Chlorates(VII) are used as non-complexing inert electrolytes. All of the chlorates are oxidizing species

$[ClO_4]^- \rightarrow Cl^-$ + 0.56 volts
$[ClO_3]^- \rightarrow Cl^-$ + 0.63 volts
$[ClO_2]^- \rightarrow Cl^-$ + 0.78 volts
$[ClO]^- \rightarrow Cl^-$ + 0.89 volts

(all in acid solution). In all cases the sodium salt is readily available; $KClO_4$ is sparingly soluble. U.S. production 1999 797 kilotonnes.

chlordane, $C_{10}H_6Cl_8$. Organochlorine derivatve. Formerly used as an insecticide but toxic to higher animals. Still used against termites.

chlorethyfos Insecticide, organophosphate derivative.

chlorex process See 2,2´-DICHLOROETHYL ETHER.

chlorfenapyr Insecticide, acaricide, pyrazole derivative.

chlorfenvinphos Insecticide, acaricide, organophosphate derivative.

chlorfluazinon Insecticide, benzylurea derivative.

chlorhexidine, $C_{22}H_{30}Cl_2N_{10}$, $[4-ClC_6H_4NHC(NH)NHC(NH)(CH_2)_3]_2$. The dihydrochloride is a white crystalline powder. m.p. about 225°C (decomp.). Used as an antiseptic and skin sterilizing agent, and as a bacteriostat in some pharmaceutical formulations.

chloric acid, $HClO_3$. See CHLORATES(V).

chloridazon Herbicide, pyridazine derivative.

chloride of lime BLEACHING POWDER.

chlorides Formally compounds of Cl(–1) formed by all elements except the noble gases ($XeCl_2$ has some stability). Chlorides of the more electropositive elements are ionic and the more electronegative elements form covalent chlorides. High oxidation state chlorides are covalent. Chlorides are formed by use of Cl_2 or HCl on the elements or by metathesis (e.g. $AgNO_3$ plus NaCl gives AgCl), anhydrous chlorides are obtained by use of ethanoyl chloride or thionyl chloride. Chloride is a good complexing agent and can act as a monodentate or bridging ligand, e.g. in $Cl_2Al(\mu-Cl)_2AlCl_2$.

chlorimuron-ethyl Herbicide, urea derivative.

chlorin The dihydroporphin ring system present in chlorophylls.

chlorinated biphenyls (diphenyls) See DIPHENYL.

chlorinated rubbers Materials containing up to 70% Cl formed by reacting rubber (solid, solution or latex) with Cl_2. The reaction is rapid to $C_{10}H_{14}Cl_2$ followed by cyclization and then formation of homogeneous materials. Stable materials used as blending agents for alkyd resins and as film formers in corrosion-resistant paints. Also used as adhesives.

HCl reacts with rubber to form hydrochlorinated rubber – also used in paints.

chlorine, Cl. Non-metallic element, at.no. 17, at.mass 35.453, m.p. –101.15°C, b.p. –34.04°C, density ρ 3.214 g l^{-1}, crustal abundance 130 p.p.m., electronic configuration [Ne] $3s^23p^5$. Occurs naturally as NaCl, $MgCl_2$ and other chlorides; sea water contains up to 1% NaCl. Cl_2 can be obtained by oxidation of NaCl with, e.g. $KMnO_4$, and is obtained commercially by electrolysis of brine or $MgCl_2$ increasingly using diaphragm cells. The economics of the electrolysis of brine depend upon the simultaneous formation of NaOH. Chlorine occurs as a greenish poisonous gas containing Cl_2 molecules. It is stored as liquid Cl_2. Chlorine is a very reactive element and combines directly with most other elements. It dissolves in water and from cold water gives chlorine hydrate, $Cl_2.7.27H_2O$ with a clathrate structure; a small proportion reacts to give Cl^- and $[OCl]^-$. Chloride is important in cellular processes for regulation of electrical activity. Chlorine is one of the basic materials of the chemical industry and is used in the production of organochlorine derivatives and indirectly in the manufacture of many other compounds which are obtained via chlorine or chloride derivatives, particularly anti-knock agents and chlorohydrocarbons (65%). Chlorine derivatives are used in water sterilization, pulp and paper manufacture (15%), solvents, polymers (principally vinyl chloride and rubbers), refrigerants and aerosol propellants. U.S. production 1999 13.4 megatonnes.

chlorine chemistry Chlorine, Group 17, is one of the halogens, electronic configuration $3s^2 3p^5$. Its typical chemistry is that of a non-metallic element in the -1 oxidation state (E° $Cl_2 \to Cl^-$ $+1.36$ volts in acid solution) forming ionic chlorides and covalent chloro derivatives. Chlorine also forms positive oxidation states – see CHLORATES – mainly with oxygen and fluorine, and with other halogens. The Cl_2^+ ion is known in the gas phase and Cl_4^+ and Cl_3^+ salts exist.

chlorine fluorides See CHLORINE HALIDES.

chlorine halides
Chlorine pentafluoride, ClF_5. M.p. -103°C, b.p. -14°C, square pyramidal. ClF_6^- is known.
Chlorine trifluoride, ClF_3. M.p. -76°C, b.p. 12°C, T-shaped, a powerful fluorinating agent.
Chlorine monofluoride, ClF. M.p. -157°C, b.p. -100°C, dissociates to the elements. Each of the chlorine fluorides is hydrolysed by water (ClF_3 explosively) and is prepared from Cl_2 and F_2.
Chlorine also forms BrCl (see BROMINE HALIDES) and ICl and ICl_3 (see IODINE CHLORIDES). See also INTERHALOGEN COMPOUNDS.

chlorine hydrate, $Cl_2.7.27H_2O$.
See CHLORINE.

chlorine oxides
Chlorine monoxide, Cl_2O. M.p. -116°C, b.p. 4°C, yellow-red gas (Cl_2 plus HgO), dissolves in water to give some HOCl. Dissociates to Cl_2 plus O_2.
Chlorine dioxide, ClO_2. M.p. -6°C, b.p. 10°C, paramagnetic yellow, explosive gas ($NaClO_3$ plus H_2SO_4). Strong oxidizing agent, gives $HClO_2$ and $HClO_3$ with water. Used as a bleach for wood-pulp and in fumigation.
Chlorine tetroxide, ClO_4, has a very short lifetime.
Dichlorine tetroxide, $ClOClO_3$, *chlorine perchlorate* ($CsClO_4$ plus chlorine fluorosulphate,, $ClOSO_2F$), *dichlorine hexoxide*, Cl_2O_6, oily red liquid (ozone plus ClO_2). All unstable.
Dichlorine heptoxide, Cl_2O_7. M.p. -91.5°C, b.p. 82°C, the most stable chlorine oxide but still explosive ($HClO_4$ plus P_2O_5), gives $HClO_4$ with water.

chlorine oxide fluorides, ClO_2F and, $ClOF_3$. Formed by hydrolysis of the fluorides, ClO_2F gives (ClO_2)$^+$ salts with e.g. BF_3.
Chlorine trioxide fluoride, ClO_3F, *perchloryl fluoride*. M.p. -148°C, b.p. -47°C ($KClO_4$ plus HF plus SbF_5 or F_2 on HSO_3F on $KClO_4$), a toxic gas, thermally stable to 500°C. It is used as a reagent to replace H in C–H-containing compounds with F – it also gives organic perchlorates, $RClO_3$ (e.g. PhLi gives $PhClO_3$). FClO, F_2ClO_2, $FClO_4$ are also known.

chlorites A type of aluminosilicate. Also the salts of $HClO_2$, CHLORATES(III).

chlormephos Insecticide, organophosphate derivative.

chlormequat chloride Plant growth regulator, quaternary ammonium salt.

chloroacetic acid See CHLOROETHANOIC ACIDS.

chloroacetone (chloropropanone), $CH_2Cl \cdot CO \cdot CH_3$. Colourless lachrymatory liquid; b.p. 119°C. Manufactured by treating propanone with bleaching powder or chlorine. It is used as a tear gas, usually mixed with the more potent bromoacetone, and in synthesis.

chloro acids Complex chloroanions are formed by most elements of the periodic table by solution of oxides or chlorides in concentrated hydrochloric acid. Potassium salts are precipitated from solution when potassium chloride is added to a solution of the chloro acid; the free acids are generally unstable.

chloroalkylamines (N-dialkylchloroalkylamines), $R_2N(CH_2)_nCl$. Formed by treating the alkanolamine with thionyl chloride. Of importance in the synthesis of pharmaceuticals.

chloroanilines (aminochlorobenzenes) Formed (mono-, di- and trichloroanilines) by the action of SO_2Cl_2 on aniline. Less basic than aniline.

chlorobenzene, C_6H_5Cl, PhCl. Colourless liquid, b.p. 132°C. Nitrates in the 2- and 4-positions. Prepared by the direct chlorination of benzene in the presence of an iron catalyst. Largely used as an intermediate in the manufacture of other chemicals, particularly phenol, DDT and aniline.

2-chlorobutadiene polymers See POLYCHLOROPRENE.

chlorocarbonic ester See ETHYL CHLORO-FORMATE.

chlorocarbons (chlorohydrocarbons)
An extremely important range of chemicals prepared by direct chlorination of hydrocarbons or by Cl_2 or HCl addition to olefins. HCl is frequently a waste product of chlorination. Subsequent treatment is by cracking. Used as cleaning fluids (tetrachloroethene, CH_3CCl_3), insecticides, polymers (vinyl chloride chloroprene), solvents (CCl_4), as chemical intermediates (CH_3Cl, C_2H_5Cl). Many organochlorine compounds present health problems.

chlorochromate(VI) salts, $MCrO_3Cl$. Obtained by treating dichromates with concentrated hydrochloric acid or by adding CrO_2Cl_2 to a saturated solution of a metal chloride. The salts contain the tetrahedral anion $[CrO_3Cl]^-$; the free acid is unknown. The solutions decompose on boiling to chlorine and chromium(III) salts. The salts are powerful oxidizing agents, the pyridinium salt is used for oxidizing alcohols to aldehydes. Chlorochromates(V) $M_2{}^1CrOCl_5$ result from the action of concentrated HCl and alkali metal chloride on CrO_3 at 0°C.

chloroethane, CH_3CH_2Cl. Colourless liquid with an ethereal odour; burns with a green-edged flame, b.p. 12.5°C. Manufactured by reacting hydrogen chloride with ethene at 40°C in the presence of $AlCl_3$ using dichloroethane as solvent, or by the catalytic or photochemical chlorination of ethane. Reacts with ammonia at 100°C to give ethylamine hydrochloride; converted to ethanol by heating with KOH. Its principal use was in the manufacture of lead tetraethyl. Having a low b.p. it is useful for the production of local anaesthesia; it is applied in the form of a fine spray, and its rapid evaporation freezes the part to which it is applied.

chloroethanoic acids (chloroacetic acids)
Monochloroethanoic acid, $CH_2Cl\cdot COOH$. B.p. 189°C. Manufactured by heating trichloroethene with sulphuric acid, or by treating glacial ethanoic acid with chlorine in the presence of red phosphorus, sulphur or ethanoic anhydride. Reacts with ammonia to give glycine, and with many types of organic compounds. It is largely used as a chemical intermediate, particularly in the manufacture of the chlorophenoxyacetic acid weedkillers, thiocyanate insecticides and various pharmaceuticals.

Dichloroethanoic acid, $CHCl_2COOH$. Low-melting solid, m.p. 5–6°C, b.p. 194°C. Prepared by the action of copper powder on trichloroethanoic acid or by the action of sodium cyanide on chloral hydrate.

Trichloroethanoic acid, CCl_3COOH. A crystalline solid which rapidly absorbs water vapour; m.p. 58°C, b.p. 196.5°C. Manufactured by the action of chlorine on ethanoic acid at 160°C in the presence of red phosphorus, sulphur or iodine. It is decomposed into chloroform and carbon dioxide by boiling water. It is a much stronger acid than either the mono- or the dichloro-acids. The sodium salt is used as a selective weedkiller.

2-chloroethyl alcohol (ethylene chlorohydrin), $ClCH_2\cdot CH_2OH$. Colourless liquid with a faint ethereal odour; b.p. 129°C. Manufactured by passing ethene into dilute chlorine water or from ethylene oxide plus HCl. Reacts with solutions of sodium hydrogen carbonate to give ethylene glycol and with solid sodium hydroxide to give ethylene oxide; with concentrated sulphuric acid at 100°C to give $\beta\beta'$-dichloroethyl ether; with ammonia and amines to give aminoethyl alcohols and with sodium salts of organic acids to give glycol esters. It is used in the preparation of these compounds and generally in synthesis.

chloroform See TRICHLOROMETHANE.

chloroformic ester See ETHYL CHLOROFORMATE.

chlorohydrins Organic compounds containing the $C(OH)\cdot CCl$ group. Formed by treating compounds containing a double bond with chlorine water, or by treating glycols or epoxides with hydrochloric acid. Converted to glycols when heated with weak alkalis such as sodium hydrogen carbonate.

chlorohydrocarbons See CHLOROCARBONS.

chlorohydroxypropane See PROPYLENE CHLOROHYDRINS.

chloroisopropyl alcohol See PROPYLENE CHLOROHYDRINS.

chloromethane (methyl chloride), CH_3Cl. Colourless gas with a pleasant ethereal odour, b.p. –24°C. Manufactured principally by the reaction of methanol and hydrogen chloride in the presence of a catalyst, either in the vapour or liquid phase. Formed naturally in coastal salt marshes and (with CH_3Br and CH_3I) in rice fields. Reacts with chlorine to give methylene chloride, chloroform and carbon tetrachloride. Forms methanol when treated with $Ca(OH)_2/H_2O$ under pressure. The chief use of methyl chloride is in the production of methylsilicones. It is also used in the manufacture of lead tetramethyl (anti-knock additive), butyl rubber, methyl cellulose and for quaternizing organic bases, e.g. Paraquat.

chloromethylation The introduction of the –CH₂Cl group into aromatic, especially activated aromatic compounds using hydrogen chloride, methanal and anhydrous zinc chloride. See MANNICH REACTION.

$$R - H + CH_2O + HCl \rightarrow R - CH_2Cl + H_2O$$

chloromethyl methyl ether, ClCH₂OCH₃. Colourless liquid; b.p. 60°C. Made by passing hydrogen chloride into a mixture of formalin and methanol. Used for CHLOROMETHYLATION, preparation of methoxymethyl ethers (used as protecting groups) and with anhydrous aluminium chloride as catalyst for introducing the formyl (–CHO) group into aromatic compounds. Carcinogen.

chloronaphthalenes
1-chloronaphthalene, α-chloronaphthalene, C₁₀H₇Cl. Colourless, b.p. 263°C. Prepared by direct chlorination of naphthalene in the presence of FeCl₃. Used as an industrial solvent.
   Other chlorinated naphthalenes. The other monochloronaphthalene, the ten theoretically possible dichloronaphthalenes, and the fourteen trichloronaphthalenes have all been prepared, generally from the corresponding amino-derivatives by diazotization and treatment with CuCl. They are of little industrial importance.

3-chloroperbenzoic acid (*m*-chloroperbenzoic acid), 3-ClC₆H₄C(O)OOH. White solid, m.p. 92°C. Prepared by treating *m*-chlorobenzoyl chloride with hydrogen peroxide and sodium carbonate. This PER-ACID is a reasonably stable, selective, active oxidizing agent.

chlorophacinone Rodenticide, indane derivative.

chlorophenols Relatively acidic materials produced by direct chlorination of the aromatic ring of phenols. Used widely in phenol-resins (with methanol) the compounds are often effective antiseptics, disinfectants, germicides, insecticides, herbicides, wood preservatives. Also used in dyestuffs.

chlorophyll The green colouring matter of plants. Chlorophyll is present in all plants, and is essential to the life of the plant, as it acts as a catalyst in the photosynthesis of carbohydrates from carbon dioxide and water. Two chlorophylls are present in plants, *chlorophyll a*, C₅₅H₇₂MgN₄O₅, which is obtained as a blue-black powder, m.p. 150–153°C, and *chlorophyll*

b, C₅₅H₇₀MgN₄O₆, dark green powder, m.p. 120–130°C. Both are soluble in alcohol and ether. The methyl group marked with an asterisk is replaced by an aldehyde group in chlorophyll *b*. The ratio of chlorophyll *a* to chlorophyll *b* in green leaves is about 3: 1. Treatment of chlorophyll *a* with acids removes the Mg, replacing it by two H atoms, and gives

R=C₂₀H₃₉ (Phytol)

an olive-brown solid, phaeophytin a. Hydrolysis of this splits off phytol, and gives phaeophorbide a. Similar compounds are obtained from chlorophyll b. Alkaline degradation of chlorophyll yields a series of phyllins – magnesium porphyrin compounds – the final member of the series being aetiophyllin, and treatment of the phyllins with acids gives porphyrins, resembling, but not identical with, those obtained from the animal blood pigments.
   Crude chlorophyll is prepared commercially from alfalfa meal or nettles by extraction with alcohol and partition into benzene. It is used as a colouring matter, particularly for foods, pharmaceutical products and as a deodorant.

chloropicrin (nitrotrichloromethane), CCl₃NO₂. A colourless, lachrymatory, toxic liquid, b.p. 112°C. Manufactured by treating sodium picrate with chlorine, or calcium picrate with bleaching powder. Used as an insecticide, nematocide, for disinfecting cereals and in synthesis.

chloroprene (2-chlorobutadiene), C₄H₅Cl, CH₂:CCl·CH:CH₂. Colourless liquid, b.p. 59°C. Manufactured by treating vinylacetylene with hydrochloric acid at 30°C in the presence of a copper ammonium chloride catalyst. Used to prepare oil and ozone synthetic rubbers. Probably carcinogenic. See POLY-CHLOROPRENE.

2-chloropropane (isopropyl chloride), C_3H_7Cl, $CH_3 \cdot CHCl \cdot CH_3$. Colourless liquid with a pleasant odour, b.p. 36.5°C. Manufactured by treating isopropyl alcohol with hydrochloric acid in the presence of zinc chloride, or by passing the vapour of the alcohol and hydrochloric acid gas over a heated metallic chloride, such as magnesium chloride. It is used as a fat solvent.

2-chloropropyl alcohol See PROPYLENE CHLOROHYDRINS.

3-chloropropylene glycol See GLYCEROL MONOCHLOROHYDRINS.

chloropyriphos (chloropyriphos-methyl) Insecticides, organophosphate derivatives.

chloroquin Important anti-malarial drug.

chlorosulfuron Herbicide, sulphonyl urea derivative.

chlorosulphonic acid See CHLOROSULPHURIC ACID.

chlorosulphuric acid (chlorosulphonic acid), $HOS(Cl)O_2$. Colourless liquid, m.p. −80°C, b.p. 158°C (SO_3 plus HCl or H_2SO_4 plus PCl_5 or $POCl_3$). Hydrolysed by water. Used as a chlorinating and sulphonating agent.

chlorothalonil Fungicide, naphthalonitrile derivative.

chlorotoluenes, C_7H_7Cl. *2-chlorotoluene*, b.p. 159°C; *4-chlorotoluene*, b.p. 162°C. Prepared by direct chlorination of toluene in the presence of a catalyst. 2-chlorotoluene can be oxidized to 2-chlorobenzaldehyde. Used as solvents and dyestuff intermediates.

chlorotoluron Herbicide, urea derivative.

chlorous acid, $HClO_2$. Unknown in the free state but CHLORATES(III), chlorites, well known and used as oxidizing agents and as bleaches.

chloroxylenol (4-chloro-3,5-dimethylphenol) Anti-microbial, preservative, also used as a chemical intermediate.

chlorpromazine (3-chloro-10-(2-dimethylaminopropyl)phenothiazine), $C_{17}H_{19}ClN_2S \cdot HCl$. White or cream coloured powder, m.p. 194–197°C. It is a major tranquillizer, also an anti-emetic. *Promazine*,

$C_{17}H_{20}N_2S \cdot HCl$. Widely used as a tranquilliser and as an adjunct to anaesthesia.

chlorthal-methyl Herbicide, chlorobenzoic acid derivative.

chlorthunel Herbicide, chlorobenzoic acid derivative.

chlozolinate Fungicide, carboximide derivative.

cholane ring system The C_{24} skeleton of the

COMMON BILE ACIDS, numbered as shown. The hydrocarbon depicted is 5β-cholane, · m.p. 90°C, which has the *cis*-ring junction as found in the major natural bile acids.

cholesteric See LIQUID CRYSTALS.

cholesterol (5-cholesten-3β-ol), $C_{27}H_{46}O$. M.p. 149°C. The principal sterol of animals, it is

found in free and esterified forms in all parts of the body. Cholesterol is obtained commercially from beef spinal cords or from wool 'wax'. The sterol is synthesized by animals, from ethanoate units via MEVALONIC ACID, SQUALENE and LANOSTEROL and can be synthesized chemically. Besides its own biological roles (e.g. as a constituent of membranes) it acts as a precursor of other steroids required in metabolism: notably the bile acids, sex hormones and adrenocortical hormones. Used in pharmaceuticals for delivery and in cosmetics.

cholic acid (3α,7α,12α-trihydroxy-5β-cholanic acid), $C_{24}H_{46}O_5$. M.p. 195°C. Bile acid – occurs as sodium salt.

choline (trimethyl-2-hydroxyethyl-ammonium hydroxide),
$[(CH_3)_3NCH_2 \cdot CH_2OH]^+OH^-$. Colourless syrup crystallizing with difficulty to a hygroscopic mass. Strongly alkaline. It is present as a constituent of lecithin in all animal and vegetable tissues, and less commonly as the free base.

Takes part in methylating reactions. Sometimes regarded as a member of the vitamin B group.

choline esterase See ACETYLCHOLINE. Hydrolytic enzyme between neuro-muscular junctions which degrades acetylcholine to choline and ethanoate. Involved in energy transfer.

chondrites The most abundant type of meteorites – small rock species.

chondroitin The matrix of cartilage. A polymer of β-D-glucuronide, 1,3-N-acetyl-D-galactosamine joined in β-1,4-linkages. It differs from HYALURONIC ACID only in the presence of galactosamine rather than glucosamine. Usually found as sulphates. Dermatan sulphate is very similar to chondroitin sulphate except that it contains L-iduronic acid instead of D-glucuronic acid. Used as a dietary supplement.

chorionic gonadotropin A glycoprotein hormone produced by the placenta and excreted in the urine during pregnancy. Its presence has been used as a method of detecting pregnancy.

chorismic acid, $C_{10}H_{10}O_6$. Monohydrate m.p. 148–149°C. An intermediate in aromatic biosynthesis via the 'shikimic acid pathway'. From SHIKIMIC ACID the sequence is through 5-phosphoshikimic acid, 3-enolpyruvylshikimic acid 5-phosphate, to chorismic acid.

chroman A strongly refractive liquid, b.p. 214–215°C which smells like peppermint. Contains the γ-pyrone skeleton (flavone, xanthone and the tocopherols (vitamin E) are chromans).

chromates Formally any oxygen-containing chromium species (see below). Normally used to describe chromates(VI).

Chromates(VI). Derivative of CrO_3 which forms the tetrahedral $(CrO_4)^{2-}$ ion and polymeric anions, e.g. $Cr_2O_7^{2-}$, $[O_3Cr-(\mu-O)CrO_3]^{2-}$, dichromate, contains oxygen bridges, in acid solution. The CrO_4^{2-} ion is yellow and $Cr_2O_7^{2-}$ red.

Free H_2CrO_4 is not known. Potassium salts are non-deliquescent and anhydrous; sodium salts are $Na_2CrO_4,10H_2O$ and $Na_2Cr_2O_7.2H_2O$. The chromates of the heavy metals are in general insoluble and are often used as pigments. Chromates(VI) are obtained from CHROMITE by heating with Na_2CO_3 in air at 1100°C followed by leaching and crystallization. Chromates(VI) are strong oxidizing agents ($E°$ $Cr_2O_7^{2-} \rightarrow Cr^{3+}$ in acid +0.33 volts) and are used as such and also in tanning, photographic processing, printing and in pigments.

Chromates(V), $(CrO_4)^{3-}$. Dark green species which hydrolyse and disproportionate to Cr(III) and Cr(VI) (K_2CrO_4 plus KOH in melt).

Chromates(IV). Mixed oxides, e.g. $M_2{}^{II}CrO_4$ prepared by solid-state reactions.

Chromates(III). Mixed oxides, e.g. $FeCr_2O_4$, often having spinel structures and prepared by solid-state reactions.

chromatography (chromatographic analysis) A series of closely related techniques for separation of components of a mixture because of the distribution of the components between a liquid or gaseous phase and a solid phase (stationary phase, often silica or alumina) which usually has a large surface area. If the stationary phase is in a tube the technique is column chromatography; when paper or thin layers of stationary phase are used the technique is a sheet technique. The active solid stationary phase is the absorbent, an inert support is the support. The whole process of separation of the sample components into separate bands or zones is the development of the chromatogram and is carried out by elution – the mobile (liquid or gas) phase being the eluting agent. The whole process gives a chromatogram.

The chromatogram can finally be used as the series of bands or zones of components or the components can be eluted successively and then detected by various means (e.g. thermal conductivity, flame ionization, electron capture detectors, or the bands can be examined chemically). If the detection is non-destructive, preparative-scale chromatography can separate measurable and useful quantities of components. The final detection stage can be coupled to a mass spectrometer (GCMS) to FTIR and to a computer for final identification.

Among the various chromatographic methods are adsorption column chromatography (use of a liquid phase in a solid column of adsorbent), partition column chromatography (distribution between two liquids in a column), thin-layer chromatography (partition on an open thin sheet), paper chromatography (use of a paper sheet as stationary phase), high-performance liquid chromatography, hplc, (partition column chromatography under high pressure), ion chromatography (IC), ion-exchange chromatography (ion exchange), gas chromatography (distribution of a gaseous solute between a gas and a liquid or solid phase), zone electrophoresis (sheet chromatography in the presence of an electric field).

Gel permeation chromatography, size exclusion chromatography, gel filtration chromatography. Techniques for separating the components of a mixture according to molecular volume differences. A porous solid phase (a polymer, molecular sieve) is used which can physically entrap small molecules in the pores whilst large molecules pass down the column more rapidly. A solvent pressure up to 1000 psi may be used.

chrome Having a chromium-plated surface.

chrome alum, $KCr(SO_4)_2.12H_2O$. A typical ALUM used as a GELATIN HARDENER.

chrome orange See LEAD CHROMATES.

chrome yellow See LEAD CHROMATES.

chromic Outdated nomenclature for chromium(III) compounds.

chromic acid A name used incorrectly for CrO_3.

chromite, $FeCr_2O_4$. Brownish-black mineral with the spinel structure used as a source of Cr and its compounds.

chromium, Cr. Metallic element, at.no. 24, at.mass 51.996, m.p. 1907°C, b.p. 2671°C, density ρ 7180, crustal abundance 100 p.p.m., electronic configuration [Ar] $3d^5 4s^1$. The most important chromium mineral is chromite, $FeCr_2O_4$, from which Cr is obtained by fusion with alkali carbonate in the presence of air, extraction with water as chromate(VI), reduction to Cr_2O_3 by carbon and reduction to metal with Al. Pure Cr can be obtained by electrolytic reduction of $[CrO_4]^{2-}$. The metal is hard silver-white in colour, bcc. The metal is very resistant to oxidation although it dissolves in non-oxidizing acids and reacts with O_2, halogens, S, etc. at high temperatures. It becomes passive in HNO_3. It is used widely in electroplating, as an additive for steel. Cr compounds are used as pigments (Cr_2O_3 and chromates(VI)) and for colouring glass, etc., in leather tanning, textiles, as wood preservatives, rust inhibitor, as catalysts, as oxidizing agents and in refractories. Chromium compounds are toxic in large quantities

but are important trace elements. World ferrochrome demand 1998 16.9 megatonnes.

chromium, organic derivatives Chromium forms a range of organic derivatives, e.g. Cr(0) (h^6–$C_6H_6)_2Cr$; Cr(I) [(h^5–$C_5H_5)Cr(CO)_3]_2$; Cr(II) $Li_4Cr_2Me_8$; Cr(III) $CrPh_3.3THF$, Li_3CrMe_6. These are prepared by the action of LiR or RMgX on the appropriate chromium derivatives.

chromium bromides Chromium(III) bromides, $CrBr_3$, and $CrBr_2$, are very similar to the corresponding chlorides.

chromium carbonyl, $Cr(CO)_6$. White solid; m.p. 149°C. Prepared from $CrCl_3$ plus reducing agent plus CO. Forms an extensive range of substituted derivatives.

chromium chemistry Chromium is a transition element of Group 6, showing oxidation states from +6 to –4: +6 – CrO_3, $[CrO_4]^{2-}$; +5 – CrF_5, $[CrO_4]^{3-}$; +4 – $[CrF_6]^{2-}$, $Cr(OBu)_4$; +3 – $CrCl_3$; +2 – CrF_2; +1 – $[Cr(bipy)_3]^+$; 0 – $[Cr(CO)_6]$; –1 – $[Cr_2(CO)_{10}]^{2-}$; –2 – $[Cr(CO)_5]^{2-}$; –4 – $[Cr(CO)_4]^{4-}$. In its higher oxidation states the Cr compounds are predominantly covalent; Cr(III) and Cr(II) compounds are largely ionic with Cr–Cr bonding in some derivatives. Chromium(II) compounds show Jahn–Teller distortion with 4 near and 2 far neighbours. The higher oxidation states are strongly oxidizing in acid solution

$[Cr_2O_7]^{2-} \rightarrow Cr^{3+}$ $E°$ +1.33 volts in acid
$[CrO_4]^{2-} \rightarrow Cr(OH)_3$ –0.13 volts in alkali

but Cr(III) and Cr(II) are stable in solution

$Cr^{3+} \rightarrow Cr$ –0.74 volts
$Cr^{2+} \rightarrow Cr$ –0.56 volts in acid

and Cr^{2+} is thus a fairly strong reducing agent. Chromium(III) forms a very extensive range of largely octahedral, inert complexes particularly with O- and N-ligands. There is an extensive aqueous chemistry of Cr(II) (blue) and Cr(III) (green) aquo-ions and of Cr(VI) anionic species.

chromium chlorides

Chromium(VI) dioxide dichloride, CrO_2Cl_2. See CHROMYL COMPOUNDS; $[CrO_3Cl]^-$ see CHLOROCHROMATE(VI) SALTS.

Chromium(IV) chloride, $CrCl_4$. Formed $CrCl_3$ and Cl_2 at 600–700°C followed by rapid cooling. Decomp. above –80°C.

Chromium(III) chloride (*chromic chloride*), $CrCl_3$. Violet solid (Cr plus Cl_2, hydrate plus $SOCl_2$) only soluble in water in presence of Cr^{2+}. Catalyst, mordant. Forms many complexes including the hydrates $[Cr(H_2O)_6]Cl_3$ – violet, $[Cr(H_2O)_5Cl]Cl_2,H_2O$ – green, $[Cr(H_2O)_4Cl_2]Cl,2H_2O$ – green.

Chromium(II) chloride (chromous chloride), $CrCl_2$. White solid (Cr plus HCl gas) dissolving in water to give a blue solution. Forms hydrates, widely used as a reducing agent.

chromium(II) ethanoate (chromium(II) acetate)

$[Cr(O_2CCH_3)_2]_2,2H_2O$. Red insoluble compound formed from sodium ethanoate and $CrCl_2$ in aqueous solution. The most stable Cr(II) compound; contains a Cr–Cr bond.

chromium fluorides

Chromium hexafluoride, CrF_6. Yellow unstable compound (CrF_3 plus F_2 at high temperatures and pressure).

Chromyl fluoride, CrO_2F_2 and *fluorochromates (VI)*, containing the $[CrO_3F]^-$ ion are known.

Chromium pentafluoride, CrF_5. Red, m.p. 30°C (CrF_3 plus F_2 at 350–500°C).

Fluorochromates(V) containing the $[CrOF_4]^-$ ion are formed from BrF_3 and $M_2Cr_2O_7$.

Chromium tetrafluoride, CrF_4. Green, sublimes 100°C (CrF_3 plus F_2).

Hexafluorochromates(IV), M_2CrF_6. Formed by the action of F_2 on 2MF plus CrF_3.

Chromium(III) fluoride, CrF_3. Green rather insoluble material (HF plus Cr_2O_3 or $CrCl_3$). Used in polishing and as a halogenation catalyst. Forms hydrates and complexes, e.g. M_3CrF_6.

Chromium(II) fluoride, CrF_2 plus HF at 600°C.

The higher chromium fluorides, CrF_6, CrF_5, CrF_4 all extremely strong oxidizing agents, immediately hydrolysed by water.

chromium hydroxide

It seems unlikely that a true chromium(III) hydroxide exists but hydroxy complexes, e.g. $[Cr(H_2O)_5OH]^{2+}$, $[(H_2O)_4Cr(\mu\text{-}OH)_2Cr(H_2O)_4]^{4+}$, are formed by addition of base to aquo-complexes and polymeric species, and ultimately gels are formed by addition of further base.

chromium oxides

Chromium trioxide, CrO_3. Red precipitate from $[CrO_4]^{2-}$ plus conc. H_2SO_4, m.p. 198°C, loses oxygen at 420°C. CrO_3 is a powerful oxidizing agent and is used as such. Acidic, gives $[CrO_4]^{2-}$ with water.

Chromium dioxide, CrO_2 (H_2O plus O_2 on CrO_3 at high temperature). Black ferromagnetic solid with the rutile structure forming chromates(IV) in solid-state reactions. Used in magnetic tapes. Very efficient reagent for oxidizing secondary alcohols to ketones.

Chromium(III) oxide, Cr_2O_3 (heat on $(NH_4)_2Cr_2O_7$, hydrolysis and ignition of chromium(III) compounds). Green compound with the corundum structure. Gives chromates(III) in solid-state reactions, anionic species with hydroxyl ions and $[Cr(H_2O)_6]^{3+}$ with acids. Used as a green pigment (viridian) and to give green colours in glass and porcelain.

chromium oxy-salts

Chromium(III) forms a sulphate which forms alums (CHROME ALUM), hydrates and sulphate complexes. The nitrate, perchlorate and phosphate are also known; all are prepared from Cr_2O_3 and the acid, and form hydrates. Used as mordants. There are many O-complexes including acetylacetonates. Chromium(II) salts can be prepared by electrolytic reduction of the chromium(III) derivatives and include the ETHANOATE and a blue sulphate. Chromium(IV) derivative, $Cr(OBu^t)_4$ (oxidation of $Cr(NEt_2)_3$ followed by alcoholysis) is monomeric.

chromium sulphates

See CHROMIUM OXY-SALTS.

chromophore

The group mainly responsible for the overall colour of the compound, i.e. in practice for causing absorption in the visible and u.v. region of the spectrum, e.g. –C=C–, –C=O, –N=N–.

chromoproteins

See PROTEINS.

chromous

Obsolete nomenclature for chromium(II).

chromyl chloride

CrO_2Cl_2. See CHROMYL COMPOUNDS.

chromyl compounds

Derivatives containing the CrO_2 unit. Chromyl chloride (NaCl plus $K_2Cr_2O_7$ plus H_2SO_4) is the best-known derivative. Dark red liquid, m.p. –96°C, b.p. 117°C; violently hydrolysed by water. Powerful oxidizing agent for, e.g., P and S, used in ÉTARD'S REACTION (oxidation of alkyl aromatic to a ketone or aldehyde).

chronopotentiometry

Study of electrode processes from potential as current varied.

chrysanthemum carboxylic acids

The acids, esters of which are the active constituents of PYRETHRUM. In dicarboxylic acids, the side-chain –C(O)OH group may be esterified: monocarboxylic acids are known. The ring –C(O)OH group may be esterified. See also ALLETHRIN, PYRETHROIDS and CINERINS.

chrysene (1,2-benzophenanthrene), $C_{18}H_{12}$.

Crystallizes in colourless plates; m.p. 254°C, b.p. 448°C. Polycyclic aromatic hydrocarbon.

chymotrypsins

An important group of proteolytic enzymes which are secreted into the intestine from the pancreas. Chymotrypsins act specifically on peptide linkages next to phenyl-alanine or tyrosine.

CI Colour Index.

CIDNP Chemically Induced Dynamic Nuclear Polarization. An n.m.r. method of studying fast reactions – spectra have unusual n.m.r. intensities.

ciment fondu An aluminous CEMENT which sets slowly but hardens rapidly. Resistant to sea water and to sulphates in solution. Also used as a refractory cement.

cinchonidine, $C_{19}H_{22}N_2O$. Cinchona alkaloid. A stereoisomer (H, vinyl interchange) of cinchonine. M.p. 210°C.

cinchonine, $C_{19}H_{22}N_2O$. Colourless needles, m.p. 255°C. Cinchona alkaloids (+)cinchonine used for the resolution of chiral acids.

1,8-cineole See EUCALYPTOL.

cinerins Constituents of PYRETHRUM. Esters of cinerolone and chrysanthemum monocarboxylic acids, insecticides.

cinmethylin Herbicide, benzyl ether derivative.

cinnabar, HgS. The principal ore of Hg. See also VERMILION RED.

cinnamic acid (3-phenylpropenoic acid), $C_9H_8O_2$, PhCH=CHC(O)OH. Colourless crystals. Decarboxylates on prolonged heating. Oxidized by nitric acid to benzoic acid. Ordinary cinnamic acid is the *trans*-isomer, m.p. 135–136°C; on irradiation with u.v. light isomerizes to the less stable *cis*-isomer, m.p. 42°C.

Prepared by heating benzaldehyde with sodium ethanoate and ethanoic anhydride (Perkin reaction) or with ethyl ethanoate and sodium ethoxide. Occurs in storax, or liquid amber, as the ester cinnamyl cinnamate made from cinnamyl alcohol, PhCH=CH·CH$_2$OH. Cinnamic acid and its derivatives are used in flavours, perfumery, cosmetics and pharmaceuticals.

cinosulfuron Herbicide, sulphonyl urea derivative.

circular dichroism (C.D.) The production of an elliptically polarized wave when a linearly polarized wave passes through a substance. The difference in the absorption intensities for right- and left-circularly polarized radiation. The plot of C.D. against frequency is the C.D. spectrum. The magnitude of the effect is expressed by the equation

$$\varphi = \pi/\lambda(\eta_l - \eta_r)$$

where φ is the ellipticity (in radians) of the emerging beam and η_l and η_r are the absorption indices of the left- and right-handed circularly-polarized light respectively. When the ellipticity is plotted as a function of wavelength, a curve results with a maximum corresponding to the wavelength of zero angle in the optical rotatory dispersion curve. Optical isomers give circular dichroism curves which are identical except that in one case the effect is positive, i.e. φ is positive throughout, whereas for the other isomer the effect is negative. Useful for determining absolute configuration of d-electron complexes.

cis Term used in designating isomers to indicate isomer with adjacent like ligands or groups. *cf. trans*.

cisplatin Important anti-cancer drug. Platinum complex.

citraconic acid (methylmaleic acid), $C_5H_6O_4$.

$$cis\text{-}[HO(O)C]CH_3C=C(CH_3)C(O)OH$$

A colourless solid crystallizing in fine needles, m.p. 91°C (decomp.). Prepared by the addition of water to citraconic anhydride which is itself prepared by the rapid distillation of anhydrous citric acid. Forms citraconic anhydride (methylmaleic anhydride) when heated. Reduced by hydrogen to pyrotartaric acid. Electrolysis of the acid yields propyne.

citral, $(CH_3)_2C=CHCH_2CH_2C(CH_3)C=CHCHO$, $C_{10}H_{16}O$. A terpene aldehyde. A volatile oil of pleasant odour forming the main constituent of lemon-grass oil from *Cymbopogon flexuosus*, and also found in other essential oils. Lemon-grass oil is an important article of commerce. The citral found in natural products is a mixture of isomers geranial (*cis*-CH$_3$.CHO) and neral (*trans*-CH$_3$.CHO).

With dilute sulphuric acid citral forms *p*-cymene. Citral can be condensed with propanone to form a ketone, pseudoionone, $C_{13}H_{20}O$, which is technically important, as it is readily convertible into α- and β-ionone. Used in synthesis of vitamin A, flavouring and perfumery.

citric acid, $C_6H_8O_7$, $C(OH)[C(O)OH]–[CH_2 C(O)(OH)]_2$. Crystallizes from water below 37°C as the monohydrate; the anhydrous acid has m.p. 153°C. It occurs in the juice of citrus fruits and in beets, cranberries and certain other acid fruits. Manufactured by the fermentation of sugar by moulds of the *Aspergillus niger* group. The acid is tribasic, and forms three series of salts; the citrates of the alkali metals are soluble, but the neutral citrates of calcium and barium are insoluble. At 175°C it loses water to give aconitic acid: at higher temperatures aconitic and citraconic anhydrides are formed. Oxidation with potassium permanganate or heating with fuming sulphuric acid gives acetone and dicarboxylic acid. It is used extensively in the soft drinks and food industries as an acidulant and flavouring, and in resins, and as a sequestering agent.

citric acid cycle (tricarboxylic acid cycle, Krebs's cycle) A cyclic sequence of reactions in cell metabolism, by which is achieved the controlled oxidative breakdown of acetylcoenzyme A, derived from carbohydrates (see GLYCOLYSIS) and fatty acids (see β-OXIDATION). Other molecules, such as amino-acids, can be modified so that they too can be fed into the cycle. The products of the cycle are carbon dioxide together with reduced coenzymes (CoASH, NADH, NADPH, $FADH_2$) which serve as energy sources; under aerobic conditions this occurs through the respiratory chain. The citric acid cycle is of fundamental importance, being at the centre of both the production of energy and the formation of complex cell constituents. See DICTIONARY OF BIOLOGY.

citrulline (α-amino-δ-ureidovaleric acid), $C_6H_{13}N_3O_3$,$H_2N\cdot CO\cdot NH(CH_2)_3\cdot CHNH_2\cdot C(O)OH$. M.p. 222°C. Intermediate in the UREA CYCLE in the excretion of excess nitrogen from the body.

Cl Chlorine.

Claisen condensation Condensation of an ester with another ester, a ketone, or a nitrile in the presence of sodium ethoxide, sodium, or sodamide, with the elimination of an alcohol. The result is the formation of a β-ketonic ester, ketone, or nitrile respectively, e.g.

$$CH_3C(O)OC_2H_5 + HCH_2C(O)OC_2H_5 \rightarrow$$
$$CH_3COCH_2C(O)OC_2H_5 + C_2H_5OH.$$

ethyl acetoacetate

The reaction is of general application and of great importance, and a large number of syntheses have been effected by its use.

Claisen reaction Condensation of an aldehyde with another aldehyde or a ketone in the presence of sodium hydroxide with the elimination of water. Thus benzaldehyde and methanal give cinnamic aldehyde, $PhCH:CH\cdot CHO$.

Clapeyron equation A thermodynamic equation applying to any two-phase equilibrium for a pure substance. The equation gives a measure of the slope of the phase boundary:

$$\frac{dP}{dT} = \frac{\Delta_{trs}S}{\Delta_{trs}V}$$

where P is the pressure, T the absolute temperature and S the entropy of the transition. If one phase is the vapour,

$$\frac{dP}{dT} = \frac{\Delta_{vap}H}{T\Delta_{vap}V}$$

and for a solid–liquid boundary

$$P = P_o + \frac{\Delta_{fus}H}{\Delta_{fus}V} \cdot \ln\frac{T}{T_o}$$

clarification The removal from a liquid of small amounts of suspended matter with the object of obtaining a clear product. Clarification may be achieved by FILTRATION, CENTRIFUGATION, or by the use of a CLARIFIER.

clarifier Large tanks with continuous feed and outflow in which the suspended matter is allowed to settle and is removed. Clarifiers are generally similar to THICKENERS except that they handle smaller quantities of solids.

Clark electrode See OXYGEN CATHODE.

classification The process of separating a mixture of particles into two or more fractions according to size, shape, density, magnetic properties, etc.

clathrate compounds Molecular compounds formed by the inclusion of molecules of one type in holes in the lattice of another. Used in the separation of, e.g. gases.

Claude process A process similar to the LINDE PROCESS for the liquefaction of air, except that additional cooling is produced by allowing the expanding gas to do external work.

Clausius–Clapeyron equation The variation of vapour pressure with temperature.

$$\frac{d\ln P}{dT} = \frac{\Delta_{vap}H}{RT}$$

clays Natural aluminosilicates which occur as a plastic paste or can be converted to a paste by grinding or mixing with water. When dried and ground, clay particles can be suspended in water almost indefinitely; suspension is aided if a small amount of Na_2CO_3 is added. Gels may be formed in concentrated suspension.

Clays have layers of linked $(Al, Si)O_4$ tetrahedra combined with layers of $Mg(OH)_2$ or $Al(OH)_3$. Clays are very important soil constituents and are used in pottery, ceramics, as rubber, paint, plastic and paper fillers, as adsorbents and in drilling muds.

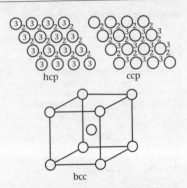

hcp   ccp

bcc

clay treatment Removal of by-products and ACID-SLUDGE from oils by adsorption on clay.

cleavage planes Directions in a crystal in which there is ready cleavage. Cleavage planes often correspond to layers of atoms or molecules in the lattice.

Clemmensen reduction Aldehydes and ketones may generally be reduced to the corresponding hydrocarbons by heating with amalgamated zinc and hydrochloric acid.

clodinafop-propargyl Herbicide, phenoxypropionate derivative.

clofencet Plant growth regulator, pyridazine derivative.

clofentezine Acaricide, tetrazine derivative.

clomazone Herbicide, oxazolidinone derivative.

clopreprop Plant growth regulator, phenoxypropionic acid derivative.

clopyralid Herbicide, pyridine derivative.

cloquintocet—mexyl Herbicide safener, quinoline derivative.

cloransulam Herbicide, triazole derivative.

close-packed structures The majority of metals, alloys and ionic inorganic compounds, and also many compounds which are appreciably covalent, have structures which can be described in terms of close-packing of the largest species, atoms or ions. In each layer of a true close-packed structure each atom has six neighbours in a hexagonal array. A similar second layer ties with its atoms above the holes of the first layer. The atoms of the third layer can be directly above the atoms of the first layer (hexagonal close-packing, hcp) or above holes in both first and second layers (cubic close-packing, ccp). In the latter case the fourth layer lies over the first. More complex sequences of layers are known. The body-centred cubic structure, bcc, does not make quite such full use of space as hcp or ccp. Many structures are described in terms of smaller cations occupying appropriate holes (octahedral Δ or tetrahedral T) between layers. For each close-packed atom there is one octahedral hole and two tetrahedral holes (in the diagrams the number n designates an atom in the nth layer, the first layer is designated O).

closo See BORANE ANIONS.

clothodin Herbicide, oxine derivative.

cloxyfonac Plant growth regulator, aryloxyalkanoic acid derivative.

clupadonic acid (docosapentaenoic acid), $C_{22}H_{34}O_2$. A straight-chain fatty acid with 5 double bonds. A major component of fish oils.

cluster compounds Species containing groups of metal atoms joined by metal–metal bonds, e.g. $[Rh_6(CO)_{16}]$, $[MO_6Cl_8]^{4+}$.

Cm Curium.

CNDO method See HÜCKEL THEORY.

Co Cobalt.

coacervation The process of separation of lyophilic (hydrated colloid) sols into two immiscible liquid phases, each of which has a different concentration of the dispersed phase.

coagel A gelatinous precipitate; e.g. rigid silica gel.

coagulation The stability of a colloidal particle is due to its surface electric charge, or its possession of a hydration sheath, or both. Any effect, chemical or physical, which removes these causes of stability produces coagulation – irreversible instability of the colloid. Thus the passage of an electric current through a colloidal sol makes the particles move to the electrodes. They are discharged and coagulated. The electrical precipitation of smokes and mists is due to a similar action.

The action of small quantities of electrolytes in coagulating hydrophobic (i.e. unhydrated) colloids is important. The colloid is coagulated by the adsorption of ions of charge opposite to that which it carries. Thus an arsenic sulphide sol (negative) can be coagulated by unipositive cations, but dipositive cations are more efficient and tripositive cations much more so, the latter producing coagulation even at great dilutions. The addition of a large excess of the coagulant may stabilize a sol of opposite charge to that of the initial colloid, by adsorption of a large excess of (in the above example, positive) ions. Other coagulating influences include ultrasonic vibrations, u.v. light and boiling (as in denaturization of proteins); the two latter effects are essentially chemical. Hydrated or hydrophilic colloids are also coagulated by electrolytes, but only in high concentrations. This effect is termed 'SALTING OUT' in contrast to the simple adsorption effect outlined above. See COLLOID, ELECTROPHORESIS, FLOCCULATION, PEPTIZATION, 'SALTING OUT'.

coal A naturally occurring carbon-containing solid fuel which exists in the form of seams at varying depths below the earth's surface. Formed by the arrested bacterial decay of vegetation, followed by chemical processes of condensation and polymerization under the influence of temperature, pressure and time. By these processes the differences between the original woody material and anthracite, the highest-ranking coal, can be explained. As the rank increases the carbon content increases from about 50 to 95%, and the oxygen content decreases from about 40 to 3%. The CALORIFIC VALUE increases (average 3.3×10^7 J kg^{-1}) through the range of coals: peat – lignites – bituminous – semi-bituminous and carbonaceous coals – to the anthracites.

The properties of the various coals in this ranking will vary considerably and many are used only for specific purposes. Coal seams frequently contain methane.

coal briquettes Blocks of coal powder mixed with a binder (coal-tar pitch or bitumen) and heated with pressing. Used as fuels, particularly domestically.

coal gas See TOWN GAS.

coalite process A process for the carbonization of coal at 600°C to give coke, coal-tar oil, hydrocarbons, carbon monoxide, ammonium compounds.

coal tar The by-product obtained from the high-temperature CARBONIZATION of coal in coke ovens and gas retorts and from the low-temperature carbonization processes for SMOKELESS FUEL production.

The composition of coal tar varies with the carbonization method but consists, largely, of mononuclear and polynuclear aromatic compounds and their derivatives. Coke-oven tars are relatively low in aliphatic and phenolic content, while low-temperature tars have much higher contents of both.

Crude tar is normally distilled in continuous plant into distillate fractions which can vary in boiling range and in name, leaving pitch as a residue.

Higher-boiling fractions supply the CREOSOTES, absorbing oils, naphthalene, anthracene, COAL TAR FUELS, TAR, etc.

coal tar fuels (CTF) Industrial fuels produced by blending distillate and residual fractions from coal tar distillation.

coal tar pigments Generally metallic lakes or insoluble metal salts of organic dyestuffs. Not now prepared from COAL TAR.

coated paper To give a fine surface for printing and for preventing ink absorption, paper is coated with a colloidal mix of e.g. kaolin, satin white, with an adhesive such as glue, casein, starch, plus a plasticizer to harden the protein and prevent frothing.

cobalamin See COBALT, ORGANOMETALLIC COMPOUNDS.

cobalt, Co. Transition metal, at.no. 27, at. mass 58.933, m.p. 1495°C, b.p. 2927°C, density ρ 8900, crustal abundance 20 p.p.m., electronic configuration [Ar] $3d^84s^2$. Obtained commercially from silver ores (arsenides and sulphides) and Ni, Cu and Pb arsenide ores (speisses). Smaltite, $CoAs_2$, and cobaltite, CoAsS, are also used as sources. The ore is roasted and Co is precipitated as the hydroxide and then reduced to Co with carbon (hcp below 417°C, ccp to m.p.). The metal is silvery white and readily polished. It dissolves in dilute acids and is slowly oxidized in air. Adsorbs hydrogen strongly. The main use of cobalt is in alloys and electroplating. $LiCoO_2$ used in lithium–cobalt batteries and in pigments. Cobalt compounds are used in paints and varnishes, catalysts. Cobalt is an essential element in the diet.

cobalt, organic derivatives Low oxidation state derivatives are well established, e.g. $MeCo(CO)_4$ ($NaCo(CO)_4$ plus MeI). Vitamin B_{12} is a naturally occurring cobalt(III) organometallic and there is an extensive chemistry of alkyl, particularly methylcobalamin (vitamin B_{12} is cyanocobalamin) derivatives. Olefin and cyclopentadienyl complexes are known, olefine complexes are intermediates in hydroformylation reactions.

cobalt alloys Up to 80% of cobalt produced is used in alloys, 20% of this for magnets. The most important cobalt alloys are stellite (up to 30% Cr, 18% W, 2½% C used in valves and cutting tools) and vitallium (20% Cr, high temperature uses). Co is also used in highspeed tools.

cobaltammines Ammonia complexes of cobalt(III), almost entirely with octahedral coordination.

cobalt bloom, $Co_3(AsO_4)_2.8H_2O$. An important source of arsenic.

cobalt bromide See COBALT HALIDES.

cobalt carbonyls Cobalt in CO at 150°C and 300 atm. gives $Co_2(CO)_8$, $[(OC)_3Co-(\mu-CO)_2-Co(CO)_3]$ with a metal–metal bond; m.p. 51°C. $Co_2(CO)_8$ gives $Co_4(CO)_{12}$ on heating. With OH^- $Co_2(CO)_8$ gives $[Co(CO)_4]^-$ which gives $[HCo(CO)_4]$ with dilute acid. $Co_2(CO)_8$ and $HCo(CO)_4$ and their derivatives are important industrial catalysts in hydroformylation reactions. Many substituted derivatives of the cobalt carbonyls are known.

cobalt chemistry Cobalt is an element of Group 9. The common oxidation states in aqueous solution are +2

$$E° \text{ for } Co^{2+} \rightarrow Co \ -0.28 \text{ volts in acid}$$

which exists as the hydrated ion, and +3 which is very unstable as the hydrate but stable as ammine and other low-spin complexes

$$E° \text{ for } [Co(NH_3)_6]^{3+} \rightarrow [Co(NH_3)_6]^{2+} +0.1 \text{ volts}$$

There are many tetrahedral and octahedral Co(II) compounds, e.g. halides and a great range of largely octahedral Co(III) complexes. Other oxidation states for Co are –1, e.g. $[Co(CO)_4]^-$; 0, $Co_2(CO)_8$; +1, $CoBr(PR_3)_3$ and $[Co(NCR)_5]^+$; +4, $(CoF_6)^{2-}$. Cobalt complexes are oxygen carriers. Cobalt can be present in enzymes and may be of great importance in their action.

cobalt fluorides See COBALT HALIDES.

cobalt halides

Cobalt(II) fluoride, CoF_2. Forms 4,3,2,0 hydrates and gives perovskites, e.g. $KCoF_3$.

Cobalt(III) fluoride, CoF_3. Brown powder (CoF_2 plus F_2) also forms a green hydrate by electrolytic oxidation. CoF_3 is widely used in the fluorination of organic derivatives. Gives complexes e.g. M_3CoF_6.

Cobalt(IV) fluoride is not known but M_2CoF_6 (Cs_2CoCl_4 plus F_2) is stable.

Cobalt chloride, $CoCl_2$. Obtained as red crystals of $CoCl_2.6H_2O$ from aqueous solution, $CoCl_2.H_2O$ and $CoCl_2$ are blue as is the $(CoCl_4)^{2-}$ ion. No higher chloride is known although cobalt haloammines, e.g. $(Co(NH_3)_5Cl)^{2+}$ are stable.

Cobalt(II) bromide and iodides are similar to the chloride.

cobalt(II) hydroxide, $Co(OH)_2$. Precipitated from cobalt(II) salts by strong base; pink or blue. Dissolves in excess base to give $[Co(OH)_4]^{2-}$ (deep blue) and $[Co(OH)_6]^{4-}$. Hydroxo complexes of cobalt(III), including hydroxy-bridged species, are known. Used in batteries and as a drier.

cobaltic Cobalt(III) derivatives.

cobaltite, CoAsS. A cobalt mineral.

cobaltous Cobalt(II) derivatives.

cobalt oxides

Cobalt(II) oxide, CoO. Olive-green solid (heat on $Co(OH)_2$ or cobalt(II) oxyacid salt in absence of air) with NaCl structure.

Tricobalt tetroxide, Co_3O_4. Black solid (ignition of CoO in air). Has spinel structure and other spinels MCo_2O_4 are also known.

Impure CoO_2 (oxidizing agent on alkaline Co(II)) and some mixed oxides of cobalt(IV) and (V), e.g. K_3CoO_4, are known.

cobalt oxyacid salts Cobalt(II) forms an extensive range of oxyacid salts, generally as hydrates, generally soluble in water. Examples are $Co(NO_3)_2.6H_2O$; $Co(ClO_4)_2.6H_2O$; $CoSO_4.7H_2O$. Few cobalt(III) oxyacid salts are known. $Co_2(SO_4)_3.18H_2O$ is formed by electrolytic oxidation and forms alums; $Co(NO_3)_3$ contains co-ordinated nitrate (CoF_3 plus N_2O_5).

cocaine (benzoylmethylecgonine),
$C_{17}H_{21}NO_4$. Colourless prisms, m.p. 98°C. Alkaloid, obtained from coca, either by direct purification, or by acid hydrolysis of the mixed alkaloids to ecgonine, which is then methylated and benzoylated. Coca consists of the dried leaves of shrubs growing in Central and South America.

Cocaine is the oldest of the local anaesthetics. Controlled substance. It is a central nervous system stimulant and is habit forming. See ECGONINE.

cochineal The dried body of the female insect *Dactylopius coccus*. The deep red colour of the cochineal is due to carminic acid. It is used as a colouring matter for foodstuffs and drugs, and as an indicator. Once used as a dyestuff.

co-current flow When two streams of material are brought into contact with each other with the object of transferring either heat or matter between them, if the streams move in the same direction flow is said to be co-current. See also COUNTER-CURRENT FLOW.

codeine (o-methylmorphine),
$C_{18}H_{21}NO_3.1H_2O$, m.p. 155°C. Prepared by methylating morphine or directly from opium. Controlled substance.

cod-liver oil Oil expressed from the fresh liver of the cod, *Gadus callarias*. It is a pale yellow liquid, slightly soluble in alcohol, miscible with organic solvents. It is a valuable source of vitamins A and D and other food factors and is used in the prevention and cure of rickets in growing children.

codon See NUCLEIC ACIDS and GENETIC CODE.

coenzyme A The most prominent acyl group transfer coenzyme in living cells. It is concerned in a variety of reactions. Structurally coenzyme A is composed of adenosine-3',5'-diphosphate and pantotheine phosphate moieties. Pantotheine is made up of PANTOTHENIC ACID, β-alanine and mercaptoethylamine. ACETYL COENZYME A and malonyl coenzyme A are the most important forms in which coenzyme A is found. Acyl derivatives of coenzyme A are usually formed in ATP-dependent reactions and are involved in many acyl group transfers. See DICTIONARY OF BIOLOGY.

coenzymes Defined as small thermostable molecules that are necessary for the action of enzymes. This term will include certain metals, but the word has come to be used only for substances which play a known part in the reaction catalysed by the enzyme. The difference between coenzymes and prosthetic groups (defined as the non-protein portions of enzymes) is one of degree rather than of kind, depending on the firmness of the combination of the protein; thus the tightly bound haematins are usually referred to as prosthetic groups and the phosphopyridine nucleotides as coenzymes, though both play essentially similar roles in biological oxidations. Coenzymes may be structurally altered in the course of the reaction but are usually regenerated in subsequent reactions.

There are various classes of coenzymes including:

Coenzymes effecting transfer of hydrogen.
Coenzymes effecting transfer of groups.
Coenzymes effecting isomerization.
See DICTIONARY OF BIOLOGY.

coherent precipitate A precipitate in a solid alloy system which is still bonded directly to the parent lattice and does not have a distinct grain boundary. Generally associated with high strength in the alloy.

coinage metals The elements Cu, Ag, Au of Group 11.

coke The dense product of the carbonization of coal in an oven. Petroleum coke is produced as a product of oil refining. Coke varies in colour from dark matter to a silvery lustrous material. Used as a fuel, as a reducing agent in blast furnaces, for producer gas formation, as electrodes.

coking Relatively severe cracking processes converting residual materials such as pitch or tar into a mixture of gases, naphtha, kerosine, gas oil and coke. The gas oil is used mainly as feedstocks for catalytic cracking.

coking coal Coal which, after carbonization, will give a marketable coke.

colemanite, $CaB_3O_4(OH)_3.H_2O$. An important source of borates.

collagen The most abundant protein in the body, containing mainly glycine, hydroxyproline and proline. Collagen is essentially insoluble but when boiled with water, the strands separate and undergo some hydrolysis, affording GELATIN. The conversion of collagen to leather involves the formation of cross linkages between collagen molecules. Collagen is found in all connective tissue, e.g. skin, cartilage, tendons, ligaments, bone.

s-collidine See 2,4,6-TRIMETHYLPYRIDINE.

colligative properties Properties of solutions, e.g. vapour pressure, osmotic pressure, boiling point, freezing point, which depend only upon the number of particles (atoms, ions, or molecules) present in solution, and are independent of the nature of the solute.

Collman's reagent, $Na_2Fe(CO)_4$ plus triphenylphosphine. Important for carbonylation of halides, i.e. $RBr \rightarrow RCHO$.

collodion A cellulose nitrate widely used as a base for lacquers, for which a ketone or ester may be used as the solvent. Used in photography, lacquers, leather, cement. Very inflammable.

colloid Disperse phase characterized by a slow rate of diffusion and by the fact that the path of a beam of light is illuminated on passage through a colloidal solution. Now defined as being in the colloidal state if its particles, of approximate size between 2 nm and 500 nm, dispersed in a continuous medium. Colloids are thus intermediate between coarse suspensions on the one hand and molecular or ionic solutions on the other. The lines of demarcation are indefinite, fine suspensions showing some colloidal effects, whilst large molecules are essentially colloidal in behaviour. Most colloids are solids in liquids, but liquids in gases (aerosols) and solids in solids (gold in glass) are also known. The commonest liquid is water but other solvents form disperse phases. Colloids are classed as lyophilic – solvent attracting – or lyophobic – solvent repelling.

It is now possible by choice of suitable conditions to prepare most compounds in this form. Many materials such as proteins, vegetable fibres, rubber, etc. are most stable or occur naturally in the colloidal state. In the colloidal state the properties of surface are all-important.

colloidal electrolyte Although the sodium salts of the lower fatty acids behave as normal weak electrolytes, with higher members of the series association occurs with onset of colloidal character. The anion is no longer a discrete fatty acid anion, but consists of an aggregate of fatty acid anions together with undissociated molecules; it is a giant polyvalent ion, dissociating at the surface and compensated by an equal number of sodium ions in close proximity to the giant ion – see MICELLES. Soaps and many dyestuffs may be classed as colloidal electrolytes.

colloid mills Devices for producing colloidal suspensions or emulsions where the particle and droplet sizes are less than one micron.

Colloid mills are used in the paint industry, and in the pharmaceutical and food-processing industries.

colour centre A crystal defect which gives rise to optical absorption (changes colour of crystal).

colour couplers Compounds used in photographic development of colour films in which the anion can react with developer to give appropriate dyestuffs.

colour development The photographic process in which developer oxidation products, formed in the emulsion layer where silver ions are reduced, are used to form dyes either by direct combination with another molecule of oxidized developer (primary colour development), or by the oxidized developer reacting with a new molecular species known as a COLOUR COUPLER to give a dye of the required absorption (secondary colour development). Multicoloured processes employ the latter system.

colour index (CI) The definitive listing of dyestuffs and pigments which includes information on commercial names, method of application, dyefastness, etc.

colour indicators INDICATORS (acid–base, oxidation–reduction) which depend for their effect on colour change. Screened indicators have an additional dye added which acts as a colour filter and makes the colour change more visible. Thus the colour change of methyl orange with pH is made more obvious by addition of some methylene blue. A universal indicator is a mixture of indicators which shows a gradual but well-marked series of colour changes over a very wide range of pH.

colour photography The process of preparation of prints or transparencies in colour. White light is split into three components – red, green and blue – by filters – yellow, magenta, cyan. The particular colour activates grains of silver halide (generally in specific layers for each colour) which contain sensitizing dyes on the grain. COLOUR DEVELOPMENT then gives rise to a coloured image.

columbite, (Fe, Mn)(Nb, Ta)$_2$O$_6$ with excess Nb. The principal ore of NIOBIUM.

columbium, Cb. Obsolete name and symbol for NIOBIUM.

combinatorial chemistry Techniques for synthesizing series of compounds and their testing – for, e.g. physiological activity or particular properties – frequently by automated methods with molecules anchored on a solid surface (resin).

combining volumes, law of See GAY-LUSSAC'S LAW.

combustion The rapid, high-temperature oxidation of fuels, converting carbon to carbon dioxide (or carbon monoxide) and hydrogen to water vapour. Any sulphur in the fuel is oxidized to the dioxide or trioxide depending on combustion conditions, while nitrogen either remains unreacted or is converted to nitrogen oxides. Most combustion reactions occur in the gas phase except for the burning of the fixed carbon in solid fuels.

The release of chemical energy during combustion of gases produces a luminous, radiating zone which is seen as the flame or FLAME FRONT.

common-ion effect In a solution of a weak electrolyte, e.g. ethanoic acid, HAc, the concentration of ions is governed by the equilibrium

$$HAc \rightleftharpoons H^+ + Ac^-$$

Addition of excess H$^+$ ions to this solution will cause the equilibrium to move towards undissociated acid thereby decreasing the concentration of Ac$^-$. This effect is known as the common-ion effect and is of considerable practical importance. Thus, e.g. in the precipitation of metal ions as insoluble sulphides, the concentration of S^{2-} in aqueous solution is controlled by the equilibrium

$$H_2S(aq) \rightleftharpoons 2H^+(aq) + S^{2-}(aq)$$

Addition of acid will reduce the concentration of S^{2-}, whilst in alkaline solution the concentration of S^{2-} will increase. Since, in order for precipitation to occur, the solubility product of the sulphide must be exceeded, i.e.

$$K_{sp} = [M^{x+}][S^{2-}]^{x/2},$$

the actual precipitation can be controlled by varying the pH of the solution.

complement The combination of an antibody and an ANTIGEN on the surface of a foreign cell leads to the 'fixation' of a group of proteins present in normal serum, collectively known as complement. Interacting serum proteins which participate immunologically in bacterial lysis and macrophage chemotaxis. The result is the activation of destructive enzymes which cause lysis of the cell. See DICTIONARY OF BIOLOGY.

complex Any compound in which the bonding is by interaction of the electrons of the donor with empty orbitals of the acceptor. In some complexes the electron flow may take place in both directions simultaneously – see BACK-BONDING. The interaction may take place between charged or uncharged species.

Where the structure is known, a complex species comprising the acceptor and its ligands is formulated within square brackets, e.g. [Co(NH$_3$)$_6$]Cl$_3$. Bridging ligands are designated μ-L, e.g. Fe$_2$(CO)$_9$ is [(OC)$_3$Fe-(μ-CO)$_3$Fe(CO)$_3$]. Hapto designates the number of ligand atoms actually bonded to the acceptor, e.g. [(h^5 – C$_5$H$_5$)Mn(CO)$_3$] has 5 carbon atoms (plus 3 carbonyls) bonded to manganese.

complex ion A complex ion is formed by the co-ordination of other ions or molecules to an ion or atom to form a stable charged entity. Thus the Co^{3+} ion and ammonia give [Co(NH$_3$)$_6$]$^{3+}$ whilst the Fe^{3+} ion and cyanide ions give the [Fe(CN)$_6$]$^{3-}$ complex. Depending upon the stability and lability of the complex a solution of a complex salt may or may not give the reactions of the individual components of the complex. Hydrated salts generally contain aquo-complexes, e.g. [Cu(H$_2$O)$_6$]$^{2+}$ and the anions of oxy-salts may be considered as being complexes formed by (mostly hypothetical) positive ions and oxide ions, e.g. the NO$_3^-$ ion as N^{5+} and 3O^{2-}.

complexometric indicator See METALLO-CHROMIC INDICATOR.

complexometric titration A TITRATION involving formation or decomposition of a complex. Thus Fe^{3+} plus SCN$^-$ is intense red; titration of the Fe^{3+} with EDTA causes decomposition of the red Fe^{3+}– SCN$^-$ complex and formation of the colourless Fe^{3+}– EDTA complex.

complexone See SEQUESTERING AGENT.

component For the PHASE RULE the number of components in a chemical system is the smallest number of chemically independent substances in terms of which the compositions of each of the phases in the system may be described separately. The number of components in a system need not necessarily equal the number of chemical species present. For example, if calcium carbonate is heated in a closed system:

$$CaCO_3 \rightleftharpoons CaO + CO_2$$

the concentration of $CaCO_3$ can be expressed in terms of those of CaO and CO_2, i.e. only two components are present.

composites High-performance materials consisting of a plastic matrix reinforced by fibres (glass, graphite, boron, etc.). Used in spacecraft, aircraft, car components, sports goods, machinery.

compound The usage of chemists in respect of the word compound is not perfectly definite; all would agree that gunpowder is a mixture and common salt a compound. Not much difficulty is found with a compound with a fixed composition such as sodium chloride, sugar or water. It is not so easy to understand the orthodox usage in connection with substances the composition of which may vary, the properties of the substance probably also varying with composition. Glass, steel, iron oxides and starch are examples of such substances; in them every atom is attracted by (bonded to) all the adjacent atoms so as to make a uniform or very nearly uniform solid; such bodies are clearly not mere mechanical mixtures; they are chemical compounds of varying or indefinite composition. The feldspars and many other silicates also have no definite composition; similarly many plastics and other polymers are completely chemically combined and must be considered as chemical compounds of varying compositions; alloys such as pewter and brass are compounds not mere mixtures. It is not useful to give a precise definition of compounds, and there are intermediate states in between the typical compound and the typical mixture.

compounding The mixing process used to produce a homogeneous mixture for use in, e.g. a plastic or rubber composition when for example rubber vulcanizing agents, fillers, antioxidants, pigments are added.

concentrated Containing a high proportion (e.g. of solute in a solution).

concentration The amount of substance in a given volume of solution or, for gases, the molar proportion in the gas. For solutions, measured as the number of mol $M\,dm^{-3}$.

concentration cell The potential of a piece of metal immersed in a solution containing its ions (see ELECTRODE POTENTIAL) depends on the concentration of the ions. Thus a cell may be set up which derives its electromotive force from the difference in concentration of solutions of the same electrolyte surrounding the two electrodes. Any cell which depends on this principle is called a concentration cell. Such a cell would require a bridge between the electrolytes in order to continue operation.

conchoidal fracture A surface resulting from the fracture of a solid which shows no regular crystal faces (cleavage planes). The fracture is generally made up of shell-like curved surfaces and is characteristic of amorphous glassy materials.

concrete Construction material resulting from the hardening of a mixture of cement, sand (or similar) and water. Can be made lighter by incorporation of, e.g. perlite or polystyrene.

condensation reactions Addition of one molecule to another with the elimination of a simple molecule such as water, ammonia, an alcohol, etc., e.g. the Claisen reaction.

condensers Heat exchangers used for condensing vapours, the heat normally being removed by cooling water.

conductance (G) The inverse of resistance. See CONDUCTIVITY, SOLIDS.

conductiometric titration A titration in which the end point is ascertained by observing the changes in conductivity of the solution.

conductive polymers Polymers, mainly organic, which conduct electricity. Examples are polyacetylene, polypyrrole, polythiophene.

conductivity The property which allows the passage of an electric current. The reciprocal of the resistance of a circuit is called the conductivity. The reciprocal of the specific resistance is the specific conductivity or conductance of a substance. Measured in siemens, m^{-1}.

For solutions

$$G = \frac{KA}{l}$$

where K is the conductivity, l the distance between the electrodes and A the cross-section area of the electrodes.

conductivity, solids Solids may conveniently be classified in terms of their electrical conductivity into electronic conductors, semiconductors and insulators. Conductance, G, is resistance^{-1} and is expressed in ohms^{-1} (formerly mho but now siemens). Metals are typical conducting solids, their conductivity increasing with decreasing temperature. Semiconductors may be classified as either intrinsic semiconductors, e.g. pure germanium or silicon and a number of transition metal oxides, or as impurity semiconductors, e.g. Al- or P-doped Si and Ge. Semiconductors may be differentiated from conducting solids in that the conductivity of the former increases markedly with increasing temperature. Refractory oxides, e.g. alumina, silica, magnesium oxide are typical insulators. Impurity-doped semiconductors find extensive use in electronic microcircuits as transistors, etc. Ionic conductors conduct by movement of ions through the solid.

Condy's fluid A disinfectant solution of calcium and potassium permanganates.

configuration The spatial arrangement of atoms or groups in molecules. Specified by placing the attached groups according to the SEQUENCE RULES and viewing the diagram or model of the molecule so that the group of lowest sequence is directed away from the viewer. Then, if the direction from the highest sequence group to the second to the third is clockwise, the molecule is designated R, if anticlockwise S. R and S do not designate absolute configuration. The absolute configuration can be determined by X-ray methods. See ISOMERISM and CONFORMATION.

conformation The term usually restricted to the potentially dynamic spatial arrangements of atoms or groups in a molecule which may be in equilibrium with other conformations. Usually no single conformation constitutes a discrete and isolable substance under usual conditions, in contrast to configurational isomers.

staggered   eclipsed

Two extreme conformations of ethane are the *eclipsed* and *staggered* forms, which are easily in-

terconvertible by rotation about the C–C bond.

Cycloalkanes present other conformational differences. For example cyclohexane molecules exist as rapidly interconverting species with *boat* or *chair* conformations, and it is not possible to separate these, although at room temperature the chair conformation is considerably more stable and comprises more than 99% of the equilibrium mixture.

chair   boat

A monosubstituted cyclohexane, e.g. methylcyclohexane, exists theoretically in two isomeric forms with a chair-form ring, and the methyl substituent either *axial* or *equatorial*. Since these rapidly interconvert through a process known

CH$_3$

axial   equatorial

as *ring inversion*, physical separation of the isomers is not possible. Physical evidence suggests that the equatorial form of methylcyclohexane predominates over the axial form. *Cis–trans*-isomers of cyclohexane derivatives have the additional possibility of conformational isomerism. Chair conformations are important in the pyranose forms of sugars. See GLUCOSE.

coniferin, $C_{16}H_{22}O_8$. The glucoside of coniferyl alcohol, present in fir trees. Used for the synthesis of vanillin by oxidation.

coniferyl alcohol, $C_{10}H_{12}O_3$. (3-CH$_3$O)(4-HO)-C$_6$H$_3$CH=CHCH$_2$OH. Prisms; m.p. 73–74°C. Occurs to a small extent in wood and is a constituent of the glycoside CONIFERIN. Oxidized to vanillin. It resinifies with mineral acids.

coning and quartering A sampling technique involving the formation of a cone, flattening, rejection of opposite quarters, repetition to obtain a sample of appropriate size.

conjugate acid The inverse of conjugate base.

conjugate base In Brønsted–Lowry acid base theory an acid is considered to dissociate into a proton and the conjugate base of the acid. Thus the conjugate base of an acid is its anion.

conjugate solutions Solutions of two substances in one another (e.g. phenol in water and water in phenol) which are in equilibrium at a particular temperature.

conjugation Alternating double (or triple) and single bonds as in

$$-\overset{|}{C} = \overset{|}{C} - \overset{|}{C} = \overset{|}{C} -$$

$$-\overset{|}{C} = \overset{|}{C} - \overset{|}{C} = O, \text{ etc.}$$

conproportionation The opposite of DIS-PROPORTIONATION, e.g. $Mo + MoF_6$ to MoF_5.

conservation of energy, law of In systems of constant mass, energy can neither be created nor destroyed. One form of energy may disappear, but another takes its place. Thus the energy possessed by a hammer may be converted into heat and sound on striking a surface. In an isolated system, the total energy of the system remains constant. Energy and mass are inter-convertible.

conservation of matter, law of Matter can neither be destroyed nor created. However, emission of radiation must be accompanied by a loss of mass, equal to E/mc^2, where E is the energy of the radiation and c the velocity of light. E/mc^2 is usually small compared with the masses of material used in ordinary chemical manipulation. There is a conservation of mass and energy considered together. See ATOMIC ENERGY.

constantan An Ni (45%) Cu (55%) alloy with high resistance and low resistance-temperature coefficient, used in resistances and thermo-couples.

constant-boiling mixture See AZEOTROPIC DISTILLATION and MIXTURES.

constant proportions, law of The statement that the composition of a pure chemical compound is independent of its method of preparation. In fact many compounds have a range of compositions – see COMPOUND, DEFECT STRUCTURES.

contact process See SULPHURIC ACID.

continuity of state As the temperature is increased towards the critical point, the properties of liquids and vapours become increasingly similar, until at the critical temperature they are identical. Although the change from liquid to vapour, or vice-versa, is normally discontinuous, a gradual transition with continuity of state is possible.

continuous counter-current decantation A method for the continuous washing of finely-divided solids to free them from impurities. A series of continuous THICKENING is used, the solids passing through them in series, counter-current to the solution.

continuous spectrum The occurrence of well defined lines (see LINE SPECTRUM), or of bands (see BAND SPECTRUM) consisting of well defined lines, arises because electronic, vibrational and rotational energy changes in a molecule can occur only in definite steps (quanta) corresponding to transitions between allowed energy levels. Certain changes, e.g. the dissociation of a molecule, are not quantized processes. Hence, if light is emitted during the occurrence of such a process, its frequency no longer has definite values, but may take any value over a continuous range of frequencies. The spectrum of such a system appears to be continuous. A continuous spectrum is characteristic of an unquantized process, such as dissociation.

contraceptive drugs Drugs used to prevent conception. The drugs are mainly hormonal steroid derivatives.

convergence frequency See CONVERGENCE LIMIT.

convergence limit The lines in a spectrum represented by a given series, and thus the energy levels, become closer and closer together as the wavelength of the lines becomes shorter, and eventually approach a limit, known as the convergence limit, which may be expressed as a frequency, or wave number, the convergence frequency.

conversion processes A general term used to describe an industrial reaction involving conversion of products, e.g. conversion of coal fuel to gaseous fuel by gasification.

converting A metallurgical oxidation process carried out by blowing air through the molten charge in a converter.

cool flames Quiescent, incomplete combustion exhibited by hydrocarbons, aldehydes and ethers. Under certain conditions the oxidation reaction can occur at low temperatures (580–800 K) emitting little heat and light and showing no FLAME FRONT or reaction zone. Also called homogeneous or slow combustion since oxidation apparently occurs simultaneously at different points in the mixture.

cooling towers Towers used to cool water for use in condensers and coolers. Cooling is generally by contact with atmospheric air.

co-ordinate bond The linkage of two atoms by a pair of electrons, both electrons being provided by one of the atoms (the donor). The co-ordinate bond is formally identical with the covalent bond. An atom capable of accepting the electrons is the acceptor, the molecule donating the electrons is the donor or ligand. Co-ordinate linkages occur widely in inorganic complexes. See VALENCY, THEORY OF.

co-ordination arrangements The arrangement of atoms or ligands around a central atom depends upon the sizes of the atoms and the electronic configuration of the central atom (lone-pairs of electrons occupy positions in space). For details of the most common co-ordination arrangements see under the individual headings. Common arrangements are (considering linked atoms only and ignoring lone-pairs):

co-ordination
  number

2	linear	$BeCl_2, CO_2$
	bent	H_2O
3	trigonal planar	BF_3
	pyramidal	NH_3
4	TETRAHEDRAL	$CH_4, [NiCl_4]^{2-}$
	square planar	$[PtCl_4]^{2-}$
5	trigonal bipyramidal	PF_5
	square pyramidal	IF_5
6	OCTAHEDRAL	$NaCl, [Co(NH_3)_6]^{3+}$
	trigonal prismatic	$NiAs$
7	pentagonal bipyramid	ReF_7^-
	face-centred octahedron	$[NbOF_6]^{3-}$
	face-centred trigonal prism	$[TaF_7]^{2-}$
8	CUBIC	$CsCl$
	SQUARE ANTIPRISM	$[Mo(CN)_8]^{4-}$
	dodecahedral	$[Mo(CN)_8]^{4-}$
12	ICOSAHEDRAL	$[Ce(NO_3)_6]^{2-}$
	CUBIC CLOSE-PACKED, perovskite	

co-ordination compound A compound containing a CO-ORDINATE BOND or bonds.

co-ordination isomerism Isomerism in co-ordination compounds in terms of the actual arrangement of the atoms, e.g. NO_2^- can bond to Co^{3+} as either Co–ONO or Co–NO_2.

co-ordination numbers The number of groups or ions surrounding a particular molecule or ion in a crystal or in solution; in complexes the total number of co-ordinated and covalently bonded groups surrounding the central atom. For elements Li to F the maximum co-ordination number for covalent compounds is 4, but 6 co-ordination can occur in ionic compounds. For elements Na to Cl the maximum co-ordination number is 6.

copolymer A complex polymer resulting from the polymerizing together of two or more monomers. A copolymer is a true compound and often has properties quite distinct from those of a physical mixture of the separately polymerized monomer components. Important industrial copolymers include those obtained from vinyl chloride–vinyl acetate and styrene–butadiene monomer mixtures. See BLOCK COPOLYMERS.

copper, Cu. Metallic element, at.no. 29, at. mass 63.546, m.p. 1084.62°C, b.p. 2562°C, density ρ 8960, crustal abundance 50 p.p.m., electronic configuration [Ar] $3d^{10}4s^1$. Coinage metal; occurs native (contains Ag, Bi, Pb, etc.) and as cuprite Cu_2O, tenorite or melaconite CuO, copper pyrites or chalcopyrite $CuFeS_2$, erubescite or bornite Cu_3FeS_3; malachite $CuCO_3 \cdot Cu(OH)_2$ (bright green) and azurite $2CuCO_3 \cdot Cu(OH)_2$ (bright blue) are naturally occurring minerals used as pigments. The sulphide ores and native copper are used as sources of copper. Copper ores are concentrated by flotation and sulphur is oxidized in converters. The crude copper is refined electrolytically. The metal is ccp and reacts with oxygen at red heat. It is attacked by halogens, oxidizing acids, ammonia and KCN solutions (the latter two in the presence of oxygen). Approx. 55% of copper is used for electrical purposes and about 15% in pipes (plumbing). Much copper is used as alloys (brasses, bronzes, gold including coins). Other copper uses include catalysis ($CuCl, CuCl_2$, CuO), semiconductors, fungicides ($CuCl_2, CuO$), printing, dyeing, pigments, paints. Copper is present in some enzymes, e.g. nitrous oxide reductase, superoxide dismutase, cytochrome oxidase, tyrosinases, and is part of the oxygen carrier in snails, crabs and some crustaceans and atacamite, $Cu_2(OH)_3Cl$ is present in the cutting edge of teeth of some worms.

copper, organic compounds Copper(I) forms alkyls and aryls and also complexes, e.g. $(h^5-C_5H_5)CuPPh_3$, $(C_6H_5Cu)_2$. Copper derivatives formed in situ are important in coupling reactions.

copper arsenate Wood preservative used as chromated copper arsenate, CCA.

copper bromides, $CuBr_2$ and $CuBr$. Very similar to copper chlorides.

copper chemistry Copper is a typical transition element of Group 11. Very low oxidation states are not stable; unstable carbonyls are formed by matrix isolation methods.

Copper(I) salts are very similar to Ag(I) salts with an insoluble chloride and iodide; they are colourless or pale yellow. Complexes are formed with phosphines and CO.

Copper(II) salts (blue in aqueous solution) are typical M(II) salts but generally have a distorted co-ordination (Jahn–Teller distortion, 4 near plus 2 far neighbours). Extensive ranges of complexes are known, particularly with *N*-ligands.

Copper(III) is known in complex oxides and fluorides and in amino-acid complexes.

Electrode potentials in acid are $Cu^+ \rightarrow Cu$ +0.34 volts, $Cu^{2+} \rightarrow Cu^+$ +0.158 volts; Cu^+ disproportionates to Cu plus Cu^{2+} in aqueous solution but is stable in many non-aqueous solvents. Many cluster compounds, e.g. $[Cu_4I_4(PMe_3)_4]$, $[Cu_6H_6(PPh_3)_6]$ are known and Cu–Cu bonding is found in many copper(II) derivatives, particularly those containing bridging ligands (e.g. $[Cu(O_2CCH_3)_2,H_2O]_2$.

copper chlorides

Copper(II) chloride, $CuCl_2$. Dark brown (Cu plus excess Cl_2) forms green $CuCl_2.2H_2O$ and many complexes, e.g. $CuCl_2$ pyridine, generally containing bridging chlorines. Complex species, e.g. $[Cu_2Cl_6]^{2-}$, $(CuCl_4)^{2-}$ yellow or green, are also known.

Copper(I) chloride, CuCl. White solid ($CuCl_2$ plus HCl plus excess copper or SO_2). Gives carbonyl and phosphine complexes.

copper chromite Mixed oxide, often with added Ba^{2+}, used as a catalyst for reduction of ketones and esters to alcohols.

copper cyanide, CuCN. White compound (Cu(II) plus KCN solution, $(CN)_2$ liberated) forms $K_3Cu(CN)_4$ with excess KCN.

copper ethanoate, $[Cu(O_2CCH_3)_2.H_2O]_2$. Dimeric compound with Cu–Cu interaction (copper carbonate plus ethanoic acid).

copper fluoride, $CuF_2.2H_2O$. Anhydrous CuF_2 and some complexes are known.

copper hydroxides Basic copper(II) salts are formed by precipitation with, e.g. NaOH. A rather indefinite copper(II) hydroxide is precipitated from copper(II) solutions. On heating converted to a hydrated oxide. Used as fungicide.

copper iodide Copper(II) iodide is not known. Addition of I^- to a solution of a copper(II) salt gives a precipitate of copper(I) iodide, CuI (I_2 liberated quantitatively).

copper naphthenates Copper(II) salts of carboxylic acids derived from crude petroleum oils and used as wood preservatives.

copper nitrate, $Cu(NO_3)_2$. Formed with $3H_2O$ (CuO, Cu, $CuCO_3$ in dil. HNO_3). Forms 9- and 6- hydrates and decomposes on heating to the oxide. Anhydrous $Cu(NO_3)_2$ is formed from Cu, liquid N_2O_4 and $C_2H_5O_2CCH_3$; it is appreciably volatile.

copper oxide chloride, $Cu_2Cl(OH)_3$. Fungicide.

copper oxides

Copper(I) oxide, Cu_2O. Red solid formed from Cu(I) salt and hydrazine or heat on CuO. Gives a cheap red glass and a cuprate, KCuO (K_2O plus Cu_2O).

Copper(II) oxide, CuO. Black solid formed by heating $Cu(OH)_2$, $Cu(NO_3)_2$, etc. Dissolves in acid to Cu(II) salts, decomposes to Cu_2O at 800°C. Polishing agent. Forms cuprates in solid-state reactions. A cuprate(III), $KCuO_2$, is also known.

copper perchlorates

Copper(II) perchlorate, $Cu(ClO_4)_2.6H_2O$. Formed from CuO and $HClO_4$.

Copper(I) perchlorate, $CuClO_4$. Formed in solutions of organic solvents (Et_2O, C_6H_6) by displacement of Ag from $AgClO_4$ by Cu.

copper sulphates

Copper(II) sulphate, $CuSO_4$. Blue crystals (blue-vitriol) of $CuSO_4.5H_2O$ from CuO and H_2SO_4; gives pale blue trihydrate, colourless monohydrate which loses most of its water at 250°C (used – blue colour – to detect water). Obtained industrially from copper ores and H_2SO_4 or Cu plus H_2SO_4 plus air. Used in agriculture, and water treatment as algicide, as a wood preservative.

Copper(I) sulphate, Cu_2SO_4. Grey solid, Cu_2O plus dimethyl sulphate.

copper sulphides

Copper(II) sulphide, CuS. Black solid, Cu plus excess S or copper(II) salt plus H_2S. Decomposes to copper(I) sulphide (Cu_2S) on heating.

co-precipitation A substance when precipitated from solution may be impure due to the presence of foreign substances adsorbed on the surface or occluded within the bulk of the precipitate particles. The contamination of the precipitate by substances which are normally soluble in the mother liquor is termed co-precipitation.

corn oil (maize oil) Food oil used in margarine manufacture. Contains glycerides of palmitic, oleic and linoleic acids.

correlation diagram A diagram of calculated energies of a molecule against assumed molecular geometry including the energies of the various molecular orbitals. Used in approaches to molecular geometry.

corresponding states (reduced variables) States for gases set up as ratios of variables to the critical constants. Used to give measures of comparison but often only approximate to real correspondence.

corrinoids Vitamin B$_{12}$ type compounds.

corrin (vitamin B$_{12}$) Composed of the nucleotide and the highly substituted, reduced, porphyrin-like corrin ring.

corroles Analogues of porphyrins prepared, e.g. by reacting pyrroles with aldehydes.

corrosion Most metals and alloys are attacked by oxygen, moisture and acids. Some are attacked by alkalis. The process of attack is called corrosion. Corrosion may result in a uniform attack which is generally not very serious; it may attack preferentially, particularly at grain boundaries, resulting in severe weakening of the metal without much visible deterioration; it may attack locally at areas where conditions are varied resulting in perforation; finally it may produce a passive oxide layer as on Cr and Al which gives protection from further corrosion. There are two main types of corrosion: atmospheric oxidation and tarnishing in dry air, which is not serious, and electrolytic corrosion under conditions of moisture or submersion.

Any two areas on a metal or alloy surface not chemically and physically identical can act as anode and cathode. Thus two phases in an alloy such as steel act as microelectrodes. The anode reaction is solution of metal or corrosion M – 2e → M^{++} for a dipositive metal. The cathode reaction is by hydrogen evolution in acid solution or oxygen reduction in neutral solution. Differential aeration is a common cause of corrosion. If one area of a metal has a higher oxygen potential than another, then it becomes cathodic

$$\tfrac{1}{2}O_2 + H_2O + 2e \rightarrow 2OH^-$$

making the area of low oxygen potential anodic with consequent corrosion. Thus corrosion is more severe under deposits of silt or in general where an area is loosely covered. Underground corrosion is caused by **1** electrolytic cells, the moist soil acting as electrolyte, **2** stray electric currents from earthing points, etc., **3** anaerobic bacteria which in the presence of iron and traces of CaSO$_4$ depolarize the microcathodes with the production of H$_2$S.

The main methods of combating corrosion are **1** treatment of water systems with inhibitors, **2** cathodic protection, **3** protection by paint, plating, galvanizing, phosphating, etc.

See also RUSTING, CATHODIC PROTECTION.

cortisol See HYDROCORTISONE.

cortisone, C$_{21}$H$_{28}$O$_5$. M.p. 215°C. A steroid

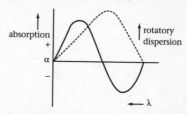

hormone produced by the adrenal glands. Anti-inflammatory drug, it can remove features of rheumatoid arthritis, but does not check the underlying disease: it is used in various diseases of the eye.

corundum, α-Al$_2$O$_3$. See ALUMINIUM OXIDE. The corundum structure has close-packed oxygen with Al in octahedral holes. Ruby, sapphire and emery are mineral varieties of corundum. Articles made from corundum are extremely inert and resistant to corrosion. Used for grinding.

COSY (correlation spectroscopy) NMR technique used to interpret complex spectra.

Cotton effect Within an absorption band, there is anomalous rotatory dispersion. In the

figure, the broken line represents an absorption band and the full line the rotatory dispersion curve. As the wavelength decreases the rotation angle increases, passing through a maximum and then decreases, passing through zero at the wavelength corresponding to the maximum of absorption. On proceeding to lower wavelengths, the angle falls until it passes through a minimum and then rises again. This phenomenon is known as the Cotton effect. This effect is general for coloured compounds as well as for colourless substances with bands in the ultraviolet. See OPTICAL ROTATORY DISPERSION, CIRCULAR DICHROISM.

coulomb The unit quantity of electricity; 96 494 coulombs, 1 Faraday, corresponds to 1 mole of electrons. Charge transported per second by a current of 1 ampere. See FARADAY'S LAWS OF ELECTROLYSIS.

coulometer An instrument or device for measuring the amount of charge (number of coulombs) passing in an electrical circuit. Thus a silver coulometer consists of 2 platinum electrodes dipping into a solution of silver nitrate. For each coulomb of charge passed 1.118×10^{-3} g of silver is deposited at the cathode. Thus, from the increase in weight of the cathode the charge passing can be calculated.

coulometry Analytical technique involving the measurement of quantities of electricity. Constant current coulometry can be used for coulometric titrations, e.g. Br_2 can be prepared electrically and used for oxidation (e.g. of olefines) *in situ* and the end point detected spectrophotometrically; Ag^+ can be generated and used to titrate Cl^-. In controlled (constant) potential coulometry one species may be oxidized or reduced and thus determined at a specific potential.

coumaphos Insecticide, organophosphate derivative.

o-coumaric acid (trans-2-hydroxy-cinnamic acid), $C_9H_8O_3$. Colourless crystals, m.p. 108°C. Obtained by boiling coumarin with sodium ethoxide. Irradiation of o-coumaric acid produces COUMARINIC ACID. The stable form of the acid is the *trans* form.

coumarin, $C_9H_6O_2$. Colourless crystals, m.p.

70°C, b.p. 290°C. Occurs in the Tonka bean. Prepared synthetically by heating salicylaldehyde with ethanoic anhydride and sodium ethanoate. It is the δ-lactone of unstable coumarinic acid. Hydrolysis with sodium hydroxide produces COUMARIC ACID. It is used in perfumery and varnishes. Coumarin derivatives form many rodenticides (e.g. Warfarin).

coumarin glycosides Widely distributed in plants. The aglucones consist of hydroxycoumarins containing 1, 2 or 3 hydroxyls or their methyl ethers.

coumarone See BENZFURAN.

coumarone and indene resins Prepared from purified naphtha of boiling range 150–200°C which is treated with H_2SO_4 to produce a resin. Used particularly in floor tiles and hot melt compositions.

coumatetryl Rodenticide, coumarin derivative.

counter-current flow When two streams of material moving in opposite direction are brought into contact for the purpose of transferring heat or matter, flow is said to be counter-current. Counter-current operation is used in preference to CO-CURRENT FLOW since it results in greater heat or mass transfer for a given set of operating conditions.

counter-ions The charge on the surface of a colloidal particle is compensated for by an equal and opposite ionic charge in the liquid in immediate contact with it. The system as a whole is electrically neutral. Those compensating ions, which occupy a diffuse layer around the particle, are termed counter-ions.

Also used to describe the oppositely charged ion balancing charge, e.g. chloride is the counter-ion to Na^+ in sodium chloride.

counters, radioactive Apparatus for the detection and quantitative estimation of radioactivity.

coupling The interaction between nuclei or electrons, particularly used in NUCLEAR MAGNETIC RESONANCE. Also reaction involving joining of two species often with elimination of a small molecule. See, e.g. HECK REACTION.

coupling constant, J See NUCLEAR MAGNETIC RESONANCE.

covalency maximum See CO-ORDINATION NUMBERS.

covalent bond The linkage of two atoms by the sharing of two electrons, one contributed by each of the atoms. The electrons are only shared equally when the atoms are identical, and in most covalent bonds the electrons are held to a greater extent by one atom than by the other (ELECTRONEGATIVITY). Covalent links are formed where electrovalent linkages would involve the interaction of small, highly charged cations with large diffuse anions. See VALENCY, THEORY OF.

covalent radius The equilibrium distance between two atoms joined by a covalent bond of a given type (i.e. single, double or triple) is found to be, within close limits, constant. Thus the C–C bond distance of 1.54 Å found in diamond is the same as that found in cyclic saturated hydrocarbons. In general the A–B covalent bond distance is close to the mean of the A–A and B–B bond distances. Thus, a set of covalent radii may be drawn up such that the sum of any two distances gives the distance between the corresponding atoms in a covalent molecule. The covalent radii are, in general, half the internuclear distances in the elements, except when the latter are very electronegative (e.g. fluorine).

Cp Cyclopentadienyl.

Cp* Methyl cyclopentadienyl.

cp Close packed.

4-CPA Plant growth regulator, phenoxyacetate derivative.

Cr Chromium.

cracked gasoline (cracked spirit) Gasoline obtained from heavier petroleum distillates by a THERMAL or, more usually, CATALYTIC cracking process. The unsaturated nature of gasoline leads to formation of gum unless inhibitors are added.

cracking The decomposition of a substance into fractions of lower molecular weight. Cracking processes, widely used in the petroleum industry, may be purely thermal or catalytic. See THERMAL CRACKING, CATALYTIC CRACKING.

cream of tartar See POTASSIUM HYDROGEN TARTRATE.

creatine (methylguanidinoethanoic acid), $C_4H_9N_3O_2,H_2NC(NH)NMeCH_2C(O)OH$. Crystallizes with $1H_2O$ lost at 100°C. Decomposes at 291°C. Creatine is present in the muscles of all vertebrates. It plays an important part in the cycle of chemical changes involving muscular contraction; phosphagen, a creatine phosphoric acid ester, breaks down to creatine and phosphoric acid, which recombine during the recovery process. Creatine is formed in the body by the methylation (by methionine) of glycocyamine. Has been used to boost muscle strength and to treat muscular diseases.

creatine phosphate (phosphocreatine, phosphagen) See CREATINE.

creatinine (1-methylglycocyanidine), $C_4H_7N_3O$. Crystallizes with $2H_2O$; m.p. 260°C (decomp.). The internal anhydride of creatine and is formed from creatine in alkaline solution. Found in human urine as a breakdown product of the body tissues.

creep The slow yielding of metals and alloys (generally at high temperatures) under a load which does not cause fracture. The process is diffusion controlled.

creosote Fractions obtained in the distillation of coal or wood tar. *Light, medium* and *heavy* fractions are classified by their boiling ranges (approx. 475–540; 540–570; 570–630 K, respectively). Creosote fractions are also referred to as naphthalene oil, wash oil, light anthracene oil, etc. Creosotes have a characteristic odour and consist of a mixture of hydrocarbons, phenols and other aromatic derivatives. They are used for timber preservation, for fluxing pitch and bitumen, and as fuels.

Medicinal creosote is a mixture of phenols, chiefly guaiacol and creosol (4-methyl-2-methoxyphenol), obtained by distillation of wood tar. B.p. 200°C. It is almost colourless with a characteristic odour and is a strong antiseptic; less toxic than phenol.

cresols (hydroxytoluenes), C_7H_8O. The cresols are colourless liquids or crystalline solids. They are volatile in steam, and are reduced to toluene by Zn dust. Present in coal tar and cracked naphtha.

2-hydroxytoluene, *o*-cresol,
          m.p. 31°C, b.p. 191°C
3-hydroxytoluene, *m*-cresol,
          m.p. 12°C, b.p. 203°C
4-hydroxytoluene, *p*-cresol,
          m.p. 35°C, b.p. 202°C

Usually prepared from the corresponding sulphonic acids by alkali fusion, methylation of phenol or from the aminotoluene by treatment with nitrous acid followed by boiling. Both 2- and 4-cresol are used as end components in azo dyes.

For uses see CRESYLIC ACIDS. World demand 40 000 tonnes p.a.

cresotic acid 2-hydroxy-3-methylbenzoic acid. 3, 4, and 5 methyl derivatives are known. Used as disinfectants and in dyestuff manufacture.

cresylic acids Methyl derivatives of phenol, e.g. C_6H_4(Me)OH are cresols; $C_6H_3Me_2$OH are xylenols. Also called tar acids as obtained from coal distillation or sweetening petroleum distillants. Used as starting materials for pesticides and herbicides and for phenolic resins, disinfectants, antioxidants and plasticizers (cresyl phosphate).

CRG process (catalytic rich gas process) A catalytic low-pressure process producing a natural gas substitute from naphtha feedstock and steam over a nickel catalyst at about 775 K. The product is a mixture of methane, carbon dioxide and some hydrogen. This may require METHANATION, carbon dioxide removal and propane blending before satisfactory properties are obtained.

cristobalite, SiO_2. The high-temperature form of SILICA, stable from 1470°C to m.p. 1710°C.

critical humidity The humidity at which the vapour pressure at the surface of a solid or solution is equal to the partial pressure of water in the atmosphere. At humidities above the critical value water will tend to be absorbed; below the critical value moisture will be lost to the atmosphere.

critical micelle concentration (CMC) Concentration of material above which micelles formed – also needs to be above Krafft temperature.

critical phenomena Above the critical temperature T_c it is impossible to liquefy a gas. The minimum pressure required to cause liquefaction at the critical temperature is the critical pressure P_c. The critical volume V_c is the volume occupied by one mole of the gas at its critical temperature and pressure.

critical solution temperature The temperature at which the mutual solubility of two phases becomes equal. For two liquids, e.g. phenol and water, the solubility of each in the other increases until at the critical solution temperature only one phase exists.

critical temperature (T_c) Temperature above or below which a specific phenomenon

occurs. Generally applied to temperature below which superconductivity occurs.

CRLS (I.R.–CRCS) Infra-red technique using Fourier Transform methods.

cross section A number denoting the effective area of a nucleus for a scattering or absorption process. See BARN. Also a measure of the tendency for two molecules to react.

crotonaldehyde See 2-BUTENAL.

crotonic acids See 2-BUTENOIC ACIDS.

crotonyl The trivial name given to the 2-butenoyl group, –C(O)CH:CHCH$_3$.

crotyl The trivial name given to the 2-butenyl group, –CH$_2$CH:CHCH$_3$.

crotyl alcohol (*cis-* and *trans-*2-buten-1-ol), C_4H_8O, CH$_3$CH:CHCH$_2$OH. Colourless liquid, b.p. 118°C. Prepared by the reduction of CROTONALDEHYDE.

crown compounds E.g. crown ethers. Macrocyclic poly derivatives containing, e.g. repeating (O–CH$_2$·CH$_2$)$_n$ units. The compound depicted would be called dibenzo-18–crown-6 (18– the total number of atoms in the ring, crown is the class name and 6 is the total number of hetero atoms in the ring), or 2,3-11,12-diphenylene-1,4,7,10,13,16-hexaoxacyclooctadecane. Simplest examples are prepared by the cyclic oligomerization of ethylene oxide. They act as complexing agents which, e.g. solubilize alkali metal ions in non-polar solvents, complex alkaline earth cations, transition metal cations and ammonium cations, e.g. 12–crown-4 is specific for the lithium cation. Used in PHASE-TRANSFER CHEMISTRY.

Thiacrown agents and derivatives containing other hetero atoms or groups also known. Sulphur compounds can extract mercury(II) from industrial waste water. See CRYPTATES.

crude oil A naturally occurring mixture consisting largely of hydrocarbons along with sulphur, nitrogen and oxygen derivatives. Crude oil is normally removed from reservoir rock in liquid form but is often associated with quantities of gas, water or solids. The properties of crude oils vary widely from light-coloured, mobile fluids to dark, viscous materials. Quantities generally measured in barrels.

Crum Brown's rule An empirical guide to substitution in benzene derivatives.

crushing and grinding Operations relating to particle size reduction. The term *crushing* is used where compressive forces are mainly employed, while *grinding* refers more to reduction by attrition and shearing actions.

cryohydric point An alternative name for the EUTECTIC POINT, usually used when one of the components is water.

cryolite, Na_3AlF_6. Mineral Na_3AlF_6, occurs native in Greenland. Made artificially from sodium aluminate, NaF, and $NaHCO_3$. Used as the electrolyte in Al manufacture and for white glass, glazes and enamels. The cryolite structure is adopted by many compounds of stoichiometry A_3BX_6. It is closely related to the ReO_3 and perovskite structures. Used as insecticide.

cryoscopic constant An alternative name for the freezing-point depression constant.

cryoscopy The determination of molecular weights by the freezing-point depression method.

cryptands Polycyclic compounds with donor atoms, generally O, N, S, situated in the bridges so that they bind metals strongly. Particular cryptands can be designed to bind most ions. Compare CROWN COMPOUNDS.

cryptates Polyheteroatom bicyclic

1:1:1 cryptate   2:2:2 cryptate

macrocycles, similar to CROWN ETHERS but possessing additional $(-XCH_2CH_2)_n$ bridges. The chemical applications of these substances are closely related to crown ethers, the [2:2:2] compound specifically complexing K^+.

crystal A discrete solid particle bounded by definite faces intersecting at definite angles, and showing certain symmetry characteristics. The external form of a crystal is due to the or-

dered arrangement of atoms or molecules or ions within the bulk of the solid.

crystal drawing Crystals are drawn in a conventional projection and a conventional position. For the cubic system the x axis does not point to the eye of the observer, but is rotated clockwise through an angle of 18° 26′, the tangent of which is ⅓; the y axis is not drawn at right angles to the vertical c axis because the eye is elevated at an angle of 9° 28′, the tangent of which is ⅙.

crystal field theory An approach to transition metal compounds which explains the u.v. and visible spectra (electronic) and magnetic properties in terms of a central ion surrounded by electron-rich ligands which split the d orbitals into a pattern determined by the geometry of the co-ordination, the distance of approach of the ligands and the tendency to covalent bonding.

crystal habit The description of the relative development of different faces and types of face on crystals which are of the same class.

crystalline state The form of the solid state with an ordered arrangement of atoms, etc. in the solid. The pattern is characteristic of the solid and generally gives rise to characteristic bounding faces of the crystal.

crystallins Water soluble proteins present in eye lenses.

crystallization The removal of a solid from solution by increasing its concentration above the saturation point in such a manner that the excess solid separates out in the form of crystals. Successive crystallization or fractional crystallization (the successive removal of the least soluble component) may be used in separation and purification.

crystallizers Equipment for producing crystals from solutions of crystallizable materials.

crystallographic axes The (arbitrary) axes defining the repeat unit (unit cell) in a crystal. See CRYSTAL STRUCTURE.

crystal nucleus A minute crystal which serves as a centre of formation for larger crystals which would not otherwise be formed. E.g. crystallization of a supersaturated solution may be induced by formation of a crystal nucleus through mechanical shock, inoculation by dust, etc.

crystals, liquid See LIQUID CRYSTALS.

crystal structure The arrangement of atoms and molecules in a crystal, as determined by X-ray, neutron and electron diffraction methods.

A characteristic of a crystal is that the atoms, molecules or ions are arranged in a regular way in three dimensions (a lattice). It follows that if we start from an arbitrary origin in the crystal and proceed in any direction we shall arrive, after travelling a certain distance, at a point with the same environment as the point of origin. We may perform this process in two other directions and find the corresponding 'repeat distances'. These three axes of reference, x, y and z, and the three repeat distances, a, b and c, define a parallelopiped, defined by the symmetry elements but generally including the smallest volume, which contains a representative portion of the crystal structure, and which if indefinitely repeated in those directions will reproduce the crystal. A characteristic parallelopiped is termed the '*unit cell*', and the repeat distances or *cell dimensions* (the sides of the unit cell) are written a, b and c, corresponding to the x, y and z axes. The angles between the axes taken in pairs, are called α, β and γ (between yz, zx and xy, respectively). The choice of axes in the crystal is not completely arbitrary, except in the triclinic system, but is determined by the positions of the symmetry elements. The shape of the unit cell, and hence the constants required to define it, depends on the symmetry of the crystal, and the following table summarizes the data required in the various systems:

Triclinic a, b, c, α, β, γ (no planes or axes of symmetry).

Monoclinic a, b, c, β, ($\alpha = \gamma = 90°$) – one 2-fold axis and/or one plane of symmetry.

Orthorhombic a, b, c, ($\alpha = \beta = \gamma = 90°$) – one 2-fold axis at the intersection of two planes or perpendicular to two other 2-fold axes.

Trigonal a ($= b$) c, ($\gamma = 120°$, $\alpha = \beta = 90°$) – one 3-fold axis perpendicular to other axes of symmetry.

Hexagonal a ($= b$) c, ($\gamma = 120°$, $\alpha = \beta = 90°$) – one 6-fold axis perpendicular to other axes of symmetry.

Tetragonal a ($= b$), c, ($\alpha = \beta = \gamma = 90°$) – one 4-fold axis perpendicular to other axes of symmetry.

Cubic a ($= b$, $= c$), ($\alpha = \beta = \gamma = 90°$) – four 3-fold axes at tetrahedral angles to one another.

The crystal structure determines not only the arrangement of atoms in the lattice but also the external form of the crystal.

crystal symmetry A term describing the regularities in the position and arrangement of faces and edges of a crystal and also of the atoms within the crystal. Such regularities are defined in terms of planes of symmetry, axes of symmetry and centres of symmetry. A plane of symmetry divides the crystal into two identical halves, each of which is the mirror image of the other. An axis of symmetry is an imaginary line drawn through the crystal in such a way that, on rotating the crystal about this line as axis, the crystal presents an identical aspect to the observer at least once in a complete rotation. If the same aspect is presented four times (as would be the case for a cube rotated about a line through the centres of opposite faces) the axis is called an axis of four-fold symmetry. Similarly, there are axes of two-, three- and six-fold symmetry. A crystal has a centre of symmetry if like faces are arranged in matched pairs in corresponding positions on opposite sides of a point. There are a limited number of possible arrangements of planes, centres and axes of symmetry, which lead to a classification of crystals according to CRYSTAL SYSTEMS. In order to describe the external form of crystals three crystallographic axes are chosen. One face, making intercepts on all the axes, is taken as the standard or parametral face to which all other faces are referred. The description of crystal symmetry given above refers to the macroscopic crystal. The symmetry elements are also present in the unit cell, where they act on the atoms in the cell, but there are also additional symmetry elements to be considered.

crystal systems Crystals are placed in one of the seven crystal systems: cubic, tetragonal, hexagonal, trigonal, orthorhombic, monoclinic and triclinic; according to their symmetry. For the characteristics of each of these systems see CRYSTAL STRUCTURE and the separate headings (e.g. CUBIC SYSTEM).

crystal violet See METHYL VIOLET.

Cs Caesium.

CS gas See MALONONITRILE.

CTAB See CETRIMIDE.

Cu Copper.

cubanes Derivatives of $(CH)_8$ hydrocarbon.

cubic close-packing (ccp) See CLOSE-PACKED STRUCTURE.

cubic co-ordination Co-ordination by eight ligands at the corners of a cube as in CsCl. A cube has 6 square faces. Compare square-anti-prism and dodecahedron (see over).

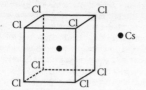

cubic system The CRYSTAL SYSTEM with three equal crystallographic axes at right angles, e.g. NaCl.

cumene (isopropylbenzene), C_9H_{12}, $PhCH(CH_3)_2$. B.p. 152°C. Obtained commercially by the reaction of benzene with propene in the presence of a catalyst such as hydrogen fluoride. Oxidation via a peroxide is a commercially valuable route to propanone and phenol. U.S. production 1999 3.14 megatonnes.

cumyl-α-hydroperoxide, $C_9H_{12}O_2$, $PhC(CH_3)_2OOH$. A commercially important intermediate obtained by passing air through liquid cumene in the presence of a trace of a basic catalyst. On treatment with acids it gives phenol and propanone.

cupferron (ammonium N-nitroso-phenylhydroxylamine), $[NH_4][ON(NO) C_6H_5]$. A reagent used for the separation of Fe/Ti and Zr which it precipitates from acid solutions. Possibly carcinogenic.

cuprammonium, $[Cu(NH_3)_4]^{2+}$. The deep blue copper(II) ammine (excess aqueous ammonia to a copper(II) salt). The solution dissolves cellulose and is used in one process of rayon manufacture.

cuprates See COPPER OXIDES.

cuprene See ACETYLENE BLACK.

cupric Copper(II) species.

cuprite (red copper ore), Cu_2O. Bright red Cu ore. The cuprite structure has linear two-fold co-ordination by oxygen.

cupron See BENZOIN.

cuprous Copper(I) species.

curie A non-S.I. unit of radioactivity. 1 Ci = 3.7 × 10^{10} disintegrations per second; the same number of disintegrations as produced by 1 g of radium. See BECQUEREL.

Curie law The molar magnetic susceptibility of a substance may be represented as:

$$\chi = C/T.$$

where C is the Curie constant, and χ contains both diamagnetic and paramagnetic contributions. Not all substances conform to this equation; a more general relationship, the Curie–Weiss law, states:

$$\chi = \frac{C}{T - \theta}.$$

θ is the Weiss constant and is characteristic of the particular substance being considered.

Curie temperature The temperature below which a paramagnetic solid undergoes a phase transition to a state where large domains of spin align co-operatively so that the substance becomes ferromagnetic. Also called Curie point.

Curie–Weiss law See CURIE LAW.

curium, Cm. Artificial actinide metal, at.no. 96, at.mass ^{244}Cm 244.06, m.p. 1345°C, b.p. 3100°C, density ρ 13 510, electronic configuration [Rn] $5f^7 6d^1 7s^2$. ^{242}Cm (162 days) and ^{244}Cm (18 years) are formed by multiple neutron irradiation of ^{239}Pu; ^{248}Cm (4.7×10^5 years) is formed by decay of ^{252}Cf, the compounds are purified by ion-exchange chromatography. The metal is formed by reduction of CmF_3 with Ba, it has the double-hexagonal cp structure. The metal is silvery in colour and is rapidly oxidized. ^{248}Cm is a potential source of higher actinides. ^{242}Cm is a power and heat source.

curium compounds These are mainly in the +3 state and include all of the trihalides.

$$Cm^{4+} \xrightarrow{\ \geq 2.8\ volt\ } Cm^{3+}$$

The +4 state is stabilized in aqueous solution by fluoride ions. CmO_2 and CmF_4 are formed by strong oxidation or the action of fluorine.

Curtius transformation An alternative to the HOFMANN TRANSFORMATION for obtaining an amine from an ester via the hydrazide, azide and isocyanate. Thus ethyl ethanoate is converted into methylamine:

$$CH_3COOC_2H_5 + NH_2NH_2 \xrightarrow{C_2H_5OH} CH_3CONH\cdot NH_2$$

$$CH_3CONH\cdot NH_2 + HNO_2 \xrightarrow{H_2O/0°C} CH_3CO\cdot N_3$$

$$CH_3CO\cdot N_3 \xrightarrow{warm} CH_3\cdot NCO + N_2$$

$$CH_3\cdot NCO + H_2O \xrightarrow{H^+} CH_3NH_2 + CO_2$$

cutback bitumen

Alternatively, the azide may be boiled with alcohol to convert it into the urethane, and the latter then hydrolysed with dilute acid. In general, the Hofmann method is rather more convenient, but the Curtius method must be employed when the compound in question is attacked by alkalis or easily brominated.

cutback bitumen A form of bituminous road binder produced by 'cutting back' the viscosity of medium-hard bitumen with volatile solvents.

cutting fluids Fluids used as coolants and lubricants in metal-cutting operations. They may be low-viscosity oils containing emulsifiers which form stable oil-in-water emulsions on dilution. Alternatively aqueous mixtures based on sodium nitrite, triethanolamine sebacate, polyglycols, etc. may be used. For heavier work straight lubricating oils are often used.

CVD Chemical vapour deposition.

cyamelide See CYANIC ACID.

cyanamide (carbodiimide), H_2NCN. Colourless, deliquescent solid; m.p. 41°C. Prepared HgO plus thiourea, CO_2 over heated $NaNH_2$, acid plus CaNCN. Normally obtained as the dimer, $H_2NC(=NH)NHCN$ or $H_2NC(=NH)N=C=NH$ (see CARBODIIMIDES). Acts as a weak acid and forms salts (see e.g. CALCIUM CYANAMIDE, a fertilizer) and as a base with strong acids. Gives urea with acids, thiourea with H_2S, guanidine with NH_3. Polymerizes to melamine, $(H_2NCN)_3$. Used as herbicide and plant growth regulator.

cyanates Salts of cyanic acid, HNCO. KNCO formed from KCN plus PbO. NH_4NCO (KNCO plus NH_4Cl) isomerizes on heating to urea. Some cyanate complexes and covalent compounds, e.g. $P(NCO)_3$ (PCl_3 plus AgNCO), are mainly N-bonded.

cyanazine Herbicide, triazine derivative.

cyanic acid (hydrogen cyanate), HNCO. Mobile volatile liquid formed by distillation of CYANURIC ACID. A weak acid, polymerizes to a solid polymer, cyamelide, above 0°C. Hydrolysed to NH_3, H_2O, CO_2 in aqueous solution. Forms CYANATES and organic isocyanates, RNCO, and cyanates, NCOR.

cyanides The salts of hydrogen cyanide, HCN. Covalent derivatives RCN are generally termed cyanides or nitriles; derivatives RNC are isocyanides, isonitriles. Cyanides are formed from HCN, by metathesis (e.g. $AgNO_3$ plus KCN

precipitates AgCN; AgCN plus ECl gives ECN), or by specific methods (see POTASSIUM CYANIDE, SODIUM CYANIDE). Cyanides are slowly hydrolysed (to CO_3^{2-} and NH_3), have solubilities similar to those of chlorides (pseudohalides) and are extremely poisonous. Many complex cyanides (cyanometallates) are known, see, e.g. CYANOFERRATES. Cyanide complexes can often be reduced to complexes in which the metal is in low oxidation states.

cyanine dyes Polymethine dyes containing $N-C(=C-C)_n=N$ and related systems. Not very fast to light, widely used as spectral sensitizers in colour photography.

cyanoacetic acid See CYANOETHANOIC ACID.

cyanoacetic ester See ETHYL CYANOETHANOATE.

2-cyanoacrylates, $CH_2=C(CN)C(O)OR$. Compounds formed by condensation of alkyl cyanoethanoates with methanal to give a polymer which can be depolymerized to the monomer. The methyl ester, b.p. 48–49°C at 2.5 mm is commercially available. Readily polymerized (through the vinyl group) by free radical or anionic polymerization and used as an adhesive by polymerization in situ.

2-cyanobenzamide, $C_8H_6N_2O$. Colourless needles, m.p. 173°C. On heating to its m.p. is converted into imidophthalimide, m.p. 203°C. Readily hydrolysed to phthalic acid. Prepared by the partial dehydration of phthalamide with ethanoic anhydride.

Converted to phthalocyanine pigments on heating with metallic salts.

cyanocobalamine See VITAMIN B_{12}.

cyanoethanoic acid (cyanoacetic acid, malonic acid mononitrile), $NC·CH_2C(O)OH$. Crystallizes in large prisms which liquefy on exposure to atmospheric moisture, m.p. 66°C. Prepared by sodium cyanide on sodium chloroethanoate followed by acidification with sulphuric acid. Decomposes at 160°C into carbon dioxide and acetonitrile. Concentrated acids or alkalis convert it to malonic acid. Condenses with aldehydes to give unsaturated α-cyano acids or unsaturated nitriles. The ethyl ester is used in organic syntheses.

cyanoethylation A particular example of a MICHAEL REACTION. Some nucleophiles, especially carbanion species, readily add across the double bond of acrylonitrile, forming a cyanoethyl derivative. E.g. propanone, in the presence of a base (e.g. benzyltrimethylammonium hydroxide or an alkali hydroxide), can undergo cyanoethylation to give mono-, di- or tricyanoethylation products.

$$CH_3COCH_3 \longrightarrow$$
$$CH_3COCH_2CH_2CH_2CN \longrightarrow$$
$$CH_3COCH(CH_2CH_2CN)_2 \longrightarrow$$
$$CH_3COC(CH_2CH_2CN)_3$$

In addition to compounds with active methylene groups, primary and secondary amines, alcohols and phenols can undergo cyanoethylation. See PROPENENITRILE.

cyanoferrates
Hexacyanoferrates(III), $[Fe(CN)_6]^{3-}$, *hexacyanoferrates(II)*, $[Fe(CN)_6]^{4-}$. Formed by the addition of alkali cyanide to a solution of an Fe(III) or Fe(II) salt. Hexacyanoferrates(III) are reddish in colour, $K_3Fe(CN)_6$ is generally obtained by oxidizing the $[Fe(CN)_6]^{4-}$ ion with Cl_2. There are many related iron cyanoferrates containing $[Fe(CN)_6]^{n-}$ and substituted species. When iron is present in two oxidation states, deeply coloured species (Prussian blue, Turnbull's blue, Berlin green) used in inks and dyestuffs are formed. In alkaline solution $[Fe(CN)_6]^{3-}$ has oxidizing properties (e.g. can be used to titrate Ce^{3+} in the presence of K_2CO_3).

Hexacyanoferrates(II) are yellow. Substituted cyanoferrates containing anions such as $[Fe(CN)_5NO]^{2-}$ (nitroprusside), $[Fe(CN)_5NO_2]^{2-}$ are known. Used in pigments and to prevent caking of salt.

cyanogen, C_2N_2, NCCN. Colourless, inflammable gas, m.p. $-28°C$, b.p. $-21°C$ with characteristic smell of bitter almonds. Very poisonous. Formed from HCN by oxidation with Cl_2 over a catalyst or NO_2, from Cu^{2+} plus CN^-, or $Hg(CN)_2$ plus $HgCl_2$. Paracyanogen, $(CN)_x$, is formed in some of these reactions. Behaves as a pseudohalogen and adds oxidatively to e.g. Pd(0) and gives CN^- and NCO^- with base.

cyanogen halides, XCN.
Cyanogen fluoride, FCN. Colourless gas (b.p. $-46°C$) prepared by pyrolysis of cyanuric fluoride. Polymerizes to $(FCN)_3$, cyanuric fluoride, at room temperature.
Cyanogen chloride, ClCN. Colourless liquid, m.p. $-7°C$, b.p. $13°C$ (aqueous CN^- plus Cl_2). Linear molecule, polymerizes to cyanuric chloride $(ClCN)_3$. Extremely poisonous.
Other halides are known.

cyanoguanidine (dicyandiamide), $C_2H_4N_4$, $H_2NC(NH)NH \cdot CN$. Colourless crystalline solid, m.p. $209°C$, soluble in water and alcohols. Prepared commercially by the dimerization of cyanamide in the presence of bases, and used in the manufacture of plastics, but mainly for the preparation of melamine, benzoguanine and diallylmelamine – all used in the formation of amino resins.

cyanohydrins Organic compounds containing the –C(OH)CN group. Formed by the action of hydrogen cyanide on aldehydes or ketones, or by the addition of sodium cyanide to a cooled suspension of the hydrogen sulphite compound of an aldehyde or ketone. Alkalis give a cyanide and the original compound; mineral acids convert them to α-hydroxycarboxylic acids.

cyanophoric glycosides Glycosides containing hydrocyanic acid which is split off on hydrolysis. Of frequent occurrence in plants. Amygdalin and glycosides of mandelonitrile are examples.

cyanophos Insecticide, organophosphate derivative.

cyanuric acid (2,4,6-triazinetriol, tricyanic acid, trihydroxycyanidine), $C_3H_3N_3O_3$. Stable solid sparingly soluble in

water. Prepared by melting urea or by hydrolysis of cyanuric chloride. *N*-esters are known. Source of cyanates, used in production of melamine, herbicides, dyestuffs.

cyanuric chloride (trichloro-s-triazene), $C_3N_3Cl_3$. M.p. $154°C$, b.p. $190°C$. Cyclic derivative prepared by polymerization of ClCN or from Cl_2, HCN and HCl. Used in the manufacture of dyestuffs, pharmaceuticals, triazine herbicides, plastics, explosives, bleaches and disinfectants. Gives cyanuric fluoride, $C_3N_3F_3$, with NaF.

cyclamate sodium, $C_6H_{11}NHSO_3Na$. Sodium salt of *N*-cyclohexylsulphamic acid, itself obtained by sulphonation of cyclohexylamine. Used as a sweetener in E.U.

cyclanilide Plant growth regulator, cyclopropylcarboxylic acid derivative.

cyclenes Cyclic hydrocarbons containing one or more double bonds.

cycle oil (cycle stock) In petroleum catalytic cracking processes the middle distillate fraction produced is recycled to the process. It may also be marketed as 'cycle oil' as a blending component of diesel oil or heating oil.

cyclic Compounds which contain a closed ring system, such as the cyclohexane ring, as opposed to the open-chain aliphatic compounds. Cyclic compounds can be subdivided into homocyclic compounds, in which the ring is composed only of carbon atoms, and heterocyclic compounds, in which the ring is composed of atoms of more than one kind.

cyclic AMP (adenosine 3´,5´-monophosphate) A phosphate nucleotide of importance in the mediation of hormone action. See DICTIONARY OF BIOLOGY.

cyclic hemiacetals See HEMIACETALS.

cyclic process A system which has undergone a series of changes and has returned to its original state and thus has completed a cycle. The whole series of changes is a cyclic process. See e.g. BORN–HABER CYCLE.

cyclic voltammetry Analytical and investigative method by which a species in solution is successively oxidized and reduced by application of a varying potential. The current is plotted against voltage.

cyclitols See INOSITOL.

cyclized rubbers Modified rubbers containing cyclic systems produced by heating rubber with catalysts ($SnCl_4$, H_2SnCl_6). They are hard, thermoplastic and non-elastic, used in corrosion-resistant paints and adhesives, particularly for rubber–metal joins.

cycloaddition Reaction of two connected groups to give a new ring, e.g. Diels–Alder addition.

cyclobarbitone, $C_{12}H_{12}N_2O_3$. Short-acting BARBITURATE used as a hypnotic.

cyclobutadiene, C_4H_4. An alkenic hydrocarbon of high reactivity and very short lifetime (<5 seconds) in the free state. It can be generated by degradation of a cyclobutadiene-metal complex, e.g. $C_4H_4Fe(CO)_3$ with Ce^{4+}, and its chemistry studied *in situ*, e.g. the addition of alkynes under Diels–Alder conditions to give 'DEWAR'

BENZENE derivatives. C_4H_4 behaves like a conjugated diene and an alkenic diradical.

cyclobutane (tetramethylene), $CH_2.CH_2.CH_2.CH_2$, C_4H_8. Colourless gas which burns with a luminous flame. Insoluble in water, soluble in organic solvents, b.p. –15°C. Prepared by treating 1,4-dibromobutane with metallic sodium. Reduced to n-butane by hydrogen at 200°C in presence of nickel catalysts.

cyclodextrans Cyclic polysugars joined through bridging oxygen atoms. Used industrially as dextrans and in polymerization of acrylates in aqueous solution, delivery vehicles for laundry softeners, odour removers.

cycloheptane See SUBERANE.

1,3,5-cycloheptatriene (tropilidene), C_7H_8. Prepared by heating norbornadiene to 450°C (also from atropine and cocaine). It is a liquid, b.p. 116–118°C, but yields solid salts of the tropylium cation by hydride abstraction, and tropolone by alkaline permanganate oxidation. Forms many metal complexes.

cyclohexane, C_6H_{12}.

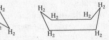

Colourless liquid, m.p. 6.5°C, b.p. 81°C. Manufactured by the reduction of benzene with hydrogen in the presence of a nickel catalyst and recovered from natural gases. It is inflammable. Boat and chair forms exist (see CONFORMATION). Used as an intermediate in the preparation of Nylon [6] and [66] via caprolactam, as a solvent for oils, fats and waxes, and as a paint remover.

cyclohexanol (hexalin, hexahydrophenol), $C_6H_{11}(OH)$, $C_6H_{12}O$. Colourless liquid, m.p. 24°C, b.p. 161°C. Manufactured by heating phenol with hydrogen under pressure in the presence of suitable catalysts. Oxidized to adipic acid (main use as intermediate for Nylon production); dehydrogenated to cyclohexanone.

Dehydrated (e.g. $AlCl_3$) to cyclohexene. Used in the manufacture of celluloid, esters (plasticizers), detergents and printing inks.

cyclohexanone ('pimelic ketone', ketohexamethylene), $C_6H_{10}O$.

Colourless liquid with a strong peppermint-like odour; b.p. 155°C. Manufactured by passing cyclohexanol vapour over a heated copper catalyst. Oxidized to adipic acid. Used in the manufacture of caprolactam, Nylon, adipic acid, nitrocellulose lacquers, celluloid, artificial leather and printing inks. Solvent for cellulose derivatives, resins, rubber.

cyclohexene, C_6H_{10}. B.p. 83°C. Widely used cyclic ethene.

cyclohexyl, $C_6H_{11}^-$. Cyclic grouping which forms many derivatives.

cyclohexylamine, $C_6H_{11}NH_2$, $C_6H_{13}N$. Colourless liquid, b.p. 134°C, miscible with water, volatile in steam. It can be extracted from aqueous solution by hydrocarbons. It is a strong base and is used as a solvent for dyestuffs, as an acid inhibitor in degreasing baths and for other industrial purposes.

cyclonite (cyclo-trimethylentrinitramine) High explosive. Formed from ethanoic acid, ethanoic anhydride and nitric acid.

1,5-cyclo-octadiene, C_8H_{12}. The *cis,cis*-isomer is obtained by a catalytic dimerization of butadiene using e.g. derivatives of nickel carbonyl. It is a colourless, mobile liquid with terpene-like odour, b.p. 151°C, m.p. –69°C. The *trans,trans*-isomer is also known.

cyclo-octatetraene (COT), C_8H_8. A golden-yellow liquid, m.p. –7°C. b.p. 142–143°C.

Prepared by the polymerization of ethyne at moderate temperature and pressure in the presence of nickel salts. The molecule is non-planar and behaves as a typical cyclic olefin, having no aromatic properties. It may be catalytically hydrogenated to cyclo-octene, but with Zn and dil. sulphuric acid gives 1,3,6-cyclooctatriene. Forms metal complexes, e.g. $(C_8H_8)_2U$, uranocene.

cyclopentadiene, C_5H_6. A colourless liquid

with a sweet, distinctive smell, b.p. 41.5–42.0°C. Insoluble in water, but soluble in all organic solvents. Obtained during the cracking of petroleum hydrocarbons, and from the distillates produced in the carbonization of coal. It polymerizes very readily on standing at room temperature to DICYCLOPENTADIENE (m.p. 32.5°C) and higher polymers, via an intermolecular Diels–Alder reaction. This is a more convenient form for handling cyclopentadiene which is easily regenerated by 'cracking' the oligomers. One hydrogen atom of the methylene group is acidic; sodium dissolves in ethereal cyclopentadiene solutions with the evolution of hydrogen, yielding $C_5H_5^-Na^+$ which is colourless to pink in the absence of air. The cyclopentadienyl anion $C_5H_5^-$ is an aromatic system. FULVENES are produced in basic media from ketones or aldehydes. CYCLOPENTADIENYLIDES are produced with many metals. Used in synthesis. It undergoes typical Diels–Alder reactions; used in the preparation of plastics and insecticides. It is the parent hydrocarbon of the metallocenes and other metal cyclopentadienyls.

cyclopentadienylides Metal derivatives of cyclopentadiene containing C_5H_5 groups. Formed by use of Grignard reagents, NaC_5H_5, LiC_5H_5, TlC_5H_5 on metal halides, carbonyls, etc. The C_5H_5 group may be pentahapto (h^5-C_5H_5) as in ferrocene (h^5-C_5H_5)$_2$Fe or monohapto, e.g. (h^5-C_5H_5)Fe(CO)$_2$.(h^1-C_5H_5). In pentahapto derivatives the C_5H_5 group is bonded symmetrically to the metal; monohapto derivatives show fluxional behaviour and each C–H may be equivalent on an n.m.r. time scale. Indene, fluorene and other cyclopentadiene derivatives form related complexes. The co-ordinated C_5H_5 group has considerable aromatic character and undergoes acylation under Friedel–Crafts conditions, etc. The C_5H_5 group can be considered to donate 5 electrons to the metal. Used as catalysts and petroleum additives.

cyclopentane, C_5H_{10}. Occurs in petroleum. Mobile flammable liquid.

cyclophanes Benzene derivatives bridged between 4- and/or 3-positions with methylene groups; hence paracyclophane, etc. It is not usual to include the ortho-derivatives since alternative names based on conventional ring systems are possible.

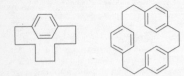

[8] paracyclophane   [2.2.2.]-paracyclophane

The value of the prefix [m] indicates the number of methylene groups in the ring, whilst the number of prefixes [m, n, etc.] refers to the benzene rings in the structure, i.e. [8]-paracyclophane, [2,2,2]-paracyclophane. Systems based upon heterocyclic molecules are also known.

cyclophosphamide, $C_7H_{15}, Cl_2N_2O_2P, H_2O$.

$$\left[\begin{array}{c} H_2 \\ C-NH \\ H_2C \qquad P \overset{O}{\underset{N(CH_2.CH_2Cl)_2}{}} \\ C-O \\ H_2 \end{array} \right].$$

Fine, white crystalline powder, m.p. 49.5–53°C. Carcinogen but has been used in the treatment of cancer.

cyclopropane (trimethylene), $CH_2:CH_2.CH_2$, C_3H_6. Colourless gas with a sweetish odour, b.p. –34.5°C. Prepared by treating 1,3-dibromopropane with zinc. It is a powerful gaseous anaesthetic, non-irritant and non-toxic to the liver and kidneys, but is a respiratory depressant.

cycloprothrin Insecticide, pyrethroid derivative.

cyclosulfamuron Herbicide, urea derivative.

cyclotrimethylene trinitramine See CYCLONITE.

cycloxydim Herbicide, oxime derivative.

cyfluthin Insecticide, pyrethroid derivative.

cyhalofop-butyl Herbicide, propionic acid derivative.

cyhalothrin Insecticide, pyrethroid derivative.

cyhexatin Acaricide, organotin derivative.

cyloate Herbicide, thiocarbamate derivative.

cymenes (isopropylmethylbenzenes), $C_{10}H_{14}$. 4-Cymene is a colourless liquid, b.p. 177°C. Insoluble in water; miscible with organic solvents. Occurs in many essential oils, such as cumin, thyme and chenopodium. Obtained from the oil in the waste liquors of the sulphite wood pulp process. Prepared by heating camphor with $ZnCl_2$, or from oil of turpentine. Oxidized by chromic acid to terephthalic acid and by nitric acid to 4-toluic acid. Used as a thinner for paints and in the manufacture of thymol. The other isomers are of little significance.

cymoxamil Fungicide, urea derivative.

cypermithrin Insecticide, pyrethroid derivative.

cyphenothrin Insecticide, pyrethroid derivative.

cyproconazole Fungicide, azole derivative.

cyprodinil Fungicide, pyrimidine derivative.

cyromazine Insecticide, triazine derivative.

cysteine (α-amino-β-thiolpropionic acid, β-mercaptoalanine), $C_3H_7NO_2S$, $HSCH_2.CHNH_2.CO_2H$. Reduction product of cystine. Amino-acid is the first step in the breakdown of cystine in the body, one molecule of cystine splitting to give two molecules of cysteine. Cysteine is soluble in water but the solution is unstable, and is reoxidized to cystine.

cystine (dicysteine), $C_6H_{12}N_2O_4S_2$.

$$[HO(O)C \cdot CH(NH_2) \cdot S]_2$$

Reduced to cysteine. Cystine is abundant in the proteins of the skeletal and connective tissues of animals and in hair and wool, from which it is readily prepared.

cytidine See NUCLEOSIDES.

cytochromes Widely distributed respiratory enzymes concerned in the electron transport chain of living cells. They are hemoproteins differing in their porphyrin groups, but some contain metals which alternate between oxidation states during reactions. Cytochrome C is present in the greatest amounts but others are also present. Cytochrome C derivatives catalyse the reduction of oxygen to water, part of the chain of reactions in extraction of energy from food. Cytochrome P_{450} introduces a hydroxyl group into molecules, e.g. in steroid biosyntheses, and drug metabolism. See DICTIONARY OF BIOLOGY.

cytosine (2-oxy-4-aminopyrimidine), $C_4H_5N_3O$.

Constituent of DNA. Can undergo 5-methylation.

cytotoxic agents Chemicals injurious to living cells. The term is applied particularly with respect to tumour formation. Many cytotoxic agents are bifunctional alkylating agents, e.g. mustard gas, nitrogen mustards.

D

D Deuterium. Also a system of designating configuration of molecules such as monosaccharides by analogy with (+)-glyceraldehyde – designated D. Molecules analogous to (–)-glyceraldehyde are designated L.

D2341 Acaricide, hydrazine carboxylate derivative.

2,4-D (2,4-dichlorophenoxyacetic acid) Herbicide.

da Dalton.

DABCO 1,4-diazabicyclo[2,2,2]octane.

Dacron A polyester.

daimuron Herbicide, urea derivative.

dalapon Herbicide, alkanoic acid derivative.

dalton (Da) Term sometimes used for the atomic mass unit.

Dalton's law of partial pressures The total pressure (P) exerted by a mixture of gases is equal to the sum of the partial pressures (p) of the components of the gas mixture. The partial pressure is defined as the pressure the gas would exert if it was contained in the same volume as that occupied by the mixture. Assumes no interaction between the gas molecules.

daminozide Herbicide, alkanoic acid derivative.

Daniell cell A Zn/Zn²⁺ ‖ Cu²⁺/Cu cell. The e.m.f. of the Daniell cell is 1.10 V and is virtually independent of temperature.

dansyl 5-Dimethylaminonaphthalene-1-sulphonyl derivatives used (e.g. chloride and fluoride) for active site investigations in serine proteinases.

DAST (diethylamino sulphur trifluoride), Et_2NSF_3. Fluorinating agent for converting, e.g. CHO to CHF_2. Similar in reactions to SF_4 but easier to use.

dative covalent bond A co-ordinate bond. See VALENCY THEORY.

dazomet Nematocide, fungicide, herbicide, insecticide; thiadiazone derivative.

Db Dubnium, element 103.

2,4-DB (4-(2,4-dichlorophenoxy) butyric acid) Herbicide.

DCIP Nematocide, haloalkyl ether.

DCO DEHYDRATED CASTOR OIL.

DDT Formerly an important insecticide. Organochlorine derivative.

d–d transition Electronic transition occurring between non-degenerate d orbitals, e.g. those which have been split by a crystal field. Generally occurs in the visible region but may also occur in the u.v. or near i.r.

deactivating collision An intermolecular collision between an 'activated' and a normal molecule, which results in the deactivation of the former without it undergoing reaction. See DEACTIVATION.

deactivation The process whereby the chemical reactivity of a substance is diminished or entirely removed. For example, the catalytic activity of metals such as platinum or nickel for various hydrocarbon reactions may be considerably reduced or completely removed by the presence of trace amounts of catalyst poisons such as sulphur or mercury, which cause deactivation of the catalyst.

Alternatively, in a photochemical reaction a molecule absorbs light. This results in it possessing more energy than an ordinary molecule and, consequently, more reactivity. However, collision of this activated molecule with another molecule may cause the former to lose its energy without reacting. The activated molecule is said to have undergone a deactivating collision and has itself become deactivated.

deactivators Additives used in gasoline to prevent the catalysis of gum formation by metals (e.g. Cu).

deaeration Removal of oxygen and other dissolved gases from solvents by physical or chemical means. Carried out on a large scale from boiler feed water with a view to reducing corrosion.

deasphalting A single-solvent precipitation process whereby petroleum residues are treated with liquid propane or similar light solvent in a counter-current column. The ASPHALTENES are precipitated and the high-quality lubricating oil stock (bright stock) dissolves in the propane and may be further refined by solvent extraction.

de Broglie equation The wavelength of particles (λ) is given by the equation:

$$\lambda = (h/p) = (h/mv)$$

where h = Planck's constant 6.6×10^{-34}, m = mass, v = velocity and p = momentum. Used particularly for electrons and protons.

Debye–Hückel theory The activity coefficient of an electrolyte depends markedly upon concentration. In dilute solutions, due to the Coulombic forces of attraction and repulsion, the ions tend to surround themselves with oppositely charged ions. Debye and Hückel showed that it was possible to explain the abnormal activity coefficients at least for very dilute solutions of electrolytes. For dilute solutions, the Debye–Hückel equation by calculations based on these Coulombic interactions is

$$\log \gamma_{\pm} = - |Z_+ - Z_-| A T^{\frac{1}{2}}$$

where γ is the activity coefficient, z the ionic charge and

$$I = \tfrac{1}{2} \Sigma z_i^2 \left(\frac{b\ i}{b\ \ominus} \right)$$

the dimensionless ionic strength of the solution.

Debye units Non-S.I. unit of dipole moments. One Debye is 3.34×10^{-30} esu cm.

decalin (decahydronaphthalene), $C_{10}H_{18}$. There are two stereoisomers of decalin corresponding to the relative positions of the hydrogens at the bridging carbons.

trans

cis

The *cis* form has m.p. –45°C, b.p. 198°C. The *trans* form has m.p. –32°C, b.p. 185°C. The *cis* form can be quantitatively converted into the *trans* form by AlCl₃. Commercial decalin is manufactured by the high-pressure hydrogenation of naphthalene at high temperatures, and contains 90% *cis* and 10% *trans*-decalin. The two forms can be separated by fractional distillation. Decalin is used as a solvent.

decanedioic acid (sebacic acid), $HO(O)C \cdot [CH_2]_8 \cdot C(O)OH$, $C_{10}H_{18}O_4$. Colourless leaflets; m.p. 134°C. Manufactured by heating castor oil with alkalis or by distillation of oleic acid. Forms an anhydride, m.p. 78°C. The esters of sebacic acid are used as plasticizers and for resins and alkyd fibres.

decanoic acid See CAPRIC ACID.

decanols (decyl alcohols), $C_{10}H_{21}OH$. Used in the manufacture of plasticizers and detergents, petroleum additives and surfactants. They are manufactured by the OXO REACTION from nonenes, CO and H_2 with reduction of the resultant decanals to decyl alcohols. The composition of the product depends on the composition of the feed, the most common commercial grade being 'iso-decanol', a mixture of 10-carbon branched-chain primary alcohols with a narrow boiling range.

n-Decyl alcohol (1-decanol) is manufactured by the telomerization of ethene in the presence of aluminium triethyl to aluminium trialkyls, which are oxidized to the alkoxides and hydrolysed by sulphuric acid to a mixture of straight-chain alcohols. From these *n*-decyl alcohol may be fractionated, though it is usual to market a mixture of alcohols. Pure *n*-decyl alcohol is a colourless liquid with a sweet fat-like odour, m.p. 7°C, b.p. 233°C. Insoluble in water, soluble in alcohol. Also used as a plant growth regulator.

decantation The removal of a liquid from a suspension or from an immiscible heavier liquid by pouring while leaving the latter in the original container.

decarbonylation Loss of CO.

decarboxylases See CARBOXYLASE.

decarboxylation Loss of CO_2.

decay constant Radioactive decay occurs according to the exponential rate law $N=N_0e^{-\lambda t}$ where N_0 is the number of atoms present at time zero and N is the number of atoms present at time t. λ is the decay constant, which is equal to $0.693/t_{1/2}$, $t_{1/2}$ being the half-life of the radioactive species, and has a characteristic value for a particular nuclide.

decoctions Pharmaceutical solutions made by boiling a drug (often naturally occurring) with water and filtering.

decomposition point In a laboratory distillation the thermometer reading coinciding with the first indications of thermal decomposition of the liquid or solid.

decomposition voltage The smallest voltage which will cause the electrolysis of an electrolyte. The decomposition voltage or potential depends both upon the nature of the electrolyte and of the electrodes.

decoupling Electronic technique used in n.m.r. spectroscopy to remove particular coupling from observation. Hence simplifies observed spectrum.

decyl alcohol See DECANOLS.

de-emulsification See DEMULSIFICATION, DEMULSIBILITY.

defect structures Structures in which there are irregularities in the lattice. Defects may be of many types but include the non-alignment of crystallites, orientational disorder (e.g. of molecules or ions), vacant sites with the migrated atom at the surface (Schottky defects), vacant sites with an interstitial atom (Frenkel defects), non-stoichiometry. Crystals exhibiting defects show anomalous physical properties, particularly density, colour, electrical conductivity. Applications of defect solids include catalysis, semiconductors, luminescent materials.

degeneracy Having the same energy; particularly applied to wave functions, e.g. in a transition element gaseous atom the five d orbitals are degenerate. In a crystal field the degeneracy is removed.

degenerate orbitals See DEGENERACY.

degree of hydrolysis The degree of hydrolysis of a salt is defined as the fraction of the total salt hydrolysed by water. Thus, if in a solution of the salt AB, 90% of it is hydrolysed to yield the base AOH and the acid BH, the degree of hydrolysis is 0.90. It may be expressed also as a percentage, i.e. 90%.

degree of polymerization Number of monomeric units in an oligomer or polymer.

degrees of freedom From the standpoint of the phase rule, the number of degrees of freedom which a system possesses is the number of variable factors, temperature, pressure and concentration of the components, which must be fixed in order to define completely the condition of the system. E.g. if the pressure and temperature of a given mass of gas are fixed, the volume of the gas must have a definite value. The system has, therefore, two degrees of freedom.

In statistical mechanics (e.g. the theory of specific heats of gases) a degree of freedom means an independent mode of absorbing energy by movement of atoms. Thus a monatomic gas has three translational degrees of freedom. Polyatomic molecules have in addition vibrational and rotational degrees of freedom.

DEHP di[2-ethyl-hexyl]phthalate See PHTHALIC ESTERS.

dehumidification The removal of condensable water vapour from a vapour–gas mixture by condensation, absorption or adsorption. The term is most frequently applied to the removal of water vapour from air.

Dehumidification may be effected by cooling. Where small quantities of dry air are required dehumidification can be carried out using chemical absorbents, e.g. calcium chloride, phosphorus pentoxide, sulphuric acid, etc.

dehydrated castor oil (DCO) Polyunsaturated carboxylic acid glyceride, typically a 9,11-octadecadienoic acid derivative, prepared by heating castor oil to 250–300°C in the presence of catalysts. Used to form sebacic and capryl derivatives (by cleavage at the C=C bond) and for paints and varnishes.

dehydration Loss of H_2O, often with formation of C=C bond.

dehydro A prefix for an organic compound indicating the presence of less hydrogen. Often synonymous with ANHYDRO-.

dehydroacetic acid (3 acetyl-6-methyl 2-H-pyron-2, 4(3H) dione), $C_8H_8O_4$. Colourless needles; m.p. 109°C, b.p. 270°C. Prepared by boiling acetoacetic ester for several hours. It is formed from keten by polymerization in the presence of $ZnBr_2$. The parent acid is unstable and always reverts to the lactone form. It is reconverted to acetoacetic ester by boiling alcoholic potassium hydroxide. Reduced by hydrogen iodide to dimethylpyrone. Used in synthesis and in plasticizers and resins.

dehydrocyclization In petroleum CRACKING or REFORMING processes alkanes undergo dehydrocyclization yielding aromatics and hydrogen, e.g. $C_7H_{16} \rightarrow PhCH_3 + H_2$.

dehydrogenases See ENZYMES.

dehydrogenation The process whereby the hydrogen content of a molecule is decreased and the degree of unsaturation is increased, e.g. cyclohexane → benzene. It is usually accomplished by catalysts such as platinum or palladium on charcoal or by heating the compound with selenium (which yields hydrogen selenide).

dehydrohalogenation Loss of HX generally to leave a π bond.

deicers Chemicals used to break up snow or ice. Urea and, e.g. NaCl or $CaCl_2$ are used for roads. Deicing fluids including propylene or ethylene glycol are used for aircraft. All deicers act by reducing the freezing point of water.

deliquescence The property of having a water vapour pressure lower than that of the surrounding air so that water is absorbed (to give a solution).

deliquescent substance A material showing deliquescence.

delocalization A description of the normal situation for molecules or ions in which a bonding electron cannot be associated only with one atom but must be considered delocalized over the whole group. Applied particularly to aromatic molecules where in, e.g. benzene, the six π electrons are delocalized over the six carbon atoms. The extra stability of the molecule over a situation with three localized double bonds is called the delocalization energy.

delphinine A plant alkaloid related to ACONITINE.

Delrin Polymethanal.

delta bonding Lateral overlap between two orbitals (e.g. d orbitals) such that there are four regions of overlap.

deltamethrin Insecticide, pyrethroid derivative.

delta value (δ) A measure of shift in NUCLEAR MAGNETIC RESONANCE.

demasking See MASKING.

demeton-s-methyl (metasystox) Insecticide, acaricide, organophosphate derivative.

demulsibility The property enabling oils to separate from oil–water emulsions. Good demulsibility is obtained by special refining processes to remove polar compounds not removed in normal refining.

demulsification Emulsions of oil and water formed in various operations may require special techniques to separate the two liquid phases. This process of demulsification can be effected by either physical or chemical means. The chemical methods depend on the opposing action of different emulsifiers on one another. Some emulsions can be broken by mechanical treatment while others can be broken electrically by a high voltage current. Some emulsions coagulate in the presence of electrolytes.

denaturants Substances added to dutiable articles to render them unfit for human consumption, e.g. denatured alcohol. See INDUSTRIAL METHYLATED SPIRITS.

denaturation The irreversible change in solubility and other properties of proteins that occurs when a protein in solution is heated, shaken or treated with acid. Good examples are the precipitation of milk with vinegar (ethanoic acid) and the scrambling and beating of eggs. See PROTEINS.

dendrimers Tree-like molecules.

dendrite A tree-like crystal formation formed because crystal growth proceeds at different rates in certain directions from the initial crystallization point. Found particularly in alloys, and ice crystals. Many salts can be induced to crystallize as dendrites (see DENDRITIC SALT). Have advantages over normal forms.

dendritic salt NaCl crystallized as dendrites. It is less dense than normal rock salt, does not cake so easily, dissolves more quickly.

denitrification The return of fixed nitrogen, generally nitrates and ammonium salts, to the atmosphere, generally as N_2 and N_2O. See NITROGEN OXIDE.

denitrogenation The removal of nitrogen compounds, particularly from gas oil cracking feedstocks, to prevent the ammonia which would be formed neutralizing the acid catalysts used. Denitrogenation can be effected catalytically.

dense media separation Separation of solids of different density by means of a liquid of intermediate density which allows a 'float and sink' separation. Liquids of suitable density can be obtained by suspending finely divided solids in water. Used for separation of mineral products such as coal, ores or aggregates.

densitometer An apparatus for measuring the intensities of e.g. lines on a photographic plate. Used in X-ray analysis and spectrometric analysis.

density (ρ) Mass per unit volume, $kg\,m^{-3}$. Relative density (specific gravity, relative to water) may be more manageable.

density-gradient column A graduated glass tube, filled with a mixture of solutions such that a density gradient is obtained throughout the length of the tube. Used for rapid determination of densities of small particles of solid which sink to a particular level in the tube.

density of states Used in thermodynamics. The number of states in an energy range divided by the width of the range.

deodorants Perfumes that mask odours when natural secretions decompose on the skin. Aluminium compounds ($Al(OH)_3$) absorb the odours.

D-2-deoxyribose (desoxyribose), $C_5H_{10}O_4$. The sugar isolated by hydrolysis of DNA. Colourless crystals, m.p. 91°C, soluble in water. See NUCLEOSIDES.

dephlegmator See PARTIAL CONDENSER.

depilatories Materials used for hair removal. The most widely used is calcium thioglycollate in an alkaline medium at pH 12.3.

depolarization ratio (ρ) Applied to Raman lines, the ratio of the intensity of the scattered light with polarization perpendicular and parallel to the incident radiation. Gives information on the symmetry of the vibration.

depot fat The reserve of fat in animals, deposited in adipose tissue and which usually undergoes metabolism slowly.

depression of freezing point When a solute is added to a pure solvent the freezing point of the solvent is lowered by an amount which is proportional to the number of solute particles in the solution. The depression caused by 1 mole of solute in 1 litre of solvent is called the cryoscopic constant, molar depression constant (K_f). K_f has a definite value for a particular solvent. Used to determine average molecular weight.

Solvent	Molecular depression constant (°C/mole)
water	1.86
benzene	4.90

deprocarb Fungicide, benzimidazole derivative.

depsides Compounds with tannin-like properties formed by the condensation of the carboxylic group of one phenolcarboxylic acid with a hydroxyl group of a similar acid.

$$HO \cdot C_6H_4 \cdot C(O)OH + HO \cdot C_6H_4 \cdot C(O)OH \rightarrow$$
$$HO \cdot C_6H_4 \cdot C(O)O \cdot C_6H_4 \cdot C(O)OH$$

Called didepsides, polydepsides, etc., depending on the number of phenol residues they contain. Obtained from lichens, present in tea.

dermatan sulphate See CHONDROITIN.

derris Obtained from the root of the Far-Eastern shrub *Derris elliptica*, and originally used as a fish poison; contains rotenone. It is applied as a dust, and is used as an insecticide.

desalination See WATER.

desensitization A process used in photography to induce a loss of sensitivity in the silver halide emulsion over part of its spectral sensitivity range. Often used during development to allow subsequent processing to be carried out in a bright light but without affecting the latent image or inhibiting development.

deshielding NMR effect that has signals at lower field (down field). Compare SHIELDING.

desiccant A material used for drying. Some desiccants can react chemically with specimens being dried. Used in a desiccator. In order of effectiveness common desiccants are $CaCl_2 < CaO < NaOH < MgO < CaSO_4 < H_2SO_4 <$ silica gel $< Mg(ClO_4)_2 < P_2O_5$.

desmedipham Herbicide, carbamate derivative.

desmetryn Herbicide, triazine derivative.

desmotropism See ISOMERISM.

desorption The reverse of adsorption.

Dess–Martin oxidation Conversion of primary and secondary alcohols to aldehydes and ketones using triacetoxyperiodinone.

destructive distillation The distillation of organic solids or liquids in which the substance being heated partly or wholly decomposes during distillation, leaving a solid or viscous liquid in the still. Volatile liquid products are collected after condensation. Typical destructive distillation processes involve the carbonization of coal, peat, wood or oil shale. See also CARBONIZATION.

desulphurization The removal of sulphur compounds from petroleum fractions using hydrogen in the presence of a catalyst. Also known as *hydrofining*, *hydrodesulphurization* and *hydrotreating*.

detergent oil Lubricating oil containing detergent or dispersant additives used in internal combustion engines in order to stabilize solid contaminants in a finely divided state, and so prevent abrasion and accumulation of deposits. Additives used include alkaline earth metal salts of petroleum sulphonic acids, alkyl phenol sulphides, etc.

detergents Water-soluble, surface-active agents capable of wetting a variety of surfaces, removing grease and oil and retaining the dirt in suspension for ease of rinsing. Soap acts as a detergent and a large range of synthetic detergents, derived from petroleum, is available. These detergents contain *hydrophilic* groups such as sulphate, sulphonate or polyether which confer water solubility. Long hydrocarbon chains – *hydrophobic* groups – enable them to dissolve oily materials.

Anionic detergents, probably the most important group, are typified by alkyl aryl sulphonates which ionize in water to produce a large anion. *Cationic* surface-active agents, which produce a large cation, are less used as detergents but find some application in the textile and dyeing industries and as adhesion and emulsifying agents. They are often quaternary ammonium compounds or derivatives of fatty acid amides. *Non-ionic* detergents generally consist of condensation products of alcohols or phenols with ethylene oxide.

detonating gas A mixture of H_2 and O_2 (molar ratio 2:1) produced by electrolysis of water and which explodes violently to reform H_2O on ignition.

detonation A type of EXPLOSION typified by a much-increased reaction rate as well as a very high velocity percussion wave and a high pressure rise. Very high localized temperatures are produced.

This type of explosion is important in engines where detonation can vary from an incipient stage, often referred to as pinking, to a more severe condition known as KNOCKING.

detoxication (detoxification) A term for the biochemical transformations which occur during the metabolism of foreign organic compounds by animals. The process does not always result in decreased toxicity.

deuterium, D. At.no.1, at.mass 2.01. An isotope of hydrogen which, because of the relatively large difference in masses, gives markedly different physical properties, including rates of reactions, to deutero compounds – bracketed figures below are for hydrogen analogues. D_2, b.p. –249.7 (–252.8°C); D_2O, b.p. 101.4 (100.0°C), m.p. 3.8 (0.0°C), ionic product at 25°C 1.1×10^{-15} (1.0×10^{-14}). Ordinary hydrogen compounds contain 1 part in 4500 D. D_2O (heavy water) is obtained by preferential evaporation of H_2O or preferential electrolysis of H_2O. Deuterium compounds are used as tracers in chemical and biological investigations; as solvents, etc., when protons are under investigation (e.g. NMR), D_2O is used as a moderator in atomic reactors.

deuteron (deuton), $^2_1D^+$. The nucleus of the deuterium atom. Used in linear accelerators for bombarding other nuclei.

Devarda's alloy 45% Al, 50% Cu, 5% Zn. Used to detect and analyse for nitrates.

developed colours Ingrain dyestuffs.

developers, photographic See AMINO-PHENOLS, colour development. Solutions which cause reduction of exposed silver halide grains to silver.

development centres The areas of a silver halide crystal in which exposure to light has induced latent images and where, on development, reduction to metallic silver will start.

devitrification The crystallization of a glass (a supercooled liquid) which then becomes opaque.

'Dewar' benzene (bicyclo[2,2,0] hexa-diene), C_6H_6. Colourless liquid with alkenic properties. Many substituted derivatives are known, the preferred method of preparation being the addition of an alkyne to a cyclobuta-diene.

'Dewar' benzene is a valence isomer of benzene, to which it reverts on heating.

Heavily substituted valence bond isomers of benzene with prism-like structures ('prismanes') are also known.

dewaxing Removal of wax from lubricating oil stock.

dexide See DEXTRAN, MUCOSE.

dextran, mucose A mucilaginous polymer of glucose made by certain bacteria. Relative mol.wt. between 30 000 and 250 000, and has considerable resistance to enzymic hydrolysis. Manufactured by the bacterial fermentation of sucrose. It can be used as a plasma substitute in blood transfusions, as an emulsifying and thickening agent, and as a stabilizer for ice creams. Used in gel filtration and chromatography (SEPHADEX) and polydexide. Sephadex 2-(diethyl) aminoether is an ion-exchange resin used medicinally.

The sodium salts of sulphuric acid esters of dextran are used as anticoagulants for the same purposes as heparin.

dextrins Intermediate products formed during the hydrolysis of starch to sugars. They are strongly dextrorotatory. Manufactured by the action of heat and hydrolytic agents on wet or dry starch and used as adhesives, for improvement of stability and masking taste or odour.

dextrorotatory Species that rotates plane-polarized light clockwise.

dextrose See GLUCOSE.

diacetin See ACETINS.

diacetone See DIACETONE ALCOHOL.

diacetone alcohol (4-hydroxy-4-methyl-pentan-2-one), $C_6H_{12}O_2$, $HOC(CH_3)_2CH_2C(O)CH_3$. A colourless, odourless liquid; b.p. 166°C. Manufactured by treating propanone with lime, or sodium or barium hydroxides. Decomposed by acids and strong alkalis. It is used as a lacquer solvent and as a dye solvent in printing, and as antifreeze in pharmaceutical and hydraulic fluids.

diacetoneamine, $C_6H_{13}NO$, $Me_2C(NH_2)$-$CH_2C(O)Me$. Colourless, strongly basic liquid; b.p. 25°C/0.2 mm. Obtained from mesityl oxide and ammonia or ammonia and propanone.

diacetyl (biacetyl), $CH_3COCOCH_3$. A greenish-yellow liquid with a powerful odour; b.p. 88°C. It occurs in bay and other essential oils, and also in butter. Prepared from butan-2-one via its isonitroso derivative hydrolysed with HCl; oxidation of 2,3-butyleneglycol with oxygen at 270°C in the presence of a Cu catalyst. Reduced to 2,3-butyleneglycol by yeast. Forms dimethylglyoxime with hydroxylamine. Hardening results from the addition of diacetyl to photographic gelatin. Used to enhance aroma of butter.

diacetylenes (dialkynes) Hydrocarbons containing two alkyne linkages in the molecule. They are mobile, colourless liquids with odours resembling that of ethyne. Their properties are similar to those of the acetylenes. Prepared by coupling copper or Grignard derivatives of monoacetylenes in the presence of $CuCl_2$. Diacetylene (buta-1,3-diyne) itself is conveniently obtained by treating 1,4-dichloro-but-2-yne with sodamide in liquid ammonia.

diafen thiuron Insecticide, thiurea derivative.

diagonal relationship A similarity (in solubility and thermal stability) between compounds of elements of the main groups of the Periodic Table related to one another diagonally (Li and Mg; Be and Al). Results from a similarity in size in combining forms.

dialin The trivial name given to each isomer (two) of dihydronaphthalene.

dialkynes See DIACETYLENES.

diallylmelamine (2-diallylamine-4,6-diamino-s-triazine) See CYANOGUANIDINE.

dialysis A process used for the purification of colloidal sols. The comparatively large particles (1–100 pm) of the sol, in contrast to ions and small molecules, cannot diffuse through a semipermeable membrane. By enclosing the impure sol within such a membrane (dialyser), which is supported in pure (preferably flowing) water, impurities of a crystalloid nature are gradually removed. The function of the kidneys in mammals is to dialyse the blood, removing impurities such as urea from the haemoglobin sol. See ELECTRODIALYSIS.

diamagnetism The property of all substances that causes repulsion of the substance from a magnetic field. Originates in the interaction of the electronic charge and the field and is an additive property of the atoms and groups present. Order of magnitude 10^{-3} that of paramagnetism, so that the effect is swamped by the presence of unpaired electrons. See MAGNETIC SUSCEPTIBILITY.

1,2-diaminoethane (ethylenediamine), $C_2H_8N_2$, $H_2NCH_2 \cdot CH_2NH_2$. Colourless liquid which fumes in air and has a strong ammoniacal odour; m.p. 11°C, b.p. 116°C. Absorbs carbon dioxide and water from the air to give ethylenediamine carbamate. Manufactured by heating 1,2-dichloroethane with ammonia under pressure in the presence of CuCl. Forms a crystalline hydrochloride, and oil-soluble soaps with fatty acids; these are used as detergents and emulsifying agents. It is also used as a solvent for certain vat dyes and for casein, shellac and resin. Forms stable complexes with transition metals, in which it acts as a chelating agent. Used in textiles, paper, coatings, films and adhesives, in emulsifying agents (after condensation with carboxylic acids), rubber formulation.

1,6-diaminohexane (1,6-hexane-diamine, hexamethylene diamine), $H_2N \cdot (CH_2)_6 \cdot NH_2$. Colourless solid when pure; m.p. 41°C, b.p. 204°C. Manufactured by the reduction of adiponitrile. Used in the production of Nylon [6, 6].

2,4-diaminotoluene A percursor of the polyurethanes. See 4-NITROTOLUENE.

diamond A crystalline form of carbon, ρ 3520. In a crystal of diamond every carbon atom is linked tetrahedrally to four others. The structure has C–C 1.54Å, equal to that found in saturated hydrocarbons.

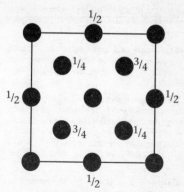

Crystal structure of diamond

This mineral, which in its transparent varieties is a most beautiful and costly gem, has been known from early times.

Important sources, e.g. in South Africa, are the basic igneous rock known as kimberlite and also alluvials. The valuable diamond may be colourless or faintly coloured, but must be transparent. The colours are mainly due to the presence of defects – an isolated nitrogen gives yellow, boron can give blue. The heavily coloured forms, known as carbonado or bort, are of no value as gems, but are used for rock drills, for lathe tools, and when powdered for cutting and polishing clear diamonds. In cutting, the natural crystalline form is obliterated and an artificial shape, which gives rise to a large amount of internal reflection producing the 'fire' of the stone, fashion. The diamond is extremely hard, and stands highest in Mohs's scale of hardness. It possesses a high refractive index, and dispersive power.

Diamond is very resistant to chemical reagents; potassium dichromate and sulphuric acid attack it with the formation of CO_2. It burns in air or oxygen at 700°C to CO_2 leaving scarcely any ash; some forms of bort may leave as much as 4.5% ash.

Very high temperature and pressure on graphite in the presence of a metal catalyst gives synthetic diamonds big enough for many industrial uses.

diamond anvil cell Device for applying high pressure to samples which are held between two diamonds. Pressures of up to 1Mbar can be obtained by the turn of a screw.

diamorphine (heroin), $C_{21}H_{23}NO_5$. Diacetylmorphine. See MORPHINE. Used in the form of its hydrochloride which is soluble in water and alcohol. M.p. 231–232°C. Controlled substance.

diaphragm cell A cell for electrolysis of brine to NaOH and Cl_2 in which the anode and cathode compartments are separated by a diaphragm.

diaphragm valve A valve for controlling the flow of liquids and gases in which a flexible diaphragm is pressed down against a ridge running across the valve body. Only the valve body and the diaphragm come into contact with the fluid and the valve may thus be made corrosion-resistant.

diaspore, α-AlO(OH). See ALUMINIUM HYDROXIDE. A major constituent of bauxite.

diastereomers Stereoisomeric structures which are not enantiomers. E.g. mesotartaric acid and either one of the optically active tartaric acids are said to be diastereomeric and differ only in the configuration at one carbon atom.

Mesotartaric acid   D- or L-Tartaric acid

Unlike enantiomers, diastereomers usually have substantially different chemical and physical properties.

diastereotopic See ENANTIOTOPIC.

diatomic molecule A molecule with two atoms, e.g. N_2, HCl.

diatomite (kieselguhr) A hydrated silica, the remains of diatoms. The particles are hollow and the material is used as a filler and an absorbent (nitroglycerine gives dynamite). Other uses are for polishing, filtration, decolorizing and for heat-insulation. Tripolite is a compact form.

Diazald A trade name for 4-ToS(O)$_2$ N(NO)CH$_3$. A white crystalline solid; m.p. 60°C. A convenient and easily stored source of diazomethane, which can be liberated by alkaline hydrolysis.

diazepine (diazepam, valium, 7-chloro-2,3-dihydro-1-methyl-5-phenyl-2H-1,4-dibenzodiazepin-2-one), $C_{16}H_{13}ClN_2$. White plates, m.p. 125°C. One of several di-azepines widely used as sedatives but diazepam is a controlled substance and is addictive.

1,2-diazine The ring system.

diazinon Insecticide, acaricide, organophosphate derivative. Use being phased out.

diazoacetic ester See ETHYL DIAZOETHANOATE.

diazoamino-compounds Prepared by condensation of a diazonium salt with a primary or secondary amine in the presence of sodium ethanoate, e.g.

$$C_6H_5N_2Cl + NH_2 \cdot C_6H_5 \rightarrow$$
$$C_6H_5 \cdot N_2 \cdot NH \cdot C_6H_5 \rightarrow$$
diazoaminobenzene

The diazoamino-compounds are usually yellow in colour, and do not dissolve in acid; they can usually be isolated and crystallized with decomposition. When treated with HNO_2 two molecules of diazonium salt are formed. Form an azo-compound when warmed with an amine and its hydrochloride.

Some diazoamino-compounds are suspect mutagenic agents.

diazo compounds Compounds containing –N=N– group attracted to carbon(s).

diazo dyestuffs Azo dyestuffs with –N$_2$– groups.

diazomethane, N_2CH_2. A yellow gas with a musty odour. It is very poisonous and highly explosive. Very soluble in ether. Prepared by treating a mixture of nitrosomethylurea and ether with cold concentrated potassium hydroxide solution, or better from DIAZALD and alkali. The resulting ethereal solution is a convenient form in which to handle diazomethane. It is a very reactive substance, generally adding :CH$_2$ groups to active hydrogens, etc. and, e.g. forms methyl chloride with hydrochloric acid, acetonitrile with hydrogen cyanide and methyl esters with organic acids. It reacts with acyl chlorides to give diazoketones, and with esters of unsaturated organic acids to give pyrazoline carboxylic acids. With aldehydes a mixture of a methyl ketone and an alkene oxide is formed; the proportions depend upon the conditions of the reaction and upon the aldehyde used. Alcohols react slowly or not at all, but phenols form methyl ethers. On photolysis it produces carbene (methylene), :CH₂, which forms cyclopropanes with alkenes. It is a useful reagent in organic syntheses. See ARNDT–EISTERT SYNTHESIS.

diazonium compounds An important class of compounds containing the RN=NX group with R aryl.

The diazonium salts are by far the most important diazo compounds. These are salts derived from the base R–N=N–OH, e.g. benzene diazonium chloride, PhN₂⁺Cl⁻.

They are prepared by the action of HNO₂ on aromatic amines. The amine is dissolved in excess of mineral acid and sodium nitrite is added slowly until a slight excess of HNO₂ is present. The reaction is usually carried out in ice-cold solution. The solution then contains the diazonium salt of the mineral acid used; anhydrous diazonium salts of varying stability may be precipitated with complex anions like PF₆⁻, SnCl₆²⁻BF₄⁻. Fluoroborates decompose to arylfluorides.

The diazonium salts usually decompose when warmed with water to give a phenol and nitrogen. When treated with CuCl, CuBr, KI, the diazo group is replaced by chlorine, bromine or iodine respectively (Sandmeyer reaction). A diazonium sulphate and hydroxylamine give an azoimide. The diazonium salt of anthranilic acid (2-aminobenzoic acid) decomposes to give BENZYNE.

The most important reaction of the diazonium salts is the condensation with phenols or aromatic amines to form the intensely coloured azo-compounds. The phenol or amine is called the secondary component, and the process of 'coupling' with a diazonium salt is the basis of manufacture of azo dyestuffs. The entering azo group goes into the 4-position of the benzene ring if this is free, otherwise it takes up the 2-position, e.g. diazotized aniline coupled with phenol gives benzeneazophenol. When only half a molecular proportion of nitrous acid is used in the diazotization of an aromatic amine a diazoamino compound is formed.

Used in reprographic processes, particularly the ozalid blue-print paper processes.

dibasic acid An acid with two replaceable hydrogen atoms which may yield two series of salts, e.g. H₂SO₄ gives NaHSO₄ and Na₂SO₄.

1,2,5,6-dibenzanthracene, C₂₂H₁₄. Crystallizes in silvery leaflets, m.p. 262°C. A polycyclic aromatic carcinogen obtained from coal tar. The 7,8 derivative, m.p. 196°C, is also known.

dibenzenechromium, C₁₂H₁₂Cr. Brown-black

crystals, m.p. 284–285°C, rapidly oxidized by air. Formed from CrCl₃, C₆H₆, AlCl₃ and reducing agents. The prototype π-arene complex with a sandwich structure, with the two benzene rings symmetrically bonded to the metal.

dibenzoyl peroxide (benzoyl peroxide), PhC(O)OOC(O)Ph. M.p. 107°C. Prepared from benzoyl chloride plus a cooled solution of sodium peroxide. Decomposes on moderate heating to give radicals and catalyses polymerization reactions, anti-Markownikoff additions (e.g. of HBr) across alkenes and radical halogenation of alkanes. Used commercially in the preparation of a wide range of styrene-based polymers. See PEROXIDES, ORGANIC.

dibenzyl See 1,2-DIPHENYLETHANE.

diborane, B₂H₆. See BORON HYDRIDES.

dibromochloropropane (1,2-dibromo-3-chloro-propane), CH₂BrCHBrCH₂Cl. B.p. 196°C. A heavy amber liquid with a mildly pungent odour prepared by the addition of bromine to propenyl chloride. It is used as a soil fumigant and as a nematocide. Can cause sterility in humans. Probably carcinogenic.

1,2-dibromoethane (ethylene dibromide), BrCH₂·CH₂Br. A colourless liquid with a sweet odour, m.p. 10°C, b.p. 132°C. Manufactured by passing ethene through bromine or bromine and water at about 20°C. Chemical properties similar to those of 1,2-dichloroethane; when heated with alkali hydroxides, vinyl bromide is formed. Used in petrols, as a fumigant for stored products and as a nematocide. Probably carcinogenic.

di-t-butyl dicarbonate, BuOC(O)OC(O)Bu. Used for the protection of amino groups.

2,6-di-t-butyl-4-methylphenol (2,6-di-tert-butyl-4-methylphenol, ionol, butylated hydroxytoluene, BHT) White solid, m.p. 69–71°C, used as an anti-oxidant especially to inhibit peroxide formation in ethers and petroleum. A potent inactivator of lipid-containing viruses. Skin irritant.

dibutyl phthalate See PHTHALIC ESTERS.

dicamba Herbicide, benzoic acid derivative.

dicarbollides See CARBORANES.

dicarboxylic acids Organic acids containing two carboxyl (–C(O)OH) groups. They form both acidic and neutral salts and esters; some give anhydrides by loss of a molecule of water between the two carboxyl groups. Many occur naturally as the free acid or an ester. They are generally prepared by the oxidation of a glycol, hydroxy-acid or hydroxyaldehyde, or by hydrolysis of a dinitrile or cyano-acid.

dichlobenil Herbicide, benzonitrile derivative.

dichlofluamid Fungicide, N-trihalomethyl derivative.

dichloramid Herbicide softener, chloramide derivative.

dichlorobenzenes, $C_6H_4Cl_2$.
1,2-dichlorobenzene, o-dichlorobenzene, b.p. 179°C.
1,3-dichlorobenzene, m-dichlorobenzene, b.p. 172°C.
1,4-dichlorobenzene, p-dichlorobenzene, m.p. 53°C, b.p. 174°C.

The 2- and 4-isomers are manufactured by the direct chlorination of benzene in the presence of iron as a catalyst, the resulting mixture being separated by fractional distillation. The 3-isomer may be obtained by isomerization of the 2- or 4-compound in the presence of a catalyst.

4-dichlorobenzene is the most important of the three compounds, being used largely as a moth repellant and air deodorant. 2-dichlorobenzene is used as a dye intermediate, insecticide, solvent and for various other purposes. The 3-isomer is not of commercial importance.

2,2-dichlorodiethyl sulphide See MUSTARD GAS.

dichlorodifluoromethane, CCl_2F_2. B.p. –30°C. Manufactured by the action of HF on CCl_4 using $SbCl_5$ as a catalyst. Known commercially as Freon-12 or Arcton-12. Was widely used as a refrigerant and aerosol propellant. It is much less toxic than carbon tetrachloride.

1,2-dichloroethane (ethylene dichloride), $CH_2Cl·CH_2Cl$. Colourless liquid with an odour like that of chloroform; b.p. 84°C. It is an excellent solvent for fats and waxes. Was first known as 'oil of Dutch chemists'. Manufactured by the vapour- or liquid-phase reaction of ethene and chlorine in the presence of a catalyst. It reacts with anhydrous ethanoates to give ethylene glycol diethanoate and with ammonia to give ethylenediamine, these reactions being employed for the manufacture of these chemicals. It burns only with difficulty and is not decomposed by boiling water.

Most of the ethylene dichloride produced is utilized for the manufacture of vinyl chloride, which may be obtained from it by pyrolysis or the action of caustic soda. Large quantities are also used in anti-knock additives for gasoline. As a solvent it has been partially displaced by TRICHLOROETHENE and TETRACHLOROETHENE. U.S. production 1999 10.3 megatonnes.

dichloroethenes (dichloroethylenes), $C_2H_2Cl_2$. There are three of these compounds.

1,1-dichloroethene (1,1-dichloroethylene, asymmetric dichloroethene), $CH_2=CCl_2$. Colourless liquid, b.p. 37°C. Prepared by heating 1,1,1- or 1,1,2-trichloroethane with excess lime at 70–80°C. Polymerizes readily to an insoluble solid.

1,2-dichloroethene (acetylene dichloride) Exists in two forms, *cis* and *trans*, both formed when acetylene tetrachloride (1,1,2,2-tetrachloroethane) is heated with water and zinc at 100°C. The product contains about 80% of the *cis*-form. When a mixture of equal volumes of ethyne and chlorine is passed over activated carbon at 40°C the *trans*-form is the chief product. *Cis*-form, b.p. 60°C; *trans*-form, b.p. 49°C. They are insoluble in water but are miscible with hydrocarbon solvents and are very similar in properties to trichloroethene; not affected by moisture or alkalis. They are used as substitutes for ether in fat extraction and as solvents, e.g. for rubber.

dichloroethylenes See DICHLOROETHENES.

2 2´-dichloroethyl ether (*bis*(2-chloroethyl) ether), $(ClCH_2CH_2)_2O$. A colourless liquid; b.p. 178°C. Manufactured by treating ethylene chlorohydrin ($CH_2Cl·CH_2OH$) with chlorine and excess ethene at 80°C; or by heating ethylene chlorohydrin with sulphuric acid at 100°C. Reacts with amines to give morpholines and with sodium tetrasulphide to give rubber-like plastics. It reacts with fused potassium hydroxide to give divinyl ether. Used as a solvent in the Chlorex process for dewaxing mineral oils, and for removing grease from textiles. Used for the manufacture of surfactants and elastomers. Probably carcinogenic.

sym-dichloro-isopropyl alcohol See GLYCEROL DICHLOROHYDRIN.

dichloromethane (methylene chloride), CH_2Cl_2. A colourless liquid with a chloroform-like odour; b.p. 41°C. Prepared by heating chloroform with zinc, alcohol and hydrochloric acid; manufactured by the direct chlorination of methane. Decomposed by water at 200°C to give methanoic and hydrochloric acids. Largely used as a solvent for polar and non-polar substances, particularly for paint removal (30%), dissolving cellulose acetate and degreasing (10%). It is more stable than carbon tetrachloride or chloroform especially towards moisture or alkali. Somewhat toxic.

dichloromethyleneammonium salts (phosgenammonium, phosgene iminium), $[Cl_2C=NR_2]^+X^-$. Prepared from Cl_2 and $[Me_2NC(S)S^-]_2$ or the corresponding thiocarbamoyl chloride. Reactive electrophiles used in synthesis, e.g. RCH_2COCl plus $[Cl_2C=NMe_2]^+$, gives $(Me_2N)ClC=C(R)COCl$.

dichlorop Herbicide, alkanoic acid derivative.

dichlorophen Acaricide, fungicide, bacteriostat, chlorophenol derivative.

dichlorophenoxy acid See 2,4-D.

1,2-dichloropropane (propylene dichloride), $CH_3CHClCH_2Cl$. Colourless liquid with a pleasant odour, b.p. 96°C. Manufactured by treating liquid chlorine with an excess of liquid propene. It is very similar in properties to 1,2-dichloroethane, and is used for similar purposes.

1,3-dichloropropane Nematocide, soil fumigant.

dichloropropanols (glycerol dichlorohydrins) *sym*-dichloroisopropyl alcohol (*glycerol α-dichlorohydrin, 1,3-dichloro-2-hydroxypropane*), $CH_2Cl·CHOH·CH_2Cl$. Colourless liquid with an ethereal odour; b.p. 174–175°C. Prepared by passing dry HCl into glycerin containing 2% ethanoic acid at 100–110°C. Converted to α-epichlorohydrin by KOH. Used as a solvent for cellulose nitrate and resins.

2,3-dichloropropanol (glycerol β-dichlorohydrin), $CH_2Cl·CHCl·CH_2OH$. Colourless liquid, b.p. 182°C. Prepared by the chlorination of propenyl alcohol. Reacts with NaOH to give epichlorohydrin.

dichlorotoluene See BENZAL CHLORIDE.

dichlorvos Insecticide, acaricide, organophosphate derivative.

dichroism PLEOCHROISM in a uniaxial crystal. See CIRCULAR DICHROISM.

dichromates Compounds containing the $[O_3CrOCrO_3]^{2-}$ anion. See CHROMATES.

diclofop-methyl Herbicide, propionic acid derivative.

diclomezine Fungicide, pyridazine derivative.

dicloran Fungicide, aniline derivative.

dicofol Acaricide, organochlorine derivative.

dicophos Insecticide, acaricide, organophosphorus derivative.

dicyanil Insecticide, pyrimidine derivative.

dicyandiamide See CYANOGUANIDINE.

dicyclohexylamine, $C_{12}H_{23}N$, $(C_6H_{11})_2NH$. M.p. 0°C. Prepared from cyclohexanone and cyclohexylamine. Industrial solvent and corrosion inhibitor.

N,N′-dicyclohexylcarbodiimide, $C_6H_{11}NC\ NC_6H_{11}$, $C_{13}H_{22}N_2$. Crystalline solid; m.p. 35–36°C, b.p. 154–156°C, prepared by oxidizing *N,N′*-dicyclohexylthiourea with HgO in carbon disulphide solution, also obtained from cyclohexylamine and phosgene at elevated temperatures. Used as a mild dehydrating agent, especially in the synthesis of peptides from amino-acids. Skin irritant.

dicyclopentadiene Colourless solid; m.p.

32°C, b.p. 170°C (decomp.), has a characteristic odour. It is the Diels–Alder product of dimerization of CYCLOPENTADIENE, the *exo*-form being formed most rapidly but the *endo*-form is thermodynamically favoured. At temperatures above 150°C a retro-Diels–Alder reaction occurs and cyclopentadiene monomer is regenerated; see DIENE REACTIONS. Forms polymer used in fabrication.

dicyclopentadienyl compounds See CYCLOPENTADIENYLIDES.

dieldrin The epoxy derivative of aldrin, now withdrawn.

dielectric constant (ε_r) The potential V of two charges q_1 and q_2 separated by r is $V = q_1q_2/4\pi\varepsilon r$ where ε is the permittivity of the medium normally expressed as relative permittivity ε_r. The dielectric constant is $\varepsilon/\varepsilon_0$ where ε_0 is the permittivity in vacuo. The polarity or polarizability of the medium affects ε_r. Ionic substances are more soluble in solvents of high ε. ε is a measure of dipole moment.

Diels–Alder reaction The Diels–Alder reaction is the 1,4-addition of an alkene or alkyne (dienophile) across a conjugated diene. An example is the addition of propenal to buta-1,3-diene to give Δ^4-tetrahydrobenzaldehyde. The

dienophile should be substituted with electron-attracting groups such as –C(O)OR, –CN, –CF$_3$, –NO$_2$, to enhance its reactivity, and the conjugated diene should have a *cis*-configuration. Its utility in synthesis arises from the stereospecific course it follows, i.e.

The addend from dimethylmaleate is also *cis*.

Diels' hydrocarbon (γ-methylcyclopentenophenanthrene), $C_{18}H_{16}$. The fundamental hydrocarbon on which the structure of the sterols and related compounds is based.

It can be obtained by heating cholesterol with selenium.

dienes Organic compounds containing two carbon-to-carbon double bonds. See DIOLEFINS.

dienochlor Acaricide, organochlorine derivative.

dienoestrol, dienestrol, $C_{18}H_{18}O_2$, [4-HOC$_6$H$_4$C(:CHMe)]$_2$. Colourless, crystalline powder; m.p. 233°C, insoluble in water, soluble in alcohol and in organic solvents. Prepared by reducing 4-hydroxypropiophenone and dehydrating the pinacol thus obtained.

It has oestrogenic properties (see OESTRADIOL) and is used for the same purposes as STILBOESTROL.

dienophile See DIELS–ALDER REACTION.

diesel fuel Fuel used for diesel (compression ignition) engines. The composition varies but is near that of GAS OIL. Fuel for diesel-engined road vehicles is commonly known as Derv.

Dieterici's equation See VAN DER WAALS' EQUATION.

diethanolamine See ETHANOLAMINES.

diethofen-carb Fungicide, carbamate derivative.

diethylamine See ETHYLAMINES.

diethyldithiocarbamic acid, Et$_2$NC(S)SH. Typical dithiocarbamic acid. Gives a brown complex with copper salts, soluble in CCl$_4$, used in the determination of copper. The Zn^{2+} and Et$_2$NH$_2^+$ salts are used as accelerators in rubber vulcanization.

1,4-diethylene dioxide See DIOXAN.

diethyleneglycol (2,2´-dihydroxydiethyl ether, digol), $C_4H_{10}O_3$, (HOCH$_2$CH$_2$)$_2$O. A colourless, hygroscopic and almost odourless liquid; b.p. 244°C. It is obtained as a by-product in the manufacture of ethylene glycol by the hydration of ethylene oxide. Manufactured by heating ethylene glycol with ethylene oxide. Solvent lubricant used in antifreezes. Used as a softening agent for textile fibres, as a solvent for certain dyes and as a moistening agent for glues, paper and tobacco. Its esters and ethers are used as solvents and plasticizers in lacquers. Forms a dinitrate used as a propellant.

diethyleneglycol diethyl ether (diethylcarbitol), EtOCH$_2$CH$_2$OCH$_2$CH$_2$OEt, $C_8H_{13}O_3$. High b.p. solvent, b.p. 188°C.

diethyleneglycol monoethylether (methyl carbitol), $C_6H_{14}O_3$, CH$_3$CH$_2$OCH$_2$ CH$_2$OCH$_2$CH$_2$OH. A colourless liquid with a pleasant odour; b.p. 193°C. Miscible with water and most organic solvents. Manufactured by heating ethylene oxide with ethylene glycol monoethyl ether under pressure. Used in the preparation of nitrocellulose lacquers and laminated glass and as solvent.

diethylenetriamine, NH$_2$CH$_2$CH$_2$NHCH$_2$CH$_2$NH$_2$. B.p. 207°C. Important polyamine with uses similar to those of diaminoethane.

diethyl ether See ETHER.

di-(2-ethylhexyl) phthalate (DEP) Plasticizer used in vinyl polymers. May be harmful to health. See PHTHALIC ESTERS.

diethyl malonate (malonic ester), $C_7H_{12}O_4$, $CH_2(COOEt)_2$. A colourless liquid with a faint aromatic odour, b.p. 199°C. Manufactured by treating sodium monochloroethanoate with sodium cyanide in alkaline solution at 60°C. The resulting sodium cyanoethanoate is hydrolysed with alcohol and sulphuric acid to give malonic ester. Converted to malonic acid and alcohol with water or dilute alkalis. Like acetoacetic ester, it reacts with sodium alkoxides or metallic sodium – a sodium atom displacing one of the hydrogen atoms of the $>CH_2$ group. This sodium derivative reacts with halogen compounds to give substituted malonic esters of the type $RCH(COOEt)_2$. With sodium followed by R'Cl disubstituted malonic esters $RR'C(COOEt)_2$ are formed. The substituted malonic esters are hydrolysed by alkali to the acids, which are readily decarboxylated to substituted ethanoic acids. Malonic ester and the esters of the substituted malonic acids react with urea to give barbituric and substituted barbituric acids, which are important drugs.

diethyl oxalate (oxalic ester), $C_6H_{10}O_4$, $(C(O)OEt)_2$. Colourless liquid; b.p. 185°C. Prepared by distilling a mixture of oxalic acid, alcohol and carbon tetrachloride. Condenses with other esters in the presence of sodium. Used in organic syntheses and as a solvent.

diethyl oxide See ETHER.

diethyl phthalate See PHTHALIC ESTERS.

diethyl succinate See ETHYL SUCCINATE.

diethyl sulphate, $(C_2H_5O)_2SO_2$. A colourless liquid with a faint ethereal odour; b.p. 208°C (decomp.). Manufactured from ethene and 100% sulphuric acid. Ethylating agent in organic syntheses. Highly toxic, carcinogen.

diethyl sulphide (ethyl sulphide), $(CH_3CH_2)_2S$. A colourless liquid with an ethereal odour when pure; usually it has a strong garlic-like odour; b.p. 92°C. Prepared by the action of KHS on ethyl chloride or potassium ethyl sulphate. When heated at 400–500°C it forms thiophene. Solvent for inorganic salts; used in plating baths.

difenacoum Rodenticide, coumarin derivative.

difenoconazole Fungicide, triazole derivative.

difenthialone
Rodenticide, coumarin derivative.

differential optical absorption spectroscopy See DOAS.

differential scanning calorimetry (DSC) See THERMAL ANALYSIS.

differential thermal analysis (DTA) See THERMAL ANALYSIS.

differential titration Analysis of a sample which contains two or more similarly reacting species which are differentiated, e.g. use of different bases or indicators.

diffraction pattern A beam of X-rays, electrons, or neutrons passed through a crystalline solid produces a diffraction pattern on a photographic plate or other recorder. The pattern is due to diffraction in a manner analogous to the diffraction of light by a diffraction grating. Reflected X-rays, electrons or neutrons from crystalline materials also give diffraction patterns. The diffraction patterns are used in structure determination.

diffusion In any gaseous mixture or liquid solution which is kept at a uniform temperature, the composition eventually becomes uniform throughout the system, no matter what was the original distribution. This is explained on the basis of the molecular motion that is diffusion. See GRAHAM'S LAW OF EFFUSION.

Graham proposed that the rate of diffusion of different gases through a porous diaphragm was inversely proportional to the square roots of their masses; this is the basis of a method of separation of gases, which has been applied successfully to the separation of hydrogen and deuterium, and of uranium isotopes using UF_6.

A diffusion mechanism is also used in DIALYSIS as a means of separating colloids from crystalloids. The rate of diffusion of molecules in gels is practically the same as in water, indicating the continuous nature of the aqueous phase. The diffusion of gases into a stream of vapour is of considerable importance in diffusion pumps.

Diffusion also takes place, although slowly, between solids. See FICK'S LAW OF DIFFUSION.

diffusion pump A non-mechanical pump used for the production of high vacua. Gas molecules are forced through a fine jet by molecular bombardment with a stream of oil or mercury vapour from a boiler. Once through the jet the heavy vapour is condensed and returned to the boiler, while the expelled gas is removed by a mechanical backing pump.

diflubenzuron Insecticide, urea derivative.

diflufenicon Herbicide, carboxamide derivative.

diflumetorim Fungicide, pyrimidine derivative.

difluoromethane, CH_2F_2. B.p. –52°C. A gas used as an azeotropic mixture with other fluorocarbons as a refrigerant. See FLUOROCARBONS.

digestion The process of ageing (recrystallization) of a precipitate to obtain it in a form more suitable for filtration.

digitalis The dried leaves of *Digitalis purpurea*, the common foxglove. It is a valuable source of the cardiac glycosides. The glycosides contain several sugars including DIGITALOSE and DIGITOXOSE, linked to steroidal aglucones. The glycosides cause a failing heart to beat more slowly, more regularly and more strongly. DIGOXIN rather than the extracts of digitalis is now used in treating cardiac failure.

digitalose (3-methyl-D-fucose), $C_7H_{14}O_5$. The sugar constituent of digitalin; m.p. 106°C.

digitoxose, $C_6H_{12}O_4$. The 2-desoxy sugar present in several of the cardiac glycosides of digitalis.

diglyme, $C_6H_{14}O_3$, $CH_3 \cdot O \cdot CH_2 \cdot CH_2 \cdot O \cdot CH_2 \cdot CH_2 \cdot O \cdot CH_3$. Dimethyl ether of diethylene glycol. A colourless liquid; b.p. 160°C. Typical ether, useful as a high-temperature solvent.

digol The trivial name for DIETHYLENEGLYCOL, $O(CH_2CH_2OH)_2$.

digoxin, $C_{41}H_{64}O_{14}$. A glycoside obtained from the leaves of *Digitalis lanata* which contains three DIGITOXOSE units and digoxigenin. Colourless crystals, m.p. 265°C. The cardiac glycoside most frequently used in treating heart failure.

dihedral angle The angle relating two parts of a molecule. In H_2O_2 the dihedral angle in the solid is 94°.

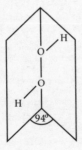

dihydroazirine See ETHYLENEIMINE.

3,4-dihydro-2H-pyran, C_5H_8O. B.p. 85°C.

Prepared by dehydration and rearrangement of tetrahydrofurfuryl alcohol over alumina. Used as a PROTECTING GROUP in organic synthesis, it forms tetrahydropyranyl ethers by reaction with alcohols under very mild acid catalysis. These ethers are cleaved easily with dilute acid after the desired reaction has been accomplished. Has also been used to protect thiol, carboxyl and certain >N–H groups.

dihydroxyacetone, $HOCH_2C(O)CH_2OH$. Crystallizes as the monomer, m.p. about 80°C, but changes to the dimer on standing, m.p. about 115°C. Obtained by the action of certain *Acetobacter* on glycerol. It is a strong reducing agent. Used in preparations for application to the skin to simulate a suntan.

1,2-dihydroxybenzene (catechol, pyrocatechol), $C_6H_6O_2$. Colourless crystals; m.p. 105°C, b.p. 240°C. Strong reducing agent.

Obtained by fusing 2-benzenedisulphonic acid with NaOH. It is used as a photographic developer, for preparation of dyes and pharmaceuticals, and as an antioxidant.

1,3-dihydroxybenzene See RESORCINOL.

1,4-dihydroxybenzene See HYDROQUINONE.

dihydroxybutanes See BUTANE DIOLS.

dihydroxychloropropanes (glycerol monochlorohydrins), $C_3H_7ClO_2$.

1,2-dihydroxy-3-chloropropane (3-chloropropylene glycol, glycerol α-monochlorohydrin), $CH_2Cl \cdot CH OH \cdot CH_2OH$. Colourless, rather viscous liquid; b.p. 139°C/18mm. Prepared by passing dry HCl into glycerin containing 2% ethanoic acid at 105–110°C. Reacts with nitric acid to give a dinitrate which is used in the manufacture of low-freezing dynamites.

1,3-dihydroxy-2-chloropropane (2-chlorotrimethylene glycol, glycerol β-monochlorohydrin), $CH_2OH \cdot CHCl \cdot CH_2OH$. Colourless liquid; b.p. 146°C/18mm. It is obtained in small amounts in the preparation of the α-chlorohydrin.

***cis*-dihydroxycyclohexadienes** Used as versatile chiral synthons.

2,2′-dihydroxydiethyl ether See DIETHYLENEGLYCOL.

1,2-dihydroxyethane (ethylene glycol), $HOCH_2 \cdot CH_2OH$. Colourless, odourless, rather viscous hygroscopic liquid having a sweet taste, b.p. 197°C. Manufactured from ethylene chlorohydrin and $NaHCO_3$ solution, or by the hydration of ethylene oxide with dilute sulphuric acid or water under pressure at 195°C. Used in antifreezes and coolants for engines (50%), in manufacture of polyester fibres (e.g. Terylene) and in the manufacture of various esters used as plasticizers. See ETHYLENEGLYCOL MONOMETHYL ETHER.

dihydroxymalonic acid See MESOXALIC ACID.

1,3-dihydroxynaphthalene See 1,3-NAPHTHALENE DIOL.

3,4-dihydroxyphenylalanine (L-Dopa), $C_9H_{11}NO_4$. M.p. 282°C (decomp.). The naturally occurring substance is laevorotatory. Aminoacid isolated from various sources, but can be synthesized. It is formed from tyrosine as the first stage in the oxidation of tyrosine to melanin. It is used in the treatment of Parkinson's disease. Forms adhesive proteins.

1,2-dihydroxypropane (propylene glycol), $CH_3 \cdot CHOH \cdot CH_2OH$. A colourless, almost odourless liquid. It has a sweet taste, but is more acrid than ethylene glycol; b.p. 187°C. Manufactured by heating propylene chlorohydrin with a solution of $NaHCO_3$ under pressure and from propylene oxide. It closely resembles dihydroxyethane in its properties, but is less toxic. Forms mono- and di-esters and ethers. Used as an anti-freeze and in the preparation of perfumes and flavouring extracts, as a solvent and in mould growth inhibition.

9,10-dihydroxystearic acid (9,10-dihydroxyoctadecanoic acid), $C_{18}H_{36}O_4$. M.p. 132°C. Used in cosmetics and toilet preparations.

diimide (diimine), $HN{=}NH$. Unstable reagent normally generated *in situ* by oxidation (Cu^{2+}/air, H_2O_2, HgO) of hydrazine, for addition to symmetrical multiple bonds (C=C, C≡C, N=N) by stereospecific *cis*-addition of hydrogen. Stable at low temperatures.

3,5-diiodotyrosine (iodogorgic acid), $C_9H_9I_2NO_3$. Pale, straw-coloured needles, m.p. 198°C. Found in marine organisms, and in the thyroid gland. Required for thryonine synthesis.

diisobutyl ketone See 2,6-DIMETHYL-4-HEPTANONE.

2,4-diisocyanatotoluene (toluene-2,4-diisocyanate, 2,4-tolylenediisocyanate, TDI) A lachrymatory liquid; b.p. 251°C. Manufactured from phosgene and 2,4-diaminotoluene. Used for preparing polyurethane foams and other elastomers by reaction with polyhydroxy compounds. Produces skin irritation and causes allergic eczema and bronchial asthma.

diisopropyl ether See ISOPROPYL ETHER.

diisopropylideneacetone See PHORONE.

dikegulac Plant growth regulator, furan derivative.

diketen (3-buteno-β-lactone), $C_4H_4O_2$. A colourless highly lachrymatory liquid; b.p. 127°C,

$$H_2C{:}C \cdot CH_2 \cdot C(O)$$
$$\underset{\displaystyle \quad \rule{1.2em}{0.4pt}\; O \;\rule{1.2em}{0.4pt}}{}$$

m.p. –6.5°C. Manufactured from propanone via keten. It reacts with alcohols and amines to give acetoacetic esters and amides respectively. Can be used as its stable propanone adduct.

diketones Organic compounds containing two keto (>C=O) groups. They are classified according to the number of carbon atoms separating the two keto groups. α- or 1,2-diketones are formed from their mono-oximes, the isonitrosoketones, by boiling with dilute sulphuric acid. The aliphatic α-diketones are yellow oils with pungent odours, while the aromatic diketones are crystalline solids. They react characteristically with 2-phenylenediamines to give quinoxalines; they form mono- and di-oximes with hydroxylamine and osazones with hydrazines. β- or 1,3-diketones $R \cdot CO \cdot CH_2 \cdot CO \cdot R$ are obtained by the reaction between an ester and a ketone in the presence of metallic sodium or sodamide. The central carbon shows acidic properties and forms metallic derivatives, many of which are insoluble in water but soluble in organic solvents. Most are chelates (see ACETYLACETONATES). These diketones can exist in keto and enol forms. They react with phenylhydrazine to give pyrazoles, and with hydroxylamine to give iso-oxazoles. γ- or 1,4-diketones ($R \cdot CO \cdot CH_2 \cdot CH_2 \cdot CO \cdot R$) are formed by treating sodium acetoacetic ester with α-bromoketones and hydrolysing the product. They readily form ring compounds.

2,5-diketopiperazine (2,5-piperazinedione, glycine anhydride), $C_4H_6N_2O_2$. Sublimes 260°C; sparingly soluble in water; hydrolysed by alkalis or mineral acids to glycylglycine.

It and substituted diketopiperazines are formed by the condensation of amino-acids, and are obtained in small quantities on the hydrolysis of proteins.

dilatancy Some pastes or suspensions (e.g. starch moistened with water) seem to become harder when they are subjected to shearing; in extreme examples, pastes, which are 'wet' enough to flow as a viscous liquid when allowed to stand, will temporarily crumble like a dry, brittle solid if suddenly stirred. This rheological phenomenon, called dilatancy, is the opposite of THIXOTROPY. It is explained by the friction of the particles against one another; when they are left undisturbed, the particles roll over one another and settle down into minimum volume, but when the medium is subjected to shear, they are forced into a less dense packing and the available fluid is insufficient to fill the interstices.

dilatometer An apparatus for measuring small changes of volume of a liquid, solution or solid immersed in a liquid. It usually consists of a cylindrical glass bulb with attached capillary tube, and change of volume of the contents of the bulb is noted by observing the movement of the meniscus in the capillary. Dilatometry is useful for determining transition temperatures, rates of reaction, polymerization, etc.

dilauryl peroxide, $(CH_3(CH_2)_{10}CO)_2O_2$. See PEROXIDES, ORGANIC.

dilute solution A dilute solution is one which contains only a small amount of solute.

dilution, infinite With electrolyte solutions the equivalent conductivity (Λ) increases as the solution is diluted. If Λ is extrapolated to zero concentration, the limiting value, Λ_0, is called the equivalent conductivity at infinite dilution.

dimedone **(5,5-dimethyl-1,3-cyclohexanedione)**, $C_8H_{12}O_2$. Greenish-yellow needles or prisms, m.p. 148–149°C, sparingly soluble in cold water. Gives crystalline condensation products with aldehydes.

dimepiperate Herbicide, thiocarbamate derivative.

dimer A compound formed by the addition polymerization of two molecules of a monomer.

dimercaprol See DITHIOGLYCEROL.

dimethachlor Herbicide, acetanilide derivative.

dimethametryn Herbicide, triazine derivative.

dimethenamide Herbicide, acetamide derivative.

dimethipin Herbicide, plant growth regulator, dithione derivative.

dimethirimol Fungicide, pyrimidine derivative.

dimethoate Insecticide, acaricide, organophosphate derivative.

dimethonaph Fungicide, morpholine derivative.

1,2-dimethoxyethane (glyme), $CH_3OCH_2 CH_2OCH_3$. A water-miscible colourless liquid, b.p. 83°C, used as a solvent.

dimethoxymethane (methylal, methylformal), $C_3H_8O_2$, $CH_2(OCH_3)_2$. Colourless liquid with a pleasant odour; b.p. 42°C. It occurs in commercial formalin, and is made by treating methylene chloride with sodium methoxide, or by treating a mixture of methyl alcohol and methanal with calcium chloride and a little hydrochloric acid. It is a good solvent, and is also used to replace methanol in many reactions.

dimethyl acetylenedicarboxylate, $C_6H_6O_4$, $CH_3O(O)C\cdot C\equiv C\cdot C(O)OCH_3$. A lachrymatory liquid; b.p. 195–198°C, which is extensively used as a dienophile in Diels–Alder reactions.

***cis*-dimethylacrylic acid** See TIGLIC ACID.

dimethylamine See METHYLAMINES.

4-dimethylaminobenzaldehyde (Ehrlich's reagent), $4\text{-}Me_2NC_6H_4CHO$. Dyestuff reagent.

3-(dimethylaminomethyl)-indole (gramine) Colourless solid; m.p. 132–134°C. An important intermediate for preparing indole derivatives, produced by treating INDOLE under MANNICH REACTION conditions with methanal and dimethylamine.

3-dimethylaminophenol, $C_8H_{11}NO$. Colourless needles; m.p. 87°C, b.p. 265–268°C. It is prepared by NaOH fusion of dimethylaniline-3-sulphonic acid, or by heating a mixture of resorcinol, dimethylamine sulphate and dimethylamine.

Other 3-(alkylamino)phenols are prepared by analogous methods. Used as intermediates in the preparation of dyestuffs.

4-dimethylaminopyridine (4-pyridinamine), $C_7H_{10}N_2$. M.p. 111–114°C. Polymerization catalyst, particularly for urethanes.

N,N-dimethylaniline, $C_8H_{11}N$, $PhNMe_2$. Colourless oil of characteristic smell; b.p. 193°C. Prepared by heating aniline with methanol and a little sulphuric acid at 215°C.

Used in determination of peroxidase and Fe^{2+} and Fe^{3+}.

dimethylanilines (xylidines) Prepared by reduction of nitro compounds. Used in dyestuff manufacture.

dimethylarsenic acid Herbicide, organoarsenic derivative.

N,N-dimethylbenzylamine, $C_6H_5CH_2NMe_2$. Used in the analysis of proteins.

2,3-dimethylbutadiene (βγ-dimethylbutadiene), C_6H_{10}, $H_2C=CMe\cdot CMe=CH_2$. Colourless liquid; b.p. 69.5°C. Manufactured from propanone by conversion to pinacone and passing its vapour over heated $KHSO_4$. Slowly changes to a rubber-like substance; the change is hastened by metallic sodium or peroxides. Used in the manufacture of artificial rubber.

5,5-dimethyl-1,3-cyclohexanedione See DIMEDONE.

dimethyl disulphide (methyl disulphide), CH_3SSCH_3. B.p. 108–110°C. Used as a solvent for sulphur in oils and in synthesis. Obtained by oxidizing methane thiol.

dimethylformamide (DMF), $HC(O)NMe_2$. Colourless liquid; b.p. 153°C, m.p. –61°C. It is an excellent solvent for a large variety of organic and inorganic materials and is widely used as a reaction medium. It can act as a catalyst in some substitution, elimination and addition reactions; synthetic applications include syntheses of aldehydes (with phosphorus oxychloride) and amines (Leuckart reactions). It is used as a solvent for lacquers, adhesives, dyes, etc.

dimethylglyoxime (diacetyl dioxime, butane-2,3-dione dioxime), $C_4H_8N_2O_2$, $CH_3C(:NOH)\cdot C(:NOH)CH_3$. Colourless needles; m.p. 240–241°C, sublimes at 215°C. Prepared by the action of hydroxylamine on diacetyl, or by treating methyl ethyl ketone with ethyl nitrite and hydrochloric acid which gives diacetyl mono-oxime which with sodium hydroxylamine monosulphonate gives dimethylglyoxime. It slowly polymerizes; condenses with 2-phenylenediamine to give quinoxaline derivatives. Quantitatively forms a dark-red crystalline nickel salt. Under appropriate conditions dimethylglyoxime can also be used for the detection and estimation of Bi, Cu, Co and Pd. Also used for rot-proofing of some fibres.

2,6-dimethyl-4-heptanone (diisobutyl ketone, isovalerone), $[(CH_3)_2CHCH_2]_2CO$. Obtained by reduction of phorone, b.p. 165°C. Very good solvent for aldehyde resins.

1,1-dimethylhydrazine (unsymdimethylhydrazine), Me_2NNH_2. M.p. –58°C, b.p. 64°C. A highly inflammable hygroscopic yellow liquid which fumes in air. Prepared from Me_2NH and NH_3 in the presence of catalysts. Used in rocket fuels. Corrosive to skin and mucous membranes.

dimethyl (methylene) ammonium iodide (Eschenmoser's salt), $(CH_3)_2C=N(CH_3)_2I$. Used in MANNICH REACTIONS and synthesis.

dimethyl phthalate See PHTHALIC ESTERS.

2,2-dimethylpropanoic acid See PIVALIC ACID.

dimethyl sulphate, $(CH_3O)_2SO_2$. A colourless and odourless liquid. Its vapour is extremely poisonous; the liquid is poisonous by absorption through the skin, b.p. 188°C. Prepared from methanol and chlorosulphonic acid. A good methylating agent, it reacts with amines and ammonia to give methylamines, with phenols and with organic acids to give methyl ethers and with organic acids to give methyl esters. Probably carcinogenic. Also methyl methane sulphonate, $CH_3S(O)_2OCH_3$.

dimethyl sulphide (methyl sulphide), CH_3SCH_3. A colourless liquid; b.p. 37°C. Prepared by methylation of methane thiol. It has a pungent odour and is used for detecting pipe leakages by mixing with the gas stream. Forms complexes with transition metals. Solvent for mineral acids.

dimethyl sulphite, $(CH_3O)_2SO$. A colourless liquid with a faint odour; b.p. 126°C. Prepared by boiling methanol with thionyl chloride. Decomposed by boiling water to methanol and SO_2. It reacts with organic acids, amines and alcohols to give methyl esters, methylamines and methyl ethers. Methylating agent in syntheses.

dimethyl sulpholane (2,4-dimethyl tetrahydrothiophene-1,1-dioxide) M.p. 280°C. Solvent for extraction processes.

dimethyl sulphone, Me_2SO_2. B.p. 238°C. Important solvent.

dimethyl sulphoxide (DMSO, methyl sulphoxide), Me₂SO. A colourless, odourless solid; m.p. 18°C, b.p. 189°C. Good solvent for a wide variety of inorganic and organic reactants. Saturated aliphatic hydrocarbons are virtually insoluble in DMSO. Having low toxicity it can be used in biology and medicine, especially for low-temperature preservation. The chemical reactions are those of a typical lower aliphatic sulphoxide. With sodium hydride a salt of the anion [CH₃SOCH₂]⁻ is produced (dimsyl sodium). The anion has several synthetic uses being a powerful nucleophile. Produced by the oxidation of dimethyl sulphide.

dimethyl terephalate See PHTHALIC ESTERS.

dimethyl vinphos Insecticide, organophosphate derivative.

dimorphism The existence of a substance in two crystal forms (see POLYMORPHISM).

dimsyl sodium See DIMETHYL SULPHOXIDE.

dineconazole Fungicide, azole derivative.

dinitramine Herbicide, nitroaniline derivative.

1,3-dinitrobenzene (m-dinitrobenzene), C₆H₄N₂O₄. Forms colourless crystals; m.p. 90°C, b.p. 302°C. Very poisonous.

It is prepared by the direct nitration of benzene or nitrobenzene, only small quantities of the 2- and 4-dinitro-compounds being produced. It is used as an intermediate in the preparation of 3-nitroaniline and 3-phenylenediamine.

4,6-dinitro-o-cresol (DNOC, 4,6-dinitro-2-hydroxytoluene), C₇H₆N₂O₅. A yellow crystalline solid; m.p. 86°C, with insecticidal, ovicidal and herbicidal properties.

dinitrogen, N₂. The term used to denote the N₂ group, particularly in complexes, e.g. (Ru(NH₃)₅N₂)Cl₂.

dinitrogen oxide, N₂O. See NITROGEN OXIDES.

dinitrogen tetroxide, N₂O₄. See NITROGEN OXIDES.

2,4-dinitrophenylhydrazine, C₆H₆N₄O₄. Violet-red crystalline solid, m.p. 199°C. Prepared from 2,4-dinitrochlorobenzene and hydrazine in boiling alcohol. It is an important reagent for aldehydes and ketones (Brady's reagent).

2,4-dinitrotoluene, C₇H₆N₂O₄. Colourless needle-shaped crystals; m.p. 71°C. Prepared by the nitration of the mixture of mono-nitrotoluenes obtained by mildly nitrating toluene. Used for the preparation of the corresponding mono- and di-amino-compounds, the latter compound being the precursor of toluene-2,4-di-isocyanate which is used in polyurethane formation.

dinobuton Acaricide, fungicide, nitrophenylcarbonate derivative.

dinocap Fungicide, acaricide, dinitrophenyl.

dinoseb See DNBP.

dinoterb Herbicide, nitrophenylcarbonate derivative.

diofenolan Insecticide, diphenyl ether derivative.

diolefins Hydrocarbons containing two double bonds. They are divided into three classes, according to the position of the double bonds. Compounds of the type –CH=C=CH– are called allenes, after the simplest member of the series. They are said to have 'cumulated' double bonds. They react in most cases normally, that is, each double bond is unaffected by the proximity of the other; they are converted by heating into the isomeric acetylenes. Compounds of the type –CH=CH–CH=CH– are said to have 'conjugated' double bonds and react somewhat differently from the other diolefins. For instance, bromine or hydrogen is often added so that a product of the type –CHBr–CH=CH–CHBr– is formed. Also, these hydrocarbons participate in the Diels–Alder reaction (see DIENE REACTIONS). They show a tendency to form rubberlike polymers. Hydrocarbons not falling into these two classes are said to have 'isolated' double bonds and each reacts normally. See ALLENES, OLEFINS, BUTADIENES and DIENES.

diorite A widely distributed igneous rock composed of plagioclase feldspar and hornblende. Used as road metal.

diosgenin, C₂₇H₄₂O₃. M.p. 204–207°C. A steroid sapogenin found abundantly in *Dioscorea* spp. (Mexican yams), and used in large amounts as one of the principal starting materials for the commercial manufacture of steroid hormones.

dioxan (1,4-diethylene dioxide) A colourless liquid with a faint and not unpleasant odour; b.p. 101°C. Manufactured by heating ethylene glycol with concentrated sulphuric acid or by passing ethylene oxide over solid NaHSO$_4$ at 120°C. Toxic. Forms complexes. Used as a solvent for cellulose acetate, resins, waxes and many organic substances.

dioxin (TCDD, 2,3,7,8-tetrachlorodibenzo-4-dioxin) A highly carcinogenic and poisonous material produced by combustion of chloroaromatics. The term is used for many related compounds.

dioxolane (1,3dioxacyclopentone), C$_3$H$_6$O$_2$. B.p.74–75°C. Solvent.

dioxygen, O$_2$. Formally the oxygen molecule but used for dioxygen complexes and dioxygenyl, (O$_2$)$^+$, salts.

dioxygenyl, (O$_2$)$^+$. See DIOXYGEN.

dipentaerythritol, C$_{10}$H$_{22}$O$_7$, [(CH$_2$OH)$_3$ CCH$_2$]$_2$O. Formed during the formation of PENTAERYTHRITOL; m.p. 222°C; generally present in pentaerythritol. Used in the preparation of drying oils, resin esters and alkyd resins; in the formation of plasticizers and fire-retardant compositions.

dipentene See LIMONENE.

diphacinone Rodenticide, indanedione derivative.

diphenamid Herbicide, amide derivative.

diphenyl, C$_{12}$H$_{10}$. Volatile, colourless solid which forms large lustrous plates, m.p. 70.5°C, b.p. 254°C. Benzene vapour passed repeatedly through an iron tube heated to 720°C produces much diphenyl and polyphenyls. Bromobenzene when heated with copper powder (Ullman reaction) and the action of CrCl$_3$ or CuCl$_2$ on

phenyl magnesium halides also provides diphenyl. Chlorinated and brominated diphenyls find extensive use in electrical components and as hydraulic fluids, but are being phased out because of health hazards. Uses include its action as a fungistat during shipment of apples and oranges, and as a heat transfer agent

(dowtherms) mixed with diphenyl ether and terphenyls.

diphenylamine, C$_{12}$H$_{11}$N, Ph$_2$NH. Colourless leaflets; m.p. 54°C, b.p. 302°C. Prepared by heating aniline and aniline hydrochloride at 200°C. It is only weakly basic, its salts with mineral acids being hydrolysed by water. It is slightly acidic, and gives an N-potassium salt. It is used as a second component in monoazo dyestuffs. Also used as a redox indicator. Diphenylamine derivatives are used as stabilizers for rocket propellants, polymers (particularly rubber), in the detection of nitrate and as fungicide.

1,2-diphenylethane (dibenzyl, bibenzyl), C$_{14}$H$_{14}$, PhCH$_2$CH$_2$Ph. Colourless crystals; m.p. 52°C, b.p. 284°C. It can be obtained by the action of sodium or copper upon benzyl chloride, by the action of AlCl$_3$ upon benzene and 1,2-dichloroethane or by heating benzoin or benzil.

diphenyl ether (phenyl ether), PhOPh. M.p. 21°C, b.p. 250°C. Prepared from NaOPh and PhCl. Used as a heat-transfer agent for perfuming soaps and as a herbicide.

diphenylethyne See TOLAN.

diphenylguanidine, C$_{13}$H$_{13}$N$_3$ PhNHC(NH)NHPh. Colourless needles; m.p. 147°C. Prepared by the action of PbO and ammonia on thiocarbanilide. Extensively used as a rubber accelerator, particularly when mixed with other accelerators such as zinc diethyl-dithiocarbamate.

diphenylpicrylhydrazyl A stable, dark violet-coloured FREE RADICAL; m.p. 137–138°C,

prepared by condensation of 1,1'-diphenylhydrazine with picryl chloride (2,4,6-trinitrochlorobenzene) and oxidation of the resulting orange hydrazine with PbO. Used as a dehydrogenating agent, an inhibitor of radical chain reactions and as an analytical reagent.

diphos Commonly used abbreviation for a diphosphine.

diphosgene, $ClCOCCl_3$. Dimer of phosgene prepared by photochlorination of methyl methanoate. B.p. 128°C. Used in synthesis.

diphosphines An important group of chelating agents including $Ph_2PCH_2CH_2PPh_2$, $Me_2PCH_2CH_2PMe_2$, etc. Prepared from R_2PNa and dihalides. Abbreviated as diphos. Used to prepare complexes as catalysts, particularly for hydrogenation.

diphosphopyridine nucleotide (DPN) See NICOTINAMIDE ADENINE DINUCLEOTIDE.

dipolar ions (zwitterions) Species formed internally, e.g. by transfer of H^+ from a carboxyl group of an amino acid to the nitrogen.

dipole moment $\rho(\mu)$ In a heteronuclear diatomic molecule, because of the difference in electronegativities of the two atoms, one atom acquires a small positive charge (q_+), the other a small negative charge (q_-). The molecule is then said to have a *dipole moment* whose magnitude $\mu = qd$, where d is the distance of separation of the charges. With polyatomic molecules the net dipole moment is the vector sum of the dipole moments of the individual bonds within the molecule. Thus symmetrical molecules, e.g. CCl_4, may contain polar bonds but possess no net dipole moment. Measurements of dipole moments (see DIELECTRIC CONSTANT) may be used to give information about the structure of complex molecules. Unit Debye D. A dipole may induce a further dipole in a neighbouring molecule and hence lead to interaction.

di-n-propyl ketone See BUTYRONE.

dipterex (dimethyl(1-hydroxy-2,2,2-trichloroethyl)phosphonate), $C_4H_8Cl_3O_4P$, $(MeO)_2P(O)CH(OH)CCl_3$. An insecticide.

dipyridyl, $C_{10}H_8N_2$. Prepared by the action of

2,2'dipyridyl   4,4'dipyridyl

sodium on pyridine with subsequent oxidation by air of the disodium salt formed. A mixture of isomers (2,2'-, 4,4'-, etc.) is formed. The quaternized 4,4' derivative (PARAQUAT) and 2,2' derivative (DIQUAT) are important HERBICIDES. The 2,2' isomer (bipy) is an important chelating agent; m.p. 70–73°C, b.p. 273°C.

Diquat (ethylene dipyridylium bromide), $C_{12}H_{12}Br_2N_2$. Herbicide which is rendered inactive on contact with the soil.

See also PARAQUAT.

diradical A molecule possessing two separate free electrons, e.g.

$$(C_6H_5)_2\dot{C}.C_6H_4.[CH_2]_4.C_6H_4.\dot{C}(C_6H_5)_2$$

direct dyes Water soluble dyestuffs, generally azo dyes although some contain carboxyl groups. Dyeing is direct from a solution of the dyestuff.

disaccharides Sugars derived from monosaccharides by the elimination of a water molecule from two monosaccharides.

disazo dyestuffs Azo dyestuffs with two $-N_2-$ groups.

discharge tube A vessel, usually glass, which contains two metal electrodes and in which the passage of electricity occurs through a gas or vapour (usually at low pressure).

disidiolide Polycyclic material that shows inhibition of protein phosphatase so that cells are no longer able to divide. Originally produced by a marine sponge.

disinfectants Materials which destroy microorganisms but not spores. Chlorine and hypochlorites (chlorates(I)) are examples.

dislocation A dislocation is normally a line or skew defect in a crystal lattice although many other types are known.

dismutation reaction A particular type of metathesis reaction, e.g. $2CH_2=CHR$ giving $CH_2=CH_2$ and $CHR=CHR$. This reaction often occurs in the presence of catalysts.

disperse dyes A class of water-insoluble dyes generally used from an aqueous suspension, the dyestuff having a high affinity for the fibre. Used particularly for nylons and synthetic fibres. The main types of disperse dyestuffs are anthraquinone and aminomonoazo compounds.

disperse phase A colloid.

dispersing agent A substance used in the production of emulsions or dispersions of immiscible liquids or liquids and solids. The dispersing agent may lower the interfacial surface tension (surface tension depressant) or increase the viscosity of the continuous phase (protective colloid).

Dispersing agents, such as polyethylene polyamide succinimides or methacrylate-type copolymers, are added to motor oils to disperse 'low-temperature sludge' formed in spark-ignition engines.

dispersion A two-phase system which is uniform on the macroscopic but not on the microscopic scale.

dispersion interaction See LONDON INTERACTION.

disproportionation A process in which a compound of one oxidation state changes to compounds of two or more oxidation states, e.g. $2Cu^+ \rightarrow Cu + Cu^{2+}$, in aqueous solution. Alternatively a redistribution of groups around a central atom, e.g.

$$2PF_4Cl \rightarrow PF_3Cl_2 + PF_5.$$

dissociation The process whereby a molecule is split into simpler fragments which may be smaller molecules, atoms, free radicals or ions. In gaseous molecules dissociation is detected by loss of vibrational structure although predissociation may distort the pattern. See BIRGE–SPONER PLOT. Particularly used in connection with photochemical, thermal and ionic (electrolytic) dissociation.

dissociation constant The measure of the extent of dissociation. The extent of dissociation is measured by the dissociation constant, K. For the process $AB \rightleftharpoons A + B$

$$K = \frac{[A][B]}{[AB]}$$

where the square brackets denote the concentrations (activities) of the species.

dissociation energy (D_e) The difference between the total energy of the dissociated species and the original species. See also BOND ENERGY.

dissymmetric See CHIRAL.

distillation The separation of a liquid from a solid or another liquid by vaporization followed by condensation. The distillation can be carried out at atmospheric or reduced pressure. Volatile material can be distilled from low-temperature baths. See FRACTIONAL DISTILLATION and RECTIFICATION.

distillation column A rectifying or fractionation column. See RECTIFICATION.

distribution law See PARTITION LAW.

disulfoton Insecticide and acaricide. Organophosphorus derivative.

disulphiram (tetraethylthiouram disulphide), $C_{10}H_{20}N_2S_4$, $Et_2NC(S)S\cdot SC(S)NEt_2$. Drug used in treating chronic alcoholism.

diterpene Unsaturated hydrocarbons, formula $C_{20}H_{32}$. May contain one or more closed carbon rings. The term is also applied to their simpler derivatives. They are mostly vegetable products. ABIETIC ACID is a diterpene derivative.

dithianone Fungicide. Dinitrile derivative.

dithiocarbamates, $M^+(R_2NCS_2)^-$. Prepared by the action of amines on CS_2 in the presence of, e.g., NaOH. Form complexes with metals (S-bonded, used in analysis, e.g. Cu^{2+}). Transition metal salts are used as fungicides and Zn salts are used in rubber vulcanization. Oxidized to THIURAM DISULPHIDES.

dithioglycerol (BAL, dimercaprol), $C_3H_8OS_2$, $CH_2SH\cdot CHSH\cdot CH_2OH$. Usually obtained as an oil, m.p. 77°C. Used in detoxication for Hg, Cu, Zn, Cd but not Pb. It acts by forming a chelate with the metal.

dithiolene ligands SCR·CRS groups chelated to metals. Known with many different R (e.g. CF_3, CN, $\frac{1}{2}C_6H_4$), the two sulphur atoms are at the correct positions for chelation. Dithiolene complexes frequently have planar and trigonal-prismatic (rather than octahedral) geometry and the complexes readily undergo oxidation and reduction, sometimes forming extended π-systems.

dithionous acid (hydrosulphurous acid, hyposulphurous acid), $H_2S_2O_4$. An acid known only in solution, unstable and a powerful reducing agent. Salts prepared by reducing sulphites. Dithionites contain the $(O_2S-SO_2)^{2-}$ ion. Eclipsed structure.

dithiothreitol (threo-2,3-di-hydroxy-1,4-dithiolbutane), $C_4H_{10}O_2S_2$, $HSCH_2CH(OH)CH(OH)CH_2SH$. A useful water-soluble reagent for protecting thiols.

dithizone (diphenylthiocarbazone), $C_{13}H_{12}N_4S$. M.p. 165–169°C. A blue-black solid. Prepared from phenylhydrazine and CS_2. Used in chloroform solution as an extraction agent for most heavy metals, particularly for the extraction and estimation of Pb.

diuron Herbicide. Urea derivative.

diverse salt effect Effect of a salt without common ions on precipitation because of the effect of the ions on the activity of the precipitating ions.

divinyl ether, C_4H_6O, $(CH_2=CH)_2O$. Colourless liquid; b.p. 28°C. Prepared from $(ClCH_2CH_2)_2O$ with fused KOH in an NH_3 atmosphere or in ethylene glycol at over 200°C. Readily oxidized by air. Slowly polymerizes to a jelly.

DME DROPPING-MERCURY ELECTRODE.

DMF, $Me_2NC(O)H$. DIMETHYL FORMAMIDE.

DMFC Direct methanol fuel cell.

DMSO, Me_2SO. Dimethyl sulphoxide.

DMT Dimethyl terephthalate. See PHTHALIC ESTERS.

DNA See NUCLEIC ACIDS.

DNBP (dinoseb, 2-*sec*-butyl-4,6-dinitrophenol), $C_{10}H_{12}N_2O$. Yellow solid; m.p. 42°C. A powerful insecticide and herbicide.

DNOC See 4,6-DINITRO-O-CRESOL. Insecticide, acaricide and herbicide.

DOAS (differential optical absorption spectroscopy) Analytical technique used particularly on gases, e.g. pollution measurement.

***cis*-13-docosenoic acid** See ERUCIC ACID.

dodecahedral co-ordination Co-ordination by eight ligands at the corner of a dodecahedron. Found in the $[Mo(CN)_8]^{4-}$ ion. A dodecahedron has twelve triangular faces and consists of

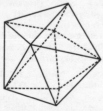

two pentagonal pyramids sharing four corners with two faces at the open edges.

dodecanedioic acid (1,10-decanedioic acid), $HO(O)C(CH_2)_{10}C(O)OH$. Plasticizer.

dodecanoic acid (lauric acid, dodecyclic acid), $C_{12}H_{24}O_2$ $CH_3.[CH_2]_{10}.C(O)OH$. Needles, m.p. 44°C, b.p. 225°C/100 mm. A fatty acid occurring as glycerides in milk, spermaceti, laurel oil, coconut oil, palm oil and other vegetable oils. The metal salts are used in detergents and shampoos. See METALLIC SOAPS.

1-dodecanol (dodecyl alcohol, lauryl alcohol), $CH_3(CH_2)_{11}OH$. Used as a wetting agent and in detergents. Prepared from C_2H_4 with $AlEt_3$ followed by hydrolysis.

dodecylphenol, $C_{18}H_{30}O$. Made from phenol and propene tetramer. Reacts with ethylene oxides to give detergents and surfactants.

dodemorph Fungicide. Morpholine derivative.

dodine Fungicide. Quinidine derivative.

Doebner–Miller reaction (αβ unsaturated carbonyl compounds) Condensation of an aromatic amine with an aldehyde or ketone in the presence of hydrochloric acid to form a quinoline derivative. A general method.

dolerite A widely distributed igneous rock composed essentially of plagioclase feldspar and augite. Used for road metal.

dolichols See POLYPRENOLS.

dolomite, $MgCO_3.CaCO_3$. The term is loosely used for all mixtures of the two carbonates in the very common 'dolomite rock' or magnesian limestone, which is used for building purposes and for the basic linings of Bessemer converters and open-hearth steel furnaces. It is one of the most important raw materials for the production of magnesium and its salts.

Donnan membrane equilibrium The distribution of ions on each side of a membrane separating two portions of a solution of an electrolyte, e.g. NaCl, in water; on one side of the membrane a polyelectrolyte, e.g. a protein, the molecules and ions of which cannot pass through the membrane. The distribution of the NaCl on the two sides of the membrane will be unequal and a membrane potential will be established. Such systems are of importance in biology.

donor See CO-ORDINATE BOND.

dopa See 3,4-DIHYDROXYPHENYLALANINE.

dopamine (4-(2-aminoethyl)-1,2-benzenediol) Anti-hypotensive and cardiotonic.

dopant Foreign species implanted into a lattice. See DOPING.

doping The name given to the practice whereby electrical, magnetic and other properties of a solid may be changed by incorporation of impurities in the host lattice. E.g. the incorporation of small amounts of Li^+ or Ga^{3+} into the nickel oxide lattice results in a respective decrease or increase in conductivity. Incorporation of e.g. B in Si forms useful SEMI-CONDUCTORS. When the dopant is not incorporated directly into the lattice, but rather forms discrete aggregations within it, the process is referred to as domain doping.

d orbitals Atomic orbitals of azimuthal quantum number 2.

double bond Atoms can share two pairs of electrons forming a double bond (two covalent bonds). Ethene has a double bond. A double bond

is stronger than a single bond although more reactive. Formally a double bond generally arises from overlap of p orbitals of two atoms which are already united by overlap of s orbitals. See PI BOND, SIGMA BOND. The atoms in such an arrangement are planar.

double layer See ELECTRICAL DOUBLE LAYER.

double refraction Having different refractive indices in different crystallographic directions.

double salt A salt formed by crystallization from a solution of two or more components that has distinct properties in the solid but in solution behaves as a mixture of the components. E.g. IRON(II) AMMONIUM SULPHATE, $FeSO_4.(NH_4)_2SO_4.6H_2O$. Double salts are either weak, easily dissociated complexes or the solid lattice has an extensive interaction, e.g. hydrogen bonding, not found in the components.

doublet Term used in n.m.r. spectroscopy to denote a signal split into an equal doublet by coupling to a further nucleus with spin ½. Also used in group theory to denote particular energy states, e.g. with s = 2.

Dow process The process for the extraction of magnesium from sea water by precipitation of $Mg(OH)_2$ by $Ca(OH)_2$ followed by solution of the hydroxide in hydrochloric acid.

dowtherm Mixtures of high-boiling organic substances used for purposes of heat exchange.

DPX-JW062, DPX-MP-062 Insecticides, oxadiazine derivatives.

drier (dryer) Plant or equipment for drying a material. Also METALLIC SOAPS and solvents added to paint, solvents and varnishes to hasten drying.

Drikold A commercial name for solid CO_2.

drilling fluids Fluids pumped through a drill bit to cool the bit, pick up cuttings and seal the wall of the borehole. Fluids vary from dirty water to very viscous solutions. Additives include, e.g, barytes (to increase density), clays, particularly bentonite (to increase viscosity), phosphates, tannins or lignite material (to reduce viscosity) and cotton seed hulls, mica or wood fibres (to seal the borehole).

droplet A small amount of liquid surrounded by its vapour and other gases.

dropping-mercury electrode (DME) An electrode consisting of a column of mercury passing through a fine capillary and emerging in the solution as fine drops to give a continuously fresh surface. Used in POLAROGRAPHY.

drug Any chemical agent that acts on biological systems or functions. Drugs can be medicines (i.e. helpful) or harmful (e.g. poisons). The term drug is also used for a substance being abused (e.g. cocaine) or used for what are considered as improper uses (e.g. muscle building). For specific information see BRITISH PHARMACOPOEIA. Drugs for medical use are categorized as to whether they are authorized for human use, have to be prescribed medically, or are freely available.

dry cleaning The process of cleaning fabrics using solvents which dissolve dirt at low temperatures but do not swell fibres. The most widely used cleaning fluids are the highly inflammable hydrocarbons and chlorinated hydrocarbons, particularly C_2Cl_4, but chlorinated species are being phased out and liquid CO_2 being used.

dry ice Solid CO_2.

drying Removal of small quantities of water from a sample by physical or chemical means. Removal of the water from a solid may be speeded up by warming the sample. Drying may also refer to the removal of solvents other than water. See also FREEZE-DRYING.

drying equipment Gases can be dried using equipment such as ABSORPTION columns, while liquids may be dried by distillation, etc.

For solids various equipment and techniques are used.

drying oils Liquids which are oxidized by air to dry, hard resins. The oils consist mainly of unsaturated triglycerides (linseed, tung, safflower oils, dehydrated castor oil (DCO) and also of unsaturated hydrocarbon polymers, e.g. polybutadiene $(-CH_2CH=CHCH_2-)_n$. Used in coatings, paints, varnishes, enamels and lacquers. Oxidation is photochemical in nature and involves formation of C–O–C cross linkages.

dry point In a laboratory distillation the thermometer reading observed when the last drop of liquid evaporates from the lowest point in the flask.

DSC Differential scanning calorimetry. See THERMAL ANALYSIS.

DTA Differential thermal analysis. See THERMAL ANALYSIS.

dubnium, Db. Element, at. no. 105. An artificial post-actimide element of Group 12 showing some chemical similarities to Nb and Ta.

dulcin (sucrol, 4-ethoxyphenylurea), $C_9H_{12}N_2O_2$. Colourless crystals; m.p. 171–172°C. Used as a sweetening agent.

dulcitol (galactitol), $C_6H_{14}O_6$. The alcohol from galactose, m.p. 198°C, of wide occurrence in plants. Can occur in the eye in cataracts.

Dulong and Petit's law The product of the atomic mass and the specific heat of a metal is of constant values approximately 6.2. Although not true for all metals at low temperatures, the exceptions and several non-metals approximate to the law at high temperatures. The molar heat capacity is approximately $25 JK^{-1}$ ($3R$ where R is the gas constant).

Dumas' method A method for determining vapour density by determining the weight of a known volume of vapour.

Also a method for determining the nitrogen content of a compound by oxidizing the compound with copper oxide and then reducing the combined nitrogen with copper to nitrogen gas, which can be measured.

duralumin An important aluminium alloy.

durene (1,2,4,5-tetramethylbenzene), C_9H_{14}. Colourless solid from coal tar, etc.; m.p. 79°C. Forms a quinone, duroquinone, a phenol, durenol acid, etc. Oxidized to pyromellitic dianhydride.

duriron An iron alloy containing Si (14%), Mn (2%), C (1%), S (0.1%). Particularly resistant to acids and used in chemical plant.

Dy DYSPROSIUM.

dyestuffs Intensely coloured compounds applied to a substrate, e.g. fibre, paper, cosmetic, hair, to give colour. React by absorption, solution, bonding. PIGMENTS retain their identity more closely on bonding to the substrate. Colours generally originate in electronic transitions; most dyestuffs are organic in nature but are frequently applied together with inorganic species. Classified in the COLOUR INDEX by chemical nature; also classified by their method of application and use: disperse (polyester, acetate); reactive (cotton, viscose); acid (nylon, wool, silk); direct (cotton, viscose); cationic (basic) (acrylics). World consumption 2001, 1.18 million tonnes.

dynamite A mixture of nitroglycerine with other substances to make the explosive safer to handle and to have specific explosive properties.

dysprosium, Dy. Lanthanide metal, at. no. 66, at. mass 162.50, m.p. 1412°C, b.p. 2567°C, density ρ 8550, crustal abundance 6 p.p.m., electronic configuration [Xe] $4f^{10}6s^2$. The metal is hcp. A Dy metal foil is used to measure neutron fluxes. Dy compounds are used in lasers, phosphors and in reactor control rods.

dysprosium compounds Most compounds are in the +3 state Dy^{3+} (f^9 yellow green) → Dy (−2.35 volts in acid) and are typical lanthanide compounds. Lower halides, e.g. DyX_2, are formed by DyX_3 plus Dy – they contain M–M bonds. Cs_3DyF_7 containing Dy(IV) is obtained using fluorine. Dicyclopentadienyldysprosium compounds reduce N_2.

E Description of stereochemistry about double bonds based on group priorities. Compare Z. See ISOMERISM.

E₁ reaction Unimolecular reaction with rate-determining step loss of leaving group.

E₂ reaction Bimolecular elimination in which base removes H^+ and leaving group departs simultaneously.

earth See ALKALINE EARTH METALS.

ebonite (hard rubber, vulcanite) A hard, black, inert material made by vulcanizing mixtures with a rubber:sulphur mixture near 2:1 and with less than 4% of the sulphur uncombined.

ebullioscopic constants See BOILING POINT, ELEVATION OF.

ebullioscopy Determination of molecular weight from rise in boiling point of solution.

ecdysone See ECDYSTEROIDS.

ecdysteroids Hormones involved in development processes in insects, marine animals and plants.

ecgonine (3-hydroxy-2-tropanecarboxylic acid), $C_9H_{15}NO_2$ Colourless

crystalline material; m.p. 203°C. The major portion of the cocaine molecule, from which it may be obtained by hydrolysis with acid. Benzoylation and methylation reconvert it to cocaine. Forms a stable hydrochloride, m.p. 246°C. See COCAINE.

eclipsed See CONFORMATION.

ecstasy (MDMA, 3,4-methylenedi-oxymethamphetamine) Used as a recreational drug but may cause permanent damage to neural cells.

Edeleanu process An extraction process utilizing liquid sulphur dioxide for the removal of aromatic hydrocarbons and polar molecules from petroleum fractions.

edifenphos Fungicide, organophosphorus derivative.

Edman degradation The sequential release of amino-acids from proteins.

EDTA ETHYLENEDIAMINETETRA-ACETIC ACID.

EDXRF See X-RAY FLUORESCENCE.

EELS (HREELS) Electron energy loss spectroscopy. Technique in which a beam of electrons is reflected from a surface. Can be used to identify down to individual atoms.

EFA ESSENTIAL FATTY ACIDS.

effective atomic number rule Applied to transition metal carbonyls and to many organo-metallic derivatives. It is frequently found that the total number of electrons available to the central atom from its own electrons, by donation (lone pairs, pairs, etc. of electrons from π-bonding systems), by covalent bonding (each covalent bond to a substituent or another metal contributes 1 electron), and taking into account charge, adds up to the total number of electrons of the next noble gas. Does not apply to paramagnetic species and there are many other exceptions to the rule. Example $ClMn(CO)_5$·Mn has 7 valence electrons, 1 from covalent bond to Cl, 5×2 electrons from carbonyls. Total 18 electrons, krypton structure.

efflorescence The loss of water by a salt hydrate arising because the pressure of water over the hydrate is greater than the partial pressure of water in the atmosphere.

effusion The passage of molecules of a gas through an orifice which has a diameter smaller than the mean free path of the gas molecules. The rate of effusion is proportional to the area of the orifice and to the mean velocity of the molecules (and hence to the reciprocal of the square root of the relative molecular mass of the gas). Used experimentally for investigating low vapour pressures and the molecular weight of gaseous species (e.g. in high-temperature equilibria); the apparatus used is called a Knudsen cell and the cavity has a small hole.

Ehrlich's reagent, $4\text{-Me}_2NC_6H_4CHO$. Prepared by formylation of $Me_2NC_6H_5$ with DMF. Used in synthesis and dyestuffs.

Eigenfunction In wave mechanics, the Schrödinger equation may be written using the Hamiltonian operator H as

$$H\psi = E\psi$$

Only certain energy values (E) will lead to solutions of this equation. The corresponding values of the wave functions are called Eigenfunctions or characteristic wave functions. Refers to a class of problems in physics and chemistry that take the form $\Omega\varphi = \lambda\varphi$ where Ω is a mathematical operator, φ the Eigenfunction and λ the Eigenvalue – an observable quantity. See DICTIONARY OF PHYSICS.

einsteinium, Es. Artificial actinide, at. no. 99, at. mass^{252}Es 252.08, m.p. 860°C, electronic configuration [Rn] $Sf^{11}7s^2$. ^{253}Es (20 days) may be formed by multiple neutron irradiation of Am, Pu and Cm; it is purified by ion exchange chromatography. Has no uses at present.

einsteinium compounds The element shows oxidation states of +2, +3 and +4, the +3 state being that of a typical tripositive actinide, EsO_2, EsO_3, $EsOCl$ and EsI_2 are known. $EsBr_3$ is reduced to $EsBr_2$ by hydrogen.

Einstein's law of photochemical equivalence Before a chemical reaction can be induced by light, the light must be absorbed by some of the reactant molecules. Einstein proposed that each reacting molecule must be excited by absorption of one quantum of light. This principle is not easy to test, however, since secondary reactions occur after the primary reaction and activated species may lose energy by processes such as fluorescence.

ejector A simple non-mechanical device used for the pumping of liquids and gases.

elaidic acid (E-9-octadecenoic), $C_{18}H_{34}O_2$, $CH_3\cdot[CH_2]_7\cdot CH\cdot CH\cdot[CH_2]_7\cdot C(O)OH$. Crystallizes in plates; m.p. 46.5°C. The *trans*-isomer of oleic acid, and is prepared from it by treatment with isomerizing agents, e.g. nitric acid.

elastane U.S. use spandex. Polymeric urethane with elastic properties.

elastase A proteinase from mammalian pancreas which hydrolyses elastin.

elastin The protein of elastic tissue, ligaments and arterial walls.

elastomers Materials with rubbery properties, i.e. capable of rapid recovery from large deformation. The internal energy of a perfect elastomer is independent of the extension. Generally synthetic materials. Useful elastomers include rubber, synthetic rubbers (poly-*cis*-1-4-isoprene and polybutadienes) and such materials as polychloroprene and butadiene copolymers.

electrical double layer An electrical double layer is generally formed at the interface between two phases or at the surface of colloid particles. One phase acquires a net positive charge and the other a net negative charge. Thus in simple terms a negatively charged dropping mercury electrode would be surrounded by an ionic sheath, in which the proportion of positive ions decreases and that of negative ions increases with increasing distance from the interface.

electric cells See BATTERIES.

electrides Compounds containing trapped electrons (formally M^-) from Group 1 elements and ethers, crown ethers or cryptanols.

electrochemical equivalent The weight of an element liberated from its ions or converted into them by unit quantity (1 coulomb) of electricity.

electrochemical series A series in order of activity of the elements. Dependent on ionization energy, solvation (hydration) energy, etc. The higher the element is in the series the greater the tendency for it to be oxidized. Similarly if a metal is added to a solution of another metal ion, which is below it in the electrochemical series, the metal ion will be displaced, e.g. zinc metal will displace copper from solutions of copper salts. The following series shows some of the more common elements in order (oxidation potentials in volts in brackets):

Li	(3.04)	Na	(2.71)	H_2	(0.00)
Cs	(2.92)	Mg	(2.38)	Cu	(−0.34)
Rb	(2.92)	H^-	(2.23)	I^-	(−0.54)
K	(2.92)	Al	(1.71)	Ag	(−0.80)
Ba	(2.90)	Mn	(1.03)	Au	(−1.42)
Sr	(2.89)	Zn	(0.76)	F^-	(−2.87)
Ca	(2.76)	Cr	(0.74)		

electrochemistry The branch of chemistry concerned with electrolysis and other similar phenomena occurring when a current is passed through a solution of an electrolyte, or concerned with the behaviour of ions in solution and the properties shown by these solutions.

electrochromatography See ELECTRO-PHORESIS.

electrocyclic reaction A term used to describe a highly stereospecific type of multicentre reaction. Reactions can involve highly specific ring opening or ring closure, e.g. *cis*-3,4-dichlorocyclobutene gives (1E, 4Z)-dichloro-1,3-butadiene. The reactions can be explained by use of Frontier orbital theory.

electrode An electro-conducting body which, when placed into an electrolyte, exhibits a certain electrical potential with respect to the bulk of the electrolyte. The material(s) used to introduce an electrical potential to a solid or solution.

electrode potential An electrode of any element in contact with a solution of its ions exhibits a potential called the electrode potential, generally in water but can apply in any solvent. The magnitude of the potential depends upon both the element and the concentration of the solution. Under conditions when no net current flows between the electrode and the solution the system is said to behave reversibly and the potential is termed the reversible electrode potential. For an element in contact with a solution of its ions at unit activity the electrode potential measured relative to the hydrogen electrode is termed the standard electrode potential ($E°$). The sign given to the electrode potential is arbitrary. Values are usually quoted as 'reduction' potentials, i.e. the electrode potential for the process $M^{n+} \rightarrow M$. A positive electrode potential indicates that the reduction process tends to occur spontaneously. A negative oxidation potential for $M^{n+} \rightarrow M$ indicates that the process $M \rightarrow M^{n+}$ tends to occur spontaneously.

If an appreciable current flows between the electrode and the solution, thus disturbing the reversible thermodynamic equilibrium conditions, the electrode is said to be polarized and the system is then operating under irreversible conditions.

electrodialysis In the DIALYSIS of colloidal sols containing electrolytes, the removal of the latter is facilitated by means of an electric field. The colloidal solution containing the unwanted electrolyte is placed between two dialysing membranes, with pure water on the other sides of the membranes. Electrodes placed in the water compartments are connected to a source of direct current. The applied e.m.f. accelerates the diffusion of the ions through the membrane and the purification of the sol becomes more rapid.

electrodispersion When an arc is struck between two metal electrodes under the surface of a liquid, particles are torn off and some remain colloidally dispersed in the liquid. A more refined electrodispersion process is obtained if the arc is maintained within a silica tube and the vaporized metal blown through a small hole in front of the arc into the dispersion medium using a stream of inert gas.

electrokinetic potential (zeta (ζ) potential) The potential difference across the diffuse part of a double layer, i.e. between the rigid solution layer and the mobile part of the solution adjacent to the bulk solution. See ELECTRICAL DOUBLE LAYER.

electrokinetics Phenomena in systems with electrical double layers at the interface between a solid and a liquid (generally aqueous) or between one liquid and another. Effects include electrophoresis, electro-osmosis, streaming potential and sedimentation potential. They all arise from a partial separation of the fixed and mobile parts of the electrical double layer.

electrolysis The process of decomposing or changing a substance, usually in solution or as a melt, by the passage of an electric current.

electrolysis, laws of See FARADAY'S LAWS OF ELECTROLYSIS.

electrolyte A solid substance which undergoes partial or complete dissociation into ions in solution, and thus acts as a conductor of electricity. Electrolytes with DISSOCIATION CONSTANTS greater than about 10^{-2} are called strong electrolytes.

electrolyte dissociation The process of the formation of ions in solution from an added solute. In the case of solid solutes with an ionic lattice (e.g. NaCl) there is complete dissociation into ions in dilute solution and conductivity is virtually independent of concentration. For other types of solute (e.g. ethanoic acid) there is an equilibrium between the ions and the undissociated molecules.

electrolytic oxidation An oxidation process effected by electrolysis.

electrolytic reduction A reduction process effected by electrolysis.

electromagnetic radiation Oscillating electric and magnetic disturbances are propagated through space as waves of electromagnetic radiation. The different forms of electromagnetic radiation, which have different wavelengths and energies, include visible, infra-red and ultra-violet radiation, radio waves, X-rays and γ-rays. All forms of electromagnetic radiation are propagated through a vacuum with the same velocity of 2.98×10^8 m sec^{-1}. Electromagnetic radiation interacts with molecules according to the frequency (energy of the radiation) u.v. – visible – near i.r. (electronic transitions), i.r. (vibrations), microwave (rotations), radio frequencies (nuclear spins), X-rays (removal of electrons).

electrometric titration Analytical method for following titrations by observing the e.m.f. of an inert electrode immersed in the solution. Of particular use in titrating coloured solutions.

electromotive force (e.m.f.) The zero-current potential of a cell balanced by an opposing potential so that no net reaction occurs.

electron Chemically the electron is the ultimate, indivisible negative charge, or particle of negative electricity, and forms an integral part of every atom. The electron has a mass of $^1/_{1837}$ that of a proton. Charge 1.602×10^{-9}c, rest mass m_e 9.109×10^{-31} kg.

electron affinity (E_{EA}) The energy released when an electron is added to a neutral atom or ion in the gaseous state, i.e.

$$A(g) + e \rightarrow A^-(g) + energy$$

Electron affinities may be estimated using a Born–Haber cycle.

electron capture See ORBITAL ELECTRON CAPTURE.

electron compounds Compounds formed by transition metals and one of the main group elements in which the structure of particular phases depends on the ratio of the total valency electrons available to the total number of atoms in the simplest empirical formula (making as-sumptions such that Fe, Co, Ni group elements have 0, Cu group 1 valence electrons). The ratios are 3:2 (β-phase bcc), 21:13 (γ-phase complex), 7:4 (ε-phase, hcp).

electron decay See BETA DECAY.

electron-deficient compounds Compounds in which a filled-shell configuration cannot be written for each atom if it is assumed that each bond between two atoms contains two electrons. The bonding in these compounds can be rationalized in terms of multi-centre bonds in which two electrons may bind many atoms. See BORON HYDRIDES.

electron density In the investigation of crystal structure by X-ray diffraction, the electron density distribution which is calculated is mapped out. Using such data, structures and bond lengths may be obtained.

In theories of bonding the term is often used to indicate the probability of finding an electron at a particular point.

electron diffraction A beam of electrons interacting with atoms or molecules is diffracted. Electron diffraction is used to study molecular structure, particularly in the gas phase and the investigation of surfaces (low-energy electron diffraction).

electron hole pair Surface electron with positive hole in lattice. Important in surface reactions.

electronegativity (χ) A measure of the tendency of an atom in a stable molecule to attract electrons within bonds. Thus in covalently bonded hydrogen chloride, HCl, the electrons are more associated with the chlorine than the hydrogen; the chlorine being the more electronegative atom. Numerical values of electronegativity have been calculated by Pauling from bond dissociation energies, ionization energies and electron affinity values. In general, electronegativity values increase from left to right across each row of the periodic table and decrease down each group. Representative values are:

H 3.1
Li 1.28 Be 1.99 B 1.83 C 2.67
         N 3.08   O 3.22 F 4.43

electron exchange ELECTRON TRANSFER without chemical change.

electronic configuration (See entry against each element.)
1 Atoms. The electronic configuration of an atom is the basis of the properties of the atom: the chemical properties are determined by the

electronic configuration of the outer shells. Each electron has four characteristic quantum numbers which are interrelated:

a. Principal quantum number, represented by n and having values 1, 2, 3 ...; this denotes the overall energy and size of the electron orbital.

b. The azimuthal quantum number, represented by l and having values 0, 1, 2 ... to $n-1$. This denotes the angular momentum and thus shape of the most probable electron distribution corresponding to the shape of the orbital. Azimuthal quantum numbers with $l = 0, 1, 2$ and 3 are also called s, p, d and f respectively.

c. A magnetic quantum number, represented by m, and having values $-l, -l+1, ... 0 ... l-1, l$. This denotes the orientation of the orbital.

d. A spin quantum number $s = \pm \frac{1}{2}$.

An s orbital is spherically symmetrical and can contain a maximum of two electrons with opposed spins. A p orbital has a solid figure-of-eight shape; there are three equivalent p orbitals for each principal quantum number; they correspond to the three axes of rectangular coordinates.

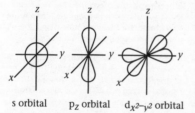

s orbital p_z orbital $d_{x^2-y^2}$ orbital

The d and f orbitals have more complex shapes; there are five equivalent d orbitals and seven equivalent f orbitals for each principal quantum number, each orbital containing a maximum of 2 electrons with opposed spins.

The above definitions must be qualified by stating that for principal quantum number 1 there are only s orbitals; for principal quantum number 2 there are only s and p orbitals; for principal quantum number 3 there are only s, p and d orbitals; for higher principal quantum numbers there are s, p, d and f orbitals.

To arrive at the electronic configuration of an atom the appropriate number of electrons are placed in the orbitals in order of energy, the orbitals of lower energy being filled first (AUFBAU PRINCIPLE), subject to the proviso that for a set of equivalent orbitals – say the three p orbitals in a set – the electrons are placed one in each orbital until all the orbitals are half filled (HUND'S RULE). The order of energy of orbitals for most

nuclei is 1s < 2s < 2p < 3s < 3p < 4s < 3d < 4p < 5s < 4d < 5p < 6s < 4f < 5d < 6p < 7s.... In any atom no two electrons may have all quantum numbers the same (PAULI EXCLUSION PRINCIPLE).

2 Molecules. The electronic configurations of molecules can be built up by direct addition of atomic orbitals (LCAO method) or by considering molecular orbitals which occupy all of the space around the atoms of the molecule (molecular orbital method). Electrons are placed in these orbitals in order of energy.

electronic transition In an atom or molecule the electrons have certain allowed energies only (orbitals). If an electron passes from one orbital to another an electronic transition occurs and there is emission or absorption of energy corresponding to the difference in energy of the two orbitals. Observed transitions may be modified by vibrational and rotational transitions.

electron microscopy The use of a beam of electrons to observe fine structure. Defects and surface detail can be observed. Under high resolution atoms and molecules can be observed.

electron pair A pair of electrons in one orbital with opposed spins; a LONE PAIR. Often the pair of electrons in a covalent or co-ordinate bond.

electron probe microanalysis (EPMA) Analytical technique.

electron spin Properties of electrons can be explained only in terms of the electrons having spin, $s = \pm \frac{1}{2}$.

electron spin resonance (esr, electron paramagnetic resonance, epr) In the absence of external fields, all spatial orientations of electron spin are equally probable. On the application of a magnetic field, however, the spins of unpaired electrons are aligned either parallel or antiparallel to the direction of the magnetic field vector. The energy difference between these two states is $g\mu H$ where g is the Landé factor, μ is the Bohr magneton and H is the strength of the applied magnetic field $g = 2.0023$. Boltzmann's law governs the distribution of spins between the two states, and at normal temperatures there is a slight excess in the lower energy state.

In an electron spin resonance spectrometer, transitions between the two states are brought about by the application of the quantum of energy $h\nu$ which is equal to $g\mu H$. The resonance condition is defined when $h\nu = g\mu H$ and this is achieved experimentally by varying H, keeping the frequency (ν) constant. The magnetic field is varied by applying microwave radiation. Free radicals give patterns depending on the number of unpaired electrons and the nuclei present. Esr spectroscopy is used extensively in chemistry in the identification and elucidation of structures of radicals.

electron transfer reactions Reactions which involve transfer of electrons from one group to another, i.e. oxidation and reduction.

electron transport chain The coupled system of enzymes and co-enzymes which is essential to the oxidation–reduction chemistry of aerobic respiration and synthesis. Involved in formation of ATP. See DICTIONARY OF BIOLOGY.

electron volt (eV) The energy acquired by an electron when accelerated through a potential of 1 volt. $1eV \sim 1.6 \times 10^{-19}$ J.

electro-osmosis If a direct current is passed through a tube of liquid which is fitted with two electrodes, one on either side of a diaphragm, the potential difference between the diaphragm and the liquid manifests itself by a movement of the liquid towards one or other of the electrodes. This phenomenon, which is akin to ELECTROPHORESIS, is termed electro-osmosis. It is modified by the presence of acids, bases and salts. The hydrogen-ion concentration at which there is no movement of liquid with respect to the membrane is termed the isoelectric point. See COLLOID.

electrophilic reagents Reagents which acquire electrons or a share in electrons from the reactant molecule; examples are the bromonium ion, Br^+ and the nitronium ion, NO_2^+. The product can frequently only accommodate the extra electrons by undergoing fission. Acidity is a special case of *electrophilicity*, the affinity for external electrons in general.

electrophilic substitution The exchange of an atom or group in a molecule for an entering electrophile according to

$$R - X + E^+ \rightarrow R - E + X^+$$

Nitration, sulphonation and Friedel–Crafts acylation of aromatic compounds (e.g. benzene) are typical examples of electrophilic aromatic substitution.

electrophoresis The migration of charged particles, colloidal particles or ions through a solution under an electric field. Variation of pH can stop movement at the ISOELECTRIC POINT. In electro-osmosis there is a constant flow of liquid relative to a stationary surface. Electrophoresis is used in analysis, particularly in biochemical applications (ionography, capillary electrophoresis, zone electrophoresis, electrochromatography) for both identification and separation. In gel electrophoresis the migration takes place through a gel – generally a cross-linked polyacrylamide gel.

electrophoretic effect The viscous drag due to the ionic atmosphere which reduces the mobility of ions etc.

electrophoretogram The separated species on, e.g., a column after electrophoresis.

electroplating The deposition of metals from solution in the form of a layer on other metals or, e.g., plastics, by passage of an electric current. A metallic article to be plated is made one electrode in a bath containing the other metal as aquo-ions or other complexes. Current density, pH, concentration, etc. all have a very marked effect on the adhesion and texture of the deposited metal. Among metals used for electroplating are Ag, Cr, Ni, Zn.

electropositive elements Elements which tend to form cations or compounds in positive oxidation states. Includes elements in Groups 1 to 13 with the actinides and lanthanides.

electrostatic precipitators Plants for the removal of fine suspended matter from a gas that depend for their action on the ionization of the gas between two highly charged electrodes. The ions so formed attach themselves to the dispersed particles, conferring a charge, with the result that the latter then migrate to the appropriate electrode.

electrovalent bond (polar bond) Bonding in compounds by electrostatic attraction.

electrovalent compounds Compounds in which the major binding force is electrostatic attraction between positive and negative ions. The lattice does not contain discrete molecules and the ions are packed together to occupy space most efficiently (generally close packing of anions). Electrovalent compounds are distinguished from covalent compounds by the conductivity of melts and solutions in polar solvents, low volatility, solubility in polar solvents.

element A substance which cannot be further divided by chemical methods. The basic substances which build up chemical compounds. An element is defined by its atomic number (nuclear charge and electronic configuration).

elementary particles The fundamental particles found in nature, from a chemical viewpoint, e.g. proton, neutron, electron.

elements, abundance of The relative abundance of elements depends on whether measurements are made on the crust, meteorites, core, etc. Crustal or atmospheric abundance p.p.m. are noted against each element.

elements of symmetry The symmetry elements, centres, axes, planes of symmetry, present in a molecule, crystal lattice or crystal. Together with the arrangement of atoms the elements of symmetry spell out the SPACE GROUP of a crystal or the point-group of a molecule.

elevation of boiling point The boiling point of a solvent is raised by the presence of a solute. For small concentrations of the solute the rise in the boiling point is proportional to the number of solute particles present in the solution. The elevation caused by 1 mole of solute in 1 litre of solvent is termed the molecular elevation constant. (See COLLIGATIVE PROPERTIES.)

elimination reaction Simultaneous loss of two groups from a molecule or intermediate.

elixirs Pharmaceutical solutions, frequently containing alcohol, used as sweetening or flavouring agents for drugs.

Ellingham diagram Graphs of Free Energy change with temperature, generally used to describe reduction of oxides.

ellipticity See CIRCULAR DICHROISM.

elutriation The process of separating a material into fractions of various sizes by allowing it to settle against an upward moving stream of fluid, usually air or water.

emanation An obsolete name for radon.

embalming The sanitization and preservation of dead bodies. Methanol is a widely used preservative.

embonic acid See PAMOIC ACID.

emerald The grass-green variety of beryl; contains some Cr^{3+}.

emerald green (Paris green, Schweinfürter green) Basic copper ethanoate arsenates(III). Insecticides.

emery Impure α-Al_2O_3 (corundum) containing some iron oxide (magnetite). Used as an abrasive and polishing agent.

emetine, $C_{29}H_{40}N_2O_4$. Alkaloid extracted from ipecacuanha or obtained by the methylation of cephaeline, another ipecacuanha alkaloid. Powerful emetic.

emission spectroscopy An analytical technique in which atoms in excited states in flames, electric arcs or in plasmas are examined by their emission spectra. Emission spectroscopy is also used to investigate the energy levels of atoms in ground and excited states.

empenthrin Insecticide, pyrethroid.

emulsification The preparation of a suspension, or emulsion, of one liquid in another.

emulsifier
1 Another name for an emulsifying agent.
2 Any machine for producing an emulsion.

emulsifying agent Dilute suspensions of oil in water behave as typical hydrophobic sols, and it is not possible to increase the concentration of oil unless a stabilizing material is added which decreases the interfacial tension. Such substances, known as emulsifying agents, are generally long-chain compounds containing a hydrophilic (carboxyl or sulphonate) group at one end of the molecule; these become orientated at the interface, the hydrophilic end projecting into the water. Large numbers of synthetic organic emulsifying agents have been developed. Many of these are sulphonates or quaternary ammonium compounds. There are also solid emulsifying agents (e.g. carbon black) which possess widely different angles of contact for the two phases.

An emulsifying agent generally produces such an emulsion that the liquid in which it is most soluble forms the external phase. Thus the alkali metal soaps and hydrophilic colloids produce O/W emulsions, oil-soluble resins of the W/O type (see EMULSION).

The commonest emulsifying agents are the soaps, which have the disadvantage of coagulating in hard waters. For edible and pharmaceutical emulsions various gums are used.

emulsion An emulsion is a disperse system (two or more phases) in which both phases are liquids; generally one of the liquids is water or an aqueous solution, and the other an oil or other water-immiscible liquid. With a given pair of liquids (e.g. an oil and water) two distinct types of emulsion are possible according to which forms the dispersion medium. Emulsions in which an oil is dispersed in water are termed oil-in-water (O/W) emulsions; those in which the water is the disperse phase are water-in-oil (W/O) emulsions. It is possible to have any phase ratio in an emulsified system. Thus naturally forming emulsions such as that of the condenser water of an engine may contain only a small fraction of oil, while semi-solid emulsion pastes may contain 95% of oil dispersed in water. In order to stabilize all but the most dilute emulsions, it is necessary to have present a third substance, the EMULSIFYING AGENT, to stabilize the system.

Emulsions have many industrial uses; they are used widely as foods and pharmaceutical preparations, cosmetics, horticultural and insecticide sprays, as oil-bound water paints, as lubricants, for spraying of roads, etc.

emulsion stabilizers Compounds which are weakly adsorbed on the grain surface of a photographic emulsion, where they displace the chemical sensitizer and prevent the appearance of fog.

enamels (vitreous enamel) A glassy coating fused to metal at high temperatures, acting as a protective and decorative surface.

enamines (vinylamines) $R_2N \cdot C = C-$ Those having a hydrogen atom on nitrogen are usually unstable and rearrange to the corresponding

$$>C=C-\underset{R}{\overset{H}{\underset{|}{N}}} \rightleftharpoons -\overset{|}{\underset{|}{C}}-\underset{H}{\overset{|}{C}}=N-R$$

imine. These are usually easily prepared from ketones with an α-hydrogen and a secondary amine, e.g. pyrrolidine, with an acid catalyst. Since many undergo C-alkylation smoothly, and the substituted enamine can be cleaved easily back to a ketone, a facile substitution of the α-position of ketones is possible.

enantiomeric excess (purity) The proportion of particular molecules produced in a reaction.

enantiomers Isomers differing in their configuration at a chiral atom.

enantiomorphic Possessing neither a plane of symmetry nor a centre of symmetry, i.e. chiral. Such a molecule cannot be brought into coincidence with its mirror image. The two enantiomorphs (the molecule and its mirror image) rotate the plane of polarized light in opposite directions. The interrelation between the two isomers is enantiomorphism.

enantioselective Preferential formation of one enantiomer.

enantiotopic The two identical ligands *a* attached to a PROCHIRAL centre in [C*aabc*] are said to be enantiotopic if their separate replacement by a group different from *a*, *b* or *c* gives rise to a pair of enantiomers [C*adbc*], and to be diastereotopic if diastereomers are obtained on replacement. E.g.

Enantiopic ligands at a prochiral centre — enantiomers

Diastereotopic ligands at a prochiral centre — diastereoisomers

enantiotropy The reversible transformation of one form or one allotrope into another at a definite transition temperature, e.g. red HgI_2 ⇌ yellow HgI_2 at 126°C.

enclosure compounds See CLATHRATE COMPOUNDS.

endergonic Having positive free energy change, i.e. needs work to drive reaction. Cf. exergonic.

endo- A prefix used to describe the orientation of atoms or groups with respect to the rest of the molecule, especially with polycyclic molecules, but not restricted to these classes. It indicates

endo-norbornyl chloride

exo-norbornyl chloride

that the substituent is facing the inside (concave side) of the molecule. In certain instances such identification may be doubtful. The stereochemically related isomer is given the prefix *exo-*.

endohedral Cage having an encapsulated heteroatom, e.g. $C_{60}Ar$.

endopeptidases A class of peptide-cleaving enzymes which act at points in the interior of polypeptide chains rather than at the terminal groups. Important examples are trypsin, chymotrypsin and pepsin – all concerned in digestion. See also EXOPEPTIDASES.

ENDOR (electron nuclear double resonance) Examination of local structure by use of radio frequencies in a magnetic field.

endosulfan Insecticide, acaricide, organochlorine derivative.

endothal Herbicide, algicide, plant growth regulator, carboxylic acid derivative.

endothermic reaction A reaction in which heat is absorbed. A spontaneous endothermic reaction must be accompanied by an increase in entropy of the system and surroundings. Increased temperature favours the products.

end point The stage in a titration when the indicator (in its widest sense) undergoes maximum change (colour, etc.) for a small amount of added titrant. The end point corresponds to a definite hydrogen or other ion concentration or redox potential. It does not necessarily correspond to the equivalence point.

enediols Organic compounds containing one carbon–carbon double bond and two hydroxyl groups.

enediynes Natural bacterial products with antibiotic and anti-tumour properties.

energy Basic unit the Joule, energy necessary to push against a force of 1 Newton for 1 metre. $1 erg = 10^{-7} J$. 1 Kilowatt hour $= 3.6 \times 10^6 J$. The calorie is still used in chemical literature. 1 cal = 4.184J.

energy levels Allowed energies of atoms or molecules of solids. Energy taken up by a molecule, say by the absorption of light, may cause an electronic transition; it may increase the energy of rotation of the molecule as a whole, or may increase the energy of vibration of the nuclei of the constituent atoms relative to one another. The changes take place only in definite steps. Every electronic orbital is associated with a specific energy value, or energy level. So the rotational and vibrational states have only certain restricted energy values, or energy levels. Diagrammatically, these are usually represented as a series of horizontal, parallel straight lines with separation proportional to the differences in energy of the various states.

energy of activation See ACTIVATION ENERGY.

enkephalins Pentapeptides, which occur naturally in the brain and have similar analgesic and other properties to MORPHINE, CODEINE.

enols (enolic compounds) The tautomeric forms of certain ketones, particularly those with an α-CH group and containing the >C=C–OH grouping. This grouping is also present in phenols, and enols resemble the phenols in several of their reactions. The hydrogen atom of the –OH group is acidic and is replaceable by sodium and other metals. The copper compounds of 1,3-diketones (acetylacetonates) and β-keto esters are of this type. The sodium compounds react with alkyl and acyl chlorides to give derivatives of the type C(R)C(O)–.

ENT 8184 Insecticide, isoindole derivative.

enthalpy (H) A thermodynamic state function defined as $H = E + PV$, where E is the internal energy and P and V are the pressure and volume respectively. For any process which occurs at constant pressure the heat absorbed or evolved is equal to the enthalpy change if the only work is pressure/volume work, i.e. $\Delta H = \Delta E + P\Delta V$ for a process at constant pressure. In chemical reactions the enthalpy change (ΔH) is related to the changes in free energy (ΔG) and entropy (ΔS) by the equation $\Delta G = \Delta H - T\Delta S$. Enthalpy is usually expressed in units of kilojoules per mole. For representative values see ATKINS or KAYE AND LABY. Enthalpy changes between standard states are designated ΔH°.

Enthalpy change

Activation $\Delta_{act}H$	Formation of activated complex
Atomization $\Delta_{at}H$	Species to gaseoust atoms (often referred to as dissociation)
Combustion $\Delta_c H$	Reaction with excess O_2 at constant volume
Electron gain $\Delta_{eg}H$	Formation of gaseous anions (electron affinity)
Formation $\Delta_f H$	Formation from elements or standard states
Fusion $\Delta_{fus}H$	Solid to liquid
Hydration $\Delta_{hyd}H$	Solvated aqueous ions from gaseous ions
Ionization $\Delta_{ion}H$	Formation of gaseous cations from atoms (ionization energy-potential)
Lattice $\Delta_{lat}H$	Formation of gaseous ions from solid
Mixing $\Delta_{mix}H$	Formation of mixtures
Reaction $\Delta_r H$	Formation of product
Solution $\Delta_{sol}H$	Solute to solution
Sublimation $\Delta_{sub}H$	Solid to gaseous atoms

Transition $\Delta_{trs}H$ Phase change
Vaporization $\Delta_{vap}H$ Liquid to gaseous atoms

Heats generally measured per mole. Should be measured from and to standard specified states.

entrainment The carrying forward by a stream of gas or vapour of fine liquid droplets.

entropy (S) A thermodynamic quantity which is a measure of the degree of disorder within any system

$$\Delta S = \int_i^f \frac{dq}{T} \text{ rev}$$

the heat supplied divided by the temperature of the addition. The greater the degree of disorder the higher the entropy. Thus, for a given substance

$$S_{gas} > S_{liquid} > S_{solid}$$

Any change taking place which results in an increase in entropy has a positive entropy change (ΔS). Most spontaneous thermodynamic processes are accompanied by an increase in entropy. Entropy has units of joules per degree K per mole. For values see ATKINS or KAYE AND LABY. The entropy of a monatomic gas is given by the Sackur–Tetrode equation.

enyl complexes Complexes of organic groupings in which an odd number of electrons must be considered to be donated to the metal, e.g.

$$\begin{array}{c} H_2C \\ HC \!-\! M \\ H_2C \end{array}$$

a π-allyl complex with 3 electrons donated (trihapto, 3 atoms bonded).

enzymes Proteins which catalyse reactions with a high degree of specificity and efficiency, e.g. orotidine 5^1-monophosphate catalyses the biosynthesis of uridine monophosphate by a factor of 10^{17}. Enzymes are present in all living organisms and are responsible for catalysing most of the reactions which take place in a cell, e.g. the hydrolysis of fats, sugars and proteins and their re-synthesis. They also catalyse the many forms of oxidation and reduction which provide energy for the cell and thus for the body.

Enzymes are obtained from plants, animals and micro-organisms by extraction with a suitable solvent, preferably after the cell structure has been destroyed by drying or grinding. They can be purified by precipitation and resolution and by fractional absorption and elution. Many enzymes have been obtained crystalline and structures determined, although much structural data can be obtained by n.m.r.

Enzymes often need for their activity the presence of a non-protein portion, which may be closely combined with the protein, in which case it is called a prosthetic group, or more loosely associated, in which case it is a coenzyme. Certain metals may be part of the enzyme, such as copper in ascorbic oxidase and selenium in glutathione peroxidase. Often the presence of other metals in solution, such as magnesium, is necessary for the action of particular enzymes.

Most enzymes are highly specific. Those that act on carbohydrates are particularly so, the slightest change in the stereochemical configuration of the molecule being sufficient to make a particular enzyme incompatible and unable to effect hydrolysis.

Enzymes act by lowering the overall activation energy of a reaction sequence by involving a series of intermediates, or a mechanism, different from the spontaneous uncatalysed reaction.

Most enzymes work best within a narrow pH range and are susceptible to a wide variety of compounds which inhibit or sometimes promote the activity. The majority of enzymes work most efficiently at around 40°C and at higher temperatures are rapidly destroyed.

The enzymes are classified into divisions based on the type of reaction catalysed and the particular substrate. The suffix '-ase' is added and recommended trivial names and systematic names for classification are usually given when quoting a particular enzyme. Any one particular enzyme has a specific code number based upon this classification.

Six main divisions of enzymes are recognized. **1** *Oxidoreductases*. Enzymes catalysing redox reactions. This group includes the dehydrogenases, oxidases, reductases, peroxidases, hydrogenases and hydroxylases. **2** *Transferases*. Enzymes which transfer a group, possibly a methyl group or a glycosyl group, from one compound to another. Examples are the aminotransferases and the methyltransferases. **3** *Hydrolases*. Enzymes catalysing the hydrolytic cleavage of C–O, C–N and C–C and other bonds. Examples are esterases, glucosidases, peptidases, proteinases, phospholipases, sulphatases and phosphatases. **4** *Lyases*. These enzymes cleave C–C, C–O, C–N and other bonds by elimination leaving double bond, or conversely add groups to double bonds. This group includes decarboxylases, hydratases, dehydratases and some carboxylases. **5** *Isomerases*. These catalyse the structural or geometric changes within a molecule. The division includes racemases, epimerases, *cis-trans*-isomerases, tautomerases and mutases. **6** *Ligases* (*synthetases*). Enzymes catalysing the joining together of two molecules coupled with the hydrolysis of a pyrophosphate bond in ADP or a similar triphosphate. They include some carboxylases and many enzymes known as synthetases.

Often there is more than one enzyme catalysing the same reaction.

Enzymes are used in starch, baking, cheese manufacture, wine-making, brewing and distillation, foodstuffs, pharmaceuticals, leather tanning, paper manufacture, adhesives, sewage disposal, animal feeds and in detergents.

Industrial enzymes are extracted from natural products or modified bacteria and fungi. See DICTIONARY OF BIOLOGY.

ephedrine, $C_{10}H_{15}NO$, $PhCH(OH)CH(NHMe)Me$. Colourless crystals containing about $\frac{1}{2}H_2O$; m.p. 40°C, b.p; 225°C. It is obtained from various species of *Ephedra*, or may be synthesized. Its pharmacological action resembles that of adrenaline.

epichlorohydrin (3-chloropropylene oxide, 3-chloro-1,2-epoxypropane),

$$CH_2.CH \cdot CH_2Cl$$
$$\diagdown \diagup$$
$$O$$

C_3H_5ClO. Colourless liquid with an odour resembling that of chloroform; b.p. 115–117°C. Prepared by treating glycerol dichlorohydrins with solid NaOH at 25–30°C, allyl chloride plus chlorine water, or chlorination of acrolein. Reduced by sodium amalgam to propenol, oxidized by nitric acid to β-chlorolactic acid. Reacts with alcohols in the presence of potassium hydroxide to give diethers of glycerol. Used in the manufacture of glycerol, glycerol ethers, triethyl oxonium salts and epoxy resins, and as a solvent.

epimerism A type of isomerism shown by substances which contain several asymmetric centres but differ in the configuration of one only of these.

In carbohydrate chemistry the term is restricted to sugars or their derivatives which differ only in the orientation of the groups attached to the carbon atom next to the potential aldehyde group of the sugar or to the corresponding group in the sugar derivative. Thus D-gluconic acid,

$$HOH_2C-\overset{\overset{\displaystyle H}{|}}{\underset{\underset{\displaystyle OH}{|}}{C}}-\overset{\overset{\displaystyle H}{|}}{\underset{\underset{\displaystyle OH}{|}}{C}}-\overset{\overset{\displaystyle OH}{|}}{\underset{\underset{\displaystyle H}{|}}{C}}-\overset{\overset{\displaystyle H}{|}}{\underset{\underset{\displaystyle OH}{|}}{C^*}}-COOH$$

and D-mannonic acid

$$HOH_2C-\overset{\overset{\displaystyle H}{|}}{\underset{\underset{\displaystyle OH}{|}}{C}}-\overset{\overset{\displaystyle H}{|}}{\underset{\underset{\displaystyle OH}{|}}{C}}-\overset{\overset{\displaystyle OH}{|}}{\underset{\underset{\displaystyle H}{|}}{C}}-\overset{\overset{\displaystyle OH}{|}}{\underset{\underset{\displaystyle H}{|}}{C^*}}-COOH$$

are epimeric at the starred atoms and each may be partially converted into the other by heating in pyridine or quinoline, such a process being called epimerization. The sugars themselves are epimerized to some extent when treated with dilute alkalis, but side reactions also occur.

epimerization See EPIMERISM.

epinephrine See ADRENALINE.

EPN Insecticide, acaricide, organophosphorus derivative.

EPO See ERYTHROPOIETIN.

epoxidation The addition of an oxygen bridge across an alkene bond, achieved in special cases by oxygen and a catalyst, but more usually by a PER-ACID such as perbenzoic acid or peracetic acid, or substituted analogues of these acids.

1,2-epoxide polymers Polymers of ethylene oxide, oxirane,

$$CH_2 - CH_2$$
$$\diagdown \diagup$$
$$O$$

and its derivatives by oxyalkylation in the presence of Lewis acids or bases or by catalytic polymerization over catalysts. The polymers are polyethylene glycols and range from liquids (lubricants, binders, solvents) to very high mol. wt. solids (for sheet, film, etc., thickeners, binders). Other epoxides, e.g. $CF_2 \cdot CF_2 \cdot O$ can also form useful polymers. See EPOXY RESINS.

epoxides Compounds containing the $C-C-O$ unit, oxirane ring. See EPOXIDATION. Very reactive, particularly under acid catalysis. Used as reactants and to form polymers.

epoxy A prefix indicating a

group in a molecule, e.g.

$$H_2C - CH_2O,$$

epoxyethane, oxiranes.

1,2-epoxypropane See PROPYLENE OXIDE.

epoxy resins Polyethers formed by condensation of, e.g. epichlorohydrin with polyols, such as bisphenol A, or by epoxidation of Diels–Alder adducts with peroxy compounds. The polyethers need curing agents to convert to resins. The resins are thermosetting, tough, adhesive, chemically resistant and have favourable electrical properties. They are used as adhesives (16%), protective coatings (48%) and in composites (22%).

epr See ELECTRON SPIN RESONANCE.

Epsom salts, $MgSO_4.7H_2O$. See MAGNESIUM SULPHATE.

EPTC Herbicide, thiocarbamate derivative.

equation of state The relation between pressure, volume and temperature of a system. See VAN DER WAALS' EQUATION.

equatorial See CONFORMATION.

equilenin, $C_{18}H_{18}O_2$. Hydroxyestra-1,3,5,7,9-pentaene. M.p. 259°C (decomp.). A steroid hormone.

equilibrium constant According to the law of mass action, for any reversible chemical reaction $aA + bB \rightleftharpoons cC + dD$, the equilibrium constant (K) is defined as

$$K = \frac{[C]^c[D]^d}{[A]^a[B]^b}$$

where [A] ..., etc. are the active masses of A ..., etc., more accurately the thermodynamic activities.

In thermodynamic terms the equilibrium constant is related to the standard chemical potential by the equation

$$\Delta\mu° = -RT\ln K,$$

where R is the gas constant.

For gas reactions where the gases are assumed to follow ideal behaviour this equation be-comes $\Delta G° = -RT\ln Kp$, where Kp is defined in terms of the partial pressures (ideally activities) of reactants and products. Thus for the general reaction above,

$$K_p = \left(\frac{P_C{}^c.P_D{}^d}{P_A{}^a.P_B{}^b}\right)$$

equilibrium diagram A simplified boiling- or melting-point diagram which shows graphically, for a liquid mixture, the composition of the vapour or solid which is in equilibrium with the liquid. Widely used in metallography to show the equilibrium between liquid and solid alloys.

equilibrium, metastable If a system, capable thermodynamically of undergoing a spontaneous change, persists without change, it is said to be in a state of metastable equilibrium. The change requires initiation, e.g. a mixture of hydrogen and oxygen at 298 K is capable of undergoing spontaneous reaction, but reaction occurs only when initiated by a spark or addition of a catalyst.

equilibrium state A system such that infinitely small changes in either direction result in opposite changes to the original position.

equilin (3-hydroxy-1,3,5(10), 7-estra-tetraen-17-one), $C_{18}H_{20}O_2$. M.p. 238–240°C. Female sex hormone found in the urine of pregnant mares.

equipartition theorem In a system at equilibrium all degrees of freedom have the same average energy. Further the average energy associated with each quadratic term (square of velocity) is $\frac{1}{2}kT$, where k is the Boltzmann constant.

equivalence point The stage in a titration when the reactants and products are present in equivalent amounts according to the stoichiometry of the reaction. This may not correspond exactly to the END POINT.

equivalent, chemical See CHEMICAL EQUIVALENT.

equivalent conductivity The specific conductance multiplied by the volume (ml) which contains 1 g equivalent of the electrolyte.

equivalent proportions, law of Substances (elements and compounds) react together in the mass ratio of their equivalents. The law does not have wide application in its simple form because of variable oxidation states and non-stoichiometries.

Er Erbium.

erbium, Er. Lanthanide metal, at.no. 68, at.mass 167.26, m.p. 1529°C, b.p. 2868°C, density ρ 9066, crustal abundance 3.8 p.p.m., electronic configuration [Xe] $4f^{12}6s^2$. The metal is hcp. Used as an additive to improve the workability of vanadium; the oxide is used in colouring glass.

erbium compounds Erbium forms a single series of typical LANTHANIDE COMPOUNDS in the +3 state E^{3+} (f^{11} rose) $\rightarrow$ Er(-2.30 volts in acid).

ergocalciferol Rodenticide, steroid derivative.

ergometrine, $C_{19}H_{23}N_3O_2$. Water-soluble alkaloid of ERGOT; m.p. 195–197°C (decomp.).

ergot Sclerotium of the fungus *Claviceps purpurea*, which develops in the ovary and replaces the grain of rye. Ergot owes its pharmacological activity almost entirely to the related alkaloids ERGOTOXINE, ERGOMETRINE and ERGOTAMINE which contain lyserigic acid residues.

ergotamine, $C_{33}H_{35}N_5O_5$. One of the ERGOT ALKALOIDS.

eriochrome black T Azo compound. A complexometric reagent for Ca^{2+} and Mg^{2+} (used for determining water hardness).

erubescite, Cu_3FeS_3. Cu ore.

erucic acid (*cis*-13-docosenoic acid), $C_{22}H_{42}O_2$, $CH_3\cdot[CH_2]_7\cdot CH{:}CH\cdot[CH_2]_{11}\cdot C(O)OH$. M.p. 35°C. An unsaturated fatty acid belonging to the oleic acid series, occurring as glycerides in rape oil and other vegetable oils. It is the *cis*-isomer, the *trans*-isomer being brassidic acid.

erythritol, $C_4H_{10}O_4$. All four tetrose alcohols (tetritols) are known, but the only one to occur naturally is erythritol or mesoerythritol which has the configuration

$$HOH_2C - \underset{\underset{OH}{|}}{\overset{\overset{H}{|}}{C}} - \underset{\underset{H}{|}}{\overset{\overset{OH}{|}}{C}} - CH_2OH.$$

It is found in lichens and in some algae. It has m.p. 120°C, is very soluble in water and is about twice as sweet as sucrose. It is a reference compound upon which the *erythro* nomenclature is based.

erythro- A prefix used to distinguish one diastereomer from the other *three* form, which is based upon an analogy to erythrose and threose. The erythro diastereomer is defined as the one in which at least two sets of similar substituents on adjacent asymmetric carbons are in the eclipsed configuration.

Erythro-form   *Threo*-form

A Newman projection of the above *erythro* form is

erythromycin, $C_{37}H_{67}NO_{13}$. Macrolide ANTIBIOTIC, produced by some species of *Streptomyces*.

erythropoietin (EPO) A peptide hormone which affects red blood cell production.

erythrose, $C_4H_8O_4$. A tetrose sugar, $(HO)H_2C\cdot C(OH)H\cdot C(OH)H\cdot CHO$, both the D- and L- forms of which have been prepared. It has been obtained as a liquid, very soluble in water and alcohol.

Es Einsteinium.

ESCA (electron spectroscopy for chemical analysis) See PHOTO-ELECTRON SPECTROSCOPY.

Eschenmoser's salt See DIMETHYL (METHYLENE) AMMONIUM IODIDE.

esfenvalerate Insecticide, pyrethroid.

esprocarb Herbicide, thiocarbamate derivative.

ESR Electron spin resonance.

essential fatty acids (EFA) A group of unsaturated acids which must be in the diet for healthy growth. A deficiency of EFA leads to lack of growth, skin lesions and necrosis. Essential fatty acids must possess an all-*cis* arrangement of double bonds in a 'methylene-interrupted' polyene system. The principal EFA are LINOLEIC ACID, ARACHIDONIC ACID, linolenic acid and γ-LINOLENIC ACID. These acids are essential as the body cannot synthesize them fast enough to meet needs. EFA are abundant in the vegetable oils and to a lesser extent in fats of some animals or of marine origin.

essential oils More or less volatile oils extracted from plants. Many essential oils are refined, separated and purified and used as sources of compounds. See OILS.

esterases See ENZYMES.

esterification The process of formation of esters by the combination of an acid with an alcohol.

esters Organic compounds formed by the union of an acid and an alcohol with elimination of water. They are volatile liquids or low-melting solids and are usually insoluble in water but soluble in alcohol or ether. Many esters have characteristic fruity odours and occur naturally in fruit. As the reaction between an acid and an alcohol is reversible, it cannot be carried to completion unless the water produced is removed. In some cases it is sufficient to have a large excess of alcohol present to secure a good yield of ester; in other cases it is usual to add a liquid such as benzene or carbon tetrachloride to the mixture and to distil off the water in the form of a low-boiling mixture. The reaction between alcohol and acid is very slow at ordinary temperatures, but the ester is formed much more rapidly at higher temperatures and in the presence of a small amount of an acid such as sulphuric, hydrochloric or benzenesulphonic acid. Esters are also produced by the action of an alcohol on an acid chloride or anhydride; amides also react with alcohols to give esters in the presence of boron trifluoride. Some esters may be obtained by the action of an aldehyde on aluminium ethoxide or isopropoxide. Methyl and ethyl esters are obtained by treating the sodium salt of the acid with methyl or ethyl sulphates. Diazomethane reacts with organic acids to give methyl esters. Esters can also be prepared by heating an alcohol with the methyl or ethyl ester of the acid; in this case an exchange of alcohols occurs and methanol or ethanol is formed. Esters are completely hydrolysed to acid and alcohol by heating with dilute sodium hydroxide solution. They react with ammonia to give amides; with ketones in the presence of sodamide to give 1,3-diketones and with Grignard reagents to give tertiary alcohols. They are reduced by a mixture of sodium and ethanol to the alcohol corresponding to the acid of the ester. Used as solvents, flavouring essences and perfumes; they are also used in many chemical processes.

estradiol (oestradiol, 1,3,5(10)-oestra-triene-3,17β-diol), $C_{18}H_{24}O_2$. M.p. 174°C. Found in the urine of pregnant mares and the ovaries of pigs. Estradiol appears to act at a cellular level, by binding to the nucleus of cells in its target organ, where it seems to stimulate the synthesis of RNA.

estriol (oestriol, 1,3,5(10)-oestra-triene-3,16α,17β-triol), $C_{18}H_{24}O_3$. M.p. 283°C. An estrogen present in the urine of pregnant women. It possesses female sex hormone activity but is not as active as estrone.

estrogens (oestrogens) The follicular female sex hormones. During pregnancy large quantities are excreted. They are found also in the urine of male members of the equine species. Estradiol is the most potent of the naturally occurring oestrogens, of which estrone, estriol, EQUILIN and equilinin are metabolic products. Stilboestrol and HEXOESTROL are synthetic products with estrogenic activity that are much used medicinally, as they can be administered by mouth and are much cheaper than the naturally occurring estrogens.

estrone (oestrone, 3-hydroxy-1,3,5(10)-oestratrien-17-one), $C_{18}H_{22}O_2$. M.p. 258°C. A female sex hormone.

Et, Ethyl C_2H_5-.

Ètard's reaction A direct method for the production of aromatic aldehydes by the oxidation of the methylated homologues of benzene by chromyl chloride, CrO_2Cl_2. When benzene homologues with longer side chains are treated with CrO_2Cl_2 a ketone is the usual product, but with ethylbenzene a mixture of phenylacetaldehyde and acetophenone is obtained.

ethalfluvalin Herbicide, nitroaniline derivative.

ethambutol An important anti-tuberculosis drug.

ethametsulfuron-methyl Herbicide, sulphonylurea derivative.

ethanal (acetaldehyde), $CH_3 \cdot CHO$. Colourless liquid with a characteristic odour; b.p. 20.8°C. Miscible with water, alcohol and ether; insoluble in concentrated calcium chloride solutions. Manufactured by **1** vapour-phase oxidation or dehydrogenation of ethanol over a catalyst; **2** vapour-phase oxidation of butane (with methanal and methanol as by-products); **3** direct oxidation of ethene over a copper(II) chloride/palladium(II) chloride catalyst

(WACKER PROCESS). Prepared on the small scale by the oxidation of ethanol with potassium dichromate and sulphuric acid. Compounds are formed with many metallic salts. It is oxidized to ethanoic acid and reduced to ethanol. Polymerization readily occurs to give paraldehyde or metaldehyde: when heated with alkalis, a brown resin is formed. Aldol is formed by the interaction of two molecules of ethanal. Ethanal is used almost exclusively as a chemical intermediate: three compounds, ethanoic acid, ethanoic anhydride and *n*-butanol accounting for the bulk of its consumption. Other industrial chemicals made from it include 2-ethylhexyl alcohol, pentaerythritol, chloral and dyestuffs.

ethanal diethanoate See ETHYLIDENE DIACETATE.

ethanamide (acetamide), $CH_3CO \cdot NH_2$. Crystallizes in long white needles which absorb water and finally liquefy. The crude substance has a strong odour of mice; m.p. 82°C, b.p. 222°C. Prepared by the dry distillation of ammonium ethanoate or by the action of ammonia on ethyl ethanoate. Weakly basic. Solvent and solubilizer for resins, etc.

ethane, $CH_3 \cdot CH_3$. Colourless, odourless gas; forms an explosive mixture with air. B.p. –89°C. It occurs in natural gas. Solid ethane has been detected on Pluto and gaseous ethane in comets. Gaseous ethane has a staggered conformation. May be prepared by reduction of ethene or ethyne by hydrogen under pressure in the presence of a nickel catalyst, or by the electrolysis of a solution of potassium ethanoate. It has the general properties of the paraffins. Used in low-temperature refrigeration plant.

ethanedial (glyoxal, biformyl), $CH(O) \cdot CHO$. Crystallizes in yellow prisms; its vapour is green and burns with a violet flame; m.p. 15°C, b.p. 51°C. Readily polymerizes on standing in the presence of moisture; the aqueous solution contains the dihydrate. It can be manufactured by oxidation of ethylene glycol with air using a copper oxide catalyst. Used to harden photographic gelatin.

ethanedioic acid See OXALIC ACID.

ethane dithiol (1,2-ethylene dimercaptan), $HSCH_2CH_2SH$. Prepared from ethanol, thiourea and $C_2H_4Br_2$. B.p. 146°C. Synthetic reagent.

ethanenitrile See ACETONITRILE.

ethanoates (acetates) Salts or esters of ETHANOIC ACID (acetic acid). The ethanoate group is often designated Ac.

ethanoic acid (acetic acid, glacial acetic acid), CH_3COOH. Colourless liquid with a pungent, irritating odour; m.p. 16.6°C, b.p. 119°C.

Manufactured by the liquid-phase oxidation of ethanal at 60°C by oxygen or air under pressure in the presence of manganese(II) ethanoate, the latter preventing the formation of perethanoic acid. Another important route is the liquid-phase oxidation of butane by air at 50 atm. and 150–250°C in the presence of a metal ethanoate. Ethanoic acid is produced by the catalytic oxidation of ethanol and by carbonylation of methanol in the presence of an Rh or Ir catalyst. Fermentation processes are used only for the production of vinegar.

Ethanoic acid will attack most metals and can form acidic, basic and normal salts. It can, however, be handled in e.g. stainless steel equipment. About half the ethanoic acid produced is used as ethanoic anhydride for the manufacture of cellulose ethanoate. Large quantities are also used for the manufacture of vinyl ethanoate and various solvents. Used as a solvent. Because a number of manufacturing processes yield large amounts of dilute acid its recovery is a problem of considerable importance. Simple rectification is expensive and hence various other processes such as azeotropic and extractive distillation and liquid–liquid extraction are used.

ethanoic anhydride (acetic anhydride), $[CH_3C(O)]_2O$. Colourless liquid with a pungent odour; b.p. 139.5°C. Hydrolysed to ethanoic acid by boiling water. Manufactured by bubbling air through a mixture of ethanal and ethanoic acid in the presence of a catalyst; or by reacting keten, derived by the cracking of ethanoic acid or propanone with ethanoic acid and by carbonylation of methane (Reppe process). It reacts with compounds containing an –OH, – SH or – NH group to give acetyl (ethanoyl) derivatives which are useful for characterization. Largely used for the production of cellulose ethanoate, also used for the manufacture of vinyl ethanoate and aspirin and other ethanoates.

ethanol (ethyl alcohol, alcohol, spirits of wine), CH_3CH_2OH. Colourless liquid with a pleasant odour, b.p. 78.3°C. Miscible with water with evolution of heat and contraction in volume; pure ethanol absorbs water vapour.

Many gases are more soluble in it than in water. Some inorganic salts and many organic compounds are soluble in ethanol. It occurs only rarely in nature, except as a result of the fermentation of sugary plant juices by yeasts, and less often by certain bacteria and moulds. Alcohol was formerly manufactured almost exclusively by the fermentation of materials containing starch and sugars, but this method is now relatively unimportant except for alcoholic beverages. It can be concentrated by distillation. Most is at present made by the catalytic hydration of ethene, or by the hydrolysis of the mono- and diethyl sulphates formed by absorbing ethene in 90% sulphuric acid. The ethene is obtained from refinery gases or other petroleum fractions by cracking. Because ethanol forms an azeotrope with water it is not possible to obtain a product containing more than 95.6% alcohol by weight by straightforward fractionation of an aqueous solution. For the manufacture of 100% ethanol AZEOTROPIC DISTILLATION is employed. Ethanol is oxidized to ethanal or ethanoic acid; with nitric acid a variety of products, including glycollic and oxalic acids, are formed. Ethanolates (ethoxides) are formed by the action of sodium, calcium, aluminium and some other metals on ethanol. It reacts with acids to give esters. With sulphuric acid it yields ether, ethene or ethyl hydrogen sulphate. Bleaching powder converts it to chloroform, while chlorine gives chloral. Ethanol is used as a starting point for the manufacture of other chemicals, principally ethanal, in foodstuffs and as a solvent as a petroleum additive. Denatured alcohol has added materials to make it less potable. Its pharmacological effects are basically those of a central depressant, low doses having an apparently stimulant effect due to the removal of normal inhibitory influences.

ethanolamines The three ethanolamines are low-melting, colourless solids which very readily absorb water and form viscous liquids; they have distinct ammoniacal odours and are strong bases.

Monoethanolamine (2-aminoethyl alcohol, 2-hydroxyethylamine), $HOCH_2CH_2NH_2$. M.p. 10.5°C, b.p. 171°C.

Diethanolamine (di-(2-hydroxyethyl)amine), $(HOCH_2CH_2)_2NH$. M.p. 28°C, b.p. 217°C/150mm.

Triethanolamine (tri-(2-hydroxyethyl)amine), $(HOCH_2CH_2)_3N$. M.p. 21°C, b.p. 277°C/150mm.

All these compounds are manufactured by heating ethylene oxide under pressure with concentrated aqueous ammonia. A mixture of the three is obtained, the proportion of each depending on the ammonia/ethylene oxide ratio used, and this is separated by fractional distillation.

The ethanolamines form soaps with fatty acids which are almost neutral in reaction and soluble in benzene. These are of great commercial importance, being used as detergents, emulsifying agents and in the preparation of cosmetics, toiletries, bactericidal and herbicidal products. Monoethanolamine and diethanolamine are widely used for removing acid constituents such as carbon dioxide and hydrogen sulphide from natural gas streams. These acid gases are then removed by steam stripping.

ethanoylation See ACETYLATION.

ethanoyl chloride (acetyl chloride), CH_3COCl. Colourless liquid with a pungent odour; fumes in moist air, producing ethanoic and hydrochloric acids. B.p. 55°C. Reacts with water and other hydroxyl-containing compounds. Prepared by the distillation of a mixture of ethanoic acid and PCl_3 or $POCl_3$. It is used to prepare ethanoyl (acetyl) derivatives of hydroxy- and amino-compounds.

ethene (ethylene), $CH_2=CH_2$. Colourless gas with a faint ethereal odour which occurs in natural gas, crude oil and coal gas. M.p. −169°C, b.p. −105°C. It is manufactured in large quantities, most of it being derived from the vapour-phase cracking of ethane, various petroleum fractions and crude oil. It is also obtained from ethanol by vapour-phase dehydration using an activated alumina catalyst at 350°C. May be prepared by dropping ethanol on to syrupy phosphoric acid heated to 220°C. It is inflammable and forms an explosive mixture with air. Readily absorbed by concentrated sulphuric acid to give ethyl hydrogen sulphate; if the reaction is carried out under pressure, diethyl sulphate is largely formed. Absorbed by solutions of potassium permanganate to give ethylene glycol: by dilute chlorine water to give ethylene chlorohydrin. Reduced by hydrogen to ethane. Reacts with hydrogen bromide and hydrogen iodide at 100°C to give ethyl bromide and iodide; hydrogen chloride does not react under these conditions. With ammonia under pressure the ethylamines are formed. Reacts with water at 450°C, or at lower temperatures in the presence of catalysts, to give ethanol. Reacts with oxygen and a palladium catalyst to give ethanal (see WACKER PROCESS).

It is given off in minute amounts by ripe tomatoes and apples and has the property of accelerating the ripening of fruit. Used in the manufacture of glycol (20%), ethylene oxide (15%), ethanol, styrene (14%), dichloroethane, ethyl chloride, vinyl esters, e.g. vinyl ethanoate and vinyl ethers, vinyl chloride (10%). When polymerized under high pressure, or at lower pressures in the presence of a catalyst, it gives ethene polymers (40%). Global capacity 1999 94 megatonnes.

ethene polymers (polyethene, polyethylene, polythene) Probably the most important polymeric substances. Polymerization is effected by high-pressure free radical processes over metal-oxide catalysts, low-pressure polymerization using Ziegler catalysts. The polymers can be produced crystalline (high density) and non-crystalline (low density). Used in fabrication (houseware, toys, etc.), pipe, wire coating, process machinery, textiles. Low density materials have a high degree of branching and are resistant to cracking. High-density materials are more rigid. Chlorination gives materials used for flooring, and electrical insulation. Chlorosulphonation (introduction of SO_2Cl groups) gives very durable materials.

ethenyl (vinyl) The group $CH_2=CH–$.

ethephon Insecticide, organophosphate derivative.

ether (diethyl ether, ethyl ether, diethyl oxide), $(CH_3CH_2)_2O$. Colourless liquid with a pleasant, characteristic odour; very volatile, its vapour forms an explosive mixture with air; b.p. 34.5°C. It is manufactured by passing ethanol vapour into a mixture of 92% ethanol and 78% sulphuric acid at 128°C. May also be produced as a by-product in the manufacture of alcohol from ethene. Commercial ether usually contains small amounts of water, ethanol, ethanal and the explosive peroxide. It is a comparatively inert compound and is an excellent solvent for a great many organic substances. Oxidized by nitric acid to ethanoic acid. Reacts with strong sulphuric acid to give ethyl hydrogen sulphate and with hydrogen iodide to give ethyl iodide. Chlorine forms various chloroethers at low temperatures. Medical and pharmaceutical uses (anaesthetic) account for a relatively small proportion of ether consumption, and it is mainly used as a chemical intermediate and as a solvent for oils, fats, waxes and alkaloids.

etherimol Fungicide, pyrimidine derivative.

ethers Organic compounds of the type R–O–R′, where R and R′ are alkyl or aryl. Formed by heating the sodium derivative of a hydroxy-compound with an alkyl or aryl halide; or by treating the hydroxy-compound with an alkyl halide in presence of silver oxide. Methyl and ethyl ethers are conveniently prepared by the action of dimethyl and diethyl sulphates on a hydroxy-compound in presence of sodium hydroxide solution. Diazomethane will react with phenols in ethereal solution to give methyl ethers. The simpler aliphatic ethers are manufactured by the action of sulphuric acid on the appropriate alcohol or olefin. They are usually liquids with pleasant odours, but some aromatic and the higher aliphatic ethers are crystalline solids. They are insoluble in water but soluble in alcohol and diethyl ether. The commonest ether, diethyl ether, is usually called simply 'ether'.

ethidium bromide (homidium bromide) Used in separation and detection of nucleic acids. Mutagen. Phenanthrene derivative.

ethinylation The reaction between ethyne and organic substrates to produce substances having an intact ethynic bond; e.g. the reaction of methanal and ethyne to give but-3-yne-1-4-diol, $HOCH_2–C≡C–CH_2OH$.

ethiofen carb Insecticide, carbamate derivative.

ethion Acaricide, insecticide, organophosphate derivative.

ethisterone (ethinyltestosterone), $C_{21}H_{28}O_2$. Steroid derivatives used in contraceptive pills.

ethofurnestat Herbicide, furan derivative.

ethoprophos Nematocide, insecticide, organophosphate derivative.

ethoxy The $CH_3CH_2O–$ group; EtO–.

ethoxycarboxylates Surfactants, foam stabilizers.

ethoxyl The group $CH_3CH_2O–$; also written EtO–.

ethoxylates Surfactants, foam stabilizers.

ethoxysulfuron Herbicide, sulphonylurea derivative.

ethychlozate (indolyl ethanoate) Plant growth regulator, imidazolyl derivative.

ethyl The group $CH_3·CH_2–$; often written $C_2H_5–$ and Et–.

ethyl acetate See ETHYL ETHANOATE.

ethyl acetoacetate (acetoacetic ester, ethyl 3-oxobutanoate),

$CH_3COCH_2C(O) OC_2H_5 \rightleftharpoons$ $CH_3C(OH){:}CH(O)OC_2H_5$. Colourless, mobile liquid with a pleasant odour; b.p. 181–182°C. Prepared by the action of sodium or sodium ethoxide on ethyl ethanoate, or of ethanol on diketen. It is the classical example of *keto-enol tautomerism*, and the ester normally contains 93% keto and 7% enol form. The proportion of the two forms is altered by change of temperature or by solution in solvents. Separation of the keto and enol forms has been effected by distillation from quartz or Pyrex apparatus: the pure components revert to the original equilibrium mixture on standing. Metallic derivatives are formed with sodium and other metals; these are of the type $CH_3C(ONa){:} CHC(O)OC_2H_5$; they react with alkyl halides to give alkyl acetoacetates $CH_3CO{\cdot}CHR{\cdot}C(O)OC_2H_5$. These also form sodium derivatives, reacting with alkyl halides in a similar manner to give di-alkyl esters. Ethyl acetoacetate and its alkyl derivatives react with strong alkalis to give ethanoic and alkylethanoic acids respectively. Dilute alkalis or acids react to give ketones. With many nitrogen compounds it reacts to form nitrogen-containing rings: thus urea gives methyl uracil; hydrazines give methyl pyrazolones; aniline gives methyl quinoline. Prolonged boiling causes decomposition with loss of alcohol and formation of dehydroacetic acid. Used widely in synthesis.

ethyl acrylate (ethyl propenoate), $C_8H_8O_2$,

$CH_2{:}CHC(O)OCH_2CH_3$. Colourless liquid; b.p. 101°C. Manufactured by treating ethylene chlorohydrin with sodium cyanide and heating the β-hydroxypropionitrile so formed with ethanol and sulphuric acid. Forms a colourless resin on standing. Used in the manufacture of synthetic resins, but the methyl ester is more commonly employed.

ethyl alcohol See ETHANOL.

ethylamines Organic compounds in which one or more of the hydrogen atoms of ammonia are replaced by ethyl groups. Colourless liquids with strongly ammoniacal odours; they are inflammable and burn with a yellow flame. Manufactured by passing ethene or ethanol vapour with ammonia under pressure over heated catalysts. The relative amounts of the three amines formed can be controlled by varying the proportions of ammonia to ethene or alcohol. They are all three strongly basic in character, form hydrochlorides and metal complexes and show the typical properties of aliphatic amines.

Ethylamine (*monoethylamine*), $CH_3CH_2NH_2$. B.p. 19°C. Prepared by reduction of acetonitrile or by heating ethyl chloride with alcoholic ammonia under pressure. It is a strong base and will displace ammonia from ammonium salts. Used in resins as stabilizer, dyestuffs and oil refining.

Diethylamine, $(CH_3CH_2)_2NH$. B.p. 55.5°C. Forms a crystalline ½ hydrate. Prepared by the action of a boiling solution of sodium hydroxide on nitrosodiethylaniline. Used in flotation agents.

Triethylamine, $(CH_3CH_2)_3N$. Oily liquid; b.p. 89°C. Prepared by heating ethylamine with an alcoholic solution of ethyl chloride under pressure. Readily oxidized by potassium permanganate.

ethyl amyl ketone (5-methyl heptanone)

B.p. 157–162°C. Solvent for resins.

ethylation Processes by which an ethyl group is added to a compound. In aliphatic chemistry this involves the substitution of the hydrogen atom of a hydroxyl, amino or imino group, and produces an ether or a secondary or tertiary amine respectively. In aromatic chemistry it may also mean the substitution of one of the hydrogen atoms of the ring by the ethyl group; this is carried out by the Friedel–Crafts reaction.

ethylbenzene, $PhCH_2CH_3$. B.p. 136°C. Recovered from mixed xylenes from petrochemical processes, prepared by addition of ethene to benzene, or, on a laboratory scale by the Friedel–Crafts reaction between C_2H_5Cl and benzene. Catalytically dehydrogenated, particularly in the presence of steam, to styrene. U.S. production 1999 5.90 megatonnes.

ethyl benzoate, $PhC(O)OEt$. Used in perfumery and fruit essences.

ethyl-n-butyl ketone See HEPTAN-3-ONE.

ethyl carbamate See URETHANE.

ethyl chloride See CHLOROETHANE.

ethyl chlorocarbonate See ETHYL CHLOROFORMATE.

ethyl chloroformate (chloroformic ester, chlorocarbonic ester, ethyl chlorocarbonate), $ClC(O)OCH_2CH_3$. Volatile liquid with an unpleasant odour; b.p. 94–95°C. Prepared by the addition of ethanol to cooled phosgene. It is very reactive and unites with many organic compounds containing –OH groups to give carbethoxy derivatives. Reacts with ethanol to give diethyl carbonate and with

ammonia to form urethane. Used in organic syntheses.

ethyl citrate (triethyl citrate), $C_{12}H_{20}O_7$. Colourless oil with a bitter taste; b.p. 185°C/17mm. Typical ester. Used as a fixative in perfumery.

ethyl cyanoacetate See ETHYL CYANO-ETHANOATE.

ethyl cyanoethanoate (cyanoacetic ester, ethyl cyanoacetate), $C_5H_7NO_2$, NC·CH$_2$C(O)OCH$_2$CH$_3$. Colourless liquid; b.p. 207°C. Prepared by boiling cyanoacetic acid with ethanol and sulphuric acid. Like acetoacetic ester, it will react with sodium and the product will combine with halogen compounds such as alkyl bromides. It may be used to replace malonic ester or acetoacetic ester in many syntheses.

ethyl diazoethanoate (ethyl diazoacetate, diazoacetic ester), $C_4H_6N_2O_2$, N$_2$CHC(O)OCH$_2$CH$_3$. Yellow oil; b.p. 84°C/61 mm. Prepared by treating a solution of ethyl glycinate hydrochloride with sodium nitrite and sulphuric acid at 0°C. Decomposes explosively if distilled at atmospheric pressure, or if treated with concentrated hydrochloric or sulphuric acid. When treated with alkalis it forms nitrogen-containing ring compounds. Forms glycollic acid on boiling with water. Reacts with alcohols to give alkoxy-ethanoic esters, and with aldehydes to give β-keto-esters. Esters of unsaturated acids react with it, forming pyrazoline carboxylic esters; these lose nitrogen on heating to give derivatives of cyclopropane. Used in organic syntheses.

ethylene See ETHENE.

ethylene chlorohydrin See 2-CHLOROETHYL ALCOHOL.

ethylenediamine See 1,2-DIAMINOETHANE.

ethylenediaminetetra-acetic acid (EDTA), (HO$_2$CCH$_2$)$_2$N·CH$_2$·CH$_2$·N(CH$_2$CO$_2$H)$_2$. Sequestering agent. A multidentate chelating agent. Forms complexes with most elements. With derivatives used in detergents and textile processing and delivery of trace elements.

ethylene dibromide See 1,2-DIBROMOETHANE.

ethylene dichloride See 1,2-DICHLOROETHANE.

ethylene dinitrate (ethylene glycol dinitrate), $C_2H_4N_2O_6$, O$_2$NOCH$_2$·CH$_2$ONO$_2$. A colourless liquid with a sweetish taste and no appreciable odour; b.p. 105°C/19mm; explodes when rapidly heated. Manufactured by treating ethylene glycol with a mixture of fuming nitric and sulphuric acids at a low temperature or by passing ethene into a cold mixture of nitric and sulphuric acids. Used as an explosive of the dynamite type; it is mixed with nitroglycerin to reduce the freezing point of dynamite.

ethylene glycol See 1,2-DIHYDROXYETHANE.

ethylene glycol dinitrate See ETHYLENE DINITRATE.

ethylene glycol monobutyl ether (butyl Cellosolve), $C_6H_{14}O_2$, CH$_3$CH$_2$CH$_2$CH$_2$OCH$_2$ CH$_2$OH. Colourless liquid with a pleasant odour; b.p. 171°C. Manufactured by heating ethylene oxide with 1-butanol in the presence of nickel sulphate as a catalyst. Used as a solvent in brushing lacquers.

ethylene glycol monoethyl ether (2-ethoxyethanol, Cellosolve), CH$_3$CH$_2$OCH$_2$CH$_2$OH. A colourless liquid with a pleasant odour; b.p. 156°C. Manufactured by heating ethylene oxide with ethanol and a catalyst, or by treating ethylene glycol with diethyl sulphate and sodium hydroxide. Used extensively as a solvent in nitrocellulose lacquers.

ethylene glycol monoethyl ether acetate (Cellosolve acetate), $C_6H_{12}O_3$, CH$_3$CH$_2$OCH$_2$CH$_2$O(O)CCH$_3$. Colourless liquid with a pleasant ethereal odour; b.p. 156°C. Manufactured by heating ethylene glycol monoethyl ether with ethanoic acid. Used as a solvent in nitrocellulose lacquers.

ethylene glycol monomethyl ether (methyl cellosolve), $C_3H_8O_2$, CH$_3$OCH$_2$CH$_2$OH. Colourless liquid with a pleasant odour; b.p. 124°C. Manufactured by heating ethylene oxide with methanol, either under pressure or in the presence of a catalyst. Also obtained by treating ethylene glycol with dimethyl sulphate and sodium hydroxide. Used as a solvent for cellulose acetate lacquers and for certain dyes and resins.

ethylene imine (dihydroazirine, aziridine, ethyleneimine), C_2H_5N.

$$\underline{CH_2 \cdot CH_2} \cdot NH$$

Colourless liquid with strong ammoniacal smell; b.p. 56°C. Miscible with water and strongly basic. Prepared commercially from 2-aminoethanol. Pure dry aziridine is comparatively stable but polymerizes explosively in the presence of traces of water.

Used industrially to cross-link hydroxylic polymers. Polyethyleneimine is a hygroscopic liquid used in paper manufacture to confer wet strength, and in textiles. *N*-alkylated derivatives also form useful polymers. Carcinogenic.

ethylene oxide (1,2-epoxyethane, oxirane)

$$\overline{CH_2 \cdot CH_2 O}$$

Colourless gas, probably carcinogenic; b.p. 10.5°C. Manufactured by heating ethylene chlorohydrin with $Ca(OH)_2$ or NaOH solution; or by the direct oxidation of ethene at 250–300°C using a silver catalyst. It forms an explosive mixture with air. Reacts with water in the presence of sulphuric acid to give ethylene glycol and polyethylene glycols; with alcohols and phenols to give ethers of glycol; and with hydrochloric acid to give ethylene chlorohydrin. It reacts with many primary and secondary amines to give ethanolamines; with organic acids to give mono-esters of ethylene glycol, and with acid anhydrides to give diesters. It is reduced by hydrogen to ethanol, and is converted to ethanal by heating at 200–300°C in the presence of alumina. Its principal use is for polymerization to 1,2-EPOXIDE POLYMERS. Also used as an intermediate in the manufacture of ethylene glycol, polyethylene glycols, glycol ethers, ethanolamines and similar compounds. It is also used as a fumigant. U.S. production 1999 4.0 megatonnes.

ethyl ethanoate (ethyl acetate, acetic ester, acetic ether), $CH_3C(O)OC_2H_5$. Colourless liquid; b.p. 77°C, with a pleasant fruity odour. Manufactured from ethanol directly or via ethanoic acid and concentrated sulphuric acid, or from ethyne via ethanal with an aluminium alkoxide as catalyst. It is an extremely useful solvent, especially for cellulose-type varnishes and adhesives; also used for cosmetics and as an artificial essence.

Used for the synthesis of ETHYL ACETOACETATE.

ethyl fluid See ANTI-KNOCK ADDITIVES.

ethyl formate See ETHYL METHANOATE.

2-ethylhexanol (2-ethylhexyl alcohol, octyl alcohol), $C_8H_{18}O$, $CH_3(CH_2)_3$ CHEtCH$_2$OH. B.p. 181°C. Manufactured by heating 1-butanol with KOH and boric oxide at 270–300°C. Used as an anti-foaming agent, for mercerizing textiles, as a solvent, as a dispersing agent in the pigment and ceramic industries and as a means of introducing the 2-ethylhexyl

group into other products. Its esters, e.g. with phthalic, stearic and phosphoric acids, are used as plasticizers. The acrylate on copolymerization with other monomers gives plastics which are 'internally' plasticized. Other esters are used as lubricants, bactericides, fungicides and insecticides. U.S. production 1999 4.0 megatonnes.

ethyl hydrogen sulphate (ethylsulphuric acid), $C_2H_6O_4S$, $(EtO)(HO)SO_2$. Oily acidic liquid. Soluble in water and slowly hydrolysed by it to ethanol and sulphuric acid. Prepared by passing ethene into concentrated sulphuric acid or by heating ethanol and sulphuric acid. Gives ethene when heated alone, and diethyl sulphate when heated with ethanol at 140°C. Forms crystalline metallic salts which are soluble in water.

ethylidene The group

$$CH_3 \cdot CH\Big\langle$$

ethyl lactate, $C_8H_{10}O_3$, $CH_3CH(OH)C(O)OEt$. A colourless liquid with a pleasant odour, b.p. 154°C. Manufactured by distilling a mixture of (±)-lactic acid, ethanol and benzene in the presence of a little sulphuric or benzenesulphonic acid. It is a solvent for cellulose nitrate and acetate and also for various resins. Used as a lacquer solvent and as a fungicide.

ethyl mercaptan (ethane thiol), C_2H_5SH. Prepared from ethanol plus H_2S, b.p. 35°C. Present in natural gas. Foul odour.

ethyl methanoate (ethyl formate), $HC(O)OCH_2CH_3$. Colourless liquid with the odour of peach kernels; b.p. 54°C. Prepared by boiling ethanol and methanoic acid in the presence of a little sulphuric acid. Used as a fumigant and larvicide for dried fruits, tobacco and foodstuffs and in flavouring. It is also used in the synthesis of aldehydes.

ethyl n-butyl ketone See HEPTAN-3-ONE.

ethyl silicate (tetraethyl silicate, 'silicon ester'), $Si(OC_2H_5)_4$. Thin mobile liquid, with a mild ester-like odour; m.p. –77°C, b.p. 168°C. Made by reacting alcohol with silicon tetrachloride. Slowly decomposed by water to give ethanol and silicic acid. It is used for waterproofing stonework, for making precision castings and for bonding.

ethyl succinate (diethyl succinate, diethyl butane dicarboxylate), $C_8H_{14}O_4$, $EtO_2CCH_2 \cdot CH_2CO_2Et$. Colourless liquid with a faint odour; b.p. 218°C. Manufactured by heating succinic acid with alcohol and sulphuric acid. Typical ester. Used as a fixative in perfumery.

ethylsulphuric acid See ETHYL HYDROGEN SULPHATE.

ethyl vinyl ether, $CH_3CH_2OCH=CH_2$, C_4H_8O. B.p. 35°C. Prepared by addition of ethyne to ethanol in the presence of catalysts. Readily polymerized, high mol.wt. polymers are used as adhesives, coatings, films; low mol.wt. polymers are used as plasticizers and resin-modifiers. Used in synthesis.

ethyne (acetylene), C_2H_2, $HC \equiv CH$. Colourless gas which when pure has a pleasant smell. M.p. –84.7°C, b.p. –80.7°C.

The compound has very wide explosive limits in air (2.3% and 80% ethyne by volume) and apart from this the liquid, and also the gas if the pressure exceeds 2 atm., are explosive even in the absence of air. For storage and transportation steel cylinders containing propanone and a porous material such as diatomaceous earth are used.

The original method for the manufacture of ethyne, the action of water on calcium carbide, is still of very great importance, but newer methods include the pyrolysis of the lower paraffins in the presence of steam, the partial oxidation of natural gas (methane) and the cracking of hydrocarbons in an electric arc.

Ethyne is the starting point for the manufacture of a wide range of chemicals, amongst which the most important are acrylonitrile, vinyl chloride, vinyl acetate, ethanal, ethanoic acid, tri- and perchloroethylene, neoprene and polyvinyl alcohol. Processes such as vinylation, ethinylation, carbonylation, oligomerization and Reppe processes offer the possibility of producing various organic chemicals cheaply. Used in oxy-acetylene welding. Polyethyne, polyacetylene can be doped with Br_2 or I_2 to become a metallic conductor.

Like all terminal alkynes, the ethynic hydrogen atoms are acidic in the sense that they are removed by suitable strong bases such as sodamide or a Grignard reagent to give $RC \equiv CNa$ or $RC \equiv CMgX$. (For ethyne R=H, although a second mole of base will also replace this hydrogen atom.) Many metals will give salts (see ACETYLIDES) which may be used for purification purposes, e.g. silver acetylide, $AgC \equiv CH$, a white shock-sensitive solid, insoluble in water, regenerates ethyne on treatment with mineral acids.

etobenzamide Herbicide, benzanilide derivative.

etofenprox Insecticide, pyrethroid.

etrimfos Insecticide, acaricide, organophosphate derivative.

Eu Europium.

eucalyptol (1,8-cineole, 1,3,3-trimethyl-2-oxa-bicyclo[2,2,2]octane), $C_{10}H_{18}O$. A very widely distributed constituent of essential oils; often considered as the active constituent of medicinal eucalyptus oil. Also present in wormseed oil, cajuput oil and in various eucalyptus oils. It is a colourless, viscous oil of characteristic camphor-like smell. M.p. –1°C, b.p. 174.4°C. Used in flavourings.

eugenol (4-allyl-2-methoxyphenol), $C_{10}H_{12}O_2$. Colourless liquid; b.p. 254°C. The chief constituent of oil of cloves, and also in other natural oils. Used in dentistry as a strong antiseptic and local anaesthetic. Used in perfumery, manufacture of vanillin, isoeugenol.

europium, Eu. Lanthanide metal, at. no. 63, at.mass 151.96, m.p. 822°C, b.p. 1529°C, density ρ 5244, crustal abundance 2.1 p.p.m., electronic configuration [Xe] $4f^7 6s^2$. The metal is bcc and is used as a neutron absorber in reactor technology. Eu compounds are used as red phosphors in colour television and as n.m.r. shift reagents.

europium compounds Europium forms compounds in the +3 and +2 oxidation states, Eu^{3+} (f^6 pale pink) → Eu –2.41 volts in acid. Europium(III) salts are typical lanthanide derivatives. Europium(II) salts are pale yellow in colour and are strong reducing agents but stable in water. EuX_2 are prepared from EuX_3 + Eu (X = Cl, Br, I) or EuF_3 + Ca; $EuCl_2$ forms a dihydrate. $EuSO_4$ is prepared by electrolytic reduction of Eu(III) in sulphuric acid. Eu(II) is probably the most stable +2 state of the lanthanides

$$Eu^{3+} \rightarrow Eu^{2+} (-0.43 \text{ volts})$$

eutectic When a mechanical mixture of two substances shows a sharp melting point, the mixture is said to be a eutectic mixture. The temperature at which this occurs is called the eutectic temperature.

eutectic point The point on the phase diagram of a mixture of substances which represents the melting point and the composition of a eutectic mixture. A eutectic melts without change in composition. The presence of eutectic points in mixtures of substances may be taken to indicate the formation of compounds between the components of the mixture in many, although by no means all, cases.

eutrophication Excessive growth of algae and higher plants on natural waters because of increased nutrients, particularly phosphates, etc., from sewage, detergents and fertilizers.

EVA plastics Copolymers of ethene and vinyl acetate (vinyl ethanoate).

evaporation All liquids and solids have a characteristic vapour pressure depending on the temperature. In a closed vessel, after a sufficient time has elapsed for equilibrium to be established, as many molecules leave the liquid surface to form vapour as return to it from the vapour phase to form liquid. In an open vessel, however, no such equilibrium is set up and if the molecules of vapour immediately over the liquid surface are removed, further liquid molecules vaporize to take their place. In this way the liquid bulk is continuously diminished and the process of evaporation occurs. In the chemical and process industries the term normally refers to the removal of water from a solution.

evaporators Equipment for the concentration of solutions by evaporation. Many types are available.

EXAFS Extended X-ray absorption fine structure spectroscopy. A spectroscopic technique which can determine structure very precisely. Uses high-intensity X-rays (e.g. from a synchrotron).

exchange reaction A reaction which occurs without chemical change. Usually applied to reactions involving isotopic replacement, e.g.

$$^{16}O_2 + 2H_2^{18}O \rightleftharpoons ^{18}O_2 + 2H_2^{16}O$$

or $H_2 + D_2 \rightleftharpoons 2HD$. Because of differences in zero point energies caused by the differing masses of the isotopes, the equilibrium constants in such reactions do not have a value of unity. Use of this fact is made in exchange reactions for the separation of isotopes.

excimer An eciplex LASER.

excipients Inert additives (pigments, solvents, filters) added to fine chemicals and made into dosage forms. Performance chemicals (e.g. heat stabilizers) are also added.

exciplex A molecular charge transfer complex formed by the complexing of a donor (or acceptor) in an excited state with an acceptor (or donor) in its ground state. An exciplex is stable in only an energetically excited state. Exciplexes

formed between atoms, e.g. Xe and Cl, can act as lasers.

excited state A state of an atom, molecule, etc., when the species has absorbed energy and become excited to a higher energy state as compared with the ground state. The excitation may be electronic, vibrational, rotational, etc.

exclusion principle See PAULI EXCLUSION PRINCIPLE.

exergonic With negative free energy change at constant T and P, cf. endergonic.

exo See ENDO.

exopeptidases A class of peptide-cleaving enzymes which act on the terminal amino-acid residues. They are further divided into carboxypeptidases acting on the carboxyl end, and aminopeptidases acting on the amino end of the peptide. See also ENDOPEPTIDASES.

exothermic reaction A reaction in which heat is evolved.

expanded plastics See CELLULAR PLASTICS.

explosion The process of an explosion. There is rapid increase in rate with rise in temperature. Reaction may lead to chain branching. Rapid combustion in homogeneous fuel/air mixtures with the FLAME FRONT passing through the mixture from the source of ignition. Explosions may occur in constant pressure or constant volume situations. In the former there are only local pressure increases at the flame front where the reaction is completed but there are no significant pressure increases in the passageway or shaft. In constant volume explosions the flame front spreads very rapidly across the container or cylinder and high turbulence creates rapid flame propagation, giving an irregular flame front and high pressure increases.

explosive cladding A coating process in which detonation imports a high velocity to the coating material which then collides with the material to be coated under such conditions that bonding, e.g. metallic bonding, occurs. Used, e.g. in the cladding of the interior surfaces of autoclaves.

explosives Substances or mixtures which, when submitted to shock, friction, sparks, etc., undergo rapid decomposition with the production of a considerable quantity of heat and large volumes of gas. The three main classes of explosives are:

Propellants. Explosives which burn at a steady speed and can be detonated only under extreme conditions.

Initiators. Explosives such as mercury fulminate and lead azide and lead styphrate which are extremely sensitive to mechanical shock, and are accordingly used in small quantities in detonators to initiate the explosion of larger masses of less sensitive material.

High explosives. Those explosives which normally burn without undue violence when ignited in an open space, but which can be detonated by a sufficiently large sudden mechanical or explosive shock.

expression The compression of a two-phase solid–liquid system in order to separate the liquid from the solid which is retained in the pressing equipment. Expression may be used instead of filtration where slurries are too viscous for pumping, or as an alternative to solvent extraction in oil production from vegetable seeds. Various forms of batch or continuous press or roller mills are used industrially.

In biochemistry the molecular signals which permit a gene to be translated (i.e. expressed). See DICTIONARY OF BIOLOGY.

extender A compounding ingredient, frequently only a filler, for rubber.

external indicators A virtually obsolete series of indicators used outside the reaction vessel; the indicator is used by transferring a drop of solution from the vessel to a drop of the indicator.

extinction coefficient (ε) A quantity which is characteristic of a medium which absorbs light. For intensity τ, $\tau = \tau_o 10^{\overline{Jl}}$ where l is the polar absorption coefficient–extinction coefficient and J is the molar concentration of absorbing species and ε the molar absorption coefficient and l the length of the sample. The extinction and absorption coefficients (K) are related; $ε = 0.4343$ K. See BEER'S LAW.

extract In solvent extraction a portion of the feed is preferentially dissolved by the solvent and recovered by distilling off the solvent. This constitutes the extract.

extraction The removal of soluble material from a solid mixture by means of a solvent, or the removal of one or more components from a liquid mixture by use of a solvent with which the liquid is immiscible or nearly so. See LEACHING, LIQUID–LIQUID EXTRACTION.

extractive distillation If two substances have boiling points close together, or form an azeotrope, separation by normal fractionation is not feasible. Extractive distillation is a method whereby a third substance is added which decreases the volatility of one compound relative to the other, and makes separation by distillation possible.

extranuclear electrons See ELECTRONIC CONFIGURATION.

F

F Fluorine.

F centre An anionic site in a crystal occupied only by an electron, e.g. NaCl plus Na vapour gives blue $Na_{1+x}Cl$ containing F centres.

face-centred cubic The CUBIC CLOSE-PACKED LATTICE (CCP). See FACE-CENTRED LATTICE.

face-centred lattice A lattice having an atom at the face centre of each unit cell.

FAD Flavin-adenine dinucleotide.

fahl ore, Cu_3SbS_3. Tetrahedrite.

Fajans' method The titration of Cl^- with Ag^+ using fluorescein as an ADSORPTION INDICATOR. At the end point the precipitate becomes red.

Fajans' rules Set of empirical rules. Ionic compounds are most readily formed by:
1 ions with low charge (e.g. Na^+ more than Al^{3+});
2 by large cations (Cs^+ more than Na^+);
3 by small anions (Cl^- more than I^-).

famoxadone Fungicide, oxazolidine derivative.

famphur Insecticide, organophosphate derivative.

Faraday (F) The electrical charge of one mole of electrons. 1 Faraday = 96 485 coulombs $mole^{-1}$.

Faraday effect (Faraday rotation) The rotation of plane-polarized light when passed through a solid or liquid, which is placed in a magnetic field. See DICTIONARY OF PHYSICS.

Faraday's laws of electrolysis These state: **1** the amount of decomposition during an electrolysis is proportional to the quantity of current passed; **2** for the same quantity of electricity passed through different solutions the extent of decomposition is proportional to the chemical equivalent of the element or group liberated.

farnesol, $C_{15}H_{26}O$. 3,7,11-trimethyldodeca-trans-2-trans-6,10-trienol. Sesquiterpene alcohol found in many essential oils. B.p.120°C/0.3 mm. Has a characteristic smell and is used in perfumery. Oxidation gives the aldehyde farnesal.

Farnesol pyrophosphate is an immediate precursor of SQUALENE, the key intermediate in steroid and triterpenoid biogenesis.

Farnesol with nerolidol, bioirritant pheromone.

fats,
$R^1C(O)O)CH_2 \cdot CH(O(O)CR^2)CH_2(O(O)CR^3$.
The fats are esters of fatty acids with glycerol with the general formula shown; R^1, R^2 and R^3 may be the same fatty acid residue, but in general the fats are mixed glycerides, each fatty acid being different. The fatty acids present in the greatest quantity in fats are oleic, palmitic and stearic acids. The term 'oil' is usually applied to those glycerides which are liquid at 20°C, and the term 'fat' to those that are solid at that temperature.

Fats are hydrolysed to glycerol and fatty acids by boiling with acids and alkalis, by superheated steam and by the action of lipases. If alkalis are used for hydrolysis, the fatty acids react with the alkalis to form soaps. Alkaline hydrolysis is therefore sometimes called saponification.

When the fats are heated above 250°C they decompose with the production of acrolein, the intense smell of which is one of the best methods for detecting fats. The extraction of fats from tissues is most conveniently carried out by extraction with ether or some other solvent.

The fats are essential constituents of the food of animals, and are the major lipid store in the body. Conversion of carbohydrates to fats in the animal body does occur. They are partially absorbed from the gut as fats to the lymphatic system and partially hydrolysed by lipases and absorbed as fatty acids which are carried direct to the liver. They are split up in the body principally by oxidation at the β-carbon atom.

fatty acids The fatty acids are monobasic acids containing only the elements carbon, hydrogen and oxygen, and consisting of an alkyl radical attached to the carboxyl group. The saturated fatty acids have the general formula $C_nH_{2n}O_2$. Methanoic and ethanoic acid are the two lowest members of this series, which includes palmitic acid and stearic acid. There are various series of unsaturated fatty acids:

The oleic acid series, $C_nH_{2n-2}O_2$, with one double bond, of which acrylic acid is the lowest member.

The linoleic acid series, $C_nH_{2n-4}O_2$, with two double bonds.

The linolenic acid series, $C_nH_{2n-6}O_2$, with three double bonds.

There also exist natural fatty acids with four or more double bonds, fatty acids with hydroxy groups in the molecule, and certain cyclic fatty acids.

The lower members of the series are liquids soluble in water and volatile in steam. As the number of carbon atoms in the molecule increases, the m.p. and b.p. rise and the acids become less soluble in water and less volatile. The higher fatty acids are solids, insoluble in water and soluble in organic solvents.

The fatty acids occur in nature chiefly as glycerides (see FATS), which constitute the most important part of the fats and oils, and as esters of other alcohols, the waxes. The naturally occurring fatty acids are mostly the normal straight-chain acids with an even number of carbon atoms. Manufactured by hydrolysis of naturally occurring glycerides (e.g. tallose, palm oil). Used in personal care products, lubricants, cleaners, coatings, adhesives, fabric conditioners. U.S. use 1997 850 megatonnes.

Fe Iron.

Fehling's solution A solution of copper sulphate, sodium potassium tartrate and NaOH used for estimating and detecting reducing sugars.

feldspars (felspars) A group of ALUMINO-SILICATES with framework structures with Al or Si in tetrahedral co-ordination. The most common constituents of igneous rocks. Subdivided into two groups depending on the detailed structural types. Used in the glass, ceramic and enamelling industries.

felspars See FELDSPARS.

femtochemistry See FEMTOSECOND SPECTROSCOPY.

femtosecond spectroscopy The study of the movement and development of atoms in molecules during reactions (10^{-15}s). Uses pulsed lasers.

fenacimol Fungicide, pyrimidine derivative.

fenamiphos Nematocide, organophosphate derivative.

fenazaquin Acaricide, quinazoline derivative.

fenbuconazole Fungicide, triazole derivative.

fenbutatin oxide Acaricide, organotin derivative.

fenchenes, $C_{10}H_{16}$. A group of generally bicyclic terpenes derived chemically from fenchone, in most cases with intramolecular rearrangement, of such a character that their carbon skeletons differ from those of fenchone.

fenchone (D-1,3,3-trimethyl-2-norbornanone), $C_{10}H_{16}O$. A dicyclic ketone of camphor-like smell, the (+)-form of which is an important constituent of fennel oil and of certain lavender oils; the (–)-form is found in thuja oil. Used in perfumery and food flavours.

fenclorim Herbicide, pyrimidine derivative.

fenfuram Fungicide, carboxamide derivative.

fennitrothiin Insecticide, organophosphate derivative.

fenobucarb Insecticide, carbamate derivative.

fenothiocarb Acaricide, thiocarbamate derivative.

fenotin Fungicide, organotin derivative.

fenoxycarb Insecticide, carbamate derivative.

fenpropamorph Fungicide, morpholine derivative.

fenpropathin Acaricide, insecticide, pyrethroid.

fenpropidin Fungicide, piperidine derivative.

fenpyroximate Acaricide, pyrazole derivative.

fenthio Insecticide, organophosphate derivative.

fentin Fungicide, acaricide, molluscide, organotin derivative.

fention Insecticide, organophosphate derivative.

Fenton's reagent An aqueous solution of $FeSO_4$ or other Fe^{2+} salt and hydrogen peroxide used for oxidizing polyhydric alcohols.

fenvalerate Insecticide, acaricide, pyrethroid.

fenvuron Herbicide, urea derivative.

FEP plastics Copolymers of tetrafluoro-ethene and hexafluoropropene. Inert and corrosion resistant as TEFLON but can be processed by melt techniques.

ferbam Fungicide, dithiocarbamate derivative.

ferimzone Fungicide, pyrimidine derivative.

fermentation Controlled microbial action to give useful products. Ethanol, lactic acid, ethanoic acid, gluconic acid, glutamic acid and many other amino-acids are produced by fermentation. Many pharmaceuticals, e.g. sterols, are produced by partial fermentation.

Fermi level The energy of occupied states in a solid at zero K.

fermions Particles with half-integral spin, e.g. electrons, protons; cf. boson.

Fermi resonance The near coincidence of the energy of an overtone(s) of a fundamental vibration with another fundamental of the same species, resulting in enhanced intensity.

fermium, Fm. Actinide element, at.no. 100, at.mass ^{257}Fm 257.1, m.p. 1527°C, electronic configuration [Rh] $5f^{12}7s^2$. Formed by multiple neutron irradiation of Am and Cm, it is purified by ion exchange chromatography.

fermium compounds No solid fermium compounds are known. The element exhibits +2 and +3 oxidation states in solution. Fermium(II) has been obtained by reduction of Fm(III) with Mg metal and stabilized in an $SmCl_2$ lattice.

ferrates Formally oxyanions of iron.

Ferrates(VI), $[FeO_4]^{2-}$. Red-purple salts ($Fe(OH)_3$ in alkali plus Cl_2; anodic oxidation of Fe in alkali). Stable in basic solution but strongly oxidizing in acid solution. Na and K salts are soluble in water.

Ferrates(IV), e.g. Sr_2FeO_4. Mixed metal oxides.

Ferrates(III), e.g. M^IFeO_2, $M^{II}Fe_2O_4$ (spinels). Mixed metal oxides. Mixed metal oxide ferrates can contain Fe(III) and Fe(IV). Ferrates(III) are of great importance. See FERRITES. Hexahydroxyferrates, $M_3^{II}(Fe(OH)_6)_2$ are formed from $M(OH)_2$ and iron(III) salts.

Ferrates(II), e.g. $Na_4[Fe(OH)_6]$ are formed in alkaline solutions.

ferredoxins Non-heme iron–sulphur proteins isolated from bacteria and plants. Contain Fe_2 and Fe_4 clusters. Involved in, e.g. nitrogen and carbon dioxide fixation and photosynthesis.

ferric compounds Iron(III) compounds.

ferrichromes Growth-promoting iron chelates.

ferricyanides See CYANOFERRATES(III).

ferrimagnetism The ordering of magnetic domains in a solid leading to non-zero magnetic susceptibility.

ferrite Bcc iron.

ferrites Chemically incorrect nomenclature. Mixed metal oxides of Fe(III). Used in electronics industry for any useful magnetic iron oxide.

ferritin A soluble iron-containing protein which is important in the absorption and storage of iron in the body.

ferrocene (di-π-cyclopentadienyl iron, bis(h⁵-cyclopentadienyl) iron), $C_{10}H_{10}$Fe.

M.p. 174°C, b.p. 249°C. The prototype of metallocenes, a typical CYCLOPENTADIENYLIDE. A sandwich compound. Oxidized to the blue ferricinium cation ((h^5-$C_5H_5)_2$Fe)⁺. Derivatives used in anti-knocks.

ferrocyanides See CYANOFERRATES(II).

ferroelectrics Compounds which show dielectric hysteresis, i.e. a reversible spontaneous polarization due to non-cancellation of the elementary dipoles in a crystal. Examples include the low-temperature $BaTiO_3$ and KH_2PO_4.

ferromagnetism Magnetic behaviour which does not follow the Curie–Weiss law and is field strength dependent. Has large positive magnetic susceptibility (χ) and the magnetic moment (M) does not fall to zero when the applied magnetic field is removed. All the magnetic moments within a crystallite can be considered parallel and the orientations in different crystallites become aligned in a magnetic field. Above the Curie point (temperature) becomes paramagnetic.

ferrosilicon Used in steels and cast iron. World use 2000 1.8 megatonnes.

ferrous Iron(II) compounds.

fertilizers Although any substance which increases production when added to the soil could be called a fertilizer, usually contain N, P or K. Ammonium sulphate is the most important nitrogenous fertilizer; sodium nitrate, ammonium nitrate, ammonia and urea are also used. A mixture of ammonium nitrate and calcium carbonate is sold under the name of Nitro-chalk. The most important fertilizers containing phosphorus are the SUPERPHOSPHATES. Total world use 135 megatonnes p.a.

ferulic acid Phenolic derivatives from plants used in food preservatives and cosmetics.

fibre reactive dyestuffs Reactive dyestuffs which form a direct chemical link to the OH groups of cellulose fibres and thus form very resistant colours.

fibres Materials which may be spun and are used in textiles, etc. Natural fibres include wool, cotton, etc. Inorganic fibres including Al_2O_3, $Al_2O_3 \cdot B_2O_3$, SiO_2, glass-wool and other materials used for insulation and packing are now available. Synthetic fibres are based either on modified natural materials – cellulosics, e.g. rayon, cellulose acetate – or on completely synthetic derivatives – polyesters, nylons, glass fibres, acrylics, polyolefins. Total U.S. production 1999 45.9 megatonnes.

Fick's law of diffusion Particle flux is proportional to concentration gradient. Flux of matter is measured as molecules per m^2 per second.

fictile Molecules which rearrange very easily. See FLUXIONAL. Eg. $Fe_3(CO)_{12}$.

field-ionization microscopy (FIM) Study of surfaces using the ionization of a noble-gas atom.

Fieser's solution An aqueous alkaline solution of sodium anthraquinone 2-sulphonate (silver salt) reduced with sodium dithionite, $Na_2S_2O_4$, and used as a scrubbing solution for partially removing O_2 from, e.g., N_2.

filament wound plastics See REINFORCED PLASTICS.

fillers Materials incorporated into polymeric compounds (plastics and rubbers) to improve the general properties, to introduce particular characteristics, or to reduce the cost even though certain physical properties may be somewhat impaired. In rigid materials, fibrous fillers (woodflour, woodpulp, cotton, glass fibre, calcium carbonate) are incorporated to improve the impact strength. Other fillers are used to increase hardness (slate powder), to improve electrical properties (mica) or to change the density ($BaSO_4$). In synthetic rubbers, fillers are classified as reinforcing fillers, inert fillers or diluents. In flexible thermoplastics, smaller proportions of inert fillers are used to reduce the cost and sometimes to aid processing.

Fillers are also used in the preparation of paper to improve the properties of the material and are added to bituminous materials such as plastics and road-surfacing materials.

films Formally any material deposited, used or studied in a thin section or in layer form. Film is generally deposited whereas sheet is rolled. Vapour deposition techniques include sputtering, evaporation. Films can be of metals, oxides, etc., but are most generally thermoplastic resins, cellophane, polyethene, polypropene, polystyrene, polyvinyl chloride, etc. Films of such polymeric organics are used as free film for wrapping, photographic images, etc.

filter Apparatus or plant for filtering operations.

filter-aids Materials such as kieselguhr which are added to a slurry to increase the rate of a filtration operation.

filter press A relatively simple and widely-used type of plant for filtration under pressure. The *chamber press* has a number of recessed plates which are covered with filter cloths, the plates then being clamped together. Feed enters through a central channel and a solid cake is built up in each of the chambers, filtrate being discharged through an outlet in the body of the press.

filtration The process of separating a solid from a liquid or gas by use of a membrane or medium (e.g. filter paper, glass sinter) which will allow the fluid to pass through but not the solid. Industrial filtration uses sand beds or perforated surfaces covered by cloth or wire gauze.

FIM (field ionization) Spectroscopic technique.

fine chemicals Pure single substances used on the basis of their chemical identity and properties as drugs, for synthesis, etc.

fine structure See MULTIPLET.

fipronil Insecticide, pyrethroid.

FIR Far infrared spectroscopy at frequencies lower than about 400 cm^{-1}.

fireclay A refractory clay consisting mainly of aluminium silicate, suitable for the manufacture of furnace bricks, crucibles, etc., and melting at over 1600°C.

fire extinguishers Materials which smother and extinguish, or at least prevent propagation of flame. Water is the most widely used extinguisher. Other important extinguishers are $NaHCO_3$ solution, CO_2 foam formed from a carbonate and acid, liquids such as $ClBrCH_2$, CH_3Br, CCl_4 (Br-containing materials being phased out) and solids such as $NaHCO_3$, $KHCO_3$, $NH_4H_2PO_4$.

fire retardant An additive which promotes resistance to burning caused by contact with a high-energy ignition source. See FLAME RESISTANT MATERIALS.

first law of thermodynamics See THERMODYNAMICS, FIRST LAW OF.

first order spectra See NUCLEAR MAGNETIC RESONANCE.

Fischer projection A method of representing three-dimensional structures in two-dimensional drawings in which the chiral atom(s) lies in the plane of the paper. The two enantiomeric forms of glyceraldehyde are represented as

$$\begin{array}{cc} CHO & CHO \\ H\!-\!\!|\!-\!OH \text{ and } HO\!-\!\!|\!-\!H \\ CH_2OH & CH_2OH \end{array}$$

See SAWHORSE PROJECTIONS. Have two horizontal bonds representing atoms in front and vertical bonds representing atoms below.

Fischer–Speier esterification The catalytic use of a strong mineral acid (HCl, H_2SO_4) to effect ester formation between an alcohol and an organic acid.

Fischer–Tropsch reaction The catalytic reaction of hydrogen and carbon monoxide (SYNTHESIS GAS) to produce high-molecular weight hydrocarbons alcohols, etc.

The reaction produces a mixture of gaseous, liquid and solid paraffinic hydrocarbons.

fission See ATOMIC ENERGY.

fission products The products of the fission of heavy nuclei. Fission is generally an asymmetric process, maximum yields occurring for products of mass 90 and mass 140, the distribution of products depending upon the energy of the bombarding particle. Fission products are generally intensely radioactive.

fixation The removal of unwanted silver halide from a photographic material by solubilization after development. The most widely used fixers are thiosulphates, which form water-soluble complex salts with silver halides. Other inorganic fixers include thiocyanates, cyanides, ammonia and concentrated potassium iodide. Organic fixers include thiourea and allylthiourea (thiosinamine). Fixing baths may also include hardeners (chromium alum, potassium alum) to toughen the final gelatin layer.

flame calorimeter Calorimeter which passes reactants etc. through flame with measurement of rise in temperature.

flame emission spectroscopy Emission spectroscopy in which the emitting atom is present in a flame.

flame front The region between the luminous zone and the dark zone of unburned gases which exists in all gas phase combustion reactions. Movement of the flame front is used to measure the BURNING VELOCITY of gaseous mixtures.

flame-resistant materials (flame retardants, flameproofing materials) Compounds added to fabric to reduce tendency to flame and to burn by glowing. The materials must be resistant to laundering and cleaning. Borates, phosphates, nitrogen (e.g. diammonium phosphate), bromine, chlorine compounds, MgO, Sb_2O_3, $Al_2O_3.3H_2O$ all have good flame resistant effects.

flame speed See BURNING VELOCITY.

flammability limits A mixture of combustible gas and air will ignite only within certain limits of composition and will not support combustion if there is too little or too much combustible present.

flamprop-M Insecticide, alanine derivative.

flash-back A situation in gas burners in which the flame speed is too high so that there will be flash-back of the flame on to the gas injector tube. This can be caused by an increase of air/fuel ratio.

flash distillation A distillation technique involving very rapid removal of solvent. Used in desalination of sea-water.

flash photolysis A technique used to study very fast reactions which, usually, involve atoms or radicals in the gas phase. A powerful flash of radiation, generally from a laser, of very high energy (usually $c.$ 10^5 joules) is passed through the reaction medium for a few ns or ps thus causing dissociation, etc. Intermediates produced in the reaction may be studied by following changes in absorption of a monochromatic light source with time. This technique has the advantage that relatively high concentrations of reaction intermediates are produced.

flash point The lowest temperature at which the application of a small flame causes the vapour above a flammable liquid to ignite when the liquid is heated under standard conditions.

flavanoids Polyphenolic anti-oxidants. Materials which inhibit oxidation of, e.g. lipids.

flavin-adenine dinucleotide (FAD) The active enzyme or prosthetic group of a large group of flavoprotein enzymes. These enzymes take part in a variety of metabolic oxidation–reduction reactions.

flavin mononucleotide (FMN) Riboflavin-5′-phosphate, the coenzyme component of some flavoprotein enzymes. It is the prosthetic group of Warburg's 'old yellow enzyme' which catalyses the oxidation of NADPH.

flavone (2-phenylchromone, 2-phenyl-benzopyrone), $C_{15}H_{10}O_2$. Colourless needles; m.p. 97°C. Insoluble in water, soluble in organic solvents. Flavone occurs naturally as dust on the flowers and leaves of primulas. It has been prepared from 2-hydroxyacetophenone and benzaldehyde.
The -2,3- reduction product formed by adding 2H is *flavanone*.

flavone glycosides Widely distributed in plants. The sugar residue is either glucose or rhamnose, occasionally a biose sugar. Further isomerization is possible according to which hydroxyl is concerned in the attachment of the sugar. They are, in general, colourless. They include apiin, acaciin, diosmin, lutusin, orobosin.

flavones A group of yellow plant pigments chemically related to flavone.

flavonol glycosides Contain derivatives of flavonol attached to sugars. They include fustin, galangin, kaempferitrin, robinin, datisein, quercitrin, quercetin, incarnatrin, quercimeritrin, serotin, etc. Very widely distributed in plants.

flavoproteins A group of proteins containing flavin in their prosthetic groups. Important in electron transport systems.

flazasulfuron Herbicide, fluoropyrimidine derivative.

flint A compact massive quartz. Used in porcelain manufacture.

flocculation Aggregation of colloid particles arising from the zeta potential at high ionic strength. Can be reversible. Cf. coagulation.

flocoumafen Rodenticide, fluorocoumarin derivative.

Florisil Trade name of a magnesium silicate used for CHROMATOGRAPHY, especially of fluorine-containing compounds.

flotation A method of separation of ores in aqueous suspension by formation of a moderately stable foam by addition of a small quantity of a suitable reagent followed by aeration. The particles of ore collect in the liquid–air interfaces of the bubbles, the foam may then be removed from the top of the cell, allowed to break down and the ore concentrate recovered.

flozasulfuron Herbicide, sulphonylurea derivative.

fluazifop-butyl Herbicides, fluorophenoxy propanoate.

fluazinam Fungicide, fluoropyrimidinamino derivative.

fluazinon Anti-tick, fluorourea derivative.

fluchloralin Herbicide, fluorobenzeneamine derivative.

flucycloxuron Insecticide, acaricide, fluorobenzamide derivative.

flucythrinate Insecticide, fluoropyrethride derivative.

fludioxanil Fungicide, fluoropyrrole derivative.

flufenoxuron Insecticide, acaricide, fluorobenzylurea derivative.

flufenprox Insecticide, fluorophenoxyether derivative.

fluidization The suspension of a mass of solid particles in an upward-flowing stream of gas or liquid. The resulting mixture assumes many of the properties of a liquid, e.g. it can transmit hydrostatic forces and solid objects of smaller density will float in it.

fluidized bed A bed or mass of particles maintained in a state of FLUIDIZATION. Fluidized beds are of importance industrially, with a wide range of applications in various forms of gas–solid contacting.

flumeturon Herbicide, fluoroarylurea derivative.

flumethrin Insecticide, fluoropyrethroid.

flumetralin Plant growth regulator, fluoronitroaniline derivative.

flumetsulam Herbicide, fluorosulphonilide derivative.

flumiclorac-pentyl Herbicide, fluorophthalimide derivative.

flumioxazin Herbicide, fluorophthalimide derivative.

fluorene, $C_{13}H_{10}$. Colourless shining flakes usually showing a violet fluorescence; m.p. 116°C, b.p. 294–295°C.

Prepared synthetically by reduction of diphenylene ketone and from ethyne and hydrogen at high temperature and obtained from coal tar.

Fluorene can be nitrated, sulphonated and chlorinated. Oxidation gives fluorenone (9 O).

fluorescein, $C_{20}H_{12}O_5$. Red crystals with green iridescence. M.p. 314–316°C (decomp.). It is made by heating phthalic anhydride with resorcinol. In alkaline solution it shows intense green fluorescence. It is a dyestuff and is used for colouring liquids in various instruments and by Water Authorities for determining the route of water supplies, leakages, etc. Used in drugs and cosmetics.

fluorescence The process of slow energy emission following absorption of electromagnetic radiation. Part of the energy thus absorbed is re-emitted as radiation of longer wavelength (lower energy). The absorption of the original radiation excites a molecule or atom and some energy may be lost to vibrations. A solvent can quench fluorescence. This excited state then returns to its original level in a series of steps, each corresponding to an intermediate energy level.

X-ray fluorescence is extensively used in the analysis of geological rock specimens. Fluorescence is used as a conventional analytical technique – see FLUORIMETRY. Compare phosphorescence.

fluorescence indicator An INDICATOR which alters the intensity of, or colour of, its fluorescence according to, e.g. the pH or redox potential of the solution. Used in the titration of deeply coloured or turbid solutions.

fluorescent brightening agents (optical brighteners) Additives for detergents, paper or textiles which absorb white light and re-emit in the violet-blue. This re-emission counteracts the faint yellow colour of bleached materials.

fluoridation The addition of small quantities of fluorides to drinking water to reduce dental decay. Can also be effected through toothpaste. NaF, SnF_2, Na_2PO_3F and Na_2SiF_6 can be used.

fluorides The salts of HF. See also FLUORINE CHEMISTRY.

fluorimetry An analytical technique using the fluorescent properties of the substance to be estimated.

fluorinating agent A substance used to introduce F or replace e.g. H by F. Examples are F_2, KF, HF, SF_4.

fluorine, F. Gaseous element, at.no. 9, at.mass 18.998, m.p. –219.67°C, b.p. –118.12°C, relative density ρ 1.696 g l^{-1}, crustal abundance 950 p.p.m., electronic configuration $[He]2s^2 2p^5$. The lightest and most reactive of the halogens. The chief commercial ore is FLUORITE, CaF_2, but many silicates contain some fluorine. The element is obtained by electrolysis of a KF–HF melt using a carbon anode; hydrogen is liberated at the cathode and must be kept rigidly separated from the fluorine. Fluorine can be handled in metal (copper, nickel, monel) and fluorocarbon apparatus and stored as a gas in cylinders or as liquid fluorine. Elementary fluorine consists of F_2 molecules but they have a relatively low dissociation energy and react with most elements (metals tend to be coated by a coherent film of fluoride). Fluorine gas, HF and fluorides are toxic in large amounts but fluorides are essential to life and beneficial in the prevention of dental caries. Fluorine compounds are used in the preparation and handling of metals (e.g. U, Al), in inert plastics, refrigerants, aerosol propellants, fluorine-containing drugs, toothpaste, water treatment. Much of fluorine is used as HF.

fluorine chemistry Fluorine is the lightest halogen, electronic configuration $1s^2 2s^2 2p^5$. It forms a single series of compounds in the -1 oxidation state ($E° - 2.82V$). High oxidation state compounds are covalent (e.g. WF_6, SF_6) low oxidation state compounds are ionic (e.g. NaF, CaF_2). In covalent fluorides the co-ordination number of F is 1 but higher co-ordination numbers are known in ionic fluorides. The F^- ion is a good complexing agent and can bridge (the bridging angle frequently approaches 180°). Fluorine forms compounds with all of the elements except He, Ne, Ar (compounds are formed here in excited states). It tends to bring out the highest oxidation states of the element and to reduce metal–metal bonding.

fluorine-containing drugs The presence of fluorine has considerable, mainly favourable, effects on the biological activity of molecules, e.g. fluorouracil. See also the many insecticides with names starting FLU.

fluorine-containing polymers Inclusion of fluorine in polymers affects the properties of the polymers greatly, particularly in conferring stability and favourable surface properties. POLYTETRAFLUOROETHENE $(-CF_2-)_n$, Teflon, PTFE is a highly crystalline solid prepared by free-radical polymerization of C_2F_4. It does not flow and fabrication uses powder metallurgy techniques. It is stable to above 250°C and is used where toughness, or good electrical insulation is required. Polyhexafluoropropene $(-CF(CF_3)CF_2-)_n$ is much more costly but is more readily fabricated. Polychlorotrifluoroethene $(-CF_2CFCl-)_n$ is almost as inert as PTFE but can be fabricated by moulding and extrusion; electrical properties are inferior to those of PTFE. Vinylidene fluoride polymers $(-CF_2CH_2-)_n$ and particularly copolymers with $CF_3CF=CF_2$ (Viton) and $CClF=CF_2$ (Kel-F) are still resistant. Polyvinylfluoride $(-CH_2CHF-)_n$ is used in films and coatings particularly in the construction industries. Perfluoropolyethers have applications as inert fluids. Inert elastomers, e.g. vinylidene fluoride and hexafluoropropene copolymers are known.

fluorine halides See HALOGEN FLUORIDES (under Cl, Br, I).

fluorine oxides See OXYGEN FLUORIDES.

fluorite (fluorspar), Mineral CaF_2. The white

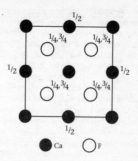

Ca ◯ F

cubic crystals are often coloured yellow, blue, green or violet and have been used for making ornaments (blue-john). It is the chief commercial source of fluorine and fluorides (e.g. HF) and is also used in glasses, enamels and as a metallurgical flux. The crystal structure of fluorite has a cubic close-packed array of Ca^{2+} with each F^- in tetrahedral holes. This structure is adopted by many oxides, fluorides and alloys, e.g. CeO_2 and $PbMg_2$. Also a pyrochlore-like material, e.g. Er_2ZrO_7, with considerable radiation resistance.

fluoroacetamide Rodenticide.

fluoroacetates See FLUOROETHANOATES.

fluoro acids Complex fluoro salts are formed by most elements but the free acids are, in general, not known (e.g. HF and BF_3 combine only in the presence of a base, e.g. NH_3). Salts are prepared by the action of F_2 on a mixture of alkali and other metal halides, by the action of HF, BrF_3, ClF_3, on appropriate mixtures of salts. The complex halide salts decompose on heating. The distinction between complex fluorides, which definitely contain complex anions, and mixed metal fluorides, e.g. $KMgF_3$, K_2UF_6, which do not contain discrete fluoro-anions is not clear-cut structurally.

fluoroapatite, $Ca_5(PO_4)_3F$. Naturally occurring phosphate rock. Of importance in teeth, where it resists dental decay, and in the production of fertilizers.

fluoroborates Salts of FLUOROBORIC ACID.

fluoroboric acid, HBF_4. Formed, with $B(OH)_3$, from BF_3 plus water. Fluoroborates contain the tetrahedral $(BF_4)^-$ ion which is only very weakly basic.

fluorocarbon polymers See FLUORINE-CONTAINING POLYMERS.

fluorocarbons A class of organic compounds in which some or all of the hydrogen atoms have been replaced by fluorine. They have very similar freezing and boiling points to the corresponding hydrocarbons but are denser and have high viscosities and low surface tensions. They are often unreactive and have great thermal stability. Fluorocarbons can be made by fluorination of hydrocarbons with fluorine or inorganic fluorinating agents such as AgF_2 and CoF_3 or by the replacement of another halogen, e.g. chlorine, by the action of an alkali halide in a solvent of high dielectric constant. They may also be prepared by fluorination in an electrolytic cell using anhydrous hydrogen fluoride and the substance to be fluorinated as electrolyte. Derivatives of hexafluorobenzene, C_6F_6, show many of the typical properties of aromatic compounds. Fluorocarbon oils, greases and inert dielectrics are used under conditions in which ordinary materials would be attacked. Perfluoro and perfluorochloro polymers are used as plastics (see FLUORINE-CONTAINING POLYMERS). Chlorofluorocarbons were formerly used as refrigerants and aerosol agents and propellants but destroy the ozone layer. Many fluorine-containing molecules are valuable surfactants. Incorporation of fluorine has major effects on physiological activity and many fluorine-containing products are used in medicine.

1-fluoro-2,4-dinitrobenzene (2,4-dinitrofluorobenzene, DNF, Sanger's reagent), $C_6H_3FN_2O$. Yellow liquid; b.p. 137°C/2mm. Reagent introduced by Sanger for labelling the terminal amino group of peptides. Probably carcinogenic.

fluoroethanoic acid, CH_2FCO_2H. Poisonous, blocks Krebs cycle. Some salts occur naturally in plants. Used as rodenticides.

fluoroform The trivial name for CF_3H, trifluoromethane.

fluoroglycofen-ethyl Herbicide, diphenyl ether derivative.

fluoroimide Fungicide, fluoromaleimide derivative.

fluorophosphonates See FLUOROPHOSPHORIC ACIDS.

fluorophosphoric acids A series of anions $(PF_6)^-$, $(PO_2F_2)^{2-}$, $(PO_3F)^{2-}$ are formed from PF_5 and HF and by hydrolysis of PF_5, etc. Hexafluorophosphates, M^IPF_6, have solubilities similar to perchlorates (BrF_3 on MF plus P_2O_5 or from hydrogen fluoride and phosphorus oxyacids).

Monofluorophosphates, (fluorophosphonates), $M^I_2PO_3F$ (H_3PO_4 in conc. HF) resemble sulphates. Esters (Ag_2PO_3F plus RX) exceedingly toxic and inhibit cholinesterase action causing muscular constriction. Derivatives used as insecticides and nerve gases. Salts of $PO_2F_2^{2-}$ used in toothpaste.

fluorosilicates Salts containing $(SiF_6)^{2-}$ (octahedral) and $(SiF_5)^-$ (trigonal bipyramidal) species; the free acid is unknown but solutions formally containing H_2SiF_6 are known (SiO_2 plus aqueous HF) and are formed in superphosphate manufacture from fluorine-containing phosphates.

fluorosulphates, MSO_3F. See FLUOROSULPHURIC ACID.

fluorosulphuric acid (fluorosulphonic acid), HSO_3F. B.p. 163°C. Stable acid (distillation of KHF_2 and oleum). Fluorosulphates are formed from the acid or by the action of peroxydisulphuryl difluoride, $FS(O)_2OOS(O)_2F$ (F_2 on SO_3), on the elements. They are similar to perchlorates. HSO_3F is a major constituent of various strongly acidic mixtures, e.g. $HSO_3F·SbF_5$, super acids, magic acid. Used to catalyse alkylation.

fluorous biphasic catalysis Use of fluorinated solvents with catalysed reactions being carried out in one phase and then moved to other phase.

fluorspar See FLUORITE.

Fluothane A trade name for HALOTHANE.

flupoxam Herbicide, fluorocarboxamide derivative.

flupropanate Herbicide, fluoropropanoate derivative.

flupyrsulfonmethyl-sodium Herbicide, fluorosulphonylurea derivative.

fluquinconazole Fungicide, fluoroazole derivative.

flurazole Herbicide, fluorothiazole derivative.

flurenol Herbicide, fluorofluorene derivative.

fluridone Herbicide, fluoropyridinone derivative.

flurochloridone Herbicide, fluoropyridinone derivative.

fluroxypyr   Herbicide, fluorooxyethanoic acid derivative.

flurprimidol   Plant growth regulator, fluoro-pyrimidylcarbinol derivative.

flurtamone   Herbicide, fluorofuranone derivative.

flusilazole   Fungicide, fluorosiliconazole derivative.

flusulfamide   Fungicide, fluorosulphon-amide derivative.

fluthiacet-methyl   Herbicide, fluorothio-acetal derivative.

flutolanil   Fungicide, fluorocarboxamide derivative.

flutriaful   Fungicide, fluorotriazole deriva-tive.

tau-fluvalinate   Insecticide, acaricide, fluoropyrethrin.

flux   1 An additive assisting in fusion – e.g. $CaCO_3$ in iron-smelting – although the lime-stone also helps with slag formation.
2 A substance enabling two pieces of metal to be joined without oxidation.
3 Flux of matter. See FICK'S LAW OF DIFFUS-ION.

fluxional   Molecules which rearrange so easily on a particular time scale (e.g. n.m.r.) that the rearrangement process can be detected are termed fluxional or stereochemically non-rigid. Example PF_5, which although it is a trigonal bipyramid shows only one type of fluorine by ^{19}F n.m.r. spectroscopy. The rearrangements can often be shown and studied by reduction of temperature. Molecules which are fluxional over all available tempera-tures are termed FICTILE.

fluxofenim   Herbicide, fluoro-oxime deriva-tive.

fly ash, Na_2CO_3. Sodium carbonate.

Fm   Fermium. Actinide element, at.no. 100.

FMN   See FLAVIN MONONUCLEOTIDE.

foamed plastics   See CELLULAR PLASTICS.

foams   A foam is a coarse dispersion of a gas in a liquid, most of the phase medium being gas with the liquid in thin laminar sheets between the gas bubbles. Foaming does not occur with pure substances and is developed by agitation of liquids and gases in the presence of stabiliz-ing agents. These are generally surface-active

substances (e.g. lauryl alcohol) which greatly lower the surface tension of the solution and which are themselves strongly adsorbed at the gas/solution interface. This adsorption process must take place very rapidly if break up of the foam is to be avoided. The persistence of differ-ent foams ranges from a few seconds (e.g. amyl alcohol solution) to hours and even months (soap and protein solutions).

Foams are used domestically and industrially and are important in rubber preparations (foamed latex) and in fire fighting. The foam floats as a continuous layer across the burning surface, so preventing the evolution of inflam-mable vapours. Foams are also used in gas ab-sorption and in the separation of proteins from biological fluids. See ANTI-FOAMING AGENTS.

folic acid   (pteroyl-L-glutamic acid, vitamin B₉, vitamin M)   Pterin-p-amino-benzoic acid – glutamic acid. Folic acid and its derivatives (mostly the tri- and heptaglutamyl peptides) are widespread in nature. It is a specif-ic growth factor for certain micro-organisms, but in animals the intestinal bacteria provide the small quantities needed for growth. Added to food. Green leaves are especially rich in this vitamin.

folpet   Fungicide, N-trihalomethylthio derivative.

fomesafen   Herbicide, fluorodiphenyl ether.

fonofos   Insecticide, organophosphate derivative.

food additives   Chemicals added to food-stuffs to preserve, improve flavour and appear-ance. Amongst types of food additives are anticaking agents (e.g. calcium phosphates), antifoaming agents (silicones), bleaching agents (peroxide), colouring agents, emulsi-fying, gelling and thickening agents (e.g. sea-weed colloids), enzymes (e.g. carbohydrases), firming agents (calcium salts), flavouring agents and enhancers (e.g. sodium glutamate), nutrients (iodide, fluoride, vitamins), pH adjusters (acids, bases), preservatives (anti-microbials, e.g. benzoic acid, SO_2, etc.), release agents (acetylated monoglycerides to prevent sticking to containers), starch modifiers (e.g. ethanoic anhydride).

force (F)   Important in chemistry in connec-tion with volume and pressure changes, surface extension and effects of electric force on rate of change of momentum, acceleration.

force constant (k) A measure of the restoring force to vibration in a bond. The stronger the bond the greater the force constant.

forchlorfenuron Plant growth regulator, urea derivative.

formaldehyde, HCHO. See METHANAL. Fungicide, bacteriostat.

formamide, $HCONH_2$. See METHANAMIDE.

formates See METHANOATES.

formation constant Stability constant.

formetinate Acaricide, insecticide, carbamate derivative.

formic acid, HCOOH. See METHANOIC ACID.

formol titration A method for estimating carboxyl groups in amino-acids and free carboxyl groups in proteins. Methanal reacts with the amino groups of most amino-acids to give neutral methylene derivatives. Thus in the presence of methanal the carboxyl groups can be estimated by titration with alkali.

formothion Acaricide, insecticide, organophosphate derivative.

formoxy The group –O·CHO.

formyl The group –C(O)H. Systematically the methanoyl group.

formylation The introduction of a formyl, methanoyl –C(O)H, group. The reaction allows the introduction of an aldehyde group. Reactions due to Gattermann, GATTERMANN–KOCH, Reimer–Tiemann (chloroform heated with sodium phenoxide) and Vilsmeier (aromatic compound, phosphorus oxychloride and N-methylformanilide) are suitable methods of formylation. Ethyl methanoate or ethyl orthoformate react with carbanions to give formyl-derivatives. Many reactions proceed through metal formyl complexes derived from carbonyl groups.

fosamine Herbicide, organophosphate derivative.

fosetyl-aluminium Fungicide, organophosphate derivative.

fosthiazole Nematocide, organophosphate derivative.

fourier transform analysis (FT analysis) The use of the mathematical technique of fourier transform analysis to obtain information from the interaction of electromagnetic radiation with scatterers, e.g. X-rays with solids to give structure, i.r. or other radiation in an interferometer to give an i.r. spectrum; n.m.r. techniques.

Fr Francium. Element at.no. 87.

fractional crystallization The process of separating the components of a mixture containing two or more materials by making use of changes in solubility with temperature.

fractional distillation The process of separating a mixture into a series of fractions of different volatilities by means of distillation.

fractionating column See RECTIFICATION.

fractionation Separation by distillation or crystallization.

francium, Fr. Radioactive alkali metal, at.no. 87, at. mass ^{223}Fr 223.02, m.p. 27°C, electronic configuration [Rn] $7s^1$. Known mainly from tracer studies. It forms a single series of Fr^+ compounds.

Franck–Condon principle The time required for an electronic transition in a molecule is very much less than the period of vibration of the constituent nuclei of the molecule. Consequently, it may be assumed that during the electronic transition the nuclei do not change their positions or momenta. This principle is of great importance in discussing the energy changes and spectra of molecules.

frankincense Gummy material from trees used in embalming. Main constituents pentacyclic terpenoids.

Frasch process A process for obtaining sulphur by passing superheated water down a shaft to liquefy subterranean natural sulphur which is blown to the surface with compressed air.

free energy A thermodynamic state function represented by G (after Willard Gibbs). The free energy change (ΔG) in any system is related to the enthalpy and entropy changes by the equation:

$$G = H - TS.$$

ΔG is a measure of the maximum amount of *useful* work which may be obtained from the change under consideration. In any system the value of the free energy change determines the position of equilibrium in that at equilibrium the free energy $\Delta G = 0$. For substances in their standard states, i.e. 1 mole at 1 atmosphere pressure and a specified temperature, the change in free energy is called the *standard free energy change*, ΔG°_f. This latter quantity is of great importance in that it determines the thermodynamic feasibility of a reaction. For a reaction to be thermodynamically feasible and thus for more products than reactants to be formed at equilibrium, the value of $\Delta G°$ must be negative, i.e. there must be a decrease in standard free energy from reactants to products. G, ΔG and $\Delta G°$ are usually expressed in kilojoules per mole for chemical reactions. The Helmholz Free Energy A is the energy change at constant pressure.

free radicals Molecules or ions with unpaired electrons and hence generally exceedingly reactive. The most stable free radicals are NO, NO_2 and molecular oxygen, O_2 (a diradical). Organic free radicals have a complete range of stability, those of relatively long life like triphenylmethyl, which exist in equilibrium with their dimeric forms in solution; and those of relatively short life like methyl, which demand special techniques for their study. Many aryl-substituted ethanes show dissociation to free radicals, and the degree of their dissociation is found to depend on (a) the nature of the solvent, (b) the concentration and temperature of the solution, and (c) the size and nature of the substituent groups.

The free radicals which have only a transient existence, like $\cdot CH_3$, $\cdot C_2H_5$ or $\cdot OH$, and are therefore met with only as intermediates in chemical reactions, can be prepared and studied directly only at low pressures of the order of 1 mm, when they may be transported from the place of preparation in a rapidly streaming inert gas without suffering too many collisions with other molecules. They may be frozen and stabilized in glasses and other matrices. Free radicals may be detected and studied by electron spin resonance spectroscopy.

freeze-drying Removal of water (or other solvent) from the frozen solid at e.g. −40° to 10°C at low pressure. Costly but avoids damage to sensitive materials; used for biological specimens and some foodstuffs.

freezing mixtures A mixture used for small-scale refrigeration consisting of salt, ice and water which freezes well below 0°C. Dry ice or liquid nitrogen and organic solvents are also used to prepare freezing mixtures. See LOW TEMPERATURE BATHS.

freezing point depression Depression of freezing point caused by a solute. The depression caused by dissolving 1 mole of the solute in 1 litre of solvent is termed the molecular depression or cryoscopic constant of the solvent. See COLLIGATIVE PROPERTIES.

Fremy's salt (potassium nitroso disulphonate), $(KSO_3)_2NO$. Oxidizing agent for phenols and anilines to quinines.

Frenkel defect See DEFECT STRUCTURES.

Freons Trade name for a group of chlorofluorocarbons. They can be made by treating a suitable chlorinated hydrocarbon with hydrogen fluoride in the presence of an inorganic chloride such as $SbCl_5$ or with SbF_3/SbF_5. They are inert substances which have been used extensively as solvents, refrigerants, blowing agents and inert dispersing agents in aerosols. Individual members are distinguished by numbers; the more important Freons include the following: Freon 12, dichlorodifluoromethane CCl_2F_2 (b.p. −30°C); Freon 21, dichlorofluoromethane $CHCl_2F$ (b.p. +9°C); Freon 142, 1-chloro-1,1-difluoroethane CH_3CClF_2. Also called Geons, Genetrons, or Arctons. Use is curtailed because of long-term effects on the upper atmosphere. See OZONE.

frequency (v Hz or $\bar{v}$) The number of wavelengths passing a fixed observer per second. Thus frequency depends on the velocity of the wave motion and upon its wavelength. A frequency (wave number) used in spectroscopy is the number of complete waves per cm (cm^{-1}), the reciprocal of the wavelength. Frequency is proportional to energy. For electromagnetic radiation the velocity c is 3×10^8 m s^{-1}.

Freundlich isotherm See ADSORPTION ISOTHERMS.

Friedel–Crafts reaction In its broadest sense this reaction covers the acylation and alkylation of aromatic compounds by an electrophilic reagent. E.g. an acid chloride RCOCl and benzene will react together in the presence of a Lewis acid to give a ketone $RCOC_6H_5$. Alkyl halides, alcohols, and under certain conditions, alkenes, may be used in the place of the acyl halide and yield the alkyl-substituted aromatic system. The Lewis acid is usually anhydrous $AlCl_3$, but HF, BF_3, $FeCl_3$, $SnCl_4$ may also be used; these serve to generate the acylium ion $(RCO)^+$ or alkyl carbenium ion R^+ which is the electrophile. Alkyl group rearrangement may occur during reaction, e.g. n-propyl bromide may yield an isopropyl substituent. The reaction is useful for the synthesis of hydrocarbons and ketones, and is used industrially.

frontier orbital symmetry The site and rate of reaction depend on geometries and relative energies of the highest occupied molecular orbital

(HOMO) of one reactant and the lowest unoccupied molecular orbital (LUMO) of the other.

froth flotation See FLOTATION.

D-fructose, $C_6H_{12}O_6$. Saccharide. Crystallizes in large needles; m.p. 102–104°C. The most common ketose sugar. Combined with glucose it occurs in sucrose and raffinose; mixed with glucose it is present in fruit juices, honey and other products; inulin and levan are built of fructose residues only. In natural products it is always in the furanose form, but it crystallizes in the pyranose form. It is very soluble in water, is twice as sweet as glucose and in many respects behaves as does glucose. Used particularly in ice-cream foodstuffs.

pyranose form

furanose form

FT-IR Fourier-transform infra-red spectroscopy. Used both to obtain the normal transmission i.r. spectrum and to study surfaces. Based on interferometric techniques.

fuberidazole Fungicide, benzimidazole derivative.

L-fucose (deoxy-D-galactose)

M.p. 145°C. The methyl pentose of gum tragacanth, of the blood group polysaccharides, and of seaweed, where its polymer fucoidin constitutes the cell walls.

fuel cell An electrical cell, which generates electricity directly by the electrochemical conversion of gaseous or liquid fuels fed to the cell as required. Fuel and oxidant are supplied to the respective electrodes, which are often porous and activated by catalysts.

Fuels which have been used include hydrogen, hydrazine, methanol and ammonia, while oxidants are usually oxygen or air. Electrolytes comprise alkali solutions, molten carbonates, solid oxides, ion-exchange resins, etc.

Batteries of fuel cells have been used in spacecraft but, in general, the high potential of the fuel cell has not been realized industrially.

fuel oils A general term applied to petroleum oil used for the production of power or heat. Fuel oils are normally divided into two categories: category 1 for engine fuels and category 2 for burner fuels. Motor and aviation gasolines are normally classified separately.

fuels Materials used for producing energy, particularly heat. They include fuel oils, coal, natural gas, synthetic gases, rocket fuels (propellants), uranium and plutonium (nuclear fuels).

fugacity (f) Used in the measure of effective pressure dependence versus the chemical potential of a gas.

fullerenes Forms of carbon generally with closed polyhedral shells, although an extensive chemistry and some interstitial chemistry (e.g. with He, Sc_2C_2) is known. Formed by ablation from graphite by an electric current or a laser. C_{60}, buckminsterfullerene, soccer-ball structure, C_{70} and other species are known and fullerene structures are related to those of nanotubes.

fuller's earth Formerly any detergent clay for fulling cloth; now a clay used mainly for its adsorptive properties in decolorizing oils, fats and waxes, e.g. for lubricants, soaps and margarine.

fulminates See FULMINIC ACID.

fulminic acid (carbonyl oxime (misnomer), carbyloxime), C=NOH. Unstable volatile compound with odour resembling HCN; polymerizes to metafulminuric acid. Hg(I) and Ag(I) salts are formed from the metal plus excess HNO_3 and added C_2H_5OH. These fulminates are explosive and shock sensitive and have been used as detonators. Fulminic acid is an isomer of cyanic acid.

fulvenes Methylene cyclopentadienes.

fumarase The enzyme catalysing the reaction between fumaric acid and water to give (–)-malic acid in the CITRIC ACID CYCLE.

fumaric acid (E-2-butenedioic acid), $C_4H_4O_4$. Crystallizes in colourless needles or prisms;

HO(O)C·CH
‖
HC·C(O)OH

m.p. 300–302°C in sealed tubes; sublimes above 200°C in open vessels. It can be manufactured by the isomerization of maleic acid by heating in the presence of catalysts, or made from carbohydrate sources by the action of moulds. When heated to 230°C it is converted to maleic anhydride; heated in sealed vessels with water at 150–170°C, it forms (±)-malic acid. Used as a food acid.

fumaryl The group $trans$–(O)C·CH=CHC(O)–

fuming sulphuric acid See SULPHURIC ACID, FUMING.

fundamental equation of thermodynamics $dU = Tds - PdV$, combining the First and Second Laws.

fundamental frequencies The characteristic vibrational modes observed in i.r. and Raman spectroscopy. The modes are those in which all the allowed vibrations can be expressed.

fungicides Chemicals that inhibit fungal attack, e.g. in agriculture, on wood, plastics, etc. Important fungicides include sulphur, polysulphides and sulphur-containing chemicals (e.g. dithiocarbonates), heavy metal (Cu, Sn, Hg, Ni) compounds.

furalaxyl Fungicide, phenylamide derivative.

furaldehyde See FURFURAL.

furan (furfuran, tetrole, oxole), C_4H_4O. Colourless liquid; b.p. 32°C. Prepared by heating furoic acid at its boiling point. Forms resins in the presence of mineral acids, but is unaffected by alkalis. Reduced by hydrogen to tetrahydrofuran. Many substituted derivatives known. Can behave as conjugated dienes.

furanose Those forms of the sugars which

contain a ring of 4 carbons and oxygen and generally, in contrast with the PYRANOSE SUGARS, how instability. They exist in α and β forms.

furathiocarb Insecticide, carbamate derivative.

furazans (1,2,5-oxadiazoles) Formed by the

action of NaOH on the dioximes of α-diketones.

furfural (furfuraldehyde, 2-furaldehyde), $C_5H_4O_2$. B.p. 162°C. Occurs in many essential oils and in fusel oil. Manufactured by heating corncobs, oat hulls or other pentose-containing material with steam under pressure at 180°C. Undergoes the Cannizzaro reaction with alkalis to give furoic acid and furfuryl alcohol. Reacts with ammonia to give furfuramide. Oxidized by sodium chlorate(V) in the presence of V_2O_5 to give fumaric acid. Forms resins with phenol, aniline and propanone. Used as a solvent, in the solvent extraction of mineral oils, in synthesis and as plasticizer.

furfuraldehyde See FURFURAL.

furfuran See FURAN.

furfuryl- The 2-CH$_2$ furyl group.

furfuryl alcohol (2-furanmethanol), $C_5H_6O_2$. Colourless liquid; b.p. 170–171°C. Prepared by the reduction of furfural or, with furoic acid, by the action of 30% NaOH solution on furfural in the cold. Forms resins with mineral acids. Solvent, wetting agent. It is poisonous.

furfurylidene The 2-furyl CH≡ group.

furilazole Herbicide safener, oxazolidine derivative.

furoic acid (2-furoic acid, furancarboxylic acid, pyromucic acid), $C_5H_4O_3$. B.p. 230–232°C. Prepared, together with furfuryl alcohol, by the action of 30% NaOH solution on furfural in the cold.

furyl- The 2-furyl or 3-furyl groups. The 2-configuration is the more common.

fusel oil A mixture of alcohols, fatty acids and esters obtained as a high-boiling fraction during the distillation of fermentation alcohol. Used for the preparation of pentanol. Toxic.

fusidic acid, $C_{31}H_{48}O_6$. An anti-bacterial steroid derivative, m.p. 192–195°C.

G

G Free energy.

g value, g factor The deviation from ideality for e.s.r. or n.m.r. spectra. Ideally 2.0023. The proportionality factor between spin magnetic moment and angular momentum. See G FACTOR.

Ga Gallium.

Gabriel's reaction The conversion of a halogen compound into the corresponding amino compound by treatment with potassium phthalimide and subsequent hydrolysis of the intermediate phthalimido compound. Thus chloroethanoic acid gives glycine. Gives a pure primary amine.

gadolinium, Gd. Lanthanide metal, at.no. 64, at.mass 157.25, m.p. 1313°C, b.p. 3273°C, density ρ 7901, crustal abundance 7.7 p.p.m., electronic configuration [Xe] $4f^7 5d^1 6s^2$. The metal is hcp (to 1262°C) and bcc (to m.p.). The largest use of Gd is in ferrites and as the selenide for electronic equipment. Gd compounds are extensively used as catalysts, particularly olefin polymerization catalysts. Gd complexes are used as magnetic relaxation agents in magnetic imaging.

gadolinium compounds Gd forms a single series of typical LANTHANIDE COMPOUNDS in the +3 state Gd^{3+} (f^7 colourless) $Gd^{3+} \rightarrow Gd$ (−2.40 volts in acid). GdI_2 (GdI_3+Gd) is $Gd^{3+}(I^-)_2 e$. Gd_2Cl_3 and GdCl ($GdCl_3$ plus Gd) have metal–metal bonding.

galactans Polysaccharides which on hydrolysis give galactose. They occur in wood and in many algae. The most important galactan is AGAR.

D-galactose, $C_6H_{12}O_6$. Crystallizes in the pyranose form; m.p. 165.5°C (anhydrous). An isomer of glucose which is fairly widely distributed in plants and animals. It is the sugar found in the brain. Chemically it is very similar to glucose. Forms galactans.

galena, PbS. A common lead ore, black or bluish-grey with a metallic lustre. It has the NaCl structure.

gallates Anionic Ga-containing species formed by dissolving Ga, Ga_2O_3, $Ga(OH)_3$ in excess base. Mixed oxides are also known.

gallic acid (3,4,5-trihydroxybenzoic acid), $C_7H_6O_5$. Colourless crystals with one molecule of water, m.p. 258°C, sparingly soluble in water and alcohol. It occurs free in woody tissue, in gall-nuts and in tea, and is a constituent of the tannins, from which it can be obtained by fermentation or by acid hydrolysis. It gives a blue-black colour with Fe^{3+} and is used in the manufacture of inks. On heating it gives pyrogallol.

Gallic acid has use as a dyestuffs intermediate and in photographic developers and tanning.

gallium, Ga. Semi-metallic element, at.no. 31, at.mass 69.723, m.p. 29.8°C, b.p. 2204°C, density ρ 6080, crustal abundance 18 p.p.m., electronic configuration [Ar] $3d^{10} 4s^2 4p^1$. Found in traces in Zn and Al minerals and in germanite (0.7% Ga). The element, metallic in appearance, has a complex structure: it is obtained by electrolysis of Ga salts. It dissolves in acids (and alkalis) and becomes coated with a thin coating of oxide on heating in air. Used in semiconductors (doping and as GaAs), light emitting diodes (GaAs; GaP; GaN), electrical devices, phosphors. World use 1999 150 tonnes.

gallium arsenide, GaAs. Important semiconductor material. Used in light-emitting diodes.

gallium chemistry Gallium is a Group 14 element. Its main oxidation state is +3 (E° for $Ga^{3+} \rightarrow Ga$ −0.56 volts in acid) and there is an extensive aqueous chemistry of the Ga^{3+} ion (generally octahedrally co-ordinated). Halides, e.g. $(GaCl_3)_2$, are dimeric and predominantly covalent. Hydride species, e.g. $GaHCl_2$, are useful reducing agents. Organic derivatives, GaR_3, are monomeric and $[R_2Ga]^+$ complexed species are stable even in aqueous solution, e.g. $[Me_2Ga(H_2O)_2]^+$, $[Me_2GaOH]_4$. Complexes are readily formed. Gallium(I) compounds are relatively unstable and occur in $GaCl_2$, GaS, etc.; contain Ga_2 units.

gallium halides Gallium(III) fluoride, GaF_3 is ionic and forms $GaF_3.H_2O$. Complexes, e.g. $(NH_4)_3GaF_6$, are formed in aqueous HF solution (decomposes to GaF_3). Gallium(III) chloride, $[Cl_2Ga(\mu\text{-}Cl)_2GaCl_2]$ is covalent and gives GaOCl with water. The bromide and iodide are similar to the chloride. $GaCl_3$ plus Ga gives $Ga^I[Ga^{III}Cl_4]$ (also Br and I) which are strong reducing agents.

gallium oxide, Ga_2O_3. Formed by heating the nitrate, sulphate, hydroxide; has various forms similar to those of Al_2O_3. Forms a hydroxide (NH_4OH on Ga^{3+} solution). Both oxide and hydroxide are amphoteric, dissolving in acids and bases (to give GALLATES).

gallium oxyacid salts These are all gallium(III) salts (Ga_2O_3 plus acid) which decompose on heating to Ga_2O_3. Known salts include $Ga(NO_3)_3.9H_2O$, $Ga(NO_3)_3.3H_2O$, $Ga(ClO_4)_3.6H_2O$, $Ga_2(SO_4)_3.18H_2O$.

gallium sulphides Ga_2S_3 (Ga plus S), GaS (Ga_2S_3 plus H_2, contains Ga_2 units) and Ga_2S (Ga plus H_2S) are known.

galvanic cell A cell that produces electricity through a chemical reaction.

galvanizing The coating of steel with Zn for protection by dipping into molten Zn or by electrodeposition (cold galvanizing). Corrosion occurs preferentially at the Zn.

gamma rays (γ-rays) Electromagnetic radiation of very short wavelength, 10^{-10} to 10^{-13}m; that is very hard X-rays. γ-Rays have great penetrating power and are emitted in most nuclear disintegrations.

gammexane See BENZENE HEXACHLORIDE.

gangliosides Sphingolipids composed of sphingosine, fatty acid, at least one sugar and sialic acid. Found in brain, nervous tissues and erythrocytes.

gangue The clay and other silicate material associated with ores.

garnets A group of silicates, $M_3^{II}M_2^{III}(SiO_4)_3$ (M^{II} = Ca, Mg, Fe; M^{III} = Al, Cr, Fe), containing discrete SiO_4^{4-} groups. Some garnets are used as gemstones with colours ranging from yellow-red to deep red and also emerald green. Garnet sand is used as an abrasive. World use 1999 245 000 tonnes.

gas The gaseous state is the most diffuse state of matter, in which molecules have almost unrestricted motion. A substance for which the volume increases continuously and without limit as the pressure is continuously reduced. A vapour is distinguished from a gas in that a gas is above the critical temperature of the substance.

gas absorption The solution of a component of a gaseous mixture in a liquid or the absorption onto a surface. Selective absorption is employed to separate components of mixtures of gases or vapours, e.g. the removal of NH_3 from NH_3–air mixtures by absorption in water.

gas adsorption Adsorption of components of a gaseous mixture in a liquid or absorption of a gas on a surface. Of great importance in catalysis. See ADSORPTION ISOTHERMS.

gas analysis A mixture of gases may be analysed by, e.g.: **1** Absorbing each constituent preferentially by means of a reagent, e.g. CO_2 in KOH solution, CO in acid or alkaline CuCl. **2** Exploding or burning the gas with oxygen or air and measuring both the change in volume and amount of waste gases formed by absorption. **3** Titration as in volumetric analysis, e.g. NH_3 and H_2S. **4** Absorbing the gas on a substance which can be weighed both before and after absorption. **5** Changes in thermal conductivity, e.g. carbon dioxide in flue gas. **6** Measurement of i.r. or u.v. spectra. **7** Vapour-phase chromatography. **8** Measurement of magnetic susceptibility.

gas calorimeter Equipment for the determination of the CALORIFIC VALUE of a fuel gas by burning at a known rate or, for small amounts of combustible gas, combustion or explosion of a known volume of gas.

gas chromatography (gas–liquid chromatography (GLC), vapour-phase chromatography (VPC)) See CHROMATOGRAPHY.

gas chromatography mass spectroscopy (GCMS) Gas chromatography with direct injection of the separated components into a mass spectrometer.

gas constant Boyle's and Charles's laws give a general equation of the form $Pv = kT$, where P is the pressure, v the volume and T the temperature of the gas, while k is a constant depending on the amount of gas used, and the units employed for expressing P and v. When 1 mole of gas is considered, k may be replaced by R, and the equation becomes $Pv = RT$. R is called the gas constant. $R = 8.314$ J mol^{-1} K^{-1}.

gas hydrates Clathrates formed by water with atomic or low-molecular weight gases and some small organic molecules. Solid water forms open cages with the gases in holes. E.g. $Cl_2, 7 \cdot 3H_2O$.

gasification Conversion of a solid or liquid feedstock to a lower molecular weight gaseous fuel of lower carbon-to-hydrogen ratio. See GASIFICATION OF SOLID FUELS, GASIFICATION OF OIL.

gasification of oil Petroleum feedstocks can be gasified to produce TOWN GAS, SYNTHESIS GAS or SUBSTITUTE NATURAL GAS (SNG). Depending on the feedstock the complete process may involve the CRACKING of large molecules followed by further gasification involving a reforming reaction, which reduces the carbon-to-hydrogen ratio of the feedstock. This may be effected by direct addition of hydrogen (hydrogasification) or by hydrogen generation from steam within the system (steam reforming). Alternatively the carbon content may be reduced by forming CO_2, coke or char which can be removed from the system.

gasification of solid fuels The conversion of the combustible matter in solid fuels to fuel gases. Various processes have been employed, ranging from the production of PRODUCER GAS, WATER GAS, to the more sophisticated LURGI PROCESS.

The main reactions involved are the exothermic producer-gas reactions, which are basically

$$C + O_2 \rightarrow CO_2$$
$$CO_2 + C \rightleftharpoons 2CO$$

and the endothermic water-gas reaction

$$C + H_2O \rightleftharpoons CO + H_2$$

If air (or oxygen) and steam are both passed through a high-temperature bed of coal or coke these reactions can be balanced, thus controlling the bed temperature and the fusion of the ash. In the higher pressure Lurgi process the gas obtained is high in methane, formed in reactions such as

$$C + 2H_2 \rightleftharpoons CH_4$$
$$CO + 3H_2 \rightleftharpoons CH_4 + H_2O$$

gas laws The laws which describe the behaviour of gases. These are Boyle's law, Charles's law, Gay-Lussac's law of gaseous volumes and Avogadro's hypothesis.

gas oil A petroleum distillate fraction intermediate between kerosine and light lubricating oil. It is used as a domestic and industrial fuel oil and as a feedstock for cracking processes. Slightly more viscous grades are used as diesel fuels.

gasoline A complex blend of hydrocarbon fractions used as a fuel for motor and aviation purposes. The overall composition may vary, depending on the processes available, but the main component is generally a straight-run NAPHTHA from the primary distillation of crude oil. This may be blended with stabilized natural gasoline, CRACKED GASOLINE from catalytic cracking of heavier fractions or from the by-products of chemical processes, components from conversion processes such as isomerization and catalytic reforming. Aviation gasoline in particular may also contain branched-chain alkanes from alkylation processes.

The various additives used to obtain the required properties include anti-knock compounds, anti-oxidants, metal deactivators, anti-icing additives, etc.

North American usage for petrol.

Gattermann–Koch reaction FORMYLATION of an aromatic hydrocarbon to yield the corresponding aldehyde by treatment with CO, HCI and $AlCl_3$; at atmospheric pressure CuCl is also required.

Gattermann's reaction A variation of the SANDMEYER REACTION; copper powder and/or copper(I) or (II) species and hydrogen halide are allowed to react with the diazonium salt solution and halogen is introduced into the aromatic nucleus in place of an amino group.

Gattermann synthesis A method for the synthesis of aromatic hydroxyaldehydes or alkoxyaldehydes. E.g. $AlCl_3$ is used to bring about the condensation of phenol with a mixture of gaseous hydrochloric acid and hydrogen cyanide.

gauche An isomer with non-eclipsed groups,

e.g. with lone pairs unaligned but not *trans*, e.g. hydrazine.

gaultherin (methyl salicylate-2-primeveroside), $C_{18}H_{26}O_{12}$. The parent glycoside of methyl salicylate, it is widely distributed in plants.

Gay-Lussac's law When gases combine they do so in volumes which are in a simple ratio to each other, and to that of the product if it is also gaseous. Thus 1 vol of nitrogen combines with 3 vol of hydrogen to form 2 vol of ammonia. The law is only approximately true and only strictly holds for ideal gases.

Charles's law, which is a quite different law, is also sometimes referred to as Gay-Lussac's law.

GC–IR Gas chromatography–infra-red spectroscopy.

GC–MS GAS CHROMATOGRAPHY–MASS SPECTROSCOPY. See CHROMATOGRAPHY.

GC–NMR Gas chromatography–n.m.r. spectroscopy.

Gd Gadolinium.

Ge Germanium.

Geiger counter A device used to detect and measure amounts of radioactivity.

gel Hydrophilic colloids are capable under certain conditions, such as lowering of temperature, of partially coagulating to a mass of intertwining filaments which may enclose the whole of the dispersion medium to produce a pseudo-solid but easily deformable mass or jelly. Such gels may, in some instances, preserve a rigidity when they contain as little as 1% of disperse phase. Gels are sometimes classified as hydrogels, alcogels, etc., according to the dispersion medium.

Inorganic gelatinous precipitates have much the same type of structure as coherent gels, and can often be made to form true gels on careful choice of conditions. Gels are sometimes divided into elastic and rigid gels. Gelatin is an example of the first type, silica gel of the second, but this classification is not strict, as even silica has a certain amount of elasticity. Gelatin gel may be converted back to the sol on heating and is thus a reversible gel; silica will not liquefy to a sol on any simple treatment, and is said to be irreversible. See COLLOIDS.

gelatin hardeners Materials used in hardening gelatin, generally in photography.

Photographic material containing gelatin can be hardened during manufacture; the process involves cross-linking between the gelatin polypeptide chains induced by the hardener.

gelatins Boiling COLLAGEN in dilute acids yields gelatins, which are roughly the constituent peptides of collagen. Gelatins swell in cold water, but are insoluble in it. They dissolve in hot water to give very viscous solutions. A solution of gelatins containing more than 1% gelatin solidifies to a jelly on cooling. They are rich in the amino-acids glycine, proline, lysine, hydroxyproline and hydroxylysine. Gelatins are manufactured from bones and hides, and are used in foodstuffs, rubbers, adhesives, and in photography.

gel chromatography See CHROMATOGRAPHY.

gel electrophoresis See ELECTROPHORESIS.

gel filtration A technique in biochemistry for separating molecules according to size. Small molecules diffuse into the pores of a gel (e.g. Sephadex), while large molecules are excluded and so are quickly eluted through a column of the gel with a continuous flow. The size of the molecule (and therefore the approximate molecular weight) can be determined by using gels of the appropriate cross-linkage. Gel filtration is commonly used to determine the molecular weight of proteins or in studies of drug-protein binding. It is a powerful tool in protein purification. Also termed gel-permeation chromatography or size-exclusion chromatography.

gelignite A form of dynamite used as an explosive.

gel-permeation chromatography (GPC) See GEL FILTRATION.

gem- A prefix indicating that particular substituents are attached to the same atom, e.g. $F_3SiSiCl_2SiF_3$, gem-dichlorohexafluorotrisilane.

gemstones Materials which because of their appearance and variety are used in jewellery and decoration. Synthetic stones simulate natural gemstones. α-Alumina (corundum) gives ruby (with Cr^{3+}) and sapphire (Fe plus Ti). Spinel, titania, $SrTiO_3$ are also used but are virtually all artificial. Diamonds can be synthesized but are then mainly used for cutting.

genetic code The relationship between the sequence of amino-acids in a protein and the sequence of nucleotides in messenger RNA which is the 'blueprint' for protein synthesis. Each amino-acid is coded by a triplet of nucleotides, e.g. phenylalanine is coded for by the pyrimidine bases uracil–uracil–uracil in messenger RNA and lysine by adenine–adenine–adenine. Most amino-acids are coded by more than one triplet (a codon) and in these cases the code is said to be 'degenerate', although it is in fact very useful in combating the effects of mutation. See also NUCLEIC ACIDS and RIBOSOMES. See DICTIONARY OF BIOLOGY.

Genetron See FREONS.

genins See STEROLS.

genome The total genetic material within a cell or individual. See DICTIONARY OF BIOLOGY.

gentian violet See METHYL VIOLET.

gentiobiose, $C_{12}H_{22}O_{11}$. The sugar of amygdalin and a number of other glycosides, including gentianose, 6-[β-D-glucopyranosido]-D-glucose. M.p. 190–195°C.

geometrical isomerism See ISOMERISM.

Geons See FREONS.

gerade See UNGERADE.

geranial See CITRAL.

geraniol (2,6-dimethyl-*trans*-2,6-octadien-1-ol) A terpene alcohol; b.p. 229–230°C. Found in a very large number of essential oils. It can also be manufactured from turpentine. It is a colourless oil, smelling of roses, and is unstable in air. It forms citral-a on oxidation. By treatment with gaseous HCl geraniol is converted into limonene. Used in perfumery and as insect attractant.

germanates Oxo compounds of germanium(IV) showing resemblances to simple silicates. E.g. $SrGeO_3$ (cyclic $[Ge_3O_9]^{6-}$) and Mg_2GeO_4 (orthogermanate). Hydrated GeO_2 dissolves in excess base to give $[Ge(OH)_6]^{2-}$ and other oxide-hydroxide anions and cations.

germanes (germanium hydrides) Monogermane, GeH_4, b.p. –88°C; Ge_2H_6, b.p. –29°C, and Ge_3H_8 are formed from GeO_2 and $LiAlH_4$ or $NaBH_4$. Higher germanes, to Ge_9H_{20}, are formed by electric discharge on GeH_4. Germanes are oxidized in air. An extensive chemistry of GeH_3^- and related groups is known. Germanes have potential as sources of very pure Ge.

germanium, Ge. Non-metallic element, at. no. 32, at.mass 72.61, m.p. 938.25°C, b.p. 2833°C, density ρ 5323, crustal abundance 1.8 p.p.m., electronic configuration [Ar] $3d^{10}4s^24p^2$. Germanium occurs in sulphides, particularly zinc ores and is obtained from cer-

tain flue dusts. It is converted to volatile $GeCl_4$ which is hydrolysed with water and GeO_2 and reduced to the element with H_2 at 600°C. It has a structure close to that of diamond; it is oxidized by heating in air and combines with chlorine on heating. Its main use is in electronics and in phosphors. Forms alloys which expand on freezing and which are used in precision casting. Used in special glasses.

germanium chemistry Germanium is a Group 14 element. The normal oxidation state is +4 in covalent $GeCl_4$, GeH_4 and ionic GeO_2. Complexes, e.g. $[GeF_6]^{2-}$, are known mainly with octahedral co-ordination. The +2 state is quite strongly reducing. Organogermanium compounds of Ge(IV) are similar to the corresponding Si or Sn compounds.

germanium halides
Germanium fluorides, GeF_4. M.p. –15°C (heat on $BaGeF_6$, Ge plus F_2) is hydrolysed by water, forms $[GeF_6]^{2-}$ in aqueous HF. GeF_2 is polymeric (Ge plus GeF_4 at 100–300°C).
Germanium chlorides, $GeCl_4$. M.p. –49°C, b.p. 86°C (Ge plus Cl_2) is hydrolysed by water. $GeCl_2$, colourless reducing agent ($GeCl_4$ plus Ge).
Bromides and iodides are similar to the chlorides.

germanium hydrides See GERMANES.

germanium oxides *Germanium(IV) oxide*, GeO_2. White solid (Ge plus O_2), forms GERMANATES. The hydrated oxide is not a simple hydroxide. *Germanium(II) oxide*, GeO. A yellow indefinite material formed by hydrolysis of $GeCl_2$.

germanium sulphides GeS_2 and GeS are known.

getter A chemical for removing impurities, most often from gases. KF acts as a getter to remove HF. Li will remove O, S, etc. from Cu and Cu-alloys.

g factor (Landé factor) The proportionality factor in the relation between the magnetic moment μ and the number of unpaired electrons

$$\mu = g\sqrt{s(s+1)}$$

For a free electron $g = 2.003$. Compare gyromagnetic ratio.

gibberellic acid (gibberellins) Plant-growth regulators, metabolites of *Gibberella fujikuroi*. Gibberellic acid is the most important member of a family of hormones called gibberellins.

gibberellins See GIBBERELLIC ACID.

Gibbs' energy (G) $G = H - TS$.

Gibbs' energy of solvation See BORN EQUATION.

Gibbs' free energy See FREE ENERGY.

Gibbs–Helmholtz equation The relation between the heats and free-energy changes which occur during a chemical reaction. For a reaction carried out at constant pressure

$$\Delta G = \Delta H - T\Delta S$$

where ΔG is the Gibbs free-energy change, ΔH is the enthalpy change and ΔS, the entropy change.

$$\left(\frac{\delta}{\delta T}\left(\frac{G}{H}\right)\right)_p = \frac{-H}{T^2}$$

Using this equation it is possible to calculate heats of reaction from the variation of ΔG with temperature.

Gibbs' isotherm The relation between the chemical potential at an interface and the change in surface tension.

Gibbs' isotherm of surface concentration The relation between the surface tension (γ) of a solution and the amount (Γ) of the solute adsorbed at unit area of the surface.

Qualitatively the equation shows that solutes which lower the surface tension have a positive surface concentration, e.g. soaps in water or amyl alcohol in water. Conversely solutes which increase the surface tension have a negative surface concentration. If the concentration dependence is known, the surface excess may be calculated, also the area per surfactant molecule.

gibbsite, γ-Al(OH)$_3$. See ALUMINIUM HYDROXIDES.

Girard's reagents Quaternary ammonium salts of the type Me$_3$NCH$_2$CONHNH$_2^+$X$^-$ which form water-soluble hydrazones with aldehydes and ketones, and are therefore separable from other neutral compounds; the aldehyde or ketone may be subsequently regenerated after separation.

glacial acetic acid See ETHANOIC ACID.

glass In general a supercooled liquid which forms a non-crystalline solid without a regular lattice. Specifically a hard, brittle, amorphous

material which is usually transparent or translucent, and resistant to chemical attack. The common lime–soda glass used for bottles and jars is a supercooled mixture of sodium and calcium silicates, prepared by fusing a mixture of sand, Na$_2$CO$_3$, lime or CaCO$_3$. Numerous special varieties of glass are now made, in which the silica is partly or (infrequently) wholly replaced by another acidic oxide, such as B$_2$O$_3$ or P$_2$O$_5$, and the Na is replaced by K (potash glass), Li, an alkaline earth metal, or Pb. 'Crown glass' contains K$_2$O or BaO as the basic constituent; 'flint glass', used largely for optical purposes, contains PbO. When glass is heated to the softening point for a prolonged period the constituents begin to crystallize and the glass 'devitrifies', i.e. there is regularity in the structure, becomes opaque and more brittle. Coloured glasses are obtained by adding small quantities of certain metallic oxides (or occasionally other compounds) to the melt. Glass can be toughened by rapid cooling of the surface or by chemical treatment of the surface. Flat sheets prepared in molten tin-float process.

glass electrode See ION-SELECTIVE ELECTRODES. Used for determination of particular ions including pH. An electrode for pH determination contains a buffer and when placed in a solution gives a potential dependent on the pH of that solution.

glass fibre Glass filaments in the form of filaments, fabric, chopped fibre, added to thermoset resins to confer strength. Used in building and for insulation.

glass transition temperature That temperature at which an amorphous solid transforms from the viscous state to a hard and brittle state.

Glauber's salt SODIUM SULPHATE.

GLC (gas–liquid chromatography) See CHROMATOGRAPHY.

globin A protein, belonging to the histone class. Combined with heme it forms the respiratory pigment haemoglobin.

globular proteins Proteins arranged so that the lipophilic sides are turned inward to face each other, whilst the hydrophilic groups are outside.

globulins See PROTEINS.

glove box (inert atmosphere box) An enclosed space, generally with a glass or plastic viewing port and gloves, used for the manipulation of moisture, oxygen sensitive or poisonous or hazardous materials. The purging may be static, e.g. by a tray of P_2O_5; or by circulation through a series of adsorbants.

glucans Polymers of D-glucopyranose; β-glucans being polymers of β-D-glucopyranose. Includes cellulose, lichenin and other materials that are constituents of cell walls.

glucitol See SORBITOL.

gluconic acid (D-gluconic acid, pentahydroxycaproic acid, dextronic acid), $C_6H_{12}O_7$. Colourless crystals; m.p. 125°C, soluble in water and alcohol. In aqueous solution forms equilibrium with its lactones. Gluconic acid is made by the oxidation of glucose by halogens, by electrolysis, by various moulds or by bacteria of the *Acetobacter* groups. Used as gluconolactone. Potassium salt used in electrolysis.

gluconolactone (D-gluconic acid δ-lactose), $C_6H_{10}O_6$. Prepared by oxidation of glucose with aqueous bromine. Component of many cleaning agents, particularly in dairies and in brewing and in textile printing.

D-glucose (dextrose), $C_6H_{12}O_6$. The most common hexose sugar. It is present in many plants, and is the sugar of the blood. Constituent of, e.g. di- and oligosaccharides, e.g. starch, cellulose, glycogen, sucrose and many glycosides, from all of which it can be obtained by hydrolysis with acids or enzymes.

Like all hexoses it can exist in a number of forms.

1 *aldehyde*

It exists in this form only in solution, though stable derivatives of the aldehyde structure are known.

2 *α- and β-glucopyranose*

Carbon atom 1 in this formula is asymmetric and two stereoisomers therefore exist, depending on whether the OH group is below (α) or

above (β) relative to the carbon atoms.

3 *α- and β-glucofuranose*

These are unstable and known only in solution, the β has the 1-OH on the same side as the $CH(OH) \cdot CH_2OH$. The ethyl glucosides can be obtained crystalline and other derivatives are known.

Ordinary glucose is α-glucopyranose monohydrate; m.p. 80–85°C, dextrorotatory (glucose). In solution the pyranose ring has the chair conformation

with the form (β-D-glucopyranose) with all substituents equatorial. The α form has axial CH_2OH and 1-OH groups. The L form is also known.

Glucose is manufactured from starch by hydrolysis with mineral acids, purification and crystallization, and is widely used in the confectionery and other food industries. It is about 70% as sweet as sucrose.

glucosidase See MALTASE.

glucosides See GLYCOSIDES.

glucuronic acid, $C_6H_{10}O_7$. M.p. 165°C. An oxidation product of glucose in which the primary alcohol group is oxidized to carboxyl; it contains the pyranose ring, and exists in α and β forms. The animal organism combines some toxic substances with glucuronic acid to excrete them in the urine. It is an important constituent of hemicelluloses and plant gums.

glue A colloidal mixture of proteins, built up from amino-acids prepared from animal or fish waste (skins, bones, tendons, etc. containing collagen) treated first with milk of lime, acidified and then heated with water at about 60°C. Glue is related to gelatine. An aqueous solution is used in flocculation, recovery of suspended particles. Hide glue (80%) forms a gel in glycerol which can be used in print rollers and in shock-absorbing equipment. On addition of methanol the gels become stable to oil and grease.

glufosinate-ammonium Herbicide, organophosphate derivative.

L-glutamic acid (α-aminoglutaric acid), $C_5H_9NO_4$, $HO_2C \cdot CH_2 \cdot CH_2 \cdot CH(NH_2) \cdot C(O)OH$. M.p. 211–213°C. Glutamic acid is one of the acidic amino-acids and is present in large quantities in cheeses. It can be produced from wheat gluten or beet sugar molasses by hydrolysis, but is usually manufactured by the fermentation of a carbohydrate in the presence of ammonium salts. Its monosodium salt has a meaty flavour and is used as a flavouring agent.

L-glutamine, $C_5H_{20}N_2O_3$, $H_2N(O)C \cdot CH_2 \cdot CH_2 \cdot CH(NH_2) \cdot C(O)OH$. Needles; m.p. 184–185°C. The monoamide of glutamic acid, widely distributed in plants, especially in seedlings of the *Cruciferae* and *Caryophyllaceae*, and in the roots of the beet, the carrot and the radish.

glutaraldehyde (pentane-1,5-dial), $H(O)CCH_2CH_2CH_2CHO$. Cross-linking agent for proteins and, commercially, for polyhydroxy resins. Also used for sterilization. Formed by hydrolysis of 3,4-dihydro-2-ethoxy-2H-pyran (from acrolein and ethyl vinyl ether).

glutaric acid (pentane-1,5-dioic acid), $C_5H_8O_4$, $HO(O)CCH_2CH_2CH_2C(O)OH$. M.p. 97–98°C, b.p. 302–304°C. Prepared by treating 1,3-dichloropropane with sodium cyanide and heating the product with NaOH. Forms an anhydride on heating at 230–280°C.

glutathione (glutamylcysteinylglycine, GSH), $C_{10}H_{17}N_3O_6S$. M.p. 190–192°C (decomp.). A tripeptide, which is very soluble in water and stable to heating. Glutathione is an important constituent of most cells, in which it is usually the major non-protein thiol. Many enzymatic reactions specifically involve glutathione either as a substrate or as a product. Glutathione can act as an oxygen carrier. The enzyme glutathione reductase reduces the oxidized form GSSG back to the reduced form (GSH). Glutathione is also involved in some S-transferase reactions and in GLYOXYLASE action.

glutelins See PROTEINS.

gluten A mixture of proteins obtained from wheat dough by washing out the starch. It contains mostly the two proteins gliadin and glutelin. Causes intestinal problems in some persons. Used in making bread.

glyceraldehyde (glyceric aldehyde, 2,3-dihydroxypropanal), $C_3H_6O_3$, $OHC \cdot CH(OH) \cdot CH_2OH$. Optically active. D-glyceraldehyde is a colourless syrup. May be prepared by mild oxidation of glycerol or by hydrolysis of glyceraldehyde acetal (prepared by oxidation of acrolein acetol). DL-glyceraldehyde forms colourless dimers, m.p. 138.5°C. Converted to methyl glyoxal by warm dilute sulphuric acid. The enantiomers are the reference compounds for the configurations of all derived longer chain sugars.

L form   D form

glyceric acid (2,3-dihydroxypropanoic acid), $C_3H_6O_4$, $HO(O)C \cdot CH(OH) \cdot CH_2OH$. An uncrystallizable syrup; optically active. Prepared by oxidation of glycerin with nitric acid.

glyceric aldehyde See GLYCERALDEHYDE.

glycerides Esters of glycerol. Classified as mono-, di- and triglycerides according to the number of acid radicals combined with the three hydroxyl groups. The triglycerides occur naturally in animal and vegetable fats and oils. See FATS.

glycerin See GLYCEROL.

glycerol (glycerin, 1,2,3-trihydroxypropane, propan-1,2,3-triol), $C_3H_8O_3$, $CH_2OH \cdot CH(OH) \cdot CH_2OH$. Normally obtained as a colourless, odourless, viscous liquid with a very sweet taste. M.p. 20°C, b.p. 182°C/20mm. It absorbs up to 50% of its weight of water vapour. It occurs (glycerides) in combination with various fatty acids in all animal and vegetable fats and oils and lipids.

Glycerin is obtained commercially as a by-product in the manufacture of soap, and by various synthetic routes starting with propene. One proceeds via propenyl chloride, dichlorohydrin, epichlorohydrin to glycerin, another via propenyl alcohol which is oxidized by hydrogen peroxide to glycerol. Some glycerol is also obtained by fermentation of sugars. It is a good solvent for many organic and inorganic compounds. It reacts with hydrochloric acid to form chlorohydrins, and with nitric acid to give nitroglycerin. With sulphuric acid or potassium hydrogen sulphate, acrolein (propenal) is formed. It is oxidized to a variety of products including glyceraldehyde, dihydroxyacetone, glyceric acid and oxalic acid. Glycerol is used in the manufacture of synthetic resins and ester gums, as a moistening agent for tobacco, in the manufacture of explosives and cellulose films, in lubricants, as a fermentation lubricant, and has many other uses.

glycerol dichlorohydrins See DICHLORO-PROPANOLS.

glycerol monochlorohydrins See DIHY-DROXYCHLOROPROPANES.

glycerophosphoric acid (3-phospho-glyceric acid) Phosphate esters of glycerol. LECITHINS are fatty acid esters of glycerophosphoric acid derivatives. Commercially glycerophosphoric acid is used to prepare the medicinal glycerophosphate salts, e.g. the calcium salt.

glyceryl The group $-OCH_2 \cdot CH(O \cdot)(-)CH_2O$.

glyceryl trinitrate See NITROGLYCERINE.

glycin See AMINOPHENOLS.

glycine (aminoethanoic acid, glycocoll), $H_2N \cdot CH_2 \cdot COOH$. Crystallizes in colourless prisms; m.p. 260°C (decomp.), turning brown at 228°C. It has a sweet taste. Glycine is the simplest of the amino-acids and is a hydrolysis product of proteins; it can be made synthetically by the action of ammonia on chloroethanoic acid and from proteins, such as gelatin, by hydrolysis with acids. Glycine can be synthesized in the animal body. Its sulphate is used in electrolysis.

glycocholic acid (cholylglycine), $C_{26}H_{43}NO_6$. See BILE SALTS.

glycogen, $(C_6H_{10}O_5)_x$. The reserve carbohydrate of the cell. The molecule is built up of a large number of short chains of α-glucose units joined by 1–4 links, the chains being crosslinked by α-1-6 glucoside links. Molecular mass about 4 million. It is a white amorphous powder with no reducing properties. It is broken down in the digestive system by glucosidases to glucose, but in the cells it is broken down to and built up from glucose-1-phosphate by means of phosphorylases.

glycol See 1,2-DIHYDROXYETHANE.

glycollic acid (hydroxyethanoic acid, hydroxyacetic acid), $CH_2OH \cdot COOH$. Colourless crystals, m.p. 80°C. Occurs in the juice of the sugar cane and beets. Prepared by boiling a concentrated aqueous solution of sodium monochloroethanoate. Also produced by electrolytic reduction of oxalic acid. Forms an anhydride when heated at 100°C. Used in textile and leather processing and cleaning (metals and dairy sanitation).

glycols Dihydric alcohols derived from aliphatic hydrocarbons by replacement of two hydrogen atoms by hydroxyl groups. They are colourless liquids. 1,2-glycols are obtained by oxidation of olefins with potassium permanganate or lead tetra-acetate; by heating olefin chlorohydrins with weak alkalis; or by heating alkane dihalides with sodium hydroxide. Of great importance as polymers, e.g. polyethylene glycol.

glycolysis The metabolic breakdown of carbohydrates in living organisms. Occurs in the presence of oxygen ('aerobic glycolysis') or in its absence ('anaerobic glycolysis'). Anaerobic glycolysis of glucose ultimately yields two moles of ATP per mole of glucose, together with lactate, which can undergo further energy-yielding degradation in the presence of oxygen. Aerobic glycolysis yields pyruvate, which in turn may give rise to acetyl-coenzyme A for the CITRIC ACID CYCLE. Pyruvate may also undergo other metabolic transformations, e.g. amination to give alanine. See DICTIONARY OF BIOLOGY.

glycoproteins See PROTEINS.

glycosidases Enzymes which split off glucoses (glucosidases) or other sugars from glycosides.

glycosides Sugar derivatives in which the hydroxyl group attached to carbon 1 is substituted by an alcoholic, phenolic or other group. The term glucoside is used for those glycosides which contain glucose as the sugar, while glycoside refers to all compounds whatever the constituent sugar. The non-sugar portion of the molecule is termed the aglycone. The simplest glucoside is methylglucoside, $C_6H_{11}O_5 \cdot OCH_3$. As the 1 carbon atom is asymmetric there are two isomers, distinguished as α and β. The natural glucosides nearly all belong to the β-series and are hydrolysed by β-glucosidases. Glycosides as a class are colourless, crystalline, bitter substances.

glycylglycine (diglycine), $C_4H_8N_2O_8$, $NH_2 \cdot CH_2 \cdot CO \cdot NH \cdot CH_2 \cdot CO_2H$. Decomposes at 260–262°C. The simplest of the dipeptides, it is formed by the action of glycyl chloride on glycine or by hydrolysing diketopiperazine with hydrochloric acid.

glyme See 1,2-DIMETHOXYETHANE. Di- or polyethers used as solvents.

glyoxal See ETHANEDIAL.

glyoxaline See IMIDAZOLE.

glyoxydiureide See ALLANTOIN.

glyoxylate cycle An ANAPLEROTIC SEQUENCE supplementing the carboxylic acid cycle in micro-organisms and in plants. Isocitric acid is cleaved to succinate and glyoxylate, the latter being used to form malic acid by combination with acetyl-coenzyme A. See DICTIONARY OF BIOLOGY.

glyoxylic acid, CH(O)·COOH. A thick syrup, rather difficult to crystallize; m.p. 98°C. Widely distributed in plant and animal tissues. Prepared by electrolytic reduction of oxalic acid. Forms salts of the type $(HO)_2CHCOOM$. Also reacts as an aldehyde. Condenses with urea to give allantoin. Reduced to glycollic acid.

glyphosphate Herbicide, N-(phosphono-methyl)-glycine.

glyptals ALKYD RESINS.

Gmelin (Handbuch der anorganische Chemie) The definitive reference series in inorganic chemistry, although the complete periodic table is not covered.

gold, Au. Metallic element, at.no. 79, at.mass 196.97, m.p. 1064°C, b.p. 2856°C, density ρ 17 310, crustal abundance 0.0011 p.p.m., electronic configuration [Xe] $5d^{10}6s^1$. Gold occurs native, typically as small particles in quartz but also as deposits after erosion of the quartz and sometimes as nuggets. Some plants, particularly members of the mustard family, accumulate significant amounts of gold. Extraction is with potassium cyanide in the presence of peroxide or air to give the $[Au(CN)_2]^-$ ion. Gold is also found in the slime after the electrolytic purification of copper. Gold is a bright yellow ccp metal. It is a good conductor of heat and electricity. Colloidal gold is formed by reducing gold chloride solution with, e.g., hydrazine. The colour of the colloid depends on the particle size. Colloidal gold is used in staining microscope slides and in electron microscope imaging. Reduction with a mixture of $SnCl_2$ and $SnCl_4$ gives Purple of Cassius, a colloidal gold oxide with adsorbed Au used in the manufacture of ruby glass. Gold is not attacked by oxygen or single acids (except H_2SeO_4). It dissolves in aqua regia and in halogen solutions and is attacked by fluorine. The main use of gold is in monetary systems but it is also used in jewellery, as an i.r. reflector, in electrical contacts and conductor systems, and its compounds in the treatment of rheumatoid arthritis. Gold–sulphur, colouring porcelain, conductor systems and catalysis. ^{198}Au is used as a radiation source (gold grains). World use 2000 3146 tonnes.

gold, organometallic compounds Gold(I) compounds RAuL (L = sulphide, phosphine, isocyanide) and gold(III) alkyls, e.g. Me_3AuPPh_3 and $[Me_2AuBr_2]^-$ are known.

gold, standard Pure gold is too soft for use as ornaments or for coinage, and is alloyed with copper or silver, or both. The fineness is expressed either in parts per thousand, or in carats. Pure gold is 24 carat fine, and the five standard alloys of 22, 18, 15, 12 and 9 carats, i.e. parts of gold in 24 of alloy, are legalized.

gold chemistry Gold is an element of Group 11. It shows oxidation states +5, +3, +2, +1, –1. The +5 state is found only in AuF_5 and hexafluoroaurates(V). The common states are +3 (generally square planar) and +1 (generally linear). Both +3 and +1 states form extensive ranges of complexes, particularly with soft bases. Organometallic compounds are readily formed. Gold(II) complexes are formed with S-ligands but many compounds which are apparently Au(II) derivatives contain Au(I) and Au(III). Gold forms stable metal–metal bonds with other metals and cluster compounds, e.g. $Au_{11}I_3L_{7 \text{ or } 8}$ (L = phosphines) are formed on reduction. CsAu contains Au(–1).

gold cyanides Important in the extraction of the metal. Gold(I) cyanide, AuCN (HCN on AuOH) gives $KAu(CN)_2$. Gold(III) cyanide, $Au(CN)_3$, (acid on $KAu(CN)_4$) forms the $[Au(CN)_4]^-$ ion ($AuCl_3$ plus KCN).

gold grains ^{198}Au encapsulated in platinum used as a radiation source.

gold halides Gold(III) fluoride (BrF_3 on Au;

decompose $AuF_3 \cdot BrF_3$ by heating) is readily hydrolysed by water. Fluoroaurates(III), $MAuF_4$, are formed by the action of BrF_3 on a mixture of MF and Au. Fluoroaurates(V) containing $[AuF_6]^-$ species are formed by the action of strong fluorinating agents, e.g. KrF_2. Red-brown AuF_5 is formed by heating $KrF_2 \cdot AuF_5$.

The remaining gold(III) halides AuX_3 ($X = Cl$, Br, I) (Au plus Cl, or Br_2, $AuCl_3$ plus KI) have dimeric structures with planar co-ordination about Au. All form haloaurates(III), $MAuX_4$, with excess halide and HX. Gold(I) halides, AuX ($X = Cl$, Br, I), are formed by gently heating the trihalides. They are insoluble in water but decompose in water to Au plus AuX_3. Gold(I) halides generally form linear complexes, particularly with phosphines and CO; complex halides $[AuX_2]^-$ ($X = Cl$, Br) are formed by reduction of $[AuX_4]^-$. AuF has been detected from laser irradiation of gold foil in the presence of SF_6 or CF_3I.

gold number A number used to define the efficiency of operation of protective colloids based upon the amount of protective colloid, preventing the colour change of a gold sol from red to blue on coagulation by electrolytes.

gold oxides Gold(I) oxide, Au_2O, is obtained from $[AuBr_2]^-$ and alkali. It decomposes above 200°C. An ill-defined gold(III) hydroxide is soluble in excess of alkali to give the $[Au(OH)_4]^-$ ion.

Goldschmidt process The preparation of sodium methanoate by pressurizing CO and NaOH at approx. 200°C. Pyrolysis then gives sodium oxalate.

Goldschmidt reaction The use of Al powder for the reduction of metal oxides.

gold sulphides Grey Au_2S and black Au_2S_3 are formed by the action of H_2S on gold derivatives. Both decompose to the metal on heating.

gonadotropins (gonadotropic hormones) Hormones influencing the activity of the sex organs. The term is not applied to the steroid sex hormones. Applied to a variety of peptide hormones, produced in the anterior lobe of the pituitary gland and elsewhere. See DICTIONARY OF BIOLOGY.

gossyplure Insecticide, pheromone.

Gouy balance A balance for the determination of magnetic susceptibility. The sample is weighed in and out of a magnetic field and the susceptibility is calculated from the difference in weights.

GPC (gel-permeation chromatography) See GEL FILTRATION.

graft copolymers See BLOCK COPOLYMERIZATION.

Graham's law of diffusion The rates at which two gases diffuse are inversely proportional to the square roots of their densities, i.e.

$$\frac{\text{rate A}}{\text{rate B}} = \sqrt{\frac{\rho \text{B}}{\rho \text{A}}}$$

See DIFFUSION. Only strictly true when the mean free path of the issuing gas is much greater than the diameter of the holes through which diffusion occurs.

Graham's salt, $(NaPO_3)_x$. A polymeric metaphosphate. See PHOSPHORUS OXYACIDS. Used in water softening, detergents, tanning.

gram atom The quantity of an element numerically equal to the relative atomic mass expressed in grams.

gram equivalent The equivalent weight of a substance expressed in grams.

gramicidin Family of antibiotics produced by soil bacteria.

gramine See 3-(DIMETHYLAMINOMETHYL)-INDOLE.

gram molecular volume The volume occupied by the gram mole of an element or compound in the gaseous state. According to Avogadro's hypothesis, under the same conditions of temperature and pressure, all gases have the same gram molecular volume. At stp the gram molecular volume is equal to 22.414 litres.

gram molecular weight The molar mass in grams.

gram molecule The quantity of a compound, or element, equal numerically to the relative molecular mass expressed in grams. The weight in grams of one mole of a substance.

granite An igneous rock extensively used as a building stone and road metal.

granulation A general term for any process for producing granules. Granulation is frequently carried out by compacting a fine material and then crushing it, and thus employs techniques of both size enlargement and size reduction.

graphene sheets The individual component sheets of graphite. Converted into nanotubes.

graphite (plumbago, black lead) A form of carbon occurring and formed in crystalline

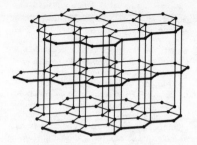

and microcrystalline forms. The structure has parallel hexagonal sheets of carbon atoms. The C–C distance within the sheets is 1.42Å, the distance found in aromatic rings. The C–C distance between sheets is 3.40Å and there is high conductivity along sheets, easy cleavage and ready incorporation of other substances between the layers (graphite compounds, lamellar compounds, intercalation compounds). Used as a lubricant, refractory, pencils, etc., electrical components, paints, carbonizing steels. Very pure graphite (coke from crude oil plus a pitch binder heated to 2800°C) is used as a moderator and reflector in nuclear reactors.

graphite compounds Compounds formed by penetration of reactants between the layers of graphite. Graphite oxide, approx. C_2O (graphite plus HNO_3 or $KMnO_4$) and graphite fluoride (graphite plus F_2) are non-conducting because of breakup of the aromatic systems of the sheets (graphite fluoride CF_x $0.5<x<1.3$ is an efficient lubricant, also used with Li in batteries). In the conducting intercalation compounds various species are inserted between the sheets. The reactivity of the intercalated compound is profoundly modified, e.g. hydrolysis of fluorides is reduced.

gravimetric analysis Methods of analysis involving precipitation, etc. which end up with a final weighing of a stable derivative.

gray (Gy) Unit of radiation dose. 1 joule Kg^{-1}. 100 rad = 1 Gy. Replaces rad.

greases See LUBRICATING GREASES.

'green' acids The name given to sulphonic acids which are present in the acid sludge phase following treatment of lubricating oil fractions with oleum or sulphuric acid. The sulphonic acids in the oil phase are termed 'MAHOGANY ACIDS' used as emulsifiers, rustproofing agents and lubricant additives. The water-soluble green acids are not used.

greenhouse effect The absorption of solar radiation by gases such as CO_2 and CH_4 leading to atmospheric warming.

greenhouse gases Gases which absorb infra-red radiation and may contribute to atmospheric warming.

green vitriol Iron(II) sulphate.

Grignard reaction Alkyl and aryl halides, particularly bromides and iodides, react with magnesium in the presence of dry ether to form compounds of the type RMgX where R represents the alkyl or aryl radical and X the halide. These are known as Grignard reagents RMgX, which are always associated with one or two molecules of ether which are co-ordinated to the magnesium. Species present in solution include R_2Mg, RMgX, MgX_2 and polymers. They react with alkyl and aryl halides to give hydrocarbons and with metal halides to give organometallics. Esters are formed when the reagents react with orthoformic, methanoic esters; carboxylic acids by combination with solid carbon dioxide. Aldehydes give secondary alcohols, while ketones give tertiary alcohols. Amides and nitriles give ketones. Water and dilute acids react to give hydrocarbons. The immediate result of the reaction is often a complex magnesium compound which is decomposed by dilute acids to give the final product.

Grignard reagents See GRIGNARD REACTION.

grinding See CRUSHING AND GRINDING.

griseoviridine Antibiotic produced by soil organisms.

ground state The lowest energy electronic, vibrational, or rotational state of an atom, molecule or ion.

group A vertical group of elements in the normal form of the PERIODIC TABLE. Within a group there are distinct chemical resemblances and gradations in properties.

group theory A mathematical method which considers the effect of a group of operators (e.g. symmetry elements, crystal field) on properties. Used in calculations on structure, spectra, magnetic susceptibility, etc.

growth hormone (somatropin) A protein hormone isolated from the anterior lobe of the pituitary gland.

guaiacol (guaic alcohol, 2-methoxyphenol), $C_7H_8O_2$. M.p. 32°C, b.p. 205°C. A constituent of guaiacum resin, and occurs in beechwood tar.

Guaiacol has a very characteristic odour and burning taste; its medicinal properties are identical with those of creosote. Used in the preparation of vanillin and paparvarin and for denaturing alcohol.

guaiazulene (S-guaiazulene, 1,4-dimethyl-7-isopropylazulene), $C_{15}H_8$. Blue plates, m.p. 31.5°C. Obtained by the dehydrogenation of guaiol with sulphur. Expectorant. If the dehydrogenation is carried out at a higher temperature using selenium an isomeric product, 2,4-dimethyl-7-isopropylazulene (Se-guaiazulene), is obtained.

guaiol, $C_{15}H_{26}O$. A crystalline alcohol; m.p. 93°C, obtained from the wood oil of *Bulnesia sarmienti*, Lorenz. On heating with sulphur it gives guaiazulene.

guanidine (iminourea), CH_5N_3, $(H_2N)_2C=NH$. Deliquescent crystals. Strongly basic, forms many salts, e.g. nitrate, m.p. 214°C. It can be prepared by oxidizing guanine or arginine; the usual method of preparation is by heating ammonium thiocyanate to give guanidine thiocyanate and obtaining the base from this. Guanidine derivatives are used in pharmaceutical manufacture, dyestuffs, explosives and resins.

guanine (6-hydroxy-2-aminopurine), $C_5H_5N_5O$.

An amorphous powder, soluble in acids and alkalis. It is present in all animal and vegetable tissues as a constituent of the nucleic acid portion of nucleoproteins.

guanosine Constituent of nucleic acids.

guazatine Fungicide, guanidine derivative.

gulose A hexose sugar.

gum acacia (gum arabic) Obtained as an exudate from acacia trees. It is the calcium salt of arabin, a complex polysaccharide built of glucuronic acid, arabinose, rhamnose and galactose units. It is used in pharmacy as an emulsifying and suspending agent and in the manufacture of plasters, in the food industry, and as an adhesive.

gums The true plant gums, e.g. gum acacia, gum tragacanth, are the dried exudates from various plants, obtained when the bark is cut or other injury is suffered. They are soluble in water to give very viscous colloidal solutions, sometimes called MUCILAGES, and are insoluble in organic solvents. They are complex polysaccharides, each containing several different sugar molecules and uronic acid groups. Used as thickening agents in the food and pharmaceutical industries etc.

In the petroleum industry the term gum refers to the dark-coloured polymer formed by the oxidation of certain unsaturated compounds of cracked or reformed gasolines.

guncotton See CELLULOSE NITRATE.

gunpowder The oldest and best-known explosive. As a propellant it has been replaced by modern smokeless powders, but it is still an important explosive used in primers, fuses and pyrotechnics. Ordinary black gunpowder consists of about 75% KNO_3, 15% charcoal and 10% sulphur. It is granulated during manufacture, grains of different sizes with different burning rates being produced, and these are polished by rotating in closed drums. Relatively insensitive to shock but is very easily ignited.

gutta-percha A naturally occurring polymeric material, isomeric with rubber but having the *trans* configuration. Obtained from the latex of various tropical trees. Insulator, dental cement.

Gy The gray. Unit of radiation.

GY-81 Fungicide, insecticide, nematocide, thiopercarbonate derivative.

gypsum Mineral $CaSO_4.2H_2O$ also known as selenite. Used in the production of plaster of Paris, to control the rate of setting of Portland cement and as a filler.

gyromagnetic ratio (γ) For a nucleus the gyromagnetic ratio is equal to the ratio of the magnitude of the magnetic moment (μ) and the angular momentum (I), i.e.

$$\mu = \gamma I \quad \text{where} \quad I = h \sqrt{I(I+1)},$$

I being the nuclear spin quantum number.

h See HAPTO.

H Hydrogen.

Ha Hahnium.

Haber process The industrial process for the direct synthesis of AMMONIA from N_2 and H_2 over a catalyst.

haem (haematin, haemin, haemocyanin, haemoglobin) See HEME etc.

haematite, α-Fe_2O_3. An important iron ore, normally blood-red in colour but forms black crystals (specular iron). Has the corundum structure with c.p. O^{2-} and octahedrally co-ordinated Fe.

hafnium, Hf. Metal, at.no. 72, at.mass 178.49, m.p. 2222°C, b.p. 4603°C, density ρ 13 310, crustal abundance 5.3 p.p.m., electronic configuration [Xe] $5d^26s^2$. An element of Group 4. Occurs with Zr in BADDELEYITE and ZIRCON. The metal is hcp. Has very similar properties to Zr and is separated by ion exchange. In contrast to Zr it is a strong neutron absorber and is used as a control rod material for nuclear reactors. Used in alloys with, e.g. W and Ta.

hafnium compounds Hafnium compounds are very similar to ZIRCONIUM COMPOUNDS and the chemistry is mainly that of the +4 state.

hahnium, Ha. Element 105. See POST-ACTINIDE ELEMENTS.

halazone, 4-HO(O)CC$_6$H$_4$SO$_2$NCl$_2$. White powder; m.p. 213°C. Used for the sterilization of drinking water.

halfenprox Acaricide, pyrethroid.

half-life (half-value period), $t_{\frac{1}{2}}$ The time taken for the concentration of a substance to fall to half its initial value. For radioactive elements the half-life is $0.69 \times 1/\lambda$ where λ is the decay constant. The term is also used for transient species (e.g. free radicals) and first-order chemical reactions.

half reaction Most generally used in redox reactions; the half reaction refers to the stoichiometry of the reaction of one of the species involved, including the electrons involved, e.g.

$$MnO_4^- + 8H^+ + 5e \rightarrow Mn^{2+} + 4H_2O$$

half-value period See HALF-LIFE.

half wave potential, $E_{\frac{1}{2}}$ The midpoint of the polarographic wave, i vs. E, of an electrochemically reversible reaction. $E_{\frac{1}{2}}$ is characteristic of the species under investigation.

halides Fluorides, chlorides, bromides and iodides; derivatives of the halogens(-1).

haloamines Halogen derivatives of ammonia. Chloramine, $ClNH_2$, is formed from NaOCl and NH_3; m.p. –66°C. Difluoroamine, F_2NH; b.p. 24°C, $H_2NC(O)NF_2$ (F_2 plus aqueous urea then H_2SO_4), and H_2NF (low-yield product from the electrolysis of fused NH_4HF_2) are NF_3 derivatives. All of the haloamines are explosive.

halofenozide Insecticide, hydrazine derivative.

haloform reaction A reaction used for the chemical recognition of methylcarbonyl (acetyl) and methylcarbinol structures, which give chloroform with NaOCl or the more easily recognized yellow iodoform with NaOI. See IODOFORM REACTION.

haloforms The trihalogen derivatives of methane, CHF_3, $CHCl_3$, $CHBr_3$ and CHI_3.

haloform test A colour reaction for detecting any of the haloforms or dihalocarbenes. A haloform or a haloform precursor is added to a mixture of 10% sodium hydroxide and pyridine. Haloforms cause the mixture to turn pink to bright blue-red at room temperature.

halogenated rubbers See CHLORINATED RUBBERS. Analogues of bromine and iodine have little stability.

halogenation A process resulting in the incorporation of halogens by addition or substitution. Among organic compounds, alkenes add halogens, interhalogens and hydrogen halides across the C=C bond; some aromatic compounds react with halogens by substitution for hydrogen atoms in the presence of Lewis acid catalysts like $AlBr_3$, $SnCl_4$, BF_3, $FeBr_3$. See also FLUORINATING AGENT.

halogens The elements fluorine, chlorine, bromine, iodine and astatine of Group 17.

halohydrins Compounds formed by addition of a halogen and hydroxyl to adjacent carbons joined by a double bond.

halonium ions Disubstituted halogen cations, RX^+R', where R, R' may be alkyl or aryl groups; intermediates in electrophilic halogen additions. May be isolated as, e.g., SbF_6^- salts from media of low nucleophilicity. They are good alkylating (and arylating) agents.

halosulfuron-methyl Herbicide, sulphonylurea derivative.

halothane (Fluothane), $CHBrCl·CF_3$. Colourless, mobile, heavy liquid, with an odour resembling chloroform and a sweet, burning taste. Non-inflammable; b.p. 49–51°C. Halothane is a general anaesthetic administered in oxygen, often along with dinitrogen oxide, N_2O. It possesses little acute toxicity, but habitual exposure to its vapour may, in rare cases, lead to liver damage.

haloxyfop Herbicide, fluorinated propionic acid derivative.

hamiltonian H An operator which carries out a mathematical operation on a function.

Hammett equation A correlation between electronic and other structure and reactivity in the side-chain derivatives of aromatic compounds and many other compounds. Derived from equilibrium constants, or other functions of molecules (e.g. the i.r. frequencies, n.m.r. shifts). For example the dissociation constants K_a of a series of 4-substituted (O_2N-, MeO–, Cl–, etc.) benzoic acids correlate with the rate constant k for the alkaline hydrolysis of 4-substituted benzyl chlorides. Similar results are obtained for 3-substituted derivatives.

The equation

$$\log k = \rho \log K + C$$

expresses this relationship, and when the ring substituent is hydrogen, k_0 is the rate of hydrolysis of benzyl chloride and K_0 is the dissociation constant of benzoic acid.

The equation

$$\log \frac{k}{k_0} = \rho\sigma$$

may be derived and is known as the Hammett equation. It is usual to maintain the term $\log K/K_0$ as referring to the dissociation of benzoic acids in water at 25°C, and to relate other measurable entities to it. The term σ is called the substituent constant since the strength of the benzoic acid depends upon the nature of the substituent. A more positive value of σ infers a more electron-attracting nature of the substituent. Negative σ values indicate electron-donating substituents.

The term ρ is a reaction constant. The slope of the straight lines is ρ, and reflects the sensitivity of the reaction under study to effects of substituents. ρ is affected by temperature, solvent changes, etc.

The equation does not hold without exceptions even for 3- and 4-substituents, especially when resonance interactions from substituents are possible.

Hantzsch synthesis The formation of pyridine and pyrrole derivatives by the condensation of ethyl acetoacetate with ammonia and an aldehyde.

hapto The number of atoms in a ligand bonded to the acceptor. E.g. ferrocene, $(C_5H_5)_2Fe$ contains symmetrically bonded pentahapto C_5 rings, designated (h^5-C_5H_5) or (η^5-C_5H_5).

hard and soft acids and bases Bases may be divided into hard, non-deformable, e.g. F^-, and soft, polarizable, potentially π-bonded, e.g. PPh_3. Acids may be divided similarly into hard, non-polarizable, e.g. Na^+, and soft, polarizable, potentially π-bonded, e.g. Pt^{2+}. Hard and soft acid–base theory (HSAB) postulates greatest stability for hard acid–hard base and soft acid–soft base pairs, e.g. Si^{4+} forms its most stable halide complexes with F^-; Pt^{4+} with I^-.

hardness The resistance of a material to pressure applied to a small area, i.e. resistance to crushing, abrasion, indentation or stretching.

Hardness is assessed on an empirical ten-point scale; diamond (10) is the hardest substance known, talc (1) is soft.

hardness of water The property conferred on water by the presence of alkaline earth salts which prevent formation of a lather with soaps. Temporary hardness, soluble Ca and Mg hydrogen carbonates, is removed by boiling; permanent hardness, other Ca and Mg salts, is removed by ion exchange (permutite process) or by detergents.

Hartree Unit of heat and energy generally derived from calculation. 1 Hartree = 627.5 kcal mol^{-1}.

Hartree–Fock method Iterative method for solving a Schrödinger Equation, i.e. the energy and electron density in a molecule.

Hartree–Fock orbital Relatively accurately calculated orbital shapes. See HARTREE–FOCK METHOD.

hassium, Hs. Element 108 produced by ^{26}Mg on ^{248}Cm. Group 8 element, forms gaseous oxide HsO_4.

HC 252 Herbicide, diphenyl ether derivative.

HCH hexachlorocyclohexane. See BHC.

hcp Hexagonal close-packed. See CLOSE-PACKED STRUCTURES.

He Helium.

heat exchangers Units where heat is transferred from one fluid stream to a second fluid stream, resulting in one being heated and the other being cooled, without change of state in either case. Heat exchangers account for a large part of chemical plant investment.

heating oil A general term describing fuel oils used in domestic and industrial heating. Often applied to kerosines or gas oils.

heats of atomization etc. See ENTHALPIES.

heat transfer media Fluids used to convey heat from the place of generation to a location where it is required. The heat may be transferred as sensible heat, latent heat or both.

Steam is by far the most widely used medium, useful up to about 475 K. Up to about 700 K organic liquids such as the DOWTHERMS and mineral oil may be used. Mercury and molten salts, such as the EUTECTIC mixture of sodium nitrite, sodium nitrate and potassium nitrate may be used up to 875 K, while above this temperature air and flue gases must be used.

heat transmission oils Thermally stable oils, usually medium viscosity distillates, employed as circulated heat transfer media for industrial purposes.

heavy hydrogen Deuterium, $^2_1$H.

heavy spar, $BaSO_4$. See BARIUM SULPHATE.

heavy water (deuterium oxide), D_2O. See DEUTERIUM. Used to form deutero derivatives for use as tracers, to determine reaction mechanisms, for the preparation of ^{2}H n.m.r. solvents and as a moderator in reactors.

Heck reaction Stereospecific coupling of alkenes etc. with RX or triflates in the presence of a palladium catalyst.

Heisenberg uncertainty principle For small particles which possess both wave and particle properties, it is not possible to specify both the position and momentum of the particle simultaneously. Mathematically the uncertainty in the position Δv and momentum Δp are related by the equation

$$\Delta p \times \Delta v = h/2\pi \ (h = \text{Planck's constant})$$

helium, He. Group 18 element. Noble gas, at.no. 2, at.mass 4.0026, m.p. –268.93°C, b.p. –267.96°C, density ρ 7.26 g l^{-1}, atmos. abundance 5.2 p.p.m. by vol., electronic configuration 1s^2. Up to 7% in some natural gases. Found in some radioactive minerals as a product of radioactive decay (the α-particle is an He^{2+} ion) (up to 7%). Separated from natural gas by liquefaction of the other gases. Used as an inert atmosphere for arc welding and for Ti, Zr, Si, Ge production, and controlled atmospheres, cryogenics, as a coolant in nuclear reactors, with 20% O_2 as an atmosphere for divers, for pressurizing liquid-fuel rockets, for filling balloons and in gas lasers. Helium has no normal chemistry although excited species containing bound He are formed in discharge tubes, and in matrices at low temperatures. Liquid He$_{II}$ has no triple point and cannot be solidified at atmospheric pressure. Liquid He$_{II}$ is a liquid which exhibits superconductivity and can flow over the edge of vessels. It changes to He$_I$ at 2.2 K. U.S. market 2003 3.4 billion cu ft.

Helmholtz free energy (A) The maximum amount of energy available to do work resulting from changes in a system at constant temperature and volume. Spontaneous changes correspond to increases in entropy. See FREE ENERGY and GIBBS–HELMHOLTZ EQUATION.

hematin See HEME and HEMIN.

hematoporphyrin See PORPHYRINS.

heme, $C_{34}H_{32}FeN_4O_4$. The prosthetic group of heme derivatives. The combination of protoporphyrin IX and iron. Structurally it resembles hemin but with the iron in the Fe(II) state. Prosthetic group of HEMOGLOBIN, CYTOCHROMES and some PEROXIDASES. The iron may be oxidized to Fe(III) during reactions involving cytochromes or peroxidases, i.e. it becomes hematin.

hemicelluloses Polysaccharide found in association with lignins in walls of plants and seaweeds.

hemihedral forms Those forms in any crystal system which show the full number of faces required by the symmetry of the system are called holohedral forms. When only half the number of faces found in the holohedral form are present, the form is said to be hemihedral.

hemiketals See HEMIACETALS.

hemimorphite, $(OH)_2Zn_4Si_2O_7.H_2O$. An important zinc ore containing $(Si_2O_7)^{6-}$ units. Formerly called calamine or electric calamine (particularly U.S. usage).

hemin, $C_{34}H_{32}ClFeN_4O_4$. The Fe(III) equivalent of HEME. It possesses one residual positive charge

and so the isolated complex contains an anion such as chloride (shown above). In alkaline solution the chloride is replaced by hydroxide to give hematin.

hemocyanin The copper-containing respiratory protein of molluscs and crustaceans.

hemoglobin The respiratory pigment of the blood of vertebrates. It is a conjugated protein consisting of the iron-porphyrin compound heme, combined with pendant globin proteins. Combines loosely with oxygen to form oxyhemoglobin, which readily loses oxygen again on exposure to a vacuum or an atmosphere devoid of oxygen. Hemoglobin combines with oxygen in the lungs, and carries it in the arteries to the tissues, where the oxygen pressure is low, and

where it is reduced, returning to the lungs via the veins. Oxyhemoglobin is scarlet in colour: reduced hemoglobin is of a purplish colour. Hemoglobin also combines very easily with carbon monoxide, forming CARBOXYHEMOGLOBIN; hemoglobin also plays a part in regulating the acidity of blood and in the carriage of carbon dioxide; muscle hemoglobin acts as a respiratory catalyst.

hendecane See UNDECANE, $C_{11}H_{24}$.

Henderson–Hasselbach equation A simplified version of the relationships used in calculations on buffer solutions.

$$pH = pK_a + \log\frac{\text{fraction neutralized}}{\text{fraction unneutralized}}$$

Henry's constant Measure of partitioning of a substance between, e.g. air and water.

Henry's law The mass of gas which is dissolved by a given volume of a liquid at constant temperature is directly proportional to the pressure of the gas. The law is obeyed only provided there is no chemical reaction between the gas and the liquid. In ideal solutions solute and solvent obey RAOULT'S LAW. The vapour pressure of the solution is proportional to the mole fractions present, using Henry's constant.

heparin A mixture of sulphonated polysaccharides that inhibits the clotting of blood. Its chief use is in treating thrombosis.

heptachlor Insecticide no longer used.

heptamethyldisilazane, $Me_3SiNMeSiMe_3$. Used to introduce =NMe groups in place of halides.

heptanal (heptaldehyde) Prepared from castor oil. Used in manufacture of 1-heptanol, 2-heptanone, methyl amyl ketone. Industrial solvent.

n-heptane, C_7H_{16}, $CH_3[CH_2]_5CH_3$. Colourless inflammable liquid; b.p. 98°C. Occurs, together with other isomeric hydrocarbons, in petroleum. Obtained by distillation of petroleum. Eight other paraffin hydrocarbons of the formula C_7H_{16} are possible. Has the general properties of the paraffins. Used with isooctane in defining the knock rating of petrols.

n-heptanoic acid (enanthic acid, oenanthic acid), $C_7H_{14}O_2$. An oily liquid, m.p. –9°C, b.p. 115–116°C/11 mm. Prepared by the oxidation of heptanal, and has growth-inhibiting properties against micro-organisms. Present in some natural oils and waxes.

heptanol (enanthic alcohol, heptyl alcohol), $C_7H_{15}OH$. 1-heptanol, b.p. 176°C. Prepared by reduction of heptanone. 2-heptanol, b.p. 158–160°C.

4-heptanone (butyrone, di-*n*-propyl-ketone), C_7H_4O, $(CH_3CH_2CH_2)_2CO$. A colourless, odorous liquid; b.p. 144°C. Prepared from butanoic acid over $CaCO_3$ at 450°C. Used as a solvent for resins, particularly the glyptal and vinyloid resins, and lacquers.

3-heptanone (ethyl *n*-butyl ketone), $C_2H_5CO(CH_2)_3CH_3$. Made by dehydrogenation of 3-heptanol; b.p. 148°C. Used as a solvent for lacquers and synthetic resins.

heptenophos Insecticide, organophosphate derivative.

heptose A carbohydrate with seven carbon atoms.

herbicides Weedkillers, often selective.

heroin See DIAMORPHINE, MORPHINE.

Hertz (Hz) The unit of frequency, s^{-1}.

Hess's law The law of constant heat summation; the total heat change accompanying a chemical reaction is independent of the route taken in reactants becoming products. Hess's law is an application of the first law of thermodynamics to chemical reactions.

heteroauxin See INDOLE-3-ETHANOIC ACID.

heterocyclic Compounds which contain a closed ring system in which the atoms are of more than one kind, for example, pyridine, thiophen and furan.

heterogeneous catalysis A catalysed reaction where the catalyst is in a different phase from the reactants, e.g. a solid catalyst in a reaction between liquid or gaseous reactants. Such reactions are generally considered to involve adsorption and reaction of the reactants at the catalyst surface.

heterogeneous reaction A reaction which occurs between substances in different phases, e.g. between a gas and a liquid.

heteroleptic Having more than one substituent, e.g. SF_5Cl.

heterolytic reaction A reaction occurring with bond breaking so that the electrons forming the broken bond are shared unequally between the fragments formed, e.g.

$$A - B \rightarrow A^+ + :B^-$$

Alternatively a reaction between a species with a pair of electrons and a species with a vacant orbital to form a covalent bond.

heteronuclear molecule See HOMONUCLEAR MOLECULE.

heteropoly acids Formally acids giving polymeric oxide- and hydroxide-bridged anions containing different metal and non-metal species. The free acids often do not exist. Examples are the phosphomolybdates.

HETP See THEORETICAL PLATE.

hexaborane (10), B_6H_{10}; **hexaborane (12)**, B_6H_{12}. See BORON HYDRIDES.

hexachlorobenzene, C_6Cl_6. Colourless crystals; m.p. 227°C, b.p. 326°C. Prepared by exhaustive chlorination of benzene with Cl_2 in the presence of $FeCl_3$ and commercially by the action of Cl_2 on hexachlorocyclohexane in C_2Cl_6. Used in preparative work (C_6Cl_5OH); for the preparation of C_6F_6 and its derivatives; as fungicides.

hexachlorobut-1,3-diene, C_4Cl_6, $Cl_2C=CCl–CCl=CCl_2$. Colourless viscous liquid; b.p. 210–211°C. Principally used as a dispersant material for obtaining i.r. spectra of solids in mulls. Skin irritant, suspected carcinogen.

hexachlorocyclopentadiene, C_5Cl_6. Liquid; b.p. 244°C, reacts with norbornadiene ([2:2:1]-bicycloheptadiene) to give the Diels–Alder adduct ALDRIN.

hexachloroethane (perchloroethane), C_2Cl_6. Colourless solid; m.p. 187°C, sublimes on heating. Manufactured by chlorination of *s*-tetrachloroethane in the presence of $AlCl_3$. Used in refining metals, particularly aluminium, in flameproofing, generation and production of artificial diamonds.

hexaconazole Fungicide, triazole derivative.

hexadecane (cetane), $C_{16}H_{34}$, $CH_3·[CH_2]_{14}·CH_3$. M.p. 18°C. A straight-chain alkane.

hexadecanoic acid See PALMITIC ACID.

hexadecanol (cetyl alcohol), $C_{16}H_{34}O$, $CH_3·[CH_2]_{14}·CH_2OH$. Colourless crystals, m.p. 49°C. It occurs as esters in various waxes, including cetyl palmitate, which is the chief component of spermaceti. Used extensively in the pharmaceutical and cosmetic industries and in gel stabilizers for greases.

(E,E)-2-4-hexadienoic acid (sorbic acid)
Mould and yeast inhibitor used in drying oils, resins.

hexaflumuron Fungicide, fluorinated benzylurea, used against termites.

hexafluorobenzene, C_6F_6. The simplest of a series of benzene derivatives having fluorine atoms in place of hydrogen. A colourless mobile liquid; m.p. 5.2°C, b.p. 80°C, thermally stable at over 500°C; undergoes nucleophilic substitution to give pentafluorophenyl derivatives. Prepared C_6Cl_6 plus KF in a polar solvent or benzene over CaF_2 at a high temperature. Readily substituted, e.g. KNH_2 gives $C_6F_5NH_2$. Used as a solvent and in preparation of fluoroaromatic derivatives.

hexafluorometallates Derivatives containing $[MF_6]^{n-}$ ions. Formed by many elements in the +3, +4, +5 oxidation states. Some derivatives, e.g. $[PF_6]^-$, $[SiF_6]^{2-}$, $[GeF_6]^{2-}$, are stable in aqueous solution and are available as solutions of strong acids. The majority of hexafluorometallates are immediately hydrolysed by water. Prepared by use of HF (aqueous or anhydrous), F_2, BrF_3.

hexafluoropropene, C_3F_6, $CF_3CF=CF_2$. B.p. –29°C. Monomer used to form polymers or copolymers with C_2F_4 and vinylidene fluoride, $CH_2=CF_2$, etc. Prepared by thermal decomposition of $CF_3CF_2CF_2CO_2Na$ or commercially by low-pressure pyrolysis of C_2F_4.

hexagonal close-packing See CLOSE-PACKED STRUCTURES.

hexagonal system The CRYSTAL SYSTEM with a 6-fold axis as principal axis of symmetry. The unit cell is taken as in the TRIGONAL SYSTEM. E.g. NiAs.

hexahydrocresols See METHYLCYCLOHEXANOL.

hexahydrophenol See CYCLOHEXANOL.

hexalin See CYCLOHEXANOL.

hexametaphosphates Derivatives of Graham's salt $(NaPO_3)_n$. See PHOSPHORUS OXYACIDS.

hexamethonium bromide, $[-(CH_2)_3N(CH)_3Br]_2H_2O$. Anti-hypertensive.

hexamethylbenzene, $C_{12}H_{18}$, $C_6(CH_3)_6$. Colourless crystalline solid; m.p. 164°C, b.p. 264°C. Prepared by the reaction of methyl chloride with benzene, toluene, or preferably pentamethylbenzene in the presence of $AlCl_3$.

It is oxidized by potassium permanganate to mellitic acid.

hexamethyldisilazane, $(CH_3)_3SiNHSi(CH_3)_3$. B.p. 125.5°C. Prepared from trimethylchlorosilane and ammonia. Used for forming trimethylsilyl derivatives of carboxyl, hydroxyl, amino and thiol groups and especially useful for protecting these groups since the Me_3Si- group is cleaved by water. Used in peptide synthesis. A convenient method of forming volatile, nonpolar derivatives of e.g. amino-acids which are then amenable to gas–liquid chromatography. The reagent is also used for siliconizing glc columns by direct injection into the gas stream. Used to prepare =NH derivatives from halides.

hexamethylenediamine See 1,6-DIAMINO-HEXANE.

hexamethylenetetramine (hexamine, methenamine, 1,3,5,7-tetra-aza-adamantane, urotropine), $C_6H_{12}N_4$, $(CH_2)_6N_4$. White crystals, sublime at 263°C with partial decomposition. Synthesized from methanal and ammonia. Hexamine is used as a starter fuel for camping stoves, as an accelerator in vulcanizing rubber and in the manufacture of resins. Anti-microbial.

hexamethyl phosphoramide See HEXAMETHYLPHOSPHORIC TRIAMIDE.

hexamethylphosphoric triamide (hexamethyl phosphoramide, HMPA, HMPT), $[(CH_3)_2N]_3PO$. M.p. 4°C, b.p. 232°C. Prepared from dimethylamine and phosphorus oxychloride. Used as an aprotic solvent, similar to liquid ammonia in solvent power but easier to handle. Solvent for organolithium compounds, Grignard reagents and the metals lithium, sodium and potassium, polymers and polymerization. Probably carcinogenic.

At temperatures of 220–240°C it functions as an efficient, neutral dehydrating agent, amides yielding nitriles and alcohols yielding alkenes.

hexamine See HEXAMETHYLENETETRAMINE.

hexan-2,5-dione (acetonylacetone), $CH_3COCH_2CH_2COCH_3$. Colourless liquid which becomes yellow on standing; b.p. 191°C. Obtained by boiling 2,5-dimethylfuran with dilute sulphuric acid. It readily condenses with a variety of substances to give derivatives of furan, thiophen and pyrrole, and is a solvent for cellulose acetate.

hexanes, C_6H_{14}. Five isomeric hexanes are known, b.p. 60–80°C. n-Hexane, $CH_3[CH_2]_4CH_3$, a colourless liquid, b.p. 69°C, is the most important isomer, and is used principally as a solvent, particularly for oil-seed extraction and in thermometers.

hexanoic acid See CAPROIC ACID.

hexanols, $C_6H_{11}OH$.

n-Hexanol, b.p. 156°C, fatty alcohol used in plasticizers. Prepared from ethanal by an aldol condensation or via hexyl$_3$Al.

4-Methyl-2-pentanol, b.p. 132°C, a solvent for nitrocellulose, urea–formaldehyde and alkyd resins. It is also used in ore flotation.

2-Ethyl-1-butanol, h.p. 147°C, is used in synthesis and as a solvent for printing inks and surface coatings.

hexazonine Herbicide, fluorobenzoyl urea derivative.

hexenes, C_6H_{12}. Various isomers depending on position of double bond. Used in synthesis.

hexestrol, $C_{18}H_{22}O_2$. Colourless crystals, m.p. 185–188°C. Shows estrogenic activity.

hexobarbitone, $C_{12}H_{16}N_2O_3$. Colourless crystals, m.p. 146°C. A rapidly absorbed hypnotic. The water-soluble sodium salt – hexabarbitone soluble – is injected intravenously to produce general anaesthesia of short duration. Habit forming.

hexone Methyl isobutyl ketone.

hexose, $C_6H_{12}O_6$. A carbohydrate with six carbon atoms. The hexoses are by far the most important of the simple sugars, and almost all the polysaccharides are built up of hexose units. Three aldohexoses, D-GLUCOSE, D-MANNOSE and D-GALACTOSE, are common in plants, either in the free state or as components of polysaccharide molecules. Other less important aldohexoses are D-idose, D-gulose, D-talose, D-allose and D-altrose.

The four ketohexoses are FRUCTOSE, SORBOSE, allulose and tagatose. Fructose diphospate is called hexose.

D-lyxo-hexulose (D(-)tagatose) A hexose sugar.

hexyl The group $C_6H_{13}^-$.

hexyleneglycol (2-methyl-2,4-pentanediol), $C_6H_4O_2$. B.p. 198°C. Used in cosmetics and hydraulic brake fluids.

hexythiazox Acaricide, thiazolidine derivative.

Heyrovsky–Ilkovic equation The relation between the HALF WAVE POTENTIAL, $E_{½}$, for a po-

larogram, the current, i, for the potential at the dropping mercury electrode, E_{DME}, the number of electrons, n, involved in the electrochemical reaction (id is the diffusion current).

$$E_{DME} = E_{½} + \frac{0.059}{n}\log\frac{id-i}{i}$$

Hf Hafnium.

Hg Mercury.

high-alumina cement See ALUMINOUS CEMENT.

highfield See N.M.R. SPECTRUM. Low chemical shift.

high performance liquid chromatography (HPLC) See CHROMATOGRAPHY.

high spin state A term used in transition-metal chemistry to denote that the compound has the maximum number of unpaired electrons consistent with the electronic configuration and stereochemistry. Most commonly used for octahedral complexes where, depending upon the crystal field splitting, two spin states are possible for, e.g., d^6 (Co^{3+}, Fe^{2+})

low spin –	high spin –
zero unpaired	four unpaired
electrons	electrons

hippuric acid (*N*-benzoylglycine), $C_9H_9NO_3$, $PhC(O)NHCH_2CO_2H$. M.p. 187°C; prepared from benzoyl chloride and glycine. Excreted in small quantities in the urine of mammals as a means of elimination of the toxic benzoic acids.

histamine (1H-imidazole-4-ethanamine), $C_5H_9N_3$. A base, formed by the bacterial degradation of histidine, and present in ergot and in many animal tissues, where it is liberated in response to injury and to ANTIGEN–ANTIBODY reactions. If injected it causes a condition of shock with dilatation of blood vessels, loss of plasma from the capillaries to the tissues and a rapid fall in blood pressure. Normally prepared from protein degradation products.

histidine (2-amino-3-(4-imidazolyl)propionic acid), $C_6H_9N_3O_2$. M.p. 277°C. The naturally occurring substance is laevorotatory. Histidine is one of the basic amino-acids occurring in proteins, the protamines and histones. It is an essential constituent of the food of animals.

histones See PROTEINS.

HMPA (HMPT) See HEXAMETHYLPHOSPHOR-IC TRIAMIDE.

HMX cyclotetramethylenetetranitramine A nitro-amine high explosive.

Ho Holmium.

Hofmann (exhaustive methylation) Formation of an olefin by pyrolysis of a quaternary ammonium salt.

Hofmann degradation of amides Amides react with solutions of chlorine or bromine in excess sodium hydroxide (NaOX) to give primary amines containing one carbon atom fewer than the original amide. Overall $RCONH_2 \rightarrow RNH_2$.

Hofmann isonitrile synthesis Formation of isonitriles from amines, chloroform in alkali.

Hofmeister series See LYOTROPIC SERIES.

holmium, Ho. Lanthanide metal, at.no. 67, at.mass 164.93, m.p. 1474°C, b.p. 2700°C, density ρ 8795, crustal abundance 1.4 p.p.m., electronic configuration [Xe] $4f^{11}6s^2$. The metal is hcp. Used in phosphors.

holmium compounds Holmium forms a single series of typical LANTHANIDE COMPOUNDS in the +3 state $Ho^{3+}(f^{10}$ brown-yellow$) \rightarrow$ Ho (−2.32 volts in acid). Lower halides contain M–M bonds.

holohedral forms See HEMIHEDRAL FORMS.

HOMO See FRONTIER ORBITAL SYMMETRY.

homo- Used as a prefix in organic chemistry it indicates a difference of $-CH_2-$ in an otherwise similar structure. It is applicable to aliphatic, alicyclic and aromatic compounds.

phthalic acid   homophthalic acid

homocyclic Compounds which contain a closed ring system in which all the atoms are the same. As the atoms are nearly always carbon atoms, the term carbocyclic is often used. These compounds can be subdivided into aromatic or benzenoid compounds and alicyclic compounds, such as cyclohexane.

homogeneous catalysis The process which occurs when the catalyst is in the same phase as the reactants, e.g. the acid-catalysed hydrolysis of an ester. Processes where the catalyst and one of the reactants are in the same phase, e.g. the hydrogenation of liquid olefins catalysed by solutions of transition metal complexes.

homogeneous combustion See COOL FLAMES.

homogeneous reaction Reactions which occur between substances in the same phase, e.g. between gases or liquids. A reaction between two solids is not usually regarded as a homogeneous reaction.

homogenizer A type of colloid mill.

homogentisic acid (2,5-dihydroxyphenylethanoic acid), $C_8H_8O_4$. M.p. 152–154°C. Occurs in plants. Occurs in the metabolism of tyrosine and phenylalanine in animals. Found in the urine of patients with alkaptonuria.

homoleptic Having only one type of substituent, e.g. MoF_6, $Ti(CH_3)_4$, $Cr(CO)_6$.

homolytic reaction A reaction such that the electrons forming the bond broken are shared equally between the fragments formed, or when species containing odd numbers of electrons (e.g. free radicals) react with each other forming a covalent bond. E.g

$$Br_2 \rightarrow 2Br\cdot$$
$$Br\cdot + CH_3CH_2\cdot \rightarrow CH_3CH_2Br$$

homonuclear molecule A molecule consisting of identical atoms or groups of atoms, e.g. O_2, N_2 or Cl_2. A heteronuclear molecule contains different atoms, e.g. HCl, NO, CO.

homopolar bond A COVALENT BOND.

homopolar crystal A crystal with all bonds homopolar (covalent), e.g. diamond.

hormones Hormones are molecules secreted directly into the bloodstream by ductless glands. They are then carried to other organs where they exert physiological action. Hormones influence the rate at which reactions proceed. The tissues which specialize in the secretion of hormones are called endocrine glands. The thyroid gland-secreting thyroxine is one example. Some hormones, however, are secreted by tissues which have other functions, e.g. the sex hormones secreted by the testes and ovaries and insulin which is secreted by parts of the pancreas. Hormones also often affect the functioning of other hormone-producing tissues so that effects can be complex and integrated. Hormones do not fall into any one particular chemical class, are water soluble and may be steroids, peptides or relatively simple compounds like adrenaline. Substances resembling the animal hormones are also produced by plants, e.g. the auxins. See DICTIONARY OF BIOLOGY.

hot carbonate processes Processes for the removal of carbon dioxide from gases by extraction with hot solutions of sodium or potassium carbonate. A significant amount of hydrogen sulphide may be removed at the same time.

hot working Deformation of metals and alloys carried out above the recrystallization temperature. There is no hardening and reductions in section can be made by processes such as hot rolling or extrusion.

HPLC High performance liquid chromatography.

HPV chemicals High production volume chemicals (broadly chemicals produced at rate of 500 tonnes p.a. or more).

HREELS See EELS.

HRMS (high resolution mass spectroscopy) Used to establish empirical formulae.

Hs Hassium. Element number 108.

Hückel approximation Method of calculating the energy of conjugated molecules considering only the resonance integral between neighbouring bonds. CNDO method.

Hückel's rule Planar monocyclic rings with $(4n + 2)$ delocalized π-electrons are aromatic.

Hudson's isorotation rule In the D-series the more dextrorotatory is named α-D and the other β-D. In the L-series the one laevorotatory is named α-L and the other β-L.

human growth hormone (HGH) Polypeptide which promotes physical development.

humectant A substance, which because of its HYGROSCOPIC nature, can be used to preserve the moisture content of materials, e.g. glycerol or sorbitol in confectionery or food, glycerol in tobacco.

Hume-Rothery's rule The statement that the phase of many alloys is determined by the ratios of total valency electrons to the number of atoms in the empirical formula. See ELECTRON COMPOUNDS.

humic acids A group of polymeric phenolic compounds present in soil, and can also be obtained from peat and brown coal or after refining sugar beet. The colouring matter of most surface water, chelate metals. Used in pigments, inks, fertilizers.

humidification The evaporation of a liquid into a gas is known as humidification and the gas is said to become humidified.

humidity The degree of wetness of moist air. Humidity may be expressed in several ways:

The *absolute humidity* is the mass of water vapour present per unit volume of moist air.

The *percentage humidity* is the ratio of the amount of water vapour present per unit mass of dry air to the amount the air could hold if saturated at the same temperature, expressed as a percentage.

The *relative humidity* is the ratio of the PARTIAL PRESSURE of the water vapour in the air to the partial pressure of water vapour in the air when saturated at the same temperature. This ratio is usually expressed as a percentage.

The humidities for systems other than air and water are defined analogously. See DICTIONARY OF PHYSICS.

humus The characteristic organic constituent of the soil. It is a dark-coloured amorphous material and is formed by the microbiological decomposition of plant materials, chiefly lignin and proteins. It occurs in the soil in company with clay, forming a colloidal clay–humus complex.

Hund's rules Rules which describe the electronic configuration of degenerate orbitals in the ground state. The electronic configuration will have the maximum number of unpaired electrons, i.e. p^3, ↑ ↑ ↑ rather than ↑↓ ↑ __ . The ground state will have the maximum orbital angular momentum.

hyaluronic acid Mucopolysaccharides composed of units of D-glucuronic acid and N-acetyl-D-glucosamine. Used in surgery and ophthalmology.

hybridization The description whereby atomic orbitals of differing type but similar energies are considered to combine to yield a set of equivalent hybrid orbitals. Hybridization is a mathematical operation in the formation of the molecular orbitals by combining the atomic orbitals of the constituent atoms of a molecule. E.g. in methane, which contains four equivalent C–H bonds. It may be considered that the

C-atom 2s and 2p orbitals are hybridized to give four equivalent sp³ orbitals which are then each combined with an H-atom 1s orbital.

hydantoin (glycolylurea, 2,4-imidazolidine dione), $C_3H_4N_2O_2$. Colourless needles; m.p. 220°C. Prepared by the condensation of glycine with potassium cyanate and boiling the hydantoic acid so formed with hydrochloric acid. It is present in beet molasses. Many substituted hydantoins known.

hydnocarpic acid ((R)-2-cyclopentane-1-undecanoic acid), $C_{16}H_{28}O_2$. M.p. 59–60°C. A fatty acid occurring as glycerides in chaulmoogra oil and other vegetable oils. Antibacterial.

hydracrylic acid See 2-HYDROXYPROPIONIC ACID.

hydramethylnon Insecticide, hydrazone derivative.

hydrates Many compounds have crystallized water additional to that required for a simple stoichiometry. Water can be bonded to cations by co-ordinate bonds from the oxygen or to anions by hydrogen bonding. The gross structure of many materials is determined by the bonding of the water of hydration.

hydration Ions (and many other species) in aqueous solutions are solvated by water and are said to be hydrated. The proton in aqueous solution is generally written as $[H_3O]^+_{aq}$ and three or four other water molecules are associated with the proton. Hydration occurs either by interaction of the lone-pairs of electrons in water with a cation or by hydrogen bonding with anions. Secondary hydration spheres also exist. Also addition of water to an olefin to give an alcohol.

hydraulic cement See CEMENT.

hydraulic conveying The conveying of particulate solids (e.g. coal, china clay, wood pulp) as a slurry through a pipeline.

hydraulic fluids Fluids used to transmit power and pressure. Most hydraulic fluids are based on low-viscosity mineral oils but for reasons of economy or safety aqueous-based fluids are also used.

hydrazides, $MNHNH_2$. The sodium and other alkali metal derivatives are formed from M, MNH_2, or MH and hydrazine, $NaNHNH_2$. Cleaves a C=C bond and, e.g., PhCH=CHMe gives PhMe and MeCH=NNH₂ (a hydrazone).

More generally derivatives of hydrazine. N_2H_4 and acyl halides give $RCONHNH_2$.

hydrazine, N_2H_4. M.p. 1.4°C, b.p. 114°C. Has structure $H_2N \cdot NH_2$ in the gauche form. Manufactured from NH_3 or urea and NaOCl or Cl_2 in the presence of a ketone and gelatin (Raschig process). Forms an azeotrope with water, anhydrous N_2H_4 is obtained by distillation over NaOH or precipitation of the sulphate, which reacts with liquid NH_3 to give $(NH_4)_2SO_4$ and N_2H_4. Hydrazine is a weak base giving hydrazinium salts, e.g. $(N_2H_5)Cl$ with strong acids. Aqueous solutions can be oxidizing giving $(NH_4)^+$ in acid with Ti^{3+}($E° + 1.27$ volt) in slow reactions, but are more generally reducing giving N_2($E°$ acid + 0.23 volt; $E°$ alkaline + 1.15 volt). Forms complexes but generally acts only as a monodentate ligand. Burns in oxygen, reacts with halogens. Used for removing O_2 from boiler-feed water, etc. and in the manufacture of hydrazides. Organic derivatives have many uses including use as high-energy fuels, blowing agents for foam plastics, antioxidants, herbicides.

hydrazinium salts See HYDRAZINE.

2-hydrazinoethanol Plant growth regulator.

hydrazobenzene (1,2-diphenylhydrazine), $C_{12}H_{12}N_2$, PhNH·NHPh. Colourless plates; m.p. 131°C. In moist air or in alcoholic solution it oxidizes spontaneously to azobenzene. With acids is converted into benzidine by an intramolecular change. It gives two molecules of aniline when treated with a strong reducing agent.

It is prepared by reduction of nitrobenzene with iron and NaOH and by electrolytic reduction of nitrobenzene. It is widely used for the preparation of benzidine.

hydrazoic acid (azoimide), N_3H, NNNH. M.p. –80°C, b.p. 37°C. Prepared from sodium azide and acid or $(N_2H_5)^+$ plus nitrous acid, HNO_2. Heavy-metal salts, azides, are used as detonators, alkali metal salts are stable and azides are used synthetically in organic chemistry.

hydrazones, $RR'C=NNH_2$. The derivatives formed by condensation between an aldehyde or ketone and hydrazine. Substituted hydrazones have been used for preparing crystalline derivatives for identification of keto-compounds.

hydrides There are several distinct types of hydrides.

1 *Salt-like*. These are the hydrides of the most electropositive elements (e.g. Na) and contain ions.

2 *Covalent.* Formed by most of the non-metals and transition metals. This class includes such diverse compounds as methane, CH_4 and iron carbonyl hydride, $H_2Fe(CO)_4$. In many compounds the hydrogen atoms act as bridges, sometimes electron deficient. Where there is more than one hydride site there is often hydrogen exchange between the sites. Hydrogens may be inside metal clusters.

3 *Complexes.* These derivatives contain complex anions which may be considered as derived from co-ordination of an H^- ion to a metal or non-metal. Examples are the BH_4^- and ReH_9^{2-} ions.

4 *Transition metal hydrides.* These are formed by hydrogen uptake by the metal. The phases are often non-stoichiometric. The oxide hydride $LaSrO_3H_{0.7}$ has long-range magnetic order.

Hydrides frequently function as hydrogenation catalysts.

hydriodic acid, HI. Aqueous solution of HY-DROGEN IODIDE.

hydroaminoalkylation   Olefin plus CO plus amine or NH_3 over catalyst to give an amine.

hydroboration   The *cis*- addition of B–H bonds across the double bonds of olefins. Thus diborane, B_2H_6, reacts with ethene to give $B(C_2H_5)_3$. Breakdown of the alkyl borane with acid gives alkanes and with hydrogen peroxide gives alcohols (the orientation is the opposite to that which would result from the direct addition of water across the double bond). Catecholborane is often used in the presence of an organolanthanide complex. Precursors for the Suzuki reaction. See ORGANOBORANES.

hydrobromic acid, HBr.   An aqueous solution of HYDROGEN BROMIDE.

hydrocarbon resins   Thermoplastic polymers of mol.wt. less than 2000 obtained by cracking petroleum and from turpentine. Used in drying oils, with rubber and as plasticizers. Class includes coumarone–indene resins, petroleum resins, cyclopentadiene resins, terpene resins.

hydrocarbons   This term includes all compounds of carbon and hydrogen only. They are subdivided into aliphatic and cyclic hydrocarbons according to the arrangement of the carbon atoms in the molecule. The aliphatic hydrocarbons are again subdivided into paraffins, olefins, diolefins, etc., according to the number of double bonds in the molecule. The cyclic hydrocarbons are subdivided into aromatic hydrocarbons and cycloparaffins.

hydrochloric acid (muriatic acid), HCl. An aqueous solution of HYDROGEN CHLORIDE. A saturated solution of HCl contains about 43% HCl and gives a constant-boiling mixture. Hydrochloric acid is used widely in the chemical industry, in the food industry (glucose, monosodium glutamate, cleaning), in metallurgy and in the oil industry. Extremely corrosive and handled in glass or plastic equipment or in apparatus using special alloys (Ta, Ni–Mo). U.S. production 1999 4.6 megatonnes.

hydrochlorides   Salts formed by organic bases with HCl (or weak bases prepared in nonbasic solvents, e.g. toluene). Also salts containing the $[ClHCl]^-$ anion.

hydrochlorinated rubber   Material used in films produced by passing HCl gas into a solution of rubber in benzene. Contains *c.* 30% Cl.

hydrocortisone (17-hydroxycorticosterone), $C_{21}H_{30}O_5$.   A CORTISONE with an 11-OH group (reduction of C=O group). White crystals; m.p. 217–220°C. Made synthetically from naturally occurring steroids, such as diosgenin or hecogenin, the 11-hydroxy group being introduced where necessary by microbiological means. Used as its 21-esters for local treatment of inflammatory and allergic conditions.

hydrocracking   Processes which involve both the catalytic cracking and hydrogenation of petroleum fractions, yielding high quality gasolines.

hydrocyanic acid, HCN.   See HYDROGEN CYANIDE.

hydrodealkylation   A catalytic process generally used for the production of benzene from alkyl aromatics, toluene being the most common feedstock. Methane is also formed.

hydrodesulphurization   See DESULPHUR-IZATION.

hydrodynamic radius of ions   The effective radius of an ion in solution.

hydrofining   See DESULPHURIZATION. Used, under milder conditions, as a SWEETENING PROCESS or to remove alkenes for improvement of stability in light petroleum fractions.

hydrofluoric acid, HF.   Aqueous solution of hydrogen fluoride. The system has a maximum boiling point at about 36% HF. Used widely in industry as a fluorinating agent and for anhydrous

HF, e.g. cryolite (Al production), UF_4 and UF_6 production; etching glass, cleaning stainless steel, etc. Extremely corrosive and handled in Monel, Teflon or platinum equipment.

hydroforming See CATALYTIC REFORMING.

hydroformylation Carbonylation reaction for reaction of an alkene with carbon monoxide and hydrogen with a Co or Rh catalyst to produce an aldehyde or a primary alcohol, depending upon conditions. See OXO REACTION.

hydrogasification The process of converting crude oil or oil fractions to SUBSTITUTE NATURAL GAS (SNG).

hydrogen, H. Gas, at.no. 1, at.mass 1.0079, m.p. $-259.34°C$, b.p. $-252.87°C$, density liquid 70.8 g l^{-1}, crustal abundance 1520 p.p.m., electronic configuration $1s^1$. There are three isotopes ^1H, ^2H (deuterium), ^3H (tritium – radioactive). The lightest element, occurs free as H_2 in some natural gases and is produced naturally by some green algae. H_2 and H_3^+ are present in the atmospheres of giant planets and hydrogen is a widespread constituent of water, minerals and natural organic materials. The element is prepared by electrolysis of water, or industrially by steam reforming hydrocarbons (C_nH_{2n+2} plus H_2O over catalyst to give CO plus H_2; then further catalyst on CO plus H_2O to give H_2 plus CO_2 followed by purification by scrubbing CO_2) and also results during catalytic reforming of petroleum. Occurs as H_2 (see ORTHO-HYDROGEN); explodes over wide concentration ranges with oxygen on sparking and reacts with halogens. Combines with most elements to give HYDRIDES. Used industrially in welding (37%), ammonia synthesis, hydrogenation of fats and oils, methanol synthesis, hydrodealkylation, etc. U.S. production 1999 460 kilotonnes.

Anti-hydrogen from a positron and negative proton has been formed.

hydrogenation A specific method of reduction in which hydrogen is added to a substance by the direct use of gaseous hydrogen, although other sources, e.g. CH_3OH, can be used. The process is normally carried out with the aid of a catalyst and proceeds more rapidly at high pressures. Hydrogenation is of great industrial importance and is widely used in petroleum refining and in the petroleum chemical industry. It is also used in the refining of fatty oils and for many other chemical processes. Also used with coal to give hydrocarbons. Many hydrogenations are known to proceed through M–H bonds.

hydrogen bond Hydrogen attached to an electronegative element, e.g. O,F, gives a highly polarized bond. If this bond is directed towards another atom, E, with a lone-pair of electrons, a bond is formed, known as a hydrogen bond, which may be represented O–H–E. Such bonds cause the self-association of water and alcohols, and form the basis of ions such as $[FHF]^-$. Hydrogen bonds are readily formed between hydroxyl and amino groups and oxygen, nitrogen and halogen atoms; they may be intermolecular or intramolecular. They are important in determining the arrangement of polypeptide and carbohydrate chains in proteins and cellulose, in adhesion and in hydrates. Also found in metal hydride derivatives.

hydrogen bromide, HBr. M.p. $-88°C$, b.p. $-67°C$. Colourless gas obtained from elements in presence of a catalyst (charcoal or Pt) or from Br_2, red phosphorus, and H_2O (HBr is the only volatile product). Gives hydrobromic acid in water (also prepared Br_2 plus H_2O with SO_2 or H_2S). Used in preparation of bromides, and as a brominating and reducing agent.

hydrogen chloride, HCl. M.p. $-115°C$, b.p. $-85°C$. Hydrogen chloride is formed as a by-product in chlorination of hydrocarbons (90%), from H_2 plus Cl_2 and NaCl plus H_2SO_4; much is used as its aqueous solution HYDROCHLORIC ACID. Hydrogen chloride is a colourless gas with an irritating smell. Forms hydrates ($1H_2O$ and $2H_2O$) and reacts with many metals (forms lower oxidation state chlorides, see CHLORINE). Used in the production of organic chlorides and as a condensing agent in organic chemistry. U.S. production 1997 1920 kilotonnes.

hydrogen cyanide (hydrocyanic acid, prussic acid), HCN. M.p. $-13°C$, b.p. $26°C$. Colourless, completely miscible with water, burns in air, smell of bitter almonds. Manufactured from methane, air, NH_3 over a catalyst at $1000°C$; also formed from metal cyanides and H_2SO_4, dehydration of formamide. Polymerizes on standing. Weak acid, gives salts, CYANIDES; forms cyanides, nitriles RCN and isocyanides, isonitriles RNC. Used in organic and synthetic fibre synthesis and as a rodenticide and insecticide. HCN is a very powerful poison.

hydrogen electrode The standard electrode against which other electrode potentials are measured. The electrode consists of a piece of platinum foil, covered with a layer of active platinum black, which is immersed in a solution containing hydrogen ions whilst hydrogen gas is bubbled over the surface. For solutions in which the concentration of hydrogen ions is equal to unit activity and for a gas pressure of 1 atm. the electrode potential is said to be normal and is arbitrarily assigned a value of zero. Although other electrode potentials are quoted relative to the standard hydrogen electrode, the latter is seldom used in practice for this purpose. Usually electrode potentials are measured against a standard calomel electrode or, e.g. ferrocene, which is then referred to the 'hydrogen scale'.

hydrogen fluoride, HF. M.p. -83°C, b.p. 19.5°C. Colourless, strongly fuming, associated liquid prepared from KHF_2 or CaF_2 and H_2SO_4. Gives HYDROFLUORIC ACID with water. Anhydrous HF is a good solvent for both inorganic salts and organic materials; it must be handled in metal (mild steel or Monel) or Teflon or Kel-F apparatus. Used as a fluorinating agent for producing chlorofluorocarbons, uranium processing and as an alkylation catalyst in the petrochemical industry. Gives fluorides as salts. The hydrogen difluoride, $[HF_2]^-$, ion is also formed. Used for aluminium production (32%), fluorocarbons (30%), alkylation (5%), uranium processing (5%), steel manufacture (5%). Forms adducts with pyridine and other bases used as fluorinating agents.

hydrogen halides, HF, HCl, HBr and HI. Hydrogen fluoride stands apart from the other hydrogen halides because of its strong hydrogen bonding and consequent association. HI is the strongest acid in water and the most readily dissociated. All form halide salts but the solubility pattern of the fluorides is frequently different from that of the other halides (e.g. AgF soluble, AgX insoluble; AlF_3 insoluble, AlX_3 soluble).

hydrogen iodide, HI. M.p. -54°C, b.p. -36°C. Colourless gas fairly readily dissociated to H_2 and I_2. Very soluble in water to give an acid solution – forms 2, 3, $4H_2O$ hydrates. Prepared from H_2 and I_2 with catalyst or I_2 plus red phosphorus plus H_2O. Aqueous hydriodic acid solution prepared I_2 plus H_2O plus H_2S. Used as a disinfectant.

hydrogen-ion concentration Measured as the number of gram ions of hydrogen ions present per litre of solution. Since these concentrations are usually small, the concentration is generally expressed as the pH of the solution; the pH being the logarithm of the reciprocal of the hydrogen-ion concentration, i.e. pH = -log[H⁺] (strictly the hydrogen ion activity).

The pH can be measured using a glass electrode or, less accurately, by coloured indicators.

hydrogen ions Cbemically free hydrogen ions are rarely formed and dissociation to H⁺ occurs only when the resultant ion can be complexed. Thus HCl gas is undissociated but gives $[H(H_2O)_n]^+Cl^-$ and $[NH_4]^+Cl^-$ with H_2O or NH_3.

hydrogenolysis The cleaving of a chemical bond by hydrogen, usually in the presence of a hydrogenation catalyst.

hydrogen overvoltage See OVERVOLTAGE.

hydrogen peroxide, H_2O_2. M.p. -0.4°C, b.p. 150°C, ρ = 1450. Colourless syrupy liquid prepared by oxidation of an anthraquinol with oxygen and subsequent reduction of the quinone with hydrogen; also prepared electrolytically via persulphates. Occurs on the surface of Europa, one of Jupiter's moons. H_2O_2 found *in vivo* implicated in various diseases. The pure liquid is stable but aqueous solutions are unstable and alkalis and dust catalyses decomposition to oxygen and water. Often used as an aqueous solution. Concentration expressed in volume of O_2 that can be liberated: 6% = 20 vol. Can act as a powerful oxidizing agent ($E°$ acid solution + 1.77 volts) when converted to H_2O and as a reducing agent ($E°$ + 0.68 volts) when converted to O_2 (e.g. by $KMnO_4$). The parent compound of peroxides. Forms adducts, perhydrates, with some salts. Used in chemicals (33%), pollution control (19%), textiles (17%), paper (13%); free H_2O_2 is used as a propellant. U.S. use 1995 360 kilotonnes.

hydrogen persulphides (sulphanes), H_2S_x ($x = 2-6$). Formed from polysulphides and acid followed by cracking. Decompose to H_2S and S.

hydrogen spectrum The atomic spectra from H atoms. See BALMER SERIES.

hydrogen sulphide (sulphuretted hydrogen), H_2S. M.p. -85°C, b.p. -61°C. Colourless, very poisonous gas produced by decaying animal or vegetable matter and present in some mineral waters. Prepared FeS plus dilute acid or *in situ* by hydrolysis of thiols. Forms sulphides with most metals, particularly in solution where it acts as a weak acid.

hydrolith, CaH_2. CALCIUM HYDRIDE.

hydrolysed product Formed as a result of hydrolysis.

hydrolysis A term used to signify reactions due to the presence of the hydrogen or hydroxyl ions of water. E.g. when salts are dissolved in water the resulting solution is often either slightly acidic or alkaline due to hydrolysis. Thus, sodium ethanoate yields an alkaline aqueous solution due to hydrolysis; some of the ethanoate ions combine with the hydrogen ions of the water to form the undissociated ethanoic acid, thereby leaving an excess of hydroxyl ions in solution. The reaction may be represented as

$$Na^+ + CH_3COO^- + H_2O \rightleftharpoons$$
$$Na^+ + OH^- + CH_3COOH$$

and the equilibrium constant, K, for this reaction is given by

$$K_1 = \frac{[OH^-][CH_3COOH]}{[CH_3COO^-][H_2O]}$$

However, the water is present in large excess, its concentration may be regarded as being constant and hence

$$K_2 = \frac{[OH^-][CH_3COOH]}{[CH_3COO^-]}$$

K_2 is called the hydrolysis constant for sodium ethanoate. Hydrolysis occurs when salts involving weak acids or bases are dissolved in water. It is often also found with metal ions in solution. The ion $[M(H_2O)_n]^{x+}$ dissociates to the hydroxy species $[M(H_2O)_{n-1}(OH)]^{(x-1)+}$. Hydrolysed species may crystallize as basic salts.

In organic chemistry the term is used to describe the conversion of an ester to an acid and an alcohol (saponification), the addition of the elements of water to a molecule, e.g. the conversion of a nitrile to an amide

$$R \cdot CN + H_2O \rightarrow RCONH_2$$

and the displacement of a group or radical by hydroxyl, e.g.

$$RCH_2Cl + H_2O \rightarrow RCH_2OH + HCl.$$

All hydrolysis reactions are catalysed by acid, or base, or both.

hydrometer A simple device for measuring the density or specific gravity of a liquid, consisting of a cylindrical float whose lower end is conical and weighted and whose upper end terminates in a narrow tube. When placed in a liquid the instrument floats upright with the tube partially immersed, the immersion depending on the liquid density; the latter may be read off from a scale on the tube.

Because of their widespread use for simple process control, hydrometers are frequently calibrated, not in specific gravity, but in some units

related to it, which bear (or bore at one time) some relationship to the concentration being so measured.

hydroperoxides A class of compounds containing the hydroperoxyl group, $HO \cdot O-$, the simplest of which is hydrogen peroxide. Organic hydroperoxides are produced by aerial oxidation of many organic compounds, and decompose, frequently with extreme ease, to give alcohols, carbonyl compounds and other products.

hydrophilic colloid Colloidal sols of proteins, carbohydrates, soaps and other complex organic materials are marked by considerable stability and are not coagulated by small concentrations of electrolyte such as are sufficient to destroy hydrophobic or suspensoid sols such as those of metals or their sulphides. This stability is due to the existence of a sheath of dispersion medium (water) which protects the hydrophilic (water-loving) particles. The essential difference between hydrophilic and hydrophobic colloids is that the former are not coagulated at the isoelectric point.

Hydrophilic sols are characterized by a high viscosity, and often set to gels on cooling; it is possible to prepare them containing a high proportion of the dispersed colloid. They are either soluble monomolecular substances (e.g. proteins, polymers) or monomolecular substances which aggregate in solution to form colloidal micelles (e.g. soaps, dyes).

hydrophobic bonding The association of non-polar molecules or groups in aqueous media, resulting from the tendency of water molecules to exclude non-polar species.

hydrophobic colloids Sols of, e.g. gold, arsenic, are stable owing to the mutual repulsion arising from the like charges on all the colloidal particles. Such sols are not hydrated. On the addition of small concentrations of electrolytes, the charges on the particles are neutralized and coagulation takes place. Such sols are termed hydrophobic (water-hating), and have a viscosity equal to, or only slightly greater than, water. Hydrophobic sols are coagulated at the isoelectric point. It is normally only possible to prepare hydrophobic sols in low concentration.

hydroprene Insecticide, juvenile hormone type.

hydroquinone, $C_6H_6O_2$, $4-C_6H_4(OH)_2$. Colourless prisms; m.p. 170°C, b.p. 285°C/730 mm. With $FeCl_3$ it is oxidized to quinone. It is best prepared from quinone by reduction with sulphur dioxide. It is used as a photographic developer and antioxidant.

hydrosilylation The addition of an Si–H bond across an olefinic bond to form an organosilane (the reaction is frequently catalysed by Pt salts) which may be used in silicone production.

hydrosulphurous acid, $H_2S_2O_4$. See DITHIONOUS ACID.

hydrotropic salts Salts mainly of long-chain organic acids which render organic materials soluble in water.

hydrovinylation Conversion of C_2H_4 etc. to vinyl derivative.

hydroxides Compounds containing inorganic OH– groups. Alkali hydroxides are strongly basic.

hydroxonium ions, $[H(H_2O)_n]^+$. See HYDRATION OF IONS.

hydroxy- A prefix used in organic chemistry to denote that a substance contains a hydroxyl (–OH) group.

hydroxyacetone See HYDROXYPROPANONE.

hydroxyapatite, $Ca_5(OH)(PO_4)_3$. The major (70%) constituent of bone and tooth enamel. Collagen protein fibres are also present in bone to give extra strength. Isostructural with APATITE.

4-hydroxybenzoic acid, $C_7H_6O_3$. (M.p. 213–214°C.) Prepared from 4-bromophenol and potassium ethyl carbonate. Dyestuff intermediate and anti-fungal agent. Methyl ester used as preservative for food, beverages.

2-hydroxybenzylalcohol (salicyl alcohol), $C_7H_8O_2$. Prepared from phenol plus CH_2Cl_2 plus NaOH. Local anaesthetic.

4-hydroxydiphenylmethane, $C_{13}H_{12}O$. Prepared from phenol plus benzyl chloride. Germicide antiseptic, preservative.

hydroxyethanoic acid See GLYCOLLIC ACID.

hydroxyethylamines See ETHANOLAMINES.

hydroxyhalides Basic metallic halides of the metals. See BASIC SALTS.

hydroxylamine, H_2NOH. Colourless solid which explodes on heating. M.p. 33°C. Weak base prepared by reduction of NO_3^- or NO_2^- electrolytically, with SO_2, or with H_2 and catalyst. The solid decomposes slowly on keeping but salts, e.g. (hydroxylamine hydrochloride,

$NH_2OH \cdot HCl$) are fairly stable. Forms oximes. Acts as a reducing agent and as an oxidizing agent. Used in formation of caprolactam, in photography as antioxidant, and in tanning.

hydroxylation The introduction of hydroxyl groups into an organic compound. Reagents include CrO_3, ozone, persulphates and perborates.

hydroxyl ion OH⁻ and radical, OH. The anion formed by ionic dissociation of water and present in hydroxides. Combines with H^+ to give water. Present in solutions of bases. The radical –OH oxidizes CO, hydrocarbons and is an active pollutant. Source of tropospheric ozone.

4-hydroxy-4-methyl-2-pentanone, $C_6H_{12}O_2$. Prepared by action of $Ba(OH)_2$ or $Ca(OH)_2$ on propane. B.p. 168°C. Solvent also used as anti-freeze agent.

3-(hydroxymethyl)pyridine (nicotinyl alcohol) Solvent for cellulose, esters, salts.

hydroxyproline (5-hydroxypyrrolidine-2-carboxylic acid), $C_5H_9NO_3$. M.p. 270°C. The laevorotatory form is a constituent of connective-tissue proteins, e.g. collagen and elastin.

2-hydroxypropionic acid (hydracrylic acid, β-lactic acid), $CH_2OH \cdot CH_2C(O)OH$. An uncrystallizable syrup. Manufactured by treating ethylene cyanhydrin with NaOH. Converted to propenoic acid on heating. Used in the manufacture of propenoic esters.

3-hydroxypropionitrile (ethylene cyanohydrin) Prepared from chlorohydrin and, e.g. NaCN. B.p. 228°C. Solvent for cellulose, esters and inorganic salts.

8-hydroxyquinoline (oxine), C_9H_7ON. Light brown needles, m.p. 75–76°C. Forms insoluble complexes with metals. Widely used in the analysis of metals. Many oxinates are extracted and the metal is estimated spectrophotometrically. Derivatives, e.g. 2-methyl, tend to be specific for, e.g., Mg^{2+}. Copper derivatives are used as fungicides. Used as a chelating agent and in anti-perspirants and deodorants.

hygroscopic Substances which absorb water from moist air and form either a moist solid or solution, e.g. $MgCl_2$, P_2O_5.

hymexazol Fungicide, isoxazole derivative.

hyoscine See SCOPOLAMINE.

hyoscyamine, $C_{17}H_{23}NO_3$. Colourless needles, m.p. 108.5°C. L-Form of ATROPINE.

hyperchromic An increase in absorption intensity. The opposite is hypochromic.

hyperconjugation Quantum-mechanical effect of transfer of electrons from occupied orbital or double bond, e.g. to an unoccupied orbital. Specifically it indicates C–H conjugation. Postulated to account for the stabilization of aliphatic carbonium ions, ease of certain elimination reactions and staggered conformation of ethane.

hyperfiltration Reverse OSMOSIS.

hyperfine structure The splitting of individual spectral bands into (many) components.

hypo, $Na_2S_2O_3$. SODIUM THIOSULPHATE. Used in photography.

hypobromates, M[BrO]. Salts of HOBr. See BROMATES.

hypobromous acid, HOBr. See BROMATES.

hypochlorites, M[ClO]. Salts of HOCl. See CHLORATES.

hypochlorous acid, HOCl. See CHLORATES. Used as powerful oxidizing and bleaching agent. See BLEACHING POWDER.

hypochromic See HYPERCHROMIC.

hypofluorites Covalent compounds containing –OF groups (e.g. F_2O, CF_3OF, SF_5OF). Prepared by the action of fluorine on an oxygen-containing compound. Ionic hypofluorites, containing OF⁻, are not known as stable compounds. HOF is a very unstable substance formed from reaction of F_2 with ice.

hypoiodates, M[OI]. See IODATES.

hypophosphoric acid, $H_4P_2O_6$. Prepared from $NaClO_2$ and red phosphorus. Salts contain $[O_3P–PO_3]^{4-}$ ions. Stable in aqueous alkali but converted to phosphate by fused NaOH. A tetrabasic acid. Forms tetra-alkyl esters.

hypophosphorous acid, H_3PO_2, $H_2P(O)OH$. Crystalline solid; m.p. 26°C. The sodium salt, NaH_2PO_2, is formed from yellow P and NaOH. The acid and its salts are powerful reducing agents giving hydrides or low oxidation state compounds. The acid is monobasic and liberates H_2 and PH_3 on heating with NaOH.

hyposulphurous acid, $H_2S_2O_4$. See DITHIONOUS ACID.

hypoxanthine (6-oxypurine), $C_5H_4N_4O$. M.p. 150°C (decomp.). Sparingly soluble in cold water, more so in hot. A breakdown product of adenine in animals.

hypoxic Natural oxygen depletion caused, e.g., by algal blooms.

hypsochromic Shift of absorption to shorter wavelength, higher frequency, higher energy. Cf. bathochromic.

hysteresis Processes which occur mainly in solids in which the response obtained on increasing the variable differs from that on reducing the variable. Occurs, e.g. during the magnetization of iron, during the absorption of moisture by textiles.

I Iodine.

ice The solid form of water. The transition point 273.15K between liquid and solid water at one atmosphere pressure is defined as 0° on the Centigrade scale, also equal to 0° Celsius. This freezing temperature is lowered by increasing the pressure. By crystallization of water under greatly increased pressure it is possible to obtain forms of solid water which are physically distinct from ordinary ice (Ice I). The molecules in ice are held together by hydrogen bonds.

ICIA 0858 Fungicide, pyridine derivative.

icosahedral Having the symmetry of an icosahedron, a solid figure with twelve apices; each face is triangular. Icosahedral units are found in many boron derivatives, e.g. $B_{12}H_{12}^{2-}$.

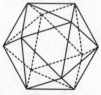

ideal gas A gas which obeys exactly the gas laws, i.e. $PV = nRT$. In practice no gas shows ideal behaviour, although helium, hydrogen and nitrogen approximate to ideality at high temperatures and low pressures.

ideal solution A solution, liquid or solid, in which the thermodynamic activity of each component is proportional to its mole fraction, i.e. a solution obeying Raoult's law. Such a solution would be formed from its components with zero heat of mixing, zero volume change and the ideal entropy of mixing. Only mixtures containing closely similar substances, e.g. isotopic elements, form ideal solutions, although many mixtures show behaviour which approximates to ideality (e.g. benzene and toluene).

idose A HEXOSE.

ignition limits See FLAMMABILITY LIMITS.

ignition temperature The minimum temperature at which a fuel can self-ignite. The temperature obtained will depend on the technique used and on the air–fuel ratio of the mixture.

ilmenite, $FeTiO_3$. A black Ti ore occurring in massive form and in sands. The structure has cp oxygen with Fe and Ti in octahedral co-ordination.

imazamethabenz-methyl Herbicide, imidazole derivative.

imazamox Herbicide, imidazole derivative.

imazapyr Herbicide, imidazole derivative.

imazaquin Herbicide, imidazole derivative.

imazathapyr Herbicide, imidazolinone derivative.

imazlil Fungicide, azole derivative.

imazosulfuron Herbicide, sulphonylurea derivative.

imibenconazol Fungicide, azole derivative.

imidacloprid Insecticide, imidazolidine derivative.

imidazole (glyoxaline, iminazole), $C_3H_4N_2$. The ring is present in POLYBENZIMIDAZOLES.

imides In organic chemistry imides are nitrogen-containing ring compounds containing the group –C(O)–NH–C(O)–. They are formed by heating dibasic acids or their anhydrides with ammonia. The hydrogen atom of the > NH group is acidic and can be replaced by a metal. Mild hydrolysis breaks the ring to give the half amide of the acid. See SUCCINIMIDE and PHTHALIMIDE. N-substituted imides can form very resistant polymers – see KEVLAR.

In inorganic chemistry the term refers to compounds containing the NH_2^- or >NH groups. These are prepared by the action of heat on amides or by metathetical reactions in liquid ammonia. The heavy metal imides are explosive.

imines Compounds containing an >NH group. May be cyclic (e.g. dihydroazirine) or linear.

iminium salts Salts containing cations in which the charge delocalization includes a nitrogen atom.

imino The name given to the group >NH where the nitrogen atom is part of a ring system or is united by both bonds to a single atom. Such compounds as R·CO·NHR are usually regarded as substituted amides and not as imino compounds.

iminoctadine Fungicide, guanidine derivative.

imiprothrin Insecticide, pyrethroid.

improper rotation (S_n) A symmetry element consisting of rotation about an axis through $2\pi/n$ followed by reflection through a plane perpendicular to the axis of rotation.

IMS Industrial methylated spirits.

In Indium.

inabenfid Plant growth regulator, pyridine derivative.

inconel See NICKEL ALLOYS.

incongruent melting Melting into its components without formation of a liquid phase, e.g. Na_2K.

indan (2,3-dihydroindene) B.p. 176°C. The first reduction product of indene. Occurs in coal tar.

indanthrene dyestuffs An important series of dyestuffs based on amino derivatives of anthraquinone.

indene (indonaphthene), C_9H_8. Colourless

liquid, b.p. 182°C, obtained from petroleum fractions. It is an acidic hydrocarbon and yields a sodium salt. The reactions of the 5-membered ring resemble those of cyclopentadiene.

indene resins See COUMARONE.

indican (indoxyl-3-glucoside), $C_{14}H_{17}NO$. See INDOXYL. Also the potassium salt of indoxyl-3-sulphuric acid, the form in which toxic indoxyl is excreted as one of the ethereal sulphates.

indicator Any substance which allows the progress of a chemical change to be followed, usually by a change of colour, or fluorescence of a solution, or by the alteration in the nature of a precipitate. E.g. methyl orange, an indicator used in the neutralization of acids with bases, is red at pH less than 3.1 and gradually changes through orange to a full yellow colour as the pH increases to 4.4.

indigo See INDIGOTIN.

indigoid dyes Vat dyes and pigments containing indigo-like groupings and giving similar colours. Contain indole or thionaphthene systems.

indigotin (indigo), $C_{10}H_{10}N_2O_2$. A very important and long-known dyestuff (woad). Originally obtained from indican (the glucoside of

indoxyl) but now prepared from anthranilic acid and chloroethanoic acid to give phenylglycine-*o*-carboxylic acid which, with KOH and NaNH$_2$, gives indoxylic acid, decarboxylated to give indoxyl and oxidized to indigotin by air. The dark blue indigotin is reduced by sodium dithionite to the soluble leuco-compound, indigotin white, which is used in dyeing and then reoxidized to indigotin. Sulphonated indigo dyestuffs are also known.

indium, In. Metal, at.no. 49, at.mass 114.82, m.p. 157°C, b.p. 2070°C, density ρ 7310, crustal abundance 0.049 p.p.m., electronic configuration [Kr] $3d^{10}5s^25p^1$. Occurs in some specimens of zinc blende; precipitated from solution electrolytically or with Zn. The metal is tetragonal with a distorted cp structure. The metal is oxidized on strong heating in air and dissolves in acids. Used in low-melting alloys. InP is used in electric motors and InGa, InAs, InS, In_4Te_3 are used in electronic components and semiconductors. Indium compounds are toxic.

indium chemistry Indium is a Group 13 element, electronic configuration $5s^2 5p^1$. It shows the normal group oxidation state of +3 ($E°In^{3+} \to In$ in acid solution –0.34 volts) and an unstable +1 state ($E°In^+ \to In$ in acid solution –0.25 volts) which disproportionates in aqueous solution. The chemistry of the +3 state is similar to that of Al(III) and Ga(III). Forms Zintl phases with Group 1 elements.

indium halides Indium forms InX_3 (X = F, Cl, Br, I), (X_2 on In). InF_3, $InCl_3$ and $InBr_3$ are fairly ionic. InI_3 is $I_2In(\mu$-$I)_2InI_2$. $InF_3.3H_2O$ and $InCl_3.4H_2O$ are known. Complex species $[InX_6]^{3-}$ are formed with F and Cl. InX (X = Cl, Br, I) (InX_3 plus In) and intermediate halides $InCl_2$ ($In^+[In^{III}Cl_4]$), In_2Cl_3 ($In^I_3[In^{III}Cl_6]$) and In_7Cl_9 are also known.

indium oxides In_2O_3 (In plus O_2) is formed by dehydration of $In(OH)_3$ (In^{3+} plus OH^-). In_2O (heat on In_2O_3), yellow when hot, red when cold, is readily oxidized to yellow In_2O_3.

indium oxyacid salts $In(NO_3)_3.5H_2O$, $In(ClO_4)_3.11H_2O$, $In_2(SO_4)_3$ are formed from $In(OH)_3$ and the appropriate acid. The sulphate forms alums.

indole, C_8H_7N. M.p. 52°C, b.p. 253–254°C.

Occurs in coal tar, in various plants and in faeces by the action of the intestinal bacteria on tryptophan. It can be prepared by the action of acid on the phenylhydrazone of pyruvic acid to give indole-2-carboxylate which can be decarboxylated to indole. Used in perfumery in highly dilute form.

indole-3-ethanoic acid (indol-3-yl acetic acid), $C_{10}H_9NO_2$. M.p. 164–165°C. Plant growth regulator.

indolizine (pyrrocoline) ring system The system is numbered as shown.

indoxyl, C_8H_7NO. Yellow crystals, m.p. 85°C;

soluble in water and alcohol. Occurs in woad as the glucoside indican, and in mammalian urine, combined with sulphuric acid, as an ester, also called indican. Arises in the body from decomposition of tryptophan.

induced dipole moment (μ) Distortion of a molecule by an electric field.

inductive effect A substituent effect on an organic compound due to the permanent polarity or polarizability of groups is called an inductive effect. It may be visualized as the inducement of charge either towards or away from a group or atom, with a resulting dipole. Accounts for, e.g. the increased acidity of trichloroethanoic acid over ethanoic acid.

industrial methylated spirits (IMS) Ethanol rendered non-potable by mixing 95% alcohol with methanol.

inelastic neutron scattering Study of the motion of molecules in liquids from passage of a beam of neutrons through a sample.

inert atmosphere box A container for carrying out chemical reactions in which the atmosphere has been purged of water, oxygen and nitrogen by recycling over active absorbants for the particular gases to be removed. Used for handling very sensitive materials; modifications can be used for handling reactive metals, e.g. Ti at high temperatures.

inert complex A complex which participates in reactions, particularly ligand-exchange reactions, either very slowly or sufficiently slowly that the rates of reaction can be readily followed. $[Co(NH_3)_6]^{3+}$ is an inert complex. Compare labile complex.

inert gases Semi-obsolete name for the NOBLE GASES.

inert gas formalism The OCTET RULE; the effective ATOMIC NUMBER RULE.

inert pair effect An apparent correlation between the existence of oxidation states n and $n − 2$ and the s^2 configuration for the heavy p block elements (n corresponds to the group-oxidation state). E.g. Pb shows oxidation states

Pb(IV) and Pb(II). Also used to describe the geometrical effects of a lone pair.

infra-red radiation (i.r.) Radiation of frequency less than about 10^{13} per sec ($10\,000$ cm^{-1}), i.e. of wavelengths greater than about 8000 Å. It is invisible to the human eye, but possesses considerable penetrating power. Photographic plates may be sensitized towards it, so that it finds application in long-distance photography. Heat-seeking devices and devices for night viewing rely on detection of i.r. radiation. Many forms of detector are known. Interatomic vibrations in molecules cause absorption in the i.r. region. See INFRA-RED SPECTROSCOPY.

infra-red spectroscopy Vibrations within molecules show characteristic absorption bands in the infra-red region of the spectrum. Infra-red spectroscopy provides valuable information about structure, and gives information on groups present and on molecular symmetry. The technique has been highly developed and is widely used as a routine tool in analysis and research. See also FOURIER TRANSFORM INFRA-RED SPECTROSCOPY.

infusions Dilute solutions containing the water-soluble components of vegetable drugs prepared by aqueous extraction of the drugs.

ingrain dyestuffs Insoluble dyestuffs formed directly on the fibre from soluble components.

inhibitor A general term for any compound which will inhibit, i.e. slow down or stop, a reaction, generally by preventing propagation of chain reactions or by passivation of surfaces, etc. The term is often used to describe any additive which will prevent a particular tendency during working operations, e.g. antioxidants, anti-corrosion agents, etc.

inisprothrin Insecticide, pyrethroid.

initial boiling point In a laboratory distillation the thermometer reading observed when the first drop of condensate falls from the lower end of the condenser tube.

initiators See EXPLOSIVES.

inks Colouring matter dissolved or dispersed in a solvent or carrier; on drying the colouring matter is bonded (by an additive) to the substrate.

inner sphere mechanism A mode of redox reaction in which complexes form an intermediate with a shared ligand.

inorganic chemistry The chemistry of all the elements other than carbon. Simple carbon compounds, e.g. carbides, carbonates, carbonyls and cyanides are often regarded as inorganic. Organic derivatives of metals and metalloids are considered as ORGANOMETALLIC.

inosinic acid (hypoxanthine riboside-5-phosphate), $C_{10}H_{13}N_4O_8P$. One of the first

ribose phosphate

purine nucleotides to be isolated. It is the biosynthetic precursor of adenylic acid (AMP) and guanylic acid (GMP). Inosinic acid may be prepared by enzymic deamination of adenylic acid. Salts are used as flavour enhancers.

inositol, $C_6H_{12}O_6$. cis-1,2,3,5-trans-4,6-hexahydroxycyclohexane. Nine stereoisomers are possible (cyclitols), but only one, mesa- or i-inositol, is widely distributed. It is present in yeast, in plants as phytic acid ester and as a constituent of both plant and animal phosphatides. M.p. 225–226°C, with 2H₂O m.p. 218°C. Has a sweet taste. i-Inositol is an essential growth factor for certain micro-organisms and is classified in the vitamin B group. It is important in cell messaging and receptor stimulation. It can most easily be obtained from corn steep liquor.

insecticides Materials used to control insects by poisoning (chrysanthemic acids, contact poisons, systemic poisons) by use of attractants or repellants (fumigation–acrylonitrile, CS₂, etc.).

insect repellants Materials that may have a repellant effect on insects.

insertion reaction A common type of reaction in which an incoming molecule is inserted between previously bonded atoms, e.g. $[(OC)_5MnCH_3]$ plus CO gives $[(OC)_5MnC(O)CH_3]$. The inserted molecule contains a multiple bond or a lone pair. The term insertion is also used in connection with enzymes. See DICTIONARY OF BIOLOGY.

insoluble azo dyes See INGRAIN DYESTUFFS.

instability constant The reverse of a STABILITY CONSTANT; the measure of dissociation.

insulator A substance with a very low electrical conductivity. In insulators there is a wide bandgap between completely filled and completely empty electronic energy levels. Most pure solid ionic substances are insulators although impurities or imperfections may introduce SEMI-CONDUCTIVITY.

insulin The hormone, secreted by the Islets of Langerhans in the pancreas, which maintains the balance of glucose metabolism. The absence of insulin leads to the disease *diabetes mellitus*.

integral tripack material Colour films and papers in which the three separate emulsions (for violet-blue, green and red) are coated on the same support in layers and, after exposure and COLOUR DEVELOPMENT, give dyes of the complementary colour (yellow, magenta and cyan).

interatomic distances The distances between the nuclei of atoms, bonded or non-bonded, in a molecule or crystal as determined by X-ray or electron diffraction, spectroscopic or other physical methods.

intercalation compounds Derivatives, particularly of graphite, in which molecules are accommodated in holes or between layers in the lattice. The resulting compounds are easily handled and stored (e.g. graphite-SbF_5, used as a fluorinating agent; graphite-$FeCl_3$: stable to water).

interfacial angles, constancy of The angle between any particular pair of faces of a crystal is constant and characteristic of the substance of the crystal. It does not depend on the size of the crystal.

interferons A group of proteins which prevent viral replication of cells. See DICTIONARY OF BIOLOGY.

interhalogen compounds Compounds formed by the halogens with one another. The known compounds are: ClF, BrF, ClBr, ICl, IBr, ClF_3, BrF_3, IF_3, ICl_3, ClF_5, BrF_5, IF_5 and IF_7. All these compounds are formed by direct combination of the two halogens. They are readily volatile and exceedingly reactive, especially if the compound in question contains fluorine. No interhalogen compounds are yet known which contain more than two different halogen atoms. Oxide derivatives are also known, e.g. IOF_5, ClO_3F. The interhalogens form related anions and cations, e.g. BrF_3 gives $[BrF_2]^+$ (with SbF_5) and $[BrF_4]^-$ (with KF). Polyhalide anions

can contain more than two halogens, e.g. $[IBrCl]^-$.

interleukins Lymphocyte communication factors. See DICTIONARY OF BIOLOGY.

intermediate A transient species occurring between reactants and products.

internal compensation A type of optical activity in which a molecule contains two asymmetric centres which produce equal and opposite rotation of the plane of polarization, with the result that the substance is optically inactive.

internal energy (U) The total energy of a system. It includes all forms of energy other than those arising from the position of the system in space. The internal energy (U) is related to the enthalpy (H) by the equation $H = U + PV$ where P is the pressure and V the volume. Expressed in joules mol^{-1}.

internal pressure Variation of internal energy with volume at constant temperature. See VAN DER WAALS EQUATION.

internal property Property that does not depend on the amount of substance, e.g. temperature, pressure.

interstitial compounds Compounds in which the structure may be considered to be derived by inserting small species (H, B, C, N) into the holes of a cp metallic lattice (particularly of transition elements).

invar Nickel alloy.

invariant system A system which, when at equilibrium, possesses no degrees of freedom. See PHASE RULE.

inverse spinel See SPINELS.

inversion See STEREOSPECIFIC REACTIONS; WALDEN INVERSION; INVERTASE. The symmetry operation, i; reflection through a point.

inversion temperature Most gases when rapidly expanded experience a cooling effect (the Joule–Thomson effect). Below 193 K hydrogen also shows this effect, but above this temperature it has a negative Joule–Thomson coefficient and is warmed by expansion. This temperature of 193 K is called the inversion temperature. Helium shows similar behaviour to hydrogen with an inversion temperature of 40 K.

invertase (sucrase) The enzyme which hydrolyses sucrose to a mixture of glucose and fructose. Used in making syrups for the confectionery trade.

invert sugar The slightly laevorotatory mixture of glucose and fructose obtained by hydrolysing sucrose.

iodates Salts containing iodine oxyanions. *Iodates*(I), M[OI], hypoiodates. Formed I_2 plus HgO; the acid is very unstable; iodates(I) are similar to chlorates(I). Iodates(III) are not known. *Iodates*(V) MIO_3. The free acid, HIO_3, the only known free halic acid (I_2 plus conc. HNO_3) dehydrates to I_2O_5. Strong oxidizing agents (IO_3^- plus I^- gives I_2). The $[IO_3]^-$ ion is pyramidal. *Iodates*(VII), periodates, $[IO_4]^-$ from HIO_4, $[IO_6]^{5-}$ from H_5IO_6, formed by electrolysis of HIO_3 which gives $[IO_6]^{5-}$ derivatives, e.g. $NaH_4IO_6.H_2O$; with base gives MIO_4. All strong oxidizing agents (potential $[IO]^- \rightarrow I^-$ +0.49 volts; $[IO_3]^- \rightarrow I^-$ +1.085 volts; $H_5IO_6 \rightarrow [IO_3]^-$ +1.7 volts). Weaker acids than perchlorates(VII). Organic derivatives, e.g. iodoxybenzoic acid, are useful reagents in synthesis.

iodic acids The acids giving rise to IODATE anions. Only HIO_3 is known in the free form. All are strong oxidizing agents.

iodides Compounds containing the I^- ion or covalent species RI. Many polyiodides, compounds between iodides and iodine, are known.

iodimetry The use of iodine in titrations, generally titrated with thiosulphate and converted to I^- using starch (blue to colourless) as indicator. Used indirectly in estimating oxidizing agents which liberate I_2 from acid KI solution.

iodine, I. Non-metallic element, at.no. 53, at.mass 126.9, m.p. 113.6°C, b.p. 185.2°C, density ρ 4930, crustal abundance 0.14 p.p.m., electronic configuration [Kr] $5s^25p^5$. Occurs as I^- in brines, as iodates(V) in caliche and in sea-water and seaweeds. Liberated by oxidation (of I^-) or reduction (of $(IO_3)^-$). The element is a black solid giving a violet vapour (I_2 molecules). Simple solutions (CCl_4) are violet, solutions in which there is appreciable charge transfer ($C_2H_5OH/KI/H_2O$) are brown. The element reacts with other halogens (to give interhalogen compounds) and bases (to give iodates). Used in disinfectants, medically (e.g. for hyperthyroidism), synthesis, photography, food supplements, electric light bulbs (quartz-halogen lamps). Iodine is an essential element in the diet and used to reduce friction between surfaces. $Na^{131}I$ is used diagnostically.

iodine chemistry Iodine is a halogen element, electronic configuration $5s^25p^5$. Its most stable oxidation states are diamagnetic –1 ($\frac{1}{2}I_2 \rightarrow I^-$ + 0.5V); +1 (ICl, also complexed I^+ cations (e.g. $[Ipy_2]^+NO_3$); $AgNO_3$ plus I_2 plus pyridine); +3 (ICl_3); +5 (IF_5, HIO_3); +7 (IF_7, KIO_4). Iodine thus shows considerably more cationic character than Cl or Br and forms some oxysalts (e.g. $(IO)_2SO_4$ (I_2O_5 plus I_2 plus H_2SO_4), $I(O_2CCH_3)_3$ (I_2 plus $(CH_3CO)_2O$ plus fuming nitric acid). I_3^+ and I_5^+ ionic species are known.

iodine chlorides Iodine monochloride, ICl, b.p. 97°C, and iodine trichloride, ICl_3, yellow-orange solid dissociated to ICl and Cl_2 above room temperature. See INTERHALOGEN COMPOUNDS.

iodine fluorides Iodine trifluoride, IF_3, yellow solid. Iodine pentafluoride, IF_5, m.p. 9°C, b.p. 100°C, colourless liquid used as mild fluormating agent and solvent, square pyramid. Iodine heptafluoride, IF_7, b.p. 4°C. See INTERHALOGEN COMPOUNDS.

iodine oxides Iodine pentoxide, I_2O_5, is the most important oxide (heat on HIO_3) and regenerates HIO_3 with water. It is white, polymeric and a strong oxidizing agent (e.g. CO gives CO_2, I_2 being liberated quantitatively). The other oxides I_2O_4, I_4O_9, I_2O_7 are also polymeric.

iodine oxyacids See IODATES.

iodoalkanes (RI) Used in synthesis, e.g. CH_3I, b.p. 42°C; C_2H_5I, b.p. 69–73°C; 1-iodopropane, b.p. 101–102°C; 2-iodopropane, b.p. 88–89°C; 1-iodobutane, b.p. 130–131°C. See IODOMETHANE.

iodobenzene, PhI, C_6H_5I. Colourless liquid, b.p. 188°C. The iodine atom is very reactive and is easily removed by metals. With chlorine an oxidized compound, phenyliodonium dichloride ($C_6H_5ICl_2$), is formed.

Iodobenzene is prepared by direct iodination of benzene with iodine and nitric acid under reflux or from diazotized aniline and potassium iodide.

iodoethane See IODOALKANES.

iodoform See TRIIODOMETHANE.

iodoform reaction A test used to detect the presence of a $C(O)\cdot CH_3$ group (or groups which may be converted to this under the reaction conditions). Iodine in alkali converts the acetyl group into iodoform.

iodogorgic acid See 3,5-DIIODOTYROSINE.

iodomethane (methyl iodide), CH_3I. Colourless liquid with characteristic odour; b.p. 42°C. Prepared by heating methanol and iodine with red phosphorus. Used in organic chemistry for methylation and for preparing methyl lithium and methyl Grignard reagents.

iodometric methods Analytical methods involving IODIMETRY.

ion An atom or group of atoms which has gained or lost one or more electrons, and thus carries a negative or a positive charge. Ions are present in solutions of electrolytes in solvents of high dielectric constant. They are generally complexed by the solvent. They are also present in melts and in gaseous ionic materials. Ions are present in gases which have been subjected to suitable electric disturbances, such as a high-tension discharge or a beam of electrons.

The word is also used to denote a unit in a solid crystal of an electrovalent compound such as NaCl in which each Na^+ is electrically attracted by the surrounding six Cl^- and each Cl^- is electrically attracted by the surrounding six Na^+. The structure of such crystals is termed ionic to indicate that the crystal is not an aggregate of independent molecules.

ion exchange When an IONOGENIC SURFACE is in contact with water or other ionizing solvent an ELECTRICAL DOUBLE LAYER is set up. The counter-ions are partially free and can readily be exchanged for others of the same sign supplied by soluble electrolytes. Thus the sodium in certain silicates can be replaced by potassium by exhaustively washing the material with a solution of KCl (see BASE EXCHANGE). It is exploited particularly with ion-exchange resins. These are synthetic insoluble cross-linked polymers carrying acidic or basic side-groups which have high exchange capacity and have many applications, including water treatment, extraction, separation, analysis and catalysis. Using a 'mixed bed' of anion- and cation-exchange resins the electrolytes can be removed from a water solution and very pure water obtained.

ion-exchange chromatography See CHROMATOGRAPHY.

ion-exchange resin See ION EXCHANGE.

ion exclusion The medium inside the pores of an ion-exchange resin is subject to the Donnan membrane equilibrium. It therefore contains a lower concentration than does the external solution of any electrolytes that may be present in the latter. This is 'ion exclusion'. On the other hand, non-electrolytes are not subject to the Donnan effect. Hence, a partial separation of electrolytes from non-electrolytes can be obtained; by using a chromatographic technique complete separation can he effected in favourable cases.

ionic atmosphere The Debye–Hückel theory of interionic attraction shows that in solutions ions of one charge type are surrounded by an atmosphere of ions predominantly of the opposite charge. The energy and chemical potential of an ion in solution are lowered as a result of the electrostatic interaction between an ion and its atmosphere. The effect felt by the central ion is normally the same in all directions. When an electric field is applied in the solution, however, the central ion moves in one direction whereas the ionic atmosphere moves in the opposite sense, giving rise to an asymmetric distribution of ions about the central ion. This effect is important in studies of electrical conductivities of solutions.

ionic liquids Low melting ionic solids containing large ions that can act as useful, non-polluting, reaction media. Example: 1-octyl-3-methylimidazolium hexafluorophosphate.

ionic product The product of the concentrations of the ions of a substance in solution. This factor is a constant for weakly ionized substances because the concentration of undissociated material may be taken to be constant if it is present in a large excess. See SOLUBILITY PRODUCT.

ionic radii The effective radius of ions in crystal lattices.

ionic strength For an electrolyte solution the ionic strength is given by

$$I = \tfrac{1}{2}\Sigma_i m_i z_i^2$$

where z is the charge on the ionic species and m is the molal concentration, the summation being continued over all the different ionic species in the solution. The concept is used in calculating activities particularly at low concentrations.

ionization, heat of The energy required to split one mole of an electrolyte into its constituent ions. For water the heat of ionization is 57.3 kJ mol^{-1}. See ENTHALPY.

ionization chamber A device for measuring the absolute intensity of a beam of X-rays or ionizing particles.

ionization energies (ionization potentials, enthalpies of ionization) The energy required to remove an electron from a free atom or ion in the gaseous state. Normally measured in electron volts or kilojoules per mole.

ionization of water   Pure water has a very low conductivity and is only very slightly ionized. The concentrations of hydrogen and hydroxyl ions, which are equal, are 10^{-7} g ions 1^{-1}. The ionic product for water is $[H^+][OH^-] = 10^{-14}$ at 25°C, but increases rapidly with temperature. The pH of pure water is clearly 7. Any solution with pH > 7 contains excess hydroxyl ions and is alkaline; those with pH < 7 are acidic, containing excess hydrogen ions.

ionization potential   See IONIZATION ENERGIES.

ionogenic surface   The stability of a colloid particle is due to the charge it carries. This charge is ionic in nature and if the stabilizing ions on the surface are due to the ionization of the particle wall material, the surface is said to be ionogenic.

ionol   See 2,6-DI-t-BUTYL-4-METHYLPHENOL.

ionone   (E-4-(2,6,6-trimethylcyclo-1-hexenyl)-3-butene-2-ol(α) and -2-butene-2-ol(β)), $C_{13}H_2O$.   Pseudoionone (see CITRAL) treated with acid, gives a mixture of α- and β-ionone. These products are powerful odorants, smelling strongly of violets.

ion-pair partition   See PHASE-TRANSFER CHEMISTRY.

ion-selective electrodes   An electrode which responds specifically to one ion present in solution and develops a potential according to the concentration of that ion. Used for metals and for anions. The electrode contains a standard solution or other specific species and a membrane or other conductor placed in contact with the solution to be examined. Glass electrodes are widely used as ion-selective electrodes.

ions, hydration   See HYDRATION.

ioxynil   Herbicide, benzonitrile derivative.

ipconazole   Fungicide, azole derivative.

iproben   Fungicide, organophosphate derivative.

iprodione   Fungicide, dicarboxamide derivative.

Ir   Iridium.

i.r.   Infra-red.

iridates   See IRIDIUM OXIDES.

iridium, Ir.   Platinum metal, at.no. 77, at.mass 192.22, m.p. 2443°C, b.p. 4428°C, density ρ 22 420, crustal abundance 3×10^{-6} p.p.m., electronic configuration [Xe] $5d^76s^2$. For separation see RUTHENIUM. It occurs in osmiridium and as a native Pt alloy. The metal is ccp with high-temperature strength and good anti-corrosion properties, but has an inclination to brittle fracture. The very high melting point is associated with good mechanical properties above 1600°C. Used to encapsulate radioactive fuels in space, electrical and in catalysts (including automotive catalysts) and in Cl_2 and NaOH production. Main uses electrochemical 25%, catalysts 8%, crucibles 11%. World demand 2001 86 000 oz.

iridium chemistry   Iridium is the heaviest member of Group 9. It shows oxidation states from +6 (IrF_6) to –1 $[Ir(CO)_4]^-$. The +6 and +5 states are strongly oxidizing and the +4 and +3 states the most stable; the +4 state for Ir is more stable than for Rh. Ir(III) and (IV) ammines and other N-bonded complexes are very stable. The +2 state is uncommon but the +1 state (generally square planar), often with phosphine or carbonyl ligands, is well known (e.g. VASKA'S COMPOUND) but undergoes oxidative addition to give Ir(III). Ir(0) exists as carbonyls, e.g. $Ir_2(CO)_8$ and phosphine complexes. Many organometallic and hydride species known.

iridium halides   The known fluorides are yellow IrF_6 (Ir plus F_2), m.p. 44°C, b.p. 53°C; yellow ($IrF_5)_4$ (Ir plus F_2); red-brown IrF_4 (Ir plus IrF_6); and black IrF_3 (heat on IrF_5). Fluoroiridates containing IrF_6^-, IrF_6^{2-} and IrF_5^{2-} species are known. $IrCl_3$ (Ir plus Cl_2 at 600°C) is the only known anhydrous chloride; $IrBr_3$, IrI_3 and hydrates are known. Chloroiridates(IV) $[IrCl_6]^{2-}$ (Ir compounds plus Cl_2 plus HNO_3) are reduced to $[IrCl_6]^{3-}$.

iridium oxides   Black IrO_2 is formed by heating Ir in air. A hydrated form and hydrated Ir_2O_3 are precipitated from Ir(IV) or Ir(III) solutions with alkali; the latter is readily oxidized to IrO_2. Mixed oxides, e.g. $CaIrO_3$, are formed by solid-state reactions; iridates are not stable in solution.

iron, Fe.   Metal, at.no. 26, at.mass 55.845, m.p. 1538°C, b.p. 2861°C, density ρ 7874, crustal abundance 41 000 p.p.m., electronic configuration [Ar] $3d^64s^2$. Iron was known in Egypt in 3000 B.C. and was used in Europe about 1000 B.C. Fe is found in most clays, sandstones and granites; meteoritic iron contains Co and Ni. The chief ores are haematite (Fe_2O_3), brown haematite (hydrated Fe_2O_3), magnetite (Fe_3O_4), siderite or spathic iron ore ($FeCO_3$), iron pyrites (FeS_2) and chalcopyrite ($CuFeS_2$). The metal is bcc below 960°C, ccp to 1401°C, bcc to 1535°C. Iron is used extensively as a metal for fabrication, either as iron or as steel.

Iron is usually prepared from oxide or carbonate ores, from which S, As, etc. have been removed by roasting in air, by reduction with carbon. The ore is mixed with coke and $CaCO_3$ and heated in a blast furnace, the maximum temperature of which is about 1300°C. The major acidic impurities are removed as slag (calcium silicate, aluminate, etc.) and the molten crude metal run off into pigs; pig iron contains 2–4% of carbon with a little P, S and Si. If the Si content is high, the C is present almost entirely as graphite; on remelting and casting, such an alloy gives grey cast iron. If the Si content is low, the carbon is present as cementite, Fe_3C, and gives white cast iron on casting. The sulphur content governs that of Mn, since excessive sulphur forms MnS, which is appreciably soluble in the slag. The cast irons are too brittle for many purposes. Wrought iron is made by melting cast iron with scrap-iron in a reverberatory furnace (puddling furnace) lined with Fe_2O_3, which oxidizes C, Si, etc., most of which are then removed by rolling. Cast iron melts at about 1200°C. Wrought-iron, which is tough, fibrous and malleable, melts at above 1500°C, though it softens at about 1000°C. It is not hardened by quenching.

Steels contain *c.* 1.5% carbon and other alloying elements; from pig-iron the C content must be reduced, from wrought-iron the C content must be increased. Above 900°C steel is a solid solution of C in γ-Fe (austenite). On slow cooling, austenite gives cementite Fe_3C and an Fe–C solution; below 690°C the γ-Fe gives a eutectic of ferrite (*c.* 0.06% C in α-Fe) and cementite. The eutectic is known as pearlite and is soft. Quenching austenite from 900°C to 150°C gives martensite (supersaturated C in α-Fe) which is very hard but may be tempered (reduced hardness but still tough) by reheating to 200–300°C.

In case hardening C and N are introduced into the iron by heating in a N_2 – NH_3 atmosphere or by immersing the steel in molten $NaCN/Na_2CO_3$ at about 900°C.

Iron is converted into steel by various processes. Iron can be rendered passive by contact with concentrated HNO_3. World steel production forecast to reach 883 megatonnes by 2003. Iron compounds are used in water as catalysts and mordants and purification. Important in proteins.

iron, organic derivatives Iron carbonyls readily form π-bonded olefin complexes by displacement of CO, and alkynes form both σ- and π-bonded derivatives. Simple σ-bonded iron organometallics are relatively unstable but h^5-cyclopentadienyl derivatives, e.g. FERROCENE, and h^4-cyclobutadiene derivatives are stable.

iron acetates See IRON ETHANOATES.

iron alums, $MFe^{III}(SO_4)_2 \cdot 12H_2O$. See ALUMS. Iron alum (ferric alum) is the ammonium salt.

iron ammonium sulphate (Mohr's salt), $(NH_4)_2Fe(SO_4)_2 \cdot 6H_2O$. Pale green crystalline solid used as a standard reducing agent (e.g. for $KMnO_4$). More stable than $FeSO_4$. Used in photography and blueprints.

iron bromides *Iron(II) bromide*, $FeBr_2$. Occurs as hydrates (Fe plus HBr). *Iron(III) bromide*, $FeBr_3$. Formed anhydrous from Fe plus Br_2 as dark red crystals. $FeBr_3 \cdot 6H_2O$ is dark green.

iron buff Hydrated iron(III) oxide. Important pigment used for black dyeing of silk and for khaki dyeing of cotton, in paints and enamels.

iron carbides Of importance in metallurgy – see IRON.

iron carbonate, $FeCO_3$. Only the iron(II) derivative known (white powder from Fe^{2+} plus Na_2CO_3 in absence of air), very readily oxidized. Occurs naturally as siderite, spathic iron ore, chalybite.

iron carbonyls Iron forms $Fe(CO)_5$ (trigonal bipyramidal), $Fe_2(CO)_9$ and $Fe_3(CO)_{12}$. $Fe(CO)_5$ is formed from Fe plus CO at 200°C and 300 atm., m.p. –20°C, b.p. 102°C. Light gives orange $Fe_2(CO)_9$ and heat gives dark green $Fe_3(CO)_{12}$. These are parent compounds of wide ranges of carbonyl derivatives in which CO is replaced by other ligands, e.g. PPh_3, butadiene, etc. They react with alkynes to give products of telomerization of the C_2 groups and insertion of CO and Fe. $Fe(CO)_5$ gives $[Fe(CO)_4]^{2-}$ with base, which reacts with acids to give $H_2Fe(CO)_4$. Halide, nitrosyl and organo derivatives are also known. Carbonyls are used in formation of finely divided iron for use in electronic equipment.

iron chemistry Iron is a first-row transition element of group 8 which shows a wide variety of oxidation states and co-ordination numbers. The most stable oxidation state is +2 (ferrous) (in e.g. $FeCl_2$, $K_4Fe(CN)_5$) although +3 (ferric)

(e.g. $FeCl_3$, $K_3Fe(CN)_6$) is also important. Many complexes are known for these states but these are labile and frequently difficult to prepare (except low-spin compounds, e.g. cyanoferrates).

$$E°Fe^{3+} \rightarrow Fe^{2+} +0.77 \text{ volts,}$$

$E°Fe^{2+} \rightarrow Fe -0.09$ volts (acid solution). Other oxidation states include -2 $[Fe(CO)_4]^{2-}$; 0 $Fe(CO)_5$; $+1$ $[Fe(H_2O)_5\text{-}NO]^{2+}$; $+4$ $[Fe(bipy)_3]_4^+$; $+5$ $[FeO_4]^{3-}$ (very unstable); $+6$ $[FeO_4]^{2-}$. The latter state is strongly oxidizing. The coordination can increase over 6 in the $+2$ and $+3$ states (e.g. Fe(III)-EDTA complex). Very important in living systems – enzymes, proteins, nitrogenases – see iron sulphides.

iron chlorides

Iron(III) chloride, $FeCl_3$. Dark red crystals (Fe plus Cl_2). Forms hydrates but hydrolysed to brown derivatives in water. Forms complex chlorides, e.g. $K_2FeCl_5.H_2O$ and $(NH_4)_2FeCl_5.H_2O$ (ammonio-chloride of iron).

Iron(II) chloride, $FeCl_2$. White solid (Fe plus HCl gas). Forms many complexes including $FeCl_2.6NH_3$, $FeCl_2.4H_2O$ (also 6, 2 and $1H_2O$), $KFeCl_3.2H_2O$.

iron, complex cyanides

See CYANOFER-RATES. Used as pigments and to prevent caking of salt. Derivatives of cyanoferrates, e.g. penta-cyanonitrosylferrates, nitroprussides, containing $[Fe(CN)_5 NO]^{2-}$, are also known.

iron ethanoates (iron acetates)

Iron(II) ethanoate, $Fe(O_2CCH_3)_2$. Pale green crystals. *Iron(III)* forms the $[Fe_3O(CH_3CO_2)_6(H_2O)_3]^+$ cation and some other hydrolysed species.

iron fluorides

Iron(III) fluoride, FeF_3. Only slightly soluble in water. White salt (HF on Fe_2O_3). Forms hydrates and complex hexafluoroferrates containing $[FeF_6]^{3-}$ ions.

Iron(II) fluoride, FeF_2. Prepared Fe plus HF gas, forms $4H_2O$ hydrate and, in the absence of oxygen, perovskites, e.g. $KFeF_3$.

iron hydrides

The $[FeH_6]^{4-}$ ion is formed from Grignard reagents and iron halides. There is further interaction with MgX species.

iron hydroxides

See IRON OXIDES.

iron iodide (iron(II) iodide), FeI_2, 9, 6, 5, 4 and $1H_2O$.

Prepared from Fe and I_2. Soluble in water, oxidized to I_2 and Fe(III) derivatives in air.

iron nitrates

Iron(III) nitrate, $Fe(NO_3)_3.9H_2O$ also $6H_2O$. The pure salt is pale amethyst but impure salts are first colourless and then brown because of hydrolysis.

Iron(II) nitrate, $Fe(NO_3)_2.6H_2O$. Unstable material (Fe plus HNO_3). Also forms $9H_2O$ hydrate.

iron oxides and hydroxides

These substances form a complex, interrelated system. They are of importance as polishes, dyes and pigments (e.g. iron buff, sienna, umber, rouge, ochre). Oxides all show non-stoichiometry. Fe_2O_3, dark brown, has two forms: one haematite with the corundum structure, the other (γ-form) with a structure related to the spinel structure and closely related to Fe_3O_4, magnetite (inverse spinel, $Fe^{3+}(Fe^{2+}O,Fe^{3+})O_4$) and FeO. $(Fe_{1-x}O)$ has a cp oxide structure (NaCl structure). Fe_2O_3 is only soluble in acids with difficulty and becomes almost insoluble if strongly ignited. FeO is difficult to prepare (heat on FeC_2O_4) and decomposes to Fe_3O_4 and Fe below 570°C. It dissolves readily in acids.

The hydroxide, $Fe(OH)_3$ (Fe^{3+} plus OH^-), has definite existence and there are many ill-defined hydrates used as pigments, particularly in paints, plastics, cements. FeOOH has two forms, goethite and lepidocrocite. Colloidal $Fe(OH)_3$ is easily obtained as a deep red sol. Many Fe(III) hydroxy complexes are known. $Fe(OH)_2$ may be formed from Fe^{2+} and OH^- in the absence of O_2 but it is very readily oxidized.

Iron oxides react with other metal oxides to give FERRATES (mixed metal oxides). Used in electrical equipment.

iron(III) perchlorate, $Fe(ClO_4)_3.6H_2O$.

iron phosphates

Iron(III) phosphate, $FePO_4.2H_2O$, is precipitated from solution but is soluble in dilute acids; complex phosphates are known (the complexing of Fe by PO_4^{3-} is used to affect redox potentials during Fe^{2+}/Fe^{3+} titrations). Medicinal iron phosphate is a dark blue powder containing iron(II and III) phosphates and iron(III) hydroxides. Food additives.

iron sulphates

Iron(III) sulphate, $Fe_2(SO_4)_3$, 12, 10, 9, 7, 6, 3 and $0 H_2O$. Formed from Fe(III) salts plus $(NH_4)_2SO_4$. Decomposes to Fe_2O_3 and SO_3 on heating. Sulphato complexes and alums are also known.

Iron(II) sulphate, $FeSO_4.7H_2O$. Ferrous sulphate, green vitriol, copperas. Forms 5 and 1 hydrates and a white anhydrous salt. Obtained from the waste liquor resulting from pickling steel in H_2SO_4. Slowly oxidized in air; forms more stable double sulphates, e.g. $M_2Fe(SO_4)_2.6H_2O$, e.g. iron ammonium sulphate.

iron sulphides FeS_2 occurs naturally as pyrites and marcasite, both containing S_2^{2-} anions. Grey FeS (Fe^{2+} plus H_2S or $(NH_4)_2S$, Fe plus S) has the nickel arsenide structure. Pyrrhotite, Fe_7S_8, is an Fe-deficient version of FeS. It seems unlikely that Fe_2S_3 exists as such. Some complex sulphides appear to be formed. Iron sulphide clusters are of great biological importance in, e.g. nitrogenases, electron transfer.

iron thiocyanates Iron(III) thiocyanate complexes are deep red species formed from Fe^{3+} and NCS^- and are used for the detection of either constituent. $Fe(NCS)_2 \cdot 3H_2O$ is known but is unimportant.

IRS Internal reflection spectroscopy. Technique used to study the i.r. spectra of solids.

Irving–Williams order The stabilities of series of complexes with different metals fall into the sequence $Mn^{2+} < Fe^{2+} < Co^{2+} < Ni^{2+} < Cu^{2+} < Zn^{2+}$. This is known as the Irving–Williams order and is general for many ligands.

isatin (indole-2,3-dione), $C_8H_5NO_2$. Yellowish-red prisms, m.p. 200–201°C. Prepared by heating α-isatinanilide with dilute mineral acid. It is used for the preparation of vat dyes by condensation with indoxyl (to give indirubin) or thioindoxyl.

isazofos Nematocide, insecticide, organophosphate derivative.

isinglass Fish glue, chiefly gelatine. Used in food stuffs, clarifying and colloid protection.

islands of nuclear stability Reasonable nuclear stability for nuclei of a group of very heavy elements.

isoamyl The group $Me_2CH \cdot CH_2CH_2-$.

isoamyl alcohol See PENTANOLS.

isoamyl ether See ISOPENTYL ETHER.

isobars Atomic species of approximately the same mass number (i.e. protons plus neutrons) but different chemical properties (i.e. numbers of protons or electrons).

isoborneol, $C_{10}H_{18}O$. M.p. 214°C. See BORNEOL.

isobornyl chloride See BORNYL CHLORIDE.

isobutane (2-methylpropane), C_4H_{10}, $Me_2CH \cdot CH_3$. Colourless gas; m.p. –145°C, b.p. –10°C. Occurs in natural gas and is produced in large amounts in the cracking of petroleum. It is used in refrigeration plant. For chemical properties see PARAFFINS.

isobutanol See 2-METHYL-1-PROPANOL.

isobutyl- (2-methylpropyl) The group $Me_2CH_2CH_2-$.

isobutyryl The 2-methylpropanoyl group, Me_2CHCO-.

isocrotonic acid See BUTENOIC ACIDS.

isocyanates Compounds containing the group $-N=C=O$. Alkyl isocyanates are prepared from dialkyl sulphates and potassium cyanate. They readily polymerize to isocyanuric esters and are formed as intermediates in the Hofmann decomposition of amides and the Curtius decomposition of azides. Aryl isocyanates are best prepared by the reaction of phosgene and an aromatic amine. PhNCO reacts with alcohols to give urethanes, with ammonia to give phenylurea and with aniline to give diphenylurea. Converted on heating to carbodiimides.

The polycondensation of di-isocyanates with polyhydric alcohols gives a wide range of polyurethanes which are used as artificial rubbers and lightweight foams, and have other important properties. Isocyanates are also used as modifiers in ALKYD RESINS.

isocyanides See ISONITRILES.

isodispersion Sols in which the dispersed particles, of colloidal dimensions, are all of approximately the same size. Hemoglobin forms an isodispersion of particles all of mol. mass 68 000.

isodurene, 1,2,3,5-tetramethylbenzene.

isoelectric point The point at which the net charge on the dispersed particles is zero – positive and negative ions being adsorbed to an equal extent. At the isoelectric point the electrophoretic velocity is zero and the substance tends to flocculate. With hydrophilic substances which are protected from coagulation by possession of a sheath of water, the stability is minimum at the isoelectric point (see COLLOID, COAGULATION, ZWITTERION).

isoelectronic Molecules or ions which have the same number of electrons, e.g. CO and N_2.

isoeugenol (2-methoxy-4(1-propenyl)-phenol) Used in manufacture of vanillin. Prepared from eugenol.

isofenphos Insecticide, organophosphate derivative.

isoleptic Having all ligands the same, e.g. $Fe(CO)_5$, $[Co(NH_3)_6]Cl_3$.

4-isoleucine (2-amino-3-methylvaleric acid), $C_6H_{13}NO_2$,
$CH_3.CH_2.CHMe.CHNH_2.C(O)OH$. Colourless crystals, m.p. 284°C (decomp.). The naturally occurring substance is dextrorotatory. An amino-acid, occurring with leucine as a product of protein hydrolysis.

isomerases Enzymes catalysing isomerization.

isomerism Compounds possessing the same composition and the same relative molecular weight, but differing in their chemical structure, are said to be isomeric, and each is an isomer, or isomeride, of the others. Isomers have differing chemical, physical and physiological properties. The isomerism may be of several types.

1 *Structural isomerism*, due to differences in the order in which the atoms are joined together.

a. Chain or nuclear isomerism, due to differences in the arrangement of, e.g. carbon atoms in the molecule as in $CH_3CH_2CH_2CH_3$ and $CH_3CH(CH_3)_2$.

b. Position isomerism, due to differences in the position of some group or atom with regard to the carbon chain or ring, as in $CH_3CH_2CH_2Cl$ and $CH_3CHClCH_3$.

c. Functional group isomerism or metamerism, due to differences in the type of compound, as in CH_3COCH_3 and CH_3CH_2CHO.

Tautomerism, or dynamic isomerism, is the special case of structural isomerism in which two isomers are directly interconvertible. The reversibility of the change is due to the mobility of a group or atom, which can move from one position to another in the molecule, often with rearrangement of one or more double bonds. Common structures showing tautomerism are the *keto-enol type*, as in acetoacetic ester.

$$> \overset{\overset{\displaystyle H}{|}}{C} - \underset{|}{C} = O \rightleftharpoons\; > C = \underset{|}{C} - C - OH$$

These isomers are known as tautomers, or tautomerides, and exist in equilibrium in the liquid state or in solution. The speed of conversion of one form to the other varies widely in different cases, and may be so slow that it is possible to isolate both forms in a relatively pure state. The term *desmotrope* is applied to those tautomers which can be separated from each other. The different tautomers can often be detected by n.m.r. spectroscopy.

2 *Space isomerism or stereoisomerism*, due to differences only in the spatial orientation of the atoms in the molecule.

a. Optical isomerism due to asymmetry. Such isomers differ in the direction in which they rotate the plane of polarized light. See optical activity.

b. Geometrical or *cis–trans* isomerism, due to different arrangements of dissimilar atoms or groups attached to two atoms joined by a double bond or forming part of certain ring structures. Thus the isomeric fumaric and maleic acids are:

$$\begin{array}{cc} H - \underset{\parallel}{C} - C(O)OH & H - \underset{\parallel}{C} - C(O)OH \\ HO(O)C - C - H & H - C - C(O)OH \end{array}$$

and

respectively. The presence of the double bond restricts the free rotation of the C atoms joined by it and permits the existence of the two forms. Similarly, in certain ring structures which need not contain double bonds, the free rotation of the atoms is restricted and other groups or atoms attached to them may lie above or below the plane of the ring and give rise to geometrical isomers. This isomerism is not confined to the >C=C< bond but also exists in compounds having >C=N- or -N=N- bonds. In the case of the >C=C< double bond or a carbon ring, the isomers are distinguished by the prefixes *cis*- and *trans*- according to whether two given groups or atoms lie on the same or on opposite sides of the plane of the double bond or ring. In the other cases the prefixes *syn*- and *anti*- are used; thus the *syn*-aldoximes have the structure:

$$\begin{array}{c} R - \underset{\parallel}{C} - H \\ N - OH \end{array}$$

For ketoximes the group which is *syn* or *anti* to the hydroxyl group must be specified. Difficulty in assigning the terms *cis*, and *trans*, arises when there is no pair of identical, or similar, groups substituting an alkene. Then the terms (E) [German *entgegen* = across] and (Z) [German *zusammen* = together] replace the terms *trans/cis*. In many, but not all, cases the term *cis* corresponds to (Z) and *trans* to (E). Under this system the two ligands attached to each olefinic carbon atom are put in order of precedence by application of SEQUENCE RULES and the symbols (E) and (Z) are then used to describe those isomers in which the ligands of higher precedence are on the opposite side (E), etc. A similar nomenclature can be applied to oximes.

Stereoisomers are also found in inorganic complexes, e.g.:

$$cis \quad Ph_3P \diagdown Pt \diagup Cl \qquad trans \quad Ph_3P \diagdown Pt \diagup Cl$$
$$Ph_3P \diagup \quad \diagdown Cl \qquad\qquad Cl \diagup \quad \diagdown PPh_3$$

which have square planar or octahedral co-ordination about the central platinum atom.

Other types of isomerism are:

Ionization isomerism, e.g. $[Pt(NH_3)_4Cl_2]Br_2$ and $[Pt(NH_3)_4Br_2]Cl_2$.

Mer and *fac* isomerism in, e.g. CoL_3X_3, arising because of different arrangements about the central atom (Co).

Linkage isomerism, e.g. the nitrite ion, can co-ordinate through either the oxygen or nitrogen atoms.

Since co-ordination numbers in inorganic chemistry vary widely from the limited range adopted by carbon, the possibilities for isomerism in inorganic derivatives are greater than in organic chemistry. Chelation can give rise to isomers.

See CONFORMATION, CHELATE COMPOUND.

isomerism, nuclear The existence of atomic nuclei with the same atomic and mass number but having different nuclear energies and radioactive decay properties. E.g. ^{124}Sb has three isomers, ^{124}Sb, ^{124m}Sb, $^{124m_2}Sb$.

isomerization The conversion of a compound to an isomer of that compound.

isomorphism Having the same crystal structure (isostructural). Isomorphous substances will often induce crystallization of each other and may even crystallize on each other (e.g. violet chromium alum will crystallize on colourless potassium alum).

isoniazide (4-pyridine-carboxylic hydrazide) Important anti-bacterial agent.

isonitriles (isocyanides, carbylamines) Organic compounds, R–NC. They are colourless, toxic liquids, having a highly disagreeable odour. Prepared by treating alkyl halides with silver cyanide, but more conveniently by dehydration of formamides (RNH·CHO) with phosphorus oxychloride in the presence of base; the reaction between a primary amine, chloroform and alcoholic KOH generates a carbylamine which is a useful test, but of limited preparative value. They are reactive substances; acids cause decomposition to a primary amine and methanoic acid. Heating to 100–200°C causes isomerization to cyanides. They are oxidized by sulphur to *iso*thiocyanates and by HgO to *iso*cyanates. Reaction with carbonyl compounds and amines gives derivatives of α-amino-acids. Their stable complexes with transition metals are analogous to the metal carbonyls, e.g. $Ni(CNPh)_4$ and $Ni(CO)_4$.

isonitrosoketones Tautomeric monoximes of α-diketones; formed by treating diketones with amyl nitrite and hydrochloric acid. Colourless crystalline solids, that dissolve in alkalis to give salts having an intense yellow colour. Form α-diketones when treated with H_2SO_4 or HNO_2.

isononanol Alcohol used as plasticizer. Di-isononal phthalate is used in food.

isooctanes See OCTANES.

isoparaffins Aliphatic hydrocarbons having a branched carbon chain structure. They usually have higher anti-knock value than the corresponding straight-chain normal paraffin. See OCTANES.

isopentyl ether, $[(CH_3)_2CHCH_2CH_2]_2O$. Generally a mixture of isomers. Good solvent, particularly for Grignard reactions.

isophorone (3,5,5-trimethylcyclohexen-1-one) Solvent used in adhesives.

isophthalic acid (1,3-benzene dicarboxylic acid) Colourless crystals, m.p. 345–347°C with sublimation. Obtained by oxidation of m-xylene.

isopiestic Solutions of salts, etc. that have the same partial pressure of solvent are termed isopiestic solutions.

isopolyacids Formally acids which give polymeric oxide- and hydroxide-bridged anions containing one metal only. Formed by, e.g. Mo (VI) $[Mo_6O_{19}]^{2-}$. The free acids often do not exist.

isopolymorphism Substances which have more than one crystalline form isomorphous with the crystalline forms of another substance are said to show isopolymorphism. Thus the octahedral and rhombic forms of As_2O_3 are isomorphous with the corresponding forms of SbO_3. These oxides are isodimorphous.

isoprene (2-methylbuta-1,3-diene), C_5H_8, $H_2C=CMe\cdot CH=CH_2$. Colourless liquid with a penetrating odour; b.p. 34°C. Obtained by the distillation of raw rubber or by passing turpentine oil vapour through a tube heated to redness. Manufactured from butanone and methanal in the presence of KOH followed by reduction and dehydration, also from propanone and ethyne in an ether solvent in the presence of powdered KOH. The resultant acetylenic alcohol is catalytically reduced to 1,1-dimethylallyl alcohol which is dehydrated by passing the vapour over alumina. It is converted to synthetic rubber by the action of catalysts (see ISOPRENE POLYMERS). Probably carcinogenic.

isoprene polymers Stereoscopic polymerization of isoprene to *cis*-polyisoprene, the synthetic analogue of natural rubber, occurs with Ziegler–Natta catalysts or lithium alkyls. Used as an elastomer and replacement for natural rubber (as polyisoprene) and in copolymers. *Trans*-polyisoprene is used in golf-ball covers.

isoprene rule This proposes that the carbon skeletons of terpenes are built up of isoprene units, so that C^4 of one unit $C^1C^2(C)C^3C^4$ is attached to C^1 of the next. The rare exceptions to this rule can be accommodated by assuming that a methyl group migration occurs at some stage in the biogenesis. The actual precursor isopentyl pyrophosphate and the closely related compound, MEVALONIC ACID, are incorporated in high yield into carotenoids, squalene, cholesterol and other terpenoid compounds.

isoprocarb Insecticide, carbamate derivative.

isopropanol See 2-PROPANOL.

isopropanolamines See 2-PROPANOL-AMINES.

isopropenyl ethanoate (isopropenyl acetate), $C_5H_8O_2$, $CH_2=C(CH_3)\cdot O\cdot CO\cdot CH_3$. Colourless liquid; b.p. 96–97°C. Prepared from propanone and keten in the presence of sulphuric acid. Acetylating agent. Gives ethanoates (acetates) with alcohols and amines, enol acetates with aldehydes and ketones, and acid anhydrides with carboxylic acids.

Isopropenyl acetate gives spinnable fibres on copolymerization with vinyl chloride.

isopropyl The group $(CH_3)_2CH-$.

isopropyl alcohol See 2-PROPANOL.

isopropylbenzene See CUMENE.

isopropyl chloride See 2-CHLOROPROPANE.

isopropyl ethanoate (isopropyl acetate), $C_5H_{10}O_2$, $CH_3C(O)OCH(CH_3)_2$. Colourless liquid with a fragrant odour; b.p. 85°C. Manufactured by leading propene into hot ethanoic acid containing sulphuric acid, or by heating isopropyl alcohol with ethanoic and sulphuric acids. Used as a solvent for cellulose nitrate and various gums and as a fuel additive.

isopropyl ether, C_6H_4O, $Me_2CHOCHMe_2$. Colourless liquid with sweet, camphor-like odour; b.p. 68°C. Made by heating isopropanol with sulphuric acid. Used as an industrial solvent particularly for dewaxing lubricating oils and for the extraction of fats and oils.

isopropylidene The group $Me_2C=$.

4,4′-isopropylidenediphenol See BISPHENOL A.

isopropylmethylbenzenes See CYMENES.

isopropylnaphthalenes Important solvents.

isoprothiolane Fungicide, dithiomalonate derivative.

isoproturon Herbicide, urea derivative.

isoquinoline, C_9H_7N. Crystallizes in tablets;

m.p. 24°C, b.p. 242°C. Odour like that of quinoline. Used in dyestuff synthesis, anti-malarials and as rubber accelerator.

isosbestic point Points on the spectra, particularly of reacting transition metal complexes, at which extinction coefficients are equal. Used in spectrophotometric studies of complex formation or of kinetics of reactions in solution to investigate processes such as isomenzation.

isostructural Having the same lattice type and crystal structure. E.g. K_2PtCl_6 and Rb_2PtCl_6.

isotactic polymers Stereospecific polymers of, e.g., $(-CH_2CHR-)_n$ with all the substituted carbons having the same stereo-configuration and, with a hypothetical straight chain, all the substituents on the same side of the chain. Prepared by polymerization using specific catalysts. Compare SYNTACTIC POLYMERS.

isothermal change A change in a system which occurs whilst the system is maintained at constant temperature on, e.g. a constant temperature bath.

isothiocyanates Derivatives of the type RNCS. See THIOCYANIC ACID.

isotones Atomic nuclei having the same number of neutrons but different mass numbers, e.g. ^{132}Xe and ^{133}Cs.

isotonic solutions Solutions which have the same osmotic pressure.

isotope Elements (or atoms) which have the same atomic number and, therefore, essentially the same chemical properties but different atomic masses are called isotopes, e.g. $^{16}_{8}O$ and $^{18}_{8}O$. In addition to stable isotopes, very many radioactive isotopes of elements are known. Although isotopes of the same element have similar chemical properties, their slight difference in mass causes slight differences in physical properties. Use of this is made in isotopic separation processes using techniques such as fractional distillation, exchange reactions, diffusion, electrolysis and electromagnetic methods.

isotope decay series The isotope sequences arising by natural decay. Can be used for dating.

isotope effect Kinetic and equilibrium effects due to slight changes in energies of species containing different isotopes, e.g. reactions involving the breaking of a C–H bond are slower when the hydrogen is replaced by deuterium or tritium and are distinguished by a 'primary kinetic isotope effect'. This difference in rate of breaking of a C–H bond compared to a C–D bond is due to the greater vibrational energy at room temperature of a C–H bond by comparison with that of the C–D bond. Consequently less energy is required to break the C–H bond.

isotopomers Species that are the same chemically but differ in their isotopic compositions, e.g. CH_4, CD_4.

isotropic A substance which possesses identical physical properties in all directions in its crystal. Cubic substances and some non-crystalline glasses are isotropic.

isouron Herbicide, urea derivative.

isovaleraldehyde (3-methylbutanol), $C_5H_{10}O$. Used in flavours and perfumes.

isovaleric acid (3-methylbutanoic acid), $C_5H_{10}O_2$. Used in flavouring and perfumery.

isoxaflutole Herbicide, fluoroketone derivative.

isoxamen Herbicide, amide derivative.

isoxathion Insecticide, organophosphate derivative.

itaconic acid (methylenesuccinic acid), $HO(O)C(CH_2)CH_2C(O)OH$, $C_5H_6O_4$. White crystals; m.p. 162–164°C. It can be prepared by the fermentation of sugar with the mould *Aspergillus terreus* or by heating citraconic anhydride with water at 150°C. Electrolysis of the potassium salt in solution gives 1,2-propadiene. Itaconic acid is used as a comonomer in plastics; its esters are polymerized to lubricating oils and plasticizers.

J

J The measure of the coupling constant in nuclear magnetic resonance. Unit Hertz, Hz. See JOULE.

J JOULE unit of energy.

Jacobson's catalyst Chiral [salenMn (III)] complex used in epoxidation.

Jahn–Teller effect The Jahn–Teller theorem states that, when any electronic state contains a number of electrons such that the degenerate orbitals are not completely filled, the geometry of the species will change so as to produce non-degenerate orbitals. Particularly applied to six co-ordinate transition metal compounds where the d^9 state is $d_\gamma{}^6 d_\epsilon{}^3$. Cu(II) generally has the distorted octahedral co-ordination typical of the Jahn–Teller effect. The deviation from a regular geometry is known as Jahn–Teller distortion. Where a species is not distorted the vibrational spectra frequently show anomalous intensities if subject to the Jahn–Teller theorem.

jj coupling See RUSSELL–SAUNDERS COUPLING.

Jones reductor A tube containing zinc amalgam. Used for reduction of solutions (e.g. $Fe^{3+} \rightarrow Fe^{2+}$) prior to titration or estimation.

Joule (J) The SI unit of energy, $kg\ m^2\ s^{-2}$.

Joule's law The internal energy of an ideal gas depends only on its temperature (being independent of its pressure and volume). Like the other gas laws, it is only approximately true. At high pressures it is invalidated by the existence of intermolecular forces.

The heat produced by an electric current I flowing through a resistance R for time t is I^2Rt.

Joule–Thomson effect (Joule–Kelvin effect) Most gases (except H and He) undergo cooling when they are expanded adiabatically through a porous barrier. This is called the Joule–Thomson effect. The change in temperature is proportional to the drop in pressure; the Joule–Thomson coefficient is the change of temperature per unit change of pressure. The effect, which is used for liquefying air, making solid carbon dioxide, etc. is shown by all gases at appropriate temperatures. H and He, which show a warming effect at ordinary temperatures, also show cooling if the experiment is conducted below their inversion temperatures.

juvenile hormones A widely distributed group of compounds which regulate larval development or reproduction of insects. Hormone activity is shown by some simple terpenoids such as the two *trans*-(Δ^6)-farnesols, their methyl ethers and methyl farnesoate dihydrochloride. Knowledge of such compounds has important implications for insect control.

K

K Potassium.

kainite, $MgSO_4,KCl.3H_2O$. Occurs as beds in the Stassfurt salt deposits. Used as a source of K salts and as fertilizer.

kairomone A compound or mixture acting between different species which adaptively favours the receiving species, e.g. a chemical by which a predator recognizes its prey or by which an insect is attracted to its food plant. See also ALLOMONE, PHEROMONE.

kaolin See CHINA CLAY.

kaolinite, $Al_2(OH)_4Si_2O_5$. The most important constituent of china clay.

Kapustinskii equation Expression for lattice energy, U, assuming hard spheres.

Karl Fischer titration The measurement of water content by use of a mixture of I_2 and SO_2 dissolved in pyridine–MeOH. HI is liberated and the pH determined with a meter.

Karplus equation The relationship between the n.m.r. coupling constant J for vic-H atoms and the dihedral angle between the C–H bonds.

K capture See RADIOACTIVITY.

Kel-F Trade name for a polymeric chlorotrifluoroethene, often copolymerized. May be a liquid or solid. Inert to chemical attack and a thermoplastic.

Kelvin (K) The fraction $^1/_{273.16}$ of the thermodynamic temperature of the triple point of water.

kephalin See CEPHALINS.

kepone An obsolete insecticide and fungicide also poisonous to mammals.

keratins An insoluble group of proteins of the scleroprotein class, found in the skin, hair, nails, horns, hoofs and feathers of animals. Keratins are insoluble in the usual protein solvents and are not attacked by pepsin or trypsin.

kernite (rasorite) A hydrated sodium borate, $Na_2B_4O_7.4H_2O$, which occurs in California and is a major source of boron compounds. See BORATES, BORAX.

kerogen Polycyclic hydrocarbon; probably has biological origin.

kerosine A petroleum fuel of approximate boiling range 410–570 K.

ketals Ketone acetals of the general formula:

$$\begin{array}{c} R \\ \diagdown \\ R' \end{array} C \begin{array}{c} OR'' \\ \diagup \\ \diagdown \\ OR''' \end{array}$$

They are colourless liquids with characteristic odours, and are prepared by the condensation of ketones with alkyl orthoformates in the presence of alcohols, or by the reaction of alkynes with alcohols in presence of HgO and BF_3. In some cases trichloroethanoic acid is used as the catalyst. They lose alcohol when heated and form vinyl ethers. Exchange of alcohol groups occurs when the ketals of the lower alcohols are boiled with alcohols of greater molecular weight. See ACETALS.

ketenes Organic compounds containing the group $>C=C=O$. They are formed by the action of zinc powder on the acid bromide or chloride of an α-bromo- or α-chloro-fatty acid, or by heating the anhydride of a dibasic acid. Keten itself and other ketens containing the group –CH=C=O are colourless substances. The remaining ketenes are highly coloured. Both types react readily with water to give acids, with alcohols to give esters and with halogens and halogen acids.

keten (ethenal), $CH_2:CO$. A colourless gas. Manufactured by passing propanone rapidly through metallic tubes heated at 550–800°C, or by heating ethanoic acid at 700–1000°C. It is very unstable and dimerizes spontaneously to diketen, 3-buteno-β-lactone. Reacts with water to form ethanoic acid, with alcohols to give ethanoates, with ethanoic acid to give ethanoic anhydride and with amines to give acetyl derivatives. Used as an acetylating agent, particularly in the manufacture of cellulose acetate (ethanoate).

keto- A prefix used to denote that the substance in question contains a carbon atom attached to an oxygen atom by a double bond and to two other carbon atoms by single bonds.

ketols Organic compounds containing both a keto- and an alcohol group. They are formed by oxidation of glycols or by condensation between two molecules of a ketone. Have the typical properties of both alcohols and ketones.

ketomalonic acid See MESOXALIC ACID.

ketone bodies (acetone bodies) Substances such as propanone (acetone) ($CH_3.CO.CH_3$), acetoacetic acid ($CH_3.CO.CH_2.COOH$) and β-hydroxybutyric acid ($CH_3.CHOH.CH_2.COOH$) produced in the liver for use by other tissues as fuel – see DICTIONARY OF BIOLOGY.

ketones Organic compounds containing the C·C(O)·C group. They include aliphatic, aromatic, cyclic and mixed ketones. The cyclic ketones have the carbon of the carbonyl group as part of a ring. They are formed by the dry distillation of the calcium or barium salts of acids, or by passing the acids over ThO_2 at 400°C. Ketones may be obtained by oxidation of secondary alcohols. (See OPPENAUER OXIDATION.) Aromatic and mixed ketones are usually prepared by Friedel–Crafts reaction using acyl halides. Aliphatic ketones are usually liquids and aromatic ketones solids. They have ethereal or aromatic odours. They are reduced to secondary alcohols, but are resistant to oxidation and give mixtures of acids and other products when strongly oxidized. They are oxidised to lactones or esters by peroxyacids or H_2O_2. They form oximes with hydroxylamine, hydrazones with hydrazine and substituted hydrazines, and semicarbazones with semicarbazide. Isonitrosoketones are formed by the action of sodium nitrite.

ketose A sugar containing a potential keto-(CO) group. The presence of the group may be obscured by its inclusion in a ring system. Called ketopentoses, ketohexoses, etc., according to the number of carbon atoms they contain.

ketoximes Organic substances containing the group >C=NOH. They are formed by treating ketones with hydroxylamine. They are similar in properties to the aldoximes, but undergo the Beckmann rearrangement when treated with sulphuric acid. They are reduced to primary amines.

ketyls E.g. $K^+(OCPh_2)^-$. Salts of radical anions of ketones, intensely coloured; dimerize readily.

Kevlar An aramide fibre.

kieselguhr See DIATOMITE.

kilogram (kg) The SI unit of mass.

kinase Protein catalysing the transfer of phosphate from ATP to the indicated substrate, e.g. hexokinase (transfer to hexose). Kinases have been implicated in a wide range of diseases.

kinetic energy Energy of a particle as a result of motion. $E_K = \frac{1}{2}mv^2$.

kinetics, chemical The branch of chemistry which is concerned with the explanation of observed characteristics of chemical reactions (e.g. the variation of reaction velocities with pressure, temperature or concentration).

kinetic theory of gases The theory in which the properties of gases are derived by applying the laws of probability and of particle dynamics to a system in which the gas is assumed to consist of elastic particles possessing temperature-dependent, random motion.

kinetin (6-furfurylaminopurine) Plant growth regulator. Brand name of hyaluronidase polysaccharides used as spreading agents.

Kipp's apparatus Laboratory equipment for the production of a gas by interaction of a liquid and a solid. It consists of three receptacles. The top is a reservoir for the liquid and is connected to the bottom. The middle contains the solid and a tap for the gas. When gas is released the liquid rises and reacts with the solid. When the tap is closed the back pressure returns the liquid to the lower reservoir and reaction ceases. Once widely used for the production of H_2S (HCl and FeS) and CO_2 (HCl and $CaCO_3$).

Kirchhoff's equation If Q is the heat evolved when a process (physical or chemical) is carried out at temperature T, the heat which would be evolved if the same process were carried out at a different temperature can be calculated with the aid of Kirchhoff's equation:

$$dQ/dT = C_1 - C_2$$

where C_1 and C_2 are the total heat capacities of the system before and after the process respectively (e.g. reactants and products in a chemical reaction). Kirchhoff's equation is a direct consequence of the first law of thermodynamics.

Kjeldahl method An analytical method for the determination of nitrogen, particularly in organic materials. The N is converted to NH_4^+ with conc. H_2SO_4 and catalysts. After neutralization the NH_3 is distilled off and estimated by titration after absorption.

knocking Both spark-ignition and compression-ignition engines may show a reduced power output accompanied by sharp knocking sounds because of explosive detonation. This is normally a function of both fuel composition and compression ratio. See ANTI-KNOCK ADDITIVES, KNOCK RATING.

knock rating The tendency for gasolines and diesel fuels to produce KNOCKING in service is designated by an OCTANE NUMBER and a cetane number.

Knudsen cell See EFFUSION.

Kohlrausch equation This equation, which describes the behaviour of strong electrolytes on dilution, states that:

$$\Lambda_\infty - \Lambda_v = kC^{\frac{1}{2}}$$

where Λ_∞ is the equivalent conductivity at infinite dilution, Λ_v that at volume v, k is a constant and C is the concentration of electrolyte. The equation is valid only at high dilutions. Λ_∞ has contributions from each ion.

kojic acid (2-hydroxymethyl-5-hydroxy-4-pyrrole) Food additive and flavour enhancer.

Kolbe reaction The preparation of saturated or unsaturated hydrocarbons by the electrolysis of solutions of the alkali salts of aliphatic carboxylic acids. Thus, ethanoic acid gives ethane,

$$2CH_3CO_2^- \rightarrow CH_3 \cdot CH_3 + 2CO_2$$

Kr Krypton.

Krafft temperature See CRITICAL MICELLE CONCENTRATION.

Krebs's cycle See CITRIC ACID CYCLE.

k resin A styrene–butadiene co-polymer.

kresox-methyl Fungicide, arylethanoic acid derivative.

Krilium The trade name of a soil conditioner. Generally has as active ingredient, a copolymer of about equal molar proportions of vinyl ethanoate and the partial methyl ester of maleic acid. Formulated with lime, bentonite, etc. Other polymers are also used.

krypton, Kr. Noble gas, at. no. 36, at. mass 83.6, m.p. –157.38°C, b.p. –153.22°C, density ρ 3.739 g l^{-1}, atmospheric abundance 1.14 p.p.m. by volume, electronic configuration [Ar] $3d^{10}$ $4s^2 4p^6$, Group 18. Obtained pure by fractional distillation of liquid air. Used (with Ar) for fluorescent lights, in photographic flash lights and to measure surface areas. Kr forms a limited range of chemical compounds and some clathrates. Linear KrF_2 (Kr plus F_2 in presence of an electrical discharge) forms adducts with Lewis acid fluorides, e.g. $KrF_2.2SbF_5$. KrF_2 is used as a strong fluorinating agent.

KTU 3616 Fungicide, carboxamide derivative.

kurchatovium Element 104. Proposed name, now called RUTHERFORDIUM Rf.

kurrol salt, $NaPO_3$. A long-chain metaphosphate formed by annealing Graham's salt. See PHOSPHORUS OXYACIDS.

KWG 4168 Fungicide, methanamine derivative.

kynurenine (3-anthraniloyl-L-alanine), $C_{10}H_{12}N_2O_3$. An intermediate in the metabolic breakdown of tryptophan.

L

L Designation of configuration. See **D**.

L (litre) Measure of volume.

L The Avogadro constant, 6.022×10^{23} mol^{-1}.

La Lanthanum.

labile complex A complex which participates in very fast reactions, particularly ligand exchange reactions, generally within the time of mixing. E.g. $[Fe(H_2O)_6]^{3+}$, $[Co(H_2O)_6]^{2+}$. Compare **INERT COMPLEX**.

lability A measure of the ease of replacement of attached groups in complexes and molecules. Cr^{3+}, Co^{3+} complexes are inert, many other complexes are relatively labile.

lactams Amino-acids when heated lose water to form lactams. The water is eliminated internally between the carboxyl group and the amino-group and a cyclic compound is formed. Also obtained by reduction of the imides of dicarboxylic acids or by isomerization of the oximes of cyclic ketones. Rings containing five, six and eight atoms can be obtained. The lactams are decomposed by heating with alkalis and the amino-acid is formed. They are colourless solids and are extremely poisonous. β-Lactams used as anti-bacterials. See **CAPROLACTAM**.

β-lactic acid 2-Hydroxypropanoic acid.

lactic acids (D,L-2-hydroxypropanoic acids), $C_3H_6O_3$, $CH_3 \cdot CHOH \cdot C(O)OH$. The acids are colourless, syrupy liquids which readily absorb moisture, and are formed by the fermentation of sugars by the *lactobacilli* and some moulds. When slowly distilled they lose water and form lactide. L-Lactic acid, sarcolactic acid, occurs in muscle, being formed by the breakdown of carbohydrate; m.p. 25–26°C. D-Lactic acid; m.p. 18°C, b.p. 122°C/14mm. Lactic acid is manufactured by the fermentation of pure sugars or of various sugar-containing materials such as starch hydrolysates, and by the hydrolysis of lactonitrile formed by reacting ethanal with HCN. The major use is in the food and beverage industries, where it is used as an acidulant, for cheese manufacture, and as a bread additive. It is also used as a chemical intermediate, in textile finishing and in leather tanning.

lactide, $C_6H_8O_4$. Colourless crystals. Prepared by the slow distillation of concentrated solutions of lactic acids. L-Lactide, structure as **LACTIDES** with two methyl groups; m.p. 95°C, b.p. 150°C/25mm. Prepared from L-lactic acid. It is partially converted to lactic acid by water. D-Lactide is similar. DL-Lactide crystallizes in colourless needles; m.p. 124.5°C, b.p. 142°C/8 mm. Obtained from DL-lactic acid.

lactides When α-hydroxy-fatty acids are heated they lose water and form lactides. These contain the group:

$$\begin{array}{c} C(O)\text{---}O\text{---}CH\text{---} \\ \text{---}CH\text{---}O\text{---}C(O) \end{array}$$

and are formed by the interaction between two molecules of the hydroxy-acid by elimination of water. They are decomposed by heating with water to regenerate the original acid. Important biodegradable solids.

lactofen Herbicide, diphenyl ether derivative.

lactones Anhydrides formed by intramolecular elimination of water between the hydroxyl and carboxyl groups of hydroxyacids. The ease with which they are formed depends upon the number of atoms in the ring to be formed. The commonest lactones are those of γ- and δ-hydroxyacids, since these have rings containing five and six atoms respectively. They are usually formed spontaneously in concentrated solutions of the acids, or by heating such solutions with small amounts of sulphuric acid or from ketones by oxidation with H_2O_2 or peroxy acids. They are usually crystalline solids which are partially decomposed by water with regeneration of the hydroxyacid.

lactose, $C_{12}H_{22}O_{11}$. Milk sugar occurs in the milk of all animals. Human milk contains 6%, cow's milk, 4%. It is manufactured by the evaporation of whey. Two glucose-4-β-galactosides. Like glucose it gives rise to two series of isomeric derivatives. Lactose is hydrolysed by lactase. Used in baking powder, chromatography.

laevorotatory See OPTICAL ACTIVITY.

laevulinic acid (levulinic acid, laevulic acid, 4-oxopentanoic acid), $C_5H_8O_3$, $CH_3COCH_2CH_2C(O)OH$. Colourless crystals; m.p. 33–35°C, b.p. 245–246°C. Prepared by heating cane sugar or starch with concentrated hydrochloric acid. Reduced to γ-valerolactone. Also reacts as a ketone. Used in cotton printing.

LAH Lithium aluminium hydride.

lake asphalts Naturally occurring mixtures of bitumen with mineral and organic matter.

lakes Insoluble pigments obtained by precipitating natural and artificial colouring matters (dyestuffs, dye-wood extracts, cochineal, etc.) on to suitable bases in the presence of Al^{3+}, Mg^{2+}, Zn^{2+}, Sn^{4+} or other metals, and compounds like tannic acid commonly used as mordants in dyeing processes.

lambda point (λ point) A phase transition which is not first order but the heat capacity becomes infinite. Includes the fluid–superfluid transition of liquid helium.

Lambert's law Layers of equal thickness of a homogeneous material absorb equal proportions of radiation:

$$I = I_0 e^{-Kd},$$

where I is the intensity of the transmitted light, I_0 that of the incident light, d the thickness of the layer and K a constant, characteristic of the substance, known as the absorption coefficient; K also depends on the wavelength of the light employed. When solutions are considered, it is clearly desirable to modify this expression to include the concentration of the absorbing molecules. This modification is embodied in BEER'S LAW.

lamellar micelles Flat biomolecular micelles with hydrophilic groups on the outside in aqueous solution.

laminarin Polysaccharide extracted from brown seaweeds. Used as anti-coagulant.

laminates See REINFORCED PLASTICS.

lamp black A soft black carbon pigment obtained by incomplete combustion of natural gas or petroleum. A soft grade of CARBON BLACK used extensively in inks and paints.

Landé g factor See G FACTOR.

langmuir An arbitrary unit used as a measure of adsorption. 10^{-6} torr adsorbant exposed to a surface for 1 second.

Langmuir adsorption isotherm See ADSORPTION ISOTHERM.

lanolin Crude preparation of cholesterol and its esters obtained from wool fat. It is a pale yellow, tenacious substance melting at about 37°C. Lanolin itself is anhydrous lanolin mixed with about 30% of water. In medicine it is used either alone or mixed with soft paraffin, lard or other fat as a base for many ointments in which lipid-soluble drugs are administered to the skin. It readily emulsifies with water, making it suitable also as a base for creams.

lanosterol (isocholesterol), $C_{30}H_{50}O$. M.p. 138–139°C. A triterpenoid or trimethylsterol 4,4,14α-trimethyl-5α-cholesta-8,24-dien-3β-ol is the precursor in animals and fungi of other sterols such as CHOLESTEROL.

lanthanide compounds The normal oxidation state is M(III) and hydrides, oxides, hydroxides and halides MX_3 (MF_3 are insoluble in water) are formed by direct reaction of the elements or from solution. Salts of oxyacids are generally hydrates with high co-ordination numbers for M. O-bonded ligands, particularly chelating ligands, form stable complexes (see SHIFT REAGENTS) and some N-bonded and halide complexes are known. Cyclopentadienyls, MCp_3 are stable; alkyls and aryls are relatively unstable. Cp_3Ln are one-electron reductants. Dihydrides, e.g. $LnH_2(THF)_2$, are known for Sm and Eu. Other compounds are useful catalysts, e.g. to induce addition of amine to alkenes. Ce, Pr and Tb form M(IV) compounds particularly MF_4 and MO_2. Some other lanthanides also form +4 complex fluorides. Sm, Eu and Yb form ionic MX_2 and M^{2+}, which exist in aqueous solution. Other lower halides contain M–M bonds although the iodides MI_3 (M = La, Ce, Gd, Pr) are $M^{3+}(I^-)_2e$.

lanthanide contraction There is a regular decrease in ionic and atomic radii across the lanthanides. This means that radii of elements after the lanthanides are very similar to radii of corresponding elements before the series (see also RELATIVISTIC EFFECTS). Thus properties depending on ionic size (e.g. structure, solubility) of elements before and after the lanthanides are very similar, e.g. Mo and W.

lanthanides The elements from lanthanum to lutetium, in which the 4f orbitals are being filled. The elements lanthanum, cerium, praseodymium, neodymium, promethium, samarium, europium, gadolinium, terbium, dysprosium, holmium, erbium, thulium, ytterbium and lutetium. Except for Pm the elements are not uncommon – the normal source is monazite which contains 90% La, Ce, Pr, Nd in its lanthanide content. The elements are separated by chromatography, ion exchange or solvent extraction. The metals are prepared by MF_3 + Ca at 1450–1700°C but for Sm, Eu and Yb the reaction La + M_2O_3 with volatilization of M is used. The metals are shiny and are attacked by water and acids. They burn in oxygen or air to M_2O_3 or MO_2 and react in hydrogen to MH_2 and MH_3. Trace amounts of lanthanides added to Ni and Co-based alloys control the diffusion mechanism in surface oxide films so that they adhere more strongly to the underlying metal – used in jet engines. Used in magnets, catalysts, glasses, glazes, superconductors.

lanthanum, La. Lanthanide metal, at.no. 57, at.mass 138.91, m.p. 918°C, b.p. 3464°C, density ρ 6145, crustal abundance 32 p.p.m., electronic configuration $[Xe]\ 5d^16s^2$. The metal is used in steels and in Mg and Al alloys. La_2O_3 is used in glasses, as a glass polish, as a high-temperature refractory and as a host matrix for fluorescent phosphors.

lanthanum compounds Lanthanum forms a single series of typical LANTHANIDE COMPOUNDS in the +3 state La^{3+} (f° colourless) $\rightarrow$ La (–2.52 volts in acid). $La(NO_3)_3.6H_2O$ is used as an analytical reagent for ethanoate ions. LaI_2 is La^{3+}, $2I^-$,e^-.

lapis lazuli Mineral ULTRAMARINE. A semi-precious stone known from antiquity.

laser A device which produces beams of monochromatic light of very great intensity in which the waves are coherent. The name is derived from Light Amplification by Stimulated Emission of Radiation.

It is essential that the atom in the excited state has a greater population in the upper excited state than the lower state. In a laser the excited atom is struck by a photon of exactly the same energy as the one which would be emitted spontaneously. The excited atom is stimulated to emit a photon and return to the ground condition with the result that two photons of precisely the same wavelength are produced and a narrow coherent beam of collimated radiation is emitted. The process can now be repeated throughout the system.

An example of a solid laser is the ruby crystal (Al_2O_3 containing about 0.05% chromium). Glasses containing Nd^{3+} are used for high-output lasers; gas lasers using helium–neon mixtures, caesium vapour and other gases have also been made. The most efficient lasers are those using semi-conductors with, for example, gallium arsenide phosphide as the active crystal. See LED, LIGHT EMITTING DIODE. Solutions of dyestuffs can act as lasers giving wide ranges of frequency.

Due to the very high intensity of the laser beams and their coherent nature they may be used in a variety of ways where controlled energy is required. Lasers are used chemically for excitation with a specific energy, e.g. in Raman spectroscopy or isotope separation. See DICTIONARY OF PHYSICS.

laser ablation The volatilization and subsequent deposition of materials to protect and modify surfaces and produce new species (e.g. nanotubes).

Lassaigne's test A general test for the presence of nitrogen, halogens or sulphur in an organic compound, a little of which is heated with a pellet of sodium in a hard-glass test tube; the hot tube is plunged into distilled water and the product ground. The presence of nitrogen is shown by the formation of Prussian blue on heating with a solution of iron(II) sulphate containing a trace of Fe(III) and hydrochloric acid. With halogen-containing substances sodium halide is formed, detectable with silver nitrate; the presence of sulphur can be shown with sodium nitroprusside or lead ethanoate.

latent heat The heat absorbed or evolved during a phase change. The heat effect in such changes is reversible. The heat absorbed when a solid is converted to liquid at the m.p. is called the latent heat of fusion, that absorbed when a liquid is converted to a gas at the b.p. and 1 atmosphere pressure is termed the latent heat of vaporization. Latent heats are usually measured in joules per mole. Latent heats are entropies of phase changes. See TROUTON'S RULE.

latent image The change occurring in silver halide grains after exposure to light which makes them developable.

latex A stable aqueous dispersion of a polymer. Originally applied to natural rubber latex from the bark of certain trees. Many synthetic rubbers and polymers (e.g. PVC, polyacrylates) are produced as latexes. Latexes may be used for direct manufacture of rubber and plastic goods, by dipping, moulding, spreading, electrodepositing and impregnating (e.g. textiles, adhesives, paints).

lattice The regular three-dimensional array of atoms in a crystal. A lattice is built up from UNIT CELLS. Lattice constants are the unit cell dimensions and angles. The dimensions are generally in Å (Angstrom) units, 10^{-8} cm, or nanometres, 10^{-9} m.

lattice energy The energy required to break down one mole of an ionic lattice into its constituent ions in the gaseous state and infinitely separate them from each other, i.e. the enthalpy change associated with the process

$$M^+X_n^-(\text{solid}) \rightarrow M^+(g) + nX^-(g)$$

Lattice energies may be derived from the BORN–HABER CYCLE or calculated using the KAPUSTINSKII or BORN–LANDÉ EQUATIONS.

Laue pattern The symmetrical array of spots obtained on a photographic plate exposed to a non-homogeneous beam of X-rays after its passage through a crystal.

lauric acid See DODECANOIC ACID.

laurite Natural $(Ru,Os)S_2$; an important source of Ru and Os.

lauryl alcohol See 1-DODECANOL.

laves phases Alloys, e.g. $MgZn_2$, where structure is determined by size rather than by electronic configurations.

law of conservation of energy See CONSERVATION OF ENERGY, LAW OF.

law of conservation of matter See CONSERVATION OF MATTER, LAW OF.

law of constancy of interfacial angles See INTERFACIAL ANGLES, CONSTANCY OF.

law of equivalent proportions See EQUIVALENT PROPORTIONS, LAW OF.

law of mass action See MASS ACTION, LAW OF.

law of multiple proportions See MULTIPLE PROPORTIONS, LAW OF.

law of rationality of indices See RATIONALITY OF INDICES, LAW OF.

law of rationality of intercepts See RATIONALITY OF INTERCEPTS, LAW OF.

lawrencium, Lr. Actinide metal, at.no. 103, at.mass ^{262}Lr 262.11, m.p. 1627°C, relative density ρ 11350, electronic configuration $[Rn]5f^{14}6d^17s^2$. Lawrencium isotopes have been formed by bombardment of Cf targets with ^{10}B and ^{11}B nuclei in a cyclotron or linear accelerator. ^{262}Lr of half life 216 min is the stablest isotope. It is known only in the +3 oxidation state; no solid compounds are known.

layer lattice Crystal structures in which well-defined layers, either simple or composite, of atoms may be distinguished. The forces between the atoms within the layers are stronger than those holding the layers together, leading to good cleavage parallel to the layers. Examples are graphite, boric acid, cadmium iodide and chloride, and molybdenum sulphide. Uses frequently depend on this cleavage.

LCAO method A method of calculation of molecular orbitals based upon the concept that the molecular orbital can be expressed as a linear combination of the atomic orbitals.

LCD Liquid crystal display.

LD$_{50}$ Measure of toxicity – likely to cause death in 50% of species. Test now obsolete.

leaching The extraction of a soluble material from an insoluble solid by dissolution in a suitable solvent. The soluble material may be either liquid or solid.

lead, Pb. Metal, at.no. 82, at.mass 207.2 (variable), m.p. 327.46°C, b.p. 1749°C, density ρ 11 350, crustal abundance 14 p.p.m., electronic configuration $[Xe]4f^{14}5d^{10}6s^26p^2$. Lead isotopes are the ultimate product of the various radioactive-decay series. Occurs naturally as galena, PbS, and cerussite, $PbCO_3$, which are roasted in air to PbO and reduced to Pb either by the reaction between PbO and PbS (to give Pb plus SO_2) or with carbon or electrolytically. The silvery-white metal is ccp. Lead is used extensively in alloys, in storage batteries (25%), for covering electrical cable, in water and noise proofing, for lead alkyls, electronic devices, ceramics, pigments, radiation and sound shields, in paints, high-quality glass. Lead compounds are toxic and hazardous to health.

lead, of pencil Writing material. Graphite-clay mixture. Organic derivatives.

lead, organic derivatives Derivatives of lead(IV) made generally by alkylation of Pb(II) compounds e.g. R_3PbX and other derivatives are known. Tetramethyl lead, Me_4Pb, and tetraethyl lead, Et_4Pb, TEL, were important antiknock additives for petroleum (gasoline) but now being phased out; prepared from Na/Pb plus RCl, electrolysis of $Na(AlEt_4)$ with a lead anode and Hg cathode; electrolysis of RMgX with lead anodes. Organolead compounds are toxic.

lead accumulator or battery A battery which consists of a number of plates of Pb; each alternate plate is connected to one wire and the remaining plates to another wire. The plates are covered with a mixture of Pb oxides and Pb sulphate, and are placed in dilute sulphuric acid. When a current is passed through the accumulator PbO_2 is formed on one set of plates and Pb on the other, and some H_2SO_4 is produced. When the accumulator is used to generate electricity H_2SO_4 acts on the Pb, the PbO and PbO_2 forming $PbSO_4$ and water to give a potential. Can provide a high current for a short time.

lead acetates See LEAD ETHANOATES.

lead–acid batteries See LEAD ACCUMULATOR.

lead azide, $Pb(N_3)_2$. An initiator ($Pb(O_2CCH_3)_2$ and NaN_3), but use being phased out. See EXPLOSIVES.

lead bromide, $PbBr_2$. M.p. 373°C, b.p. 916°C. Precipitated from cold water and soluble in hot. No lead(IV) bromide is known.

lead carbonate, $PbCO_3$. Occurs naturally as cerussite ($Pb(O_2CCH_3)_2$ plus $(NH_4)_2CO_3$). Forms basic carbonates, e.g. $Pb_3(OH)_2(CO_3)_2$, white lead, prepared from lead, air, CO_2, steam, ethanoic acid or electrolytically; used in paints but phased out in U.S.A.

lead chemistry An electropositive element of Group 14. Lead shows true cationic character in the +2 state in, e.g. PbF_2, but the +4 state is largely covalent and is quite strongly oxidizing. The +4 state forms series of complexes and some complexes, e.g. chlorides, and oxo-hydroxide complexes are formed by Pb(II). The lone pair of electrons in Pb(II) has little stereochemical effect. Lead(IV) organic derivatives are stable and compounds with Pb–Pb bonds are formed. PbH_4 (acid plus Mg/Pb) is unstable.

lead chlorides
Lead(II) chloride, $PbCl_2$. Insoluble cold water, sparingly soluble hot. Matlockite is PbFCl. $PbCl_2$ forms many complex chlorides and also basic chlorides. Cassel yellow (approx. $PbCl_2 \cdot 7PbO$) is prepared by fusion of the constituents and is used as a pigment.

Lead(IV) chloride, $PbCl_4$. Is a yellow liquid, m.p. –15°C, decomposes explosively at 100°C (H_2SO_4 on $(NH_4)_2PbCl_6$). Yellow hexachloroplumbates(IV), e.g. $(NH_4)_2PbCl_6$, are precipitated from the yellow solution resulting from PbO_2 in cold HCl.

lead chromates $PbCrO_4$ is precipitated from solutions of Pb^{2+} and CrO_4^{2-} or $Cr_2O_7^{2-}$. In basic solution the precipitate is more orange (chrome yellow, chrome orange). Primrose chrome contains some $PbSO_4$. Cologne yellow is prepared by heating $PbSO_4$ with a $K_2Cr_2O_7$ solution. Used in pigments.

lead ethanoates
Lead(II) ethanoate, $Pb(O_2CCH_3)_2 \cdot 3H_2O$. Sugar of lead. Soluble in water (ethanoic acid and ethanoic anhydride plus Pb_3O_4). Used as a mordant.

Lead(IV) ethanoate, $Pb(O_2CCH_3)_4$. (Pb(II) ethanoate plus Cl_2) is a powerful oxidizing agent which will convert vicinal glycols to aldehydes or ketones and 1,2-dicarboxylic acids into alkenes. Primary amides give ketones and amines give nitriles.

lead fluorides
Lead(II) fluoride, PbF_2. Rutile structure, precipitated from aqueous solution.

Lead(IV) fluoride, PbF_4. Strong fluorinating agent (PbF_2 plus F_2). Forms complexes containing the $[PbF_6]^{2-}$ ion.

lead hydroxides See LEAD OXIDES.

lead iodide, PbI_2. Golden-yellow, precipitated from solution. Forms complex iodides. No lead(IV) iodide is known. Used as mosaic gold in bronzing.

lead nitrate, $Pb(NO_3)_2$. Colourless crystals (PbO plus HNO_3). Forms basic nitrates. Used in calico printing, as a mordant in dyeing and in the manufacture of chrome yellow pigment.

lead oxides
Lead(II) oxide, PbO. Exists in two forms as orange-red litharge and yellow massicot. Made by oxidation of Pb followed by rapid cooling (to avoid formation of Pb_3O_4). Used in accumulators and also in ceramics, pigments and insecticides. A normal hydroxide is not known but hydrolysis of lead(II) oxyacid salts gives polymeric cationic species, e.g. $[Pb_6O(OH)_6]^{4+}$ and plumbates are formed with excess base.

Trilead tetroxide, Pb_3O_4. Red lead and other intermediate phases, e.g. Pb_7O_{11} and Pb_2O_3 are formed by heating PbO or PbO_2. Uses are as for PbO but extensively used in corrosion-resistant paints, glasses.

Lead(IV) oxide, PbO_2. Chocolate brown (electrolytic oxidation of Pb(II) salts). Used as an oxidizing agent and in accumulators.

Plumbates(IV), e.g. K_2PbO_3. Formed by solid state reactions and plumbates containing $[Pb(OH)_6]^{2-}$ ions are formed from aqueous solution.

lead silicates Silicates with low lead concentrations are used in glazing pottery and glass manufacture.

lead sulphate, $PbSO_4$. Occurs naturally as anglesite, precipitated from Pb(II) solutions. Basic salts are known and are used as white pigments. Used in batteries and varnishes.

lead sulphide, PbS. Black compound occurring naturally as galena, precipitated from aqueous solution. Used as rectifier.

lead tetraacetate See LEAD ETHANOATES.

lead tetraethyl (TEL), Et_4Pb. See LEAD, ORGANIC DERIVATIVES.

lead tetramethyl, Me_4Pb. See LEAD, ORGANIC DERIVATIVES.

lean gas A fuel gas defined as having a CALORIFIC VALUE of 100–450Btu/ft^3 (3.7–16.8 MJ m^3). Obtained from liquid and solid hydrocarbons by steam refining in the presence of a catalyst at about 1075 K to yield mixtures of CO and H_2O, or by partial oxidation at 1350–1650 K to give a mixture of CO, H_2 and CO_2 plus N_2 if air is the oxidant. See LURGI COAL GASIFICATION PROCESS, PRODUCER GAS, WATER GAS.

leather Animal and other skins treated by TANNING. Has a characteristic porous composition. Polymers which fairly closely simulate leather are now available.

Leblanc process Virtually obsolete process for Na_2CO_3. NaCl with H_2SO_4 gives Na_2SO_4 (HCl gas by-product); $[SO_4]^{2-}$ plus coke gives Na_2S which with $CaCO_3$ gives Na_2CO_3 plus CaS.

Le Chatelier's theorem or principle When a constraint e.g. pressure or heat is applied to a system in equilibrium, the equilibrium will tend to move in such a way as to neutralize the effect of the constraint. Thus, e.g. in a reaction such as

$$N_2O_4 \rightleftharpoons 2NO_2$$

increasing the total pressure will cause the equilibrium to move towards the formation of more N_2O_4 thereby decreasing the total number of molecules in the system.

lecithin A generic name for substances of the

$$CH_2 - OCOR^1$$
$$CH - OCOR^2$$
$$CH_2O.\underset{O}{P} - OCH_2CH_2N^+(CH_3)_3$$

type shown, where R^1 and R^2 are fatty acid residues. Usually one acid is saturated and one unsaturated. Lecithins are found in every animal and plant cell and are an essential constituent of cells, especially membranes. They swell up with water to give slimy emulsions or colloidal solutions. They melt at about 60°C.

Commercial lecithin is a mixture of phosphatides and glycerides obtained in the manufacture of soya bean oil. It gives a thick yellow emulsion with water. It is widely used in the food and other industries. See also PHOSPHOLIPIDS.

Leclanché cell See BATTERIES.

lectins Proteins which agglutinate erythrocytes and other cells.

LED (light-emitting diode) Emission caused by recombination of electrons and holes under an electric current.

LEED See LOW ENERGY ELECTRON DIFFRACTION.

LEEM (low energy electron microscope) Device for study of surfaces.

lel See CHELATE EFFECT.

LEP Light-emitting polymer.

leucine (L-α-aminoisocaproic acid), $C_6H_{13}NO_2$, $Me_2CHCH_2CH(NH_2)C(O)OH$. Colourless plates, m.p. 293–295°C (decomp.). Optically active. Leucine is one of the most common of the amino-acids obtained from proteins.

leucinol (mimosine, 3-hydroxy-4-oxo-1(4)-pyridine) From mimosa. Used as depilatory.

Leuckart reaction The conversion of ketones and aromatic aldehydes to primary amines by reaction with ammonium methanoate at a high temperature.

leuco compounds Colourless products from dyestuffs or prepared during dyestuff synthesis. E.g. reduction of triphenylmethane dyes containing $[Ar_3C]^+$ ions gives colourless Ar_3CH. Leuco bases, e.g. Ar_3COH, and other derivatives are known.

levorotating See LAEVOROTATING.

levulinic acid (4-oxopentanoic acid), $C_5H_8O_3$. From starch or cane sugar. Used in synthesis, rubbers and plastics.

Lewis acid See ACID.

Lewis base See BASE.

lewisite (2-chlorovinyldichloroarsine, 1-chloro-2-dichloroarsino-ethene), $ClC_2H_2AsCl_2$, $ClCH:CH\cdot AsCl_2$. A pale yellow liquid, m.p. $-13°C$, b.p. $190°C$. It has a strong smell resembling that of geraniums. War gas, systemic poison. It is hydrolysed by water and also destroyed by alkalis and by oxidizing agents. It can be manufactured by bubbling ethyne through a mixture of anhydrous arsenic trichloride and aluminium chloride.

Li Lithium.

libration In a crystalline lattice the rotational motions of molecules within the lattice are often restricted due to bonding with neighbouring atoms. The resulting oscillations of the whole molecule are called librational modes.

librium (benzodiazepine) Tranquillizer.

LIBS Laser ablation.

licanic acid, $C_{18}H_{28}O_3$, $CH_3\cdot[CH_2]_3\cdot[CH:CH]_3\cdot[CH_2]_4\cdot CO\cdot[CH_2]_2\cdot C(O)OH$. Occurs in oiticica and other oils which were previously used in the protective-coating industry.

Liebermann's reaction A colour-test for $-NO$ or $-OH$ groups. In concentrated sulphuric acid and with a crystal of phenol added characteristic colours are formed.

lift-off In gas burners having a stationary flame it is essential that the flame speed of the combustible mixture should be balanced by the velocity of the mixture in the burner tube or flame ports. If the flame speed is too low or the mixture velocity too high the flame can lift off the burner mouth or flame ports.

ligand A complexing group in co-ordination chemistry. Generally the entity from which electrons are donated. NH_3 is a ligand in $[Co(NH_3)_6]Cl_3$.
  Peripheral groups in proteins which can cause effects at distant sites.

ligand field theory An extension of crystal-field theory which takes account of electron transfer.

light-emitting diode See LED.

light–heavy selectivity In solvent extraction, the phenomenon in which the extractive power of the solvent is dependent on the molecular weight of the component extracted.

light scattering The scattering of light by suspended particles. Can be used to estimate molecular weights of macromolecules and to detect end points when a precipitate is formed in the reaction. See BROWNIAN MOVEMENTS.

light stabilizers See ULTRA-VIOLET ABSORBERS.

lignans A group of natural products obtained by ethereal or alcoholic extraction from the wood or exuded resin of the *Coniferae* and other plants. Characterized by the presence in the molecule of 4-hydroxyphenylpropene linked through oxygen. Related to LIGNIN.

lignin A highly polymeric substance occurring with cellulose in lignified plant tissues. Facilitates water and nutrient transport. It is largely responsible for the strength of wood, which contains 25–30%, and it is extractable from wood pulp by the action of SO_2 and limewater: it occurs (up to 6%) in sulphite waste liquors from paper mills. Lignin is an aromatic–propene polymer of variable mol. wt. It is of commercial value as a source of vanillin, phenols and other aromatic chemicals. Used as a filler for plastics and as a dispersant, emulsifier, etc.

lignite (brown coal) Immature coals intermediate in composition between peat and bituminous coals. They occur in thick seams, often quite near the earth's surface.

lignoceric acid (tetracosanoic acid), $C_{24}H_{48}O_2$. M.p. $84°C$. Fatty acid present free and combined in many oils, fats and waxes but principally in TALL OIL.

ligroine PETROLEUM ETHER; that fraction of refined naphtha with b.p. generally 130–$145°C$.

lime See CALCIUM HYDROXIDE and CALCIUM OXIDE.

limestone Rocks of sedimentary origin containing the remains of marine organisms or chemically precipitated or transported calcium carbonate. The pure mineral consists of $CaCO_3$ in the form of calcite, but is rarely found. Commercial limestone contains iron oxide, alumina, magnesia, silica and sulphur, with a CaO content of 22–56% and a MgO content of up to 21%. It is used as such as a fertilizer and for many other purposes, or is calcined to calcium oxide.

limewater See CALCIUM HYDROXIDE. Used as a mild antacid. Passage of CO_2 gives a precipitate of $CaCO_3$.

limiting density (of a gas) The density value of this when pressure approaches zero is called the limiting density of the gas. It is the density the gas would possess if it was an ideal gas, i.e. a gas which obeys the gas laws perfectly.

limonene (citrene, carvene, p-mentha-1,8-diene), $C_{10}H_{10}$. B.p. 176°C. An optically active terpene. Its racemic form is known as dipentene. Very widely distributed. Used as a solvent and in resins and wetting and dispersing agents.

linalool (3,7-dimethyl-1,6-octadien-3-ol), $C_{10}H_{18}O$. B.p. 198–199°C. Present in many essential oils. It is readily converted into geraniol by acid reagents. Used in perfumery.

lindane See gamma BHC.

Linde process A high-pressure process for the production of liquid air, oxygen and nitrogen by compression of air to about 200 bar (20 MN m^{-2}) followed by expansion, refrigeration and fractionation in a double column.

linkage BOND.

linoleic acid (cis,cis,cis-9,12-octadeca-dienoic acid), $C_{18}H_{32}O_2$. A fatty acid which is easily oxidized in air. It occurs widely in the form of glycerides in vegetable oils and in mammalian lipids. Cholesteryl linoleate is an important constituent of blood. The acid also occurs in lecithins. ESSENTIAL FATTY ACID of human diet. Used in paints, coatings, emulsifiers.

linolenic acid (cis,cis,cis-9,12,15-octa-decatrienoic acid), $C_{18}H_{30}O_2$. A fatty acid which occurs in glycerides of linseed amongst other oils. It also occurs widely in lecithin and serum triglycerides. γ-Linolenic acid, cis,cis,cis-6,9,12-octadecatrienoic acid, has been isolated from the seed oil of *Oenothera biennis*. See ESSENTIAL FATTY ACIDS.

linseed oil See LINOLENIC ACID. Oil of vegetable origin used as a drying oil in alkyd resins and in margarine.

linuron Herbicide, urea derivative.

lipases Esterases that act upon the fatty acid esters of glycerol (triglycerides, diglycerides and monoglycerides). Class of hydrolases. See also PHOSPHOLIPASES. Used in detergents, food processing.

lipids These are natural substances of a fat-like nature. The exact definition, however, varies. Strictly they are fatty acids or derivatives which are soluble in organic solvents and insoluble in water, e.g. the simple fats and waxes, and the phospholipids and cerebrosides. Many would also consider such compounds as sterols and squalene to be lipids.

lipophilic groups Non-polar groups in molecules that can cluster together in micelles.

lipoproteins See PROTEINS.

liposomes Amphoteric lipid molecules that can self-assemble in aqueous media into closed spherical structures.

liquefied natural gas (LNG) Natural gas liquefied at atmospheric pressure by cooling to about –160°C, normally for purposes of transportation.

liquefied petroleum gas (LPG) Mixtures of C_3 and C_4 hydrocarbons obtained from petroleum refining operations. Commercially two mixtures are available; *commercial butane* containing about 85% butanes with some propane and pentanes, and *commercial propane* containing about 92% propane along with some ethane and butanes.

These fuels are used for domestic and industrial purposes and are stored and transported as liquids in thick-walled spheres or cylinders. Liquefaction is obtained by compression or refrigeration.

Gross calorific values are high, being about 100 000 kJ/m^3. See CALOR GAS.

liquid crystals On heating a number of substances, e.g. 4-azoxyanisole and ammonium oleate, a cloudy liquid is first formed that changes at higher temperatures to a clear liquid, each transition occurring at a fixed temperature. The cloudy liquid has definite ordered structure and it is called a liquid crystal or, more generally, is said to be in a mesomorphic state; its viscosity, etc., are those of liquids. Substances which form liquid crystals are composed of molecules which possess a high degree of asymmetry – long thin molecules or flat planar ones – which tend to allow little rotation in the liquid state at low temperature. Liquid crystals exhibiting two-dimensional order are said to be in the cholesteric or smectic phases; those showing only one-dimensional order are nematic. Chiral smectics can show ferroelectric effects and be used in electronic applications. Application of a voltage causes disorder, the basis of the extensive use of liquid crystals in liquid crystal displays (LCD). Cholesteric liquids scatter coloured, polarized light and can be made the basis of thermometric units. Spectroscopic studies in the presence of liquid crystals give much additional information over more simple techniques.

liquid–liquid extraction A method of extracting a desired component from a liquid mixture by bringing the solution into contact with a second liquid, the *solvent*, in which the component is also soluble, and which is immiscible with the first liquid or nearly so. Some of the component enters the solvent, forming an *extract*, while the solution that is left, the *raffinate*, is depleted in the component by this amount.

liquid oxygen explosives (LOX explosives) Explosives based on liquid oxygen and a fuel, generally carbon black.

liquid paraffin A highly refined white oil used medicinally as a powerful laxative. See PETROLATUM.

liquids, structure of The structure of a liquid is intermediate between the perfect disorder of molecules in a gas and the highly perfect three-dimensional order in a solid. The immediate environment of an atom or molecule in a liquid is not very different from that in a solid and the density of packing is only slightly lower, but the order is very local and no long-range order occurs. Liquids tend to minimize their surface area.

liquidus curve A plot of freezing points of a molten mixture of substances with composition of the mixture. Such mixtures usually freeze over a range of temperature. If the temperature at which the last traces of liquid just solidify (assuming that sufficient time has been allowed for equilibrium to be established) is plotted against composition the resulting line is called a solidus curve.

litharge, PbO. See LEAD OXIDE.

lithia, Li_2O. See LITHIUM OXIDE.

lithium, Li. At.no. 3, at.mass 6.941, m.p. 180.50°C, b.p. 1342°C, density ρ 534, crustal abundance 20 p.p.m., electronic configuration [He] $2s^1$. The lightest alkali metal. The most important Li minerals are lepidolite, spodumene, petalite and amblygonite (all aluminosilicates). The metal is obtained by electrolysis of the fused chloride. It is a silver-white metal, bcc, ccp and other forms at high pressure, which tarnishes rapidly in air and reacts with water, halogens, nitrogen, hydrogen. Lithium is used in low-melting alloys as a heat transfer medium.

Lithium compounds are used extensively in organic synthesis (organolithiums), in ceramics and glasses (Li_2CO_3), in fluxes (LiF), in lubricating greases (LiOH), in air-conditioning plant (LiCl and LiBr), in fungicides, in foodstuffs, in storage batteries, in Al production and in treatment of depression. U.S. use 1997 3100 tonnes.

lithium, organic derivatives Of great use in industrial and laboratory synthesis. Prepared RCl in benzene or petroleum ether plus Li and also from: R'Li on RH; R'Li plus RBr; Li plus R_2Hg. Solutions, of, e.g. Bu^nLi, are commercially available. Often polymeric, e.g. $(LiMe)_4$. With chelating amines, e.g. $Me_2NCH_2CH_2NMe_2$, form monomeric derivatives which are very active. In addition to synthesis lithium organic derivatives are used in alkene polymerization.

lithium alkyls and aryls (LIR) Formed from Li plus RCl in ether, benzene, etc. Polymeric and mixed species, e.g. $(CH_3Li)_4$, are known. Used in synthesis. See LITHIUM ORGANIC DERIVATIVES.

lithium aluminium hydride (lithium tetrahydroaluminate, LAH), $LiAlH_4$. A specific reducing agent widely used in organic chemistry and as an agent for the formation of aluminium derivatives and hydrides (e.g. of B, Si). See ALUMINIUM HYDRIDE.

lithium batteries Rechargeable batteries using oxides such as $LiCoO_2$ as cathodes and graphite or Li–Sn alloys as anodes. Discharged Li intercalates into the anode. See BATTERIES.

lithium borohydride, $LiBH_4$. Prepared from diborane plus lithium ethyl. Strong specific reducing agent.

lithium carbonate, Li_2CO_3. Sparingly soluble in water (only alkali carbonate to be so); forms soluble $Li(HCO_3)_2$ with excess CO_2, loses CO_2 on strong heating. Used in the manufacture of other lithium compounds.

lithium chemistry Lithium is an alkali metal of Group 1 forming a single series of Li(I) compounds ($E°$ $Li^+ \rightarrow$ Li –3.04 volts in acid). Li is exceptional amongst the alkali metals in forming an insoluble carbonate and fluoride (similar to Be^{2+}). Li compounds are predominantly ionic but covalent and polymeric derivatives, particularly lithium alkyls, are known.

lithium chloride, LiCl. Very hygroscopic (Li_2CO_3 plus HCl) forms hydrates. Used in welding aluminium.

lithium fluoride, LiF. Relatively insoluble in water, precipitated from solution. Forms glasses.

lithium hydride, LiH. Formed from Li plus H_2 at 500°C. Stable crystalline compound, gives LiOH and H_2 with water (may ignite in moist atmosphere). Used as a source of H_2, as a reducing agent and for the preparation of hydrides.

lithium hydroxide, $LiOH,H_2O$. Strong base (Li salt or ore plus $Ca(OH)_2$). Used in greases, storage batteries and absorbing CO_2.

lithium oxide (lithia), Li_2O. White solid (Li plus O_2). Gives LiOH with water.

lithium tetrahydroaluminate, $LiAlH_4$. See LITHIUM ALUMINIUM HYDRIDE.

lithography The printing process involving the use of a photographically or otherwise prepared surface, on which the image areas are oleophilic (non-aqueous, ink receptive) and the non-image areas are hydrophilic (water receptive, ink repellant). The printing plate is generally made of grained Al or Zn, on which is coated the light-sensitive material. Also used for printing circuits on a silicon surface (microlithography).

lithopone (Charlton white, Orr's white) A mixture of $BaSO_4$ (70%) and ZnS (30%) prepared from BaS and $ZnSO_4$ solution and used after calcination as a pigment.

litmus A natural colouring matter obtained from lichens after oxidation in the presence of NH_3. Used as indicator (red acid, blue alkaline) largely in the form of litmus paper.

litre (L) Measure of volume, dm^3.

LNG LIQUEFIED NATURAL GAS.

London interactions (dispersion interactions) The interactions of non-polar molecules from their transient dipoles.

lone pair A pair of electrons in a molecule which is not shared by two of the constituent atoms, i.e. does not take part in the direct bonding. Lone pairs can generally form co-ordinate bonds. They generally influence the stereochemistry of the molecule. Thus CH_4, NH_3 (one lone pair) and OH_2 (two lone pairs) all have structures based on a tetrahedral distribution of bonding and lone pairs about the central atom.

Loschmidt's constant The number of molecules of an ideal gas in unit volume at stp. German usage for AVOGADRO CONSTANT.

low-energy electron diffraction (LEED) A technique used to investigate the surface structure of solids. The surface under investigation is bombarded with electrons of low energy, typically between 6 and 200eV, and the electrons diffracted by the surface atoms are examined. Using this method it is possible to examine both surface structure and rearrangements which may occur following chemisorption and surface reaction.

low-spin state See HIGH-SPIN STATE. The minimum number of unpaired electrons.

low-temperature baths These are used for freezing and distilling gases and low-boiling liquids. Liquid N_2 (–196°C), liquid O_2 (–183°C – very hazardous with any organic material), solid CO_2 (–78.5°C) are used together with various slush baths (liquid plus liquid N_2 or CO_2) –160°C isopentane; –140°C, 30–60 petroleum ether; –130°C pentane; –126°C methylcyclohexane; –116°C diethyl ether; –111°C CCl_3F; –95°C toluene; –60°C $CHCl=CCl_2$; –45°C C_6H_5Cl; –23°C CCl_4.

low-temperature carbonization (LTC) See CARBONIZATION.

LOX explosives See LIQUID OXYGEN EXPLOSIVES.

LPG See LIQUEFIED PETROLEUM GAS, CALOR GAS.

Lr Lawrencium, element 103.

LS coupling RUSSELL–SAUNDERS COUPLING.

LSD Lysergic acid diethylamide. A hallucinogen.

LTC LOW-TEMPERATURE CARBONIZATION.

Lu Lutetium.

lubricant Material used for the purposes of lubrication. Primarily mineral oils derived from crude petroleum, although some vegetable oils are used in 'compounded' oils for special purposes. More recently synthetic fluids have been used, particularly for aviation turbines.

The main synthetic fluids used as special lubricants are esters, polyglycols, silicones, halogenated hydrocarbons and polyphenyl ethers. See LUBRICATING GREASES.

lubricating greases Solid or semi-fluid lubricants made by the incorporation of a thickening agent in a liquid lubricant. This structure gives resistance to shear forces and to temperature.

The base lubricant is usually a petroleum

oil. The thickener usually consists of a soap or soap mixture, soaps of lithium, calcium, sodium, aluminium, etc. In addition grease may contain small amounts of free alkali, free fatty acid, glycerine, anti-oxidant, extreme-pressure agent, graphite or molybdenum disulphide.

Non-soap greases using finely divided solids as thickeners are useful as lubricants at elevated temperatures. Materials used include organoclays such as dimethyldioctyl decylammonium bentonite (Bentone greases) or selected dyestuffs which produce brightly coloured greases.

luciferins Substances responsible for bioluminescence. Enzymatic oxidation is responsible for the characteristic luminescence of the firefly. The oxidation product is formed in an excited state which liberates excess energy as light rather than heat.

lufenuron Insecticide, acaricide, benzylurea derivative.

luminescence See PHOSPHORESCENCE.

luminol (5-aminophthalylhydrazide) Yellow crystals, m.p. 329–332°C. Reaction with some oxidizing agents in the presence of alkali to emit a blue light (chemiluminescence). Used in detection of Cu, Fe and peroxides.

luminous paints Compounded from solid materials (such as calcium sulphide) which, by use of suitable methods of preparation and by addition of traces of heavy metals, can be made to exhibit phosphorescence.

lumisterol Member of vitamin D group.

LUMO See FRONTIER ORBITAL SYMMETRY.

lunar caustic Fused silver nitrate. 97% $AgNO_3$ plus AgCl. Used as a caustic.

Lurgi coal gasification process A process involving the gasification of coal under pressure with mixtures of steam and oxygen. The reactions produce CO and H_2.

lutein (luteol, xanthophyll), $C_{40}H_{56}O_2$. Carotenoid (see CAROTENE). The most common of the xanthophyll pigments, it is present in all green leaves, in blossoms and in various animal sources. It crystallizes in violet prisms with one molecule of methanol; m.p. 193°C, soluble in organic solvents giving yellow solutions. It is related to α-carotene in the same way as zeaxanthin is to β-carotene.

lutetium, Lu. Metal, at.no. 71, at.mass 174.97, m.p. 1663°C, b.p. 3402°C, density ρ 9841, crystal abundance 0.51 p.p.m., electronic configuration [Xe] $4f^{14}5d^16s^2$. The last of the LANTHANIDE(4f) elements. The metal is hcp.

lutetium compounds Lutetium forms a single series of typical LANTHANIDE COMPOUNDS in the +3 state:

$$Lu^{3+} (f^{14} \text{ colourless}) \rightarrow Lu \ (-2.25 \text{ volts in acid}).$$

lutidine, C_7H_9N. The lutidines are dimethylpyridines. The, best known is 2,6-lutidine, which is isolated from the basic fraction of coal tar and synthesized from ethyl acetoacetate, methanol and ammonia. It is an oily liquid, b.p. 144°C.

lycopene, $C_{40}H_{56}$. The red carotenoid pigment of tomatoes, rose hips and many other berries. M.p. 175°C.

lycopodium powder Spores of club moss. Used in coating pills, and in explosives, fireworks and foundry work.

Lyman series See BALMER SERIES.

lyogels See XEROGELS.

lyophilic Colloids which are solvent attracting. See COLLOIDS.

lyophobic Colloids which are solvent repelling. See COLLOIDS.

lyotropic series Although hydrophilic sols are not affected by small concentrations of electrolytes, they may be 'salted out' by the addition of certain salts. Citrates, tartrates and sulphates are very efficient in this connection; iodides and thiocyanates tend to disperse rather than coagulate. Thus silk or cellulose will peptize to colloidal sols in strong thiocyanate solutions but are reprecipitated by the addition of sulphates to the sols. The arrangement of the different anions in order of their salting-out efficiency is termed the lyotropic or Hofmeister series. This also has a connection with the coagulating power of salts on hydrophobic sols.

lysergic acid (9,10-didehydro-6-methylergoline-8-carboxylic acid), $C_{16}H_{16}N_2O_2$. M.p. 238°C (decomp.). A product obtained by hydrolysing the ergot alkaloids. It is reduced with sodium and amyl alcohol to the dihydro compound. The diethylamide is a strong hallucinogen and is a controlled substance.

lysine (α,ε-diaminocaproic acid), $C_6H_{14}N_2O_2$, $H_2N \cdot CH_2 \cdot [CH_2]_3 \cdot CH(NH_2) \cdot COOH$. Colourless needles, m.p. 224°C (decomp.), very soluble in water, insoluble in alcohol. L-(+)-Lysine is one of the basic amino-acids occurring in particularly large quantities in the protamine and histone classes of proteins. It is an essential amino-acid, which cannot be synthesized by the body and must be present in food for proper growth. It can be manufactured by various fermentation processes or by synthesis. Food additive.

lysol A soapy solution containing 50% isomeric cresols. Used as a disinfectant.

lysosomes Particles occurring in cells that contain hydrolytic enzymes. The sites of diges-tion in cells. Damage to the lysosomal mem-brane can lead to release of those enzymes which attack the rest of the cell, as occurs in dy-ing or damaged tissue.

lysozyme An enzyme which occurs especially in nasal mucosa, egg white, plant latex and var-ious animal tissues. It has the ability to rupture the walls of bacteria.

D-lyxo-hexulose (D(-)tagatose) A hexose sugar.

M

macrolides Compounds with antibiotic activity made by *Streptomyces* spp. and characterized by having a macrocyclic ring and often containing sugars. The macrolides, which include ERYTHROMYCIN, are most active against gram-positive bacteria.

macromolecule Molecules of high molecular weight (usually greater than 10 000). Examples are proteins and some polymers. The term macromolecular is also used to describe structures such as that of diamond, which may be regarded as being one large molecule.

Maddrell salt, $NaPO_3$. A long-chain phosphate obtained by heating $Na_2H_2P_4O_7$ to 230–300°C. See PHOSPHORUS OXYACIDS.

Madelung constant See BORN–LANDÉ EQUATION.

magic acid See ANTIMONY FLUORIDES.

magic angle spinning (MAS) Nuclear magnetic resonance technique for solids.

magnesia Magnesium oxide.

magnesite, $MgCO_3$. A mineral form of magnesium carbonate used as a source of magnesium compounds and in the manufacture of refractories.

magnesium, Mg. Metal, at.no. 12, at.mass 24.305, m.p. 650°C, b.p. 1090°C, density ρ 1738, crustal abundance 23000 p.p.m., electronic configuration [Ne] $3s^2$. Alkaline earth metal of Group 2. The principal useful ores are dolomite $(Ca,Mg)CO_3$, carnallite $(KMgCl_3.6H_2O)$, kainite $(MgSO_4.3Na_2SO_4)$, magnesite $(MgCO_3)$, schönite $(K_2Mg(SO_4)_2.6H_2O)$. Obtained from one of the above ores or from sea water (see DOW PROCESS) or brines (precipitated as $Mg(OH)_2$). The metal is obtained by electrolysis of an $MgCl_2$-halide melt or by reduction of fused dolomite with ferrosilicon. The metal is silverwhite, very light, hcp. Burns in air after ignition, reacts slowly with moist air to become covered with a film of oxide. Essential element. Used in alloys (43%) and castings (20%), for deoxidizing and desulphurizing metals, in batteries, sacrificial corrosion and flash bulbs. Used in glass, ceramics, fillers $(MgCO_3, MgCl_2)$, in flocculating agents, catalysts, medicine, sugar refining $(Mg(OH)_2)$, refractories, cement, paper manufacture (MgO and mixed magnesium oxides), in tanning, sizing, etc. $(MgSO_4)$.

magnesium, organic derivatives, RMgX, R_2Mg. See GRIGNARD REAGENTS.

magnesium alloys Used on account of their light weight (e.g. aircraft industry) and for casting, forging, extrusion, etc.

magnesium carbonate, $MgCO_3$. Occurs naturally as magnesite and in dolomite. Addition of CO_3^{2-} to Mg^{2+} solutions gives basic carbonates which dissolve in excess CO_2 to give $Mg(HCO_3)_2$. Heating the solution to 50°C gives crystals of $MgCO_3.3H_2O$; the 5 and 1 hydrates are also known. Basic carbonates of approximate composition $3MgCO_3.Mg(OH)_2.4H_2O$ are used medicinally as antacids and mild laxatives.

magnesium chemistry Magnesium is an alkaline earth element of Group 2. It shows a single +2 oxidation state ($E°$ $Mg^{2+} \rightarrow Mg$ –2.38 volts in acid solution). Compounds are largely ionic although there is major covalent character in Grignard reagents and in some other derivatives. Mg compounds generally have 6-coordinate Mg; $[Mg(H_2O)_6]^{2+}$ is not hydrolysed. Many complexes, particularly with O- and N-ligands, are known. Mg complexes are of great importance in photosynthesis in plants (chlorophyll contains a MgN_4 group but there is additional interaction with other O atoms).

magnesium chloride, $MgCl_2$. Obtained from carnallite, $KCl.MgCl_2.6H_2O$ by fusion when the KCl is precipitated; cooling gives carnallite and $MgCl_2.6H_2O$. Dehydrated by heating in HCl gas. $MgCl_2$ has m.p. 708°C, b.p. 1412°C. The hygroscopic hexahydrate is used in moistening cotton threads in spinning and for fireproofing. Oxide chlorides (heat on hydrates) are used in cements. Fused $MgCl_2$ is electrolysed commercially to give Mg and Cl_2. $MgCl_2$ forms complexes including anionic chlorocomplexes, e.g. $(Et_4N)_2[MgCl_4]$.

magnesium halides Magnesium fluoride, MgF_2, is precipitated from aqueous solution. It has the rutile structure. Perovskites, e.g. $KMgF_3$, are known. The other halides are similar to the CHLORIDE although complexes are not known. They have layer lattices. Used in ceramics and glass.

magnesium hydroxide, $Mg(OH)_2$. Formed by hydration of MgO or precipitation from an Mg^{2+} solution with OH^-. Only slightly soluble in water to give an alkaline solution. Occurs naturally as brucite. $Mg(OH)_2$ layers occur in silicates – e.g. the chondrodites.

magnesium nitrate, $Mg(NO_3)_2$. Stable salt. Forms 9, 6 and 2 hydrates. Hydrates dehydrate to basic salts but $Mg(NO_3)_2$ can be obtained by heating hydrates in nitric acid vapour.

magnesium nitride, Mg_3N_2. Formed from elements above 300°C (used to remove N_2). Hydrolysed to NH_3 and $Mg(OH)_2$. The amide is used as a polymerization catalyst.

magnesium oxide (magnesia), MgO. Occurs naturally as periclase. Prepared (m.p. 2640°C) by combustion of Mg in O_2 or ignition of $Mg(OH)_2$, $MgCO_3$ or $Mg(NO_3)_2$. Has NaCl structure. Hydrated (rate depends on whether previously ignited) to $Mg(OH)_2$. Used as a refractory, for heat insulation and as an antacid. Forms a peroxide which can give H_2O_2.

magnesium perchlorate, $Mg(ClO_4)_2 \cdot 6H_2O$. Formed from solution. The anhydrous material (anhydrone) is used for drying gases.

magnesium phosphates, $Mg_3(PO_4)_2 \cdot 8H_2O$. Precipitated from aqueous solutions. $MgNH_4PO_4$ is precipitated from solutions containing ammonium salts; used in estimation of Mg^{2+} and PO_4^{3-}.

magnesium silicates Many silicates contain Mg^{2+}. Used medicinally to reduce stomach acidity and (after hydrolysis to active SiO_2) to remove toxins.

magnesium sulphate, $MgSO_4$. Obtained (with $7H_2O$ – Epsom salts) from $MgCO_3$ and H_2SO_4; occurs naturally as kieserite or reichardite, $MgSO_4 \cdot H_2O$. Hydrates give $MgSO_4$ at 200°C. Epsom salts are used as a purgative, as a dressing for cotton goods and in dyeing.

magneson (4[4-nitrophenylazo]-1,3-benzenediol), $C_{12}H_9N_3O_4$. A brownish-red powder, soluble in sodium hydroxide. Used for the detection and estimation of magnesium with which it forms a blue lake in alkaline solutions.

magnet Body possessing magnetism, either temporary or permanent.

magnetic moment Property of a particle arising from its spin. For paramagnetic substances the magnetic moment (μ) is related to the magnetic susceptibility (χ) by the equation

$$\mu = 2.84\sqrt{\chi_m(T - \theta)}$$

where χ_m is the molar susceptibility, T the absolute temperature and θ is a constant for the particular substance (called the Curie temperature). Measurements are of importance in investigation of electronic configurations.

magnetic optical rotatory dispersion See MAGNETIC POLARIZATION OF LIGHT.

magnetic polarization of light When a beam of plane polarized light (radiation) is passed through a transparent medium which is placed in a magnetic field, the plane of polarization is rotated. This phenomenon is known as magnetic rotation, magnetic optical rotatory dispersion, the Faraday effect. The beam must be travelling in the same direction as the magnetic lines of force. Magnetic rotation is not connected with normal optical activity. The effect is shown by transparent substances whether or not they possess a centre of asymmetry. See DICTIONARY OF PHYSICS.

magnetic quantum number See ELECTRONIC CONFIGURATION.

magnetic resonance See NUCLEAR MAGNETIC RESONANCE, ELECTRON SPIN RESONANCE.

magnetic separation Separation techniques involving magnetic properties. Materials, e.g. minerals, can be separated into ferromagnetic, magnetic, diamagnetic and nonmagnetic fractions. Separation is generally carried out on crushed material of small particle size.

magnetic susceptibility Atoms or molecules may have a permanent magnetic moment or an induced moment due to the influence of an electric magnetic field. In general, the extent of magnetization of any substance is a function of the field in which that substance is placed. The magnetic susceptibility per unit volume, κ, is defined as the intensity of magnetization induced (I) divided by the field strength (H):

$$\kappa = I/H.$$

The molar magnetic susceptibility, χ_m, is given by the expression:

$$\chi_m = \frac{M}{\rho} \cdot \kappa$$

where M is the molecular mass and ρ the density of the substance. Magnetic susceptibilities are usually determined using a Gouy balance or a SQUID but nuclear magnetic resonance methods may also be used. For a paramagnetic substance $1/\chi$ is proportional to the absolute temperature (Curie–Weiss law). See MAGNETIC MOMENT and DICTIONARY OF PHYSICS.

magnetic tape Used for recording information (sound, video, computer output, instrumental output). Consists of magnetic particles (generally γ-Fe_2O_3 but other materials, e.g. CrO_2, also used). Information is imprinted via electrical signals converted to magnetic fields, which in turn affect the magnetization of the tape particles.

magnetite, Fe_3O_4. The mineral form of black Fe_3O_4 with the spinel structure. Used as an iron ore, a flux, a pigment for glazes and occasionally as a refractory material for lining furnaces for producing iron.

magnetogyric ratio (γ) The proportionality factor between orbital angular momentum and magnetic moment.

magneton (Bohr magneton) (μ) The fundamental quantum of magnetic moment.

Magnus's green salt, $[Pt(NH_3)_4][PtCl_4]$.

mahogany acids The sulphonic acids remaining in the oil phase following the refining of lubricating oil by oleum or sulphuric acid. See GREEN ACIDS.

maize oil See CORN OIL.

malachite, $Cu_2(OH)_2CO_3$. A native green hydrated carbonate of copper used as a decorative stone as it takes a high polish. Produced artificially (green verditer) from Na_2CO_3 and a solution of $CuSO_4$ and used as an artist's colour.

malathion Insecticide, acaricide, organophosphate derivative.

maleamic acid, $C_4H_5NO_3$, $H_2NC(O)CH=CHC(O)OH$ (or cyclized). The half amide of maleic acid; m.p. 172–173°C (decomp.). Its ammonium salt, which is made by treating maleic anhydride with ammonia, is used to make KRILIUM.

maleic acid (z-butenedioic acid), $C_4H_4O_4$. Colourless prisms; m.p. 130°C. Manufactured by treating maleic anhydride with water. It is converted to the anhydride by heating at 140°C. By prolonged heating at 150°C or by heating with water under pressure at 200°C, it is converted to the isomeric (*trans*) fumaric acid. Reduced by hydrogen to succinic acid. Oxidized by alkaline solutions of potassium permanganate to mesotartaric acid. When heated with solutions of sodium hydroxide at 100°C, sodium(±)-malate is formed. Used in the preparation of (±)-malic acid and in polymers.

maleic anhydride, $C_4H_2O_3$. Colourless; m.p. 53°C, b.p. 200°C. Manufactured by the oxidation

$$\begin{array}{c}HC{-}CO \\ \parallel \qquad \quad O \\ HC{-}CO\end{array}$$

of benzene by air at 400–450°C over a vanadium pentoxide catalyst; furfural, crotonaldehyde and butenes may also be used as starting materials. Reacts with hot water to give maleic acid. Reduced by hydrogen in presence of catalysts to succinic anhydride. Reacts as a Diels–Alder dienophile. Forms resinous substances with terpenes. Used in the production of polyester resins and copolymers. It is also employed in the manufacture of alkyd resins for varnishes, drying oils, agricultural chemicals and fumaric acid and to retard rancidity in oils and fats.

maleic hydrazide Plant growth regulator.

maleimide (2,5-pyrroledione), $C_4H_3NO_2$. Forms spin-labels with free radicals.

malic acid (L-hydroxysuccinic acid), $C_4H_6O_5$. $HO(O)CCH_2 \cdot CH(OH)C(O)OH$. (L-)-Malic acid crystallizes in colourless needles; m.p. 100°C. It occurs in many acid fruits, such as grapes, apples and gooseberries. It can be prepared by microbiological processes using various moulds or from (+)-bromosuccinic acid by the action of NaOH. Used as a chemical intermediate and in perfumery and flavouring.

malleability The property which enables a body – particularly a metal – to be extended in all directions by hammering or rolling. The degree of malleability is gauged by the thinness of leaf or foil which it is possible to produce.

malonic acid (propanedioic acid), $C_3H_4O_4$, $CH_2(C(O)OH)_2$. Colourless crystals; m.p. 136°C. Occurs in the mixed calcium salts obtained during the processing of sugar beet. Prepared from sodium cyanoethanoate with NaOH. Decomposes above 140°C to give ethanoic acid. Reacts with aldehydes in presence of primary and secondary bases to give unsaturated substituted acids; some of these very readily lose CO_2 to give $\alpha\beta$-unsaturated fatty acids. Its esters are used in organic syntheses.

malonic ester See DIETHYL MALONATE.

malononitrile, $CH_2(CN)_2$. Prepared from NaCN and sodium chloroethanoate, b.p. 223°C, polymerizes above 120°C. Used to prepare 2-chlorobenzal malononitrile, $ClC_6H_4C=C(CN)_2$, CS gas – a riot-control gas. Also used in the synthesis of vitamin B, and as a polar-additive for lubricating oils.

malt Cereal grains which have been allowed to sprout and then are heated. The sprouting activates the enzyme systems. Barley, wheat, rye, oats, sorghum, corn and rice are converted to malt. Malt is used in brewing, in food additives and as animal feed.

maltase The enzyme found in malt that is specific in splitting maltose into two molecules of glucose. Also found in the digestive systems of animals and in yeast.

maltose, $C_{12}H_{22}O_{11}$. A disaccharide which is present free in small quantities in barley grains and some other plants but is more commonly produced by the action of amylase on starch or glycogen. It is 4-[α-D-glucopyranosido]-D-glucopyranose; the α-form crystallizes as the monohydrate in colourless needles; m.p. 102–103°C. Used in brewing, soft drinks and foods.

mancopper Fungicide; thiocarbamate of Mn and Cu.

mancozeb Fungicide; dithiocarbamate of Mn and Zn.

mandelic acid (l-hydroxy(1-phenyl) ethanoic acid), $C_8H_8O_3$. Colourless prisms; (±) m.p. 118°C, (+) or (–) m.p. 133°C. Occurs combined in the glucoside, amygdalin. Prepared by hydrolysis of mandelonitrile (benzaldehyde cyanohydrin). It is administered in large doses in the treatment of urinary infections.

mandelonitrile See MANDELIC ACID, AMYGDALIN.

maneb Fungicide, dithiocarbamate of Mn.

manganates Double oxides containing manganese.
Manganates(VII), $[MnO_4]^-$. Permanganates. Dark purple tetrahedral anion (electrolytic oxidation of $[MnO_4]^{2-}$). Powerful oxidizing agent. $E°$ $(MnO_4)^-$ to MnO_2 in basic solution +1.68 volts; $E°$ $[MnO_4]^-$ to Mn^{2+} in acid solution +1.49 volts; used in volumetric analysis. See PERMANGANATE TITRATIONS. The free acids $HMnO_4$ and $HMnO_4.2H_2O$ are obtained at –75°C with $(Ba(MnO_4)_2$ plus $H_2SO_4)$, both are violent oxidizing agents; these are the acids of the oxide Mn_2O_7. Used in bleaching, washing, photography, tanning.
Manganates(VI), $[MnO_4]^{2-}$. Deep green tetrahedral ion (MnO_2 plus fused KOH with KNO_3, air, etc.). Only stable in basic solution.
Manganates(V) (hypomanganates), $[MnO_4]^{3-}$. Deep blue ion (MnO_2 in conc. KOH; $[MnO_4]^-$ plus excess $[SO_3]^{2-}$). The salts are rapidly hydrolysed.
Manganates(IV) (manganites) Mixed-metal oxides containing Mn(IV). Prepared by solid-state reactions.
Manganates(III) Mixed-metal oxides; present in the spinel Mn_3O_4, $Mn^{II}Mn^{III}_2O_4$.

manganese, Mn. Transition metal, at.no. 25, at.mass 54.938, m.p. 1246°C, b.p. 2061°C, density ρ 7200–7400, crustal abundance 950 p.p.m., electronic configuration [Ar] $3d^54s^2$. A transition element of Group 7. The principal ore is pyrolusite (MnO_2) but higher oxides (Mn_2O_3, Mn_3O_4) and the carbonate are also known. Occurs in nodules on the seabed. After roasting, manganese ores are reduced with Al or C; pure Mn is formed by electrolytic reduction. Mn metal is a soft grey metal with various low-temperature forms which do not correspond to those adopted by other metals. Mn is a reactive metal combining with oxygen on heating in air, reacts slowly with water, combines with halogens, N_2, P, Si, S and C on heating. Essential element. The major use of Mn is in steel (98%) as a deoxidizing and desulphurizing agent. Also used in alloys. Manganese compounds are used in dyes, paints, batteries (MnO_2), chemical processes, fertilizers, herbicides, fungicides. World production of Mn metal 1998 7 million tonnes.

manganese, organic derivatives Simple manganese alkyls are not very stable but carbonyl derivatives, e.g. $CH_3Mn(CO)_5$ (CH_3I plus $NaMn(CO)_5$), are well known. Cyclopentadienyl derivatives, e.g. $[(h^5-C_5H_5)Mn(CO)_3]$ and $[(h^5-C_5H_5)_2Mn]$, and π-arene complexes, e.g. $[(h^6-C_6H_6)Mn(CO)_3]^+$, are stable.

manganese alums Derivatives of manganese(III) sulphate prepared by electrolytic oxidation of Mn(II). See ALUMS.

manganese borates Rather indefinite materials precipitated from Mn^{2+} solutions by borates and used as driers for linseed oil (catalysts for oxidation of the oil).

manganese bromide, $MnBr_2$. Forms hydrates (anhydrous by heating hydrates in stream of HBr gas). Very soluble in water.

manganese carbonate, $MnCO_3$. Occurs naturally as the mineral rhodochrosite, precipitated from Mn^{2+} solutions by use of $NaHCO_3$ solution saturated with CO_2. Normally precipitated as basic carbonates. Rapidly oxidized in air. Pigment.

manganese carbonyl (dimanganese decacarbonyl), $Mn_2(CO)_{10}$. Formed by action of CO under pressure on a Mn(II) salt in the presence of a strong reducing agent. Golden-yellow crystals, structure $(OC)_5Mn–Mn(CO)_5$. Parent compound of series of manganese carbonyl derivatives, e.g. $Na[Mn^{-1}(CO)_5]$ ($Mn_2(CO)_{10}$ plus Na); $[Mn(CO)_5Cl]$ ($Mn_2(CO)_{10}$ plus Cl_2); $[(h^5-C_5H_5)Mn(CO)_3]$ (NaC_5H_5 plus $Mn_2(CO)_{10}$). Forms catalysts and anti-knock agents.

manganese chemistry Manganese is an electropositive transition element of Group 7. It shows a great range of oxidation states. Mn(II) shows a range of cationic chemistry

$$E° \; Mn^{2+} \rightarrow Mn \; -1.03 \text{ volts in acid}$$

but in basic solution hydroxide or oxide species are formed and the tendency to oxo species increases with the higher oxidation states. Mn(II) and Mn(III) form extensive ranges of complexes. Mn(II) compounds are predominantly very pale pink in colour; most are spin-free (except cyanides). Manganese(II) alkyls and aryls are known. Manganese(III) compounds (brown) show Jahn–Teller distortion

$$(E° \; Mn^{3+} \rightarrow Mn^{2+} \text{ in aq. sol. } +1.51 \text{ volts}).$$

Manganese(IV) is confined to MnO_2, MnF_4 and some complexes. Mn(IV) is strongly oxidizing as are manganese(V), (VI) and (VIII) – see MAN-

GANATES. These compounds generally have tetrahedral co-ordination. Lower oxidation state compounds are formed with soft ligands. Cyanide forms complexes with Mn(III), Mn(II). Mn(I) ($K_5Mn(CN)_6$ formed by reduction of $K_4Mn(CN)_6$) and Mn(0); ($K_6Mn(CN)_6$ formed by reduction of $K_5Mn(CN)_6$ with K in liquid NH_3) and CO in $Mn^0{}_2(CO)_{10}$ and $[Mn^{-1}(CO)_5]^-$. There is some metal–metal bonding but very little in comparison with Re. Manganese compounds are of importance in photosynthesis and enzyme systems.

manganese chlorides The only stable chloride is manganese(II) chloride, $MnCl_2$. Forms hydrates. Forms range of complex chlorides, e.g. $KMnCl_3$ (perovskite), M_2MnCl_4 (tetrahedral – green, octahedral – pink), and K_4MnCl_6. Manganese(III) chloride, $MnCl_3$, is black and decomposes at –40°C (Mn(III) ethanoate plus HCl at –100°C). Red–brown complexes, e.g. K_2MnCl_5 (MnO_2 in HCl plus KCl) are known. Chloromanganates(IV), e.g. Rb_2MnCl_6, are deep red compounds (MnO_2 in cold HCl plus MCl). Very unstable manganese(VII) and (VI) oxide chlorides, MnO_3Cl, $MnOCl_3$ (green) and MnO_2Cl_2 (brown) are formed from Mn_2O_7 and chlorosulphuric acid.

manganese cyanides Simple cyanides are not known but complex cyanides of manganese(III) (orange $[Mn(CN)_6]^{3-}$); manganese(II) (violet or yellow, e.g. $[Mn(CN)_6]^{4-}$); manganese(I) ($[Mn(CN)_6]^{5-}$); and manganese(0) ($[Mn(CN)_6]^{6-}$) are formed by the appropriate oxidation or reduction in the presence of cyanide. These are all spin-paired compounds.

manganese ethanoates Manganese(II) ethanoate, $Mn(O_2CCH_3)_2.4H_2O$, pink crystals. Manganese(III) ethanoate, $[Mn_3O(O_2CCH_3)_6]^{+-}$ $[O_2CCH_3]^-$, red–brown (Mn(II) ethanoate plus $KMnO_4$ in CH_3CO_2H). Useful starting point for manganese(III) compounds; oxidizing agent and catalyst for decarboxylation. Used as mordants and paint driers.

manganese fluorides

Manganese(II) fluoride, MnF_2. Forms hydrates (MnF_2 obtained by heating NH_4MnF_3; tetrahydrate from aqueous solution). Solutions are hydrolysed. Insoluble perovskites, e.g. $KMnF_3$, formed from solution.

Manganese(III) fluoride, MnF_3. Formed from F_2 on, e.g. $MnCl_2$. Purple–red; forms complexes, e.g. $K_2MnF_5.H_2O$; K_3MnF_6, hydrolysed by water. Used as fluorinating agent.

Blue *manganese(IV) fluoride*, MnF_4 (Mn plus F_2). Is immediately hydrolysed by water;

complex hexafluoromanganates containing $[MnF_6]^{2-}$ ions are yellow (electrolytic oxidation of lower fluorides or use of BrF_3).

Green *permanganyl fluoride*, MnO_3F. Is an explosive liquid ($KMnO_4$ plus HSO_3F).

manganese hydroxides Relatively indefinite compounds formed by precipitation. $Mn(OH)_2$ occurs as pyrochroite and has a layer lattice. Readily oxidized in air. $MnO \cdot OH$ occurs as groutite and manganite with two forms as $AlO \cdot OH$ and $FeO \cdot OH$; an important source of manganese(III) compounds. Hydrated MANGANESE(IV) OXIDES are known but are not simple hydroxides.

manganese iodide, MnI_2. Forms hydrates.

manganese nitrate, $Mn(NO_3)_2$. Formed anhydrous from Mn plus N_2O_4. Forms hydrates. Decomposes on heating to MnO_2.

manganese oxides The lower oxides are similar to those of iron; non-stoichiometric phases are formed. The higher oxides form MANGANATES. MnO is stable up to $MnO_{1.13}$ and has the sodium chloride structure. Grey or green, formed by reduction of higher oxides or heating $Mn(OH)_2$ or $MnCO_3$ in the absence of air. Mn_3O_4 is the stable oxide in air above 940°C. It is black with a distorted spinel structure. It occurs naturally as haussmannite. Mn_2O_3 has two forms (one related to the spinel structure of Mn_3O_4). Occurs naturally as braunite; formed by heating MnO_2 in air at 550–900°C. There are related hydroxy compounds, $MnO \cdot OH$ (see MANGANESE HYDROXIDES). Mn_5O_8 is $Mn_2^{II}Mn_3^{IV}O_8$. MnO_2 occurs as pyrolusite, polianite and ramsdellite. The structural chemistry is extremely complex with MnO_6 octahedra sharing corners and some Mn^{4+} ions replaced by Mn^{2+} or related low oxidation state ions. Formed fairly pure by heating $Mn(NO_3)_2$. Powerful oxidizing agent (organic materials and Cl_2: Weldon process) and catalyst (e.g. decomposition of $KClO_3$). Used in dry cells, glass (to give amethyst tint). Mn_2O_7, oily molecular oxide formed from $KMnO_4$ and concentrated H_2SO_4. Very strong oxidizing agent; can explode.

manganese phosphates Manganese(II) phosphates are formed from aqueous solution – often by precipitation – e.g. $Mn_3(PO_4)_2.7H_2O$, $NH_4MnPO_4.H_2O$ (latter used for estimation of Mn, weighed as $Mn_2P_2O_7$ after ignition).

manganese sulphates
Manganese(II) sulphate, $MnSO_4$, forms hydrates. Used in dyeing and glazing. Manganese(III) sulphate, $Mn_2(SO_4)_3$, formed from MnO_2 and H_2SO_4 or by electrolytic oxidation of $MnSO_4$ solution. Green mass, forms violet aq. sol., hydrolysed by excess water. Forms alums and certainly contains sulphato complexes. Manganese(IV) sulphate, $Mn(SO_4)_2$, black crystals ($MnSO_4$ in H_2SO_4 plus $KMnO_4$) readily hydrolysed.

manganic salts Generally manganese(III) salts.

manganites Generally applied to MANGANATES(IV).

manganous salts Manganese(II) derivatives.

mannans Polysaccharides made up of a chain of mannose units. Occur in some cell walls.

Mannich reaction The replacement of active hydrogen atoms in organic compounds (e.g. a hydrogen of a ketone or a hydrogen on the aromatic ring of a phenol) by aminomethyl or substituted aminomethyl groups. Process is also called aminomethylation, and may be accomplished by reacting 37% methanal with the desired amine, e.g. $(CH_3)_2NH$, and the compound RH with active hydrogen atom(s).

$$(CH_3)_2NH + CH_2O + RH \rightarrow$$
$$R \cdot CH_2 . N(CH_3)_2 + H_2O$$

D-mannitol (mannite), $C_6H_{14}O_6$. Mannitol,

$$CH_2(OH) \begin{matrix} H & H & OHOH \\ C & C & C & C \\ OH & OH & H & H \end{matrix} CH_2(OH)$$

the alcohol corresponding to mannose, is widely distributed in plants and fungi. White crystalline powder with a sweet taste; m.p. 168°C. It can be made from glucose by electrolytic reduction or isolated from natural sources, e.g. seaweed or manna ash. Used (often with boric acid) as dry electrolyte, also in foods, lubricants.

D-mannose, $C_6H_{12}O_6$. Obtained by hydrolysis of mannans. Oxidized ring form of mannitol.

mannuronic acid See ALGIN.

manometers Instruments used to measure pressure differences between two points (one may be the atmosphere) in a liquid or a gas. In its simplest form consists of an upright transparent U-tube partially filled with liquid (often Hg); the difference in levels gives the pressure difference directly. Mechanical manometers (using the deflection of a diaphragm) and other devices are more costly but more precise methods of pressure measurement.

manufactured gas A general term for fuel gas obtained by a conversion process from solid or liquid hydrocarbons. See TOWN GAS.

marble, $CaCO_3$. A dense form of CALCIUM CARBONATE often used decoratively.

margaric acid (heptadecanoic acid), $C_{17}H_{34}O_2$.

margarine Manufactured food product used as spread. Produced from vegetable oils; contains some polyunsaturated fats. About 20% is in aqueous phase (milk or water) together with food additives, vitamins and colour.

Markovnikov's rule In the addition of hydrogen halides to an ethylenic double bond, the halogen attaches itself to the carbon atom united to the smaller number of hydrogen atoms. This rule may generally be relied on to predict the major product of such an addition and may be easily understood by considering the relative stabilities of the alternative carbenium ions produced by protonation of the alkene; in some cases some of the alternative compound is formed. The rule usually breaks down for hydrogen bromide addition reactions if traces of peroxides are present (anti-Markovnikov addition).

marsh gas METHANE. Almost obsolete name.

Marsh's test for arsenic The arsenic-containing specimen is converted to volatile AsH_3 which is decomposed to a brown stain on heating. Estimation is by comparison of stains. Sb reacts similarly but the Sb stain is not soluble in NaOCl.

martensitic transformation Transformations involving lattice shear and no diffusion.

maser A device which involves the laser principle but produces microwave radiation.

masking An analytical technique involving addition of a masking agent which stops interfering species participating in a reaction. Demasking is the reverse process, e.g. breakdown of a cyanide complex with acid.

mass action, law of The rate at which a substance reacts is proportional to its active mass (concentration), and hence the velocity of a chemical reaction is proportional to the products of the concentrations of the reactants. Thus, in the reaction

$$A + B \rightleftharpoons C + D$$

the velocity of the forward reaction is

$$V_f = k_1[A][B]$$

where k_1 is a constant, and the square brackets denote activities (concentrations). Similarly, for the backward reaction,

$$V_b = k_2[C][D]$$

At equilibrium, $V_f = V_b$

or $$k_1[A][B] = k_2[C][D]$$

hence $$\frac{k_1}{k_2} = \frac{[C][D]}{[A][B]} = K$$

The constants k_1 and k_2 are called velocity constants, and K is called the equilibrium constant.

mass defect See PACKING FRACTION.

massicot, PbO. See LEAD OXIDES.

mass number The total number of major particles (protons and neutrons) in the nucleus.

mass spectrograph See MASS SPECTROMETER.

mass spectrometer A device in which molecules are ionized and split and the accelerated ions are separated according to their mass-to-charge ratio. In the mass spectrometer the relative abundance of isotopes or other ionized species is determined by measuring the positive or negative ion currents arriving at a fixed focus, with various controlled magnetic fields and accelerating potentials. Resolution may be to 1 part in 50 000 and accuracy to better than 1 part in 10^6. Used in identification, to determine molecular weight and to investigate breakdown.

mass spectrum The results obtained from a mass spectrometer are called mass spectra. They consist of ions of varying intensity at different (m/e) ratios and are recorded graphically or as computer output. For elements the individual lines correspond to different isotopes. For compounds individual lines may be due to the presence of isotopic species, fragmentation of the molecule or ionization of the parent species in the mass spectrometer.

masurium Obsolete name for TECHNETIUM.

match Practical method of producing flame. The usual method of flame production is a friction-initiated reaction between potassium chlorate and red phosphorus. The active materials are in a glue matrix and a glass powder is often used in the striker. The match may be wood or cardboard.

matlockite, PbFCl. See LEAD CHLORIDES.

matrix isolation The formation and study of unstable species by reaction between two or more substrates, generally in the atomic or small molecular form, often produced in a KNUDSEN CELL or, e.g., in a discharge and in which the reactive species are condensed out on a cold surface in the presence of a great excess of inert material (generally argon). Species such as metal clusters, actinide carbonyls, (M plus CO), PdC_6H_5 and uncomplexed $MgCl_2$ have been studied. Characterization is generally by spectroscopic methods.

matte The indefinite mixture of artificial sulphides produced by a smelting operation.

mauveine The first synthetic dye, synthesized in 1856.

maximum co-ordination number See CO-ORDINATION NUMBERS.

maximum covalency See CO-ORDINATION NUMBERS.

maximum multiplicity, law of See HUND'S RULES.

MB599 Insecticide, synergist, phenylbut-2-ynyl ether.

MCPA Herbicide, aryloxyalkanoic acid derivative. Also MCPA-ethyl.

MCPB Herbicide, aryloxyethanoic acid derivative.

Md Mendelevium. Element 101.

MDI See METHYLENE DI(4-PHENYL)ISOCYANATE.

MDMA See ECSTASY. Drug of abuse.

Me Methyl, CH_3-.

mean free path The average distance traversed by a molecule between collisions with another molecule. At stp the mean free path for hydrogen molecules is 70 nm.

mecarbam Insecticide, organophosphate derivative.

mechanism of reaction The manner in which a reaction occurs. It is expressed in terms of a series of chemical equations. Derived from observation and hence only a hypothesis. Thus the mechanism of the reaction of hydrogen peroxide with hydriodic acid may be written as

$$HI + H_2O_2 \rightarrow H_2O + HOI$$
$$HOI + HI \rightarrow H_2O + I_2$$

mecoprop Herbicide, aryloxyalkanoic acid derivative.

medfluidide Herbicide, sulphonamide derivative.

medicinal paraffin oil See LIQUID PARAFFIN.

medicinal white oil See LIQUID PARAFFIN, PETROLATUM.

medronic acid See METHYLENE DIPHOSPHORIC ACID.

meitnerium, Mt. Element, atomic number 109.

MEK Methyl ethyl ketone. See BUTANONE.

melamine (cyanuramide,2,4,6-triamino-1,3,5-triazine), $C_3H_6N_6$. Colourless; m.p. 354°C. It is manufactured by heating dicyandiamide, $H_2N\cdot C(NH)\cdot NH\cdot CN$, either alone or in the presence of ammonia or other alkalis, in various solvents. Important material in the plastics industry. Condensed with methanal and other substances it gives thermosetting resins that are remarkably stable to heat and light.

melamine resins See MELAMINE.

melanins The brown pigments of hair, skin and eyes. It is a polymer arising by oxidation of tyrosine by the action of tyrosinase.

melatonin (N-acetyl-5-methoxytryptamine) Hormone from pineal gland.

melissic acid (triacontanoic acid), $C_{30}H_{60}O_2$, $CH_3\cdot[CH_2]_{28}\cdot C(O)OH$. A fatty acid occurring in beeswax, m.p. 94°C.

mellisyl alcohol (myricyl alcohol, triacontanol), $C_{30}H_{62}O$, $CH_3\cdot[CH_2]_{28}\cdot CH_2OH$. Colourless crystals; m.p. 87°C, soluble in organic solvents. Present as melissyl palmitate in beeswax. Plant growth regulator.

mellitic acid (benzenehexacarboxylic acid), $C_{12}H_6O_{12}$. Colourless needles; m.p. 286–288°C. When heated it decomposes into pyromellitic anhydride, water and CO_2.

Occurs as the aluminium salt (honeystone) in some lignite beds. Prepared by oxidation of charcoal with concentrated nitric acid.

It condenses with resorcinol and aminophenols to give phthalein and rhodamine dyestuffs respectively. Esters are used in the formation of POLYIMIDES.

melting point The temperature at which a solid changes into a liquid at a specified pressure. Pure solids usually have sharp well-defined melting points. Mixtures usually melt over a range of temperatures and have, therefore, ill-defined melting points.

membrane Porous materials which allow passage of solvent but impede passage of solids and larger molecules. Generally made of polymers. Used for purification cells, osmosis and separation.

membrane cell An electrolytic cell for the production of NaOH and Cl_2 from brine in which the anode and cathode are separated by a membrane. See CASTNER–KELLNER CELL, MERCURY CELL, DIAPHRAGM CELL.

membrane equilibrium See DONNAN MEMBRANE EQUILIBRIUM.

membrane hydrolysis When a colloidal electrolyte consisting, e.g. of giant anions and normal cations, is separated by a membrane from pure water, the cations are able to diffuse out but the colloidal anions are retained. For each positive ion which penetrates the membrane a molecule of water will dissociate to provide a compensating OH^- ion, the corresponding H^+ ion tending to diffuse through the membrane to replace the original cation of the colloidal electrolyte. The liquid inside the dialyser thus becomes acid, that outside the dialyser alkaline. This process is termed membrane hydrolysis. See also COLLOIDS, COLLOIDAL ELECTROLYTE, MEMBRANE EQUILIBRIUM.

membrane separation Chemical separation through a membrane.

menaquirines, menadione Part of the vitamin A series.

mendelevium, Md. Post-actinide metal, at. no. 101, at.mass ^{258}Md 258.1, m.p. 827°C, electronic configuration [Rn] $5f^{13}7s^2$. ^{236}Md (75 minutes) has been formed by bombardment of Es with ^{4}He particles in a cyclotron or linear accelerator. The Md^{2+} ion is fairly stable in aqueous solution

$$Md^{3+} \xrightarrow{+0.2 V} Md^{2+}$$

No solid compounds are known.

p-mentha-1,5-diene (1-$\Delta^{2:8(9)}$-menthadiene) , $C_{10}H_{16}$. The most important menthadiene, an optically active monocyclic terpene found in chenopodium oil. Used in the manufacture of p-cymene.

menthol (1α,2β,5α)-5-methyl-2-(1-methylethyl)-cyclohexanol), $C_{10}H_{20}O$. An optically active monocyclic terpene alcohol.

A constituent of many oils. (–)-Menthol, m.p. 43°C, b.p. 216°C. (–)-Menthol and its stereoisomerides are obtainable by chemical reactions from a large number of terpene derivatives, e.g. by the reduction of menthone, isomenthone or piperitone. (–)-Menthol can be converted into 4-cymene and can be oxidized to menthone. It is used externally as an analgesic in rheumatism and by inhalation in the alleviation of nasal congestion and sinusitis. Used in lacquers, confectionery, cigarettes.

menthone ((2,5-trans)-5-methyl-2-(1-methylethyl)cyclohexanone), $C_{10}H_{18}O$. An optically active monocyclic ketone. Obtained by oxidizing menthol. (–)-Menthone is present in some oils. When (–)-menthone is dissolved in strong sulphuric acid and the mixture cooled, it is partially converted into (+)-isomenthone. Used in perfumery and flavouring.

mepanipyrin Fungicide, pyrimidine derivative.

mepiquat Herbicide, quaternary ammonium derivative.

mepiquat chloride Plant growth regulator, quaternary ammonium derivative.

mepronil Fungicide, carboxamide derivative.

mer isomer The isomer of octahedral Ma_3b_3 in which the 3a ligands occupy one edge of the octahedron. In the fac isomer the 3a ligands occupy one face.

mercarbam Insecticide, acaricide, organophosphate derivative.

mercaptals, $RR'C(SR'')_2$. Sulphur analogues of acetals. Oily liquids with unpleasant odours. Formed by treating mercaptans with aldehydes or ketones. They are frequently employed in synthetic organic chemistry as PROTECTING GROUPS for the carbonyl group.

mercaptans (thiols, RSH) Organic compounds containing an –SH group directly united to a carbon atom. The sulphur analogues of alcohols. They are liquids with strong and unpleasant odours. Found in crude petroleum. Prepared by the action of alkyl or aryl halides on KHS or by the reduction of the chlorides of sulphonic acids. Give complexes and metal derivatives, e.g. $Pb(SR)_2$, mercaptides. Oxidized by air or mild oxidizing agents to disulphides (RS·SR); nitric acid oxidizes them to sulphonic acids. React with aldehydes and ketones to form mercaptals and mercaptols. Long-chain derivatives oxidized to give surfactants. Used in polymerization systems to control the length of the polymer chain. Also used as odorants, particularly in natural gas and LPG, rubber vulcanizers, flotation agents and in the manufacture of various agricultural chemicals.

mercaptides See MERCAPTANS.

mercaptoacetic acid See THIOGLYCOLLIC ACID.

mercaptobenzthiazole, $C_7H_5NS_2$. Yellow crystalline powder, m.p. 174–179°C. Prepared by treatment of thiocarbanilide with sulphur, or by heating aniline, carbon disulphide and sulphur. It is an important rubber accelerator. On oxidation gives dibenzthiazyl disulphide, also a rubber accelerator. Salts used as fungicides.

mercaptoethanoic acid See THIOGLYCOLLIC ACID.

Merck Index Comprehensive listing of organic and inorganic compounds with their preparations and uses.

mercuration The formation of Hg(II) derivatives of aromatic compounds by direct substitution, e.g. $Hg(O_2CCH_3)_2$ reacts with C_6H_6 to give $C_6H_5HgO_2CCH_3$. The resulting organomercurials are useful synthetic intermediates, e.g. for preparing halogeno derivatives.

mercuric Mercury(II) compounds.

mercurous Mercury(I) compounds.

mercury, Hg. Liquid metal, at.no. 80, at.mass 200.59, m.p. −38.83°C, b.p. 356.73°C, density ρ 13546, crustal abundance 0.05 p.p.m., electronic configuration [Xe] $4f^{14}5d^{10}5s^2$. In Hg extraction CINNABAR (HgS) is roasted in air to give Hg vapour. Mercury is a silver-white liquid metal with a rhombohedral structure derived from hcp. The metal is unattacked by dilute acids but dissolves in hot oxidizing acids. Hg is used in electrical equipment – switches, Hg-vapour lamps and also in pumps and thermometers and in $NaOH–Cl_2$ production. Uses of Hg compounds include pesticides, dental fillings, batteries, catalysis. Hg and many Hg compounds are toxic and tend to accumulate in higher animals although some strains of *E. coli* can metabolize Hg compounds. U.S. use 2001 220 tonnes and falling.

mercury, organic derivatives Organomercury(II) derivatives, e.g. R_2Hg and RHgX are well established (e.g. RMgX plus mercury halide; C_6H_6 plus $Hg(O_2CCH_3)_2$ gives $C_6H_5HgO_2CCH_3$) but mercury(I) derivatives are

unknown. The CH_3Hg^+ ion has severe effects on the nervous system and is the agent for environmental toxicity of Hg. Olefine and alkyne complexes are formed and are intermediates in reactions involving Hg as a catalyst. Organomercury compounds are used extensively to prepare other organometallic derivatives; they have a limited use as bactericides.

mercury amalgams Intermetallic compounds of mercury. Some have definite compositions, e.g. $NaHg_2$, and are effectively alloys. Hg-rich amalgams are liquid. Fe does not form amalgams and is used for containers for mercury. Used in dental stoppings; cadmium amalgam is the cathode in the standard WESTON CELL.

mercury chemistry Mercury is the heaviest Group 12 element. The normal oxidation state is +2 ($E°$ Hg^{2+} → Hg +0.796 volts in acid solution). The Hg^{2+} ion is large and many compounds have high covalent character; 2-co-ordination, linear, is common but higher co-ordination is known. Mercury(I) ($E°$ $½Hg_2^{2+}$ → Hg +0.798 volts in acid solution), exists as the (Hg–Hg)$^{2+}$ ion. Other polymercury cations, e.g. Hg_3^{2+} (Hg plus AsF_5), Hg_4^{2+}, etc., are also known. All positive oxidation state compounds of Hg are readily reduced to the metal.

mercury chlorides
Mercury(II) chloride, $HgCl_2$. Corrosive sublimate, m.p. 280°C, b.p. 302°C. Essentially covalent material (Hg plus Cl_2; Hg plus aqua regia). Forms complex halide ions, e.g. $(HgCl_4)^{2-}$, $(HgCl_3)^-$, in excess HCl and forms complexes. Very poisonous. Used as fungicide, disinfectant, depolarizer for batteries.

Mercury(I) chloride, Hg_2Cl_2. Calomel, white solid precipitated from Hg_2^{2+} solutions by Cl^- or from Hg plus $HgCl_2$. Used as an insecticide and fungicide and in the calomel electrode.

mercury fulminate, $Hg(ONC)_2$. Grey crystals prepared by dissolving Hg in concentrated HNO_3 and pouring solution into C_2H_5OH. Used as a detonator. Safe if stored under water but relatively unstable in hot climates.

mercury iodides
Mercury(II) iodide, HgI_2. Scarlet (to 126°C) or yellow substance ($HgCl_2$ solution plus KI or Hg plus I_2). Forms complex iodides with excess iodide (NESSLER'S REAGENT).

Mercury(I) iodide, Hg_2I_2. Pale green, is precipitated from a solution of a Hg_2^{2+} salt by I^-.

mercury nitrates
Mercury(II) nitrate, $Hg(NO_3)_2$. Forms hydrates

with 8, 2, 1 and $\frac{1}{2}H_2O$ (Hg in concentrated HNO_3). Hydrolysed by water – forms basic nitrates. Used in felt manufacture.

Mercury(I) nitrate, $Hg_2(NO_3)_2$. Forms dihydrate from aqueous solution (Hg plus cold dilute HNO_3). Useful source of mercury(I) compounds.

mercury oxides *Mercury(II) oxide*, HgO. Yellow–orange powder (Hg^{2+} solution plus excess NaOH). Decomposes to Hg and O_2 on heating. Many derivatives containing an oxygen co-ordinated to three or four mercury atoms are known. HgO is fungicide, pigment, and depolarizer in batteries. Mercury(I) oxide appears not to exist.

mercury sulphates

Mercury(II) sulphate, $HgSO_4.H_2O$. Colourless solid (Hg plus excess hot H_2SO_4). Readily hydrolysed to Hg_2OSO_4 (turpeth mineral).

Mercury(I) sulphate, Hg_2SO_4. Formed by precipitation or excess Hg plus H_2SO_4. Hydrolysed to basic salts.

mercury sulphide, HgS. Occurs naturally as red cinnabar but a metastable black form (zinc blende structure) may be formed by precipitation or by grinding Hg plus S. Dissolves in alkali sulphide solutions to give thiomercurates. Used as a red pigment – see VERMILION RED.

merocyanine dyes Polymethine dyestuffs comprising a nitrogen-containing heterocyclic system linked through a conjugated chain to an acidic ketomethylene heterocycle. Used as dyestuffs and as spectral sensitizers in photography.

mesaconic acid (methylfumaric acid), $C_5H_6O_4$. HO(O)CCH·C(CH_3)C(O)OH. *Trans* carboxyl groups. Colourless crystals; m.p. 204°C. Prepared by photochemical isomerization of citraconic acid in the presence of bromine.

mescaline (β-(3,4,5-trimethoxyphenyl)-ethanamine), $C_{11}H_{17}NO_3$. An alkaloid obtained from *Lophophora williamsii*. M.p. 35°C, b.p. 180°C/12 mm. It is a central nervous system depressant and causes visual hallucinations. Mescaline is the hallucinatory principle of peyote. Controlled substance.

mesitite See BREUNNERITE.

mesitylene (1,3,5-trimethylbenzene), C_9H_{12}. Colourless liquid; b.p. 165°C. It occurs in crude petroleum. Prepared by adding concentrated sulphuric acid to cooled propanone, and then heating the mixture after standing.

mesityl oxide (isopropylidene acetone, 4-methyl-3-penten-2-one), $C_6H_{10}O$, Me_2C=CHCOCH$_3$. Colourless liquid; b.p. 129°C, with a strong peppermint-like odour. Prepared by distilling diacetone alcohol in the presence of a trace of iodine. Converted to phorone by heating in propanone with dehydrating agents such as sulphuric acid. It is a solvent for resins and lacquers. Used in synthesis of methyl isobutyl ketone.

meso compounds Compounds which contain tetrahedral atoms with four different groups but for which the whole molecule is achiral.

meso-ionic A compound is defined as meso-ionic if it is a heterocycle which cannot be represented satisfactorily by any one covalent or polar structure, and possesses a sextet of electrons in association with all the atoms comprising the ring. e.g. SYDNONES and NITRON.

mesomerism Assumed in molecules which may be formulated (in the conventional manner) in two or more ways that have the same spatial arrangement of the atoms and nearly the same potential energy, e.g. in the Kekulé forms of benzene. In such cases the actual distribution of the electrons which go to make up the bonds is a weighted mean of the distributions corresponding to the various possible formulae. There is, however, no tautomerism or oscillation between the various forms. The electron distribution corresponds to an intermediate state. Termed mesomerism or resonance. It leads to an increase in stability, changes in structure and reactivity.

mesomorph Orderly arrangement of micelles.

mesomorphic state See LIQUID CRYSTALS.

mesophase (mesogen) A phase intermediate between a liquid and a solid. See LIQUID CRYSTALS.

mesoxalic acid (ketomalonic acid, oxo-propanedioic acid), $C_3H_2O_5$. Occurs with $1H_2O$ probably as dihydroxymalonic acid, $(HO)_2C(CO_2H)_2$; m.p. 121°C. Prepared from beet molasses.

mesoporous solids Regular solids, e.g. zeolites.

mesyl The methane sulphonyl group; hence mesylate.

meta In meta-cresol (*m*-cresol, 3-hydroxytoluene), the CH_3 and OH groups are said to be in the meta position to each other. The same prefix is found in metaphosphoric acid and metabisulphites, but in these cases it is always written in full.

For the substance 'meta', see METALDEHYDE.

metabolism Chemical (enzymatic) changes that occur in a living animal or plant. Metabolism is sub-divided into anabolism, i.e. synthetic reactions such as the manufacture of proteins and fats, and catabolism, i.e. destructive reactions such as the breakdown of sugar to carbon dioxide and water, which release large amounts of energy.

metal A general term characterizing certain elements, e.g. gold, silver, copper, mercury, sodium, which have a characteristic lustrous appearance, which are good conductors of heat and electricity and generally enter chemical reactions as positive ions, or cations. There are certain elements, such as, for example, tellurium, which have the physical properties of a metal and the chemical properties of a nonmetal. The distinction between the two groups is not sharp. The typical metallic structures are the body-centred cubic, bcc, the face-centred cubic (cubic close-packed), ccp, and the hexagonal close-packed structures, hcp. The following table of examples shows the widespread occurrence of these structures.

Structure	Metal
body-centred cubic (bcc)	Li, Na, K, Rb, Cs, Ba, β-Zr, V, Nb, Ta, α-Cr, Mo, α-W, α-Fe
face-centred cubic (cubic close-packing, ccp)	Cu, Ag, Au, Ca (below 450°C), Sr, Al, β-La, β-Tl, Th, Pb, γ-Fe, β-Co, β-Ni, Rh, Pd, Ir, Pt
hexagonal close packed (hcp)	Be, Mg, Ca (above 450°C), Y, α-La, α-Ti, Ti, α-Zr, Hf, β-Cr, Re, α-Co, α-Ni, Ru, Os.

A few metals have more complicated structures (noted under the elements concerned) whilst the structures of the metalloids (semi-metals) are intermediate between those of true metals and homopolar compounds. E.g. germanium and grey tin have the diamond structure; arsenic, antimony and bismuth have layer structures, and selenium and tellurium have chain

structures in which each atom has only two neighbours (compare the twelve equidistant neighbours in close-packed structures). Equally, certain alloys and compounds have all the properties of a metallic element. Bonding in metals is generally considered as being by loss of one or more electrons which form a macroscopic molecular orbital surrounding the close-packed metal ions. See BAND THEORY.

metalation The removal of a relatively acidic proton from a compound and replacement by a metal atom, such as lithium, sodium or potassium (see also MERCURATION). The reaction is facilitated by the addition of chelating bases like N,N,N',N'-tetramethylethylenediamine. The products are used in synthesis.

metal atom deposition The coating of a surface with a metallic layer. See also MATRIX ISOLATION.

metal carbonyls Co-ordination derivatives between CO and a metal in which the carbon is generally bonded to the metal. Most transition metal and actinide elements form carbonyls, $M(CO)_x$ ($x = 1, 2, 3$, etc.) using matrix isolation techniques involving metal atoms and CO. These derivatives are only stable in matrices at low temperature. Most of the transition metals form stable carbonyls (MCl_n, $Macac_m$, etc., plus reducing agent plus CO under pressure). The carbonyl groups can also be bridging.

Simple stable carbonyls, except $V(CO)_6$, have an electronic configuration corresponding to the next noble gas. Carbonyl groups can be substituted by other uncharged ligands (e.g. PPh_3, pyridine) organic groups (olefins, aromatic derivatives) and also form other derivatives, e.g. halides, hydrides, sulphides.

metal cluster compounds Compounds containing clusters of metal atoms linked together by covalent (or co-ordinate) and multicentre bands.

metaldehyde, $(C_2H_4O)_n$, ($n = 4$ or 6). A solid crystalline substance, sublimes without melting at 112–115°C; stable when pure; readily formed when ethanal is left in the presence of a catalyst at low temperatures, but has unpredictable stability and will revert to the monomer. Molluscide and fuel.

metallabenzenes See BENZENE.

metallic conduction Conduction in metals arises from the movement of electrons within a partly filled band and does not cause decomposition. The conductivity of a metal may be explained by assuming that the 'sea' of electrons behaves as gas molecules observing the kinetic laws. In metallic conductors conductivity decreases with temperature.

metallic soaps Water-insoluble, alkaline earth, heavy metal, Li salts of long-chain carboxylic acids used as DRIERS, in greases and heavy-duty lubricating oils, fungicides (Cu and Zn salts), rubber, waterproofing, cosmetics and pharmaceuticals (Zn and Mg stearates). Napalm is an Al soap ($Al(OH)R_2$) used for gelling gasoline and as an incendiary.

metallizable dyes (mordant dyes) Dyes that have a metal atom added *in situ* (chelated) after application to the fibre.

metallocenes The *bis*-(h^5-cyclopentadienyl-metal) compounds, $(C_5H_5)_2M$, formed by reaction between the cyclopentadienyl anion and suitable derivatives of the transition metals, e.g. M=Fe, ferrocene; M=Co, cobaltocene; M=Ni, nickelocene. Known for, e.g., Ti, green; Cr, red, Fe, orange; Co, purple; Os, yellow. Only those metallocenes of the iron group (Fe, Ru, Os) are diamagnetic, the others exhibiting varying degrees of paramagnetism, e.g. Cp_2Co, one unpaired electron; Cp_2Mn, five unpaired electrons. Only the iron group metallocenes exhibit aromatic behaviour (see FERROCENE) and are not particularly sensitive to oxidation. Many metallocenes may be oxidized to cations, e.g. $[(C_5H_5)_2Fe^+]$, ferricinium; $[(C_5H_5)_2Co^+]$, cobalticinium. Metallocene derivatives are readily formed either by reaction of the metal or by use of substituted cyclopentadiene derivatives. Used as polymerization catalysts for olefins.

metallochromic indicator A compound which forms a complex with a markedly different colour from that of the free indicator, e.g. NCS^- used with Fe^{3+}, eriochrome black T used with Mg^{2+}.

metalloids Elements which have properties between the obvious metals and obvious non-metals. The classification is arbitrary and often based on the structure and properties of the free element. E.g. As, Sb, Bi. See METALS.

metallurgy The science of the extraction, working and use of metals and their alloys.

metal–metal bonds Covalent or co-ordinate bonds between metals in compounds, e.g. $(OC)_5Mn–Mn(CO)_5$. Often inferred from a low magnetic susceptibility, Raman spectroscopy or electron counting (EFFECTIVE ATOMIC NUMBER RULE) but only established by X-ray structure determinations. Bond orders can be up to 4.

metal passivators In catalytic operations deposition of trace amounts of certain metals may passivate or poison the catalyst. Typical metal passivators are V and Ni present in petroleum fractions.

metal surface treatments All used metal objects are surface treated at some time. Oxides are removed chemically. Dirt and grease are removed with organic solvents particularly chlorocarbons or alkaline detergents. Surfaces are then pickled in acids (generally phosphoric). Surfaces may be treated with other metals, e.g. zinc, oxidized with O or F or treated with an organic substrate. See UNIMOLECULAR FILMS.

metam Fungicide, nematocide, herbicide, insecticide, methyl isothiocyanate derivative.

metamerism See ISOMERISM.

metamitron Herbicide, triazinone derivative.

metaphosphates PHOSPHATES containing cyclic or chain-linked PO_4^{3-} groups.

metastable state See EQUILIBRIUM, METASTABLE.

metasystox See OXYDEMETON METHYL. Pesticide, organophosphate derivative.

metathetical reaction An exchange reaction, e.g.:

$$CF_3I + NaMn(CO)_5 \rightarrow CF_3Mn(CO)_5 + NaI$$

or

$$2AgNO_3(aq) + BaCl_2(aq) \rightarrow 2AgCl + Ba(NO_3)_2.$$

The driving force is generally the insolubility of one of the products in the reaction medium or the volatility of one of the products.

metazachlor Herbicide, chloroacetamide derivative.

metconazole Fungicide, azole derivative.

methaben thiazuron Herbicide, urea derivative.

methacrifos Insecticide, acaricide, organophosphate derivative.

methacrylic acid (α-methylacrylic acid, 2-methylpropen-2-oic acid), $C_4H_6O_2$, $CH_2=CMeC(O)OH$. Colourless prisms; m.p. 15–16°C, b.p. 160.5°C. Manufactured by treating propanone cyanohydrin with dilute sulphuric acid. Polymerizes when distilled or when heated with hydrochloric acid under pressure, see ACRYLIC ACID POLYMERS. Used in the preparation of synthetic acrylate resins; the methyl and ethyl esters form important glasslike polymers.

methadone (amidone, 6-dimethylamino-4,4-diphenyl-3-heptanone), $C_{21}H_{27}NO$, $Me_2NCH(CH_3)CH_2CPh_2C(O)Et$. Prepared from 2-chloro-1-dimethylaminopropane by treating with diphenylmethyl cyanide in the presence of sodamide, and converting to a mixture of ketones from which the methadone is separated. It is converted to its hydrochloride, which is a colourless crystalline powder, soluble in water and in alcohol, m.p. 235°C. It is topologically similar to morphine. It is a powerful opiate, which also depresses the cough centre. Controlled substance.

methamidophos   Insecticide, acaricide, organophosphate derivative.

methanal (formaldehyde), HCHO. Colourless gas with a characteristic and pungent odour. Soluble in water to 52%: the commercial solution – formalin – contains about 37% methanal. It is manufactured by passing methanol vapour and air over a heated metal or metal oxide catalyst, and to a lesser extent by the oxidation of paraffinic hydrocarbons. It reacts with water to give stable hydrates, called polymethylene glycols, of the type $(CH_2O)_n.H_2O$. In dilute solutions most of the formaldehyde is present as methylene glycol, $CH_2(OH)_2$ or $CH_2O.H_2O$. In more concentrated solutions, trimethylene glycol $(CH_2O)_3.H_2O$ is also present. When formalin is allowed to stand, flocculent masses of paraformaldehyde are precipitated. Its principal use is in the manufacture of urea- (30%), phenol- (20%), melamine-, acetal- (10%) and polyformaldehyde resins. Large quantities are also used in the manufacture of ethylene glycol, pentaerythritol and hexamethylenetetramine.

It is a powerful germicide, whether in solution or vapour. Irritant, may be carcinogenic. Dilute solutions are used for sterilizing surgical instruments, has applications in certain skin diseases and for storing pathological specimens.

methanamide (formamide), $HCONH_2$. A colourless, rather viscous, odourless liquid; it absorbs water vapour; m.p. 2.5°C, b.p. 210°C (decomp.). Manufactured by direct reaction of carbon monoxide and ammonia under pressure, or by distillation of ammonium methanoate. It is converted to methanoic acid or a methanoate by concentrated acids or alkalis. Forms compounds with metals and with acids. It and its derivatives are good solvents for many organic and inorganic compounds.

methanation   The process of converting CO or CO_2 to methane by a catalytic reaction with hydrogen under pressure. Methanation is used industrially in several ways including **1** for removal of CO from the gas stream in ammonia synthesis, **2** the purification of hydrogen by removing CO and **3** conversion of CO and hydrogen to methane to increase the calorific value in TOWN GAS or SNG production.

methane (marsh gas), CH_4. A colourless, odourless gas; m.p. –184°C, b.p. –164°C. Liquid at –11°C under pressure. It is the chief constituent of natural gas associated with oil, formed by the decay of vegetable matter; it occurs in coal mines, where explosive mixtures of it with air are known as fire damp, and in coal gas, marine and deep rock sediments, waste tips, and in atmospheres of outer planets. It is manufactured by passing a mixture of one volume of CO and three volumes of hydrogen at atmospheric pressure over a nickel catalyst heated to 230–250°C. The gas mixture is obtained from WATER GAS, also obtained from the anaerobic digestion of sewage. Methane is chemically fairly inert but reacts explosively with chlorine at ordinary temperatures; at low temperatures chloromethane is formed. Forms hydrates. Used in the manufacture of HCN, ethyne, methanol chloro- and dichloromethane, hydrogen, ammonia, carbon monoxide, the production of carbon black. Used as a fuel.

methanesulphonic acid, CH_3SO_2OH. M.p. 20°C, b.p. 167°C/10 mm, obtained by oxidation of dimethyl disulphide or by the catalytic reaction between methane and sulphur trioxide. Polymerization catalyst and used in methylation and esterification. See SULPHONIC ACIDS (ALIPHATIC).

Converted by, e.g., thionyl chloride, to methanesulphonyl chloride (mesyl chloride) which is useful for characterizing alcohols, amines, etc. as methanesulphonyl (mesyl) derivatives.

methanethiol, CH_3SH_8. B.p. 6°C; used in jet fuels. Found in natural gas; prepared from methanol plus H_2S, sodium methyl sulphonate and KHS. Used as odorant in gas supplies.

methanoates (formates) Esters and salts of methanoic acid.

methanoic acid (formic acid), H·COOH. A colourless liquid which fumes slightly and has a penetrating odour; m.p. 8.4°C, b.p. 100.5°C. Occurs in sweat and urine and in stinging nettles. Prepared by heating oxalic acid with glycerol, when formic acid distils off. Manufactured by decomposing sodium methanoate, from the reaction of carbon monoxide and sodium hydroxide with sulphuric acid; also made by the hydrolysis of methyl methanoate obtained by reacting carbon monoxide with methanol. It is used in textile dyeing and finishing, in leather tanning and as an intermediate for other chemicals. It is a good solvent for many organic and inorganic compounds and is a strong acid.

methanol (methyl alcohol, wood spirit, wood naphtha), CH_3OH. Colourless liquid with a spirituous odour; it is poisonous, small doses causing blindness; b.p. 64.5°C. Occurs as esters in various plant oils, such as wintergreen. It is manufactured by reacting together hydrogen and CO or CO_2, the reaction being carried out over a catalyst at 50–350 bar and 250–400°C. Some methanol is also produced by the partial oxidation of hydrocarbons. Steam reforming of methanol, i.e. reaction with H_2O over a catalyst, can produce H_2 and CO_2. Some CO can be formed but over suitable catalysts this is greatly reduced and the reform gas can be used in fuel cells. It is readily oxidized by air in presence of nickel or platinum to methanal. Reacts with sulphuric acid to give methyl hydrogen sulphate, dimethyl sulphate and dimethyl ether. Forms chloromethane with hydrogen chloride in the presence of zinc chloride. When heated with soda-lime it is converted to sodium methanoate and hydrogen. Reacts with sodium to give sodium methoxide. It is a good solvent for many inorganic salts and organic compounds. Used in the manufacture of methanal (45%), methanoic acid (10%), chloromethane (5%), MTBE (5%) and many other organic compounds, as a solvent and as a denaturant for ethyl alcohol and anti-freeze. World demand 2000: 30 megatonnes.

methenamine See HEXAMINE.

methidathion Insecticide, acaricide, organophosphate derivative.

methiocarb Insecticide, acaricide, molluscide, carbamate derivative.

methionic acid See METHYLENEDISULPHONIC ACID.

methionine (2-amino-4-(methylthio) butanoic acid), $C_5H_{11}NO_2S$, $CH_3SCH_2CH_2CH(NH_2)CO_2H$. M.p. 283°C (decomp.). Soluble in water and alcohol. The naturally occurring substance is laevorotatory. Methionine is one of the natural sulphur-containing amino-acids. It is an essential constituent of the food of mammals and is particularly important in that it and choline are the only compounds in the diet known to take part in methylating reactions. Used in animal feed.

methomyl Insecticide, acaricide, carbamate derivative.

methoprene Insecticide, juvenile hormone mimic.

methoxy- A prefix used to denote that the substance contains a methoxyl, CH_3O-, group.

methoxychlor Insecticide, methoxy benzene derivative.

5′-methoxyhydnocarpine An inhibitor that reduces multi-drug resistance in bacteria. Isolated from barberry plants.

methyl The group CH_3-, Me–. The CH_3· radical has been observed in planetary atmospheres.

methyl acrylate (methyl propenoate), $C_4H_6O_2$, CH_2:CHC(O)OCH$_3$. Colourless liquid; b.p. 80°C, insoluble in water, soluble in organic solvents. It is manufactured **1** by treating ethylene chlorohydrin with sodium cyanide and heating the β-hydroxypropionitrile so formed with methanol and sulphuric acid, **2** by the pyrolysis of the ethanoate derived from methyl lactate, or **3** by the reaction of ethyne, CO and methanol in the presence of nickel carbonyl. It readily polymerizes to give colourless rubber-like poly(methyl acrylate). See ACRYLATE RESINS.

methylal See DIMETHOXYMETHANE.

methyl alcohol See METHANOL.

methylaluminoxane Methylaluminium oxide polymer, used as a polymerization catalyst.

methylamines Compounds of ammonia in which 1, 2 or 3 of the hydrogen atoms have been replaced by methyl groups. They are manufactured by the vapour-phase reaction of methanol and ammonia at 350–450°C under pressure over an alumina catalyst. The product consists of a mixture of all three compounds; the proportion of the two higher amines may be increased by recycling the monomethylamine or reducing the ammonia/methanol ratio; the proportion of monomethylamine may be increased by diluting the reaction mixture with water or trimethylamine. All are basic and form alkaline solutions in water and complexes with acceptors. Basicity increases with the number of methyl groups.

Monomethylamine (*methylamine, methanamine*), CH_3NH_2. Colourless, inflammable gas with an ammoniacal odour, very soluble in water, m.p. $-92.5°C$, b.p. $-60°C$. Occurs naturally in some plants, in herring brine and in crude bone oil. It may be prepared in the laboratory by heating methanal with ammonium chloride, or from CH_4, NH_4Cl and $ZnCl_2$. Monomethylamine is largely employed in the manufacture of herbicides, fungicides, surface-active agents and in tanning.

Dimethylamine, C_2H_7N, $(CH_3)_2NH$. Colourless, inflammable liquid with an ammoniacal odour, m.p. $-96°C$, b.p. $7°C$. Occurs naturally in herring brine. Prepared in the laboratory by treating nitrosodimethylaniline with a hot solution of sodium hydroxide. Dimethylamine is largely used in the manufacture of other chemicals. These include the solvents dimethylacetamide and dimethylformamide, the rocket propellant unsymmetrical dimethylhydrazine, surface-active agents, herbicides, fungicides and rubber accelerators.

Trimethylamine, C_3H_9N, $(CH_3)_3N$. Colourless liquid with a strong fishy odour, miscible with water, m.p. $-124°C$, b.p. $3.5°C$. It occurs naturally in plants, herring brine, bone oil and urine. It reacts with hydrogen peroxide to give trimethylamine oxide and with ethylene oxide to give choline. Used in manufacture of quaternary ammonium compounds (bacteriostats).

4-methylamino phenol See AMINOPHE-NOLS.

methyl n-amyl ketone (heptan-2-one), $CH_3CO(CH_2)_4CH_3$. B.p. $152°C$. Occurs in oil of cloves, manufactured by dehydrogenation of 2-heptanol. Used as a solvent for lacquers and resins.

methyl arsonic acid, $CH_3As(O)(OH)_2$. Herbicide, also salts.

methylated spirits See INDUSTRIAL METH-YLATED SPIRITS.

methylation The name given to a process by which a methyl group is added to a compound. In aliphatic chemistry this involves substitution of the hydrogen atom or addition of a CH_2 carbene of, e.g., a hydroxyl, amino or imino group, and produces an ether or a secondary or tertiary amine respectively. In aromatic chemistry it may also mean the substitution of one of the hydrogen atoms of the ring by the methyl group; this is carried out by the Friedel–Crafts reaction. Amines and amino-compounds may be methylated by heating with methanal in methanoic acid solution, or by heating with methyl iodide or dimethyl sulphate. For methylation of hydroxy compounds see ETHERS.

2-methyl benzophenone   Fixative in perfumery.

methyl benzylamine (α-phenylethylamine), $Ph\cdot CH(CH_3)NH_2$. Chiral agent used in synthesis.

methyl bromide   See BROMOMETHANE. Fumigant.

3-methyl-1-butanol (isopentyl alcohol) Obtained from fusel ore; good solvent.

3-methyl-2-butanone (methyl isopropyl ketone) Solvent, b.p. $94°C$.

methyl *t*-butyl ether (MTBE), $CH_3OC(CH_3)_3$. Important solvent.

methyl butynol (2-methyl-3-butyn-2-ol), $HC\equiv CC(Me)OHMe$. Colourless liquid, b.p. $104°C$, obtained from ethyne and propanone. It is used as a corrosion inhibitor and as a chemical intermediate.

methyl cellosolve   See ETHYLENE GLYCOL MONOMETHYL ETHER.

methyl cellulose Prepared from wood pulp or cotton and methanol. Used in greases.

methyl chloride See CHLOROMETHANE.

methyl chloroethanoate (methyl chloracetate), $ClCH_2C(O)OCH_3$. Solvent, b.p. $130°C$.

methyl chloroform See 1,1,1-TRICHLORO-ETHANE.

methyl cyanide See ACETONITRILE.

methylcyclohexane, C_7H_{14}, $C_6H_{11}CH_3$. Colourless liquid, b.p. $101°C$, obtained by hydrogenating toluene and used as a solvent.

methylcyclohexanol (methylhexalin, hexahydrocresol), $C_7H_{14}O$. The commercial product is a mixture of the 1,2;1,3;1,4 isomers: It is a colourless, rather viscous liquid, with a penetrating odour, b.p. $165-180°C$. Manufactured by reduction of cresols with hydrogen in the presence of nickel catalysts. Oxidized to methylcyclohexanones. Used as a solvent for fats, oils, gums and waxes, and in nitrocellulose lacquers. Soaps containing it are important detergents. The esters are used as plasticizers in lacquers. The individual methylcyclohexanols are prepared in a similar manner from the pure cresol, and exist in cis and trans modifications. They are volatile in steam.

methylcyclohexanone, $C_7H_{12}O$. The commercial product is a mixture of the three isomers corresponding to methylcyclohexanols. It is a colourless liquid with a penetrating odour, b.p. 164–172°C. Manufactured by passing the vapour of commercial methylcyclohexanol over a heated copper catalyst. Used as solvent for resins and gums in the manufacture of lacquers.

methyl cyclopentadienyl manganese tricarbonyl See MMT.

methyldemeton METASYSTOX.

methyl dymron Herbicide, urea derivative.

methylene The divalent group $=CH_2$. The $:CH_2$ group is the simplest carbene.

methylene blue (Basic Blue 8) (3,7-*bis* (dimethylamino)-phenothiazin-5-ium chloride) Dyestuff prepared from dimethylaniline and thiosulphuric acid. It is used in dyeing fast fibres and in calico printing, as a staining material in bacteriological work and microscopy and as a mild antiseptic.

methylene chloride See DICHLOROMETHANE.

4,4´-methylene dianiline Curing agent for resins. Used in preparation of polymethylene amides.

methylene diphosphonic acid (medronic acid), $CH_2[P(O)(OH)_2]_2$. $^{99}Tc^{111}$ complexes used for radiotherapy.

methylenedisulphonic acid (methionic acid), $CH_4O_6S_2$, $H_2C(SO_3H)_2$. A colourless, crystalline solid which readily absorbs water vapour; decomposes on distillation. The potassium salt is prepared by heating methylene chloride with an aqueous solution of potassium sulphite under pressure at 150–160°C. The free acid is obtained by decomposing the sparingly soluble barium salt with sulphuric acid. The aryl esters are very stable, but the alkyl esters decompose on heating to give ethers. Resembles malonic acid in some of its reactions. Aluminium salt used as anti-perspirant.

methylene di(4-phenylisocyanate) Intermediate used in manufacture of urethane foams.

methyl ethyl ketone (MEK) See BUTANONE.

methyl fluorosulphate (methyl fluorosulphonate), $FS(O)_2OCH_3$. Colourless liquid, b.p. 94°C. Functions as a powerful methylating agent, even for amides and nitriles which are not attacked by conventional alkylating agents like dialkyl sulphates.

methyl formal See DIMETHOXYMETHANE.

methyl-α-D-glucopyranoside (α-methyl glucoside) Formed from glucose and methanol. Used in drying oils, esters, varnishes, plasticizers.

methyl glyoxal (pyruvaldehyde, 2-oxopropanol, pyruvic aldehyde), $CH_3C(O)CHO$. Yellow liquid, b.p. 72°C, with pungent odour. The liquid readily polymerizes. Obtained from dihydroxyacetone and $CaCO_3$, fermentation of sugar.

methyl green Dyestuff. Used in dyeing and printing and as a biological stain.

methylhexalin See METHYLCYCLOHEXANOL.

methyl hydrazine, $H_3CNH\cdot NH_2$. B.p. 87.5°C. Rocket fuel.

methyl (2-hydroxybenzoate) (methyl salicylate), $C_8H_8O_3$. Colourless liquid, m.p. –8.6°C, b.p. 223°C. Exists in an almost pure state in the essential oils of wintergreen and sweet birch, but is mostly prepared by esterifying salicylic acid. It has a pleasant characteristic odour and is readily absorbed through the skin. It is used in dyestuffs manufacture, perfumery and as a flavouring agent in food, drinks, dentifrices and cosmetics. It has the general medicinal properties of the salicylates, and is applied either alone or with other analgesics for easing the pain of lumbago, rheumatism and sciatica.

methyl hydroxybenzoate, methylparaken Prepared from 4-hydroxybenzoic acid and methanol. Used as food preservative and in cosmetics.

3-methyl indole See SKATOLE.

methyl iodide See IODOMETHANE.

methyl isobutyl ketone (MIBK, 4-methyl-2-pentanone, hexone), $CH_3COCH_2CHMe_2$. B.p. 117–118°C. Ketone used in solvent extraction, e.g. for tantalum fluorides and mineral oils, and as a solvent for many polymers, e.g. acrylic esters, alkyds, polyvinyl acetate. Prepared by selective hydrogenation of mesityl oxide.

methyl isocyanate (MIC), H_3CNCO. Formed from methylamine and phosgene. Used industrially in polymers but potentially explosive.

methyl lithium, H_3CLi. See LITHIUM ORGANIC DERIVATIVES.

methyl mercaptan See METHANE THIOL.

methylmercury compounds See MERCURY, ORGANIC COMPOUNDS.

methyl methacrylate (methyl α-methylacrylate, methyl 2-methylpropenoate methyl ester, MMA), $CH_2=C(CH_3)C(O)OMe$, $C_5H_8O_2$. Colourless liquid; b.p. 100°C. Manufactured by heating acetone cyanohydrin with methanol and sulphuric acid. It is usually supplied containing dissolved polymerization inhibitor, on removal of which it is readily polymerized to a glass-like polymer. See ACRYLATE RESINS.

methyl methane sulphonate (methyl sulphonic acid-methyl ester), $H_3CSO_2(OCH_3)$. Mutagen, teratogen, b.p. 203°C.

methyl methanoate (methyl formate), $HC(O)OCH_3$. Solvent.

methylnaphthalenes Found extensively in crude oil and reformed petroleum; the various mono- and dimethyl derivatives are all known and readily interconverted. Oxidized to naphthalene carboxylic acids which are used as dyestuff intermediates.

N-methyl-N′-nitro-N-nitrosoguanidine, $C_2H_5N_5O_3$. M.p. 118°C. Once used as a precursor for the production of DIAZOMETHANE. Now accepted as carcinogenic.

methyl orange (4[4′(dimethylamino)-phenylazo]benzene sulphonic acid, sodium salt), $C_{14}H_{14}N_3NaO_3S$. Indicator, pH range 3.1 (red) to 4.4 (yellow).

2-methyl-2-4-pentanediol See HEXYLENE GLYCOL.

4-methyl-2-pentanone See METHYL ISOBUTYL KETONE.

2-methyl-1-propanol (isobutanol) Obtained from carbohydrates, b.p. 108°C. Paint and varnish solvent, fruit flavour.

2-methyl propen-2-oic acid See METHACRYLIC ACID.

3-methyl-1H-pyrazole Antidote to methanol poisoning.

methyl pyridines See PICOLINES.

1-methyl-2-pyrrolidinone Industrial solvent used particularly for purification of gases.

N-methyl-2-pyrrolidone Solvent particularly for resins.

methyl red (2-carboxybenzeneazo-dimethyl-aniline), $C_{15}H_{15}N_3O_2$. M.p. 181–182°C. Indicator, pH range 4.4 (red) to 6.0 (yellow).

methyl salicylate See METHYL(2-HYDROXY)-BENZOATE.

methyl succinic acid See PYROTARTARIC ACID.

methyl sulphide See DIMETHYL SULPHIDE.

methyl sulphone, $(CH_3)_2SO_2$. B.p. 238°C. Solvent.

methyl sulphoxide See DIMETHYL SULPHOXIDE.

methyl vinyl ketone (3-buten-2-one), $H_2C=CHCOCH_3$. B.p. 81°C. Prepared from methanol and propanone.

methyl violet (Basic Violet 1, gentian violet benzoate) A violet dye obtained by the oxidation of dimethylaniline with $CuCl_2$. It consists of a mixture of the hydrochlorides of tetra-, penta- and hexamethyl-4-rosanilines. It is used in dyeing jute, for colouring methylated spirits, as a bacteriological stain and as an indicator.

metiram Fungicide, dithiocarbamate derivative.

metnan Fungicide, dithiocarbamate derivative.

metobenzucon Herbicide, urea derivative.

metobromium Herbicide, urea derivative.

metol See AMINOPHENOLS.

metolachlor Herbicide, chloroacetanilide derivative.

metolcarb Insecticide, carbamate derivative.

metosulam Herbicide, urea derivative.

metribuzin Herbicide, triazinone derivative.

metronidazole (2-methyl-5-nitro-1-imidazole ethanol) Anti-bacterial and antiprotozoal.

metsulfuron-methyl Herbicide, sulphonyl-urea derivative.

mevalonic acid, $C_6H_{12}O_4$. D-Mevalonic acid, the 3(R) form of 3,5-dihydroxy-3-methyl-pentanoic acid·

D-Mevalonic acid is the fundamental intermediate in the biosynthesis of the terpenoids and steroids, together classed as polyisoprenoids. The biogenetic isoprene unit is isopentenyl pyrophosphate, which arises by enzymic decarboxylation–dehydration of mevalonic acid-5-pyrophosphate.

mevinphos Insecticide, acaricide, organophosphate derivative.

Mg Magnesium.

MIBK Methyl isobutyl ketone.

micas Aluminosilicates with layers of linked (Si, Al)O_4 tetrahedra similar to those of clays or talc. Readily split into thin insulating sheets; used as fillers and to impart lustre to wallpapers and paints. The most important micas are phlogopite, $KMg_3(OH)_2Si_3AlO_{10}$, and muscovite, $KAl_2(OH)_2Si_3AlO_{10}$. If there are hydrated ions between the layers (instead of K) 'hydrated micas' montmorillonite, vermiculite, normally classified as clays, are found. Fluorophlogopite, $K_2Mg_6Al_2Si_6O_{20}F_4$ with other micas is widely used in the electrical industry. Used in pearlescent pigments, coated with metal oxides.

micelle A term used to describe the colloidal particles in aggregation colloids, e.g. soaps and dyes. A micelle is a submicroscopic aggregate of molecules or polyelectrolyte particles with non-polar groups on the inside and hydrophilic groups on the outside. Micelles form only above the critical micelle concentration CMC.

Michaelis constant An experimentally determined parameter inversely indicative of the affinity of an enzyme for its substrate. For a constant enzyme concentration, the Michaelis constant is that substrate concentration at which the rate of reaction is half its maximum rate. In general, the Michaelis constant is equivalent to the dissociation constant of the enzyme–substrate complex. See DICTIONARY OF BIOLOGY.

Michael reaction Originally applied in a restricted sense to the base catalysed addition of active methylene compounds (e.g. malonic ester) to activated unsaturated systems (e.g. α:β unsaturated ketones), now used in a broader sense to define the conjugate additions of nucleophiles across double bonds.

Michler's hydrol (bis(4-dimethylamino-phenyl)methanol) See MICHLER'S KETONE.

Michler's ketone (bis(4-dimethylamino-phenyl)ketone), $C_{17}H_{20}N_2O$. Glistening leaflets; m.p. 172°C. Prepared by the action of phosgene on dimethylaniline at 100°C. Reduction with sodium amalgam in alcohol gives Michler's hydrol. Treatment with PCl_3 gives the dichloride for the preparation of triphenylmethane dyes. Probably carcinogenic.

microbalance A balance capable of detecting and measuring changes in weight of the order of 10^{-6}g or less. For ordinary quantitative analysis on the small scale, the microbalances in use generally weigh up to about 10g, and are sensitive to a change of 10^{-6}g. For special purposes balances have been constructed which are sensitive to as little as 10^{-11}g, but the maximum load is correspondingly smaller than that of the ordinary analytical microbalance. Electromicrobalances are systems where the displacement is nullified electrically. Such balances are easier to operate and much more robust than conventional microbalances.

microcosmic salt, $NaNH_4HPO_4.4H_2O$. See SODIUM AMMONIUM HYDROGEN PHOSPHATE.

microcrystalline wax See PETROLEUM WAX.

micron (μ) One micron = 10^{-6} m (semi-obsolete).

microscope Magnifying instrument using visible light with relatively low magnification. Used chemically for identification, particularly of minerals (often by study of crystal form) and explosives. For higher magnification an electron microscope is used.

microwaves Electromagnetic radiation of frequency 10^{10} to 10^{12} Hz – frequencies 300 to 3 GHz. Absorption produces rotation in molecules and MICROWAVE SPECTRA are used to determine precise molecular dimensions. Physical effects – heating used for food – polymerization and reaction can be induced by microwaves.

microwave spectroscopy Electromagnetic frequencies 300–3 GHz. Spectra are obtained by passing a microwave beam, produced by a Klystron maser or magnetron. The resolution of such a microwave spectrometer is vastly superior to the best infra-red instruments and wavelengths are measured to seven figures. The spectra correspond to transitions between rotational levels. Microwave spectroscopy is used for the identification of interstellar molecules.

milbemectin Acaricide, insecticide, natural product.

milk Obtained from mammals but particularly cows. Constitution: fat 4%, solids 12%, protein 3%, lactose 4.5%, vitamins, essential fatty acids, calcium, phosphate, iron.

milk of lime A suspension of $Ca(OH)_2$ in water.

Miller indices The labels applied to planes in a crystal lattice. The reciprocal of the intercept on the axes.

millimetre of mercury (mm Hg) Non S.I. unit of pressure. 133.3 Pa. Approximately 1 torr.

millimicron (mμ) One millimicron = 10^{-9}m.

mineral colours Inorganic pigments (prussian blue, chrome yellow, iron buff, etc.), once used extensively when precipitated within fibres, now mainly used for colouring paints, papers, plastics.

mineral dressing (ore dressing) Concentration of the valuable constituents of a mineral by the application of purely physical processes. Techniques in use include dense media separation, froth flotation, electrostatic and magnetic separation, and jigging.

mineral solvents See WHITE SPIRITS.

minium, Pb_3O_4. See LEAD OXIDES.

mirex Former insecticide, carcinogenic, no longer used.

miscibility A term used to denote the extent of mixing. Gases mix in all proportions and are said to be completely miscible. However liquids may be completely miscible, e.g. alcohol and water, only partially miscible, i.e. will only dissolve in each other to a limited extent, e.g. aniline and water; almost completely immiscible, e.g. water and mercury. The miscibility of liquids depends upon their chemical and physical similarities, in particular on their internal pressures. Liquids which have approximately equal internal pressures are miscible in all proportions.

mispickel (arsenopyrite), FeAsS. A whitish common mineral. A principal source of As.

mixed crystals Crystals deposited together from a solution containing two substances (which are normally isomorphous). Also termed solid solutions.

mixed indicator The use of a mixture of two or more INDICATORS in order to sharpen the colour change or to restrict the pH range (etc.) over which the colour change occurs.

mixed metal oxides Compounds which are formally derived from oxides but contain two or more metal species often in an arbitrary ratio. Generally formed by heating mixtures of the appropriate oxy-salts. The structures are determined largely by close packing of the oxide ions and the metal atoms occupy octahedral or tetrahedral holes between the oxygen atoms. Very few mixed metal oxides contain discrete oxyanions.

MMT (methylcyclopentadienyl manganese tricarbonyl) Fuel additive to enhance octane control but little used.

Mn Manganese.

Mo Molybdenum.

MO MOLECULAR ORBITAL (theory).

mobility, ionic In an electric field an ion in solution is accelerated towards an electrode and is slowed by solvent friction and ions of opposed charge so that the ion reaches a terminal velocity. The drift speed is proportional to the strength of the applied field $u\varepsilon$ where u is the mobility of the ion. Ionic mobilities are related to the molar conductivity $\lambda = ZuF$ (Z is charge, F is Faraday constant).

Mohr method Titration of Cl^- with Ag^+ in the presence of added CrO_4^{2-}. A red precipitate forms at the end point.

Mohr's salt, $(NH_4)_2SO_4 \cdot FeSO_4 \cdot 6H_2O$. Iron ammonium sulphate. Used in analysis.

molality The molar concentration of a solution expressed in terms of the number of moles of solute dissolved in unit mass of solvent. Not to be confused with molarity.

molar absorptivity The proportionality factor relating absorbance of radiation (light) to molar concentration and path length.

molar conductivity The specific conductance of a solution of an electrolyte multiplied by the volume in ml containing 1 mole of the electrolyte. Conductivity per unit molar concentration siemens m^2 mol^{-1}. Varies slightly with concentration even for strong electrolytes.

molar heat The amount of heat required to raise the temperature of 1 mole of a substance by 1°C, at either constant pressure or constant volume.

molarity The concentration of a solution expressed as the number of moles of the solute in 1 litre of solution.

molar mass (M) The relative mass per mole of its atoms, molecules, or formula weights. Atomic weight and molecular weight were used extensively for M.

mole S.I. unit. The amount of a substance which contains one gram formula weight of the substance. One mole of any substance contains 6.0223×10^{23} (Avogadro's number) molecules or atoms. Formerly called one gram molecule, as many particles as there are in 0.012 kg of ^{12}C.

molecular beam A beam of molecules in which all the molecules have velocities, and therefore energies, which fall into a very narrow range. Such beams may be used to measure molecular velocity distributions, molecular collision cross-sections and in kinetic studies.

molecular diameters Although very few molecules are in fact spherical, for the purpose of chemical kinetics they are assumed to be so. Some typical values are: H_2, 2.38×10^{-8}cm; O_2, 3.19×10^{-8}cm; NH_3, 3.9×10^{-8}cm; and C_6H_6, 6.6×10^{-8}cm.

molecular distillation A distillation process carried out at very low pressures, 1.3 N m^{-2} or less. Under these conditions the mean free path of the molecules is of the same order as the distance between the surface of the liquid being distilled (the *distilland*) and the surface of the condensate (*distillate*), and they can therefore travel between the two with relatively few collisions. Because of this, distillation can proceed at a fast rate and the risk of thermal decomposition is minimized.

Molecular distillation is used in the separation and purification of vitamins and other natural products, and for the distillation of high-boiling synthetic organic compounds.

molecular formula The number of each kind of atom in a molecule, e.g. for benzene C_6H_6 the empirical formula is CH.

molecular ion The cation corresponding to the molecule less one electron as observed in a mass spectrometer. Also used for anions when these are studied.

molecularity The number of molecules involved in a proposed mechanism.

molecular orbitals (MO) The electronic orbitals which belong to a group of atoms forming a molecule. The orbitals may be bonding, anti-bonding or non-bonding according to whether the presence of an electron tends to hold the molecule together, to cause disruption of the molecule or to have no bonding effect on the molecule. The total number of molecular orbitals must be equal to the total number of atomic orbitals used in the bonding. In general only outer electrons need be considered for molecular orbitals.

molecular refractivity See SPECIFIC REFRACTIVITY.

molecular sieve ZEOLITES containing channels which can be used for the specific absorption of, e.g., water or gases. Sieves containing different pore sizes will absorb different molecules. Used for complexing, stabilizing, drying.

molecular spectrum See BAND SPECTRUM and CONTINUOUS SPECTRUM.

molecular weight The relative weight of one molecule referred to the standard of $^{12}C = 12.000$ amu. The molecular weight is equal to the sum of the atomic weights of the constituent nuclei. The molecular weight expressed in grams is the gram molecular weight. Determined from the volume occupied by a known weight at a particular pressure, by mass spectroscopy, ultracentrifuge, light scattering, viscosity, crystallography.

molecule The smallest particle of matter which can exist in a free state (see ATOM). In the case of ionic substances such as sodium chloride, the molecule is considered as NaCl, which exists as an ion-pair in the gas phase, although the solid consists of an ordered arrangement of Na^+ and Cl^- ions.

molecule, mass of The actual mass of a molecule can be obtained by dividing the gram molecular weight by Avogadro's number. Thus the mass of a nitrogen molecule is

$$\frac{28}{6.02 \times 10^{23}} = 4.65 \times 10^{-23}g$$

mole fraction The mole fraction (x_A) of a compound A in a mixture containing in addition only compounds B, C and D is the ratio of the number of molecules of A to the total number of molecules present. In practice, the concentrations of A, B, C and D are expressed as fractions of the number of moles of each present. The sum of the mole fractions of the components of the system, i.e. $x_A + x_B + x_C + x_D$ is equal to unity.

molinate Herbicide, thiocarbamate derivative.

molybdates Molybdates are very similar to TUNGSTATES. Isopolymolybdates $[Mo_7O_{24}]^{6-}$ and $[Mo_8O_{26}]^{4-}$ are well established. Heteropolymolybdates are similar to the corresponding tungstates. Used as corrosion inhibitors for metals, particularly in antifreezes.

molybdenite, MoS_2. The principal ore of Mo. Has a layer lattice and is used as a lubricant.

molybdenum, Mo. Metal of Group 6, at.no. 42, at.mass 95.94, m.p. 2623°C, b.p. 4639°C, density ρ 10.22, crustal abundance 1.5 p.p.m., electronic configuration [Kr] $4d^55s^1$. Principal ore molybdenite, MoS_2, roasted to MoO_3, converted to ammonium molybdate, purified, heated to MoO_3 and reduced to Mo with hydrogen. The metal is bcc. The massive metal is lustrous and attacked only by HNO_3 – HF or fused KNO_3 – NaOH or Na_2O_2. Mo metal combines with oxygen at red heat and slowly assumes a blue patina in air. Used extensively in steels, particularly cutting steels for high-temperature use, and as a filament material. MoS_2 is an important solid lubricant and Mo compounds are used in pigments. World production Mo 2000 130 000 tonnes. Mo is a necessary trace element in animal diets and the active enzymes of nitrogen-fixing bacteria contain Mo and Fe.

molybdenum, organometallic derivatives A few relatively stable molybdenum alkyls are known (e.g. $MoCl_3Me.Et_2O$ [$MoCl_4$ + $ZnMe_2$], $Li_4[Mo_2Me_8].4Et_2O(Mo(O)_2(CH_3)_2$ + Li Me)). Cyclopentadienyl derivatives, e.g. [($h^5-C_5H_5)Mo(CO)_3]_2$ and many π-acetylene complexes are known and stable.

molybdenum blue See MOLYBDENUM OXIDES.

molybdenum carbides Very hard materials of potential importance as catalysts.

molybdenum carbonyl derivatives The parent carbonyl, $Mo(CO)_6$, is a colourless sublimable solid formed from CO + reducing agent + $MoCl_5$. Carbonyl groups can be substituted by many other ligands to give, e.g. $Mo(CO)_4(PPh_3)_2$.

molybdenum chemistry Molybdenum shows oxidation states from +6 to –2. There is little aqueous chemistry except that of complex oxyanions and some complex halides. MoF_6 and $MoCl_6$ are molecular and lower oxidation state halides, polymeric with some Mo–Mo bonding. Carbonyl and phosphine derivatives are typical low oxidation state compounds. Complexes are formed with most donor atoms but particularly with O– and S– ligands in higher oxidation states. Cyanide and thiocyanate (N-bonded) complexes are well established. N_2 complexes are known and analogues are formed during nitrogen fixation.

molybdenum halides The known fluorides are MoF_6 (Mo + F_2), b.p. 35°C; tetrameric MoF_5 (Mo + MoF_6), Mo_2F_9 (Mo$(CO)_6$ + F_2), MoF_4 (heat on Mo_2F_9) and MoF_3 (Mo + MoF_6). Complex fluorides containing Mo(VI), (V) and (III) are known containing ions such as $[MoF_6]^-$. The known chlorides are grey $MoCl_6$ (MoO_3 + $SOCl_2$), dimeric $MoCl_5$ (Mo + Cl_2), $MoCl_4$ ($MoCl_5$ + C_2Cl_4), $MoCl_3$ and $MoCl_2$ (reduction of $MoCl_5$). Both $MoCl_3$ (Mo_2 pairs) and $MoCl_2$ ($[Mo_6Cl_8]Cl_4$) contain metal–metal bonds. Complex chlorides containing, e.g. $[MoCl_6]^-$, are known. Oxide halides are formed by hydrolysis of halides or by halogenation of oxides. The halides and oxide halides all form complexes with a range of ligands. Bromides and iodides, MoX_4, MoX_3 and MoX_2 are similar to chlorides.

molybdenum oxides Two oxides exist, white MoO_3 (heat on the hydrated oxide formed by acidification of a molybdate) and brown–violet MoO_2 (MoO_3 + H_2). MoO_2 contains Mo–Mo bonds and non-stoichiometric phases are also known. Mild reduction of hydrated MoO_3 gives a blue hydrous oxide, molybdenum blue. MoO_3 forms mixed oxides on fusing with other oxides – these contain linked MoO_6 octahedra rather than discrete anions. MoO_3 is formally the acid oxide of MOLYBDATES(VI). Used in glazes.

molybdenum sulphides, MoS_3, MoS_2, Mo_2S_3. MoS_2 has a layer structure which leads to its use as a solid lubricant and in catalysts, Li batteries and nanotubes.

molybic acids See MOLYBDATES.

monazite A mixed phosphate (Ce,La,Nd,Pr) PO_4 containing some thorium silicate (ThO_2 1–18%). The principal lanthanide ore.

Mond process A process for the extraction of nickel from nickel ores. See NICKEL.

monel A nickel alloy.

monensin Polyether antibiotic used as a feed addition for poultry.

monlinuron Herbicide, urea derivative.

monoacetin See ACETINS.

monobasic acid An acid which has formally only one replaceable hydrogen and forms only one series of salts, e.g. HCl and H_3PO_2 $(HOP(O)_2H_2)$.

monochloroethane, C_2H_5Cl. See CHLORO-ETHANE.

monochloroethylene, $CH_2=CHCl$. See VINYL CHLORIDE.

monochloromethane, CH_3Cl. See CHLORO-METHANE.

monochromatic radiation Radiation of a single frequency (wavelength). See MONO-CHROMATOR.

monochromator A device, prism, grating, single crystal for selecting a narrow band of wavelengths from a wide wavelength range of radiation or particles. Used in spectrophotom-eters and in X-ray diffraction equipment.

monoclinic system The CRYSTAL SYSTEM with one 2-fold axis and/or one plane of sym-metry. A unit cell has three unequal axes, one perpendicular to the other two which are ob-liquely inclined.

monoclonal antibody An engineered ver-sion of an antibody used medicinally.

monocrotophos Insecticide, acaricide, organophosphate derivative.

monodentate ligand A ligand with only one co-ordinately active lone pair of electrons which interact through a single atom, e.g. NH_3.

monodisperse Having a single, or close to single, molecular mass.

monolayers See UNIMOLECULAR FILMS.

monomer A simple molecule which can po-tentially be polymerized or telomerized.

monosaccharides Sugars with the formula $C_nH_{2n}O_n$ ($n = 5$ or 6).

monosodium glutamate See GLUTAMIC ACID.

monotropy The thermodynamic stability of only one form of a substance. All other forms are metastable and tend to change to the stable form. Phosphorus has only the violet form stable; the common yellow form is metastable.

monovinylacetylene (buta-1-ene-3-yne), $CH_2:CH·C:CH$. Colourless gas with a sweet odour; b.p. 5°C. Manufactured by the controlled low-temperature telomerization of ethyne in the presence of an aqueous solution of CuCl and NH_4Cl. Reduced by hydrogen to butadiene and, finally, butane. Reacts with water in the presence of $HgSO_4$ to give methyl vinyl ketone. Forms salts. Forms 2-chloro-buta-diene (chloroprene) with hydrochloric acid and certain metallic chlorides.

Monte Carlo methods Computer simula-tion methods used in the study of solvation.

montmorillonite A clay mineral, an alu-minosilicate with variable composition. Has a high cation-exchange capacity. Two varieties are known, one swelling and giving gels in water, the other having marked absorptive properties. BENTONITE FULLER'S EARTH consist mainly of montmorillonite.

MOPS (3-N-morpholinepropane sul-phonic acid) Biological buffer.

mordant A substance used to fix or develop the colour in dyeing.

mordant dyes See METALLIZABLE DYES.

morphine, $C_{17}H_{19}NO_3$. The principal alkal-oid of opium. It acts both as a base and as a phe-nol and reacts to form methylmorphine (CO-DEINE) and diacetylmorphine (DIAMORPHINE or heroin). Benzylisoquinoline derivative.

Morphine and its salts are valuable analgesic drugs but are highly addictive.

morpholine, C_4H_9NO. A colourless liquid with a strong ammoniacal odour; b.p. 129°C.

Manufactured by heating 2,2'-dichlorodiethyl ether with ammonia, or by heating diethanol-amine with 70% sulphuric acid. It is a moderate-ly strong base, forming soaps with fatty acids. Absorbs water and carbon dioxide from the air to give morpholine carbamate, a colourless, crystalline solid. It has a high solvent power for a very wide range of substances. Used as a solvent and for decreasing corrosion in boilers. Its soaps are used as emulsifying agents.

Morse potential Form of potential energy curve for interaction between atoms that allows for dissociation.

mosaic gold, SnS_2. See TIN SULPHIDES.

Mössbauer effect The resonance fluorescence by γ-radiation of an atomic nucleus, returning from an excited state to the ground state. The resonance energy is characteristic of the chemical environment of the nucleus and Mössbauer spectroscopy may be used to yield information about this chemical environment. Used particularly in the study of Fe, Sn and Sb compounds.

motor spirit Fuel with a boiling range of approximately 310–450 K used for internal combustion engines. See GASOLINE.

moulding The making of shaped objects in metals, plastics or other materials. Moulding can involve use of liquid material or may involve fashioning under pressure.

MS Mass spectrometry. See MASS SPECTROMETER.

MSG Sodium glutamate. See GLUTAMIC ACID.

Mt (meitnerium) Element number 109.

MTBE Methyl *t*-butyl ether. Used as a gasoline additive but use is being curtailed.

mu- (μ-) Designation of a bridging species. E.g. gaseous aluminium chloride is $[Cl_2Al (\mu-Cl)_2AlCl_2]$.

mucic acid (galactaric acid, 2,3,4,5-tetrahydroxyhexanedioic acid), $C_6H_{10}O_8$.

$$(HO_2C) \overset{H}{\underset{OH}{C}} \overset{OH}{\underset{H}{C}} \overset{OH}{\underset{H}{C}} \overset{H}{\underset{OH}{C}} (CO_2H)$$

Colourless crystals. Manufactured by the oxidation of lactose or the galactans from wood with nitric acid. When heated with water it forms a soluble lactone. Converted to furoic acid by heat and to allomucic acid, an optical isomer, by heating with pyridine. Distillation of the ammonium salt gives pyrrole. Used in the manufacture of pyrrole and other heterocycles.

mucilages Polysaccharides, generally containing galacturonic acid, xylose and arabinose residues which swell in water. See GUMS.

mucins Glycoproteins found in saliva, intestinal juices, etc.

mucochloric acid (2,3-dichloro-4-oxo-2-butenoic acid), $C_4H_2Cl_2O_3$. Colourless plates; m.p. 127°C. More correctly represented as a LACTONE, it is prepared from furfuraldehyde and chlorine in aqueous solution. Skin irritant. Used to harden gelatine.

mucopeptides Glycoproteins, structural components of the cell walls of bacteria. They are insoluble polymers of *N*-acetylglucosamine and *N*-acetylmuramic acid.

mucoproteins See PROTEINS.

mucose See DEXTRAN.

muffle furnaces A furnace designed so that the charge does not come into contact with the hot combustion gases. This is effected by putting the charge in a retort or 'box' and heating the outside walls.

mull A thick paste formed by grinding an insoluble solid with an inert liquid and used for studying spectra of the solid.

Mulliken symbols The designators, arising from group theory, of electronic states of an ion often in a crystal field. A and B are singly degenerate, E doubly degenerate, T triply degenerate states. The D state of a free ion shows E and T states in an octahedral field.

mullite, $Al_6Si_2O_{13}$. The stablest aluminium silicate, formed by heating other aluminosilicates; used in refractories and glasses.

multicentre bond A bond formed by a bonding molecular orbital between three or more atoms containing a single pair of electrons. Such bonds occur in, e.g. the boron hydrides.

multidentate ligand (polydentate ligand) A ligand which has two or more sites at which it can co-ordinate. Such a ligand can form a bridge or can form a chelate compound. An example is $H_2NCH_2CH_2NH_2$.

multiple bonding Bonding between atoms in which there is overlap of orbitals in more than one region in space. The electrons may come equally from each atom or be derived from only one of the atoms (co-ordinate bond, back bonding). Generally the primary bond is a σ-bond with extra interaction by π- or δ-bonding.

multiple-effect evaporator In a simple evaporator, containing only one evaporator unit or *effect*, approximately 1 kg of steam is required to evaporate 1 kg of water. A multiple-effect evaporator contains several evaporator units in series, and the steam generated in one effect is fed to the calandria of the next, where it acts as a heating medium and produces further evaporation. Thus 1 kg of steam is able to evaporate, very approximately, N kg of water, where N is the number of effects.

multiple proportions, law of Originally proposed by Dalton in 1804, the law states that when two elements A and B combine to form more than one compound, the weights of B which combine with a fixed weight of A are in the proportion of small whole numbers. E.g. in dinitrogen oxide, nitrogen oxide and dinitrogen tetroxide the amounts of oxygen which combine with one part of nitrogen are in the proportion 1:2:4. This law is of very limited value as very many examples of non-stoichiometric compounds are known.

multiplet In a spectrum the individual lines often consist of two or more closely grouped fine lines, which make up the fine structure of the original line. Each such group, which constitutes a line in the ordinary spectrum, is called a multiplet. In an electronic spectrum this fine structure arises from the slightly differing amounts of energy an electron possesses due to its spin. The multiplicity is the maximum possible number of such energy values which an atom in a specified state can possess by virtue of the electron spin. A multiplet which arises from an electronic transition where there are two spin states is called a *doublet*; where there are three spin states a *triplet* is found.

In addition to this electron spin fine structure there are often still finer lines present. These are known as the *hyperfine* structure, arising from the different weights of the isotopes of an element or from the spin of the nucleus.

In nuclear spectroscopy fine structure arises from coupling between nuclear spins.

muriatic acid An old name for hydrochloric acid, still occasionally used in industry. Potassium chloride is sometimes called 'muriate of potash'.

muscalone Insecticide, pheromone.

muscone (3-methylcyclopentadecanone), $C_{16}H_{30}O$. A yellow liquid; b.p. 330°C. The perfume base obtained from the scent glands of the Tibetan musk deer; also available by synthesis.

musk, artificial Fixatives for perfumes. Musk itself is used but many other species, e.g. nitroaromatics, tetralones, are also used.

mustard gas (sulphur mustard, bis (2-chloroethyl) sulphide), $C_4H_8Cl_2S$, $(CH_2ClCH_2)_2S$. Colourless oily liquid with a faint garlic-like odour; m.p. 13–14°C, b.p. 215–217°C. Manufactured by treating S_2Cl_2 with ethene at 30–35°C. Decomposed violently by bleaching powder. It is a powerful vesicant and poison, and causes conjunctivitis and temporary blindness. Agent in chemical warfare. Blistering agent.

mustard oil See PROPENYL ISOTHIOCYANATE.

mutagens Chemical substances or physical agents which raise the frequency of mutation greatly above the normal level. Many chemical mutagens are carcinogens.

mutarotation The change of optical activity of an optically active substance when it is dissolved in water, or some other solvent, and undergoes chemical change.

Frequently applied to changes observed in sugar solutions, e.g. of β- to α- glucopyranose.

mutases See ISOMERASES and ENZYMES.

mycobutanil Fungicide, triazole derivative.

mycolic acids Naturally occurring fatty acids $R^2CH(OH)\cdot CHR^1C(O)OH$, where R^1 is in range C_{20-24}, isolated from mycobacteria.

myoglobin The oxygen-carrying protein of vertebrate muscle. Myoglobin is used by muscle as a rapidly available source of oxygen.

myosin See ACTOMYOSIN.

myrcene, $C_{10}H_{16}$, $Me_2C=CHCH_2CH_2C(=CH_2)CH=CH_2$. An acyclic monoterpene, b.p. 166–168°C, found in many essential oils. Used in perfumery.

myricyl alcohol See MELISSYL ALCOHOL.

myristic acid (n-tetradecanoic acid), $C_{14}H_{28}O_2$, $CH_3\cdot[CH_2]_{12}\cdot C(O)OH$. M.p. 58°C, b.p. 250°C/100mm. See TRIMYRISTIN. A fatty acid occurring as glycerides in milk and in large quantities in certain vegetable oils. Used in surfactants.

myristyl alcohol (tetradecanol), $C_{14}H_{30}O$. Emollient.

N The newton – the unit of force.

N Nitrogen.

N$_A$ The Avogadro constant, L.

Na Sodium.

NAA Neutron Activation Analysis.

nabam Fungicide, algicide, dithiocarbamate derivative.

nacrite See KAOLINITE.

NAD See NICOTINAMIDE ADENINE DINUCLEO-TIDE.

NADP See NICOTINAMIDE ADENINE DINUCLEO-TIDE PHOSPHATE.

naled Insecticide, acaricide, organophosphate derivative.

nandrolone Nortestosterone, steroid.

nanoparticles Small ordered particles formed by self-assembly.

nanoscale materials Materials of physical size of a few nanometres. Have exceptional and potentially useful properties, e.g. films of CaF$_2$ and BaF$_2$ of less than 20 nm have ionic conductivities up to 1000 times greater than in bulk.

nanotubes Originally only a form of carbon but now applied to any species forming tube-like structures from vapourized materials and from solution. Structures based on folded and rolled sheets of graphite-like atoms. Have useful electronic and structural properties. Nano structures in which one or more dimensions is smaller than 100 nm. Fibrous forms can be obtained using aqueous surfactants, vapourized materials or from solution.

napalm The aluminium soap of NAPHTHENIC and PALMITIC acids used to gel gasoline. The thickened gasoline has been used for military flame throwers and incendiary bombs. See METALLIC SOAPS.

naphtha A term used to describe SPECIAL BOILING POINT spirits (SBP) but also used as a description of light distillate feedstock for gas or petrochemical manufacture. The boiling range is generally about 40–150°C. Used as solvent.

naphthalene, C$_{10}$H$_8$. Very volatile in steam,

readily sublimes at low temperatures; m.p. 80°C, b.p. 218°C. White crystalline solid with a penetrating, tarry smell; it burns with a very smoky flame. Obtained commercially from coal tar and petroleum fractions by demethylation of methylnaphthalenes with H$_2$ at 750°C and 10–70atm.

It is a typically aromatic compound and gives addition and substitution reactions more readily than benzene. Can be reduced to a series of compounds containing 2–10 additional hydrogen atoms (e.g. tetralin, decalin), which are liquids of value as solvents. Exhaustive chlorination gives rise to wax-like compounds. It gives rise to two series of monosubstitution products depending upon whether the substituent is in the 1- or 2-position. Readily nitrates and sulphonates to give valuable dyestuffs intermediates, the substituent entering chiefly into the 1-position except in the case of sulphonation at a high temperature (150°C), which gives a mixture containing 80% of naphthalene-2-sulphonic acid. With alkali metals gives a radical anion which is a good reducing agent.

Most naphthalene produced is utilized in the manufacture of phthalic anhydride, for plasticizers, alkyd resins and polyesters. It is also used in the manufacture of 2-naphthol and insecticides.

Naphthalene derivatives are of importance, particularly as dyestuff intermediates.

1,3-naphthalene diol Used as a reagent for sugars, oils, etc.

naphthalene sulphonic acids Naphthalene monosulphonic acids are obtained by direct sulphonation of naphthalene with concentrated sulphuric acid. A mixture of the two possible isomers is always obtained in proportions varying with the temperature of reaction. Fusion with NaOH gives the appropriate naphthol.

Naphthalene-1-sulphonic acid crystallizes with $2H_2O$, m.p. 90°C. Used for the preparation of 1,8- and 1,5-nitronaphthalene sulphonic acids and 1-naphthol.

Naphthalene-2-sulphonic acid crystallizes with $3H_2O$, m.p. 83°C. It is also used for the preparation of nitro-derivatives.

Naphthalene disulphonic acids are prepared by more prolonged sulphonation than in the preparation of the monosulphonic acids. Four isomeric acids are obtained. The 2,7-derivative is used in dyestuff manufacture.

Naphthalene trisulphonic acids can be obtained by more drastic sulphonation of naphthalene or its mono- and disulphonic acids. Only the 1,3,5-, 1,3,6- and 1,3,7-acids are obtained. The most important is the 1,3,6-acid which is used as a dyestuff intermediate.

naphthenes Cyclo-alkanes; saturated hydrocarbons containing at least one closed ring of carbon atoms, e.g. cyclohexane. Also naphthalene.

naphthenic acids Carboxylic acids obtained from crude petroleum via the water-soluble sodium salts. Many are derivatives of cyclopentane. The sodium salts are important detergents and emulsifying agents, while many other salts have importance – see METALLIC SOAPS. Fungicides.

naphthoic acids Naphthalene carboxylic acids. Prepared by oxidization of methylnaphthalenes or from bromo derivatives via Grignard reagents.

1-naphthol (α-naphthol), $C_{10}H_8O$. Colourless; m.p. 94°C, b.p. 278–280°C. Usually prepared on the large scale by NaOH fusion of sodium naphthalene-1-sulphonate, but can also be obtained by high-temperature alkaline digestion of 1-chloro-naphthalene. Oxidized by acid permanganate or nitric acid to phthalic acid. Gives nitrosocompounds when treated with nitrous acid. Important dyestuff intermediate.

2-naphthol (β-naphthol), $C_{10}H_8O$. White crystals, slightly pink when impure, m.p. 122°C, b.p. 285–286°C. Prepared commercially by NaOH fusion of sodium naphthalene-2-sulphonate. With nitrous acid it gives 1-nitroso-2-naphthol. It can also be chlorinated and sulphonated. Oxidized ultimately to phthalic acid on prolonged oxidation. Used chiefly for the manufacture of dyes and antioxidants.

2-(1-naphthylacetamide) Plant growth regulator.

(2-naphthyl) acetic acid Plant growth regulator.

2-(1-naphthyl) acetic acid Plant growth regulator.

1-naphthylamine (α-naphthylamine), $C_{10}H_9N$. Colourless crystals; m.p. 50°C, b.p. 301°C. Basic, forming sparingly soluble salts with mineral acids. Prepared by the reduction of 1-nitronaphthalene with iron and a trace of hydrochloric acid or by the action of ammonia upon 1-naphthol at a high temperature and pressure. Readily diazotizes and couples to aromatic hydroxylic or basic compounds. Used as a component in a number of important monoazo dyes, but its use has been severely curtailed because of its potent carcinogenicity.

2-naphthylamine (β-naphthylamine), $C_{10}H_9N$. Crystallizes in lustrous leaflets; m.p. 112°C, b.p. 294°C. It is prepared by heating 2-naphthol with a strong solution of ammonium sulphite and ammonia at 150°C. It was used as an end component in a few azo-dyes, but this use has been discontinued because of its carcinogenic character.

1-naphthyl-*N*-methylcarbamate Insecticide. See CARBARYL.

(z-naphthyloxy) acetic acid Plant growth regulator.

naproanilide Herbicide, alkanamide derivative.

napropamide Herbicide, alkanoate derivative.

nascent hydrogen A term formerly used for hydrogen produced by electrolysis or chemical reaction and used *in situ* for reduction. The powerful reducing action is due to reaction at surfaces or to the formation of hydride species.

natamycin Fungicide, natural product.

natrolite, $Na_2(Al_2Si_3O_{10}).2H_2O$. See ZEOLITES.

natron Mixture of Na_2CO_3, $NaHCO_3$, $NaCl$ used in mummification.

natural gas Gas obtained from underground accumulations which may or may not be directly associated with crude oil. Such gas would normally contain at least 90% methane and other low-boiling hydrocarbons, the remainder being nitrogen and carbon dioxide. When natural gas is associated with crude oil it frequently contains appreciable amounts of higher hydrocarbons, pentanes, hexanes, etc., and is referred to as wet gas. Used as a fuel and raw material.

Nb Niobium.

NBS *N*-bromosuccinimide.

Ne Neon.

near infrared Region of the electromagnetic spectrum corresponding to electronic and vibrational transitions.

neburon Herbicide, urea derivative.

Néel point The temperature above which magnetic susceptibility becomes normal. See ANTIFERROMAGNETISM and FERROMAGNETISM.

negative adsorption When adsorption at a surface occurs from a solution the concentration of some components may be higher at the surface than in the bulk, whilst with other components the concentration will necessarily be less at the surface than in the bulk. These latter components are said to undergo negative adsorption. According to the Gibbs' adsorption isotherm any solute which increases the interfacial tension of the system will be negatively adsorbed, e.g. the negative adsorption of sodium chloride at the air/water interface, which has been confirmed experimentally.

neighbouring group effect Effect on reactivity or spectra attributed to neighbouring groups; either steric or electronic.

nematic liquid crystals See LIQUID CRYSTALS.

nematocide A substance for reducing numbers of worm-like species.

neodymium, Nd. Lanthanide metal, at.no. 60, at.mass 144.24, m.p. 1021°C, b.p. 3074°C, density ρ 7008, crustal abundance 38 p.p.m., electronic configuration [Xe] $4f^46s^2$. A typical LANTHANIDE. Nd is used in glass and in capacitors, Nd–Fe–B magnets.

neodymium compounds Stable Nd compounds are typical lanthanide compounds in the +3 state

$$Nd^{3+} (f^3 \text{ red}) \rightarrow Nd \ (-2.44 \text{ volts in acid})$$

A +4 fluoride, Cs_3NdF_7, has been prepared by the action of F_2 on $CsCl + NdCl_3$. $NdCl_2$ ($NdCl_3$ + Nd) and NdI_2 are known and contain M–M bonds.

neon, Ne. Noble gas, at.no. 10, at.mass 20.180, m.p. –248.59°C, b.p. –246.08°C, density ρ 1247g cm^3, crustal abundance 7 x 10^{-5} p.p.m., electronic configuration [He] $2s^22p^6$. A relatively abundant noble gas (1.82×10^{-3}% of air). Obtained pure by fractionation of liquid air. Widely used in fluorescent tubes (red neon lights), lighting and Geiger–Muller tubes, electrical equipment and in gas lasers. Ne forms no normal chemical compounds.

neopentyl The group Me_3CCH_2-. Derivatives, e.g. neopentyl glycol $(CH_3)_2C(CH_2OH)_2$, are used in resins.

neophyl The group $PhMe_2CCH_2-$.

neoprene See POLYCHLOROPRENE.

nephelauxetic effect The changes in the electronic (d → d, etc.) spectra of complexes due to varying degrees of covalent character in the metal–ligand bonds. Measured as the ratio, β, of the Racah parameter B in the complex as compared with B in the free ion. The greater β, the greater the ionic character in the bond.

nepheline (syenite) Feldspar mineral.

nephelometry A method of quantitative analysis which involves the spectrophotometric estimation of the scattering of light by a colloidal suspension of a precipitate.

neptunium, Np. Actinide metal, at.no. 93, at.mass ^{237}Np 237.05, m.p. 644°C, density ρ 20 250, electronic configuration [Rn] $5f^57s^2$. Traces of Np occur naturally, being formed by neutron capture by natural uranium, but Np is normally prepared artificially by neutron irradiation of ^{238}U or ^{235}U [^{237}Np (2.2×10^6 years)]. Np is separated by selective oxidation and solvent extraction. The metal is formed by reduction of NpF_3 with lithium; there are six crystal forms. Neptunium is converted into ^{238}Pu for use as a power source.

neptunium compounds Neptunium is a typical early actinide element and forms compounds similar to those of uranium but also showing a strongly oxidizing +7 oxidation state.

$$\text{NpO}_2^{2+} \text{ (pink)} \xrightarrow{+\ 1.14} \text{NpO}_2^{+} \text{ (green)} \xrightarrow{-\ 0.74}$$

$$\text{Np}^{4+}\text{(yellow-green)} \xrightarrow{+\ 0.155} \text{Np}^{3+}\text{(blue)}$$

$$\xrightarrow{-\ 1.83} \text{Np (volts in 1M acid)}.$$

The metal dissolves in acid to give Np^{3+} oxidized by air to Np^{4+} and to NpO_2^{+} by mild oxidants. Electrolytic oxidation gives green NpO_5^{3-} stabilized by large cations. NpF_6 (NpF_4 plus fluorine, m.p. 55°C orange) and NpF_5 (reduction of NpF_6 with iodine) are the only +6 and +5 halides. NpF_4 is formed by the action of HF and oxygen on NpO_2 (NpCl_4 and NpBr_4 are also known); all NpX_3 are formed from aqueous solution. The stable binary oxide is NpO_2. Complex fluorides of Np(V) and Np(IV) and complex chlorides of Np(IV) are formed from aqueous solution. The NpO_2^{2+} and NpX_4 derivatives form many complexes particularly with oxygen-donors.

neral See CITRAL.

Nernst equation The relation between the e.m.f. of a cell to the concentrations or, more accurately, the activities of the reactants and products of the cell reaction. For a reaction- aA+bB ⇌ cC + dD The e.m.f. of the cell is given by:

$$E = E^\circ - RT \ln \left(\frac{a_C^c a_D^d}{a_A^a a_B^b} \right)$$

where E° is the standard e.m.f of the cell and the a_n are the activities of the reactants and products under a given set of conditions.

nerol (3,7-dimethyl-2,6-octadien-1-ol), $C_{10}H_{18}O$. B.p. 225–226°C. A terpenic alcohol and a constituent of neroli, petit-grain and bergamot, and other essential oils. Used in perfumery.

nerolidol, $C_{15}H_{26}O$. B.p. 276°C. A sesquiterpene alcohol.

nerve gases See FLUOROPHOSPHORIC ACIDS as examples.

Nessler's reagent An alkaline solution of HgI_2 in KI used for detecting and estimating ammonia (brown colour or precipitate formed).

Nessler tubes Cylinders of thin glass, generally graduated, used for comparing turbidities and colours of solutions.

neuraminic acid (prehemataminic acid), $C_9H_{17}NO_8$. An important amino sugar which is widely distributed in animal tissues and secretions especially in the form of the N-acyl derivatives (see SIALIC ACIDS).

neurine (trimethylvinylammonium hydroxide, trimethylethenylammonium hydroxide), $C_5H_{13}NO$, $[\text{Me}_3\text{NCH=CH}_2]^+\text{OH}^-$. A liquid forming a crystalline trihydrate. It is present free and combined in brain and other animal and vegetable products and is formed as a product of putrefaction of lecithin. It can be prepared synthetically from choline and decomposes easily to trimethylamine.

neutralization, heat of The amount of heat evolved when 1 g equivalent of an acid is neutralized by 1 g equivalent of a base. For strong acids and strong bases in dilute solution the only reaction which occurs is H$^+$ + OH$^-$ → H_2O and the heat of neutralization has a constant value of 57.35 kJ.

neutral spirits Distillate from ethanolic solutions containing >95% ethanol (>190 proof), used for blending.

neutrino An elementary particle without charge with variable mass and with spin ½. See DICTIONARY OF PHYSICS.

neutron One of the fundamental particles in an atom. It has almost the same mass as a proton but is electrically neutral. A free neutron, obtained from an atomic pile, has a mean lifetime of 914 s; it decays to yield a proton and an electron. A neutron has magnetic moment 1.91 B.M. and interacts with atoms with unpaired electrons. Neutron rest mass 1.674×10^{-27} kg, 939 MeV. Neutrons are used for the determination of molecular structure by the diffraction which occurs when a substance is subjected to a neutron flux, in neutron activation analysis, and for the formation of new nuclides by neutron capture.

neutron activation analysis (NAA) Method for determination of trace concentrations by neutron capture followed by study of induced radioactivity.

neutron diffraction Method of structure analysis using a beam of neutrons.

neutron scattering Determination of structure using a beam of neutrons.

Neville-Winther acid The trivial name given to 4-hydroxy-1-naphthalene sulphonic acid.

Newlands's law See OCTAVES, LAW OF.

Newman projections See SAWHORSE PROJECTIONS.

newtons (N) The basic unit of force 1 N = 1 kg m s^{-2}. The unit of PRESSURE in the atmosphere = 101 325 N m^{-2}.

Ni Nickel.

niacin See NICOTINIC ACID.

nichrome A nickel alloy.

nicin Small peptide used as food preservative. Also has anti-microbial properties. Vitamin cofactor.

nickel, Ni. Transition metal, at.no. 28, at.mass 58.693, m.p. 1455°C, b.p. 2913°C, density ρ 8902, crustal abundance 8 p.p.m., electronic configuration [Ar] $3d^8 4s^2$. Nickel is a transition element of Group 10 occurring naturally as mixed metal sulphide ores (e.g. millerite, pentlandite), as mixed arsenides, antimonides, sulphides, as garnierite and in the iron ore pyrrhotite, $Fe_n S_{n+1}$. The ore is roasted to NiO which is reduced to Ni with carbon. The details of the final purification depend upon the type of the ore but may involve electrolytic processes or conversion to $Ni(CO)_4$ at 50°C with CO followed by decomposition to pure Ni at 180°C (Mond process). The silver-white metal is ccp. The metal is not tarnished by air but is attacked by acids (except conc. HNO_3). It is resistant to F_2. It is used extensively in alloys, particularly in stainless steels and cast iron, as a coinage metal, in plating and batteries. Used in glass (green), in catalysts (particularly for hydrogenation). Nickel is an essential element for some enzymes. World production 1998 1.03 million tonnes.

nickel, organic derivatives The most important nickel organic derivatives are the bis-π-allyls, e.g. $(h^3-C_3H_5)_2Ni$ (C_3H_5MgBr plus $NiBr_2$) is a very active catalyst for cyclotelomerization of diolefins. Ni(0) complexes, e.g. $[Ni\{P(OEt)_3\}_4]$, catalyse olefin–diene coupling to give 1,4-hexadienes. Nickelocene, $(h^5-C_5H_5)_2Ni$, is an electron-rich compound and reacts as if it had a free olefinic bond. Simple aryls, e.g. $(R_3P)_2NiX(Ar)$, are also known.

nickel accumulator The nickel accumulator uses the reaction

$$Fe(s) + NiO·OH + H_2O \rightarrow Fe(OH)_2(s) + Ni(OH)_2$$

The reaction is carried out in an alkaline medium, usually potassium hydroxide with some lithium hydroxide. Cadmium or Cd–Fe alloys are also used for the negative plate. On charging, the $Fe(OH)_2$ is reduced to Fe and the $Ni(OH)_2$ oxidized to nickel(III) oxide. Its discharge voltage is only about 1.2 V, and as compared with the lead accumulator it has a rather low current efficiency and is rather more expensive. It has some compensating advantages in lower weight and greater mechanical strength, for the electrode materials are compressed into perforated pockets of nickel-plated steel, and the accumulator can therefore withstand mechanical shocks and high rates of charge or discharge.

nickel alloys Have corrosion and scaling resistance but are expensive and so are confined to specialized applications. Examples are monel, hastelloy, nichrome, inconel, nimonics, permalloy, invar, nilo.

nickel ammines Anhydrous Ni salts and Ni salts in the presence of aqueous ammonia form ammines, e.g. $[Ni(NH_3)_6]^{2+}$ (deep blue). Many of the ammines lose ammonia on exposure to air.

nickel ammonium sulphate, $Ni(NH_4)_2(SO_4)_2.6H_2O$. Blue-green crystalline material formed from a solution of the components. Used in electroplating.

nickel arsenide, NiAs. An important structural type, the As is hcp with the Ni in octahedral holes and the As in trigonal prismatic coordination. This structure is adopted by, e.g. CoTe, CrSb, FeS, NbS.

nickelates Nominally any anionic nickel-containing species; practically oxo species, many are mixed metal oxides. Nickelates(IV) are ill defined (NiO plus alkali plus KNO_3 or O_2). Nickelates(III), $MNiO_2$, are prepared similarly.

nickel   *bis* **(dimethylthiocarbamate)** Fungicide.

nickel bromide, $NiBr_2$. Forms hydrates.

nickel carbonates Unstable ill-defined basic materials precipitated from solution. $NiCO_3.6H_2O$ ($NaHCO_3$ solution plus CO_2) and $NiCO_3$ ($CaCO_3$ plus $NiCl_2$) are both relatively unstable. Used in ceramics and glazes.

nickel carbonyl, $Ni(CO)_4$. M.p. –25°C, b.p. 43°C. Toxic material prepared Ni plus CO at less than 100°C. Decomposes to pure Ni at higher temperatures (Mond process for purifying Ni). Forms extensive ranges of substituted derivatives including some cluster-carbonyl derivatives. Used in synthesis and coatings.

nickel chemistry Nickel is an electropositive element of Group 10. The most stable oxidation state is Ni(II) (E° $Ni^{2+} \rightarrow$ Ni –0.24 volts in acid solution) (green or yellow octahedral; red planar; blue tetrahedral) which readily forms complexes. Ni(III) and (IV) are known for fluorides and oxides and some other Ni(III) complexes are known. Ni(I), e.g. $Ni(PPh_3)_3Cl$ is fairly stable. Ni(0), carbonyl, $Ni(PPh_3)_4$ and many organic derivatives, and Ni(–1) is present in carbonylates.

nickel chloride, $NiCl_2$. Forms hydrates with octahedral nickel. Complex chlorides, e.g. blue $[NiCl_4]^{2-}$ (formed in ethanol), are known.

nickel cyanides Nickel(II) cyanide, $Ni(CN)_2$, is precipitated from a Ni^{2+} salt and CN^- in aqueous solution. Forms $Ni(CN)_2.4H_2O$ and many complexes including ammonia complexes with strong clathrating powers for aromatic hydrocarbons. Gives the $[Ni(CN)_4]^{2-}$ ion with excess CN^-; these can be reduced to red $[Ni_2(CN)_6]^{4-}$ (contains Ni–Ni bond) and $[Ni(CN)_4]^{4-}$ ions.

nickel dimethylglyoxime, $C_8H_{14}N_4NiO_4$. The normal form in which nickel is weighed in analysis. There is metal–metal bonding in the solid. The red complex is precipitated from alkaline solution.

nickel fluorides Nickel(II) fluoride, NiF_2, yellow-green, forms green $NiF_2,4H_2O$, and perovskites, e.g. $KNiF_3$, K_3NiF_6 and K_2NiF_6 are known. NiF_3 and higher fluorides are used in fluorination.

nickel hydroxide Apple-green $Ni(OH)_2$ is formed by precipitation from aqueous solutions of Ni(II) salts. It is not amphoteric. Black NiO·OH ($Ni(NO_3)_2$ solution plus Br_2 in aqueous KOH) and related compounds are useful oxidizing agents (see NICKEL ACCUMULATOR).

nickel iodide, NiI_2. Forms hexahydrate. Similar to bromide.

nickel nitrate, $Ni(NO_3)_2.6H_2O$. Green crystals from aqueous solution. $Ni(NO_3)_2$ is formed from Ni plus N_2O_4.

nickel oxides Nickel(II) oxide, NiO, has the NaCl structure (heat on oxyacid salt or $Ni(OH)_2$). Dissolves in acids. Higher nickel oxides are not known although indefinite oxide-hydroxides (e.g. NiO·OH – see HYDROXIDE) are formed.

nickel perchlorate, $Ni(ClO_4)_2.xH_2O$. Soluble nickel salt.

nickel sulphate, $NiSO_4.xH_2O$. Fairly soluble nickel salt. Forms double sulphates, e.g. nickel ammonium sulphate.

nickel vitriol, $NiSO_4.7H_2O$. Hydrated nickel sulphate.

niclosamide Molluscide, amidine derivative.

nicol prism A calcite crystal, cut in such a way that only one of the refracted rays emerges. The emerging beam is plane polarized. See DICTIONARY OF PHYSICS.

nicosulfuran Herbicide, sulphoxylurea derivative.

nicotinamide See NICOTINIC ACID.

nicotinamide adenine dinucleotide (NAD) Adenosine 5'-(trihydrogen phosphate)-P'-5 ester with 3-(amino carbonyl)-1-β-D-ribofuranyl pyridinium inner salt. An extremely important and widespread coenzyme, formerly known as diphosphopyridine nucleotide (DPN). The biologically active form of nicotinic acid; the reduced form is NADH. Involved in many oxidation–reduction reactions. See DICTIONARY OF BIOLOGY.

nicotinamide adenine dinucleotide phosphate (NADP) An important coenzyme of the pyridine nucleotide class. The structure is that of NAD except that the adenosine has an additional phosphate group at the 2'-position. Involved in oxidation–reduction reactions. See DICTIONARY OF BIOLOGY.

nicotines (3-(1-methyl-2-pyrrolidinyl)-pyridine), $C_{10}H_{14}N_2$. Pure nicotine is a colourless liquid, b.p. 247°C, but darkens on exposure to air and light. Crude nicotine contains small amounts of other alkaloids, but (–)-nicotine is the principal component. Insecticide, usually manufactured from tobacco.

nicotinic acid (niacin, 3-pyridine carboxylic acid, vitamin B_3), $C_6H_5NO_2$. M.p. 232°C. It can be prepared by oxidizing nicotine with a variety of agents, or, more cheaply, from pyridine or quinoline. Essential component of mammalian diet. It is the pellagra-preventing factor of vitamin B. The amide, nicotinamide is incorporated into NICOTINAMIDE ADENINE DINUCLEOTIDE. Precursor of NAD and NADP, acts as a coenzyme. Used in foods, fuels and pharmaceuticals.

nido Applied to open structures, e.g. $B_4C_2H_8$ (nest-like). See CARBORANES and BORANES.

ninhydrin (1,2,3-triketohydrindene hydrate), $C_9H_4O_3.H_2O$. Light brown crystals, losing water at 125–130°C, m.p. 242°C (decomp.). Ninhydrin can be prepared by oxidizing diketohydrindene with SeO_2. It gives a blue colour on heating with proteins, peptides and amino-acids and is much used for their detection, particularly as a spray reagent in paper chromatography.

niobates See NIOBIUM OXIDES.

niobite, $(Fe,Mn)(Nb,Ta)_2O_6$. See COLUMBITE.

niobium, Nb. Transition metal, at.no. 41, at. mass 92.906, m.p. 2477°C, b.p. 4744°C, density ρ 8570, crustal abundance 25 p.p.m., electronic configuration [Kr] $4d^4 5s^1$. An element of Group 5. Once known as columbium, Cb. Apart from pyrochlor, $CaNaNb_2O_6F$, all Nb minerals contain Ta although Nb is much more abundant. The important mineral is columbite–tantalite $(Fe,Mn)(Nb,Ta)_2O_6$. Ores are treated by NaOH fusion followed by acid washing to give mixed pentoxides. Nb is separated by liquid extraction from HF solution with methyl isobutylketone. The metal is prepared from Nb_2O_5 by reduction with carbon, it is bcc, remains lustrous in air but reacts with oxygen or steam at high temperatures and dissolves in HNO_3–HF. Nb is used in special steels and electronics, Nb/Ti alloys are superconductors. World use 1997 28 700 tonnes.

niobium chemistry Niobium chemistry is dominated by the +5 oxidation state although lower HALIDES, which, with the exception of the fluorides, contain metal–metal bonds, and lower oxides are known. Hydrides (e.g. $NbH_{0.6-0.8}$, NbH_{2-x}), borides (NbB, Nb_3B_2), carbides (Nb_4C_3, NbC), and nitrides (Nb_2N, Nb_4N_5, Nb_5N_6, NbN) are readily formed. Alkoxides (e.g. $Nb(OEt)_5$ from $NbCl_5$ + EtOH + NH_3) and dialkylamides (e.g. $Nb(NMe_2)_5$ from $NbCl_5$ and $LiNMe_2$) form well-established classes of compounds. A range of complexes, including complex halides, is known. Sulphides (Nb_3S_4, NbS_2) are known. Low oxidation state species such as $[Nb(CO)_6]^-$ and $[Nb(CO)_5]^{3-}$ are known.

niobium halides Known for all NbX_5 and $NbX_4.NbX_5$ (Nb + X_2); NbF_5, m.p. 72°C, b.p. 235°C; $NbCl_5$, m.p. 205°C, b.p. 254°C. Hydrolysis gives oxide halides, e.g. $NbOF_3$ and $NbOCl_3$. Complexes are formed with many ligands and complex halides, particularly complex fluorides and oxide fluorides, e.g. NbF_6^-, $NbOF_5^{2-}$, are present in acid solution. NbX_4 (NbX_5 + Nb) are dark coloured solids with (except for NbF_4) some metal–metal bonding. NbI_4 is easily formed by heating NbI_5. Adducts are formed with various ligands. Lower halides, e.g. $[Nb_6Cl_{12}]Cl_{4/2}$, are formed by further reduction. They contain metal-cluster species and there is

extensive chemistry based on $(Nb_6X_{12})^{n+}$ (n=2, 3 or 4) and $Nb_4X_{11}^-$ species. NbF_3 has an ionic structure. Lower halides, e.g. [$NbCl_3$; dimethoxy-ethane] used to promote coupling of organic molecules.

niobium organometallic chemistry Nb-alkyls and aryls, e.g. $MeNbCl_4$ and $NbMe_5(Me_2PCH_2CH_2PMe_2)$, are known although relatively unstable. Cyclopentadienyl derivatives (h^5–$C_5H_5)_2NbX_3$, (h^5–C_5H_5) $Nb(CO)_4$ are relatively stable as is $[Nb(CO)_6]^-$ ($NbCl_5$ + Na in diglyme + CO (pressure)). Tantalum compounds are very similar.

niobium oxides Nb_2O_5 is the most important oxide and is prepared by dehydration of the hydrated oxide formed by hydrolysis of halides. It is a dense, white, inert material. Reduction of Nb_2O_5 gives NbO_2 and NbO. The pentoxide forms niobates, e.g. $KNbO_3$, as mixed-metal oxides by fusion with other oxides or carbonates. Reduction gives BRONZES, and, e.g. $NaNbO_2$ with Nb–Nb bonding. Nb_2O_5 is used in electronic components. Isopolyniobates, e.g. $[H_xNb_6O_{19}]^{(8-x)-}$ (x = 0, 1, 2) are formed by fusing Nb_2O_5 with alkali and dissolving the melt in water.

NIR Near infrared.

nitempyram Insecticide, vinylidene derivative.

nithiazine Insecticide, thiazinone derivative.

nitramide, H_2NNO_2. See NITROGEN, OXY-ACIDS.

nitramine See TETRYL.

nitrapyrin Bactericide, pyridine derivative.

nitrates Salts of NITRIC ACID, HNO_3. All metallic nitrates are soluble in water, some, e.g. $UO_2(NO_3)_2$, are readily extracted into organic solvents. Formed, generally as hydrates, from nitric acid and metal oxides, hydroxides and carbonates. Anhydrous nitrates, e.g. $Cu(NO_3)_2$, formed by dissolving Cu in liquid N_2O_4 or reacting with N_2O_5; many adducts are formed with excess N_2O_4; they are soluble in organic solvents and are often volatile. Many nitrates are ionic but heavy metal nitrates and anhydrous nitrates have covalently bonded nitrate groups. On heating most nitrates decompose to the metal oxide and NO_2 and O_2, although $NaNO_3$ and KNO_3 give nitrites and oxygen and NH_4NO_3 gives N_2O and H_2O. Nitrates may be tested for by the 'brown ring' test; iron(II) sulphate is added to the cold solution to be tested, and concentrated sulphuric acid added down the side of the test tube; in the presence of a nitrate a brown or black ring is formed at the junction of the two liquids. This test is unreliable in the presence of bromide or iodide (which also give coloured rings); a sounder test and estimation of nitrate is by reduction to ammonia with Devarda's or Arnd's alloy and NaOH or by precipitating nitron nitrate. Nitric acid and the nitrates (particularly those of Na, K, Ca and NH_4) are of great commercial importance; practically all modern explosives contain high proportions of ammonium nitrate or organic nitro-compounds. Nitrates are also employed as fertilizers and as preservatives.

nitrating acid Generally a mixture of nitric and sulphuric acids used for commercial nitration, e.g. in the production of nitro compounds and NITROGLYCERIN.

nitration See NITRATING ACID, NITRO-COMPOUNDS.

nitrene group, RN=. The imino group. Nitrene derivatives, e.g. $ReCl_3(PR'_3)_2NR$ and $[WF_5NMe]^-$ are formed by condensation reactions.

nitric acid, HNO_3. M.p. –42°C, b.p. 83°C (decomp.). Extremely important acid produced industrially by oxidation of NH_3 over Pt to NO and H_2O followed by reaction of the NO with oxygen to NO_2 and further oxidation and absorption in water to give aqueous HNO_3. Forms constant-boiling mixture (c. 68% HNO_3) and 1:1 and 1:3 hydrates. Concentrated with conc. H_2SO_4. Gives NITRATES. Gaseous nitric acid is $HONO_2$ with planar nitrogen. $HNO_3.3H_2O$ is present in stratospheric clouds. Used in the production of fertilizers (NH_4NO_3), explosives, dyestuffs, etc. Behaves as a very strong acid but also as a strong oxidizing agent (reduced to various nitrogen oxides or even to NH_3), especially to non-metals and organic materials. Can cause explosions with organic compounds, e.g. alcohols. May be handled in stainless steel or Al equipment as oxide films are formed. U.S. production 1999 9.07 megatonnes.

nitric oxide, NO. See NITROGEN OXIDES.

nitrides Compounds of nitrogen with other elements prepared by the action of N_2 or NH_3 on the element or by ammonolysis of halides. Electropositive elements form ionic nitrides formally containing N^{3-} ions (e.g. Li_3N), hydrolysed to NH_3. Transition elements form compounds that structurally can be considered to have N atoms in holes in metal lattices. These compounds are very hard, inert and high melting. Many ternary nitrides are known. The less electropositive elements form covalent nitrides – formally derivatives of NH_3 but often containing rings (e.g. BN, P_3N_5). Transition metal nitrides and BN, Si_3N_4, AlN are of importance as hard materials and as ceramics.

nitriding A method of surface hardening steel by treating it in an atmosphere of NH_3 which gives a surface layer of nitride. The steel must contain specific alloying elements, e.g. Al, Cr, Mo.

nitrido complexes Complexes containing the N^{3-} ligand, e.g. $[OsO_3N]^-$, the osmiamate ion.

nitrification Oxidation of NH_4^+ to NO_2^- and NO_3^- by soil bacteria. See DICTIONARY OF BIOLOGY.

nitrile rubber A copolymer of butadiene (55–80%) with propenenitrile, acrylonitrile. Polymerized in an emulsion. An oil- and solvent-resistant rubber used in sealing rings, gaskets, hoses, etc.

nitriles Organic cyanides containing the –C≡N group. They are colourless liquids or solids, with peculiar but not unpleasant odours. They are formed by heating amides with P_2O_5 or by treating halogen compounds with NaCN. They are decomposed by acids or alkalis to give carboxylic acids containing the same number of carbon atoms as the nitrile. Reduced to primary amines. Aliphatic nitriles are valuable solvents for many metal complexes and salts and are used as chemical intermediates, fuel additives, lubricants, plasticizers, insect repellants, weed control, CO_2 removal from natural gas, electroplating additives.

nitrilotriacetic acid, $N(CH_2COOH)_3$. An important complexing (sequestering) agent. Forms particularly stable chelates with Zn^{2+}. Used in detergents, may be carcinogenic.

nitrites Salts containing the $[NO_2]^-$ ion. Salts of NITROUS ACID. Organic O-bonded derivatives RONO. Metal complexes O- or N-bonded. Alkali nitrites are used as preservatives.

nitroamines Aromatic organic compounds containing both nitro and amino groups, usually prepared by partial reduction of polynitro-compounds. They have importance as dyestuffs intermediates. See NITRO ANILINES.

2-nitroaniline (1-amino-2-nitrobenzene), $C_6H_6N_2O_2$. Prepared from 1,2-dinitrobenzene by reduction of one nitro group. Dyestuff intermediate.

3-nitroaniline (1-amino-3-nitrobenzene) Yellow needles; m.p. 114°C, b.p. 285°C. Prepared by sodium sulphide reduction of one of the nitro groups in 3-dinitrobenzene. Dyestuff intermediate.

4-nitroaniline (1-amino-4-nitrobenzene) Yellow crystals; m.p. 147°C. Prepared by heating 4-nitrochlorobenzene with concentrated aqueous ammonia in an autoclave at 170°C. It is also prepared by alkaline hydrolysis of 4-nitroacetanilide or by nitrating and hydrolysing benzylideneaniline.

It is an important dyestuffs intermediate, and also as a source of 4-phenylenediamine.

2-nitroanisole (o-nitroanisole), $C_7H_7NO_3$. Colourless oil, m.p. 9°C, b.p. 273°C. Prepared by heating 2-nitrochlorobenzene with NaOMe or from 2-nitrophenol and MeCl. Important dyestuffs intermediate.

nitroanthraquinones The direct nitration of anthraquinone results chiefly in substitution at the α-positions, 1-nitro- and 1,5- and 1,8-dinitroanthraquinones being obtained.

The compounds have limited value as dyestuffs intermediates.

nitrobenzene, $C_6H_5NO_2$, $PhNO_2$. Colourless, highly refractive liquid with characteristic smell; m.p. 6°C, b.p. 211°C.

It is manufactured by reacting benzene with a mixture of nitric and sulphuric acids. Most of the nitrobenzene produced is used to manufacture aniline, which is obtained by reduction. A considerable proportion is used as a raw material in the dyestuffs industry, either as nitrobenzene as such, or as aniline. Used in soap manufacture.

Further nitration gives 3-dinitrobenzene; sulphonation gives 3-nitrobenzene sulphonic acid. Reduction gives first azoxybenzene, then azobenzene and aniline, depending upon the conditions.

nitrobenzoic acids, $C_7H_9NO_4$. All isomers known; used as synthetic reagents.

nitrobiphenyl, $C_{12}H_9NO_2$. 2-, 3-, and 4- derivatives known. 2- used as plasticizer, fungicide. 4- is a suspected carcinogen.

nitrocellulose See CELLULOSE NITRATE.

nitrochlorobenzenes, $C_6H_4ClNO_2$.
2-Nitrochlorobenzene, needles, m.p. 32°C, b.p. 245°C.

4-Nitrochlorobenzene, prisms or leaflets, m.p. 83°C, b.p. 238°C.

A mixture of the two mononitrochlorobenzenes is prepared by nitration of chlorobenzene. Further nitration of the mixture or of either of the mononitro compounds gives 2,4-dinitrochlorobenzene, m.p. 51°C, b.p. 315°C.

The nitrochlorobenzenes are valuable dyestuffs intermediates. High toxicity.

nitro-compounds (aromatic) A large group of compounds which contain the nitro group, NO_2.

Prepared by the direct action of nitric acid. The reaction is greatly facilitated if a mixture of nitric and sulphuric acids is used.

In the nitration of benzene, 3-dinitro- and 1,3,5-trinitrobenzenes are obtained under more vigorous conditions. With naphthalene, 1-nitronaphthalene is the first product and further nitration gives a mixture of 1,5- and 1,8-dinitronaphthalenes; 2-nitronaphthalene is not obtained.

The nitrohydrocarbons are neutral substances, but when a nitro group is introduced into a phenol or amine the acidic properties are greatly increased or the basicity decreased. The presence of a nitro group makes halogen atoms in the same molecule more reactive.

On reduction, the nitro-compounds give rise to a series of products, e.g.

$$PhNO_2 \rightarrow \text{nitrosobenzene, } PhNO \rightarrow$$
$$\text{N-phenylhydroxylamine, } PhNHOH$$

and further reduction can give azoxybenzene, azobenzene, hydrazobenzene and aniline.

Used commercially for reduction to the amines for conversion to dyestuffs. This is usually done in one stage with iron and a small amount of hydrochloric acid.

Some nitro-compounds are themselves coloured and can be used as dyestuffs, e.g. picric acid. For aliphatic nitro-compounds see NITROPARAFFINS.

nitro-dyes The coloured salts of various nitrophenols or sulphonic acid derivatives. Acid dyestuffs.

nitrogen, N. Gaseous element, at.no. 7, at. mass 14.007, m.p. -210.00°C, b.p. -195.79°C, density (liquid) 1.145 g l^{-1}, crustal abundance 25 p.p.m., electronic configuration [He] $2s^2 2p^3$. Colourless gas existing as N_2 (air contains 75.5% by weight, 78.1% by volume) and also as nitrates, ammonia and in proteins in animal and vegetable tissues (about 16% N by weight). Obtained industrially by fractionation of liquid air, obtained on a laboratory scale by heating NH_4NO_2 (a mixture of $NaNO_2$ and NH_4Cl) or barium or sodium azides. N_2 is fairly inert at room temperature (see NITROGEN FIXATION) but combines with most electropositive elements on heating and with O_2 (under irradiation), H_2 (in presence of catalyst forms AMMONIA in HABER PROCESS). CaC_2 (to give calcium cyanamide), MOH plus carbon (to give cyanides). Active nitrogen is excited N_2 produced by an electric discharge. Used in ammonia and nitric acid synthesis and as an inert atmosphere. Liquid nitrogen is used as a refrigerant.

nitrogenase A complex of metal-containing proteins present in nitrogen-fixing bacteria and blue–green algae.

nitrogen chemistry Nitrogen is an electronegative element of Group 15. Its only simple ionic chemistry is that of the nitride ion N^{3-}. The normal oxidation state is +3; pyramidal (rapid inversion) in 3-co-ordinate molecules, e.g. NH_3, 4-co-ordinate tetrahedral in, e.g. $[NH_4]^+$. The +5 oxidation state does not show 5-co-ordination but is 4-co-ordinated in, e.g. $[NF_4]^+$ and 3-co-ordinate in the planar nitrate, $[NO_3]^-$ ion. Nitrogen–nitrogen bonded species include N_2O, N_3^-, N_2, N_5^+.

nitrogen chlorides Nitrogen trichloride, NCl_3, explosive oily liquid formed from Cl_2 and NH_4Cl solution. $NHCl_2$ and NH_2Cl (mono and dichloroamine) are also formed. ClNO, NITROSYL chloride, NF_2Cl and $NFCl_2$ are also known – see NITROSYL HALIDES. NCl_3 used in bread manufacture (not U.S.A.).

nitrogen-donors Tripositive nitrogen compounds, particularly ammonia and its derivatives, form extensive series of complexes. Nitrosyl complexes are N-bonded. $(NO_2)^-$ can be O- or N-bonded but the latter is more stable. Nitrates are O-bonded.

nitrogen fixation The use of atmospheric nitrogen in the manufacture of commercially important nitrogen compounds, nitric acid, ammonia and ammonium salts, mainly used as fertilizers. Conversion to ammonia is effected by the Haber process, and nitric acid is obtained by oxidation of ammonia. Bacteria present in the roots of certain plants fix atmospheric nitrogen; so do *Azotobacter*, free-living soil micro-organisms (see NITROGENASE). Some metal complexes interact with N_2 to give N_2 complexes which may be reduced to NH_3. The return of the fixed nitrogen to the atmosphere

is DENITRIFICATION which involves nitrogen oxides. See DICTIONARY OF BIOLOGY.

nitrogen fluorides *Nitrogen trifluoride*, NF_3. M.p. –206.7°C, b.p. –129°C, prepared by the electrolysis of fused ammonium hydrogen fluoride. It is a very inert substance, but the other nitrogen fluorides tend to explode, either spontaneously or in the presence of organic materials. Other nitrogen fluorides are *tetrafluorohydrazine*, F_2NNF_2, prepared by the action of Cu on NF_3, and *cis-* and *trans-difluorodiazine*, FN=NF. There is an extensive chemistry of the substituted nitrogen fluorides, e.g. nitrosodifluoroamine, F_2NNO. NF_3 is used in etching for semi-conductors and liquid crystal displays and cleaning sites for vapour deposition. Some organonitrogen fluorides are very good fluorinating agents.

nitrogen hydrides The simple hydride is ammonia, NH_3, but hydrazine, N_2H_4 and unstable diazine HN=NH are known and derivatives of these, e.g. $[NH_4]^+N_3^-$, ammonium azide.

nitrogen iodides *Nitrogen tri-iodide*, NI_3. Black powder formed from NH_3 gas over $KIBr_2$. Alcoholic solutions of I_2 and NH_3 give a black solid $NH_3 \cdot NI_3$ (NI_4 tetrahedra) which is stable when moist but shock-sensitive when dry.

nitrogen mustards Poly-(2-chloroethyl)-amines, $RN(CH_2CH_2Cl)_2$, in which R is an alkyl, alkylamine or chloroalkyl group. These have vesicant and other properties similar to those of mustard gas. Have been used in the treatment of leukaemia, Hodgkin's disease and similar conditions.

nitrogen oxides
Dinitrogen oxide (*nitrous oxide*), N_2O. Colourless gas, m.p. –91°C, b.p. –88.5°C (heat on NH_4NO_3). Decomposes to N_2 and O_2 above 500°C; can be detonated. Linear molecule NNO. Used as a mild anaesthetic.

Nitrogen oxide (*nitric oxide*), NO. Colourless gas, paramagnetic with only slight tendency to dimerization, m.p. –164°C, b.p. –152°C. Prepared in laboratory from Cu, and 8M HNO_3 and industrially by oxidation of ammonia over a catalyst. Decomposes at high pressures to N_2O and NO_2, reacts with O_2 (to NO_2) and halogens (to XNO). Removal of an electron gives the NO^+ (NITROSONIUM ion); combines with transition metal derivatives to give NITROSYLS. Used in production of nitric acid. Has great biological importance in signalling between cells.

Dinitrogen trioxide, N_2O_3. Only stable in solid state (m.p. –102°C). Pale blue solid giving deep blue liquid; the gas contains some $ONNO_2$ molecules. Prepared from NO and O_2 or NO and N_2O_4 with freezing; decomposes to NO and NO_2.

Nitrogen dioxide, NO_2 and *dinitrogen tetroxide*, N_2O_4. Are in equilibrium (NO_2 brown, paramagnetic, N_2O_4 colourless, diamagnetic) with N_2O_4 in the solid (m.p. –11°C, b.p. 21°C). NO_2 is a bent molecule ONO, N_2O_4 is largely planar O_2NNO_2. Formed from NO plus O_2 or Cu plus HNO_3. The liquid (particularly in the presence of donor molecules such as $CH_3CO_2C_2H_5$) acts as a good solvent and anhydrous nitrates, e.g. $Cu((NO_3)_2$ (Cu plus N_2O_4) may be prepared in this solvent. Used as an oxidizing agent (nitrous fumes) and as an intermediate in the formation of nitric acid.

NO_x. Mixture of nitrogen oxides formed during petroleum combustion or by lightning. Can react with volatile hydrocarbons in heat and sunlight to give ozone. Pollutant.

Dinitrogen pentoxide, N_2O_5. White solid (HNO_3 plus P_2O_5) readily decomposes to NO_2 and O_2, sublimes 32.5°C. In solid state $(NO_2)^+(NO_3)^-$; gaseous molecules are O_2NONO_2 but unstable. Used as nitrating agent.

Nitrogen trioxide, NO_3. Unstable white solid (N_2O_5 plus O_3).

nitrogen (oxyacids)

Hyponitrous acid, $H_2N_2O_2$. A solution of the sodium salt is formed from NO and Na in dimethoxyethane; $Ag_2N_2O_2$ is precipitated from solution, $H_2N_2O_2$ formed from $Ag_2N_2O_2$ and HCl in ether. The hyponitrite ion is *trans*$(O–N–N–O)^{2-}$.

Nitroxylic acid, $H_2N_2O_2$. Yellow Na_2NO_2 formed by electrolysis $NaNO_2$ in NH_3. Structure unknown; free acid unknown.

Nitramide, H_2NNO_2. A weak acid.

Nitrous acid, HNO_2. Unstable weak acid prepared by acidifying solutions of nitrites, e.g. KNO_2. The free acid is known in the gas phase (NO, NO_2 plus H_2O). Nitrites are formed by reduction of nitrates with C or Pb. Can act as oxidizing agent (to e.g. I^-, Fe^{2+}), more normally a reducing agent (converted to HNO_3). Forms both O- and N-nitrites with organic groups and in co-ordination compounds; the N-bonded nitro-derivatives are generally more stable. Organic thionitrites ONSR are also known.

Hyponitric acid, $H_2N_2O_3$. Trioxo dinitrate(II). Sodium salt from H_2NOH in MeOH with $C_2H_5NO_2$; many complexes are known; the free acid is unstable.

Peroxonitrous acid, HOONO. Isomer of nitric acid (HNO_2 plus H_2O_2).

Nitric acid, HNO_3. See separate entry.

nitrogen sulphides See SULPHUR NITROGEN DERIVATIVES.

nitrogen trichloride See NITROGEN CHLORIDES.

nitrogen trifluoride See NITROGEN FLUORIDES.

nitroglycerin(e) (glyceryl trinitrate, 1,2,3-propanetriol trinitrate), $C_3H_5N_3O_9$, $CH_2(ONO_2)·CH(ONO_2)·CH_2(ONO_2)$. Oily liquid prepared by treating glycerin with a mixture of nitric and sulphuric acids. When pure it is colourless, odourless, insoluble in water; m.p. 8°C. It is a very powerful and dangerous explosive and is never used alone due to its sensitivity. Used in propellants and dynamites. Used in medicine to give relief from chest pain in angina (provides NO).

1-nitroguanidine, $H_2N·C(:NH)·NH·NO_2$. White crystalline powder, m.p. 232°C; soluble in hot water. Usually manufactured from calcium carbide via calcium cyanamide, dicyandiamide and guanidine nitrate which is converted to nitroguanidine by the action of conc. H_2SO_4. It is used in some modern propellants to make them cooler and flashless.

nitromethane, CH_3NO_2. B.p. 100°C. Widely used solvent. The lowest NITROPARAFFIN.

nitron, $C_{20}H_{16}N_4$, PhN·N·NPh·CH·NPh. 1,4-diphenyl-3-(phenylamino)-1H-1,2,4-triazolium inner salt. Used in ethanoic acid solution as a precipitating agent for nitrates. ClO_4^-, PF_6^-; ReO_4^-, WO_4^{2-} also form precipitates.

1-nitronaphthalene, $C_{10}H_7NO_2$. Yellow needles; m.p. 61°C; b.p. 304°C. Oxidized by chromic acid or potassium permanganate to 3-nitrophthalic acid; acid reducing agents give 1-naphthylamine. Readily nitrated, sulphonated and chlorinated.

Prepared by the direct nitration of naphthalene with a mixture of nitric and sulphuric acids. Its chief use was for the preparation of 1-naphthylamine and its derivatives.

1-nitro-2-naphthol Used to prevent gum build up in petroleum.

nitronium salts Salts containing the NO_2^+ ion, e.g. $[NO_2]^+[WF_6]^-$. Prepared by the action of strong acids on concentrated nitric acid, by reactions involving nitrogen dioxide in bromine trifluoride or by NO_2 plus oxidizing agents. The nitronium ion is the active agent in the nitration of aromatic systems in mixtures of nitric acid and sulphuric acids or with nitronium fluoroborate.

nitroparaffins, $C_nN_{2n+1}NO_2$. Colourless, pleasant-smelling but rather toxic liquids, sparingly soluble in water. They may be prepared by the action of silver nitrite on alkyl halides. The lower members are manufactured by the vapour-phase reaction of propane with nitric acid at 400°C; a mixture of products results which may be separated by distillation. Nitroparaffins containing an α-hydrogen atom dissolve in aqueous alkali giving the salt of the *aci*-nitro compound, e.g.

$$CH_3NO_2 \rightarrow [CH_2 = NO_2]^-Na^+$$

which on treatment with concentrated mineral acids is hydrolysed to the corresponding aldehyde or ketone. Reduction of nitroparaffins with tin and hydrochloric acid gives the amine.

The lower nitroparaffins are used as propellants, as solvents and as chemical intermediates, e.g. nitromethane is an excellent solvent for polar materials especially metal salts.

***p*-nitrophenetole (4-ethoxynitrobenzene),** $C_8H_9NO_3$. Yellow crystals; m.p. 58°C, b.p. 283°C. Prepared by the ethylation of 4-nitrophenol with ethyl chloride. It is used for preparing 4-phenetidine.

2-nitrophenol (o-nitrophenol), $C_6H_5NO_3$. Bright yellow needles; m.p. 45°C, b.p. 214°C. Prepared together with 4-nitrophenol by careful nitration of phenol. Sodium sulphide reduces it to 2-aminophenol which is used in dyestuffs and photographic processes.

4-nitrophenol (p-nitrophenol), $C_6H_5NO_3$. Colourless needles; m.p. 114°C. Prepared as 2-nitrophenol. Reduction with iron and hydrochloric acid gives 4-aminophenol.

nitropropanes See NITROPARAFFINS.

nitroprussides Containing the $[Fe(CN)_5NO]^{2-}$ ion. See NITROSYLS.

nitrosamines Compounds having the general formula

$$\underset{R'}{\overset{R}{>}} N \cdot NO.$$

Most are yellow oils or solids and arise from the interaction of nitrous acid with secondary or tertiary amines. The great majority are carcinogenic to experimental animals.

4-nitroso-N,N-dimethylaniline, $C_8H_{10}N_2$. The free base forms dark green leaflets; m.p.

85°C. On standing it goes browner in colour. It is basic. It reduces to 4-aminodimethylaniline and gives dimethylamine with hot sodium hydroxide solution. It is prepared by the action of $NaNO_2$ in acid upon dimethylaniline at 0°C.

It is used for the preparation of azines and oxazines by condensation with 3-diamines and phenols.

4-nitrosodiphenylamine Accelerator for vulcanizing.

nitroso dyes Dyestuffs containing nitroso-, NO, groups, generally quinone oximes prepared from nitrous acid and phenols or hydroxylamine on quinones. Acid dyes.

nitrosomethylurea, $NH_2CON(NO)CH_3$. A colourless crystalline solid; m.p. 123–124°C. Unstable, carcinogenic. Reacts with cold potassium hydroxide solution to give diazomethane, but this use has now been superseded by DIAZALD.

nitrosonium hydrogen sulphate (nitrososulphuric acid, chamber crystals), $NOHSO_4$. White solid; m.p. 73°C (decomp.). Prepared SO_2 and fuming nitric acid. Used in diazotization. NO^+ salt.

nitrosonium salts Salts containing the NO^+ ion which result from the reaction of nitrosyl chloride with metallic chlorides, e.g. $[NO]^+[AlCl_4]^-$, NO plus oxidizing agents, or from reactions in BrF_3.

nitrosonium tetrafluoroborate, $NOBF_4$. White solid; NOF plus BF_3. It is very hygroscopic; reaction with water gives oxides of nitrogen, but reactions with alcohols give nitrito esters. Aromatic primary amine hydrogen chlorides give diazonium tetrafluoroborates which may be pyrolysed to give aromatic fluoro derivatives. Secondary aliphatic amines react with it to give nitrosoamines.

4-nitrosophenol, $C_6H_5NO_2$. Greyish-brown leaflets, decompose at 124°C. It gives a red sodium salt. Prepared by the action of nitrous acid upon phenol or by the action of hydroxylamine hydrochloride upon 4-benzoquinone. It can act either as a nitrosophenol or as a quinone monoxime, 4-nitrosoanisole, $CH_3O \cdot C_6H_4 \cdot NO$ is known and is colourless.

4-Nitrosophenol condenses with bases to form indophenols, which are used for the production of sulphur colours.

nitrososulphuric acid (nitrosulphuric acid, nitrosulphonic acid) Alternative names for nitrosonium hydrogen sulphate, $NOHSO_4$.

nitrosyl halides A series of covalent compounds, XNO, formally derivatives of [NO]$^+$ salts, prepared by direct interaction of X_2 and NO. Form nitrosonium salts with Lewis acid halides, act as strong oxidizing agents. Have bent molecules. FNO, m.p. $-133°C$, b.p. $-60°C$, colourless; ClNO, m.p. $-62°C$, b.p. $-6°C$, orange–yellow; BrNO, m.p. $-56°C$, b.p. 0°C. Other nitrosyl derivatives and nitrogen trifluoride oxide, F_3NO, and nitryl halides, e.g. FNO_2 are known. Nitrosyl chloride is used in the production of caprolactam from cyclohexane.

nitrosyls Compounds containing NO bonded to metals through the nitrogen. In general similar to carbonyls but the M–N bonds are more stable to substitution than M–C bonds. For electron-counting assumed to have 3-electrons donated to the metal (i.e. NO$^+$ complexes). Formed by direct action of NO or indirectly by the action of nitrosyl halides. The iron group forms very stable nitrosyls, e.g. $Fe(CO)_2(NO)_2$, $[Fe(CN)_5NO]^{2-}$ (nitroprusside ion), $[Ru(NO)Cl_5]^{2-}$. $[Cr(NO)_4]$ is the only homoleptic nitrosyl. Thionitrosyls containing MNS groups are also known.

nitrosyl sulphuric acid, $NOHSO_4$. See NITROSONIUM HYDROGEN SULPHATE.

nitrothal isopropyl Fungicide, benzene carboxylate derivative.

nitrotoluenes, $C_7H_7NO_2$. Prepared by the direct nitration of toluene. Used in dyestuff preparation.

2-nitrotoluene, b.p. 222°C.

3-nitrotoluene, b.p. 230°C.

4-nitrotoluene, colourless crystals; m.p. 51–52°C, b.p. 238°C.

Further nitration of the 2- and 4- isomers yields 2,4-dinitrotoluene, yellow crystals, m.p. 71°C. This material is reduced to 2,4-diaminotoluene treated with phosgene to give 2,4-diisocyanatotoluene, a precursor of polyurethanes.

nitrourea, $NH_2CONHNO_2$. Colourless crystalline solid; m.p. 158°C. Prepared by adding urea nitrate to concentrated sulphuric acid at 0°C and pouring the mixture on to ice. It is a strong acid, and forms salts. Not readily oxidized but can be detonated. Used in the preparation of semicarbazide.

nitrous acid See NITROGEN OXYACIDS.

nitrous fumes An NO–N_2O_4 mixture used in oxidation. Also NO_x formed in car exhausts.

nitrous oxide, N_2O. See NITROGEN OXIDES.

nitroxides Compounds containing an R^1R^2NO group. Paramagnetic (unpaired electron) and widely used to study free radicals or the effect of spin-labelled species.

nitroxylic acid, H_2NO_2. See NITROGEN (OXY-ACIDS).

nitryl halides, XNO_2. Known for F and Cl; the parent compounds of nitronium salts, formally halogen nitro derivatives. FNO_2, m.p. $-166°C$, b.p. $-72°C$, colourless, prepared NO_2Cl plus AgF or N_2O_4 plus CoF_3. $ClNO_2$, m.p. $-145°C$, b.p. $-15°C$, colourless, prepared chlorosulphuric acid plus HNO_3. Both hydrolysed by water.

n.m.r. Abbreviation for NUCLEAR MAGNETIC RESONANCE.

n.m.r. shift reagents Various paramagnetic lanthanide chelates, principally of β-diketones, useful for simplifying n.m.r. spectra because of their ability to cause large changes in chemical shifts of nuclei adjacent to an electronegative substituent in a molecule. Yb^{3+}, Pr^{3+}, Eu^{3+} particularly used.

No Nobelium.

nobelium, No. Actinide metal, at.no. 102, at. mass ^{259}No 259.10, m.p. 827°C, electronic configuration [Rn] $5f^{14}7s^2$. ^{255}No (3 min) is produced by the action of ^{10}C and ^{13}C nuclei on a curium target.

nobelium chemistry Nobelium shows only +2 and +3 oxidation states. The +2 state (f^{14}) being most stable. $No^{3+} \xrightarrow{+\ 1.45V} No^{2+} \xrightarrow{-\ 1.64V} No$

noble gases The elements helium, neon, argon, krypton, xenon and radon. All occur as minor constituents of the atmosphere and He is a component of some natural hydrocarbon gases. Separated by fractional distillation. Used (particularly He, Ar) to provide an inert atmosphere, e.g. for welding, and in electric light bulbs, valves and discharge tubes (particularly Ne). Liquid He is used for achievement of low temperatures. Kr is used as insulating material in double glazing. The amounts of He and Ar formed in minerals by radioactive decay can be used to determine the age of the specimen. Xe and to a lesser extent Kr and Rn have a chemistry; the other noble gases do not form chemical compounds although mixed species can be formed in matrices. For other uses see individual entries. See EXCIMERS.

node A region where the wave function of an atom or molecule passes through zero.

NOE See OVERHAUSER EFFECT.

nonanoic acid (pelargonic acid), $C_9H_{18}O_2$, $CH_3 \cdot [CH_2]_7 \cdot C(O)OH$. An oily liquid, with a rancid odour, m.p. 12.5°C, b.p. 253–254°C. It can be prepared by the oxidation of oleic acid or undecane. It has been found, with other fatty acids with odd numbers of carbon atoms, in small amounts in human hair. Herbicide, plant growth regulator.

non-aqueous solution A solution in which the solvent is not water but may be organic (e.g. Et_2O) or inorganic (e.g. NH_3), protonic (e.g. EtOH) or non-protonic (e.g. BrF_3). Non-aqueous solvents may be self-ionized, e.g. $2NH_3 \rightleftharpoons NH_4^+ + NH_2^-$.

non-aqueous solvents See NON-AQUEOUS SOLUTION.

non-aqueous titrations Analytical procedures carried out in organic and other non-aqueous media because of insolubility of the reactants in water, or because some acids and bases cannot be titrated in water. Thus H_2SO_4 and $HClO_4$ can be titrated as a mixture in methyl isobutyl ketone whilst amines can be titrated in 3-methylsulpholane.

non-ionic detergents See DETERGENTS.

non-polar molecule A molecule which has a zero dipole moment. Symmetrical covalent compounds, e.g. CCl_4, are non-polar.

non-soap greases See LUBRICATING GREASES.

non-stoichiometric compounds Compounds in which the ratios of the numbers of atoms present do not follow simple integers. Many transition metal oxides, silicates, etc. are non-stoichiometric and the phenomenon is common. Non-stoichiometry has great effect on physical properties, e.g. conductivity, catalysis.

nonylphenol, $C_{15}H_{20}O$. Made phenol plus nonylenes (propene trimer), contains branched chains. Used to form phenolic resins (with aldehydes) and reaction with ethylene oxide to give surfactants.

nopinene See β-PINENE.

nor- The most common use of the prefix nor- in the name of an organic compound is to indicate the loss of a CH_2 group, thus nornicotine has the formula $C_9H_{12}N_2$, while nicotine is $C_{10}H_{14}N_2$. The same convention is used with steroids. In terpene nomenclature nor- indicates the loss of all the methyl groups from the

parent compound. With amino-acids however the use of nor- turns the trivial name of a branched-chain compound to the isomeric straight-chain compound; thus norvaline is 2-amino-n-valeric acid.

noradrenaline (norepinephrine), $C_8H_{10}NO_3$. M.p. 103°C. Biosynthetically serves as a precursor for adrenaline (has N-methyl). See DICTIONARY OF BIOLOGY.

noraluron Insecticide, benzylurea derivative.

norbornadiene (bicyclo[2,2,1]hepta-2,5-diene), C_7H_8

A slightly coloured liquid, b.p. 90°C, obtained by a Diels–Alder reaction between cyclopentadiene and ethyne at approximately 150°C. Above 450°C it rearranges to cycloheptatriene (tropilidene)..

norbornene (bicyclo-[2,2,1]-hept-2-ene) M.p. –46°C, b.p. 96°C. Derivatives, e.g. -2-methanol and 2-methyl acrylate used in co-polymers and potentially in photoresists.

Nordhausen sulphuric acid See SULPHURIC ACID, FUMING.

norethisterone (17-ethinyl-19-nortestosterone) See ETHISTERONE.

norethondione ETHISTERONE derivative.

norflurazon Herbicide, pyridazinone derivative.

normality See NORMAL SOLUTION.

normal liquid A liquid in which the molecules have no tendency to associate. The simpler liquid hydrocarbons approximate to this type, while water, ethanoic acid and alcohol are partially associated.

normal solution A solution which contains 1 g equivalent of the substance in 1 litre of solution. A normal solution must have the equivalent referred to a specific reaction but this is often not done. A normal solution may be indicated by the letter N, thus, N hydrochloric acid signifies a normal solution of hydrogen chloride (for titration with base). Use of molarity is preferred.

normal temperature and pressure See STANDARD TEMPERATURE AND PRESSURE.

Np Neptunium.

NQR Nuclear quadrupole resonance.

NTA Nitrilotriacetic acid, $N(CH_2COOH)_3$.

ntp See STANDARD TEMPERATURE AND PRESSURE.

nualimol Fungicide, pyrimidinyl derivative.

nuclear isomerism See ISOMERISM, NUCLEAR.

nuclear magnetic resonance A technique for studying nuclear magnetic moments. The substance under test is placed in a stable strong magnetic field. This produces a splitting of the nuclear energy levels. The sample is then subjected to an additional, weak, oscillating magnetic field produced by passing r.f. current through a surrounding coil, and the frequency of this field is slowly 'scanned' over an appropriate range. At certain precise frequencies the nuclear magnets resonate with the field in undergoing transitions between the magnetic energy levels. The resonance is detected in a search coil and is amplified and recorded. The technique has assumed great importance for analysis of molecular structure because the resonance frequencies of a particular element are influenced by the electronic environment in which the atom is situated. N.m.r. studies can be made on nuclei with non-zero spin (see NUCLEAR PARAMAGNETISM). Three different kinds of hydrogen atom are detected in ethanol, corresponding to those in the CH_3, CH_2 and OH groups. In addition to direct information from the resonance frequencies the nuclei couple together to give fine structure in the n.m.r. spectrum. The extent and nature of the coupling provides further information on molecular structure.

Information from an n.m.r. spectrum is classified into the chemical shift, δ (the relative shift from a standard [Me_4Si for ^1H, ^{13}C; CCl_3F for ^{19}F] which is rendered independent of the field), and the coupling constants, J, which are determined directly from the spectra.

Nuclei with a large shift are described as deshielded. If the coupling constant is of the same order as the chemical shift differences the spectra are called second order; if the chemical shift differences are much greater than the coupling constant the spectra are simpler and first order.

Complex n.m.r. spectra can be simplified by the use of N.M.R. SHIFT REAGENTS or by use of higher fields or decoupling.

Nuclear magnetic resonance can be applied to solutions and solids including living material (magnetic resonance imaging). Can be used analytically, e.g. for estimation of water and oils in foodstuffs.

nuclear Overhauser effect See OVERHAUSER EFFECT.

nuclear paramagnetism The nuclei of many species have spin and the total angular momentum is given by

$$\frac{h}{2\pi}\sqrt{I(I+1)}$$

where I is the nuclear spin number ($I = 0, \frac{1}{2}, 1, \frac{3}{2} \ldots$). $I = 0$ corresponds to a nucleus without spin. Nuclei behave as magnetic dipoles and nuclear magnetic resonance techniques can be used to study nuclei with $I > 0$.

Representative nuclear moments

Isotope	Spin no. (I)	% abundance
^1H	$\frac{1}{2}$	99.98
^2H	1	0.0156
^{12}C	0	98.9
^{13}C	$\frac{1}{2}$	1.1
^{14}N	1	99.62
^{15}N	$\frac{1}{2}$	0.38
^{19}F	$\frac{1}{2}$	100
^{31}P	$\frac{1}{2}$	100

nuclear quadrupole moment The electric quadrupole moment of the nucleus. Affected by chemical environment and can be used to investigate structure (NQR spectroscopy).

nuclear spin The nucleus has a property with the characteristics of spin. The angular momentum of nuclear particles is expressed in units of $(h/2\pi)$. (h = Planck's constant). See NUCLEAR MAGNETIC RESONANCE and NUCLEAR PARAMAGNETISM.

nucleases Those enzymes which hydrolyse nucleic acids to oligonucleotides or mononucleotides. Nucleases are present in plasma, intestinal juice and the liver.

nucleic acids These are essential components of all living cells, as they carry the necessary hereditary information enabling highly specific proteins to be constructed. There are two principal types of nucleic acid; deoxyribonucleic acid, DNA, is found in the nuclei of cells; ribonucleic acid, RNA, is found mainly in the cytoplasm. They are polynucleotides; DNA is built of the nucleotides of the purines adenine and guanine and the pyrimidines cytosine and thymine (see NUCLEOSIDES). The DNA molecule consists of two helical chains coiled round the same axis, each chain consisting of alternate phosphate and sugar groups connected through carbons 3 and 5 of the pentose molecule, and with the bases attached to the pentose. The chains are generally held together by hydrogen bonds between the bases. Spatial considerations demand that of two linked bases one must be a purine and the other a pyrimidine. It is also assumed that only two specific pairs can bind, adenine with thymine and guanine with cytosine. Thus given a sequence of bases on one chain, that on the other is determined. If the chains are uncoiled and separated any new chain built on one of the originals has a copied sequence. The chains can be either right- or left-handed, apparently as criteria for molecular recognition.

In RNA the sugar is ribose and uracil replaces thymine. Different varieties of RNA exist with widely different molecular masses. Messenger RNA (transfer RNA) conveys information in the form of the sequence of bases from DNA to RNA on the ribosomes, where proteins are built up from amino-acids on the RNA template. Each sequence of three bases, called a codon, codes for a specific amino-acid. Thus the order of amino-acids on a protein is ultimately based on the order of nucleotides on the DNA molecule. See GENETIC CODE.

Transfer or soluble RNAs are specific carrier molecules for amino-acids during protein synthesis on ribosomes with ribosomal RNA as the template. There is at least one t-RNA molecule for each amino-acid.

Synthetic and natural segments of DNA and RNA transmit genetic information. See DICTIONARY OF BIOLOGY.

nucleide See NUCLIDE.

nucleon A collective name for particles of mass number 1, i.e. protons, neutrons.

nucleophilic reagents Groups which act by donating or sharing their electrons. E.g. the hydroxyl anion, halide ions, OR^- and SR^- groups, and amines. Can denote two or more pairs of electrons – see AMBIDENT ANIONS.

nucleophilic substitution Reactions proceeding in the broadest sense according to the equation

$$R - X + :Nuc \rightarrow R - Nuc + X^- \qquad \mathbf{1}$$

where R may be an alkyl, aryl, metal or metalloid group, and X and Nuc may be a wide variety of both inorganic and organic anions: in addition Nuc may be an uncharged compound with an unshared electron pair, e.g. amines, water. The mechanisms of these reactions have been

investigated, and proceed in two ways: S_n2 reactions (substitution nucleophilic bimolecular) proceed formally according to step **1** since the rate-determining step of the reaction involves two particles, i.e. the kinetics are second order: in general S_n2 reactions involve a Walden inversion and there is a 5-co-ordinate intermediate in which both X and N are bonded to the same carbon of R. S_n1 reactions (substitution nucleophilic unimolecular) are dependent upon a preliminary cleavage of the R–X bond, step **2**, in determining the rate

$$R - X \rightarrow R^+ + Nuc^- \qquad \mathbf{2}$$

of the reaction: this is slow compared to step **3**

$$R^+ + Nuc \rightarrow R - Nuc^+ \qquad \mathbf{3}$$

These reactions follow first-order kinetics and proceed with racemization if the reaction site is an optically active centre. For alkyl halides nucleophilic substitution proceeds easily: primary halides favour S_n2 mechanisms and tertiary halides favour S_n1 mechanisms. Aryl halides undergo nucleophilic substitution with difficulty and sometimes involve ARYNE intermediates.

nucleoproteins See PROTEINS.

nucleosides Glycosides of heterocyclic bases, in particular of purines and pyrimidines. They are crystalline substances, sparingly soluble in water. The nucleosides forming part of the molecule of ribonucleic acid, RNA, are the 9-β-D-ribofuranosides of adenine and guanine (adenosine and guanosine), and the 3-β-D-ribofuranosides of cytosine and uracil (cytidine and uridine). In deoxyribonucleic acid the sugar is deoxyribose and the uracil is replaced by thymine. See DICTIONARY OF BIOLOGY.

adenine, 1,H-purin-6-amine
guanine, 2-amino-1,7-dihydro-6H-purin-6-one
cytidine, 4-amino-1-β-D-ribofuranosyl-pyrimidine
uridine, 1-β-D-ribofuranosyl-uracil

nucleotides Originally applied to the phosphate esters of ribose and deoxyribose with purine or pyrimidine bases attached, which were identified as the constituents of NUCLEIC ACIDS. Now used for phosphates of glycosides of all heterocyclic bases to encompass such compounds as ADENOSINE TRIPHOSPHATE, NICOTINAMIDE ADENINE DINUCLEOTIDE and other co-enzymes, i.e. containing base–sugar–phosphate units, of importance in gene functions.

nucleus, atomic The nucleus of an atom is the small but massive internal core of the atom in which is concentrated the weight and positive charge. The weight, magnetic resonance and radioactive properties of the atom are associated with the nucleus, whereas chemical properties and u.v. and visible spectra are associated with the outer electrons. The nucleus is built up from sub-atomic particles.

nuclide A species of atom characterized by the constitution of its nucleus. Commonly used in the context of radioactive decay.

Nujol (petrolatum) A trade name for a heavy medicinal LIQUID PARAFFIN. Extensively used as a mulling agent in spectroscopy and in foodstuffs and cosmetics.

nutraceuticals Unregulated health-enhancing or disease-preventing products.

Nylon (polyamides) With $+C(O)NH+$ groupings, a class of synthetic fibres and plastics. Manufactured by condensation polymerization of α, ω-aminomonocarboxylic acids or of aliphatic diamines with aliphatic dicarboxylic acids, e.g. adipic acid and hexamethylene diamine. Also formed specifically, e.g. from caprolactam. The different Nylons are identified by reference to the carbon numbers of the diacid and diamine (e.g. Nylon 66 is from hexamethylene diamine and adipic acid). Thermoplastic materials with high m.p., insolubility, toughness, impact resistance, low friction. Used in monofilaments, textiles, cables, insulation and in packing materials.

O

O Oxygen.

OAS Optoacoustic spectroscopy.

ob See CHELATE COMPOUND.

obsidian Natural volcanic glass. Used for tool-making by pre-industrial man.

occlusion The retention of a gas or of solids by a metal, or the absorption of an electrolyte by a precipitate.

ochre Earthy iron-containing pigments, yellow to brownish red in colour.

ocimene **(3,7-dimethyl-1,3,6-octa-triene)**, $C_{10}H_{16}$. An acyclic monoterpene. Two forms. It is found in various essential oils.

octadecane derivatives Derived from $CH_3(CH_2)_{16}CH_3$, e.g. STEARIC ACID (octadecanoic acid), STEARYL ALCOHOL (octadecanol).

Z-(9)-octadecen-1-ol (oleyl alcohol) Prepared from butyl oleates or TRIOLEIN by reduction. Used to prepare sulphonic acids as surfactants, detergents, wetting agents, antifoam agents, plasticizers.

E-9-octadienoic acid See ELAIDIC ACID.

octahedral co-ordination Co-ordination of six ligands which are disposed at the corners of an

View down 3-fold axis

octahedron. SF_6 has octahedral co-ordination about sulphur. An octahedron has eight triangular faces.

octahedron See OCTAHEDRAL CO-ORDINATION.

octamethylcyclotetrasiloxane Cyclic material used for the preparation of dimethyl silicones.

octamethyltrisiloxane Linear material used to prepare silicones.

octane number Motor or aviation gasolines are designated for KNOCK RATING by the arbitrary scale of octane number. Iso-octane (2,2,4-trimethylpentane) is given a value of 100 and n-heptane a value of 0. The octane number of the gasoline is the percentage of iso-octane in a blend with n-heptane which has the same anti-knock performance as the gasoline under standard test conditions.

octanes, C_8H_{18}. Hydrocarbons of the paraffin series. There are eighteen possible paraffin hydrocarbons of this formula. They occur in petroleum and have boiling-points between 99°C and 125°C. The most important isomer is 2,2,4-trimethylpentane, $(CH_3)_3C \cdot CH_2 \cdot CH(CH_3)_2$, usually called iso-octane. Produced in large quantities by various methods from the butane–butylene fraction of the gas from the cracking of petroleum. It is a colourless liquid; b.p. 99°C. It has marked anti-knock properties and is used as a standard for determining the 'knock rating' of petrols. Used as a solvent and a thinner. Another isomer, 2-methyl-heptane, is sometimes also called iso-octane. Used in the preparation of capryl derivatives for perfumes, varnishes, oils.

n-octanoic acid (caprylic acid), $C_8H_{16}O_2$, $CH_3 \cdot [CH_2]_6 \cdot C(O)OH$. M.p. 16°C, b.p. 239°C. A fatty acid occurring free in sweat, as esters in fusel oil, from which it is most easily prepared, and as glycerides in the milk of cows and goats and in coconut and palm oils. Used in perfumery and solvents.

1-octanol Used in perfumes (as esters) and as copolymer.

2-octanol (capryl alcohol) B.p. 178°C. Occurs in glycerides from milk and in natural oils (from sodium ricinoleate). Used as foam-reducing agent in soaps, fats, and esters, as plasticizer.

octanols octyl alcohols.

octant rule Empirical rule which relates the sign and the amplitude of COTTON EFFECT curves with the geometry and substitution of cyclohexanone derivatives and has been applied especially to steroids. It may also be extended to other cyclic ketones.

octaves, law of Newlands (1863–1864) found that when the elements were arranged in order of increasing atomic weights, the first known element was similar to the eighth known, the second to the ninth, the third to the tenth and so on. He termed this relationship the 'law of octaves'. See PERIODIC LAW. The law is a natural consequence of the early part of the periodic table.

octet A group of 8 valence electrons (neglecting d electrons). Noble gases (except He) have 8 valence electrons and such arrangements have a certain stability in both ionic compounds, e.g. Na^+F^-, and covalent compounds, e.g. CCl_4. There are many compounds which do not have octets of electrons, e.g. most transition element compounds (except compounds for which the EFFECTIVE ATOMIC NUMBER RULE holds), NO (paramagnetic), BF_3 (B has 6 electrons) and SF_6 (S has 12 electrons).

octhilinone Fungicide, bactericide, thiazyl derivative.

octyl alcohols Compounds with the formula $C_8H_{17}OH$. See 2-OCTANOL.

4-*t*-octylphenol (diisobutylphenol), $C_{11}H_{13}O$, $Me_3CCH_2CMe_2C_6H_4OH$. M.p. 81–83°C, b.p. 286–288°C. Made by alkylation of phenol. Forms oil-soluble resins with methanal (salts used as oil additives) and surfactants (with ethylene oxide).

2-(octylthio)ethanol Insect repellant.

oenanthic acid See *N*-HEPTANOIC ACID.

oestrogens See ESTROGENS.

ofurace Fungicide, phenylamide derivative.

oil Naturally occurring hydrocarbons, mainly aliphatic, used as a fuel after refining and as a feedstock for the petrochemical industry. Calorific value *c.* 4.4×10^7 J kg^{-1}. World production 1998 3400 million tonnes. See CRUDE OIL.

Also used to describe oily substances obtained from natural products, e.g. oil of anise. Examples are given below. Used mainly in perfumes, soaps, flavouring, medicinals.

oil of angelica Contains phellandrene and valeric acid. Used in liquors.

oil of anise Commonly called oil of aniseed, it is obtained from star anise fruits, *Illicium verum*. It consists chiefly of ANETHOLE. Used in flavouring.

oil of bay From leaves of pimenta. Contains 50% eugenol. Used in pharmaceuticals.

oil of bitter orange 90% D-limonenen. Used in flavouring, perfumery.

oil of calamus From *Acorus calamus*. Araceael. 76% β-asarone. Used in perfumery.

oil of camphor
1 Light oil of camphor is an almost colourless fraction containing a small amount of camphor, about 30% of cineole and the remainder terpenes.
2 Dark oil of camphor consists very largely of safrole.
Used as solvent and in lacquers and detergents.

oil of caraway Obtained from *Carum carvi*. It contains about 60% of carvone. Used in lacquers.

oil of cardamom Used in flavouring.

oil of celery From celery seed. Used as flavour for soft drinks.

oil of champaca From flowers. Used in perfumery.

oil of cinnamon Volatile oil from the bark of *Cinnamomum zeylanicum*. The chief constituent of this oil is cinnamic aldehyde, $C_6H_5 \cdot CH:CH \cdot CHO$ (50%). In the U.S. the rectified oil of cassia is known as oil of cinnamon. Used in flavouring, perfumery.

oil of citronella From flowers of *Cymbopogum* (Andropogan) *nardus*. 67% geraniol. Used as perfume, insect repellant.

oil of cloves The volatile oil distilled from cloves consisting largely of eugenol. Used in confectionery and toothpaste.

oil of copaiba Contains caryophyllene and cadinene. Used in perfumery.

oil of coriander The volatile oil distilled from the ripe fruits of *Coriandrum sativum*. Its chief constituent is coriandrol, $C_{10}H_{17}OH$. Used in flavouring.

oil of cumin From *Cuminum cyminum*. Contains 30–40% cuminaldehyde. Used in food flavouring.

oil of cypress From *Cupressus sempervirens*. Contains furfural, D-pinene, D-camphene. Used in perfumery.

oil of garlic Flavouring agent.

oil of geranium From pelargonium geranium. Contains geraniol esters. Used in perfumery, ointments, manufacture of rhodinol.

oil of ginger Contains L-zingiberene. Food flavouring.

oil of juniper From berries of juniper. Contains pinene, cadinene. Used in foodstuffs and perfumery.

oil of lavender Contains linalyl esters. Used in flavouring and perfumery.

oil of lemon From lemon peel. Contains limonene, terpinene, phellandrene, pinene. Used in flavouring and perfumery.

oil of lemon grass From *Cymbopogon citratus*. Contains citral. Used as source of citral (vitamin A) and in perfumery.

oil of linaloe From a Mexican wood. Contains linalool. Used in perfumery.

oil of marjoram From *Origanum marjorana*. Contains 40% terpenes. Used in perfumes and microscopy.

oil of mustard — expressed Expressed from mustard seeds. Contains glycerides. Used in manufacture of margarine, soap, and as a lubricant.

oil of nutmeg Used (small quantities only) as a flavouring.

oil of orange From orange peel and flowers. Contains D-limonene. Used in flavouring and perfumery.

oil of origanum From wild marjoram. Contains carvacrol. Used in perfumery.

oil of parsley Used in flavouring.

oil of patchouli From *Labiatae* species. Contains patchouli alcohol. Used in perfumery.

oil of pepper From unripe fruit of black pepper. Contains L-phellandrene. Used as condiment.

oil of peppermint The oil distilled from the fresh flowering tops of *Mentha piperita*, containing about 50% of menthol together with menthyl esters and the ketone, MENTHONE. Mulled aromatic carminative. Used in flavouring.

oil of pettigrain From *Citrus vulgaris*. Contains linalyl ethanoate. Used in perfumery.

oil of rose Contains geraniol and citronallol. Used in perfumery and flavouring.

oil of rosemary From *Rosmarinus officinalis*. Contains borneol. Used in liniments and pain lotions.

oil of rue From *Ruta graveolens*. Contains methyl nonyl ketone. Used in flavouring.

oil of sweet almond Expressed from kernels of various species of prunus. Mainly glyceryl oleate. Used in perfumery, soaps, lubricants.

oil of sweet bay From laurel. Contains terpenes. Used in perfumery.

oil of turpentine (spirit of turpentine) From various pines. Contains terpenes. Used as a solvent for oils, resins, varnishes, paints.

oil of vetiver From vetiver grass. Contains vetivone. Used in soaps and perfumery.

oil of white cedar From *Thuja occidentalis*. Used in perfumery.

oil of wintergreen See METHYL SALICYLATE.

olefin complexes Co-ordination compounds in which an olefin is bonded to a metal by overlap of the π-bonding orbitals of the olefin with a suitable orbital of the metal and by back bonding from a suitable orbital of the metal into a π-anti-bonding orbital of the olefin. The olefin (or acetylene) is thus bonded sideways on to the metal. Olefin complexes are formed by many transition and non-transition metals and are intermediates during many reactions of olefins in the presence of metals. Examples

olefin polymers See POLYALKENES.

olefins See ALKENES.

oleic acid (*cis*-9-octadecenoic acid), $C_{18}H_{34}O_2$, $CH_3[CH_2]_7 \cdot CH:CH \cdot [CH_2]_7 \cdot C(O)OH$. Colourless liquid. B.p. 286°C/100 mm, insoluble in water. Oleic acid occurs naturally in larger quantities than any other fatty acid, being present as glycerides in most fats and oils. It forms one-third of the total fatty acids of cow's milk. A crude grade from inedible tallow is used in the production of lubricants, detergents, resins and other products. Isomerizes to the *trans* form elaidic acid. See VEGETABLE OILS. Herbicide, insecticide, fungicide, often as salts.

oleum See SULPHURIC ACID, FUMING.

oleyl alcohol See OCTADECEN-1-OL.

oligomer A polymer composed from only a small number of monomeric units, hence oligomerization. Thus ethene can be polymerized to polyethene or oligomerized to C_{16}–C_{20} molecules of importance for detergents, etc.

oligosaccharides Polymeric sugar derivatives. Can be synthesized by automated processes.

olive oil Cooking oil and foodstuff. Mainly mixed glycerides of oleic (83%) and palmitic acids.

olivine, Mg_2SiO_4, Fe_2SiO_4. A member of the silicates, $M_2^{II}SiO_4$ (M=Mg, Fe, Mn) containing discrete SiO_4^{2-} anions. The Mg-rich olivines are used in the manufacture of refractory bricks.

omethoate Insecticide, acaricide, organophosphate derivative.

-onium compounds Groups of compounds containing cations of the type R_xA^+ which are analogous to ammonium compounds. Ions include phosphonium R_4P^+, oxonium R_3O^+, sulphonium R_3S^+, iodonium R_2I^+.

opal Semi-precious stone; amorphous form of hydrated SiO_2. Single crystals have photonic properties.

open shell compound A species for which the total number of valency electrons does not correspond to a noble gas configuration or the OCTET rule. E.g. BF_3 (6 outer electrons), SF_6 (12 outer electrons). In the transition series, applied to compounds in which there is a partially filled d shell. Not of wide use.

opioids Natural or synthetic drugs exerting analgesic effects. See DICTIONARY OF BIOLOGY.

opium Dried latex which exudes from the unripe capsules of *Papaver somniferum*. It contains a large number of alkaloids, chief among which are MORPHINE, CODEINE and papaverine.

Oppenauer oxidation The oxidation of secondary alcohols to ketones using aluminium *tert*-butoxide propanone. Reverse of the Meerwein–Ponndorf–Varley reduction.

opsin See RHODOPSIN.

optical activity The property possessed by certain substances of rotating the plane of polarization of polarized light. The property is exhibited by certain solids and liquids (e.g. lactic acid), and may also be observed in solutions of such substances and in the vapour phase. It is associated with asymmetry either of a molecule, an ion or a crystal lattice. Molecules may be asymmetric (chiral) due to having four different groups joined to one carbon atom, or from restricted rotation about a central bond (see CHIRALITY).

Optically active substances are termed dextrorotatory or laevorotatory according to whether the plane of polarization of the light is rotated to the right or to the left with respect to the direction of incidence of the light. The prefixes *d*- and *l*- were once used to indicate which optical isomer was which, but now the correct prefixes are (+)- for dextrorotatory, and (–)- for laevorotatory compounds, with (±)- for racemic compounds. These signs are arbitrary as the rotation can change with wavelength of the light (COTTON EFFECT, OPTICAL ROTATORY DISPERSION). With carbohydrates and amino-acids the prefixes D- and L- are used to indicate configuration, not direction of rotation. The convention has been made that the configuration of D-glyceric aldehyde shall be represented by formula **1** and L-glyceric aldehyde by formula **2**.

$$\begin{array}{cc}
\text{CHO} & \text{CHO} \\
| & | \\
\text{H} - \text{C} - \text{OH} \quad & \text{OH} - \text{C} - \text{H} \\
| & | \\
\text{CH}_2\text{OH} & \text{CH}_2\text{OH} \\
\mathbf{1} & \mathbf{2}
\end{array}$$

When the formulae of other sugars are written in this way those which have the configuration

$$\begin{array}{c}
| \\
\text{H} - \text{C} - \text{OH} \\
| \\
\text{CH}_2\text{OH}
\end{array}$$

are given the prefix D-, while L- is used for those with the configuration

$$OH - \underset{|}{\overset{|}{C}} - H$$
$$CH_2OH$$

The same nomenclature has been adopted for amino-acids, the configurational family to which the α-carbon atom belongs being denoted by the prefixes D- and L-.

The R, S convention is a scheme which has largely superseded the D, L system to denote configuration about a chiral centre in a molecule. The convention allows unequivocal designation of the configuration in a description of the positions in space of ligands attached to a chiral centre in relation to an agreed standard of chirality like a right-hand helix. See CONFIG-URATION.

Groups attached to the chiral centre are given an order of priority according to the SEQUENCE RULES.

optical activity index Observed as the differential interaction of molecules with the two hands of circularly polarized light (CHIROPTIC-AL SPECTROSCOPY). The phase difference, $\Delta\theta$, between the two components is

$$\Delta\theta = (n_R - N_L) \cdot \frac{2\pi l}{\lambda}$$

where n_R and n_L are the refractive indices, l is the sample length and λ the wavelength.

optical brighteners See BLEACHING AGENTS.

optical electrons In general, only the outermost electrons in an atom are concerned in producing the emission or absorption of visible light. Such electrons are termed optical electrons.

optical exaltation An abnormal increase in the refractivity of an organic compound, over and above the additive amount to be expected from the bonds present, due to the presence of a system of conjugated double bonds.

optical purity The ratio of a substance's specific rotation to the specific rotation of the pure enantiomer.

optical rotation Rotation of the plane of polarized light generally by an optically active material. The specific rotation depends on concentration, path length, and the material causing rotation.

optical rotatory dispersion (ORD) A description of the changes in optical rotation of an optically active molecule, organic or inor-ganic, with the wavelength (frequency) of the light. The variation of specific rotation with wavelength increases as an absorption band is approached, and the rotation may pass through a maximum or minimum when the absorption band is crossed. This is the so-called COTTON EFFECT. The effect is particularly sensitive to structural features of many molecules and is used in structural and stereochemical studies. ORD may be induced by a magnetic field, see MAGNETIC POLARIZATION OF LIGHT.

optoacoustic spectrometry A spectro-photometric technique whereby modulated electromagnetic radiation falling upon the sample causes heating when absorption occurs. The heating causes a pressure change in the cell containing the sample, which is detected by a microphone. The technique can be used to examine most regions of the electromagnetic spectrum.

orbencarb Herbicide, thiocarbamate derivative.

orbital Loosely used to describe the geometrical figure which describes the most probable location of an electron. More accurately an allowed energy level for electrons. See ELEC-TRONIC CONFIGURATION.

orbital angular momentum (L) The total angular momentum of a set of electrons in an atom or ion. For filled and empty shells, L is zero.

orbital electron capture Electron capture by the nucleus in which a proton forms a neutron and the atomic number decreases by one.

orbital magnetic moment The component of the magnetic moment of an atom or ion resulting from the motion of the electron around the nucleus.

orbital splitting The raising of the degeneracy of a series of orbitals of equivalent energy, e.g. by placing in a CRYSTAL FIELD.

ORD OPTICAL ROTATORY DISPERSION.

order of reaction The total power to which the concentration terms are raised in the mathematical equation which describes the variation of the rate of a reaction with concentration of reactants at a given temperature. For example, if in the reaction of two substances A and B the rate of reaction is proportional to the concentrations of each reactant, the reaction is said to be first order in A and in B and second order overall. The order may be other than unity and in some cases may be non-integral. It is an experimentally determined quantity which does not necessarily have any relationship to the actual number of molecules involved in the reaction since many reactions proceed in a series of steps. The order refers only to the slowest step.

ore dressing See MINERAL DRESSING.

organic Until the early part of the nineteenth century chemical substances of animal or plant origin were designated as 'organic' and held to differ fundamentally from 'inorganic' substances of mineral origin in that a 'vital force' was required for their production. The 'vital force' theory was discredited in 1828 by Wöhler but the term 'organic' has remained. Organic chemistry is now the study of the compounds of carbon, whether they be isolated from natural sources or synthesized in the laboratory. Most, but not all, organic compounds contain H as well as C, while other common elements are O, N, the halogens, S and P. These and other elements are usually bound to carbon by covalent bonds. A few very simple carbon compounds, e.g. metallic carbonates, are considered as inorganic, and the large class of organometallic compounds is of interest to both inorganic and organic chemists.

organoboranes Compounds containing at least one direct carbon-to-boron bond. The commonest classes are the trialkylboranes, R_3B, and the dialkylboranes, R_2BH. Developed as extremely useful organic synthetic reagents, they are decomposed by aqueous acid to alkanes and by alkaline peroxide to alcohols: the latter reaction provides a convenient method, via hydroboration, for the effective *cis*-hydration of alkenes in an anti-Markovnikov direction. Acid hydrolysis of the hydroboration product of internal acetylenes provides a convenient synthesis of *cis*-alkenes.

Methods of producing B–C bonds include HYDROBORATION (direct addition of B–H bonds across the double bonds of olefins), nucleophilic displacement at a boron atom in BX_3 (X = halogen or $B(OR)_3$) by, e.g. a Grignard reagent, and a *pseudo*-Friedel–Crafts reaction with an aromatic hydrocarbon, BX_3, and $AlCl_3$. Ph_3B and $(Pr^0O)_3B$ are used in processes for the manufacture of adiponitrile.

organoborates Compounds formally containing BR_4^- anions although some derivatives, particularly lithium compounds, e.g. $LiBMe_4$, are polymeric and covalent. Most easily prepared from metal alkyls and trialkylboranes.

Postulated as intermediates in reactions between organoboranes and nucleophiles. The BR_4^- ions react with electrophiles in synthetically useful reactions. There may be intermolecular transfer to E^+, e.g. $RCOX$ reacts with BR'_3 to give ketones $RCOR'$, or intramolecular transfer particularly to the α-carbon, e.g.

$$NaR_3B - C \equiv CR' \xrightarrow{\text{HCl}} R_2BCR' = CHR'$$

The organoboranes formed by such reactions may be decomposed to olefins.

organochlorine insecticides Highly active insecticides, formerly widely used. They find their way into the bodies of animals and insects and are stored, principally in the fat. The application of organochlorine insecticides is now widely restricted. Organochlorine compounds are produced by some lilies as antifungal compounds.

organoelement compounds Compounds of an element containing organic residues bonded directly to the element. See ORGANOMETALLIC COMPOUNDS. Formed by all elements except for the noble gases.

organolithic Composites of organics and silicates.

organometallic compounds Compounds in which carbon atoms or organic groups are linked to metal or metalloid atoms. Metal isonitrile and carbonyl derivatives are considered as organometallic although carbides and cyanides are not generally so classified. Organometallic derivatives range from substantially ionic, e.g. $NaCH_3$, to covalent, e.g. $H_3CMn(CO)_5$. Widely used in synthesis (e.g. $RMgX$, Grignard reagents) and catalysis (e.g. Ziegler catalysts and aluminium alkyls). Cobalamins are cobalt complexes of corrinoids and are naturally occurring organometallics – include Vitamin B_{12}.

organometalloids Compounds containing a direct bond between one or more carbon atoms and one or more metalloid atoms.

organophosphorus compounds Compounds containing C–P bonds but often extended to esters and thio-esters (and so listed in this Dictionary). The most important application of organophosphorus compounds is in the field of pesticides.

There are some human medical problems associated with organophosphorus compounds.

Organophosphorus compounds are widely used as ligands in metal-complex chemistry, including some catalytic processes; triphenylphosphine is the commonest such ligand. Several classes of organophosphorus compounds are employed in organic synthesis, e.g. triphenylphosphine in the Wittig synthesis of alkenes and HMPA (hexamethylphosphoramide) as a solvent. PHOSPHATE ESTERS are used for flame-proofing cotton and synthetic materials, as gasoline additives and for solvent extraction. Used as pesticides.

organosilicon compounds R_nSiX_{4-n} but also oxides, etc. Halides formed from R_4Si and halogen. Used in production of silicones. Derivatives with SiOR bonds occur naturally; part of the structural material of plants.

organosol A collodial dispersion in which an organic dispersion medium is used. Specific names such as alcosol, benzosol may be used to indicate that the dispersion medium is alcohol or benzene. Name used for, e.g. cellulose nitrate, collodions, rubber solutions in toluene or naphtha.

organotin derivatives See TIN, ORGANIC DERIVATIVES. Used as fungicides.

Orgel diagrams Simple graphs showing the relation between the energies of various electronic states and the crystal field splitting.

ornithine (2,5-diaminovaleric acid), $C_5H_{12}N_2O_2$, $NH_2 \cdot CH_2 \cdot CH_2 \cdot CH_2 \cdot CH(NH_2) \cdot C(O)OH$. M.p. 140°C. An amino-acid occurring in the urine of some birds as dibenzoylornithine. Ornithine is a precursor of arginine in plants, animals and bacteria.

orotidine (5´-monophosphate decarboxylase) Enzyme that catalyses last steps in biosynthesis of uridine monophosphate. Accelerates this reaction by factor of 10^{17}.

ortho In the case of disubstituted derivatives of benzene the prefix ortho has a precise structural significance. Thus in ortho-cresol, commonly written o-cresol; the CH_3 and OH groups

are said to be ortho to each other. Using numbers to designate substituent position o-cresol is 2-hydroxytoluene.

The prefix is also used in orthophosphates, orthocarbonates, orthoformates and orthosilicates, which are derivatives of $PO(OH)_3$, the hypothetical $C(OH)_4$, $HC(OH)_3$ and $Si(OH)_4$ respectively; in these the word 'ortho' is always written in full.

orthoformic ester See TRIETHYL ORTHOFORMATE.

ortho-hydrogen The hydrogen molecule, H_2, exists in two forms because of differences in the relative spin of the nuclei of the two atoms. When spins are parallel we have *ortho*-hydrogen; when the spins are symmetrical, or in opposite directions, we have *para*-hydrogen. Ordinary hydrogen is a mixture of the two forms; at normal temperatures the composition being 25% *para* and 75% *ortho*. Almost pure *para*-hydrogen is prepared by cooling ordinary hydrogen to a very low temperature in contact with charcoal or over metals which act as catalysts for the conversion. The two forms have identical chemical properties but slightly different physical properties. *Ortho*-hydrogen is unstable with respect to *para*-hydrogen.

orthorhombic system The CRYSTAL SYSTEM with a 2-fold axis at the intersection of two planes of symmetry or perpendicular to two other 2-fold axes. A unit cell has three unequal axes at right angles.

Os Osmium.

osazones Organic compounds containing two (C=N–NHAr) groups on adjacent carbons. Formed by treating α-diketones, α-hydroxyaldehydes, hydroxyketones, aminoaldehydes or aminoketones with arylhydrazines. Sugars can be identified by their osazones, which have characteristic melting-points, formation times or crystal appearance.

osmates See OSMIUM OXIDES.

osmiamates See OSMIUM OXIDES.

osmic acid The name given to osmium tetroxide and its solution in water. This solution contains free OsO_4. In alkaline solution $[OsO_4(OH)_2]^{2-}$ ions are present.

osmiridium Native or synthetic alloys of osmium (15–40%) and iridium (50–80%). It is used for special purposes in which a very hard and incorrodible material is required, such as the tips of fountain pen nibs and sparking points.

osmium, Os. Transition metal, at.no. 76, at. mass 190.23, m.p. 3033°C, b.p. 5012°C, density ρ 22 590, crustal abundance 1×10^{-4} p.p.m., electronic configuration [Xe] $4f^{14}5d^66s^2$. Osmium is a platinum metal of Group 8. It occurs as a sulphide or as osmiridium (extraction see RUTHENIUM). The metal is hcp and has very similar properties to Ru. Oxygen at high temperatures gives OsO_4. Os is used for hardening purposes in alloys with other platinum metals (osmiridium) and as a catalyst in hydrogenation and ammonia production.

osmium chemistry The chemistry is similar to that of Ru but there is greater stability of the higher oxidation states (halides and oxides); oxidation states from +8 to 0 are known. $Os(CO)_5$ and some polymeric carbonyls are formed and the organometallic chemistry is similar to that of Fe. Os(III) and Os(II) (predominantly octahedral) form extensive series of complexes with N-, P- and As- ligands. Osmium nitrosyls are very stable, as are nitrido derivatives (see also OXIDES).

osmium halides Yellow OsF_7 (Os plus F_2 at 60°C and 400 atm) dissociates to OsF_6 very readily. Pale green OsF_6 (Os plus F_2) m.p. 32°C, blue tetrameric OsF_5 and yellow OsF_4 (both OsF_6 plus $W(CO)_6$) are the known fluorides. Oxide fluorides, e.g. $OsOF_5$ (Os plus F_2 plus O_2) and complex fluorides, e.g. $[OsF_6]^-$, $[OsF_6]^{2-}$ are known. $(OsCl_5)_2$ and red $OsCl_4$ (Os plus Cl_2 at 600°C) decompose on heating to green $OsCl_3$. Chloroosmates(IV) and (III), $[OsCl_6]^{2-}$ and $[OsCl_6]^{3-}$, are useful starting materials in Os chemistry. $OsBr_4$, $OsBr_3$ and OsI_3 and some oxide halides and oxide halide complexes are known.

osmium oxides Colourless OsO_4 (Os plus O_2 at high temperatures), m.p. 25°C, b.p. 100°C, is a toxic material with a penetrating odour. It is used as an oxidizing agent; alkenes give dihydroxy species. Dark-coloured OsO_2 (Os plus OsO_4) has the rutile structure.

In strongly alkaline solution OsO_4 gives red osmates(VIII), $[OsO_4(OH)_2]^{2-}$, which can be reduced to pink or blue osmates(VI), $[OsO_2(OH)_4]^{2-}$. Osmates(VII) are formed by heating mixed oxides in oxygen. Osmates(VI) undergo substitution by anions, e.g. CN^- to give $[OsO_2(CN)_4]^{2-}$.

Osmiamates, e.g. $KOsO_3N$ ($K_2OsO_4(OH)_2$ plus ammonia) are stable and related species, e.g. $[Os^{VI}NCl_4]^-$, can be formed by substitution.

osmosis When a solvent and a solution are separated by a semi-permeable membrane, such as a piece of parchment or a plant or animal membrane, the solvent tends to flow spontaneously through the membrane into the solution. This process is known as osmosis. A similar flow is observed when two solutions of different

chemical potential are separated by a semipermeable membrane.

osmotic pressure (π) The excess hydrostatic pressure which must be applied in order to counterbalance the process of osmosis. Thus a solution under this excess pressure can come to osmotic equilibrium with the same solvent on the other side of a membrane permeable only to the solvent molecules. For dilute solutions of non-electrolytes the osmotic pressure is equal to the pressure they would exert if they existed as gases under the same conditions of molecular concentration and temperature. For dilute solutions the osmotic pressure (π) is given by

$$\pi = \frac{n_B RT}{V}$$

where n_b/v is the concentration of solute, R is the gas constant and T is the absolute temperature. Osmometry may be used for the determination of molar mass.

Ostwald dilution law Variation of molar conductivity with concentration.

Ostwald ripening A process of crystal growth in which a mixture of coarse and fine crystals of a substance are left in contact with a solvent. This results in a growth of the large crystals at the expense of the fine crystals.

ounce (oz) Non-S.I. unit of weight. Used chemically for platinum metals. 1 oz = 3.110 × 10^{-2} kg.

outer sphere complexes Weak complexes in which there is minimal disturbance of the inner co-ordinated ligands. Postulated as the mechanism in many electron transfer reactions.

ovalbumin (egg albumin) The chief polypeptide constituent of white of egg. Used as foodstuff.

Overhauser effect (nuclear Overhauser effect) Transfer of energy from, e.g. ^1H, to another nucleus, e.g. ^{13}C, which enhances n.m.r. intensities.

overtones Weak vibrational absorption corresponding to multiples or adducts of fundamental frequencies.

over-voltage (over-potential) The excess potential, over and above that of the reversible electrode, that must be applied to an electrode for electrolysis to proceed at a measurable rate. E.g. at a platinized platinum electrode the evolution of hydrogen will occur at practically the reversible electrode potential. However, at a mercury electrode no hydrogen is evolved until an over-voltage of approximately 1 volt is reached. Of great practical importance in processes such as electroplating, electro-analysis, electro-reduction, polarography, etc. Overvoltage occurs because of irreversible processes at the electrode or depletion of the electrolyte close to the electrode.

oxabetrinil Herbicide safener, acetonitrile derivative.

oxadiareyl Herbicide, oxadiazyl derivative.

oxadiazon Herbicide, oxadiazyl derivative.

oxadixyl Fungicide, phenylamide derivative.

oxalates Salts or esters of oxalic acid.

oxalic acid (ethanedioic acid), $C_2H_2O_4$, $HO(O)C \cdot C(O)OH$. Crystallizes from water in large colourless prisms containing $2H_2O$. It is poisonous, causing paralysis of the nervous system; m.p. 101°C (hydrate), 189°C (anhydrous), sublimes at 157°C. It occurs as the free acid in beet leaves, and as potassium hydrogen oxalate in wood sorrel and rhubarb. Commercially oxalic acid is made from sodium methanoate. This is obtained from anhydrous NaOH with CO at 150–200°C and 7–10 atm. At lower pressure sodium oxalate formed; from the sodium salt the acid is readily liberated by sulphuric acid. Oxalic acid is also obtained as a by-product in the manufacture of citric acid and by the oxidation of carbohydrates with nitric acid in presence of V_2O_5 and by the carboxylation of alcohols in the presence of NO and a palladium salt.

When heated with sulphuric acid the acid is decomposed to CO, CO_2 and H_2O. Gives methanoic acid when heated with glycerin. Forms both acid and neutral salts. The normal potassium and ammonium salts are soluble in water; the corresponding acid salts are less soluble and the salts of other metals are only sparingly soluble. The calcium salt is very insoluble. Oxalic acid is used for metal cleaning and as a chemical intermediate. Because of its bleaching action it is widely employed for textile finishing and cleaning, and is also used for whitening leather.

A common domestic use is the removal of rust or blood stains from white materials, e.g. cotton.

oxamide, $C_2H_4N_2O_2$, $(C(O)NH_2)_2$. Colourless crystalline solid from ethyl oxalate and concentrated ammonia. Forms cyanogen when heated with P_2O_5.

oxamyl Insecticide, nematocide, oxine carbamate derivative.

oxanthrol Reduction product of anthraquinone.

oxasulfuron Herbicide, sulphonylurea derivative.

oxazole ring The ring numbered as shown.

oxazolidinones Class of antibiotics.

oxetane polymers Polymers formally derived from OXETANE $\overline{CH_2CH_2CH_2O}$, many of the oxetane derivatives being derived from pentaery-thritol so that the repeating unit is $(CH_2CR_2CH_2O)n$. Polymerization is with Lewis acids, particularly BF3.

Poly (3,3-*bis* (chloromethyl) oxetane), chlorinated polyether (R=CH_2Cl) prepared by a Williamson reaction. Widely used for injection moulding and coating.

oxetanes Derivatives of oxetane C_3H_6O.

$$CH_2CH_2CH_2O$$

Formed by an intramolecular Williamson reaction on, e.g. pentaerythritol derivatives, pyrolysis of cyclic carbonate esters or photochemical addition of aldehydes or ketones to olefins. The commercially important 3,3-*bis*(chloromethyl)oxetane, b.p. 90°C, is prepared by a Williamson reaction; oxetanes readily polymerize with Lewis acids to give useful OXETANE POLYMERS.

oxidases Enzymes that participate in oxidation reactions.

oxidation Any process by which the proportion of the electronegative constituent in a compound is increased, e.g. $Cu_2O \rightarrow CuO$. In oxidation electrons are removed from the species oxidized, e.g. $Cu^+ \rightarrow Cu^{2+} + e$, and the OXIDATION STATE is increased. In organic chemistry particularly the most obvious effect is to increase the proportion of oxygen, e.g. $C_2H_5OH \rightarrow CH_3CO_2H$. For specific selective oxidizing agents see e.g. OPPENAUER OXIDATION, SELENIUM DIOXIDE, CHROMIUM TRIOXIDE, OSMIUM TETROXIDE.

β-oxidation The iterative degradation of fatty acids in biological systems which removes a 2-carbon fragment each time.

oxidation–reduction indicator
See REDOX INDICATOR.

oxidation–reduction potential
See REDOX.

oxidation state The difference (+ve or –ve) between the number of electrons associated with an atom in a compound as compared with an atom of the element. In ions corresponds to the ionic charge, e.g. NaCl[Na (+1), Cl (–1)]; in covalent compounds to the formal charge assuming an ionic structure, taking account of electronegativity to determine charge, e.g. CCl_4 is assumed to be $C^{4+}4Cl^-$ and thus C (+4). Coordinate bonds do not affect the oxidation state, e.g. $[Co(NH_3)_6]^{3+}$ contains Co (+3) and $Fe(CO)_5$ contains Fe (0). Elements are in the zero oxidation state. The oxidation state, Roman numerals, is used in systematic nomenclature, e.g. MnO_4^- is manganate(VII).

oxidative addition Reaction in which there is direct addition to an element or multiple bond with increase in oxidation state. E.g. addition of Cl_2 to PCl_3 gives PCl_5.

oxidative phosphorylation The process occurring along the respiratory or ELECTRON TRANSPORT CHAIN by which ATP is generated through a series of coupled reactions from reduced coenzymes, such as NADH.

oxide A compound of oxygen with another element. Oxides are divided into acidic oxides, which react with bases to form salts (e.g. SO_2, P_2O_5); basic oxides which react with acids to form salts (e.g. CuO, CaO); amphoteric oxides, which exhibit both basic and acidic properties (e.g. Al_2O_3); neutral oxides (e.g. CO, NO) which do not react with either acids or gases; peroxides, derivatives of hydrogen peroxide; and superoxides containing O_2^- species. Free acids and gases cannot always be dehydrated to oxides. Many oxides (e.g. FeO) are non-stoichiometric and contain metal atoms in two or more oxidation states. Most solid-state oxides have structures with close-packed oxide ions. Non-stoichiometry may increase on heating (e.g. ZnO) and may cause marked colour changes. Fusion of two or more oxides often gives mixed metal oxides which frequently do not contain discrete oxyanions. Often conducting and may contain magnetic domains.

oxidized rubber (Rubbone) Produced by oxidation of rubber with air in the presence of a catalyst. Can be vulcanized and chlorinated. Used for impregnation, electrical insulation and as an adhesive.

oxidizing agent A material which brings about oxidation and in the process is itself reduced. E.g. MnO_4^- in acid oxidizes $Fe^{2+} \rightarrow Fe^{3+}$ and is itself reduced to Mn^{2+}.

oximes Organic compounds containing the group $=N \cdot OH$ united directly to a carbon atom. See ALDOXIMES and KETOXIMES.

oxine 8-hydroxyquinoline. Oxine copper fungicide.

oxirane The systematic name for the oxygen-

containing 3-membered heterocycle, more commonly called an epoxide.

Oxitol See CELLOSOLVE.

oxolinic acid Bactericide, quinoline derivative.

oxonium A positive ion containing a central oxygen atom. The hydroxonium ion, H_3O^+, exists in aqueous solutions of acid, and solid salts containing this ion may be prepared. The trimethyloxonium, Me_3O^+, and triphenyloxonium, Ph_3O^+, ions and their derivatives are also known. Trialkyloxonium salts are strong alkylating agents for organic and inorganic nucleophiles.

oxopyrazolines See PYRAZOLONES.

oxo reaction (hydroformylation) A reaction which involves the addition of water gas or synthesis gas ($H_2 + CO$) to an alkene under pressure in the presence of a catalyst. The latter may be mixtures of oxides of cobalt, thorium and copper on kieselguhr, a cobalt carbonyl or rhodium phosphine complexes. The products are aldehydes which may be further hydrogenated using the same catalysts to primary alcohols. A modification using carbon monoxide and water instead of water gas yields carboxylic acids. A terminal alkene $R \cdot CH = CH_2$ usually yields a mixture of terminal aldehyde $R \cdot CH_2 \cdot CH_2 \cdot CHO$ and branched aldehyde $R \cdot CH(CHO) \cdot CH_3$ depending upon catalyst and conditions. These may be converted to the mixed alcohols, especially where used for plasticizers or detergent manufacture. See CARBONYLATION, HYDROFORMYLATION.

oxyanions Anions which may formally be considered as being formed by co-ordination of oxide ions, O^{2-}, to metal or non-metal cations, e.g. the sulphate ion SO_4^{2-} is formally built up from S^{6+} and $4O^{2-}$ ions. Oxyanions are the normal species in water as complex fluoroanions are the normal species in anhydrous hydrogen fluoride.

oxyazo dyes Azo dyestuffs containing a hydroxyl group.

oxycarboxin Fungicide, amide derivative.

oxycyanogen, $(OCN)_2$. A pseudohalogen (KOCN plus Cl_2).

oxydemeton methyl, $C_6H_{15}O_4PS_2$. A systemic and contact insecticide.

oxyflurofen Herbicide, diphenyl ether derivative.

oxygen, O. Diatomic gas, at.no. 8, at.mass 15.999, m.p. -218.79°C, b.p. -182.95°C, density gas 1.429 g l^{-1}, density liquid 1.308 g l^{-1}, crustal abundance 474 000 p.p.m. (atmosphere 20.9%), electronic configuration [He] $2s^22p^4$. Colourless, odourless gas condensing to a pale-blue liquid and gas. Under high pressure forms a red polycrystalline phase and gives a superconducting metal at even higher pressures. Occurs in air (21% by volume) and by far the most abundant element (lithosphere 47% by weight, oceans 89%). Essential for most forms of life on earth; carried through body attached to the iron of heme or myoglobin. Manufactured by distillation of liquid air and can be prepared in laboratory by heating $KClO_3$ with MnO_2. Reacts with most elements on heating to form oxides. Exists as paramagnetic O_2 and as ozone, O_3. Used in steel making (65%), in synthesis (e.g. HNO_3, H_2SO_4, synthesis gas, ethylene oxide), in welding, in mining, explosives (see LIQUID OXYGEN EXPLOSIVES), in rocket fuels, sewage digestion. U.S. production 1999 697 kilotonnes.

oxygen carrier A molecule which can react directly with molecular oxygen (to form a peroxide or superoxide derivative, a complex) and can subsequently use the oxygen in an oxidation reaction – generally of an organic substrate, e.g. hemoglobin.

oxygen cathode A platinum or gold electrode used in amperometry for measuring oxygen content by measuring the rate of cathodic reduction of O_2. The cathode is used with a reference electrode. Can use enzymes immobilized on the electrode to measure O_2 uptake. A Clark electrode is a typical oxygen electrode.

oxygen chemistry Oxygen is an electronegative Group 16 element, electronic configuration $1s^22s^22p^4$. Its predominant chemistry is in the -2 oxidation state in OXIDES (covalent, or ionic), oxygen halides, alkoxides (ROM compounds) and ethers. The M–O bond order may approach 2 in compounds such as F_3PO. The oxygen commonly is 2 (bent) co-ordinate but co-ordination numbers of 3 (R_3O^+ and TiO_2), 4 (Cu_4OCl_6), 6 (MgO) are known. O–O bonding is fairly unusual but is known in e.g. O_2, O_2^-, O_3 and peroxy compounds. The di-oxygenyl $(O_2)^+$ cation (O_2 plus PtF_6 gives O_2PtF_6; O_2F_2 plus BF_3 gives O_2BF_4) shows that other oxidation states can exist. $[O_2]^-$ and $[O_2]^{2-}$ formally contain other oxidation states. Oxygen in many compounds can act as ligands.

oxygen fluorides *Oxygen difluoride*, OF_2. Prepared by passing F_2 through dilute cooled NaOH solution. M.p. -224°C, b.p. -145°C. It is only slowly hydrolysed by water at ambient temperature, and is a strong oxidizing agent when sparked.

Other oxygen fluorides, O_2F_2, O_4F_2, O_5F_2 and O_6F_2, are prepared by passing an electric spark through a mixture of O_2, F_2 and sometimes O_3.

oxymercuration Addition of OH and HgO(O)CR to a multiple bond.

oxytetracyclin Widely used antibiotic.

oxythallation Thallium(III) salts oxidize a wide variety of organic substrates under mild conditions in alcohol, dilute acid or diglyme solution. Unstable organothallium derivatives are produced as intermediates.

oxytocin A polypeptide hormone secreted by the posterior lobe of the pituitary gland.

ozalid process See BLUE PRINT PAPER.

ozokerite A naturally occurring mineral wax which on refining yields a hard white wax known as CERESIN.

ozone, O_3. Allotrope of oxygen, m.p. -193°C, b.p. -112°C, blue in colour. Present in upper atmosphere and absorbs u.v. radiation and hence protects earth's surface from excess u.v. radiation. Holes can be caused in the ozone layers because of the reaction of ozone with species formed by, e.g. CFCs. Ozone is formed at ground level by interaction of nitrogen oxides, NO_x, with hydrocarbons (from auto emissions). It is a respiratory irritant and can cause crop damage. Prepared by action of electric discharge on O_2 and can be separated from O_2 by fractionation. Has bent molecule. Used for oxidation, sterilization and purification.

ozonides, $-HC(\mu-OO)(\mu-O)CH-$. Products of the action of ozone on various classes of unsaturated organic compounds. Prepared by passing ozonized air or oxygen into the unsaturated compound dissolved in a suitable solvent. Thick oils with an unpleasant choking smell. Readily decomposed, some by water or ethanoic acid and formerly used to establish the position of the double bond in the original unsaturated compound. Inorganic ozonides, such as potassium ozonide, KO_3, also exist. These contain the O_3^- ion.

ozonizer Apparatus for the preparation of ozone by passing oxygen through an electrical discharge.

ozonolysis A procedure for the addition of ozone to an unsaturated organic compound, giving an ozonide which can be unstable and is not normally isolated but is decomposed *in situ*. E.g. an unsymmetrical alkene gives 2 ketones

$$\frac{R^1}{R^2}\!\!>\!\!C = C\!\!<\!\!\frac{R^3}{R^4}\xrightarrow[\text{decomp.}]{O_3}\frac{R^1}{R^2}\!\!>\!\!C = C + \frac{R^3}{R^4}\!\!>\!\!C = O$$

after decomposition of the ozonide with zinc and dilute aqueous acid.

P

P Phosphorus.

Pa Protactinium, pascal.

PAB See 4-AMINOBENZOIC ACID.

packed column (packed tower) Used for gas–liquid contacting, for GAS ABSORPTION, for small-scale industrial distillation operations, and for LIQUID–LIQUID EXTRACTION.

packing fraction The mass of an atom is not the simple sum of the masses of the elementary particles making up the atom, since some of the mass is converted to energy ($E = mc^2$) which is used to hold the nucleus together. The difference between the actual atomic mass and the mass number (the total mass of the protons and neutrons which constitute the nucleus) is termed the mass defect. The packing fraction is defined as the mass defect divided by the mass number. For the lightest and heaviest elements, the packing fraction is positive; intermediate elements have a negative packing fraction and hence have greater stability.

The term packing fraction is also applied to the space occupied in close-packed structures.

paclobutrazol Plant growth regulator, azole derivative.

PAH Polymeric hydrocarbon found in meteorites.

paint Covering material for wood, metal, etc., used to improve appearance and to protect the substrate against e.g. rust. The basic components of paints are pigment (decoration but also protects against sunlight), the non-volatile vehicle (drying oils, epoxy resins, etc.) which holds the pigment in place and the volatile vehicle (water, hydrocarbon solvents) which allows ease of application and eventually evaporates. For metallic paints see ALUMINIUM. U.S. production 1999 1.3 million gallons.

palladium, Pd. Metallic element, at.no. 46, at.mass 106.42, m.p. 1555°C, b.p. 2963°C, density ρ 12 020, crustal abundance 6×10^{-4} p.p.m., electronic configuration [Kr] $4d^{10}$. The least noble of the platinum metals, resembling Pt in its mechanical properties. Pd compounds are readily reduced to the metal which has the ccp structure. The metal is dissolved by conc. HNO_3 or hot H_2SO_4; it is oxidized in air at dull red heat. Used in alloys (white gold, Au plus Pd) (8%), as a catalyst (particularly for hydrogenation and exhausts) and in dental work (31%) and in electrical components (44%). World demand 2001 730 000 oz.

palladium ammines An extensive range of palladium(II) complexes, generally 4-coordinate, containing Pd–NH_3 (or amine) bonds.

palladium black A finely divided precipitate obtained by reducing solutions of Pd salts. Used as a hydrogenation catalyst.

palladium chemistry Palladium is an element of Group 10. The chemistry is similar to that of Pt although the +6 and +5 oxidation states are missing. There are extensive ranges of complexes. Organometallic derivatives and hydrides are much less stable than for Pt. Pd forms an aquoion $[Pd(H_2O)_4]^{2+}$ and salts containing this ion are known. Complexes of Pd(O), e.g. Pd(PPh$_3$)$_3$, are important catalysts.

palladium halides Palladium fluorides known are brick-red PdF$_4$ (PdF$_3$ plus F$_2$ under pressure), black PdF$_3$ (Pd^{II}Pd^{IV}F$_6$; PdCl$_2$ plus BrF$_3$) and violet PdF$_2$ (PdF$_3$ plus SeF$_4$). Fluoropalladates(IV), containing [PdF$_6$]$^{2-}$ ions, are formed from BrF$_3$ solution; all palladium fluorides are hydrolysed by water. The only known chloride is PdCl$_2$ used in photography and electroplating (Pd plus Cl$_2$); [PdCl$_6$]$^{2-}$ and [PdCl$_4$]$^{2-}$ species are formed in solutions containing hydrochloric acid. PdBr$_2$ and PdI$_2$ are also known.

Extensive series of complexes are formed with PdX_2.

palladium hydrides Palladium absorbs hydrogen up to $PtH_{0.6+x}$. Hydrido complexes, e.g. $[PtHBr(PEt_3)_2]$, are formed by reduction of the dihalides. Hydrido derivatives are presumably intermediates in hydrogenation reactions using Pd catalysts. Pd/H involved in hydrogen sensors.

palladium organic derivatives Palladium alkyls and aryls are much less stable than those of Pt. Pd–olefin complexes and allyl derivatives, e.g. $(C_3H_5PdCl)_2$, are well established. They are of major importance in catalysis, e.g. the WACKER PROCESS.

palladium oxide The only known oxide is PdO, formed by direct combination of the elements at about 800°C or by ignition of salts. It decomposes to the elements at 875°C. A hydrated form is precipitated from solution. Used as a catalyst.

palmitic acid (hexadecanoic acid), $C_{16}H_{32}O_2$, $CH_3 \cdot [CH_2]_{14}C(O)OH$. Crystallizes in needles, m.p. 63°C, b.p. 351°C. Widespread fatty acids occur as glycerides in most animal and vegetable fats and oils and as esters of alcohols other than glycerol in various waxes. The best sources are palm oil and Japan wax. It forms 40% of the total fatty acids of cows' milk. A solid mixture of palmitic and stearic acids, 'stearine', is used for making candles, and the sodium and potassium salts of palmitic and stearic acids are soaps. Used in lubricants, liniments. Palmitic acid can be obtained by hydrolysing natural fats with superheated steam when it is the first fatty acid to distil over.

palm oil Source of palmitic acid.

palygorskite A form of FULLER'S EARTH.

pamoic acid (embonic acid, 2,2´-dihy-droxy-1,1´-dinaphthyl-methane-3,3´´-dicarboxylic acid), $C_{23}H_{16}O_6$. Used to form salts with organic bases; used in medicine where insolubility in water and lack of absorption from the alimentary canal are required.

panchromatic sensitization Extension of the sensitivity of photographic silver halide emulsions to the whole of the visible spectrum by addition of sensitizing dyes.

pancreatine Mixture of peptides from pig pancreas. Converts starch into soluble carbohydrates.

pantothenic acid (vitamin B_5), $C_9H_{17}NO_5$, $HOCH_2CMe_2CHOHNHCH_2CH_2C(O)OH$. A constituent of coenzyme A. Important in fatty-acid metabolism.

papain Proteolytic enzyme from papaya. Used in tenderizing meat.

paper An interwoven mat of hydrated cellulose fibres. Most types of fibrous cellulose material can be used in its preparation. The properties of the paper depend upon the degree of hydration and sub-division of the fibres. In order to make paper less porous and partially water-repellent, it is SIZED. Some inorganic materials, e.g. $BaSO_4$, may be added to the paper as fillers. Production 1999 96.9 million tonnes (37% recycled).

paper chromatography See CHROMATOGRAPHY.

para In para-cresol, p-cresol, the CH_3 and OH

groups are in the para or 1,4 position to each other. Thus systematically 4-hydroxymethyl-benzene or 4-hydroxytoluene. This use of the prefix is confined to disubstituted benzene derivatives; in such cases as parahydrogen and paraldehyde the prefix has no uniform structural significance and is always written in full.

parabens Alkylhydroxybenzoates used as food preservatives.

paracetamol (N-(4-hydroxyphenyl)-acet-amide), $C_8H_9NO_2$. White crystalline powder; m.p. 169–172°C. Widely used as an analgesic.

parachor A parameter empirically relating [surface tension]$^{1/2}$ and molecular volume. Formerly used to indicate structure.

paracyanogen, $(CN)_x$. The white or brown polymer formed when cyanogen is heated to 500°C.

paraffin See KEROSINE.

paraffin oil See KEROSINE.

paraffins See ALKANES. Term used industrially and domestically in UK for saturated hydrocarbons.

paraffin wax See PETROLEUM WAX.

paraformaldehyde (paraform, trioxymethylene, polyoxymethylene) A mixture of polymethylene glycols of the type $(CH_2O)_n \cdot xH_2O$ where n is 6–50. It is a white, amorphous powder having the odour of methanal, m.p. 120–130°C. The commercial product contains 95% methanal and is obtained in white flocculent masses when solutions of methanal are evaporated or allowed to stand. When heated it is converted to methanal. Used as a convenient solid substitute for methanal as disinfectant, fumigant and in dentistry.

para-hydrogen See ORTHO-HYDROGEN.

paraldehyde (2,4,6-trimethyl-1,3,5-trioxinone), $(C_2H_4O)_3$. A colourless, mobile liquid,

m.p. 125°C, b.p. 124°C, slightly soluble in water. Formed when ethanal is allowed to stand in the presence of a catalyst at moderate temperatures. Habit forming.

paramagnetism If when a substance is placed in a magnetic field it causes a greater concentration of the lines of magnetic force within itself than in the surrounding magnetic field, it is said to exhibit paramagnetism. Paramagnetism is associated with the presence of unpaired electrons in an ion or molecule in a random orientation. Susceptibility falls with rise in temperature – Curie law. Magnetic susceptibility measurements of paramagnetic substances have contributed significantly to the theory of valency and knowledge of structure. Adiabatic paramagnetization is used to attain very low temperatures.

Paraquat Herbicide, bipyridinium derivative.

parathion/parathion-methyl Insecticides, acaricides, organophosphates.

Paris green See EMERALD GREEN.

parity See UNGERADE. Symmetry behaviour of orbitals.

partial condenser A type of condenser. Vapour is partially condensed by heat exchange with another process stream, and the remainder is condensed in a *final condenser* using cooling water.

partial pressure In a mixture of gases or vapours each constituent can be considered to contribute to the total pressure that pressure which the same amount of the constituent would exert if it were present alone in a vessel of the same volume as that occupied by the mixture (Dalton's law). This is called the partial pressure of the constituent.

particle size measurement Methods include passing through graded sieves, for gels or for small particles by molecular weight determination, light scattering, etc.

parting The process of separating Au from Ag. As the chemical properties of the elements are very different this is an easy matter and methods used include dissolving the Ag in conc. H_2SO_4 or HNO_3 or the preferential reaction of Ag with Cl_2 to give AgCl.

parting agents See ABHERENTS.

partition column chromatography See CHROMATOGRAPHY.

partition function The total information in the thermodynamic properties of a system.

partition law (distribution law) In a heterogeneous system of two or more phases in equilibrium, the ratio of the activities (or less accurately the concentrations) of the same molecular species in the phases is a constant at constant temperature. This constant is termed the partition coefficient.

PAS 4-aminosalicylic acid. Also photo-acoustic spectroscopy.

Pascal (Pa) The S.I. unit of pressure, $N\,m^{-2}$.

Paschen series See BALMER SERIES.

passivators See METAL PASSIVATORS.

passivity Many metals (e.g. Fe, Co, Ni, Cr) form a surface layer of oxide in contact with strong oxidizing agents (e.g. conc. HNO_3, chromic acid, H_2O_2). Further solution does not occur and many properties are altered. Other metals, e.g. Al and Mg, normally have a tough surface film of oxide which prevents further attack so that although electropositive they do not dissolve in water. Passivity can also be conferred by fluorine which forms a layer of fluoride on, e.g. Cu or Ni.

Pasteur effect The decomposition of sugar by respiration is often greater in the absence of oxygen than in its presence. See DICTIONARY OF BIOLOGY.

pasteurization The process of partial removal of some non-spore-producing organisms from milk, butter, beer, wine by application of low heat (63°C for 30 minutes, 72°C for 20 seconds, 132°C for 1 second) or radiation.

patina The decorative and corrosion-resistant surface given to bronze, iron, etc. as an oxide film formed on warming the metal.

Pauli exclusion principle In any atom no two electrons can have all four quantum numbers the same.

Pb Lead.

PCB Polychlorinated biphenyls.

PCR Polymerase chain reaction.

Pd Palladium.

peanut oil (arachis oil) Glycerides of arachidic acid. Used in manufacture of soaps, margarine, paints.

pearl ash See POTASSIUM CARBONATE.

peat A naturally occurring, brown, fibrous mass of partially decayed plant material. The calorific values of peat on a dry basis range from about 16–20 MJ/kg. Used for firing power stations, converted to peat charcoal by low-temperature CARBONIZATION and in horticulture.

pebulate Herbicide, thiocarbamate derivative.

pectic compounds Polysaccharides present in cell walls.

pectin A purified carbohydrate product obtained from the dilute acid extract of the inner portion of the rind of citrus fruits or from apple pomace. It is a mixture, the most important constituent of which is methyl pectate, which is the methyl ester of pectic acid, a high molecular weight polymer of D-galacturonic acid. Pectin also contains araban and galactan. Pectin is manufactured as a white powder, soluble in water, and used to assist the setting of jams and jellies, and for numerous other purposes.

pefurazoate Fungicide, azole derivative.

PEG Polyethylene glycol.

pelargonic acid See NONANOIC ACID.

pelargonin The anthocyan glycoside of plants. It is the 3,5-diglucoside of pelargonidin.

Other anthocyans which contain pelargonidin are **1** callistiphin, **2** monardaein or salvianin. See ANTHOCYANINS.

penconazole Fungicide, azole derivative.

pencycuron Fungicide, urea derivative.

pendimethalin Herbicide, nitroaniline derivative.

penicillin A group of BACTERICIDAL ANTIBIOTICS containing the basic ring structure of penicillanic acid.

Many semi-synthetic penicillins are made from 6-aminopenicillanic acid (6-APA, $R=NH_2$).

pentaborane (9), B_5H_9; **pentaborane (11)**, B_5H_{11}. See BORON HYDRIDES.

pentacene Five linearly fused benzene rings. Can act as a transistor and can be doped with I_2 or Br_2 to construct photovoltaic diodes.

pentachlorophenol, C_6Cl_5OH. Solid having a pungent odour when warm. Prepared by direct chlorination of phenol or polychlorophenols using a catalyst for the last stages of chlorination or by the hydrolysis of C_6Cl_6 with NaOH in methanol at 130°C. Source of chlorobenzene derivatives. Insecticide, fungicide, herbicide, but being phased out.

pentaerythritol, $C_5H_{12}O_4$, $C(CH_2OH)_4$. Colourless crystalline compound, m.p. 260°C. Made by reacting ethanal and methanal in the presence of alkali. It is used in alkyd resin production, in the manufacture of its tetranitrate (explosive), and in the production of oxetanes. See DIPENTAERYTHRITOL and TRIPENTAERYTHRITOL.

pentaerythritol tetranitrate (PETN or penthrite), $C(CH_2 \cdot O \cdot NO_2)_4$. Needles, m.p. 140–141°C. Made by nitrating pentaerythritol; an extremely powerful and violent explosive. Mixed with TNT it forms pentolite. Used medicinally in tablet form to relieve the pain of angina pectoris.

pentan-2,4-dione See ACETYLACETONE.

pentanedioic acid See GLUTARIC ACID.

1,5-pentanediol Plasticizer.

pentanes, C_5H_{12}. There are three possible isomers of this formula; they occur in the low boiling fractions of petroleum. Inflammable liquids. $CH_3 \cdot CH_2 \cdot CH_2 \cdot CH_2 \cdot CH_3$, n-pentane, b.p. 38°C, is used as a standard illuminant in photometry.

pentanochlor Herbicide, anilide derivative.

pentanoic acids (valeric acids), $C_5H_{10}O_2$. n-Pentanoic acid, $CH_3 \cdot CH_2 \cdot CH_2 \cdot CH_2 \cdot C(O)OH$, is a colourless liquid with an unpleasant odour, b.p. 186°C. Prepared from n-butyl iodide by a Grignard reaction or by oxidation of n-amyl alcohol.

Isopentanoic acid, $Me_2CHCH_2C(O)OH$, is a colourless liquid with an unpleasant odour, b.p. 177°C. Occurs in the roots of valerian and angelica together with an optically active form of methylethylethanoic acid. Prepared by oxidation of isopentanol. A mixture of acids similar to that obtained from valerian roots is prepared by oxidation of fusel oil.

pentanols, $C_5H_{12}O$. Eight alcohols of this formula are possible. The pentanol of commerce is obtained by distillation of fusel oil and is a mixture of isopentanol with from 13–60% secondary pentanol; it has b.p. 128–132°C. It is used in the preparation of amyl acetate, amyl nitrite and amylene. The most important of the pentanol alcohols are:

1 *Isopentanol alcohol, isobutyl carbinol (3-methylbutanol),* $(CH_3)_2CHCH_2CH_2OH$. A colourless liquid with an unpleasant odour; inhalation of its vapour causes violent coughing, b.p. 131°C. Obtained from fusel oil or by treating $(CH_3)_2CHCH_2CH_2Cl$ with sodium hydroxide. Oxidized by chromic acid to 3-methylbutanoic acid. **2** *Active pentanol alcohol (sec. butyl carbinol, 2-methylbutanol),* $CH_3CH_2CH(CH_3)CH_2OH$. B.p. 128°C. **3** *Normal pentanol alcohol (1-pentanol),* $CH_3CH_2CH_2CH_2CH_2OH$. B.p. 137°C, obtained by treating 1-chloropentane with sodium hydroxide. **4** *Amylene hydrate, tert. pentanol alcohol, 2-methyl-2-butanol,* $CH_3CH_2C(CH_3)_2OH$. Liquid with a penetrating camphor-like odour, b.p. 102°C. Prepared from β-isoamylene (2-methyl-2-butene) by passing into sulphuric acid and then diluting with water.

Complex mixtures of various pentanol alcohols are manufactured from chloropentanes derived from petroleum.

pentosans Hemicelluloses which give pentoses on hydrolysis. The most widely distributed are xylan and araban.

pentose A carbohydrate with five carbon atoms. Of the aldopentoses, both stereoisomers of arabinose and the D-forms of xylose and ribose occur naturally; lyxose does not occur naturally. There are four (two pairs of stereoisomers) possible ketopentoses (pentuloses).

pentose phosphate pathway A sequence of reactions of sugar phosphates that provides a source of hexoses, pentoses, heptoses and tetroses for a living cell. It is concerned in the initial stages of photosynthesis. A major function appears to be the generation of NADPH which is apparently preferred to NADH in biosynthetic reactions.

pentoxazone Herbicide, oxazolidine derivative.

pepsin (pepsinogen) The principal proteolytic enzyme in the gastric juice of vertebrates which splits links between the α-carboxyl group of a dicarboxylic amino-acid and the α-amino group of an aromatic acid.

peptidases Proteolytic enzymes, ENDOPEPTIDASES, attack the internal links of a peptide chain.

peptides Substances composed of two or more amino-acid residues, and designated as di-, tri-, oligo-, polypeptides, etc. according to the number of amino-acids linked by the peptide bond –CO·NH–. The simplest peptide is glycylglycine, $H_2N \cdot CH_2 \cdot CO \cdot NH \cdot CH_2 \cdot CO_2H$.

Most naturally occurring peptides possess important physiological activity. Peptides also arise as partial hydrolysis products of proteins. Methods for peptide synthesis are very highly developed. See also PROTEINS. Most are present as helices in the solid and some in solution. Folding is of great importance. See DICTIONARY OF BIOLOGY.

peptidoglycan The main constituent of bacterial cell walls.

peptization This term is sometimes used in a general way to imply the converse of coagulation, i.e. dispersion, especially when the process results in the formation of a colloidal sol. Peptization is, however, generally restricted to a chemical means of dispersion in which the colloidal particles are stabilized by the adsorption of charged ions. An example of peptization is the formation of silver iodide sols by shaking the neutral precipitate with a small excess of either KI or $AgNO_3$; with the former reagent the resultant sol is negative; with the latter, positive.

peptoids Peptide-like chains of N-substituted glycines. Have side chains on nitrogen.

peptones Derived proteins of low mol.wt. formed by degradation of proteins by, e.g. enzymes.

per-acids Oxyacids in which $[O-O]^{2-}$ or $[O-OH]^-$ groups have replaced oxide in oxyanions (compounds containing the superoxide ion $[O_2]^-$ are formally not per-acids but the distinction is often difficult to make). The O–O group may be bonded end on $[M-O-O]$ or symmetrically

$$\left[M \longleftarrow \begin{matrix} O \\ | \\ O \end{matrix} \right].$$

Prepared by use of H_2O_2 or by electrolytic oxidation. Persulphuric acids, perborates, are of importance. (Permanganates, perchlorates and periodates are not salts of per-acids.) Organic per-acids are prepared similarly. The more stable per-acids such as perbenzoic acid, 3-chloroperbenzoic acid, monoperphthalic acid, convert olefins to epoxides and by hydrolysis to *trans*-glycols. Perethanoic and permethanoic acids give glycol monoesters with olefins. See PEROXIDES, ORGANIC.

perborates (perborax) Sodium perborate (borax plus Na_2O_2 or H_2O_2 plus NaOH). Quite stable and used as a bleaching agent and antiseptic.

perbromates, $M[BrO_4]$. See BROMATES.

perbromic acid, $HBrO_4$. See BROMATES.

percarbonates Carbonates with varying amounts of H_2O_2 of crystallization are stable as solids and $2Na_2CO_3.3H_2O_2$ is used in laundering and for cleaning dentures.

perchlorates, $MClO_4$. See CHLORATES(VII). Organic perchlorates tend to be explosive.

perchloric acid, $HClO_4$. See CHLORATES(VII). Perchloric acid ($KClO_4$ plus H_2SO_4) is a colourless or yellowish strongly fuming liquid which can be distilled at low pressure and gives crystalline $HClO_4.H_2O$. The free acid can explode and frequently does so in the presence of organic materials. $HClO_4$ is a strong oxidizing acid and is sometimes used in analysis. Forms ionized salts.

perchloroethylene Tetrachloroethene, $Cl_2C=CCl_2$.

perchloryl fluoride, $FClO_3$. See CHLORINE OXIDE FLUORIDES.

perewskites See PEROVSKITE.

perfect gas A gas that obeys the gas laws. A gas without atomic or molecular interactions.

perfluoroalkyl derivatives Organic derivatives in which all the H atoms, except those in the functional groups, have been replaced by fluorine. Thus C_6F_6 is perfluorobenzene; $C_6F_5C(O)OH$ is perfluorobenzoic acid.

Perfluoroalkyl derivatives have important technical uses, e.g. sulphonic acids as surfactants. Introduction of perfluoroalkyl groups or F atoms confers enhanced properties on many pharmaceuticals.

perfume chemistry The broad contents of perfumes are essential oils (of plant origin), flower oils, natural extracts from resins, gums, etc., and fixatives (e.g. ambergris, civet) which render the perfume more lasting. Materials used in perfumery are both natural and synthetic.

peri A prefix sometimes used to denote the positional relationship between groups occupying the 1,8 positions on the naphthalene ring. As a special case the two groups may be part of a third ring.

l-Naphthylamine-8-sulphonic acid is sometimes referred to as peri acid.

periclase MgO either natural or formed by heating $Mg(OH)_2$ used in furnace linings.

pericyclic Reaction mechanism in which electrons move in a single synchronous step, e.g. simultaneous addition.

L(−)-perillaldehyde (4(1-methylethyl-enyl)-1-cyclohexene-1-carboxalde-hyde), $C_{10}H_{14}O$. Natural product. Oxime used as artificial sweetener in Japan.

period A series of elements in the periodic table, a period being elements from an alkali metal to a noble gas. The periods are **1** the very short period H, He, **2** two short periods Li–Ne and Na–Ar, **3** two long periods containing transition elements K–Kr and Rb–Xe, **4** three very long periods containing the rare-earth elements Cs–Rn, Fr–Lr and **5** an unfinished period.

periodates See IODATES(VII).

periodic acids See IODATES(VII).

periodic law Mendeléeff (1869), independently of Newlands (octaves law), put forward the law that 'the properties of the elements are periodic functions of their atomic weights', i.e. by arranging the elements in order of increasing atomic weights, elements having similar

Periodic Table

Group	1	2	3	4	5	6	7	8	9	10	11	12	13	14	15	16	17	18
Outer electron configuration	s^1	s^2	d^1	d^2	d^3	d^4	d^5	d^6	d^7	d^8	d^9	d^{10}	p^1	p^2	p^3	p^4	p^5	p^6
	H																	He
	Li	Be											B	C	N	O	F	Ne
	Na	Mg											Al	Si	P	S	Cl	Ar
	K	Ca	Sc	Ti	V	Cr	Mn	Fe	Co	Ni	Cu	Zn	Ga	Ge	As	Se	Br	Kr
	Rb	Sr	Y	Zr	Nb	Mo	Tc	Ru	Rh	Pd	Ag	Cd	In	Sn	Sb	Te	I	Xe
	Cs	Ba	La*	Hf	Ta	W	Re	Os	Ir	Pt	Au	Hg	Tl	Pb	Bi	Po	At	Rn
	Fr	Ra	Ac+	Rf	Db	Sg	Bh	Hs	Mt	110	111	112	(113)	(114)	(115)	(116)	(117)	(118)

Lanthanides*	La	Ce	Pr	Nd	Pm	Sm	Eu	Gd	Tb	Dy	Ho	Er	Tm	Yb	Lu
Actinides+	Ac	Th	Pa	U	Np	Pu	Am	Cm	Bk	Cf	Es	Fm	Md	No	Lr

properties occur at fixed intervals. This principle, with the modification that atomic numbers are used instead of atomic weights, forms the basis of the periodic table.

Periodic table A table of the elements arranged in order of atomic numbers, the arrangement being such as to emphasize the chemical relationships between the elements and to emphasize the relationship between elements with similar configurations. Various systems are used for numbering the vertical groups (elements with similar chemistries). See PERIODIC LAW. See table at top of this page. Elements numbered in brackets have not yet been identified. Elements numbered but not in brackets have not yet been given an agreed name.

periplanar The arrangement for the five atoms involved in an E2 reaction. The anti arrangement of incoming base and leaving group is preferred (anti-periplanar).

peritectic A type of solid–liquid phase equilibrium system, commonly found with binary alloys, in which the two components form limited ranges of solid solution in each other, although they are completely miscible in the liquid state. A peritectic system differs from a eutectic system in that the former shows no minimum in the freezing-point composition curve.

Perkin reaction A condensation between aromatic aldehydes and the sodium salts of fatty acids or their aromatic derivatives. The reaction between benzaldehyde and sodium ethanoate in the presence of ethanoic anhydride leads to sodium cinnamate

$$C_6H_5CHO + H_3C \cdot COONa \rightarrow C_6H_5CH{:}CHCOONa$$

In general, condensation takes place at the α-carbon atom, leading to simple cinnamic acids or to their α-substituted derivatives. When possible, the anhydride corresponding to the sodium salt should be the condensing agent.

perlite Originally the name of a glassy volcanic rock, but now also applied to a lightweight product produced on a considerable scale by rapidly heating to their softening point crushed and graded perlites and similar glassy, highly siliceous volcanic materials containing 2–6% of water. The particles lose water as steam, yielding a product with ρ approx. 1200.

Used as a lightweight aggregate in heat and sound insulating plasters, as a loose fill and in lightweight concrete. World demand 1998 2.2 million tonnes.

permanganates Salts containing the dark purple $[MnO_4]^-$ ion. $KMnO_4$ is the most important salt (MnO_2 (pyrolusite) plus KOH plus air at 300°C followed by treatment of $[MnO_4]^{2-}$ electrolytically or with CO_2 to give $[MnO_4]^-$ plus MnO_2). Used in metal treatment and as a bactericide. $NaMnO_4 \cdot 3H_2O$ is more soluble; a solution is used as a disinfectant (Condy's fluid). $AgMnO_4$ is fairly insoluble but is very reactive. See MANGANATES(VII).

permanganate titrations On account of its intense colour and easy purification, potassium permanganate is an important volumetric oxidizing agent which requires no indicator even in 0.01 M-solution. When used, as in most estimations, in presence of sulphuric acid, it is reduced to Mn^{2+} (5 electrons). Solutions are standardized after having been boiled (or left to stand for some days) and then filtered. The best standards are sodium oxalate and pure (electrolytic) Fe; crystallized oxalic acid and $Fe(NH_4)_2(SO_4)_2.12H_2O$ are also frequently used but are less reliable. The solution may be used for the direct titration of Fe^{2+}, H_2O_2, $[Fe(CN)_6]^{4-}$, oxalate. Suitable back titrations enable it to be used for the estimation of a great variety of oxidizing and reducing agents. In the titration of Fe^{2+}, Cl^- interferes but titration can be carried out in presence of an excess of manganese(II) sulphate. In some estimations permanganate is used in almost neutral solution, when hydrated manganese dioxide (or manganate(IV)) is formed. An important instance of this procedure is the determination of Mn(II) (the Volhard method); the titration is carried out in the presence of a Zn^{2+} salt and suspended ZnO.

permanganic acid, $HMnO_4$. See MANGANATES(VII).

permethrin Insecticide, pyrethroid.

permittivity (ε) See DIELECTRIC CONSTANT.

permonosulphuric acid (Caro's acid), $HOS(O)_2OOH$. Powerful oxidizing agent (H_2O_2 on conc. H_2SO_4 or HSO_3Cl; hydrolysis of $H_2S_2O_6$ with H_2SO_4). The pure acid can be obtained crystalline. Used widely in organic synthesis. The sodium salt is used as an antiviral agent.

permutite process The process of softening hard water by replacing Ca^{2+} and Mg^{2+} by Na^+ by passage through a ZEOLITE. The zeolite is regenerated with NaCl solution.

perovskite Mineral $CaTiO_3$.

perovskite structure A structure adopted by many ABX_3 compounds in which A and B are positive ions and X is, e.g. F^- or O^{2-}. The structure has A at the corners of a cube, B at the body centre and X (A and X are ccp) at the face centres. A has 12X neighbours, B has 6X (octahedral) neighbours. Examples are $KNiF_3$, $BaTiO_3$, $KNbO_3$. Many perovskites show distortions from the idealized cubic structure.

peroxidases The enzymes which destroy peroxides by transferring peroxide oxygen to oxidizable substances, e.g. reduced gluta-thione.

peroxides Compounds containing singly bonded pairs of oxygen atoms – derivatives of H_2O_2. They yield H_2O with dilute acids. BaO_2 and Na_2O_2 are typical peroxides. PbO_2, NO_2, MnO_2 are sometimes erroneously termed peroxides on account of their stoichiometry.

peroxides, organic Organic peroxides are used as sources of free radicals for free radical polymerization, as curing agents for resins and elastomers and as cross-linking agents for polyolefins. The most commonly used are dibenzoyl peroxide (m.p. 106°C) and dilauroyl peroxide (m.p. 55°C) prepared from the acid chloride or anhydride and Na_2O_2 or H_2O_2. Other peroxides are similar. Peroxyethanoic acid (b.p. 36°C/30 mm) is used for epoxidation and bleaching. It is prepared from ethanoic acid and H_2O_2. Other peroxyacids are similar. See PER-ACIDS.

peroxychromium compounds H_2O_2 reduces $[CrO_4]^{2-}$ to Cr^{3+} but gives blue $CrO(O_2)_2$ which can be extracted into ether (test for Cr or, after distillation of CrO_2Cl_2, for Cl^-). Other derivatives are also known and are relatively stable.

peroxyethanoic acid, $CH_3C(O)OOH$. See PEROXIDES, ORGANIC.

perrhenates Rhenates(VII). See RHENIUM OXIDES.

persistent organic pollutants (POPS) Materials, e.g. chloroaromatics, that only slowly break down in the environment.

Perspex An acrylic resin.

persulphates Salts of PERSULPHURIC ACIDS.

persulphuric acid (perdisulphuric acid), $HOS(O)_2OOS(O)_2OH$. Dibasic acid formed as salts by electrolysis of sulphates at low temperatures and high current density. The acid and persulphates are strong oxidizing agents ($E°[S_2O_8]^{2-}$ to SO_4^{2-} +2.01 volts in acid) but the reactions are often slow. Compare permonosulphuric acid.

pervaporation Technique for separation of volatile liquids by passage through a membrane, i.e. combination of permeation and evaporation. Used particularly for dehydration, e.g. of ethanol.

perxenates Xenates(VIII).

PES Photoelectron spectroscopy.

pesticides Herbicides, insecticides, nematocides, fungicides, plant growth regulators. In this Dictionary bacteriocides also included. World use 2.5 million tonnes p.a.

PET See POLYETHYLENE TEREPHTHALATE.

petalite, (Li, Na)(AlSi$_4$)O$_{10}$. Aluminosilicate containing 2–4.5% Li$_2$O, used as a source of Li.

PETN See PENTAERYTHRITOL TETRANITRATE.

petrochemicals (petroleum chemicals) Petroleum chemicals may be obtained from a variety of petroleum fractions, normally from processing of intermediates produced in primary conversion processes. The main basic raw materials are natural gas, refinery gases, liquid petroleum fractions and waxes, while the primary conversion processes include cracking, partial oxidation and pyrolysis.

petrol Motor fuel, gasoline or aviation fuel (U.K. usage). Originally a trade name.

petrolatum (petroleum jelly, paraffin jelly, Nujol, vaseline) An unctuous, salve-like residuum from non-asphaltic petroleum. It is available commercially under a wide variety of names and in various degrees of purity. Colloidal system of branched-chain hydrocarbons. Used in pharmacy, as a basis for medicaments, as lubricants and in foodstuffs. See PETROLEUM WAX.

petroleum A term used to describe oil products in general but also used for CRUDE OIL, petroleum oils.

petroleum coke The solid, carbonaceous material obtained as a by-product in THERMAL CRACKING processes for the production of gasoline. See COKE.

petroleum ether (light petroleum) Lower aliphatic hydrocarbons of high volatility and a narrow distillation range. Made in a number of grades, the two most common having distillation ranges 40–60°C and 60–80°C which contain C$_5$ and C$_6$ hydrocarbons respectively. Ligroin fraction b.p. 35–60°C.

petroleum jelly See PETROLATUM.

petroleum wax Solid, waxy hydrocarbons obtained from crude oils, especially paraffinic crudes.

Paraffin waxes have a macrocrystalline structure and consist largely of n-alkanes of formulae C$_{20}$H$_{42}$ and upwards with some iso-and cyclo-alkanes.

Microcrystalline waxes, produced from heavy lubricating oil residues, have a microcrystalline structure and consist largely of iso-and cycloalkanes with some aromatics.

Petroleum wax is used in the manufacture of candles, polishes, lubricants, cosmetics, ointments and for waterproofing purposes. Waxes are also used as a cracking feedstock for the production of 1-alkenes for conversion to detergents.

pewter A tin–lead alloy.

Ph The phenyl group, C$_6$H$_5$–.

pH See HYDROGEN-ION CONCENTRATION.

phaeophorbide See CHLOROPHYLL.

phaeophytin See CHLOROPHYLL.

phalloidins See AMANITINS.

Pharmacopoeia The legal standard for drugs and drug usage. British Pharmacopoeia and U.S.A. Pharmacopoeia most used reference systems.

phase A system, which is uniform in chemical composition and physical properties and separated from other homogeneous parts of the system by boundary surfaces. E.g. in a closed system containing ice, water and water vapour, there are three phases. Since all gases are completely miscible with each other, gases can only exhibit one phase. With liquids and solids several phases are possible.

phase diagram A diagram showing the conditions of equilibrium between various phases. In one-component systems a pressure–temperature diagram is used; in binary systems a three-dimensional diagram relating pressure, composition and temperature, or two-dimensional diagrams relating composition and temperature at constant pressure or composition and pressure at constant temperature, are used.

phase rule The phase rule relates the number of phases, P, the number of degrees of freedom (variance), F, and the number of components, C, in a chemical system according to the equation

$$P + F = C + 2$$

phase-transfer chemistry Reactions between heterogeneous systems (solid–liquid or liquid–liquid) brought about by addition of a catalytic amount of an agent which transfers a reactant, often a nucleophile, across the interface into the system where reaction can occur. Thus R$_4$N$^+$ or R$_4$P$^+$ cations are used to transfer otherwise insoluble anions into non-polar media from aqueous media. CROWN ETHERS can be used to catalyse direct solid–liquid phase transfer of salts into non-polar solvents, e.g. KOH into toluene. Also called ion-pair partition.

α-phellandrene (p-mentha-1,5-diene, 5-isopropyl-2-methyl-1,3-cyclohexadiene), $C_{10}H_{16}$. A colourless odoriferous oil, b.p. 175–176°C. A monocyclic terpene, occurs in oils. Used as a fragrance.

β-phellandrene (p-mentha-1(7)2-diene, 3-isopropyl-6-methylenecyclohexene), $C_{10}H_{16}$. A colourless, odoriferous oil; b.p. 171–172°C. A monocyclic terpene.

phenacetin (N(4-ethoxyphenyl)acetamide, N-acetyl-4-phenetidine), $C_{10}H_{13}NO_2$. White crystals, m.p. 137–138°C. Prepared from phenol, via 4-nitrophenol, 4-nitrophenetole and 4-phenetidine. Antipyretic analgesic. It has chronic toxicity towards the kidney and may be carcinogenic.

phenacyl The trivial name for the group $PhC(O)CH_2-$.

phenanthrene ring system The system numbered as shown.

2-phenanthroline (4,5-phenanthroline), $C_{12}H_8N_2$. As phenanthrene but with 4,5 nitrogens. Made by heating o-phenylenediamine with glycerol, nitrobenzene and conc. sulphuric acid. It exists as a monohydrate, m.p. 94°C, or anhydrous, m.p. 117°C. It is used as a complexing reagent, in the estimation of Fe^{2+} and in the preparation of numerous chelates of metals, generally in low oxidation states.

phenates O-Metal derivatives of phenol (and other phenols).

phenethyl alcohol (2-phenylethanol), $C_8H_{10}O$. Obtained from essential oils and prepared from ethyl phenyl ketone by reduction. Used in flavouring and perfumes and pharmaceuticals.

p-phenetidine (4-ethoxyaniline), $C_8H_{11}NO$. Colourless oil, m.p. 2–4°C, b.p. 254°C. Prepared by the reduction of 4-nitrophenetole with iron and hydrochloric acid. Dyestuffs intermediate. Its acetyl derivative is phenacetin.

phenetole (ethoxybenzene), $C_8H_{10}O$, PhOEt. A colourless, aromatic liquid, b.p. 172°C. It is prepared by heating KOPh with ethyl iodide, or by adding ethanol to a mixture of phenol and P_2O_5 at 200°C.

phenidone See 1-PHENYL-PYRAZOLID-3-ONE.

phenmodipham Herbicide, carbamate derivative.

phenobarbitone (phenylethylbarbituric acid, 5-ethyl-5-phenyl-2,4,6-(1H, 3H,5H)-pyrimidine trione), $C_{12}H_{12}N_2O_3$. White crystals, m.p. 174°C. Prepared by condensing the ethyl ester of phenylethylmalonic acid with urea. It is a more active hypnotic than barbitone. It and its sodium salt – soluble phenobarbitone – are used as sedatives.

phenol (carbolic acid, hydroxybenzene), C_6H_6O, PhOH. Colourless crystals; m.p. 43°C, b.p. 183°C. Soluble in water at room temperature.

Obtained synthetically by: fusion of sodium benzenesulphonate with NaOH to give sodium phenate; hydrolysis of chlorobenzene by dilute NaOH at 400°C and 300 atm. to give sodium phenate (Dow process); catalytic vapour-phase reaction of steam and chlorobenzene at 500°C (Raschig process); direct oxidation of cumene (isopropylbenzene) to the hydroperoxide, followed by acid cleavage to propanone and phenol; catalytic liquid-phase oxidation of toluene to benzoic acid and then phenol. Where the phenate is formed, phenol is liberated by acidification.

Phenol is acidic and forms metallic salts. It is readily halogenated, sulphonated and nitrated. Used in quantity for the manufacture of phenolic and epoxy resins (46%), caprolactam (15%), bisphenol A (20%) and alkyl phenols (6%). Thus it is a starting material for making both precursors of Nylon 66. Disinfectant, it is also used in the manufacture of various dyes, explosives, pharmaceuticals and perfumes.

phenol aldehydes Aromatic aldehydes (R'O)ArCHO. Examples anisaldehyde, ethylvanillin. Used in perfumery and flavouring, electroplating, metal complexing.

phenol ethers, ArOR. (R aliphatic or aromatic). Examples diphenyl ether, anisole. Occur in essential oils. Herbicides, used as heat-transfer media (Ph_2O) and antioxidants (with bulky substituents). Form stable resins.

phenol–formaldehyde resins Resins formed principally by phenol but also other aromatic hydroxyl compounds (resorcinol, cresols, xylenols, 4-t-butylphenol, 4-phenyl-phenol) with methanal under acid or basic conditions. Used for mouldings, coatings and bondings.

phenolic resins A class of polymers resulting from interaction of phenols and aldehydes (generally methanal). See PHENOL-FORMALDEHYDE RESINS.

phenolphthalein (3,3-bis(4-hydroxy-phenyl)-1-(3H)-monobenzofuranone), $C_{20}H_{14}O_4$. Colourless crystals; m.p. 254°C. Prepared by heating phthalic anhydride and phenol in the presence of sulphuric acid. Extensively used as an indicator with pH range 8.3 (colourless) to 10.4 (red) and as reagent for oxidases, blood, etc. It has been used medicinally as a purgative. Probably carcinogenic.

phenol red (phenolsulphonphthalein), $C_{19}H_{14}O_5S$. Sulphonic acid analogue of phenolphthalein. Indicator.

phenols Compounds containing at least one hydroxyl group attached directly to a carbon atom of an aromatic ring. Acidic, and soluble in caustic alkalis forming metallic salts, the phenates. Their solubility in water increases with the number of hydroxyl groups. Polyhydric phenols have a strong reducing action. They are widely distributed in natural products, e.g. tannins, anthocyanins, tyrosine.

phenothrin Insecticide, pyrethroid.

phenoxyacetic acid herbicides See 2,4-D, MCPA, DICHLOROPHENOXYETHANOIC ACID.

2-phenoxyethanol (ethylene glycol monophenyl ether, (1-hydroxy-2-phenoxyethane)), $C_6H_5O\cdot CH_2\cdot CH_2OH$. A colourless, viscous liquid, b.p. 117°C/7 mm. Made by heating sodium phenate with ethylene chlorohydrin. Antiseptic.

phenoxy resins (polyhydroxyethers) Linear thermoplastics formed by condensation

of epichlorhydrin, bisphenol A and NaOH in dimethylsulphoxide. Used for containers.

phenyl The group C_6H_5– or Ph. The radical occurs as an intermediate in many organic reactions in solution, e.g. in the decomposition of the benzene diazonium cation in the presence of Cu(I).

phenylacetaldehyde (benzene aceto-aldehyde), $PhCH_2CHO$. Prepared by oxidation of alcohols. Used in perfumes.

phenylacetic acid See PHENYLETHANOIC ACID.

phenylacetone, $PhCH_2COCH_3$. Used in synthesis, particularly in reactions requiring benzyl radicals.

phenylalanine (α-amino-β-phenylpropionic acid), $C_9H_{11}NO_2$, $PhCH_2CHNH_2\cdot C(O)OH$. Crystallizes in colourless leaves; m.p. 283°C. The naturally occurring substance is laevorotatory. Essential amino-acid. Involved in phenyl ketonurea.

phenylamine (aniline), C_6H_7N, $PhNH_2$. Colourless oily liquid, turning brown on oxidation. m.p. –6.2°C, b.p. 184°C. Manufactured from nitrobenzene by vapour-phase hydrogenation in the presence of a copper catalyst, or by reduction with iron and water containing a trace of hydrochloric acid. Used in the manufacture of antioxidants and vulcanization accelerators for the rubber industry, but it is also employed for the manufacture of dyes and pharmaceuticals.

On acetylation it gives acetanilide. Can be nitrated and halogenated. It is basic and gives water-soluble salts with mineral acids. Heating aniline sulphate at 190°C gives sulphanilic acid. When heated with alkyl chlorides or aliphatic alcohols mono- and di-alkyl derivatives are obtained, e.g. dimethylaniline. With glycerol and sulphuric acid (Skraup's reaction) quinoline is obtained, while quinaldine can be prepared by the reaction between aniline, paraldehyde and hydrochloric acid.

With sodium nitrite in dilute HCl solution at 0°C aniline gives benzene diazonium chloride, which loses nitrogen on warming and gives phenol. The diazonium group may also be replaced by halogeno- or cyano-groups by double decomposition with CuX or CuCN (Sandmeyer reaction); it can be coupled with phenols and amines to give coloured azo-compounds. Can be carbonylated.

Oxidized to polyaniline, a conducting polymer.

U.S. production 1999 713 000 tonnes.

4-phenyl-2-buten-2-one (benzylidene acetin), $PhCH=CHCOCH_3$. Prepared by condensation of propanol and benzaldehyde, b.p. 261°C. Used in synthesis and perfumery.

2-phenylenediamine (1,2-diaminobenzene), $C_8H_2N_2$. Brown-yellow crystals; m.p. 103–104°C, b.p. 256–258°C. Its solutions reduce Ag^+ ions and it is used as a photographic developer. It is also used as a dye-precursor, for the synthesis of phenazine derivatives and for characterizing (*inter alia*) α-diketones.

3-phenylenediamine (1,3-diaminobenzene), $C_6H_8N_2$. Colourless crystals; m.p. 63°C, b.p. 287°C. Turns brown in air. Prepared by a one-stage reduction of 3-dinitrobenzene with iron and hydrochloric acid.

Basic, forms a stable water-soluble dihydrochloride. Diazotization gives brown azo dyes (Bismarck brown) by the coupling of the partially diazotized base with the excess of diamine. Is also used as an end component of many azo-dyes, readily coupling with one or two molecules of diazo compound. Used in resins, block copolymers, fibres.

4-phenylenediamine (1,4-diaminobenzene), $C_6H_8N_2$. White crystals; m.p. 147°C, b.p. 267°C, darken rapidly in air. Prepared by reducing 4-nitroaniline or aminoazobenzene. Oxidizing agents convert it to quinone derivatives, hence it cannot be diazotized with nitric acid.

It is used for hair and fur dyeing, as a rubber accelerator and as a photographic developer and in dyestuff manufacture.

phenylethanoic acid, $PhCH_2CO_2H$. M.p. 76°C, b.p. 266°C. Occurs naturally as esters, manufactured by hydrolysis of benzyl cyanide. Used as a penicillin precursor, in amphetamine production, and in perfumes.

phenyl ether See DIPHENYL ETHER.

2-phenylethyl alcohol, $PhCH_2CH_2OH$. B.p. 220°C. Has a faint rose odour, occurs free or combined in many essential oils (e.g. orange juice). Made by reduction of ethyl phenylacetate with Na in alcohol or better by Friedel–Crafts reaction between benzene and ethylene oxide. Used extensively in perfumery (the ethanoate is also used).

phenylethyl ethanoate See PHENYLETHYL ALCOHOL.

N-phenylglycine, $PhNHCH_2C(O)OH$ $C_8H_9NO_2$. White crystals, m.p. 127°C. It can be prepared: **1** by the condensation of aniline with chloroethanoic acid. **2** By heating aniline, NaOH, methanal and KCN. **3** By heating aniline, lime and trichloroethene. An important intermediate in the indigo industry.

N-phenylglycine-4-carboxylic acid ((2-carboxyphenyl)aminoethanoic acid), $C_9H_9NO_4$. Sandy powder; m.p. 207°C (decomp.).

Prepared by the action of chloroethanoic acid upon anthranilic acid. It can also be prepared by the action of KCN and methanal on anthranilic acid, followed by hydrolysis with NaOH of the nitrile of anthranilic acid first produced.

It is used for the production of indigo.

phenylhydrazine, $C_6H_8N_2$, $PhNHNH_2$. A colourless, refractive oil; b.p. 240–241°C, with slight decomposition. Basic, oxidizes in air. It is very poisonous by skin absorption and inhalation of the vapour.

It is prepared commercially by treating benzene diazonium chloride with sodium sulphite and then reducing the mixture with zinc dust and ethanoic acid. It can also be prepared by reduction of benzene diazonium chloride with $SnCl_2$.

Strong reducing agent; it reduces Fehling's solution in the cold and aromatic nitro compounds to amines. With ketonic groups, hydrazones are formed. Compounds containing the grouping –CHOHCO–, e.g. glucose, fructose, etc., react with two molecules of phenylhydrazine to give osazones.

Condenses with acetoacetic ester to give a pyrazolone derivative which on methylation gives phenazone. The sulphonic acid similarly gives rise to the tartrazine dyestuffs. It is used to make indole derivatives by the Fischer process.

phenylhydrazine-4-sulphonic acid See PHENYLHYDRAZINE.

phenylhydrazones See PHENYLHYDRAZINE.

phenyl isocyanate (isocyanatobenzene), C_6H_5NCO, PhNCO. A pungent lachrymatory almost colourless liquid; m.p. –33°C, b.p. 162°C. Used as a dehydrating agent and for characterization of alcohols. Prepared from aniline and phosgene in the presence of hydrogen chloride.

phenylisothiocyanate, PhNCS. B.p. 221°C. Used in sequencing peptides.

phenylmagnesium halides See GRIGNARD REAGENTS.

phenylmercuric ethanoate See MERCURY, ORGANIC DERIVATIVES.

2-phenylphenol Fungicide.

phenylphenols Prepared from diphenyl ether. Used in rubber manufacture.

phenyl phosphate Normally Na_2PO_4Ph. Used in testing for completeness of pasteurization.

1-phenyl-1-propanol (α-ethylbenzyl alcohol) Prepared from benzaldehyde or benzoyl chloride. B.p. 210°C. Heat-transfer medium, used in perfumery.

1-phenylpyrazolid-3-one (phenidone), $C_9H_{10}N_2O$. Colourless crystalline solid, m.p. 121°C. Made by reacting phenylhydrazine with 3-chloropropionic acid to the hydrazide which cyclizes to the product. Important as a photographic developing agent.

phenyl salicylate (salol, 2-hydroxybenzoate, 2-hydroxybenzoic acid, phenyl ester) Formed from $POCl_3$, PhOH and salicylic acid. Used in manufacture of polishes, waxes, lacquers, adhesives and as pharmaceutical.

phenytoin (5,5-diphenylhydantoin), $C_{15}H_{12}N_2O_2$. Used as its sodium salt, which is a white hygroscopic powder. Prepared from benzil, urea and NaOH.

pheromones Substances secreted externally by an individual and which elicit a specific response from other individuals of the same species. E.g. insect sex attractants such as bombykol (hexadeca-10-*trans*, 12-*cis*-dien-1-ol) from the silk moth; trail markers of ants; and 'Queen substance' of the honey bee (9-oxodec-2-*trans*-enoic acid).

philosopher's stone See TRANSMUTATION.

phloroglucinol (1,3,5-trihydroxybenzene), $C_6H_6O_3$. Colourless crystals; $2H_2O$ from water. Water lost at 100°C. Anhydrous compound, m.p. 200–219°C.

It occurs in many natural glycosides. It can be prepared by fusing resorcinol with NaOH and is manufactured from trinitrobenzene via trinitrobenzoic acid and triaminobenzene.

It reacts chemically as if it were either 1,3,5-trihydroxybenzene or the triketone of hexahydrobenzene. Used in printing, textiles, dyeing, photography, adhesives, pharmaceuticals.

pH meter An instrument used for measuring the pH of solutions, suspensions, etc., generally with the aid of a glass electrode. Both 'null' type, which is the more accurate, and 'direct-reading' instruments are available.

phorate Insecticide, acaricide, nematocide, organophosphate derivative.

phorone (di-isopropylideneacetone, 2,6-dimethyl-2,5-heptadiene-4-one), $C_9H_{14}O$, $Me_2C=CHCOCH=CMe_2$. Yellow liquid having a camphor-like odour; m.p. 28°C, b.p. 198.5°C. Formed when propanone is saturated with HCl and allowed to stand. Resembles camphor in many of its properties and is a solvent for cellulose nitrate. Used to prepare diisobutyl ketone (reduction).

phosalone Insecticide, acaricide, organophosphate derivative.

phosgene, Cl_2CO. See CARBONYL DERIVATIVES. Stored as triphosgene, *bis*(trichloromethyl)carbonate which decomposes to phosgene on heating.

phosgenammonium salts (phosgene iminium salts) See DICHLOROMETHYLENE-AMMONIUM SALTS.

phosmet Insecticide, acaricide, organophosphate derivative.

phosphagen High-energy phosphate compounds which act as energy reservoirs in the cell (ADP, ATP).

phosphamidon Insecticide, acaricide, organophosphate derivative.

phosphatases A group of enzymes which hydrolyse and transfer phosphate groups.

phosphate coatings Commonly applied to steels and other alloys in order to give improved corrosion resistance. Such coatings provide an excellent keying surface for painting and also improve the retention of lubricants on the surface.

phosphate esters, $(RO)_3PO$. Derivatives formed by phosphoric acids – particularly H_3PO_4 – and alkyl or aryl alcohols. Tricresylphosphate, $(MeC_6H_4O)_3PO$ (a mixture of all isomers) formed Cl_3PO on technical cresol, is important as a plasticizer, flame-retardant agent for fabrics and as a hydraulic fluid. $(MeO)_2(PhO)PO$, $(BuO)_3PO$, $(MeO)_2(xylylO)PO$, $(MeO)_3PO$ are used as gasoline additives. Phosphate esters (e.g. $(BuO)_3PO$) are extensively used in solvent-extraction processes. Esters are used in plasticizers, lacquers, varnishes. The genetic substances DNA and RNA are phosphate esters and many phosphates are of great biological importance.

phosphates Formally any salt of a PHOSPHORUS OXYACID. Generally applied to salts of H_3PO_4. Widely used as fertilizers, e.g. ammonium superphosphate. U.S. production 1999 46 million tonnes.

phosphatides See PHOSPHOLIPIDS.

phosphatidic acid (diacylglycerol phosphate) The parent compounds of phosphoglycerides.

phosphazenes (phosphonitrilic derivatives) Linear or cyclic compounds containing $(P=N)_x$ groups. A group of polymers. The simplest compounds, the chlorides, are formed by heating PCl_5 with NH_4Cl. The cyclic dimers, trimers and tetramers are the most important. Linear phosphazenes are used as flame retardants, elastomers in gaskets and hoses and in medical devices. Phosphazenes with perfluoroalkoxy groups are used in pharmaceuticals and as solvents in batteries. The chlorine atoms may be substituted by many other groups. The rings have low hydrolytic stability.

phosphides Compounds of the elements with phosphorus. The non-metals form covalent materials (e.g. S). The most electropositive elements form ionic derivatives, e.g. Na_3P, rapidly hydrolysed to PH_3 by water. The transition elements form derivatives containing P_n groups (e.g. CdP_2) or hard materials with complex structures.

phosphinates Alkyl derivatives of $HOP(O)H_2$.

phosphine, PH_3. See PHOSPHORUS HYDRIDES.

phosphinites Derivatives of $R_2P(H)O$. Insecticides, rodenticides, often used as Al, Mg, Zn salts.

phosphites (phosphonates), $(RO)_3P$. Derivatives of phosphorous acid, prepared PCl_3 plus ROH or ArOH. Used as intermediates in organic synthesis; with R′X, $(RO)_3P$ undergoes the Michaelis–Arbuzov reaction to $(RO)_2P(O)R′$. Form complexes with transition metals. $(RO)_3P$ used as anti-oxidants in lubricating oils. Metal derivatives of phosphorous acid are also termed phosphites – see PHOSPHORUS OXYACIDS.

phosphocreatine See CREATINE.

phosphoglycerides Glycerophosphoric acids.

phospholipases Enzymes which cleave phospholipids either at the carboxylate or at the phosphate linkages.

phospholipids, R′C(O)OCH$_2$·CH(OC(O)R″)·CH$_2$OP(O)$_2$OR‴. Substances of a fat-like nature which are essential components of living cells. They contain phosphorus and usually nitrogen and on hydrolysis give fatty acids, phosphoric acids and bases such as choline (from R‴). The phospholipids consist of a number of different species usually with glycerol as the 'backbone'. They include lecithins, cephalins and sphingomyelins. Amphoteric and mainly exist as components of cell membranes. Known specifically as phosphatides. See DICTIONARY OF BIOLOGY.

phosphomolybdates Salts of the heterophosphomolybdic acids (see MOLYBDATES) but particularly of $H_3PMo_{12}O_{40}$. The yellow ammonium salt is used for the estimation of Mo or P.

phosphonates Derivatives such as $RP(O)(OR)_2$.

phosphonites Derivatives such as $RP(OR)_2$, a dialkyl alkylphosphonite.

phosphonitrilic derivatives See PHOSPHAZENES.

phosphonium salts White crystalline salts analogous to the ammonium halides, e.g. $[PH_4]I$. Formed by the action of dry hydrogen halides on phosphine, which functions as a weak base. Phosphonium salts dissociate into PH_3 and HX at normal pressures and temperatures. Organophosphonium salts, e.g. $[(C_6H_5)_3PH]^+[BF_4]^-$, are much more stable than the unsubstituted derivatives; the stability increases with increasing substitution of hydrogen by alkyl or aryl groups. Triphenylalkylphosphonium salts having the general formula $[(C_6H_5)_3PCHR^1R^2]^+X^-$ are used to make phosphoranes (phosphonium ylids)

$$(C_6H_5)_3PC\begin{smallmatrix} R^1 \\ \\ R_2 \end{smallmatrix}$$

in the Wittig reaction. Used in phase-transfer catalysis.

phosphoproteins See PROTEINS.

phosphoranes Derivatives generally covalent, of 5-covalent phosphorus, i.e. R_5P, Ph_5P. The Wittig intermediates $RR′C=PR″_3$ can also be included under this heading.

phosphorescence Some substances absorb radiation and as a result are induced to emit radiation themselves. When this emission ceases when the irradiation is stopped, the process is called fluorescence; when the emission continues after the source of irradiation is removed, the process is termed phosphorescence. Phosphorescence results from a process involving a triplet state in which the atoms or molecules of the absorber are capable of existing in metastable electronic states. See PHOSPHORS.

phosphoric acids Oxyacids containing phosphorus(V) - see PHOSPHORUS OXYACIDS. Prepared phosphates plus H_2SO_4 or P oxidized to P_2O_5 hydrolysed to H_3PO_4. The salts of phosphoric acids, phosphates, occur naturally (particularly calcium phosphate, apatite) and organic derivatives are essential constituents of living tissues. Phosphates, particularly SUPER-PHOSPHATES, are of importance as fertilizers. U.S. production of P_2O_5 equivalent 1999 13.9 million tonnes.

phosphorimetry The use of phosphorescence in analysis. A non-routine method used for trace concentrations.

phosphorous acids PHOSPHORUS OXYACIDS containing phosphorus(III) particularly H_3PO_3.

phosphors Substances which exhibit phosphorescence. Doped oxide lattices, e.g. Sb, Mn; $Ca_5(F,Cl)(PO_4)_3$ are used in fluorescent lamps. YVO_4:Eu is a red phosphor used in colour television tubes, on stamps to assist with automatic sorting. Zn_2SiO_4:Mn i.r.-stimulated phosphors are used for night vision. Organic phosphors are used as OPTICAL BRIGHTENERS.

phosphorus, P. Non-metal, at.no. 15, at. mass 30.974, m.p. 44.15°C, b.p. 280.5°C, density ρ 1820, crustal abundance 1000 p.p.m., electronic configuration $[Ne]3s^23p^3$. Phosphorus is a non-metal of Group 15. It occurs naturally as calcium phosphate and as APATITE and FLUOROAPATITE. Phosphorus is obtained from $Ca_3(PO_4)_2$ by fusion with sand and coke in an electric arc furnace; P distils out and is condensed under water. There are many allotropes but three main forms. The most reactive form is white phosphorus (stored under water) which contains P_4 molecules. It is purified by zone melting. Heating for a long period at a high temperature gives red phosphorus. It is a polymeric material. Heating white phosphorus under high pressure gives black phosphorus with a graphite-like structure. The vapour contains P_4 and P_3 species. White phosphorus is spontaneously inflammable, black phosphorus is almost non-inflammable. P dissolves in alkalis but only in oxidizing acids.

Phosphorus compounds are very important as fertilizers and are widely used in matches, pesticides, special glasses and chinaware, alloys (steels, phosphor bronze), and metal treating (10%), detergents (40%), electrical components (e.g. GaP), foods and drinks (10%). Phosphates are an essential constituent of living organisms occurring in bone, teeth and DNA, RNA, etc.

phosphorus chemistry Phosphorus is an element of Group 15. The stable oxidation states are +5, (5-co-ordinate covalent), e.g. PF_5, 6-co-ordinate complexes (e.g. $[PF_6^-]$), 4-co-ordinate tetrahedral, e.g. $[PCl_4]^+$, $[PO_4]^{3-}$ and +3 (mainly 3-co-ordinate pyramidal, e.g. PH_3). Catenation occurs readily and is frequently based on P_4 tetrahedra although cyclic phosphines (e.g. $[CF_3P]_{4 \text{ and } 5}$) are known. Cationic species, e.g. P_4^{2+} are formed in oxidizing media such as oleum. Phosphorus(V) compounds can accept lone pairs of electrons, phosphorus(III) compounds can act as donors.

phosphorus halides Halides of the types PX_5, PX_3, POX_3, PSX_3 and P_2X_4 are known. Pentahalides are known for fluorine, b.p. –75°C (P plus F_2); PCl_5 plus NaF or AsF_3 (trigonal bipyramidal)) Cl, decomp. 167°C, (P plus excess Cl_2 $[PCl_4]^+[PCl_6]^-$), Br, $[PBr_4]^+Br^-$. POX_3 result from partial hydrolysis. The trihalides (all X) are pyramidal and are prepared by PCl_3 plus NaF or P plus a deficiency of X_2. (PF_3, b.p. –101°C, PCl_3, b.p. 76°C); P_2X_4(F,Cl,I) by coupling reactions or use of electrical discharges. PX_3 are very weak Lewis acids but good Lewis bases (e.g. $[Ni(PF_3)_4]$) forming very stable complexes. PX_5 are good acceptors. The phosphorus halides are used in the synthesis of phosphorus compounds, particularly phosphate and phosphite esters, and the chlorides as chlorinating agents PF_5 as a catalyst. PF_5 is hydrolysed to fluorophosphoric acids.

phosphorus hydrides Phosphine, PH_3, m.p. –133.5°C, b.p. –88°C. Formed water on Ca_3P_2 or yellow phosphorus plus NaOH solution. Decomposes at 450°C. Weak base (see PHOSPHONIUM SALTS). As generally prepared spontaneously inflammable because of the presence of P_2H_4, b.p. 52°C. Alkyl and aryl phosphines can contain P–H bonds but are more stable than PH_3 (particularly aryl derivatives). P_3H_5 and higher phosphines known.

phosphorus (organophosphorus acids)

Acid	Ester, etc.
$P(OH)_3$	
phosphor*ous* → acid	phosph*ite* e.g. $(CH_3O)_3P$ trimethyl phosphite
$HP(OH)_2$	
phosphon*ous* → acid	phosphon*ite* e.g. $C_6H_5P(OC_2H_5)_2$ diethylphenyl phosphonite

$H_3CP(OH)_2$

| methyl → phosphon*ous* acid | methylphosphon*ite* |

$H_2P(OH)$

| phosphin*ous* → acid | phosphin*ite* e.g. $(CH_3)_2POEt$ ethyl dimethyl phosphinite |

$HP(O)(OH)_2$

| phosphon*ic* → acid | phosphon*ate* |

$H_3CP(O)(OH)_2$

| methyl → phosphon*ic* acid | methylphosphon*ate* e.g. $CH_3P(O)(OC_6H_5)_2$ diphenyl methyl-phosphon*ate* |

$H_2P(O)OH$

| phosphin*ic* → acid, hypophosphorous acid | phosphin*ate* e.g. $(C_6H_5)_2P(O)OCH_3$ methyl diphenyl-phosphinate |

$OP(OH)_3$

| phosphor*ic* → acid | phosph*ate* e.g. $(C_2H_5)_3P=O$ triethylphosph*ate* |

The nomenclature considers organophosphorus compounds to be derivatives of phosphorus hydrides, oxyacids and oxides. Some of the parent compounds are unstable and cannot be isolated.

phosphorus oxides Phosphorus forms a series of oxides the most important of which are P_4O_{10} and P_4O_6.

Phosphorus(V) oxide, P_4O_{10}. Phosphorus pentoxide. There are several forms (P burned in excess O_2). P_4O_{10} has a great affinity for water and is used as a drying and dehydrating agent (e.g. HNO_3 gives N_2O_5); it is itself converted to phosphoric acids (see PHOSPHORUS OXYACIDS). P_4O_{10} reacts with metal oxides to give phosphates. Its structure has a P_4 tetrahedron with O along each edge and on each P. P_4O_9, P_4O_8 and P_4O_7 are intermediate oxides (heat on P_4O_6) with similar structures to P_4O_{10}.

Phosphorus(III) oxide, P_4O_6. Phosphorus trioxide, m.p. 24°C, b.p. 174°C. A waxy material (burn P in deficiency of O_2). It burns in excess O_2 to P_2O_5, reacts with, e.g. Cl_2 to $POCl_3$ and dissolves in water to give phosphorus(III) oxyacids. The structure is similar to that of P_4O_{10} but without the terminal oxygens.

phosphorus oxyacids Because phosphorus shows oxidation states +5 and +3 and can in addition share oxygens (particularly between tetrahedral atoms) there is a range of phosphorus oxyacids, cf. phosphorus (organophosphorus acids). POH groups have ionizable hydrogen, P–H groups are non-ionizable.

Hypophosphorous acid, H_3PO_2, $H_2P(O)OH$. A monobasic acid. $Ba(H_2PO_2)_2$ is formed when white phosphorus is dissolved in $Ba(OH)_2$ solution. H_3PO_2 and its salts are strong reducing agents.

Phosphorous acid, H_3PO_3, $HP(O)(OH)_2$. A dibasic acid (PCl_3 plus cold water), strong reducing agent.

Hypophosphoric acid, $H_4P_2O_6$, $HP(O)(OH)(\mu-O)P(O)(OH)_2$. It can be tetrabasic (Na salt from NaOCl on red phosphorus). See also PHOSPHORUS (ORGANOPHOSPHORUS ACIDS).

Phosphoric acids are derivatives of P_4O_{10}. Condensed acids and anions are known. Orthophosphoric acid, H_3PO_4, $OP(OH)_3$, gives tetrahedral anions. Pyrophosphoric acid, $H_4P_2O_7$, $(HO)_2P(O)(\mu-O)P(O)(OH)_2$, a tetrabasic acid. Metaphosphates (HPO_3) are obtained by dehydration of H_3PO_4 at 300°C. Many species are known; a cyclic $(P_3O_9)^{3-}$ ion and various chain phosphates, e.g. Maddrell's salt, Kurrol's salt, Graham's salt, all $(NaPO_3)_x$, are well characterized. So-called sodium hexametaphosphate is a long-chain compound used in water softening – chelates Ca^{2+} ions. Short-chain $[P_3O_{10}]^{5-}$, $[P_4O_{13}]^{6-}$ are also known.

Polyphosphates are used as detergents (calgon) but disposal of the residual phosphate causes major problems. Phosphates are of importance as flameproofing agents. See also PHOSPHORUS.

phosphorus sulphides Yellow solids P_4S_3, P_4S_4, P_4S_5, P_4S_7 and P_4S_{10} formed from red phosphorus and sulphur. The structures are based on a P_4 tetrahedron with some bridging S atoms and, in P_4S_7 and P_4S_{10}, terminal S atoms. Used in matches; P_4S_{10} in the preparation of organosulphur compounds (particularly pesticides and oil additives).

phosphorylases Enzymes that catalyse the addition of phosphate to organic species. Often connected with energy availability.

phosphoryl halides, $P(O)X_3$. See PHOSPHORUS HALIDES.

phosphotungstates Heteropolytungstates. See TUNGSTATES. Used as reagents for alkaloids.

photo-acoustic spectroscopy (PAS) A spectroscopic technique in which a sample is irradiated with chopped monochromatic radiation and the absorption of the radiation causes a temperature fluctuation which is transmitted to the gas in contact with the sample and which is in turn transmitted to a microphone. Used to measure the optical absorption spectra of solids and liquids, including opaque materials.

photochemistry Chemical reactions brought about by the action of light. Only light which is absorbed by the system can produce chemical effects; any light which passes unchanged through the material or is scattered is necessarily ineffective. Consequently, it is essential to determine the absorption spectrum of the substances concerned and to identify the spectral regions responsible for the reaction. Absorption of light is a quantum process. The primary step in a photochemical reaction is therefore the absorption of a quantum of light energy by a particular atom, molecule or ion which is thereby raised to an excited state containing an excess quantum of energy equal to hv (where h is Planck's constant and v is the frequency of the light absorbed). Only electronic excitation can bring about chemical reaction and therefore radiation in the far i.r. region is ineffective. Photochemical reactions are generally noticed with visible or u.v. radiation, the latter being the more powerful. Radiation in the far u.v. is powerful enough to break any chemical bond: e.g. oxygen gas is dissociated into oxygen atoms. However, not all photochemical reactions involve actual dissociation of molecules into atoms; many proceed by way of electronically excited molecules. An excited molecule may return to its ground state within about 10^{-4} sec while emitting its absorbed energy as fluorescence or passing it on to neighbouring molecules as thermal energy; in these cases no photochemical reaction is produced. In some cases, however, the excited molecule is capable of undergoing reaction, the light supplying the activation energy needed to induce reaction.

After the primary step in a photochemical reaction, the secondary processes may be quite complicated, e.g. when atoms and free radicals are formed. Consequently the quantum yield, i.e. the number of molecules which are caused to react for a single quantum of light absorbed, is only exceptionally equal to exactly unity. E.g. the quantum yield of the decomposition of methyl iodide by u.v. light is only about 10^{-2} because some of the free radicals formed recombine. The quantum yield of the reaction of $H_2 + Cl_2$ is 10^4 to 10^6 (and the mixture may explode) because this is a chain reaction.

The absorbed light may act as catalyst for a spontaneous reaction, but in other cases it may supply energy to make possible a reaction which, without light, would be thermodynamically impossible. In some cases, such a reaction reverses itself by thermal reaction (e.g. if left in the dark) and, hence, during irradiation a *photostationary state* is reached.

The light producing a photochemical reaction is most commonly absorbed by one of the reactants, but many examples are known where energy absorbed by another species is passed to the reactants; this is the phenomenon of *photosensitization.*

photoconduction The phenomenon observed when certain substances, e.g. selenium, which are semi-conductors in the dark, become good conductors when exposed to light. Photoconduction arises from the absorption of light energy causing electrons to be promoted from the valence band of the solid into the conduction band. It is the basis of some commercial photocopying processes.

photodissociation The process which occurs when a molecule on absorbing a quantum energy undergoes dissociation to smaller molecules, radicals or atoms. E.g. propanone undergoes photodissociation to methyl radicals and carbon monoxide.

photoelectric cells See PHOTOELECTRIC EFFECT. Since the intensity of emission is proportional to the intensity of the incident radiation, the effect can be used for measuring light intensity.

photoelectric effect The liberation of electrons by electromagnetic radiation. Most solids are only affected by vacuum u.v. radiation although Group 1 metals are semi-conductors that emit from visible and near i.r. radiation.

photo-electron spectroscopy Bombardment of an atom with photons results in the emission of an electron of energy E, where $E = (E_i - E_b)$, E_i and E_b, being the respective energies of the photon and of binding of the ejected electron. The technique may be used for measuring binding energies of valence electrons or for inner core electrons. Inner core spectroscopy, more commonly termed ESCA (electron spectroscopy for chemical analysis) or X-ray photo-electron spectroscopy (XPS), generally uses Al($k\alpha$) 1486 eV as the photon source. Ultra-violet photo-electron spectroscopy and synchrotron radiation used for examining valence-shell electrons.

photoemission spectroscopy The application of photoelectron spectroscopy to surfaces.

photographic developers Chemical reducing agents which specifically reduce silver halide grains in the photographic plate which have been exposed to light.

photographic gelatin The medium for silver halide emulsion used in photography. It is a protective colloid for the emulsion grains but also enhances sensitivity.

photography Process of preparing an image. Chemical photography uses a photographic film containing silver halide. The halide ions generate electrons paired with a positive pole and the Ag^+ captures the electrons to give Ag atoms which cluster to form the latent image.

photoisomerization Interconversion of isomers under the influence of electromagnetic radiation.

photoluminescence A general term used to describe the emission of light as a result of an initial absorption of light. The phenomena of fluorescence and phosphorescence are examples of photoluminescence.

photolysis The decomposition or activation and reaction of a substance or substances on exposure to electromagnetic radiation.

photon A quantum of light energy. The energy of a photon is given by $E = hv$, where v is the frequency of the radiation and h is Planck's constant. Intensity is determined by number of photons.

photosensitization A process in which a photochemical reaction is induced to occur by the presence of a substance (the photosensitizer) which absorbs the light but itself is substantially unchanged at the end of the reaction, the absorbed light energy being passed on to the main reactants. E.g. when hydrogen is exposed to light of wavelength 254nm, no absorption of the light occurs, and the hydrogen remains completely unaffected. If mercury vapour is added to the hydrogen, the mercury atoms are excited. When such an excited mercury atom collides with a hydrogen molecule, it can transfer some of its energy to the hydrogen, and cause it to dissociate into atoms. The hydrogen has been made sensitive to the light which it does not absorb.

photostationary state A photostationary state is attained when the rate of removal of reactants by light is equal to the rate of recombination or reaction of the products to yield the reactants.

photosynthesis The process by which green plants build up their carbon compounds from atmospheric carbon dioxide using light as the energy source. More generally any process initiated by radiation that produces energy, e.g. radiation of CO_2 to carbohydrates with simultaneous release of O_2. Energy is absorbed using light-harvesting complexes which incorporate carotenoids.

phototropy The reversible change of colour of certain substances when exposed to light of a certain wavelength.

photovoltaic cells (PV) Convert light, typically sunlight, to energy. Made from two layers of semi-conducting material, e.g. doped silicon, one with P, one with B.

phoxin Insecticide, organophosphate derivative.

phrenosin, $C_{48}H_{93}O_9$.   M.p. 212°C. A CEREBROSIDE.

phthalamide, $C_8H_8N_2O_2$, 1,2-$C_6H_4(CONH_2)_2$. Colourless crystals; on heating to 200–210°C melts (decomp.) into phthalimide and ammonia.

Prepared by stirring phthalimide with cold concentrated ammonia solution. Hydrolysed to phthalic acid with dilute acids. Dehydration with ethanoic anhydride gives first 2-cyanobenzamide and then phthalonitrile.

phthalic acid, $C_8H_6O_4$, 1,2-$C_6H_4(C(O)OH)_2$. Colourless crystals; m.p. 190–210°C. At m.p. it decomposes into water and PHTHALIC ANHYDRIDE. Prepared by oxidation of naphthalene or 2-xylene, or by alkaline hydrolysis of phthalic anhydride.

Dibasic acid, forms stable metallic salts. Distillation with soda lime gives benzene. Its reactions are similar to phthalic anhydride, in which form it is almost invariably used.

phthalic anhydride (1,3-isobenzofurandione), $C_8H_4O_3$. Long silky needles; m.p. 130°C, b.p. 284°C. Slowly converted into phthalic acid in water.

Prepared by the oxidation of naphthalene with sulphuric acid at 270–300°C in the presence of $HgSO_4$; under these conditions the phthalic anhydride sublimes out of the reaction mixture as formed. It is manufactured by the vapour-phase oxidation of naphthalene or

2-xylene over a V_2O_5 catalyst.

Treatment with PCl_5 gives phthalyl chloride; reduction with zinc and ethanoic acid or NaOH gives phthalide. Fusion with urea gives phthalimide.

Valuable dyestuffs intermediate. It can be condensed with benzene and $AlCl_3$ to give anthraquinone. With phenols and a condensing agent, phthalic anhydride gives rise to phthalein dyestuffs, e.g. phenolphthalein, eosin. It also condenses with quinaldine derivatives to give 'quinoline' dyes. The chief commercial uses of phthalic anhydride are in the manufacture of dialkyl phthalate plasticizers and alkyd resins.

phthalic esters Because of their low vapour pressures and chemical stability, various esters of phthalic acid such as diethyl (b.p. 298°C), dibutyl (b.p. 340°C) bis(2-ethylbenzyl) and di-n-octyl (b.p. 248°C) are used as plasticizers. The dimethyl ester is used in the manufacture of polyethylene terephthalate and the diethyl ester in the manufacture of celluloid and as a solvent. The dimethyl and dibutyl esters are used as insect repellants but there are possible health hazards. They are all manufactured by reacting phthalic anhydride with the alcohol in the presence of catalytic amounts of sulphuric acid. The dimethyl ester is used in alkyd resins.

phthalide, $C_8H_6O_2$. M.p. 75°C. Formed

by reduction of phthalic anhydride with Zn dust and NaOH. Fusion with KCN and hydrolysis gives homophthalic acid $1,2-C_6H_4((CO)OH)CH_2C(O)OH$. Also fungicide, tetrachlorophthalate.

phthalimide, $C_8H_5NO_2$. Colourless plates, m.p. 230°C. It is prepared on the large scale by passing ammonia into molten phthalic anhydride. Also obtained by melting together urea and phthalic anhydride.

Phthalimide dissolves in alkalis to give N-metal derivatives, but on warming the solution the ring opens, and salts of phthalamic acid are obtained. Reduced to phthalide with zinc and NaOH solution; prolonged hydrolysis with acids or alkalis gives phthalic acid or its salts. Treatment with concentrated ammonia solution gives phthalamide. Used in the Gabriel synthesis of amino derivatives from halides.

When treated with alkaline hypochlorite solution anthranilic acid is obtained. This reaction, the first stage in the indigo synthesis, is the most important outlet for phthalimide.

phthalocyanines An important class of organic colouring matters, used as pigments and of more limited use as dyestuffs. Have great fastness and brilliance of shade; green to blue in colour.

They can be prepared by: **1** heating phthalic anhydride, urea (or ammonia) and a metallic salt; **2** heating 2-cyanobenzamide and a metallic salt; **3** heating phthalonitrile with a metal or metallic salt.

Depending upon the metallic compound used, different metallic phthalocyanine derivatives are obtained, e.g. when copper chloride is used copper phthalocyanine (Monastral Fast Blue B) is obtained.

The phthalocyanines can be dispersed as pigments or they can be sulphonated: dispersion is carried out by solution in sulphuric acid, followed by precipitation in water.

Very stable to heat, light and chemical reagents. The metal-free and heavy metal compounds sublime practically unchanged at 550–580°C.

phthalonitrile, $C_8H_4N_2$, $1,2-C_6H_4(CN)_2$. Colourless needles; m.p. 141°C, b.p. 290°C. Slowly hydrolysed to phthalic acid by dilute acids and alkalis.

Obtained by dehydration of 2-cyanobenzamide or phthalamide with ethanoic acid. On a larger scale the dehydration is carried out by means of phosgene in pyridine or from a mixture of ammonia and phthalimide over a catalyst at 350–450°C. Used to prepare PHTHALOCYANINES.

phthioic acids (phthienoic acids) See MYCOLIC ACIDS.

phyllins See CHLOROPHYLL.

physisorption See ADSORPTION.

phytic acid (1,2,3,4,5,6-cyclohexane-hexolphosphoric acid), 1,2,3,4,5,6,-$C_6H_6(OPO(OH)_2)_6$. *Myo*-inositol hexaphosphoric acid, a water-soluble syrup, sparingly soluble in organic solvents and strongly acidic Important in the body's use of iron and nutrients. Occurs as its insoluble calcium–magnesium salt, phytin, in seeds. It is also found in blood plasma phospholipids and in the erythrocytes of chicken blood. Used as a fermentation nutrient.

phytin See PHYTIC ACID.

phytochemistry The chemistry of compounds associated with plants.

phytol (2,6,10,14-tetramethylhexadec-14-en-16-ol), $C_{20}H_{40}O$. A diterpenic alcohol obtained by the action of alkalis on chlorophyll. Colourless oil; b.p. 202–204°C/10mm. On oxidation it yields a ketone $C_{18}H_{36}O$; b.p. 175°C/11mm. Used in preparation of vitamins E and K.

phytoremediation The use of plants to extract and concentrate, e.g., heavy metals.

phytotoxic Poisonous to plants.

pi bonding (π-bonding) Bonding by interaction of orbitals of such shape that there are

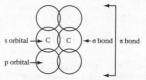

s orbital — C   C — σ bond   π bond

p orbital —

two regions of overlap. Thus ethene, C_2H_4, is considered to have a σ-bond formed by overlap of s-orbitals and a π-bond formed by overlap of p-orbitals. π-Bonding is of importance in co-ordination compounds where co-ordinate bonds may have π-character. The electrons can be donated from the donor or there may be electron transfer from the acceptor to the donor (back-bonding).

pickling The immersion of metal parts in dil. H_2SO_4 or HCl, followed by washing to obtain a clean surface which can be subsequently treated, e.g. by galvanizing.

picloram Herbicide, chlorinated pyridine carboxylic acid derivative.

picolines (methylpyridines), C_6H_7N. Found with pyridine in bone oil and coal tar. α-Picoline (2-methylpyridine), b.p. 129°C; β-picoline (3-methyl), b.p. 144°C; γ-picoline (4-methyl), b.p. 143°C. 2-Picoline is used in the manufacture of the monomer 2-vinylpyridine, 3-picoline for nicotinic acid, and 4-picoline for isoniazid. Picolines are used as solvents, dyestuffs intermediates and in insecticide manufacture.

picramic acid (2-amino-4,6-dinitrophenol), $C_6H_5N_3O_5$. Red needles; m.p. 168–169°C. Prepared by reduction of picric acid with sodium hydrogen sulphide. It is used for dyestuff synthesis.

picrates Derivatives (salts) of picric acid, i.e. $X^+[OC_6H_2(NO_2)_3]^-$. A convenient method of characterization of basic organic compounds. Some picrates, especially of metals, explode on heating. Aromatic hydrocarbons form charge transfer complexes with picric acid, also known as picrates.

picric acid (2,4,6-trinitrophenol), $C_6H_3N_3O_7$. Bright yellow crystals; m.p. 122°C. Prepared by the nitration of phenolsulphonic acid. It is used in dyeing and as an explosive. In the cast form it is known as lyddite. Used in matches, leather, textiles, batteries.

picrolonic acid (2,4-dihydro-5-methyl-4-nitro-2-(4-nitrophenyl)-3H-pyrazol-3-one), $C_{10}H_8N_4O_5$. Fine yellow needles; m.p. 124°C (decomp.). Used for the isolation and identification of organic bases, and in tests for calcium, copper and lead.

piezoelectricity The electric charge developed in anisotropic crystals when subjected to stress (e.g. pressure). Used in electrical components and in lighters.

pig iron The product from a blast furnace after cooling in sand moulds. Essentially the same as cast iron but the latter has generally been remelted and cast to a finished shape.

pigments Materials which are used to impart colour to surfaces, plastics, inks, etc., although the pigments may incidentally affect other properties of the substrate. Distinct from DYE-STUFFS which operate on a molecular level, pigments tend to be particulate and insoluble. Many pigments are inorganic in nature. White (TiO_2, basic lead carbonate and sulphate, zinc oxide and sulphide); red and brown (iron oxides, red lead, cadmium reds); yellow (iron oxides, lead, zinc and cadmium chromates); green (chromium derivatives); blue (cyanoferrates, ultramarines); black (C). Metal powders (Al, bronze, Zn) are also used in surface coatings. Many insoluble organic dyestuffs (e.g. phthalocyanines) are also used as pigments.

pimelic acid (heptanedioic acid), $HO(O)C \cdot [CH_2]_5 \cdot C(O)OH$, $C_7H_{12}O_4$. Colourless prisms; m.p. 105°C. Present in castor oil. It can be prepared from pentamethylene dibromide or cyclohexanone via the nitrile and by the reduction of salicylic acid followed by oxidation of the ring system. Also produced by oxidizing capric or oleic acids. Used in polymeric systems.

pinacol (pinacone, 2,3-dimethylbutane-2,3-diol), $C_6H_{14}O_2$. Colourless crystals; m.p. 38°C, b.p. 175°C. Forms hexahydrate, m.p. 47°C. Prepared by treating a mixture of propanone and benzene with magnesium amalgam. When heated with sulphuric acid, it forms pinacolone. 2,3-Dimethylbutadiene is formed when the vapour of pinacol is passed over alumina at 400°C.

pinacolone (pinacolin, 3,3-dimethyl-2-butanone), $C_6H_{12}O$. Colourless liquid with a camphor-like odour; b.p. 103–106°C/746mm. Prepared by heating pinacol hydrate with sulphuric acid. It is oxidized to pivalic acid.

pinacol–pinacolone rearrangement When heated with mineral or organic acids, pinacols undergo a molecular rearrangement with loss of water to give ketones, pinacolones.

pinacols, $(HO)R^1R^2C \cdot CR^3R^4(OH)$. 1,2-Glycols where R_1, R_2, etc. are alkyl or aryl groups. They are formed by reducing aromatic ketones with zinc and dilute acid, or from aliphatic ketones by reduction with magnesium amalgam.

pinacone See PINACOL.

pindone Rodenticide, indanone derivative.

α-pinene (2,6,6,-trimethylbicyclo-[3,1,1]hept-2-ene), $C_{10}H_{16}$. A dicyclic terpene; m.p. –50°C, b.p. 156°C. The most important of the terpene hydrocarbons. It is found in most essential oils derived from the *Coniferae*, and is the main constituent of turpentine oil. Contains two asymmetric carbon atoms. The (+)-form is obtained from Greek turpentine oil. When heated under pressure at 250–270°C, α-pinene is converted into dipentene. It can be reduced by hydrogen in the presence of a catalyst to form pinane. It is oxidized by air or oxygen, both wet and dry. With moist air it forms sobrerol, $C_{10}H_{18}O_2$, which forms the oxide pinol by the action of weak mineral acids.

By oxidation with permanganate it forms pinonic acid, $C_{10}H_{16}O_3$, a monobasic acid derived from cyclobutane. With strong sulphuric acid it forms a mixture of limonene, dipentene, terpinolene, terpinene, camphene and 4-cymene. Hydrogen chloride reacts with turpentine oil to give $C_{10}H_{17}Cl$, bornyl chloride, 'artificial camphor'.

β-pinene (nopinene, 6,6-dimethyl-2-methylenebicyclo[3H]heptane), $C_{10}H_{16}$. α-Pinene in nature is usually accompanied by smaller amounts of (-)-β-pinene. Natural (+)-β-pinene extracted from the ripe fruits of *Ferula galbaniflua*. B.p. 162–163°C. On oxidation with potassium permanganate it forms nopinic acid, $C_{10}H_{16}O_3$; m.p. 126–127°C.

pink salt Ammonium hexachlorostannate(IV).

pinocamphone, $C_{10}H_{16}O$. A dicyclic ketone, the (-)-form of which constitutes about 45% of oil of hyssop.

pinocarveol, $C_{10}H_{16}O$. A dicyclic terpene alcohol, the (-)-form of which is present in the oil of *Eucalyptus globulus*.

piperalin Fungicide, dichlorobenzoate derivative.

piperazine (hexahydropyrazine), $HN(CH_2CH_2)_2NH$, $C_4H_{10}N_2$. The hexahydrate, m.p. 44°C (104°C when anhydrous), b.p. 126°C. Made by the action of alcoholic ammonia on 1,2-dichloroethane, piperazine is used in human and veterinary medicine.

2,5-piperazine dione See DIKETOPIPERAZINE.

piperidine (hexahydropyridine, 2,5-piperazinedione), $C_5H_{11}N$. Colourless liquid with a characteristic ammoniacal smell; m.p. –9°C, b.p. 106°C. Miscible with water. It is present in pepper as the alkaloid piperine. It can be prepared by the reduction of pyridine, either electrolytically or by other means. Piperidine is a strong base, behaving like the aliphatic amines.

piperitol See PIPERITONE.

piperitone, $C_{10}H_{16}O$. An optically active naturally occurring terpene ketone, the oxidation product of piperitol. Formed in eucalyptus oil and in Japanese peppermint oil, and in the oil from Sudan Mahareb grass. Piperitone is of considerable technical importance, oxidized to thymol and reduced to menthene, menthols and α-phellanedrene. It is a colourless oil of a pleasant peppermint-like smell.

piperonal (3,4-methylenedioxybenz-aldehyde, 1,3-benzodioxole-5-carbox-aldehyde), $C_8H_6O_3$. White crystals; m.p. 37°C, b.p. 263°C. Occurs associated with vanillin. Obtained on oxidation of various natural products such as piperine. Used extensively in soap perfumery, flavouring, synthesis.

piperonyl butoxide Insecticide synergist, cyclic polyether derivative.

PIPES (1,4-piperazine-*bis*-(2-ethene sulphonic acid)) Zwitterionic compound. Used as biological buffer.

pipette Apparatus for obtaining a given volume of solution. The apparatus may be filled by sucking or, more safely, by applying a negative pressure or by automatic means.

piremicarb Insecticide, acaricide, pyrimidyl derivative.

pitchblende, $UO_{2-2.25}$. An important uranium ore also containing useful quantities of Ra. Th, Ce, Pb are also present.

pitch (coal tar pitch) The product left when the more volatile portions of coal tar are distilled.

pivalic acid (trimethylacetic acid, 2,2-dimethylpropanoic acid), $(CH_3)_3C \cdot C(O)OH$. Colourless solid; m.p. 35.5°C, b.p. 164°C. Prepared by the carboxylation of isobutene with carbon monoxide in sulphuric acid or by the oxidation of pinacone with NaOBr. Intermediate in the synthesis of various industrial products.

pK For any acid HA which partially dissociates in solution, the equilibrium

$$HA(solv.) \rightleftharpoons H^+(solv.) + A^-(solv.)$$

is defined by an equilibrium constant K, where

$$K = \frac{[H^+][A^-]}{[HA]}$$

Concentrations have been used in place of activities when concentrations are low. The negative logarithm of the equilibrium constant is termed the pK, i.e. pK = −log K. For acids, the pK is denoted by pK_a. Similarly for a base BOH, the pK_b is given by

$$pK_b = -\log K_b, \text{ where}$$

$$K_b = \frac{[B^+][OH^-]}{[BOH]}$$

planar complexes Complexes in which the acceptor atom is surrounded by four ligands in a plane. If predominantly covalent the hybridization is normally considered as dsp^2. E.g. $[Pt(NH_3)_4]^{2+}$, $[Ir(CO)Cl(PPh_3)_2]$ and $[PtCl_4]^{2-}$.

Planck's constant (*h*) The constant relating the energy and frequency of radiation, $E = hv$. It has a value of 6.62×10^{-34} J s.

plane of symmetry See SYMMETRY ELEMENTS.

planetary electrons Extranuclear electrons. See ELECTRONIC CONFIGURATION.

plant hormones (plant growth substances) Compounds other than nutrients which affect physiological processes in plants. E.g. auxins, gibberellins.

plasma A highly ionized gas.

plasmalogens Phospholipids analogous to lecithins and cephalins. Occur in brain and muscle.

plasmoids Small rings of DNA found in bacteria.

plaster A material used to cover and produce a flat surface on a wall or other structure. The chief materials are **1** a mixture of lime and sand or of Portland cement and sand, and **2** materials consisting essentially of plaster of Paris.

plaster of Paris An incompletely hydrated calcium sulphate, $2CaSO_4.H_2O$, produced by heating gypsum, $CaSO_4.2H_2O$. The setting of the plaster is the result of rehydration due to the added water; it is accompanied by a slight expansion.

plastic additives Materials incorporated into plastics to change properties, increase stability (remove free radicals), add flame retardancy, give light stability, pigments, antistatics.

plastic explosives Explosives which can be moulded by hand. E.g. cyclonite mixed with an oily or rubbery binder.

plasticizers Materials incorporated into plastic resins to change workability, flexibility, flow and impact resistance. Thermoplastic resins (particularly PVC and polyvinyl acetate) undergo major changes in properties in the presence of plasticizers. Plasticizers are high mol.wt. liquids or low-melting solids often carboxylic phosphate esters although hydrocarbons, halogenated hydrocarbons, ethers, polyglycols are also used.

plastics Artificial polymeric materials, usually organic in nature, which have generally been shaped by flow thermally or by moulding. Many artificial elastomers are plastics. Plastics can be classified on the basis of their thermal behaviour. Thermoplastics (e.g. ethene polymers, polyvinyl chloride) can be repeatedly softened by heating and rehardened on cooling. Thermoset or thermocured plastics (e.g. phenol-formaldehyde resins) are infusible and insoluble in their final moulded shape. Plastics are also characterized on their chemical derivation or final structure. Plastics are moulded and shaped whilst in their heat-softened stage and then cured by continued heating (thermosetting plastics) or cooling (thermoplastics). Extrusion can give pipe, rod, etc. World use 1999 33.6 million tonnes, 40% packaging, 18% construction materials.

plastoquinones A family of trisubstituted benzoquinones with polyisoprenyl groups. They occur in chloroplasts and play an essential part in photophosphorylation during photosynthesis.

plate (tray) The vapour–liquid contacting units in an absorption or FRACTIONATING COLUMN. The term 'plate' tends to be used in Britain, 'tray' in the U.S. See PLATE COLUMN, BUBBLE-CAP PLATE.

plate column (tray column) The plate column, or tray column as it is known in the U.S., is the plant most widely used for vapour–liquid contacting, and as such is employed in GAS ABSORPTION and STRIPPING and for fractional distillation. It consists of a vertical cylindrical column containing a series of horizontal *plates* or *trays* spaced up to 1 m apart and set one above the other all the way up the column. Liquid flows down the column from plate to plate and vapour passes upwards through holes in them.

Since the transfer of material between phases takes place on the plates, the degree of gas absorption, or of separation in the case of a distillation column, depends directly on their number.

plate efficiency (tray efficiency) If a plate in a distillation or absorption column were 100% efficient, perfect contacting would occur on it, and the liquid and vapour streams leaving the plate would be in equilibrium with each other. In practice the efficiency of plates is usually less than 100%, with the result that the number required to achieve a given degree of separation or absorption is greater than the theoretical number.

platforming Catalytic process using a platinum catalyst to convert petroleum fraction into high-octane fuels.

platinum, Pt. Metal, at.no. 78, at.mass 195.08, m.p. 1768°C, b.p. 3825°C, density ρ 21 450, crustal abundance 0.0001 p.p.m., electronic configuration [Xe] $4f^{14}5d^96s^1$. The most abundant of the PLATINUM METALS, occurring naturally and in traces in heavy metal sulphide ores. Pt compounds are readily reduced to the metal which has the ccp structure. The metal is very malleable and ductile. It is attacked by F_2 and Cl_2 above 300°C, dissolved by aqua regia and fused alloys. The metal is used extensively in investment jewellery, laboratory ware, thermocouples (particularly Pt/Pt–Rh), electrical contacts, and catalysis (SO_3, NH_3, hydrocarbon cracking, methanal production, exhaust catalysts). In catalysis the metal is generally impregnated on an inert support. Used in fabrication of high-quality glassware. Pt compounds have anti-tumour activity. World demand 2002 637 000 oz.

platinum, organic derivatives Stable alkyls and aryls, e.g. [*trans*-PtBrMe(PEt$_3$)$_2$] are formed by the action of Grignard reagents on the halides. Tetrameric species, e.g. [Me$_3$PtCl]$_4$, are formed by the reaction of, e.g. PtCl$_4$. Olefins and alkynes form π-bonded complexes, e.g. K[PtCl$_3$(C$_2$H$_4$)], [PtCl$_2$(C$_2$H$_4$)]$_2$, [(Ph$_3$P)$_2$Pt(PhC$_2$Ph)].

platinum ammines Ranges of Pt(II) (4-coordinate) and Pt(IV) (octahedral) complexes containing Pt–NH$_3$ (and amine) groups.

platinum black A finely divided precipitate formed from Pt(II) solutions by reducing agents. Used as a hydrogenation catalyst. Compare PLATINUM OXIDES.

platinum chemistry Platinum is the heaviest element of Group 10. It shows oxidation states of +6 and +5 (fluoride derivatives only), +4, +2, +1 and 0. Pt(II) and (IV) form extensive series of complexes generally with a square planar or octahedral co-ordination respectively. Halides, N-ligands (particularly amines and ammonium), phosphines form stable complexes with both oxidation states. Cyanide and carbonyl complexes are particularly stable with Pt(II). Some complexes are anti-tumour agents. Pt(0) complexes are stable with phosphine ligands; cluster derivatives are formed in many cases. Organic derivatives are known for Pt(II) and Pt(IV).

platinum halides Platinum fluorides known are volatile red PtF_6 (Pt plus F_2), m.p. 61°C, a very strong oxidizing agent (O_2 gives $O_2^+PtF_6^-$), red polymeric PtF_5 (Pt plus F_2), yellow brown PtF_4 (Pt plus BrF_3). Fluoroplatinates(V) and (IV) (e.g. K_2PtCl_6 plus BrF_3 gives K_2PtF_6) contain complex anions. All platinum fluoride species except $[PtF_6]^{2-}$ are hydrolysed by water. Known chlorides are red–brown $PtCl_4$, green-black $PtCl_3$ (contains Pt^{II} and Pt^{IV}) and red-black $PtCl_2$ (all formed from Pt and Cl_2). Chloroplatinates containing, e.g. $[PtCl_6]^{2-}$ and $[PtCl_4]^{2-}$ ions are stable. Bromides and iodides are similar to chlorides; the stability of the complex halides increases with the mass of the halogen. Complexes are readily formed between Pt(II) and (IV) halides and donors, e.g. phosphines, thiols.

platinum metals The elements ruthenium and osmium (iron group), rhodium and iridium (cobalt group) and palladium and platinum (nickel group). They occur together as native alloys (native platinum, osmiridium), in mixed sulphide ores and as traces in Au, Ag and Cu sulphide ores. See the individual elements.

platinum oxides Brown PtO_2, Adam's catalyst for hydrogenation, is formed by dehydration of the hydrated oxide precipitated on hydrolysis of $PtCl_6^{2-}$. PtO ($PtCl_2$ plus KNO_3) and Pt_3O_4 (Pt electrode or Pt plus pressure of O_2). Mixed oxides, e.g. $Tl_2^{III}Pt_2^{IV}O_7$, are known.

pleochroism The property of a crystal of having a different colour depending upon the direction of transmitted light through the crystal. Cyanoplatinates, e.g. $K_2Pt(CN)_4$, are strongly pleochroic.

plumbates See LEAD OXIDES.

plutonium, Pu. Actinide metal, at.no. 94, at. mass ^{239}Pu 239.05, m.p. 640°C, b.p. 3228°C, density ρ 19 840, electronic configuration [Rn] $5f^67s^2$. Traces of Pu occur naturally, being formed by neutron capture by natural uranium. Pu is normally prepared by neutron irradiation of ^{238}U [^{239}Pu (24000 years)]. ^{238}Pu (86 years) is formed from Np. Pu is separated by selective oxidation and solvent extraction. The metal is formed by reduction of PuF_4 with calcium; there are six crystal forms. ^{239}Pu is used in nuclear weapons and reactors; ^{238}Pu is used as a nuclear power source (including space exploration). Inhalation of Pu compounds is very dangerous.

plutonium compounds Plutonium chemistry is similar to that of uranium with a greater stability of the lower oxidation states but also an unstable +7 state formed by electrolytic oxidation in alkaline solution

$$PuO_2^{2+} \text{ (pink)} \xrightarrow{+0.91} PuO_2^+ \text{ (red)} \xrightarrow{+1.17}$$
$$Pu^{4+} \text{ (green)} \xrightarrow{+0.98}$$
$$Pu^{3+} \text{ (blue)} \xrightarrow{-2.05} Pu \text{ (volts in 1M acid)}$$

The metal reacts slowly with water and rapidly with dilute acids. In aqueous solutions the +3, +4, +5 and +6 oxidation states co-exist. PuF_6 (red–brown, mp. 52°C, b.p. 62°C) is formed from fluorine and PuF_4. PuF_4 is prepared from PuO_2, HF and oxygen. $PuCl_4$ is known in the gas phase. All PuX_3 are known. The stable oxide is PuO_{2+x}. Pu_2O_3 is also known. Complex fluorides are formed by Pu(V) and Pu(IV) and chlorides by Pu(IV). The PuO_2^{2+} ion and PuX_4 form complexes with oxygen donors. $Pu(C_3H_5)_3$ has been prepared.

plywood Layers of wood bonded by adhesives (generally phenol–formaldehyde, urea–formaldehyde, melamine–formaldehyde resins). A laminate.

Pm Promethium.

PMMA Polymethylmethacrylate.

pnictogens The elements nitrogen, phosphorus, arsenic, antimony and bismuth of Group 16. Pnictides contain X^{3-} species.

Po Polonium.

point group A set of symmetry elements passing through or referring to a single point. Refers to molecules. In lattices where symmetry elements may not all pass through one point the corresponding arrangement is referred to as a space group.

Poisson–Boltzmann equation Used to calculate molecular electrostatic potentials.

polar bond Covalent bond with unequal distribution of electrons.

polarimeter An instrument used to measure the amount by which the plane of polarization of plane-polarized light is rotated on passing through an optically active medium.

polarizability When a molecule is placed in an electric field the electrons tend to be drawn away from the nuclei with the result that a dipole is induced in the molecule. The strength of the dipole divided by the field strength is termed the polarizability of the molecule. Such *electronic* polarizability can be determined from measurements of refractive index at wavelengths in the visible spectral region. Similar measurements in the far i.r. give, additionally, the *atomic* polarizability. Polarizability has units of (coulomb–metre)^2 joule^{-1}.

polarization When molecules possessing dipole moments are placed in an electric field they tend to align themselves in a certain direction within the field. This effect is called orientation polarization. Simultaneously the electrons within each molecule are displaced slightly towards the positive pole of the field. This effect is called electron polarization. Further nuclei having a positive charge are slightly displaced relative to each other. This atomic polarization is of very small magnitude. The total molecular polarization is the sum of the orientation, electronic and atomic polarizations.

polar molecule As the two electrons in a covalent link are not shared equally (different electronegativities), there is a separation of charge over the bond and, if the effects are not cancelled out, over the whole molecule. Such a molecule is known as a polar molecule. Thus in covalent hydrogen chloride

$$\overset{\delta+}{H} - \overset{\delta-}{Cl}$$

the chlorine has a greater share of the bonding pair than the hydrogen and the molecule has a resultant dipole moment. Lone pairs of electrons also make molecules strongly polar.

polarography An electrochemical method of analysis, particularly suitable for dilute solutions of substances which are susceptible to electrolytic reduction (or oxidation) at a mercury cathode. i is plotted against v generally using a clean surface such as a dropping mercury electrode, although Pt electrodes can be used. This polarogram shows a step for each reducible species present in solution. This step is of characteristic potential and of height proportional to the concentration of the component and the number of electrons in the process. The method is particularly useful for determining trace amounts of metal and for investigations of complexes in solution and for studies in non-aqueous solvents.

polar solvent A solvent, the molecules of which have good solvating power. Generally, the greater the dipole of the molecules of the solvent the better the liquid as a solvent for ionic substances.

pollucite, $Cs(AlSi_2O_6).xH_2O$. The major mineral source of Cs. Contains $(Al,Si)O_4$ tetrahedra linked into a three-dimensional network.

polonium, Po. Non-metallic element, at.no. 84, at.mass ^{210}Po 209.98, m.p. 254°C, b.p. 962°C, density ρ 9340, electronic configuration [Xe] $5d^{10}6s^26p^4$. A decay product from heavier elements but prepared by neutron irradiation of ^{209}Bi. ^{210}Po has a half life of 138 days, it is an α-emitter and is a very hazardous material. Separated from Bi by electrochemical methods and sublimed. The metal is obtained by thermal decomposition of the sulphide. Used to remove static electricity and as a neutron source.

polonium chemistry Polonium is an element of Group 16, and shows many similarities to tellurium. However polonium hydroxide and dioxide are more basic than the corresponding tellurium compounds, and the dihalides are quite stable. The *halides* PoX$_4$ and PoX$_2$ are known. The tetrapositive halides are obtained by the action of the metal with the halogen and form complex halides, M_2PoX_6. Both PoCl$_2$, and PoBr$_2$ are obtained by thermal degradation of the tetrahalides. *Polonium dioxide*, PoO$_2$, is obtained from the elements at 250°C, it is basic in character; a monoxide, PoO, and a trioxide, PoO$_3$ are also known. Lead and mercuric polonides, containing the Po^{2-} ion, have been prepared by interaction of the elements. Most polonium salts are highly coloured, being red or yellow. All polonium compounds are readily hydrolysed and reduced.

poly- Polymeric derivative. See below or under particular heading. E.g. for polyacrolein see ACROLEIN POLYMERS.

polyacetaldehyde See ALDEHYDE POLYMERS.

polyacetals Polymers formed from polyols, e.g. $HO(CH_2)_nOH$, and carbonyl derivatives, e.g. methanal CH_2O, which react to give $HO[(CH_2)_nOCH_2O]_x(CH_2)_n-OH$. Ring structures are also formed. Pentaerythritol and glutonaldehyde are typical components. Used in coatings. See POLYETHYLENE GLYCOLS.

polyacrylamide, $\{CH_2CHCONH_2\}_n$. Linear polymers through the propene groups. High mol. wt. material, gels, gums, plastics soluble in water. By reaction at the amide group, can be converted into polyelectrolytes or thermosetting resins used in gel electrophoresis. Widely used polymer, particularly because of water solubility. Substituted acrylamides, e.g. methacrylamide, $CH_2{:}CMeCONH_2$, give similar polymers.

polyacrylonitrile Polymeric acrylonitrile, polymeric PROPENENITRILE, $(CH_2 \cdot CHCN)_n$. Important fibre, elastomer. Frequently used as a copolymer with, e.g. butadiene, styrene or vinyl chloride. Polymerization initiated by free radicals or by light. Used in finishings, knitwear.

polyalkenes Polymeric olefins. See, e.g. BUTENE POLYMERS, BUTYL RUBBER, ETHENE POLYMERS, PROPENE POLYMERS.

poly(alkylene oxides) See 1,2-EPOXIDE POLYMERS.

polyalkylidenes Polymers, $(CHR)_n$, formed, e.g. from the reaction of diazomethane and alcohols or hydroxylamine derivatives in the presence of boron compounds or metal complexes. Polymethylene is formally the same as polyethene and the properties of the various polymers depend upon the degree of polymerization, crystallinity and the stereochemistry.

polyamides See NYLON.

polyamines (poly(alkylene polyamines)), $(HN-R-NH-R')_n$. Hydrophilic polar substances formed by reacting alkylene polyamines or simple amines with alkylene dihalides. Used as flocculants, for, e.g. cellulose fibre and mineral ore suspensions.

polyampholyte A macromolecule with both anionic and cationic character.

polyaniline Conducting polymers.

polyanion A macromolecule with anionic character.

polybenzimidazoles Heat-resistant polymers containing benzimidazole rings. The monomers are formed by condensation of aromatic diamine and dicarboxylic acid derivatives and the polymers are formed by condensation in a solution of P_2O_5 in phosphoric acid or a high-boiling-point solvent. Used in high-strength adhesives and laminates and in heat-resistant materials.

poly blends Mixtures of structurally different polymers or copolymers.

polycarbonates A group of thermoplastics

$$H \left(O(RO)\overset{\overset{O}{\|}}{C} \right)_n OROH$$

characterized by their toughness, high softening point and clarity. Polyesters of carbonic acid with aliphatic or aromatic dihydroxy compounds. A typical polycarbonate is formed from bisphenol A and phosgene. Used in cars and in spectacle lenses.

polycation A macromolecule with cationic character.

polychloral See ALDEHYDE POLYMERS.

polychlorinated biphenyls Formerly widely used in electrical goods and as hydraulic fluids. Carcinogenic.

polychloroprene (neoprene) Synthetic rubbers obtained by emulsion polymerization of chloroprene, $CH_2=CCl \cdot CH=CH_2$. The mol.wt. is controlled by addition of sulphur or sulphur-containing additives during the polymerization. Neoprene has high tensile strength, resilience and resistance to abrasion. It is resistant to solvents and is used in automotive goods, in wire and cable sheathing, road and building construction, and as a fibre binder for e.g. paper.

polychlorotrifluoroethene See FLUORINE-CONTAINING POLYMERS.

polydentate ligand See MULTIDENTATE LIGAND.

polydextrose Manufactured from dextrose with some sorbitol and oleic acid. Used in reduced-calorie foodstuffs.

polydispersion In most colloidal sols the dispersed particle sizes vary over wide ranges unless special means are adopted to exclude particles outside certain size limits. Sols with wide ranges of particle size are said to be polydispersed.

polyelectrolyte A macromolecular compound containing many ionizable groups within the same molecule. A polyelectrolyte may be purely anionic or cationic, or may be amphoteric; the individual groups may be weak or strong acids. E.g. strongly acidic – polystyrene sulphonic acid; weakly acidic – polymethacrylic acid; strongly basic – polyvinylpyridinium bromide; weakly basic – polyvinylamine; amphoteric – polyglycine, proteins (polyampholytes). Although the term is usually used for water-soluble substances, ion-exchange resins and fibrous proteins could be considered insoluble polyelectrolytes.

polyene antibiotics See MACROLIDES.

polyenes Compounds with many carbon-to-carbon double bonds, e.g. carotenoids.

polyesters Polymers formed by condensation of polyhydric alcohols (generally glycol or propylene glycol) and polybasic acids (generally maleic or terephthalic acid), i.e.

$$-\left(CH_2CH_2OC(O)\underset{}{\bigcirc}C(O)-O\right)_n$$

Used as fibres, particularly in textiles and film. Many other polyester polymers are of importance, e.g. unsaturated polyester resins from phthalic anhydride, propylene glycol and maleic anhydride, used with reinforcement in boats, cars, etc. (alkyd resins).

polyethene (polyethylene) See ETHENE POLYMERS.

polyethers Polymers with $\{C-O-C\}_n$ units in the backbone. See e.g. ALDEHYDE POLYMERS, 1,2-EPOXIDE POLYMERS, POLY(OXYPHENYLENES), PHENOXY RESINS, POLYACETALS.

polyethylene glycols (PEG), $HO(CH_2CH_2O)_nH$. Polyether glycols having mol.wt. above 200. Prepared from ethylene oxide and water, dihydroxyethene or diethylene glycol plus base. Water soluble. Short-chain members ($n = 1, 2$, etc.) are used as solvents. Polyethylene glycols are used in pharmaceuticals, textiles, paper coatings and to form protein-resistant surfaces.

polyethyleneimine See DIHYDROAZIRINE.

polyethylene terephthalate (PET) Polymer used for bottles and containers. World use about 1 million tonnes p.a.

polyformaldehyde See ALDEHYDE POLYMERS.

polyglycols See POLYETHYLENE GLYCOLS.

polyhalides Compounds containing complex anions formally formed by addition of halogens and interhalogens to halides. Anions of types $[AB_2]^-$, $[AB_4]^-$, $[AB_6]^-$ (B may be all the same or different (e.g. $[IClBr]^-$)). Polyiodides are specific polyhalides containing species such as I_3^-, I_7^-, I_9^-. Polyhalides often dissociate, i.e. lose halogen, on heating the solid complexes.

polyhedral skeletal electron pair theory (PSEPT) Approach to the electronic structure of boranes, carbolations, metal clusters. See WADE'S RULES.

polyhexafluoropropene See FLUORINE-CONTAINING POLYMERS.

polyimides Film, plastics, wire enamels containing the imide group

$$— C(O)NC(O) —$$

derived from aromatic acids. Have good heat-resistant properties. Prepared from diamines and polycarboxylic acids, half esters, anhydrides (alkylene pyromellitimide from mellitic acid derivatives).

polyiodides See IODIDES and POLYHALIDES.

polyketides Natural organic compounds regarded as arising from poly-β-ketonic intermediates derived by condensation of ethanoate units. Biosynthesis of fatty acids, many aromatic compounds, natural oxygen heterocyclics, tetracycline antibiotics and other fungal metabolites occur via polyketide routes. Mevalonic acid is a special case, arising from only three ethanoate units.

polylactic acids (polylactide) Polymer formed from natural products via dextrose and lactic acid. Used in house furnishings. Degradable.

polymannuronic acid See ALGINIC ACID.

polymeric reagents Solid, insoluble reagents for organic synthesis, produced by introducing typical organic functional groups [e.g. Br, Li, SO_3H, CH_2Cl] onto a small proportion of benzene rings of cross-linked polystyrene. They allow selective reactivity, simplified reaction work-up (filtration) and high-purity product isolation, low volatility of noxious reagents and polymer recycling. Catalyst systems in which active groups, e.g. PR_2, co-ordinated to metals and substituted onto a polymer are also available.

polymerization Any process that results in the formation of large molecules (macromolecules), consisting of repeated structural units. These structural units (described as 'mers') are contributed by the reacting monomers, but, constitutionally, need not be identical with them.

Polymerization processes yielding polymers, whose mers are constitutionally identical to the reacting monomers, are now classified as *addition polymerizations*. Thus styrene can be converted, by addition polymerization, to polystyrene:

$$PhCH = CH_2 \longrightarrow [PhCHCH_2 -]_n$$

Such reactions can be initiated by free radicals, derived from initiators such as benzoyl peroxide, ammonium persulphate or azo*bis*isobutyronitrile by radiation, u.v. or γ-rays; or by ionic mechanisms by reactive ions, derived, for example, from BF_3 or $TiCl_4$. The process is also known in the plastics industry as 'vinyl polymerization' since the best-known examples are those involving vinyl or related monomers. Addition polymerizations are characterized by their extremely rapid rates, achieved at relatively low temperatures, especially in the case of ionic initiated polymerizations. The kinetics are typical of those expected of chain reactions. Industrially the process is important for the production of many polymers and copolymers used for the manufacture of plastics and synthetic rubbers.

In mass polymerization bulk monomer is converted to polymers. In solution polymerization the reaction is completed in the presence of a solvent. In suspension, dispersed mass, pearl or granular polymerization the monomer, containing dissolved initiator, is polymerized while dispersed in the form of fine droplets in a second non-reactive liquid (usually water). In emulsion polymerization an aqueous emulsion of the monomer in the presence of a water-soluble initiator is converted to a polymer latex (colloidal dispersion of polymer in water).

Condensation polymerization: each step in the process results in the elimination of some easily removed molecule (often water). E.g. the polyester polyethylene terephthalate (Terylene) is formed by the condensation polymerization (polycondensation) of ethylene glycol with terephthalic acid:

$$n\text{HO·}[CH_2]_2\text{·OH} + n\text{HO·CO} \bigcirc \text{C(O)OH}$$

$$\downarrow$$

$$\left(\text{O·}[CH_2]_2\text{·O·CO} \bigcirc \text{·CO} \right)_n + n\text{H}_2\text{O}$$

Here the empirical compositions of polymer and reacting monomers are different and the structural unit (mer) has no constitutional identity with the monomer components.

The kinetics of this type of polymerization are the same as for simple condensation; for this reason, the use of the term polycondensation is perhaps more appropriate. Unless kinetic evidence suggests otherwise, polymerizations involving the formation of chain polymers from

cyclic compounds, following ring scission, are classed as condensation polymerizations. Some important condensation polymers are those derived from phenol, urea, or melamine and methanal; the polyesters; the polyamides (Nylons); the polysiloxanes (silicones).

polymer-supported reagents Used for synthesis. Facilitate clean up and reaction. See POLYMERIC REAGENTS.

polymethine dyes See cyanine dyes.

polymethylene See POLYALKYLIDENES.

polymethylene glycols See METHANAL.

polymethylmethacrylate (PMMA) Important polymer. See METHYLMETHACRYLATE. Used in bone cements.

polymorphism The existence of a substance in more than one crystal form, e.g. HgI_2 yellow–orthorhombic, red–tetragonal. The different forms are referred to as polymorphs. Substances which exist in two forms are referred to as dimorphic. Polymorphism of the elements is known as ALLOTROPY. Thermodynamically unstable forms (e.g. the yellow form of HgI_2) are referred to as METASTABLE.

polymyxins A group of naturally occurring antibacterials.

polyoxins Fungicides, prepared by fermentation.

polyoxyethylene polymers From polyethylene glycol (PEG) derivatives. Fatty acids plus ethylene oxide or PEG liquids or soft solids used as emulsifiers, defoamers, wetting agents, detergents, lubricants.

polyoxymethylene See PARAFORMALDE-HYDE.

poly(oxyphenylenes) (poly(phenylene oxides), poly(phenylene ethers)) Polymers,

$$\left(\bigcirc \substack{R \\ R} \text{O} \right)_n$$

(R often Me) formed by oxidative polymerization of phenols using oxygen with copper and an amine (pyridine) as catalysts. The products are thermoplastics used in engineering applications and in electrical equipment. Have good heat-resistant properties.

poly(oxytetramethylene)glycols
See TETRAHYDROFURAN POLYMERS.

polyphenol antioxidants A group of compounds occurring in vegetables, fruit, chocolate. Of importance in inhibiting cardiovascular disease. Inhibit oxidation of cholesterol.

poly(phenylene ethers)
See POLY(OXYPHENYLENES).

poly(phenylene oxides)
See POLY(OXYPHENYLENES).

poly(phenylene sulphides)

Formed commercially by the polymerization of S, Na_2CO_3 and PhX in a sealed container at 275–370°C; substituted derivatives are known but not fully evaluated. Used as high-temperature adhesives, for laminates and in coatings.

polyphosphates See PHOSPHORUS OXYACIDS.

polyprenols Polymeric monohydroxylic alcohols with a carbon skeleton composed of isoprene units. Found in bacteria, yeasts, plants and mammals. Solanesols are found in green leaves, dolichols in bacteria and mammals and betulaprenols in silver birch wood. Polyprenols play an important role as carriers (usually as their phosphates) in the transfer of sugars from nucleotide diphosphate sugars to a wide range of acceptors including the biosynthesis of carbohydrate chains. The acceptors are usually membrane or wall-bound and include polysaccharides, glycopeptides, glycolipids, etc. Polyprenols are biosynthesized from mevalonic acid.

polypropene See PROPENE POLYMERS.

polypropylene See PROPENE POLYMERS.

polypyridine Light emitter.

polysaccharides Carbohydrates derived from monosaccharides by the removal of $n–1$ molecules of water from n molecules of monosaccharides. Higher carbohydrates are polysaccharides. Can be sequenced. Hydrolysed to monosaccharides.

polysorbates See SORBIC ACID.

polystyrene (polyvinylbenzene),
$(PhCH·CH_2)_n$. Important polymer used in plastics, synthetic rubber, resins.

polysulphide rubber (thiokol),
$CH_2·S(S)–S(S)_x$. Prepared dihaloalkanes and sodium polysulphide. Chemically resistant. Used in resins and rubbers.

polysulphides Materials containing anions such as $[S_2]^{2-}$ (e.g. marcasite FeS_2), $[S_3]^{2-}$, $[S_4]^{2-}$, $[S_5]^{2-}$, $[S_6]^{2-}$ formed as aqueous solutions from sulphides and sulphur. Also polysulphide polymers.

polyterephthalic acid See POLYESTERS.

polytetrafluoroethene (polytetrafluoroethylene, Teflon, Fluon, PTFE)
See FLUORINE-CONTAINING POLYMERS. A very tough, translucent material of high softening point (320°C), excellent chemical resistance, low coefficient of friction and good electrical insulating properties.

poly(tetramethylene oxide) See TETRAHYDROFURAN POLYMERS.

polythionic acids (polythionates), $H_2S_nO_6$.
A series of more or less unstable dibasic acids, obtained by the action of iodine on sodium sulphite, thiosulphate and mixtures of these, by the action of hydrogen sulphide on other members of the same series and of H_2S on H_2SO_3 at 0°C. The best known as alkali salts are: dithionic acid $H_2S_2O_6$, trithionic acid $H_2S_3O_6$, tetrathionic acid $H_2S_4O_6$, pentathionic acid $H_2S_5O_6$ and hexathionic acid $H_2S_6O_6$. Known for $H_2S_nO_6$ up to $n > 22$.

polyurethanes Condensation polymers formed from polyhydroxy compounds and polyisocyanates containing $[ROC(O)NHR^1NHC(O)O]_n$ repeating units. The main use is in construction and foams. See ISOCYANATES.

poly(vinyl acetals),

Resins formed from the reaction of poly(vinyl alcohol) with aldehydes. The formal derivative (from methanal) is used in wire coatings and adhesives and the butyral (from butanal) is used in metal paints, wood-sealers, adhesives and in safety glass interlayers.

polyvinyl acetate See VINYL ACETATE.

poly(vinyl alcohol), $\{CHOH·CH_2\}_n$.
Prepared generally from polyvinyl acetate (ethanoate) using methanol and base and water. Forms copolymers. Used as a size in the textile industry, in aqueous adhesives, in the

production of polyvinyl acetates (e.g. butynal) for safety glasses.

polyvinyl chloride (PVC) See VINYL CHLORIDE POLYMERS.

POPOP (1,4-bis(5-phenyloxazol-2-yl)-benzene) Used in dye laser systems.

POPS (persistent organic pollutants) Materials which persist in the environment and accumulate in animal fats.

porphin See PORPHYRINS.

porphyrinogens Biosynthetic intermediates of PORPHYRINS containing no methylene bridges between the pyrroles.

porphyrins A group of naturally occurring pigments. Haemoglobin and other animal respiratory pigments and chlorophyll, the respiratory catalyst of plants, are compounds of porphyrins with metals. They are derivatives of porphin which has the structure shown.

Important porphyrins include: *protoporphyrin*, $C_{34}H_{34}N_4O_4$, the porphyrin present in haem; *mesoporphyrin*, $C_{34}H_{38}N_4O_4$, obtained by treating hemin with hydriodic acid; *hematoporphyrin*, $C_{34}H_{38}N_4O_6$, obtained from haemin or haematin by the action of strong acids; *coproporphyrin*, $C_{36}H_{38}N_4O_8$, found in the faeces and in normal urine, in the serum of various animals, and in certain yeasts.

Portland cement A HYDRAULIC CEMENT made from $CaCO_3$ and aluminium silicates.

positive ray analysis When a beam of positive particles (rays), formed by a high potential or electric discharge on molecules, is passed through a magnetic and an electric field, the various ions are deflected to varying extents depending upon their velocities, masses and charges. Such analysis is the basis of a mass spectrometer, although negative ions can also be used.

positron (β^+) The positive counterpart of the electron. The ultimate unit of positive electricity.

positron decay See ELECTRON DECAY.

positron emission tomography (PET) Physiological study of positron emission from a species (e.g. ^{11}C) ingested into the animal under study.

post-actinide elements See PERIODIC TABLE. Elements following Lr, part of a 6d transition series. Elements have been prepared up to atomic number 109 (Rf, Db, Sg, Bh, Hs, Mt) and atoms of 110, 111, 112 described but names have not been agreed. See PERIODIC TABLE.

potash Potassium hydroxide.

potash alum See ALUM.

potassamide (potassium amide), KNH_2. White solid, base on ammonia system (K in liquid NH_3 gives a blue solution which is a good reducing agent; this liberates H_2 and KNH_2 is formed on removal of excess NH_3).

potassium, K. Alkali metal, at.no. 19, at.mass 39.098, m.p. 63.7°C, b.p. 756°C, density ρ 862, crustal abundance 21 000 p.p.m., electronic configuration [Ar] $4s^1$. Widely distributed in silicate rocks, e.g. orthoclase, $KAlSi_3O_8$, in plants as the oxalate, tartrate and in blood, milk, etc., also occurs widely in salt beds, e.g. as carnallite, $KMgCl_3.6H_2O$ (see POTASSIUM CHLORIDE) and in sea water. Is an essential element particularly in its involvement in ion-transport across all membranes (nerve impulses). The metal is prepared by reduction of KCl with Na at high temperature and low pressure or can be obtained by electrolysis. It is a soft, silver-white metal, bcc, reacts violently with water and is readily oxidized (O_2 gives KO_2). Used as a reducing agent, an Na/K alloy may be used for heat transfer. Dissolves in mercury and forms an amalgam. Potassium compounds are used extensively as fertilizers (95%); KOH is used as an electrolyte in batteries, glass, ceramics and in synthesis, ^{40}K is radioactive. Total U.S. use of potassium salts (K_2O equivalent) 1999 6.0 million tonnes.

potassium, organic derivatives Potassium forms a limited range of almost ionic derivatives, e.g. $K^+[CPh_3]^-$. The alkyls are very reactive, and can be used in metallation reactions.

potassium acetate, KO_2CCH_3. See POTASSIUM ETHANOATE.

potassium amide See POTASSAMIDE.

potassium antimonyl tartrate See ANTI-MONY POTASSIUM TARTRATE.

potassium arsenate, KH_2AsO_4. Used in textiles, tanning, paper treatment.

potassium bromate (potassium bromate(V)), $KBrO_3$. Formed from Br_2 and hot aqueous KOH. Decomposes to KBr and O_2 on heating. Used as a volumetric standard and in bread and flour. See BROMATES(V).

potassium bromide, KBr. Obtained KOH or K_2CO_3 plus HBr or Br_2 plus KOH. M.p. 728°C, b.p. 1376°C. Used in photography and as a sedative.

potassium *t*-butoxide, $KOCMe_3$. M.p. 220°C. White solid (K in Me_3COH); very strong base. Hygroscopic.

potassium carbonate (pearl ash), K_2CO_3. Deliquescent powder made by modification of Solvay process using alcoholic solutions. Used in standard alkali solutions, glass and tanning.

potassium chemistry Potassium is a typical alkali metal of Group 1. In its simple compounds it shows only the +1 oxidation state, generally 6-co-ordinate. Complexes are formed with water, NH_3 (relatively unstable) and macrocyclic ligands such as crown ethers. K^- complexes are known.

potassium chlorate (potassium chlorate(V)), $KClO_3$. Not very soluble in water, deposited from chlorate(V) solutions (Cl_2 plus $Ca(OH)_2$ or electrolysis of aqueous NaCl). On heating gives KCl and $KClO_4$ but decomposes to KCl at high temperatures. Used in explosives, fireworks, dyeing, matches, printing.

potassium chloride, KCl. M.p. 776°C, sublimes 1500°C, ρ 1 984. Occurs naturally as sylvine and in sylvinite (KCl–NaCl), carnallite ($KMgCl_3.6H_2O$), kainite ($KCl.MgSO_4.3H_2O$), 'hard salt' ($KCl–NaCl–MgSO_4$) and in many brines. Separated by fractional crystallization, soluble water and lower alcohols. Used in fertilizer production, to produce other potassium salts, and in electrolysis.

potassium chromate, K_2CrO_4. See CHROMATES(VI). Used in enamels and leather production.

potassium citrate, $K_3C_6H_5O_7 \cdot H_2O$. Colourless crystals; very soluble in water, slightly soluble in alcohol. Used medicinally as a diuretic.

potassium cyanate, KOCN. Obtained by PbO oxidation of aqueous KCN. Used to prepare cyanates.

potassium cyanide, KCN. M.p. 635°C. Prepared fused K_2CO_3 plus C plus NH_3 gas (Beilby's process). Hydrolysed in aqueous solution. Used to prepare cyanides. Very poisonous.

potassium dichromate, $K_2Cr_2O_7$. M.p. 396°C. Orange-red crystals prepared from chromite, K_2CO_3 and CaO followed by acidification. Used in the preparation of chrome pigments, bleaching, as an oxidizing agent (matches), in organic chemistry and in volumetric analysis.

potassium ethanoate (potassium acetate), KO_2CCH_3. Occurs in the sap of plants. Used as alkali and for deicing runways.

potassium ethoxide, KOC_2H_5. Base similar to potassium *t*-butoxide.

potassium ferricyanide, $K_3Fe(CN)_6$. See CYANOFERRATES(III).

potassium ferrocyanide, $K_4Fe(CN)_6$. See CYANOFERRATES(II). Used in volumetric analysis.

potassium fluoride, KF. M.p. 858°C, b.p. 1505°C. Obtained KOH or K_2CO_3 plus HF, has NaCl structure. Forms hydrates and adducts including KHF_2, the hydrogen fluoride, with HF. The latter acid salts are used as electrolytes in electrolysis to form fluorine. KF is poisonous in large quantities but is used in water fluoridation, glass frosting, fluxes, forming CF compounds.

potassium hydrogen carbonate, $KHCO_3$. Less soluble in water than K_2CO_3, precipitated from an aqueous solution of K_2CO_3 and CO_2. Used in foods (baking powder), medicine, fire extinguishers.

potassium hydrogen fluoride, KHF_2. See POTASSIUM FLUORIDE.

potassium hydrogen sulphate, $KHSO_4$. Formed K_2SO_4 plus H_2SO_4. Other acid salts, e.g. $K_2SO_4.3H_2SO_4$, are known.

potassium hydrogen tartrate (potassium bitartrate, argol, cream of tartar), $C_4H_5O_6K$. Colourless salt, soluble boiling water; occurs in grape juice, deposited as argol during fermentation. Used in baking powders (liberates CO_2 with $NaHCO_3$).

potassium hydroxide, KOH. M.p. 306°C, b.p. 1320°C. Prepared by electrolysis of KCl solution (also Ba(OH)$_2$ plus K$_2$SO$_4$ or H$_2$O/K amalgam). Forms hydrates. Strong base, used as electrolyte in batteries and as base particularly for fertilizers, mordant and soap manufacture.

potassium iodate (potassium iodate (V)), KIO$_3$. Formed I$_2$ plus hot KOH solution. Forms acid salts, e.g. KIO$_3$·HIO$_3$. Used in volumetric analysis.

potassium iodide, KI. Formed I$_2$ plus hot KOH solution. Soluble water and many organic solvents; dissolves iodine to solutions of polyiodides containing [I$_3$]$^-$, [I$_5$]$^-$, etc. Used medicinally and in table salt.

potassium metabisulphite, K$_2$S$_2$O$_5$. Used as anti-fermentation agent and bleach.

potassium methoxide, KOCH$_3$. Similar to POTASSIUM *t*-BUTOXIDE.

potassium nitrate, KNO$_3$. Prepared by fractional crystallization from a solution of NaNO$_3$ and KCl. Soluble in water, NH$_3$ and methanol. Used in gunpowder, tobacco treatment.

potassium nitrite, KNO$_2$. M.p. 440°C. Prepared KNO$_3$ plus Pb. Used in diazotization.

potassium oxalate, C$_2$O$_4$K$_2$·H$_2$O. The acid salt C$_2$O$_4$HK·H$_2$O is also known and decomposes to the quadroxalate or tetroxalate, C$_2$O$_4$HK·C$_2$O$_4$K$_2$,2H$_2$O. Potassium oxalate (salts of sorrel, salts of lemon) is used for removing ink stains and iron mould, in photography and as a mordant. All formed from KOH or K$_2$CO$_3$ and oxalic acid.

potassium oxides Potassium burns in air to give mainly the orange superoxide KO$_2$ which decomposes to K$_2$O$_2$ and K$_2$O on heating. It gives H$_2$O$_2$, KOH and O$_2$ with water. Potassium peroxide, K$_2$O$_2$, formed from O$_2$ into a solution of K in liquid NH$_3$ at –60°C or by heating KO$_2$ gives H$_2$O$_2$ with acids. Potassium monoxide, K$_2$O, formed by heating KNO$_3$ with K or KN$_3$. Reacts violently with water to give KOH.

potassium perchlorate (potassium chlorate(VII)), KClO$_4$. Formed by heating KClO$_3$ to 500°C or HClO$_4$ plus KOH. Relatively insoluble water (used for determination of K$^+$).

potassium periodate (potassium iodate(VII)), KIO$_4$. Prepared KIO$_3$ solution plus Cl$_2$ and used as an oxidizing agent. Other iodates(VII), e.g. K$_4$I$_2$O$_9$.9H$_2$O, are known. See IODATES.

potassium permanganate (potassium manganate(VII)), KMnO$_4$. See PERMAN-GANATES. Used extensively as an oxidizing agent (bleach, tanning, photography) and as a volumetric reagent.

potassium peroxydisulphate, K$_2$S$_2$O$_8$. Obtained by double decomposition between (NH$_4$)$_2$S$_2$O$_8$ and K$_2$CO$_3$ or by electrolytic oxidation of K$_2$SO$_4$. Used as bleach. The peroxymonosulphate, K$_2$SO$_5$, is also known.

potassium phosphates Many phosphates including K$_3$PO$_4$, K$_4$P$_2$O$_7$, and hydrogen phosphates are known. KH$_2$PO$_4$ forms very large crystals from which slices are used in wavelength converters for lasers. Used in detergents, surfactants, water treatment.

potassium sodium tartrate tetrahydrate (Rochelle salt, Seignette salt), C$_4$H$_4$O$_6$KNa.4H$_2$O. Used in preparing mirrors, forms piezoelectric crystals.

potassium sulphate, K$_2$SO$_4$. Occurs naturally with other sulphates as glaserite, SCHÖNITE and syngenite, and purified by crystallization. Can be prepared from KOH and H$_2$SO$_4$. Used as fertilizer especially where Cl$^-$ content must be kept down in irrigated soils. Forms acid salts, e.g. K$_2$SO$_4$.1, 3, ¾, H$_2$SO$_4$, from solutions in aqueous H$_2$SO$_4$.

potassium sulphite, K$_2$SO$_3$. Formed KOH solution plus SO$_2$. Potassium pyrosulphite, K$_2$S$_2$O$_5$, is also formed from this solution.

potassium tetrafluoroborate, KBF$_4$. Precipitated from solution.

potassium thiocyanate, KSCN. Prepared KCN plus S (the reaction with a solution of S in, e.g. C$_6$H$_6$ or propanone, is fast and quantitative). Used to prepare thiocyanates and in volumetric analysis.

potential energy (V) Energy as a result of position.

potentiometric titration At the end point in a titration there is a rapid change in the concentration of all of the reacting species. An inert electrode immersed in the solution thus shows a rapid change in e.m.f. as the end point is approached and the end point can be determined by observing the change in e.m.f. of the electrode during the titration. This method is especially useful when coloured solutions are used and the use of indicators is impractical.

potter's clay A term used for any clay or earth which can be used for making pottery, either alone or after admixture with non-plastic materials. The red colour of some pottery is due to iron compounds in the clay; these decompose on heating and liberate Fe_2O_3 or Fe(III) complexes which are red in colour. Other colours are produced by firing the kiln or oven under reducing conditions, or by adding coloured oxides to the clay.

pounds per square inch (p.s.i.) Non-S.I. unit of pressure 1 p.s.i. = 6.894 kPa.

powder diffraction See X-RAY DIFFRACTION.

powder metallurgy The techniques of sintering powders under pressure but below the melting point to produce solid objects. Used for metals, alloys and some plastics.

Pr Praseodymium. Also propyl.

prallethrin Insecticide, pyrethroid.

praseodymium, Pr. Lanthanide metal, at. no. 59, at. mass 140.91, m.p. 931 °C, b.p. 3520 °C, density ρ 6773, crustal abundance 9.5 p.p.m., electronic configuration [Xe] $4f^36s^2$. A typical LANTHANIDE. The metal is hcp (to 798°C) and bcc (to m.p.). Used in glass and in thermoelectric materials, in n.m.r. shift reagents.

praseodymium compounds Praseodymium(III) compounds are typical LANTHANIDE COMPOUNDS Pr^{3+} (f^2 pale green) $\to$ Pr (−2.47 volts in acid). PrF_4 is formed by the action of HF on Na_2PrF_6 (F_2 on NaF/PrF_3). Pr(III) oxysalts are decomposed in air to Pr_6O_{11} and O_2 under pressure and give PrO_2 at 500°C – intermediate phases are known. Lower halides contain M–M bonds except PrI_2 which is $Pr^{3+}(I)_2^-e$.

precipitation The formation of an insoluble compound from solution either by interaction of two salts, e.g.

$$NaCl + AgNO_3 \xrightarrow{H_2O} AgCl\downarrow + NaNO_3$$

or by temperature change affecting solubility. The formation of a precipitate is governed by the solubility product. In analysis, precipitation is important in gravimetric and some forms of qualitative analysis. The form and purity of the precipitate depend upon the conditions used for the precipitation. Easily filtered precipitates are frequently formed by use of compounds which slowly release the precipitating reagent.

precipitation hardening (age hardening) The hardening of certain alloys which takes place with time, when a supersaturated solution tends to decompose with partial precipitation of the solute metal as an intermetallic compound.

precipitation indicator An INDICATOR which functions by giving a coloured precipitate at the END POINT. Thus K_2CrO_4 is used as indicator in the estimation of Cl^- by Ag^+; at the end point there is a red precipitate of Ag_2CrO_4.

predissociation See DISSOCIATION.

prednisolone (11β,17α,21-trihydroxy-pregna-1,4-diene-3,20-dione), $C_{21}H_{28}O_5$. Glucocorticoid. More potent than cortisone, it is used for the same purposes. It is especially useful in the treatment of asthma.

pregnane (17β-ethylaetiocholane), $C_{21}H_{36}$.

M.p. 83.5°C. Does not occur naturally, but has been obtained synthetically. It is the basic hydrocarbon skeleton of biologically and clinically important steroids and part of the signalling pathway.

pregnanediol (5β-pregnane-3α,20α-diol), $C_{21}H_{36}O_2$. M.p. 238°C. Formed by reduction of PROGESTERONE in the liver.

prehnitene (1,2,3,4-tetramethylbenzene), $C_6H_2(CH_3)_4$.

premetallized dyes Dyestuffs similar to METALLIZABLE DYES except that the soluble metal complexes are formed prior to application to the fibre.

prephenic acid, $C_{10}H_{10}O_6$. An intermediate in the 'shikimic acid pathway' of aromatic biosynthesis, arising by rearrangement of CHORISMIC ACID.

pressure, gas A measure of the force exerted by a gas on the walls of its container. 1 pascal Pa (force m^{-2}); one Newton $metre^{-2}$. The atmosphere (atm) is 101 325 Pa. A 760 mm column of mercury exerts a pressure of 1 atm, 760 torr. 1 torr is 133.32 Pa. 1 BAR = 10^5 Pa.

preticlaclor Herbicide, actinilide derivative.

prilling Conversion of materials into forms which have desirable properties, e.g. granular to pellets.

primary carbon Only one other carbon atom bonded to it.

primeverose, $C_{11}H_{20}O_{10}$. Colourless crystals; m.p. 208°C. A disaccharide, glucose-6-β-D-xyloside, found as a constituent of the glycosides gaultherin, rhamnicosin, primeverin, genticaulin, etc.

primisulfuron-methyl Herbicide, sulphonylurea derivative.

primitive lattice A lattice having only one equivalent point in the unit cell. Often a primitive lattice is expressed with the atom or molecule corresponding to the point at the corners of the unit cell.

principal quantum number See ELECTRONIC CONFIGURATION.

printing inks Normally mixtures of pigment (e.g. carbon black) dispersed in an oil of a suitable viscosity.

prion Naturally occurring protein involved in transmission of spongiform encephalopothies. Infective material has unusual folding pattern.

prismanes See 'DEWAR' BENZENE.

probenazole Fungicide, bactericide, thiazole derivative.

procaine (diethylaminoethyl-4-aminobenzoate), $NH_2 \cdot C_6H_4 \cdot C(O)OC_2H_4 \cdot N(C_2H_5)_2$, $C_{13}H_{20}N_2O_2$. M.p. ($2H_2O$) 51°C (61°C anhydrous). Its hydrochloride has m.p. 153–156°C and is a powerful local anaesthetic.

prochiral An atom C-linked to two identical ligands and to two other different ligands (C *aa-bc*) is said to be prochiral because if one of the ligands *a* is replaced by a ligand *d* different from *a*, *b* and *c*, a chiral centre is produced in (C *abcd*). Similarly for trigonal centres.

procymidone Fungicide, dicarboximide derivative.

prodamine Herbicide, dinitroaniline derivative.

producer gas (blow gas) A mixture of CO (20%) and N_2 (75%) produced by passing air and steam through a bed of incandescent coke or coal in a *producer*. Used in ammonia synthesis.

profenofos Insecticide, acaricide, organophosphate derivative.

progesterone, $C_{21}H_{30}O_2$. Steroid derivative. Progesterone is the hormone that governs the growth and development of the uterus during pregnancy. It also exerts an anti-ovulatory effect when given during days 5 to 25 of the normal menstrual cycle: this is the basis for the use of some synthetic progestins as oral contraceptives.

Progesterone can be manufactured from a species of yam. Used in the production of cortisones.

proguanil, $C_{11}H_{16}ClN_5$. Obtained as the hydrochloride, m.p. 248°C, prepared by treating 4-chlorophenyldicyandiamide with isopropylamine hydrochloride. It is used for the prevention and treatment of malaria.

prohexadrone-calcium Plant growth regulator, cyclohexane carboxylate derivative.

prolactin (lactation-stimulating hormone) Hormones, the presence of which is necessary for the secretion of milk, a protein.

prolamines See PROTEINS.

proline (2-pyrrolidinecarboxylic acid), $C_5H_9NO_2$. Colourless crystals; m.p. 220–222°C. Not strictly an amino-acid.

promazine See CHLORPROMAZINE.

promethazine (10-(2-dimethylaminopropyl)-phenothiazine), $C_{17}H_{20}N_2S$. Used as the hydrochloride in the treatment of allergies. See PHENOTHIAZINE.

promethium, Pm. Artificial lanthanide metal, at.no. 61, at.mass ^{145}Pm 146.92, m.p. 1042°C, b.p. 3000°C, density ρ 7264, electronic configuration [Xe] $4f^56s^2$. Traces of Pm occur naturally as a fission product of uranium, ^{147}Pm (2.64 years) is obtained from nuclear reactors and used as an energy source. It is a typical LANTHANIDE. Previously called illinium. Used in luminescent paint for watch dials.

promethium compounds Promethium forms a single series of typical LANTHANIDE COMPOUNDS in the +3 state

Pm^{3+} (f^4 rose pink) → Pm (–2.42 volts in acid)

prometon Herbicide, triazine derivative.

prometryn Herbicide, triazine derivative.

prontosil An orange-red azo dye. An early anti-bacterial. Sulphanilamide derivative.

proof spirit The legal standard used in assessing the strength of alcoholic liquors for excise purposes. U.S. proof is 2 times the alcohol contents by volume. British 100° proof contains 11 parts alcohol to 10 parts water by volume. Measured from specific gravity.

propachlor Herbicide, acetanilide derivative.

1,2-propadiene (allene), $CH_2=C=CH_2$, C_3H_4. Colourless gas prepared by the electrolysis of potassium itaconate, or by the action of zinc and alcohol on 1,3-dibromopropane. It is easily isomerized to propyne (methylacetylene).

propamocarb hydrochloride Fungicide, carbamate derivative.

propanal (propionaldehyde), CH_3CH_2CHO. Colourless liquid; b.p. 48°C. Prepared by dehydrogenation of n-propanol over a catalyst, Fischer–Tropsch or oxo combination of C_2H_4, CO and H_2. Used for production of trimethylolethane by reaction with methanol (used in alkyd resins) and for hydrogenation to propanol or oxidation to propanoic acid.

propane, C_3H_8, $CH_3CH_2CH_3$. A colourless inflammable gas with a peculiar odour. M.P. –190°C, b.p. –44.5°C. It occurs in the natural gas from petroleum. Also obtained by reduction of propene. Has general properties of the paraffins. Used as a refrigerant and as a fuel.

propanedioic acid See MALONIC ACID.

1,3-propanediol, $(HO)[CH_2]_3OH$. Used in production of polytrimethylene terephthalate. Prepared from ethylene oxide and methanol followed by hydrogenation.

1,2-propanediol, $CH_3CH(OH)CH_2OH$. Prepared from glycerol or propene oxide by oxidation. B.p. 187–189°C. Used in anti-freezes, resins.

1,3-propane dithiol, $HS[CH_2]_3SH$. Liquid, b.p. 170–171°C, with an extremely disagreeable odour. It is principally used in organic chemistry for protecting a carbonyl group as a trimethylene dithioketal whilst transformations are effected at other centres of the molecule.

The protecting group can be removed by treatment with dilute acid and $HgCl_2$.

propanil Herbicide, anilide derivative.

propanoic acid (propiolic acid, propionic acid), $C_3H_6O_2$, $CH_3CH_2C(O)OH$. Colourless liquid with an odour resembling that of ethanoic acid; m.p. 24°C, b.p. 141°C. Occurs in the products of the distillation of wood. Prepared by the oxidation of propanol, propanal or by the reduction of propenoic acid. Used in production of esters and polymers, flavours. Propanoic anhydride is used similarly.

1-propanol (n-propyl alcohol), C_3H_8O, $CH_3CH_2CH_2OH$. Colourless liquid with a pleasant odour; b.p. 97°C. It occurs in fusel oil. Also obtained by the hydrogenation of propene oxide. Oxidized to propanal and propanoic acid. Forms propene when heated over alumina. Used as a solvent and for the preparation of esters (propyl acetate).

2-propanol (isopropyl alcohol), $CH_3CH_2OHCH_3$. B.p. 82°C. Manufactured by hydrolysis of propene. Used in the production of acetone (propanone) by oxidation, for the preparation of esters (e.g. the ethanoate used as a solvent), amines (diisopropylamines, etc.), glycerol, hydrogen peroxide. The alcohol is used as an important solvent for many paints, coatings, resins, aerosols, anti-freezes. U.S. production 1999 0.66 million tonnes.

2-propanolamines (isopropanolamines) Alkanolamines with at least one $NCH_2CHOHCH_3$ grouping. Important materials include monoisopropanolamine NH_2CH_2-$CHOHCH_3$, b.p. 159°C; di-isopropanolamine $NH(CH_2CHOHCH_3)_2$, b.p. 248°C; tri-isopropanolamine $N(CH_2CHOHCH_3)_3$, b.p. 300°C. Manufactured from ammonia and propylene oxide. Used as weedkillers, as stabilizers for plastics, in detergents, alkanolamine soaps, for sweetening natural gas and in synthesis.

propanone (acetone, dimethyl ketone), C_3H_6O, CH_3COCH_3. Colourless, volatile liquid with a pleasant, ethereal odour; highly inflammable, b.p. 56°C. It is largely manufactured by the dehydrogenation of 2-propanol over a copper catalyst at 500°C and 4 atm and by oxidation of cumene. Considerable quantities are also produced in the manufacture of other chemicals, notably glycerin, hydrogen peroxide and phenol. It occurs in significant amounts in blood and urine in certain pathological conditions. Ammonia reacts with boiling propanone to give diacetonamine. Sodium hydroxide causes condensation to diacetone alcohol; under more vigorous conditions mesityl oxide and phorone are produced. These are also formed, together with mesitylene, by the action of small amounts of mineral acids. Reduction of propanone gives isopropanol and pinacol. Crystalline derivatives are formed with hydroxylamine, phenylhydrazine and semicarbazide. It is detected by the formation of iodoform when treated with iodine and sodium hydroxide. Propanone is used as a solvent and for the manufacture of methyl methacrylate (30%), methyl isobutyl ketone, methyl isobutyl carbinol and various other chemicals.

propaphos Insecticide, organophosphate derivative.

propaqilafop Herbicide, arylpropionic acid derivative.

propargite Acaricide, cyclohexylpropynyl sulphite derivative.

propargyl alcohol See 2-PROPYN-1-OL.

propazine Herbicide, triazine derivative.

propellants A general description for an explosive used to propel a rocket, bullet, shot or shell.

The word 'propellant' is also used for the liquefied gas in a pressurized aerosol container.

propenal (acrolein, acraldehyde, vinyl aldehyde), $C_3H_4O_2$, $CH_2{:}CH{\cdot}CHO$. A colourless, volatile liquid, with characteristic odour. The vapour is poisonous, and intensely irritating to eyes and nose; b.p. 53°C. Prepared by the distillation of a mixture of glycerin, potassium sulphate and potassium hydrogen sulphate. It is manufactured by direct oxidation of propene or cross-condensation of ethanal with methanal.

When exposed to sunlight, it is converted to a white insoluble resin, disacryl. Oxidized by air to propenoic acid; small amounts of hydroquinone will inhibit this. Bromine forms a dibromide which is converted by barium hydroxide into DL-fructose. The acrid odour of burning fats is due to traces of propenal. It is used in the production of methionine and in controlled polymerization reactions to give ACROLEIN POLYMERS and in perfume production.

propene (propylene), C_3H_6, $CH_3CH{=}CH_2$. Colourless gas; m.p. −185°C, b.p. −48°C. Formed by cracking petroleum. Prepared by passing the vapour of 1-propanol over heated alumina or methanol plus steam over a catalyst. Reacts with sulphuric acid to give 2-propyl ether and 2-propanol, with halogens to give propene dihalides, with dilute chlorine water to give propene chlorohydrin, with hydrogen in the presence of catalysts to give propane, and with benzene to give cumene (route to phenol

and propanone). Used in the manufacture of these compounds and others derived from them but principally to form polymers (70%) (polypropene (25%), polyacrylonitrile (15%)), cumene (10%) and solvents. Me_2CHOH (10%), propylene oxide (10%). See PROPENE POLYMERS.

propenenitrile (acrylonitrile, vinyl cyanide), C_3H_3N, $CH_2{:}CH{\cdot}CN$. Volatile liquid; b.p. 78°C. Manufactured by the catalytic dehydration of ethylene cyanhydrin, by the addition of hydrogen cyanide to ethyne in the presence of CuCl or the reaction of propene, ammonia and air in the presence of a molybdenum-based catalyst.

Propenenitrile is very active, both in polymerization and in undergoing ready addition to compounds containing active hydrogen (e.g. the process of cyanoethylation). Polymers and copolymers of propenenitrile are industrially important as synthetic fibres, nitrile rubbers and as components in several thermoplastic compositions. It is an effective fumigant against stored grain insects.

propene oxide See PROPYLENE OXIDE.

propene polymers (polypropene, polypropylene) An important group of polymers used as moulding resins and in extruded forms (e.g. film, fibres, filters). Can be electroplated. Useful polymerization is by Ziegler catalysis and gives an isotactic material.

propenoic acid (acrylic acid, vinylformic acid), $C_3H_4O_2$, $CH_2{:}CH{\cdot}C(O)OH$. Colourless liquid having an odour resembling that of ethanoic acid; m.p. 13°C, b.p. 141°C. Prepared by oxidizing propenal with moist AgO or treating 3-hydroxypropionitrile with sulphuric acid. Slowly converted to a resin at ordinary temperatures. Important glass-like resins are now manufactured from methyl acrylate, see ACRYLIC RESINS. Propenoic acid itself can also be polymerized to important polymers – see ACRYLIC ACID POLYMERS.

propenol (allyl alcohol), C_3H_6O, $CH_2{=}CHCH_2OH$. Colourless liquid with a pungent odour; b.p. 97°C. Prepared by heating glycerol with oxalic acid, or by reduction of propenal. Manufacture: **1** direct chlorination of propene at 500°C followed by hydrolysis of allyl chloride; **2** reaction of propenal and isopropanol (both derived from propene) at 400°C in the presence of a catalyst. Used for synthesis and in resins and plasticizers.

propenyl (allyl) The $CH_2{=}CHCH_2-$ group.

propenyl isothiocyanate (allyl isothiocyanate, mustard oil), C_4H_5NS, $CH_2=CHCH_2NCS$. Colourless liquid with an intensely pungent odour; b.p. 151°C. From oil derived from black mustard. Manufactured from propenyl iodide and potassium thiocyanate. Reacts with ammonia to give propenyl thiourea. It is a strong vesicant.

propenyl polymers See ALLYL POLYMERS.

propenylthiourea (allylthiourea, thiosinamine, rhodallin), $C_4H_8N_2S$, $CH_2:CHCH_2NHCSNH_2$. Colourless crystalline solid with a faint garlic-like odour; m.p. 74°C. Manufactured by treating propenyl isothiocyanate with a solution of ammonia in alcohol. Used medicinally, toxic side reactions may occur. Propenyl thiourea is a chemical sensitizer for photographic silver halide emulsions, but somewhat toxic.

propetamphos Insecticide, acaricide, organophosphate derivative.

propham Herbicide, carbamate derivative.

propiconazole Fungicide, azole derivative.

propineb Fungicide, dithiocarbamate (Zn salt) derivative.

β-propiolactone, $C_3H_4O_2$, $\overline{CH_2.CH_2C(O)O}$. Colourless liquid; b.p. 162°C (decomp.). Prepared by reacting ketene with methanal. Highly reactive compound has many synthetic uses, chiefly for adding the group $-CH_2CH_2C(O)-$. It is a potent carcinogen.

propiolic acid See PROPANOIC ACID.

propiolic alcohol See 2-PROPYN-1-OL.

propionaldehyde See PROPANAL.

propionic acid See PROPANOIC ACID.

propionyl acetate (n-propyl ester of ethanoic acid) M.p. 192°C. Used in flavours, perfumes and as a solvent, particularly for resins.

propiophenone, $Ph(CO)CH_2CH_3$. Prepared from propanyl chloride, benzene and $AlCl_3$. M.p. 219°C. Used in manufacture of perfumes.

propisochlor Herbicide, acetonitrile derivative.

propoxor Insecticide, carbamate derivative.

propyl The C_3H_7- group which occurs in two isomeric forms: normal propyl, or Pr^n, 1-propyl, $CH_3.CH_2.CH_2-$, and *iso* propyl, Pr^i, 2-propyl, $(CH_3)_2CH-$.

propyl alcohols See PROPANOL.

n-propylbenzene, $PhCH_2CH_2CH_3$. Prepared ethyl sulphate on benzyl magnesium bromide. Used in textile printing and dyestuffs.

propylene See PROPENE.

propylene chlorohydrins (chlorohydroxypropanes), C_3H_7ClO. Both chlorohydrins are formed as colourless liquids by the action of dilute solutions of HOCl on propene, or by the action of HOCl on propylene oxide (1,2-epoxypropane). The product contains about 90% α-chlorohydrin and 10% β-chlorohydrin.

α-Propylene chlorohydrin (1-chloro-2-hydroxypropane), $CH_2Cl·CHOH·CH_3$. B.p. 127°C. Used in organic syntheses particularly for propylene oxide.

β-Propylene chlorohydrin (2-chloropropanol), $CH_3CHClCH_2OH$. B.p. 134°C.

Both of these chlorohydrins are converted to 1,2-dihydroxypropane by heating with solutions of sodium hydrogen carbonate; when heated with solid sodium hydroxide they give 1,2-epoxypropane.

propylene dichloride See 1,2 DICHLOROPROPANE.

propylene glycol See 1,2-DIHYDROXYPROPANE.

propylene oxide (1,2-epoxypropane, methyl oxinane), C_3H_6O, $CH_3\overline{CH·CH_2O}$. Colourless liquid; b.p. 34°C. Manufactured by heating propylene chlorohydrin with solid NaOH, or CaO oxidation of and from isobutane. Reacts with water in the presence of sulphuric acid to give propylene glycol, and with alcohols and phenols to give ethers. It resembles ethylene oxide in properties, but is somewhat less reactive. Used in chemical syntheses particularly reactions with polyols to give polyglycols, and as a solvent for foams, lubricants and surfactants.

propyl gallate (3,4,5-trihydroxybenzoic acid, propyl ester) Anti-oxidant.

propylparaben (4-hydroxybenzoic acid, propyl ester) Food preservative, antifungal agent.

propyne (allylene, methylethyne), C_3H_3, $CH_3C{\equiv}CH$. B.p. –23°C. Prepared by the action of alcoholic potassium hydroxide on 1,2-dibromopropane, or by the reaction of dimethyl sulphate on sodium acetylide in liquid ammonia. The chemical properties of propyne resemble closely those of ethyne.

2-propyn-1-ol (propargyl alcohol, propiolic alcohol), C_3H_4O, HC⦂C·CH$_2$OH. B.p. 115°C. Prepared from ethyne plus CH_3CHO or CH_3CH_2OH. Easily polymerized by heat or alkali, and used widely in organic synthesis.

prostaglandins A group of 20-carbon unsaturated, hydroxylated fatty acids containing C_5 rings, occurring in mammalian organs, tissues and secretions and in some simple animals like soft coral. The prostaglandins are anti-virals and have wide physiological activity. Unsaturated fatty acids like ARACHIDONIC ACID are biological precursors to prostaglandins. See also THROMBOXANES.

prosthetic group The non-protein group of a conjugated protein. See ENZYMES.

protactinium, Pa. Actinide element, at.no. 91, at.mass 231.04, m.p. 1572°C, b.p. *ca.* 4300°C, density ρ 15 370, crustal abundance 3.27×10^{-4} p.p.m., electronic configuration [Rn] $5f^26d^17s^2$. Originally separated from uranium ores by co-precipitation with ZrO_2. Now separated from residues from uranium extraction plants by solvent extraction. The most stable isotope is ^{231}Pa (32 340 years). The metal has been obtained by reduction of PaF$_4$ with barium at 1400°C. The silvery metal has a modified bcc structure. Protactinium compounds are very hazardous to health.

protactinium compounds In its compounds Pa shows oxidation states of +5 and +4

$$Pa^{5+} \xrightarrow{-0.1} Pa^{4+} \xrightarrow{-0.9} Pa \text{ (volts in aq. solution)}$$

the +5 state is most stable. All the halides PaX$_5$ and PaX$_4$ are known; brown PaF$_4$ is formed from Pa$_2$O$_3$, excess H$_2$ and HF; F$_2$ reacts with PaF$_4$ to give colourless PaF$_5$; PaCl$_5$ is a yellow, volatile solid (sublimes 200°C) prepared from SOCl$_2$ and Pa$_2$O$_5$ at 350–500°C; it is reduced to PaCl$_4$ by H$_2$ at 400°C. All of the halides are hydrolysed to hydrates, then oxide halides PaOX$_2$ and PaOX$_3$, and eventually to hydrated oxides. Complex halides such as K$_2$PaF$_7$ are formed from solution and also complexes, e.g. between the halides and N- and O-donors. The hydrated pentoxide on heating gives Pa$_2$O$_5$, PaO$_{2.3}$, then PaO$_2$ although there are very complex relationships in the Pa–O series. Some Pa(III) derivatives known. Pa(C$_5$H$_5$)$_4$ has been prepared from PaCl$_4$ and KC$_5$H$_3$. Pa metal gives PaH$_3$ with hydrogen.

protamines See PROTEINS.

proteases Enzymes that break amide bonds.

proteasome Enzyme that breaks down proteins.

protecting group A group which is added to one functional group in a molecule before chemical treatment of a different functional group. A protecting group may prevent unwanted reaction at the site it is protecting or it may serve to direct a reagent to another site. Protecting groups are easily attachable without causing racemization at chiral centres and are removable under conditions where the remainder of the molecule is unaffected. Protecting groups are used extensively in synthesizing peptides from amino-acids.

protective colloids Hydrophilic colloids, e.g. gelatin, being themselves unaffected by small concentrations of electrolytes, due to hydration, are able, when added in very small amounts, to protect hydrophobic sols from the coagulating influence of electrolytes. These are therefore termed protective colloids.

All hydrophilic colloids possess some degree of protective action and gelatin, starch and casein are used commercially for this purpose.

proteins The chief nitrogenous constituents of living organisms. They contain about 50% carbon, about 25% oxygen, 15% nitrogen, 7% hydrogen and some sulphur. They are formed from a mixture of AMINO-ACIDS which they give on hydrolysis. Proteins are precipitated by alcohol, propanone and by strong solutions of salts.

The proteins can be classified as follows but one particular protein may fall into more than one class.
Simple proteins:

Albumins. Soluble proteins both in water and in dilute aqueous salt solutions; found in all living tissue. Typical albumins are ovalbumin from eggs and lactalbumin from milk.

Globulins. Proteins insoluble in water, soluble in dilute salt solutions. They include such proteins as myosin from muscle, fibrinogen from blood and edestin from hemp.

Protamines. Strongly basic, low mol. wt. proteins which contain high levels of arginine and lysine, but no sulphur-containing amino-acids. They are soluble proteins, associated with nucleic acids and are obtained in large quantity from fish spermatozoa.

Histones. Small unfolded chromosomal proteins.

Prolamines. Proteins insoluble in water, but dissolving in aqueous alcohol solutions. Found in the seeds of cereals.

Glutelins. Also insoluble in water and in 7% aqueous ethanol. Soluble in acids and alkalis. Also found in cereals.

Scleroproteins. Insoluble proteins obtained from the skeletal and connective tissues of animals. Typical classes are KERATINS, COLLAGENS and ELASTINS.

Conjugated proteins:

Proteins with a PROSTHETIC GROUP:

Phosphoproteins. Contain phosphates and include casein from milk and phosvitin from egg yolk.

Chromoproteins. Compounds of proteins with heme or some similar pigments.

Nucleoproteins. The prosthetic group of the nucleoproteins is a nucleic acid, often linked through salt linkages with protamines or histones. The nucleoproteins are present in the nuclei of all cells. Chromosomes are largely nucleoproteins and some plant viruses and bacteriophages have been shown to be pure nucleoproteins.

Lipoproteins. The lipid moiety of lipoproteins is quite variable both qualitatively and quantitatively. The α-lipoprotein of serum contains glyceride, phosphatide and cholesterol to about 30–40% of the total complex. The *β*-lipoprotein of serum contains some glyceride but phosphatide and cholesterol account for nearly 75% of the total.

Glycoproteins or mucoproteins. Compounds of proteins with carbohydrates. All glycoproteins contain a hexosamine and usually sulphate, ethanoate and glucuronic acid. The carbohydrate–protein linkages are in some cases covalent and in others of the salt type. Aqueous solutions of glycoproteins are extremely viscous.

Enzymes may be classified under some of the above headings.

Proteins consist of large numbers of amino-acids joined by the peptide link –CO–NH– into –C(O)NHCHR′CONHCHR′′– chains. These chains are called peptides and may be broken into smaller chains by partial hydrolysis (see PEPTIDES). Proteins may contain more than one peptide chain.

The three-dimensional arrangement of peptides is very important in determining the properties of the protein and there is extensive hydrogen bonding. In fibrous proteins the polypeptide chain is in the form of a regular helix, called the α-helix, but other forms are known. Globular proteins consist of chains folded round themselves. Found particularly in ribosomes.

Methods of amino-acid and peptide analysis enable the complete amino-acid sequence of a number of proteins to be worked out. The grosser structure can be determined by X-ray diffraction or n.m.r. procedures. Proteins have molecular weights ranging from about 6 000 000 to 5000 (although the dividing line between a protein and a peptide is ill defined). Edible proteins can be produced from petroleum and nutrients under fermentation. Particular proteins fold in specific ways and folding in another way confers very different physiological properties, e.g. protein folding implicated in CJD and Alzheimer's Disease. See DICTIONARY OF BIOLOGY.

proteomics The science of proteins; the information in protein structures. Hence proteomes, the study of the distribution and function of proteins.

prothrombin See THROMBIN.

protocatechuic acid (3,4-dihydroxy-benzoic acid), $C_7H_6O_4$. Crystallizes with $1H_2O$; m.p. 199°C. It occurs in the free state in the onion and other plants; is a constituent of one group of tannins, and is a product of the alkaline decomposition of resins.

protogenic See AMPHIPROTIC.

proton One of the units from which all forms of matter are built. The proton is identical to the nucleus of the hydrogen atom, having a mass of one atomic unit and a unit positive charge equivalent to the negative charge on the electron. The proton is 1836 times the weight of the electron. Rest mass 1.672×10^{-27} kg, spin ½.

protonation Addition of a proton, e.g. H_2O is protonated by HCl to form $[H_3O]^+Cl^-$.

protonic acid Molecules which give H^+ by dissociation, e.g. HCl, H_2SO_4.

proton motive force See CHEMIOSMOTIC HYPOTHESIS.

protophilic See AMPHIPROTIC.

protoporphyrin See PORPHYRINS.

Prussian blue, $KFeFe(CN)_6$. See CYANO-FERRATES.

prussiates An old name for CYANIDES.

prussic acid, HCN. Old name for HYDROGEN CYANIDE.

PSEPT See POLYHEDRAL SKELETAL ELECTRON PAIR THEORY.

pseudocumene **(1,2,4-trimethyl-benzene)**, $C_6H_3Me_3$. Liquid produced by catalytic cracking of methyl benzenes. Used to prepare trimellitic anhydride.

pseudohalogens Compounds which show a resemblance to the halogens in their reactions and in the properties of their compounds. Pseudohalogens X_2 form X^-, XO^- and XO_3^- anions; AgX compounds are insoluble in water. E.g. $(CN)_2$, $(SCN)_2$, $(SeCN)_2$, $(SCSN_3)_2$.

pseudoionone See CITRAL.

pseudomorphic Having the same crystal form.

psychotomimetic drugs (psychopharmacological agents) Drugs which produce changes in thought, perception and mood. Examples are MESCALINE, LYSERGIC ACID DIETHYLAMIDE (LSD-25), psilocybin and the active constituents of marihuana, notably TETRAHYDROCANNABINOL.

Pt Platinum.

PTA 4-Terephthalic acid.

pteridines Heterocyclic compounds that take part in cell metabolism.

pterins Derivatives of pteridine which include the butterfly pigments leucopterin and xanthopterin and the vitamin, FOLIC ACID.

pteroylglutamic acid See FOLIC ACID.

PTFE POLYTETRAFLUOROETHENE. See also FLUORINE-CONTAINING POLYMERS.

Pu Plutonium.

pulp Fibrous cellulose material made from wood, cotton, linen, straw, etc., by mechanical disintegration with possibly some chemical treatment, or by chemical treatment (sodium sulphide and sodium hydroxide or sodium hydrogen sulphite or bisulphite). Pulp is used in paper, rayon, cellulose production.

pumice Volcanic material used as abrasive and in fireproofing and filtration.

pump A device, mechanical or using liquids, for moving a fluid against a pressure.

purine, $C_5H_4N_4$. A crystalline solid; m.p. 216–217°C. It can be prepared from uric acid, but is

not obtained naturally, and is not physiologically important although many of its derivatives are. It is the parent compound of a group of compounds of animal and vegetable origin, collectively called purines. The group includes adenine and guanine, which are constituents of the nucleic acid portion of nucleoproteins, their breakdown products, hypoxanthine, xanthine and uric acid, and the drugs caffeine, theobromine and theophylline.

Purple of Cassius See GOLD.

putrescine (tetramethylenediamine, 1,4-diaminobutane), $H_2N \cdot [CH_2]_4 \cdot NH_2$. M.p. 27–28°C, b.p. 158–159°C. Putrescine is found associated with cadaverine in putrefying tissue, being formed by bacterial action on the amino-acid arginine and is present to a small extent in many cells. It is present in ergot.

putty powder See TIN OXIDES.

PVC Polyvinyl chloride. See VINYL CHLORIDE POLYMERS.

pyknometer An instrument for measuring the density of a substance by determining the weight of a known volume.

pymetrozine Insecticide, triazine derivative.

pyraclofos Insecticide, organophosphate derivative.

pyraflufen-ethyl Herbicide, piperazole derivative.

pyranose The stable ring form of the sugars

which contains a C_5O ring skeleton. Glucose is structurally a 1,5-glucopyranose with a primary alcohol group – CH_2OH as a side chain. Such pyranose derivatives are stable and crystalline in contradistinction to the FURANOSE SUGARS.

pyrazinamide Important anti-tuberculosis drug.

pyrazine (1,4-diazine), $C_4H_4N_2$ (see over).

pyrazole (1,2-diazole), $C_3H_4N_2$. A colourless crystalline substance **1**, m.p. 70°C. Obtained

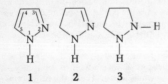

by passing ethyne into cold ethereal solutions of diazomethane. It is a weak base. Partial reduction gives pyrazoline **2**, and complete reduction leads to pyrazolidine **3**. Pyrazole is an aromatic compound, undergoing electrophilic substitution in the 4-position.

pyrazolidine See PYRAZOLE.

pyrazoline See PYRAZOLE.

pyrazolinones See PYRAZOLONES.

pyrazolones (pyrazolinones, oxopyrazolines 5-oxo-1,3-substituted pyrazolines) Used extensively as dyes and pigments, in colour photography and in pharmaceuticals. Prepared from acetoacetic ester and phenylhydrazines.

pyrazolynate Herbicide, pyrazole derivative.

pyrazophos Fungicide, organophosphate derivative.

pyrazo sulfuron-ethyl Herbicide, sulphonylurea derivative.

pyrazoxyfen Herbicide, pyrazole derivative.

pyrethrins Constituents of pyrethrum and compounds of related structure.

pyrethroids Pyrethrins.

pyrethrum Insecticide. The mixture of substances obtained by grinding or extracting the dried flowers of *Chrysanthemum cinerariaefolium*. Usually used in conjunction with a synergist such as PIPERONYL BUTOXIDE.

Pyrex A trade name for a heat-resisting borosilicate glass containing a high percentage of SiO_2 with some B, Al and alkalis. Pyrex has high mechanical strength.

pyributicarb Herbicide, fungicide, thiocarbamate derivative.

pyridaben Insecticide, acaricide, pyridazinone derivative.

pyridaphenthion Insecticide, acaricide, organophosphate derivative.

pyridate Herbicide, pyridazine derivative.

pyridine, C_5H_5N. Colourless refractive hygroscopic liquid; b.p. 115.3°C. Strong characteristic

smell. Burns with a smoky flame. Strong base, forms stable pyridinium salts (N-protonated) with mineral acids and quaternary compounds with alkyl halides, e.g. 1-methylpyridinium iodide $[C_5H_5NCH_3]I$. Forms a characteristic picrate and complexes with most metal salts. Reduced to piperidine, $C_5H_{11}N$, by sodium and alcohol, oxidized to pyridine-N-oxide by peracids. Very poisonous.

Commercially, pyridine is manufactured from ethyne and ammonia. It is used as a solvent, particularly in the plastics industry, for mineral acids and in the manufacture of nicotinic acid, various drugs and rubber chemicals.

pyridine N-oxide, C_5H_5NO. Prepared from pyridine and H_2O_2. M.p. 65°C. Used in synthesis.

pyridinium Ions formed by N-co-ordination of pyridine to a proton or cation.

pyridinium bromide (perbromide), $[pyNH]^+Br_3^-$. Prepared from pyridine, HBr and Br_2. Used in small-scale brominations.

pyridoxal See PYRIDOXINE.

pyridoxine (vitamin-B₆, 2-methyl-3-hydroxy-4,5-bis(hydroxy-methyl)pyridine), $C_8H_{11}NO_3$. Water-soluble vitamin. Precursor of pyridoxal phosphate, a prosthetic group associated with transamination. See DICTIONARY OF BIOLOGY.

pyrifenfox Fungicide, oxime derivative.

pyrimethanil Fungicide, pyrimidine derivative.

pyrimidine, $C_4H_4N_2$. Crystalline compound

with a penetrating smell, m.p. 20–22°C, b.p. 124°C. It can be prepared from barbituric acid via trichloropyrimidine.

It is the parent substance of a group of compounds which includes cytosine, thymine and uracil, which are constituents of nucleic acids and barbituric acid and its derivatives.

pyrimidipen Insecticide, acaricide, pyrimidine derivative.

pyriminobac-methyl Herbicide, pyrimidinyl derivative.

pyriproyfen Insecticide, juvenile hormone.

pyrites, FeS_2. A common mineral, brass-yellow with a metallic lustre. The structure contains S_2^{2-} species. Used for production of H_2SO_4.

pyrithiobac-sodium Herbicide, pyrimidinyl derivative.

pyro acids Acids or anions derived from (often hypothetical) acids containing two oxygroups with a bridging (μ) oxygen, e.g. $[O_3SOSO_3]^{2-}$, pyrosulphate.

pyrocatechol See 1,2-DIHYDROXYBENZENE.

pyrochlore (pyrochlorite), $NaCaNb_2O_6F$. A niobium mineral sometimes containing some Ta. A series of oxide–fluorides of stoichiometry $A_2B_2X_7$.

pyrogallol (1,2,3-trihydroxybenzene), $C_6H_6O_3$. White lustrous needles, m.p. 132°C, b.p. 210°C (decomp). Alkaline solutions rapidly absorb oxygen from the air and become dark brown in colour. It is prepared by digesting gallic acid with water at 200°C. Extensively used as a photographic developer, and in gas analysis as an absorbent for oxygen.

pyroligneous acid The crude brown liquor obtained by the distillation of wood. Contains ethanoic acid. Methanol and propanone are also present.

pyrolusite, β-MnO_2. An important manganese mineral, iron-grey in colour with the rutile structure.

pyrolysis Subjecting a material, e.g. petroleum, to high temperature, often in the absence of air.

pyromellitic acid (1,2,4,5-benzene-tetracarboxylic acid), $C_6H_2(CO_2H)_4$. Prepared by oxidation of durene. M.p. 260°C (decomp.). Readily converted to pyromellitic dianhydride, m.p. 287°C, used as a cross-linking agent for epoxy resins and also with diamines giving polyimides with excellent high-temperature resistance.

pyromucic acid See FUROIC ACID.

pyrones Compounds containing the ring systems shown which occur in nature, e.g. as part of the anthocyans.

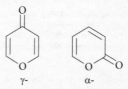

γ-   α-

pyrophoric metals Some metals when produced in a porous condition or finely divided state by reduction at low temperatures are very active and are often spontaneously inflammable in air. Thus pyrophoric iron can be obtained by reduction of iron(II) oxalate and many others by distillation of amalgams at low pressures.

Pyrophoric alloy is a Ce–Fe alloy used in lighters.

pyrophosphates Salts containing $[P_2O_7]^{4-}$ species. See PHOSPHORUS OXYACIDS and BISPHOSPHONATES.

pyroquilon Fungicide, quinoline derivative.

pyrosols The electrolysis of fused salts results in cloudy solutions called pyrosols which are generally regarded as colloidal, the free metal being the dispersed material. Pyrosols are easily produced by dissolving the metal directly in the fused salt, e.g. zinc in fused zinc chloride.

pyrosulphuric acid, $H_2S_2O_7$. Present in fuming sulphuric acid. Pyrosulphates contain $[S_2O_7]^{2-}$ ions.

pyrosulphurous acid, $H_2S_2O_5$. The free acid is unknown but pyrosulphites contain $[S_2O_5]^{2-}$ $[O_3SS(O)_2]^{2-}$ ions. Formed from sulphites, $[SO_3]^{2-}$, and SO_2.

pyrotechnics Fireworks and related substances. Combustible mixtures, including illuminating, incendiary, priming, warning and smoke-producing compositions. They contain an oxidant, such as nitrate or chlorate, and combustible substances such as charcoal, sulphur, antimony sulphide, etc. Metallic powders are incorporated to give illuminating effects and metallic salts provide coloured effects. In incendiary compositions a highly exothermic reaction occurs giving rise to intense heat, as in thermite. Smoke-producing compositions may contain phosphorus or hexachloroethane.

pyroxenes A group of silicate minerals containing simple chains of stoichiometry $(SiO_3)_n{}^{2n-}$. Example are enstatite, $MgSiO_3$ and diopside $CaMg(SiO_3)_2$.

pyrrocoline ring system See INDOLIZINE RING SYSTEM.

pyrrole (azole, imidole), C_4H_5N. Colourless oil, b.p. 130°C. Prepared by heating ammonium mucate, or from butyne-1,4-diol and ammonia in the presence of an alumina catalyst. The pyrrole molecule is aromatic in character. It is not basic and the imino-hydrogen atom can be replaced by potassium. Many pyrrole derivatives occur naturally, e.g. proline, indican, heme and chlorophyll.

pyrrolidine (tetrahydropyrrole), C_4H_9N. Almost colourless, ammonia-like liquid, b.p. 88–89°C, which fumes in air. Strong base. It occurs naturally in tobacco leaves, but is made industrially by hydrogenation of pyrrole.

2-pyrrolidinone (2-oxypyrrolidine), C_4H_7NO. B.p. 251°C, m.p. 25°C. Manufactured

from butyrolactone and ammonia. Easily hydrolysed to 4-amino-butanoic acid, its most important use is for the formation of *N*-VINYLPYRROLIDONE by reaction with ethyne and as a high boiling solvent and plasticizer.

pyruvic acid (2-oxopropanoic acid), $CH_3COCOOH$. A colourless liquid with an odour resembling that of ethanoic acid, m.p. 13°C, b.p. 65°C/10mm. It is an intermediate in the breakdown of sugars to alcohol by yeast. Prepared by distilling tartaric acid with potassium hydrogen sulphate. Tends to polymerize to a solid (m.p. 92°C). Oxidized to oxalic acid or ethanoic acid. Reduced to (±)-lactic acid.

pyruvic aldehyde See METHYL GLYOXAL.

Q

quadruple point The unique conditions of temperature and pressure at which four phases of a two-component system exist in equilibrium.

quadrupole moment The coupling of the nuclear quadrupole with the asymmetric electron fields in compounds. Measurement of quadrupole moments yields information about the electronic distribution over the atoms within in a molecule or ion.

qualitative analysis Analysis for the identification of constituents.

quantitative analysis Analysis for the estimation of constituents.

quantum In Planck's quantum theory the energy associated with electromagnetic radiation consists of discrete units, quanta, which, for radiation of given frequency, have definite energies. Thus the transfer of energy can occur only in definite parcels, or quanta, and not continuously as in a stream of fluids. Conversely, matter has wave-like properties. For radiation of frequency v, the unit of energy, the quantum, is equal to hv, where h is Planck's constant. When a quantum has particle properties it is known as a PHOTON.

quantum efficiency In radiation-induced processes the actual number of species which are decomposed or reacted per quantum of energy absorbed (quantum yield).

quantum number The extranuclear electrons in an atom or the atoms in a molecule possess energy due to type, rotation, vibration or spin. By quantum theory the energy possessed by such a particle is quantized, i.e. can have only certain definite values. Each such value for a particular form of energy is a multiple – a small whole number, or sometimes one-half – of a quantum, which is characteristic of the form of energy under consideration. This multiple, which defines the energy of the particle, is called the quantum number.

quantum yield See QUANTUM EFFICIENCY.

quartering See CONING AND QUARTERING.

quartet A group of four closely spaced lines in a spectrum (n.m.r., etc.).

quartz, SiO_2. The low-temperature form of SILICA. Quartz is used extensively for optical components. It is more transparent to u.v. radiation than is glass.

quartz glass See SILICA VITREOUS.

quasi crystals Generally alloys with local order but without total crystalline order. Used in coatings for cooking utensils and razor blades.

quaternary salts Cationic species of the right-hand side of the Periodic Table – often stabilized by co-ordination of a base. The products of reaction, e.g. tertiary amines or phosphines with alkyl halides. Ionic salts. The quaternary ammonium salts are used extensively in fabric softening (cations react with fabrics), cationic emulsions and as germicides, antistatic agents.

quenching The removal of energy from excited species (including fluorescing species) by a solvent.

quercitrin, $C_{21}H_{20}O_4$. Pale yellow crystals, m.p. 169°C, a glycoside widely distributed in plants. Hydrolysed by acids to rhamnose and quercitin (3',4'5,7-tetrahydroxyflavanol). Prepared commercially from oak bark, and is still used as a natural dyestuff (lemon flavin).

quicklime, CaO. Calcium oxide.

quicksilver An old name for mercury.

quinaldine (2-methylquinoline), $C_{10}H_9N$. Colourless oily liquid, b.p. 246–247°C. Volatile in steam. Forms stable salts and quaternary compounds. It is prepared by heating together aniline, ethanal and zinc chloride. It is used for the preparation of photosensitizing dyes.

quinalphos Insecticide, acaricide, organophosphate derivative.

quincloral Herbicide, quinoline derivative.

quinhydrone, $C_{12}H_{10}O_4$. A molecular compound, stable in the solid state, consisting of alternate quinone and hydroquinone molecules linked together by hydrogen bonds. It is prepared by mixing alcoholic solutions of quinone and hydroquinone and crystallizes in reddish-brown needles with a green lustre, m.p. 171°c. It is very sparingly soluble in cold water, and is split into its constituents by boiling water. It was used in the QUINHYDRONE ELECTRODE. Related compounds of quinone with phenols are known and these are collectively called quinhydrones.

quinhydrone electrode An electrode used in pH measurement relying on the redox reaction, hydroquinone ⇌ quinone. Can be used in non-aqueous solvents.

quinidine, $C_{20}H_{24}N_2O_2.2H_2O$. M.p. 172°C. Dextrorotatory stereoisomer of quinine. It occurs in cinchona bark and is obtained as a by-product in QUININE manufacture. It has similar medicinal properties to quinine.

quinine, $C_{20}H_{24}N_2O_2.3H_2O$. White microcrystalline powder, m.p. 57°C (177°C when anhydrous). The principal alkaloid of cinchona

bark. Quinine and its salts were used for the treatment of malaria but now many forms of malaria have developed immunity and other anti-malarials are used.

quinitol The trivial name for 1,4-cyclohexanediol.

quinizarin (1,4-dihydroxy-9,10-anthraquinone), $C_{14}H_8O_4$. Used for the preparation of various dyestuffs.

quinmerac Herbicide, quinoline derivative.

quinoclamine Herbicide, acaricide, naphthaquinone derivative.

quinol An alternative name for HYDRO-QUINONE.

quinoline (benzazine), C_9H_7N. Colourless oily refractive liquid; b.p. 238°C. Very hygroscopic, disagreeable odour. Oxidized by alkaline

permanganate to quinolinic acid (pyridine-2-dicarboxylic acid). Reduced by tin and hydrochloric acid to the tetrahydro-compound. Basic, gives stable salts with mineral acids, and quaternary ammonium compounds with alkyl halides (quinolinium salts).

Occurs in the high-boiling fraction of coal tar. Most conveniently prepared by Skraup's reaction by heating a mixture of aniline, glycerol, sulphuric acid and nitrobenzene. Used in the manufacture of dyestuffs and pharmaceuticals.

quinolinol 8-HYDROXYQUINOLINE.

quinone Another name for benzoquinone.

quinones Diketones formally derived by replacing the >CH_2 groups in a dihydroaromatic system by >C=O. They are highly coloured substances many of which occur in nature, e.g. vitamin K. They are prepared by oxidation of phenols. They give addition compounds with many types of organic compounds, especially quinols. They are readily reduced to dihydric phenols and are used as dehydrogenating agents. Used in photography and as oxidants.

quinoxyfen Fungicide, quinoline derivative.

quintet A group of five closely spaced lines in a spectrum.

quintozene Fungicide, chloroaromatic derivative.

quinuclidine, $C_7H_{13}N$. A sublimable solid;

m.p. 158–159°C. A strong, hindered base (pK_b = 11), it forms very soluble crystalline complexes with organometallic compounds of magnesium, lithium and zinc. These are useful for characterization and for conversion reactions like METALATION, e.g. this base promotes reaction of toluene with an alkyl-lithium to form benzyl-lithium.

quizalofop Herbicide, arylpropionic acid derivative.

quorn (lycoprotein) High-protein meat substitute.

R

R Designation of CONFIGURATION.

Ra Radium.

Racah parameters The parameters used to express quantitatively the interelectronic repulsion between the various energy levels of an atom. Generally expressed as B and C. The ratios between B in a compound and B in the free ion give a measure of the NEPHELAUXETIC EFFECT.

racemate (racemic compound) An equimolecular compound or mixture of the dextro and laevoratatory optically active forms of a given compound.

DL-racemic acid See TARTARIC ACID.

racemization The process in which an achiral stereoisomer is converted into a racemate. EPIMERIZATION is a special case of racemization. Heat and the action of acids or bases can encourage racemization.

rad See GRAY. 0.01 Gy.

radiation A term used to describe any form of electromagnetic radiation. Absorbed dose measured in GRAYS.

radiative decay process Loss of excitation energy as a photon. Transfers energy as heat.

radical Term usually applied to species which have one or more free valencies. See FREE RADICALS.

Formerly applied to a group of atoms, present in many compounds which can pass unchanged through a series of chemical reactions.

radii Sizes derived from distances in compounds.

radioactive decay series The series of isotopes into which a radioactive nucleus is successively transformed.

radioactivity Becquerel (1896) found that uranium salts had a pronounced effect on a photographic plate even when they were separated by thin sheets of metal. This property was called radioactivity. Three main types of radiation are concerned, α-, β- and γ-rays, which arise from the spontaneous disintegration of the nucleus of the radioactive element. The emission of an α particle corresponds to a decrease in mass of the nucleus by four units; β particle emission corresponds to an increase in atomic number of the nucleus by one unit. The emission of γ-rays corresponds to loss in energy by the nucleus; they are often accompanied by X-rays.

There are other less common types of radioactive decay. Positron emission results in a decrease by one unit in the atomic number; K capture involves the incorporation of one of the extranuclear electrons into the nucleus, the atomic number is again decreased by one unit.

All elements of atomic number greater than 83 exhibit radioactive decay; K, Rb, Ir and a few other light elements emit β particles. The heavy elements decay through various isotopes until a stable nucleus is reached. Known half-lives range very widely.

The rate of radioactive transformations cannot be altered by changing the conditions which are available in the laboratory. The process is a spontaneous one. Measured in curies (Ci) 3.7×10^{10} disintegrations per second.

Radioactive elements may be prepared artificially by bombarding the atoms of ordinary stable elements with, e.g. helium nuclei. See RADIOACTIVITY, ARTIFICIAL.

radioactivity, artificial When an element undergoes a nuclear reaction, e.g. as a result of bombardment with α-particles, protons, deuterons or neutrons or heavier ions, the resulting atoms may be stable or metastable. In the latter case, they will be radioactive, and the phenomenon is known as artificial or induced radioactivity.

For example the reaction between aluminium nuclei and α-particles is expressed by the equation

$$^{27}_{13}Al + {}^4_2He \rightarrow {}^{30}_{15}P + {}^1_0n$$

(^1_0n represents a neutron). The new phosphorus atoms disintegrate into atoms of a silicon isotope and positrons $^{30}_{15}P \rightarrow {}^{30}_{14}Si + {}^0_1e$, this process produces the observed artificial radioactivity. In the process, only a fraction of the α-particles causing disintegration actually give rise to active phosphorus, the majority following an (α, p) reaction giving $^{30}_{14}Si$ directly. The emission of positrons or other charged particles in such processes is detected experimentally by the ionization which they produce in gases; in counters this effect is used to count the number of particles emitted and hence to determine the rate of decay of the activity.

Normally, artificial radioactivity is induced only by bombardment with particles of appreciable mass – α-particles, protons, deuterons, neutrons, or ions of heavier elements, but γ-rays of very high energy are capable of disintegrating atoms with the ejection of a neutron from each nucleus and the formation of an isotope of the original element, which emits β-rays. Neutrons possess no electric charge and can penetrate the heaviest atoms without being subject to strong electrostatic forces. For this reason neutrons are particularly effective in producing artificial activity. The best sources of neutrons are nuclear reactors and the cyclotron, and by their use many radioactive isotopes have been prepared.

radiocarbon dating Carbon in living things contains a uniform proportion of radioactive ^{14}C, which is produced constantly in the atmosphere by the action of cosmic radiation. Once the host dies, the exchange of nonradioactive carbon with the atmospheric ^{14}C ceases. Thus from the amount of ^{14}C in the dead sample the age of it may be determined. ^{14}C has a half-life of 5570 years and the method is useful for samples having ages up to 30 000 years.

radium, Ra. Radioactive metal, at.no. 88, at. mass 226.03, m.p. 700°C, density ρ 5650, crystal abundance 6×10^{-7} p.p.m., electronic configuration [Rn] $7s^2$. Various isotopes are members of various radioactive decay series; ^{226}Ra is most stable, half-life 1600 years. Ra is isolated from uranium ores; the element is obtained by electrolysis using a Hg cathode. Ra is a white metal which tarnishes in air and reacts with water (at least partly radiochemically). Used in self-luminous paints, as a neutron source, in radio-

therapy (being supplemented by artificial radioisotopes), radiography of metals.

radium compounds Radium is an alkaline earth of Group 2. Its compounds are very similar to the corresponding barium derivatives although they are generally less soluble. Radium halides RaF_2, $RaCl_2$ and $RaBr_2$ are known. The carbonate, fluoroberyllate and iodate are insoluble in water.

radon, Rn. Radioactive noble gas, at.no. 86, at.mass 222.02, m.p. –71°C, b.p. –61.7°C, density (liquid) ρ 4400, present naturally in trace amounts, electronic configuration [Xe] – $4f^{14}5d^{10}6s^26p^6$. Radon is a product of radioactive decay of the heavy elements. ^{222}Rn is obtained as a gas from $RaCl_2$ solution. Rn is radioactive and has been used as a radiation source and as a gaseous tracer; it is a considerable hazard in uranium mines and some homes. Group 18 element. Rn forms compounds, particularly a fluoride and solid adducts between the fluoride and Lewis acid fluorides.

raffinate See LIQUID–LIQUID EXTRACTION.

raffinose (melitose), $C_{18}H_{32}O_{16}$. Large prisms as the pentahydrate; m.p. 80°C or m.p. 118–119°C when anhydrous. The best-known trisaccharide, it is composed of galactose, glucose and fructose. It is found in sugar beet, but the best source is cotton-seed meal, which contains 8%. It has no reducing power.

Raman effect When light of frequency v_0 is scattered by molecules of a substance, which have a vibrational frequency of v_1, the scattered light when analysed spectroscopically has lines of frequency $v_0 \pm v_1$ etc. This spectrum is called a Raman spectrum and the frequencies $v_1, v_2 ...$ correspond to the vibrational or rotational changes in the molecule. The selection rules for Raman activity are different from those for i.r. activity and the two types of spectroscopy are complementary in the study of molecular structure. Raman spectrometers generally use lasers for excitation. In the resonance Raman effect excitation at a frequency corresponding to electronic absorption causes great enhancement of the Raman spectrum.

Rotational Raman spectra give accurate information on bond lengths. Used to identify minerals, enamels, etc.

ramoplanin Potent antibiotic, macrocyclic decapeptide.

Raney nickel A form of nickel prepared by treating an Al–Ni alloy with NaOH solution.

The nickel is left in a spongy mass which is pyrophoric when dry. This form of nickel is a strong catalyst, especially for hydrogenations.

Raoult's law When a solute is dissolved in a solvent, the vapour pressure of the latter is lowered proportionally to the mole fraction of solute present. Since the lowering of vapour pressure causes an elevation of the boiling point and a depression of the freezing point, Raoult's law applies and leads to the conclusion that the elevation of boiling point or depression of freezing point is proportional to the weight of the solute and inversely proportional to its molecular weight. Raoult's law is strictly only applicable to ideal solutions since it assumes that there is no chemical interaction between the solute and solvent molecules.

rare earths Strictly the oxides of the lanthanide elements but often applied to the elements themselves. See LANTHANIDES. World demand rare earth oxides 1997 66 000 tonnes. Used in magnets, Ni–H batteries, catalysts, phosphors and electrical components.

rare gases The noble gases.

Raschig process See HYDRAZINE.

rasorite See KERNITE.

Rast's method A virtually obsolete method of determining molecular masses by measuring the depression of freezing point of a solvent by a known weight of the solute whose molecular weight is required. Camphor is commonly used as the solvent because of its high molecular depression constant. Has use for determining molecular masses of polymers.

rate-determining step The slowest step in a sequence of reactions. Generally the one that effectively determines the overall rate.

rate of reaction The rate of consumption of one of the reactants or the rate of formation of one of the products. Often proportional to the concentration (activity) to a given power of one of the species. The rates of most reactions increase with temperature. See REACTION VELOCITY.

rationality of indices, law of See RATIONALITY OF INTERCEPTS, LAW OF.

rationality of intercepts, law of The intercepts of the faces of a crystal upon the axes of the crystal bear a simple ratio to each other.

ratio of specific heats The ratio between the specific heat of a gas at constant pressure (C_p) and that at constant volume (C_v). This ratio (C_p/C_v) decreases with the number of degrees of freedom which in turn varies with the number of atoms in each molecule of the gas; monatomic gases have a value of 1.67; diatomic, 1.40; and triatomic 1.33.

rayon Artificial fibres made of regenerated cellulose formerly known as viscose, artificial silk, fibre silk, etc. See CELLULOSE ACETATE.

Rb Rubidium.

RBS See RUTHERFORD BACK-SCATTERING.

RDX (1,3,5-trinitro-1,3,5-triazacyclohexane, cyclonite) Important explosive.

Re Rhenium.

reaction mechanism A step-by-step description of the sequence of processes that together make up a full description of the mechanism.

reaction velocity The rate of a chemical reaction expressed either as the rate of disappearance of reactant molecules or the rate of formation of product molecules. See RATE OF REACTION.

reactive dyes Dyes which contain a functional group (chloro-substituted heterocycle, acrylamide, sulphonic acid derivative) capable of reacting directly with the fibre.

reactor Large-scale chemical processes are carried out in various environments, usually in tanks, tubes, towers or FLUIDIZED BEDS. These are usually referred to as chemical reactors and form the heart of the overall chemical process. Reactor design involving scale-up from laboratory reactions is an important aspect of chemical technology or chemical engineering. The term reactor is also used for a nuclear reactor.

realgar The mineral form of As_4S_4, formerly used as an orange–red pigment. Formed artificially by distilling sulphur and excess of arsenic.

rearrangement A reaction that takes place, apparently completely internally, to give a new compound, e.g. a *cis*- to *trans*- rearrangement.

reboiler See RECTIFICATION.

reciprocal proportions, law of A special case of the LAW OF EQUIVALENT PROPORTIONS stating that 'the weights of two or more substances which separately react chemically with identical weights of another substance are also the weights which react with each other or are simple multiples of them'. Thus 23g of Na react with 35.5g of Cl_2 to give NaCl; 23g of Na react with 1g of H_2 to give NaH; 35.5g of Cl_2 react with 1g of H_2 to give HCl. The law is not of completely general application owing to the existence of variable oxidation states and non-stoichiometric compounds.

recoil atom When an atomic nucleus disintegrates the fragments receive equal momentum. In the case of a heavy nucleus emitting a light fragment, the heavy remnant, termed a recoil atom, will recoil with a comparatively low velocity. The recoil atom has a sufficiently high energy to rupture bonds which it may form or have with other atoms. The resulting group may not recombine (Szilard–Chalmers effect) and it may thus be possible to separate chemically the atoms which have undergone radioactive disintegration. The 'hot atom' may recombine with its original partner or may combine with another group to form a new chemical compound.

recrystallization The process of repeated crystallization carried out with the object of removing impurities or of obtaining more satisfactory crystals of a substance which is already pure.

rectification A method of separating a mixture into its components by fractional distillation. The terms rectification and fractional distillation are frequently used synonymously, despite the fact that two other methods are covered by the latter term.

A *rectifying* or *fractionating column* consists of a long vertical cylinder containing either a series of *plates*, or a *packing* material having a large specific surface, both these arrangements ensuring good liquid–vapour contacting over the length of the unit. At the top of the column there is a *condenser* and at the bottom a vaporizing unit called a *reboiler*.

Rectification is of great industrial importance; in particular, it is the principal separation process employed in the petroleum industry.

See also PACKED COLUMN, PLATE COLUMN.

rectified spirit A solution of ethanol in water which contains 90% v/v ethyl alcohol. Rectified spirit is 57.8° over proof (OP). See PROOF SPIRIT.

rectifying column See RECTIFICATION.

recuperators (regenerators) Heat-exchangers used in furnace operations where combustion air or lean fuel gas may be preheated.

red lead, Pb_3O_4. See LEAD OXIDES.

red mercury A colloid from an amalgam of Hg and $Hg_2Sb_2O_7$. If used to dissolve a neutron emitter such as ^{252}Cf can induce nuclear fission.

redox An abbreviation for oxidation–reduction, e.g. redox potential; reduction–oxidation potential.

redox catalyst A combination of a free-radical catalyst with a reducing ion or salt which increases the rate of free-radical production. Used in polymerization reactions. A catalyst which increases the rate of a free-radical reaction – often polymerization.

redox indicator A substance which undergoes a visible colour change during its reversible oxidation or reduction. The substance has a definite, and preferably narrow, potential range over which the colour change occurs. E.g. methylene blue is colourless when reduced, blue when oxidized.

reduced mass For any system of particles of masses m and M in harmonic motion about each other, the reduced mass μ is given by

$$\mu = \frac{mM}{(m + M)}$$

reducing agent A material which brings about reduction – and in the process is itself oxidized. E.g. Sn^{2+} reduces Fe^{3+} and is itself oxidized to Sn^{4+}.

reducing sugar A sugar that shows reducing properties – contains hemiacetal or hemiketal groups.

reduction Chemical processes in which the proportion of electronegative substituent is decreased, e.g. $FeCl_3 \rightarrow FeCl_2$, or the charge on an ion is made more negative, e.g. $Sn^{4+} \rightarrow Sn^{2+}$ or $[Fe(CN)_6]^{3-} \rightarrow [Fe(CN)_6]^{4-}$, or the oxidation number is lowered. In organic chemistry the most obvious effect is to increase the proportion of hydrogen and to decrease the proportion of oxygen and to reduce the number of multiple bonds, e.g. C_2H_4 (ethene) $\rightarrow C_2H_6$ (ethane).

reductones Enediols stabilized by conjugation to a carbonyl or similar group.

reference electrode A standard electrode used in electrochemistry. Examples are the calomel electrode, Ag–AgCl, $Hg–Hg_2SO_4$ electrodes.

reference state The most stable state at a particular temperature and 1 bar (except for phosphorus where white P is designated the standard state).

refining The removal of impurities from metals or alloys in metallurgy using, e.g. oxidation, electrolysis, chemical separation, distillation, zone refining.

refining, petroleum The general process of converting crude oil into a wide range of petroleum products including fuels, lubricants, bitumens, waxes, etc. Specific refining operations include physical separation techniques such as DISTILLATION, solvent extraction, ABSORPTION, etc. Chemical refining operations involve many treatments including sulphuric acid treatment, SWEETENING, DESULPHURIZATION, etc.

reflux Liquid from partial condensation of vapour which is returned to the top of a fractionating column and allowed to flow down the column counter-current to the ascending vapour.

Reformatski reaction Aldehydes and ketones react with α-bromo-fatty acid esters in the presence of zinc powder to give β-hydroxyesters which may be dehydrated to give α-, β-unsaturated esters. α-Chloroesters will react if copper powder is used in conjunction with the zinc.

reform gas See METHANOL.

reforming The process of subjecting straight-run gasoline from primary distillation to a thermal or catalytic treatment. Straight-chain alkanes are converted to isomers while cycloalkanes are dehydrogenated to form aromatics. Thermal reforming is similar to THERMAL CRACKING but higher temperatures (500–600°C) and pressures (about 7000 kN /m⁻²) are used. The process results in gasoline of higher OCTANE NUMBER and wider boiling range. The reforming process is usually carried out in the presence of a catalyst as lower temperatures and pressures may be used and higher yields of particular gasolines are obtained. See CATALYTIC REFORMING.

refractory materials (refractories) A general term for materials which are not damaged by heating to at least 1500°C in a clean (oxidizing) atmosphere.

Acid refractory materials include fireclays, flint clays, china clays (kaolins), silica, flint, chalcedony, ganister and titanium dioxide.

Neutral refractory materials include graphite, charcoal, coke, chromite and various carbides.

Basic refractory materials include lime, magnesia, various materials composed chiefly of alumina (bauxite, diaspore, laterite, gibbsite, etc.), dolomite and most of the more expensive refractory oxides, particularly zirconia.

refrigerants Materials used for cooling, e.g. liquid nitrogen, solid CO_2. See LOW-TEMPERATURE BATHS.

refrigeration The method of achieving refrigeration, i.e. lowering the temperature of a material below that of its surroundings. The usual method is by vapour compression. The cycle consists of compressing a refrigerant vapour and condensing it, then passing the high-pressure liquid through a reducing valve. Reduction in pressure causes part of the liquid to vaporize resulting in a considerable lowering of temperature. The cold liquid passes to an evaporator where it is vaporized by heat exchange with the medium to be cooled, the vapour formed passing back to the compressor.

Many condensable vapours have been used as refrigerants, but the main ones are chlorofluorocarbons (FREONS, now being phased out), polyfluorocarbons and carbon dioxide.

regioselectivity The tendency for reactions which could proceed in two or more ways to proceed in one manner only, e.g. addition to olefin derivatives to give only 1,2-adducts.

regiospecific Regioselectivity to give a specific product.

regular system The CUBIC SYSTEM.

Reinecke salts, $(NH_4)(or K)[Cr(SCN)_4-(NH_3)_2]H_2O$.

reinforced plastics COMPOSITES between plastics, generally thermosetting resins, and strengthening materials which may run from carbon metal, and glass fibres, fabrics, paper, etc. Laminates have the thermosetting resin built up by heat and pressure.

relative atomic mass (A_r) See ATOMIC WEIGHT.

relativistic effects Chemical and spectroscopic effects arising from interaction between heavy nuclei and outer electrons. Major reason for close similarities between second- and third-row transition elements.

relaxation The process of loss of energy from an excited state to another excited state or to the ground state. Return to equilibrium – see ULTRASONICS.

release agents See ABHERENTS.

remote handling facilities Equipment used to carry out chemical reactions with intensely radioactive elements, e.g. Am, to manipulate the very active fuel elements from a nuclear reactor or to handle biologically active materials.

rennet See RENNIN.

rennin The milk-clotting enzyme from the stomach of young mammals. Under the name of rennet it is used in cheesemaking and for making rennet casein and junket.

Reppe process The carboxylation of an alcohol in the presence of cobalt or rhodium derivatives, e.g. CH_3OH to $CH_3C(O)OH$.

reserve acidity and alkalinity A buffer solution is sometimes referred to as a solution with reserve acidity or alkalinity.

resins High mol.wt. materials, often of variable but controlled mol.wt., which soften at high temperatures. See THERMOPLASTIC RESINS. Synthetic resins are formed by polymerization. Natural resins occur in vegetable products (rosin) or from insects (shellac).

resmethrin Insecticide, pyrethroid.

resolution The separation of a racemate into its two enantiomorphic forms (i.e. (R) and (S) forms).

resonance In the optical sense the term refers to the absorption of radiation (visible, i.r., microwave) by a system which it is capable of emitting. E.g. mercury vapour, when suitably excited in an electrical discharge, will emit light of wavelength λ 2536 Å. This same wavelength is very readily absorbed by Hg vapour, and the energy so absorbed may be utilized in producing photochemical reactions. The process is analogous to acoustic resonance – the coupling of two oscillators.

The term resonance is also applied in valency. The context of resonance in this sense is that if the valency electrons in a molecule are capable of several alternative arrangements which differ by only a small amount in energy and have no geometrical differences, then the actual arrangement will be a hybrid of these various alternatives. See MESOMERISM. The stabilization of such a system over the non-resonating forms is the resonance energy.

resonance ionization spectroscopy A technique allowing the selection, isolation and observation of as small an amount as a single distinct atom by using a laser to ionize that atom selectively.

resonance Raman spectroscopy See RAMAN EFFECT.

resorcinol (1,3-dihydroxybenzene, 1,3-benzenediol), $C_6H_6O_2$. Colourless needles; m.p. 110°C, b.p. 276°C.

Obtained by KOH fusion of many resins. 3-benzenedisulphonic acid with caustic soda, and NaOH fusion of 2- and 4-benzenedisulphonic acids.

It is extensively used in the preparation of dyestuffs. Combines with diazonium salts to form oxyazo colouring matters. Gives rise to fluorescein dyes on fusion with phthalic anhydride. Used for production of plasticizers, resins, adhesives.

respiratory pigments A group of pigments, of which hemoglobin is the most important member, which act as oxygen carriers in living organisms. A number of oxygen carriers, e.g. myoglobin cytochromes, are not in fact highly coloured.

retinene (retinol, vitamin-A aldehyde), $C_{20}H_{28}O$. Orange crystals; m.p. 61–64°C. Obtained by oxidation of vitamin A and β-carotene, and by various syntheses. Retinol generally refers to the all-*trans* isomer which has essentially the same bioactivity as the corresponding vitamin-A alcohol. Retinol is the intermediate in the conversion in mammals of β-carotene to vitamin A by oxidative cleavage of its central double bond. See also VITAMIN A and RHODOPSIN.

retinoids Diterpenoids, part of nuclear receptors, required for development and growth.

retrosynthesis Planning and execution of synthesis backwards through various precursors.

reverse osmosis Separation of solute from a solution by causing the solvent to flow through a membrane at pressures higher than the normal OSMOTIC PRESSURE.

Reverse osmosis is used for desalination of sea water, treatment of recycle water in chemical plants and separation of industrial wastes. The technique is applied to concentration dehydration of food products such as milk and fruit juices and also to the removal of alcohol from beverages. See ULTRAFILTRATION.

reversible process For a system in equilibrium a change (process) is said to occur reversibly when the direction of change can be made to occur at will in either direction by an infinitely small change in one of the factors controlling the position of equilibrium (pressure, temperature, concentration). For example in a reaction such as

$$N_2O_4 \rightleftharpoons 2NO_2$$

which is at equilibrium, a slight increase in pressure will shift the position of equilibrium to the N_2O_4 side. However, if the pressure is restored to its original value the equilibrium will immediately return to its former position. An irreversible change is one in which a slight change in conditions results in a complete change in equilibrium which cannot be reversed when the conditions are restored to their original values.

Rf (Rutherfordium) Element number 104.

R_f, R_s, R_r Measures of retention of solute in chromatography, defined as the ratio of the distance travelled by the solute to the distance travelled by the mobile phase.

RH-2485 Insecticide, hydrazine derivative.

Rh Rhodium.

rH The log of the reciprocal of the partial pressure of hydrogen in equilibrium with a system; thus rH = –log p. Used as a concept in redox reactions.

L-rhamnose (methylpentose), $C_6H_{12}O_5$. A constituent of many glycosides, which is widely distributed in plants, particularly in combination with flavanol derivatives and in saponins. Crystallizes with $1H_2O$, m.p. 94°C; anhydrous, m.p. 122°C. Resembles mannose in behaviour.

rhenates See RHENIUM OXIDES.

rhenium, Re. Metal, at.no. 75, at.mass 186.21, m.p. 3186°C, b.p. 5596°C, density ρ 21 220, crustal abundance 4×10^{-4} p.p.m., electronic configuration [Xe] $4f^{14}5d^56s^2$. Re is a rare metal extracted from flue dusts after roasting Mo or Cu ores. The metal is obtained by reduction of Re compounds with H_2; it has a hcp structure; is silvery grey in colour, dissolves in H_2O_2 and reacts with hot Cl_2, S or O_2. Used as an additive in W- and Mo-based alloys, in Re–W thermocouples and in catalysts. Used in electronic components and jewellery.

rhenium chemistry Rhenium is an element of Group 7. It shows oxidation states +7 to –1; the higher oxidation states are much less strongly oxidizing than those of Mn. There is no chemistry of the hydrated cations, organometallic derivatives are similar to those of Mn. Lower oxidation state compounds contain Re–

Re bonds. Extensive series of complexes are known, many containing Re=O or Re≡N groups. The $[ReH_9]^{2-}$ ion is formed by reduction of ReO_4^- with K in ethylenediamine and a series of phosphine hydride complexes, e.g. $ReH_7(PR_3)_2$, is known.

rhenium halides The only Re(VII) halide is ReF_7 (Re + F_2 3 atm. at 400°C), an orange solid, b.p. 74°C. The hexahalides are pale yellow ReF_6 (Re + F_2) m.p. 19°C, b.p. 48°C and dark green $ReCl_6$ (ReF_6 + BCl_3) m.p. 29°C. Yellow–green ReF_5 (ReF_6 + $W(CO)_6$) (disproportionates to ReF_6 and ReF_4 on heating) and $ReCl_5$ (heat on $ReCl_6$) have halogen bridges. $ReCl_4$ ($ReCl_5$ + C_2Cl_4) has Cl bridges. Red Re_3Cl_9 (heat on $ReCl_5$) has a triangle of three Re atoms. The bromides and iodides are similar: $ReBr_5$, $ReBr_4$, $ReBr_3$, ReI_4, $(ReI_3)_3$, ReI_2 and ReI are known. All of the higher halides are hydrolysed by water. Complex halides are known for Re(VI), $[ReF_7]^-$; (V), $[ReF_6]^-$ and $[ReOX_4]^-$; (IV), $[ReF_6]^{2-}$, $[ReCl_6]^{2-}$; (III), $[Re_3Cl_{12}]^{3-}$ and $[Re_2Cl_8]^{2-}$. The Re(III) species contain Re–Re bonds. Oxide halides, e.g. blue $ReOF_4$ and black $ReOF_3$, are known.

rhenium oxides Solid yellow Re_2O_7 (Re + O_2) contains linked ReO_6 and ReO_4 groups. It is a volatile material and the parent of rhenates(VII), salts. ReO_3 (Re_2O_7 + CO) has a classical ionic structure, rhenates(VI) are not known. ReO_2 is formed as a hydrate (ReO_4^- + Zn + HCl) which may be dehydrated. Colourless rhenate(VII) (perrhenate), ReO_4^-, is a mild oxidizing agent, $[ReO_6]^{5-}$ and $[ReO_5]^{3-}$ are formed in very basic solution or by fusion with peroxides.

rheopexy The rheological phenomenon whereby the gelation of some thixotropic sols is accelerated by gentle mechanical agitation. An example is gypsum–water paste which, if allowed to stand, solidifies in about 10 minutes, but when gently agitated solidifies in seconds.

rhodallin See PROPENYLTHIOUREA.

rhodanine (2-thiooxo-4-thiazoledine) Prepared sodium chloroethanoate on ammonium dithiocarbamate. Used in synthesis of phenylalanine. Rhodanine 6G used in dyelasers.

rhodium, Rh. Platinum group metal, at.no. 45, at.mass 102.91, m.p. 1964°C, b.p. 3695°C, density ρ 12 410, crustal abundance 2×10^{-4} p.p.m., electronic configuration [Kr] $4d^85s^1$. One of the platinum group metals. Found with other platinum metals and often in Ni–Cu deposits. After reduction to the mixed metals Rh is extracted from the aqua regia-insoluble material with fused $KHSO_4$, extracted with water and precipitated as hydrated Rh_2O_3 which dissolves in HCl to give $(NH_4)_3RhCl_6$. The metal can be obtained by reduction with H_2, it is ccp and is very inert. It is used as an alloying agent for Pt or Pd and as a catalyst (68%). ^{105}Rh used as a radioactive drug. World demand 2001, 574 000 oz.

rhodium chemistry Rhodium is a member of Group 9. It shows oxidation states from +6 (RhF_6) to –1 $[Rh(CO)_4]^-$. The +6, +5 and +4 states are strongly oxidizing and Rh(III) is the stablest state. The +2 state is uncommon but square planar complexes in the +1 state, often with phosphine ligands, are well established. The +1 complexes are active catalysts (WILKINSON'S CATALYST) and are readily transformed to octahedral Rh(III) compounds. Many Rh(0) compounds are carbonyl derivatives, e.g. $Rh_2(CO)_8$, $Rh_6(CO)_{16}$. Rh(III) is stable in aqueous solution as the $[Rh(H_2O)_6]^{3+}$ ion, N-bonded complexes, particularly ammines, and O-complexes, e.g. sulphates, are stable. Rh(III) is generally available as $RhCl_3$ hydrates or halide complexes.

rhodium halides The known fluorides are black RhF_6 (Rh + F_2), m.p. 70°C, red $(RhF_5)_4$ (Rh + F_2), purple-red RhF_4 ($RhBr_3$ + BrF_3) and RhF_3 (RhI_3 + F_2). Fluororhodates containing $[RhF_6]^-$, $[RhF_6]^{2-}$ ions are known. $RhCl_3$ (Rh + Cl_2 at 300°C) is the only anhydrous chloride although hydrates are known from aqueous solution. $[RhCl_6]^{2-}$, $[RhCl_6]^{3-}$ and some aquo species are formed in solution. $RhBr_3$ and RhI_3 are stable.

rhodium oxide Grey Rh_2O_3 is formed by heating Rh in air to 600°C. Hydrated forms are precipitated from Na_3RhCl_6 with alkali. Black RhO_2 with the rutile structure formed by heating Rh_2O_3 to 800°C in a high pressure of oxygen. Na_2RhO_3 and other rhodates(IV) as well as rhodates(III), $M^{II}Rh_2O_4$ and $MRhO_2$ are formed by solid-state reactions but rhodates are not stable in solution.

rhodopsin A major photosensitive pigment of the eye, composed of the protein opsin and retinol.

rhombohedral system An alternative representation of the TRIGONAL or HEXAGONAL crystal systems.

riboflavin (vitamin B_2), $C_{17}H_{20}N_4O_6$. Crystallizes in orange needles; m.p. 271°C (decomp.). Soluble in water. Riboflavin is part of the original vitamin B_2 complex. It occurs widely in nature, particularly in liver, milk and white of egg. It is the precursor of FLAVOPROTEINS.

See also FLAVIN-ADENINE DINUCLEOTIDE and FLAVIN MONONUCLEOTIDE.

ribonucleic acid Single stranded nucleic acid containing ribose and uracil. Involved in protein synthesis. See DICTIONARY OF BIOLOGY.

D-ribose, $C_3H_{10}O_5$. M.p. 87°C. The sugar of ribonucleic acid; present in all plant and animal cells. It has the furanose structure shown.

ribosomes The protein synthesizing material in living cells can act as ribozymes. Ribosomes 'translate' the code in a messenger RNA molecule to form a polypeptide chain. See DICTIONARY OF BIOLOGY.

ribozymes RNA species which act as catalysts. The protein synthesizing material in living cells can act as ribozymes.

ribulose, $C_5H_{10}O_5$. Ribulose-1,5-diphosphate

is a key intermediate in the photosynthetic fixation of atmospheric CO_2.

ricin A poisonous protein of the lectin class from the seeds of the castor bean.

ricinoleic acid, $C_{18}H_{34}O_3$. *Cis*-12-hydroxy-9-octadecenoic acid. M.p. 4–5°C, b.p. 226–228°C/10 mm. It comprises about 85% of the acids of castor oil.

The name ricinoleic acid is used in commerce for the mixture of fatty acids obtained by hydrolysing castor oil, and its salts are called ricinoleates.

Rietveld analysis Method of analysis of neutron diffraction and X-ray powder data and derivation of structure.

rifampicin Important anti-tuberculosis drug. Semi-synthetic.

rimsulfuron Herbicide, sulphonylurea derivative.

Rinmann's green, $ZnCo_2O_4$. A spinel formed when cobalt nitrate solution is placed on zinc oxide and the mixture heated to redness. The green colour forms a delicate test for Zn.

Rn Radon.

RNA See NUCLEIC ACIDS.

Rochelle salt, $C_4H_4O_6KNa,4H_2O$. Tartrate. See POTASSIUM SODIUM TARTRATE. From wine making.

rock crystal, SiO_2. See QUARTZ.

rock salt, NaCl. Natural SODIUM CHLORIDE. Occurs in extensive deposits.

rodinal (4-hydroxyaminobenzene) See AMINOPHENOLS.

röntgen The quantity of ionizing radiation. A lethal whole-body single dose of radiation is about 500 röntgens for man. Rarely used – expressed as coulombs per kg.

Röntgen rays X-rays.

rosin A solid resin which occurs in the oil from pine trees and is also obtained from TALL OIL. Mainly composed of monocarboxylic acids related to abietic acids. Used in the preparation of various compounds (including esters) and in lacquers, plasticizers, flotation agents, adhesives, paper size.

rotamers Isomers formed by restricted rotation.

rotational spectrum When a molecule absorbs energy it may increase its total rotational energy as a whole (see BAND SPECTRUM). Conversely, the transition from a state of higher to one of lower rotational energy corresponds to the emission of energy, which may be of radiation of frequency v; the energy E emitted being given by $E = hv$. The lines in the molecular spectrum corresponding to such transitions are termed rotational bands, which are characteristic of the molecule. Taken collectively the rotational bands comprise the rotational spectrum of the molecule. Rotational transitions are associated with lower energies than vibrational or electronic transitions, and hence the frequency of the emitted radiation is small. Rotational spectra are associated with the far i.r. and microwave regions of the electromagnetic spectrum. Radiation can be induced by appropriate lasers and can cause dissociation.

rotatory power, specific The specific rotatory power of a pure liquid is given by

$$[\alpha]^T_D = (\alpha/ld)$$

where $[\alpha]^T_D$ is the specific rotatory power at temperature T for the sodium D-line; α is the rotation of the plane of polarization produced by a column of liquid of length l (decimetres) and density d. For solutions the corresponding equation is

$$[\alpha]^T_D = 100\alpha/lc$$

where c is the solute concentration (g ml^{-1}).

The molecular rotation is given by

$$[M]^T_D = M[\alpha]/100$$

where M is the molecular weight. See CIRCULAR DICHROISM.

rotaxanes Compounds with rings circling a linear molecule.

rotenone Insecticide, acaricide, chromanone derivative, derris.

rouge A finely divided form of iron(III) oxide, Fe_2O_3, used for polishing purposes and in cosmetics on account of its smooth texture and freedom from gritty particles.

royal jelly The sole nutrient of queen bee larvae.

R$_s$ See R$_f$.

R,S convention See OPTICAL ACTIVITY.

Ru Ruthenium.

Ru15525 Insecticide, pyrethroid.

rubber A high mol.wt. natural or synthetic polymer which exhibits elasticity at room temperature. Natural rubber is obtained almost exclusively from the tree *Hevea Braziliensis* as a latex containing 30–36% rubber exuded from grooves cut into the bark. The latex is strained, diluted with water, coagulated with methanoic or ethanoic acids to a solid rubber. For use the rubber is VULCANIZED, which introduces a degree of cross-linking and wear and tensile strength are increased by adding a reinforcing filler (generally carbon black) in the process of compounding. Natural rubber (25% use) is z-polyisoprene. Synthetic rubbers are synthetic ISOPRENE POLYMERS, BUTYL RUBBERS, ethene–propene copolymers, various vinyl polymers, styrene–butadiene, butadiene polymers, neoprene. See also ELASTOMERS. U.S. production of synthetic rubber 1999 2.4 million tonnes.

rubber conversion products Products made from rubber. See chlorinated rubber, cyclized rubber, oxidized rubber, hydrochlorinated rubber.

rubber hydrochloride See HYDROCHLORINATED RUBBER.

rubber oxidation products See OXIDIZED RUBBER.

rubber (synthetic) See ISOPRENE POLYMERS, RUBBER, BUNA RUBBERS, NEOPRENE. Synthetic rubbers are generally more expensive than natural rubber but may be oil and chemical resistant.

rubbone See OXIDIZED RUBBER.

rubidium, Rb. Alkali metal, at.no. 37, at.mass 85.468, m.p. 39.30°C, b.p. 688°C, density ρ 1532, crustal abundance 90 p.p.m., electronic configuration [Kr] $5s^1$. Occurs in traces in salt deposits and also in the lithium mineral lepidolite. The metal, bcc, is formed from RbCl and Ca. It is silver-white and very reactive. ^{87}Rb is a weak β-emitter. The metal is used as a getter and in photocells. It may find use in ion-propulsion rocket motors. Rubidium salts used in zeolites and photoelectric cells.

rubidium chemistry Rubidium is a Group 1 element. It forms a single series of +1 compounds ($Rb^+ \rightarrow$ Rb -2.92 V), mainly ionic although some complexes with, e.g. crown ethers, are known. A series of lower oxides containing metal cluster cations are formed by burning Rb in a deficiency of oxygen.

ruby Impure α-Al_2O_3 containing some chromium. Occurs as gemstones and made by fusion of Al_2O_3 and traces of Cr_2O_3. Used as a gemstone and in lasers.

Russell–Saunders coupling (LS coupling) A system of arriving at the different energy levels associated with unfilled shells, in which electron spins are assumed to couple only with electron spins, and orbital momenta to couple only with orbital momenta. Used to describe the ions of the first row of the transition elements. The heavier elements require the use of jj-coupling in which spin-orbital interaction is assumed.

rusting The formation of a loose flaky layer of hydrated iron(III) oxide on the surface of iron or steel. The reaction proceeds in a moist atmosphere via Fe(II) ions; it is accelerated by atmospheric SO_2 or H_2SO_4 (cities) or Cl⁻ (coast). It is prevented by plating, painting or alloying (e.g. 12–18% Cr gives a protective oxide layer). Rusting is negligible in hot, dry areas as the critical humidity is too low.

ruthenates See RUTHENIUM OXIDES.

ruthenium, Ru. Platinum metal, at.no. 44, at.mass 101.07, m.p. 2334°C, b.p. 4150°C, density ρ 12 410, crustal abundance 0.001 p.p.m., electronic configuration [Kr] $4d^7 5s^1$. Occurs native as an alloy with the other platinum metals (osmiridium) and also in some sulphide ores. The elements are separated by electrolysis followed by chemical separation, ion exchange and solvent extraction to give pure Ru. The metal is obtained from its compounds by H_2 reduction and is grey–white in colour, hcp. The metal is unattacked by most acids, O_2 and F_2 react at high temperatures and the metal dissolves in fused Na_2CO_3, $KClO_3$, etc. Ru compounds are used in catalysis, hydrogenation and, e.g. for ammonia production. Has medicinal uses. Used in anode plating in chlorine production and the metal is used in alloys and in electrical applications. World demand 2001 349 000 oz.

ruthenium chemistry Ruthenium is an element of Group 8 which shows oxidation states from +8 to –2. The +8 state is more strongly oxidizing than for Os; the principal lower oxidation states are +3 and +2. There is some chemistry of the aquo-ions $[Ru(H_2O)_6]^{n+}$ ($n = 3,2$) although most anions complex with the metal. N-bonded complexes are particularly stable and many ammines (Ru(II) and Ru(III)), dinitrogen and nitrosyl complexes are known. Phosphine complexes are stable with lower oxidation states and chloride– and hydride–phosphine complexes are good hydrogenation catalysts. Many carbonyl derivatives, e.g. $[RuCl_2(CO)_2(PEt_3)_2]$, and $[Ru(CO)_5]$, are known and these and other low oxidation state compounds form organometallic derivatives similar to those of Fe.

ruthenium halides Dark brown RuF_6 (Ru + F_2), m.p. 54°C, is octahedral molecular. Dark green RuF_5 (Ru + F_2), m.p. 86°C, is tetrameric with fluorine bridges. Pink RuF_4 ($RuF_5 + I_2$) and brown RuF_3 ($RuF_5 + I_2$ at high temperatures) are the known fluorides. $RuCl_3$ (Ru + Cl_2) is the only established chloride although many aquochloro complexes of Ru(IV), Ru(III) and Ru(II) are known. $RuBr_3$ and RuI_3 are also known. Complex fluorides of Ru(V) and Ru(IV), e.g. $KRuF_6$ and K_2RuF_6, are formed by fluorination in the absence of water. Some oxide fluorides, e.g. $RuOF_4$, are formed by hydrolysis.

ruthenium oxides Yellow RuO_4 (RuO_2 + KIO_4), m.p. 25°C, b.p. 40°C, is fairly unstable and a strong oxidizing agent cleanly converting secondary alcohols to ketones. It decomposes to blue RuO_2 (Ru + O_2 at 1000°C). Green ruthenates(VII), RuO_4^- (Ru fused in KNO_3 + KOH); decompose in water to orange ruthenates(VI), RuO_4^{2-} and oxygen.

Rutherford back scattering (RBS) Technique for determining atomic profiles by depth in solids which are bombarded with an ion beam and elastic scattered particles are examined at a fixed backward angle.

rutherfordium, Rf.   Element. At.no. 104. Formerly designated kurchatovium. The first element of the fourth transition series.

rutile   Mineral TiO_2, a reddish-brown material used as a source of Ti and in ceramics. The tetragonal structure has each Ti octahedrally co-ordinated by O and each O co-ordinated by three Ti in an approximate plane. This structure is adopted by many oxides and fluorides, e.g. PbO_2 and MnO_2.

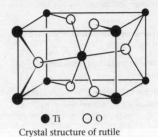

● Ti   ○ O

Crystal structure of rutile

rutin, $C_{27}H_{30}O_{16} \cdot 3H_2O$.   The 3-rhamnoglucoside of 5,7,3′,4′-tetrahydroxyflavanol.

Rydberg constant   See BALMER SERIES. 16 973 731.5 m^{-1}.

Rydberg states   Diffuse electronic states.

S Sulphur.

S Designation of CONFIGURATION. Siemens.

S 421 Insecticide synergist, chloroalkyl ether.

sabinene, $C_{10}H_{16}$. Colourless oil; b.p. 165°C.

A dicyclic monoterpene, in many essential oils.

sabinol, $C_{10}H_{16}O$. The 2-hydroxy derivative of sabinene. B.p. 208°C.

saccharates Normally the salts of saccharic acid, but the term is also used in the sugar industry for the alkali and alkaline earth compounds of sucrose: strictly these should be called sucrates.

saccharic acid The saccharic acids have the formula HO(O)C·[CHOH]$_4$·C(O)OH, and are produced by the oxidation of the corresponding aldoses with nitric acid. Isomeric forms known.

By saccharic acid is usually meant D-glucosaccharic acid, D-GLUCONIC ACID m.p. 125–126°C, obtained by the oxidation of glucose or starch.

saccharides Oligosaccharides, polymeric sugars linked through oxygen bridges.

saccharin (1,2-benzisothiazol-3(2H)-one), $C_7H_5NO_3S$. White crystals; m.p. 224°C (decomp.). It is made by the oxidation of toluene-o-sulphonamide with alkaline permanganate. Used extensively as a sweetening agent, usually in the form of the sodium salt.

Sackur–Tetrode equation Equation for the molar entropy of an ideal monatomic gas.

safflower oil An important oleic–linoleic acid oil from safflower seed used in DRYING OILS, alkyd resins, paints and as edible oil.

safrole (3,4-methylenedioxyallylbenzene), $C_{10}H_{10}O_2$. Colourless liquid; m.p. 11°C, b.p. 232°C. Obtained from oil of camphor, and is the chief constituent of other essential oils.

sal ammoniac, NH_4Cl. See AMMONIUM CHLORIDE.

salicin, $C_{13}H_{18}O_7$. The β-D-glucoside of 2-hydroxybenzyl alcohol, $C_6H_{11}O_5$·O·C_6H_4·CH_2OH. Colourless, bitter crystals, m.p. 201°C; soluble in water and alcohol. Occurs in the leaves, bark and twigs of species of willow and poplar. Used in synthesis and as analgesic.

salicyl alcohol (saligenin, 2-hydroxybenzyl alcohol), $C_7H_8O_2$. White crystals; m.p. 87°C. It is obtained by the reduction of salicylaldehyde or by hydrolysis of salicin. Local anaesthetic.

salicylaldehyde (2-hydroxybenzaldehyde), $C_7H_6O_2$. Oily liquid of aromatic odour; b.p. 196°C. It is prepared by the action of chloroform and caustic potash on phenol or by the oxidation of salicin. It is easily reduced to salicyl alcohol or oxidized to salicylic acid. Used in perfumery.

salicylamide, $C_7H_7NO_2$. 2-hydroxybenzamide. White crystalline powder, m.p. 140°C. It has antipyretic, analgesic and antirheumatic properties.

salicylic acid (2-hydroxybenzoic acid), $C_7H_6O_3$. Colourless needles; m.p. 159°C.

Occurs naturally as METHYL SALICYLATE, methyl-ester-2-hydroxybenzoate, oil of wintergreen, from which it can be obtained by hydrolysis with alcoholic KOH. It is manufactured by heating NaOPh with carbon dioxide under pressure and acidifying the sodium salicylate formed. Used in pharmaceuticals, principally aspirin, in the manufacture of dyes, liniments and rust-releasing fluids.

saligenin See SALICYL ALCOHOL.

salinity Measure of salt content – particularly of sodium chloride in irrigation water. Many crop plants cannot thrive in saline conditions.

salol See PHENYL SALICYLATE. It is used as a coating for pills, lacquers, polymers.

salt Commonly sodium chloride. More generally a salt is the substance produced by interaction of equivalent quantities of acid and base. To consider aquo-acids: if hydrogen replacement is complete the salt is said to be normal (e.g. Na_2SO_4), and if only partial, an acid salt is formed (e.g. $NaHSO_4$). In general, only the salts of strong acids with strong bases are stable in solution. If either or both constituents are weak, hydrolysis occurs and the solution becomes acidic or basic.

salt bridge In many electrochemical cells, electrical conductivity may be achieved whilst keeping the two solutions in the cell apart by means of a porous plug. The resistance of the solution in this case is very high. In practice the liquid-junction potential may be greatly reduced by means of a salt bridge, which consists of a tube containing a saturated solution of a salt (usually KCl) in an organic gel, e.g. agar. The ends of the salt bridge dip into the two half cells and the current is carried by the K^+ and Cl^- ions which have approximately the same ionic velocities.

salt hydrates CLATHRATES formed by frameworks of hydrogen-bonded water and anions with cations occupying holes in the lattice. Examples have very high water content, such as $(Pr^n_3S)F.20H_2O$.

salting out A term used for both colloidal and non-colloidal solutions. In the context of colloids, salting out refers to the coagulation of hydrophilic colloids, e.g. gelatin or soaps, by the addition of large concentrations of electrolytes. This should not be confused with the coagulation of hydrophobic sols by small quantities of electrolytes. With non-colloidal solutions the term salting out is used to describe the reduction in solubility of a non-electrolyte by addition of a strong electrolyte.

salvarsan (arsphenamine,4,4´-(1,2-di-arsendiyl)-bis(2-aminophenol)dihydro-chloride) The first anti-syphilitic drug, one of the first specific drugs.

sal volatile Ammonia, ammonium carbonate, oil of lemon and nutmeg in an alcoholic solution used as a first-aid stimulant and also in dyspepsia medicines.

SAM Self-assembled monolayer.

samarium, Sm. Lanthanide metal, at.no. 62, at.mass 150.36, m.p. 1074°C, b.p. 1794°C, density ρ 7520, crustal abundance 7.9 p.p.m., electronic configuration [Xe] $4f^66s^2$. $SmCo_5$ is used in permanent magnets. The oxide is used in optical glasses and in control rods of nuclear reactors.

samarium compounds Samarium shows +3 and +2 oxidation states. Sm(III) compounds are typical LANTHANIDE COMPOUNDS,

$$Sm^{3+} (f^5 \text{ yellow}) \rightarrow Sm (-2.41 \text{ volts in acid})$$

SmX_2 (SmX_3 + Sm) are red–brown in colour. SmI_2 can be obtained as a blue solution in tetrahydrofuran. It is a powerful reducing and complexing agent. Used in many chemical reactions. SmO is formed similarly and Sm^{2+} salts are rapidly oxidized by water.

sampling The process of obtaining a small quantity of material, e.g. for analysis, truly representative of the whole. Samples can be withdrawn by hand or using a mechanical sampler. Taking a sample of a sample is known as secondary sampling (see CONING AND QUARTERING). Heterogeneous mixtures must be handled statistically, homogeneous samples (gases or liquids) are handled simply.

sand An accumulation of minerals, mainly SiO_2, resulting from rock disintegration.

Sandmeyer's reaction The replacement of a diazonium group by a halogen or pseudohalogen atom or group. It allows the conversion of an aromatic primary amine into the corresponding halogen compound. The amine is diazotized, and the diazonium salt solution is treated with CuCl, etc., causing the evolution of nitrogen and isolation of the halobenzene. Chloride, bromide and cyanide react easily; potassium iodide is sufficient for preparation of the iodo-compound. Aromatic fluorides are prepared after isolation and thermal decomposition of the diazonium fluoroborate salt. See also GATTERMANN'S REACTION.

sandwich compounds A trivial name, originally applied to the bis-π-cyclopentadienyl-metal compounds like ferrocene but, since used variously to cover all the compounds where a metal atom is disposed between two cyclocarbon residues such as cyclopentadienyl, benzene, cyclohexadienyl, cyclobutadiene, tropylium, cyclo-octatetraene, etc. Those compounds having only one such ring are correspondingly called half sandwich compounds, e.g. cyclopentadienylmanganese tricarbonyl, $C_5H_5Mn(CO)_3$. Triple-decker derivatives are also known. See ORGANOMETALLIC COMPOUNDS, METALLOCENES, FERROCENE.

sandwich plastics Reinforced plastics.

Sanger's reagent (DNFB, 1-fluoro-2,4-dinitrobenzene) Used for labelling terminal amino acids.

α-santalol (5-(2,3-dimethyl)tricyclic [2,2′0²,⁶]-hept-3-yl-2-methyl-2-pentan-l-ol) Sesquiterpene alcohol used in perfumes, soaps, detergents.

santene, C_9H_{14}, and **santenone (π-nor-camphor)**, $C_9H_{14}O$. Have related cyclic structures. Terpene-like molecules.

sapogenin glycosides Aglucones, found in many plants; fish poisons.

saponification The alkaline HYDROLYSIS of an ester to an alcohol and the alkali-metal salt of a carboxylic acid (a soap).

saponins Glycosides found in plants, having fish poisoning and anti-fungal properties.

sapphire As a gemstone a blue form of ALUMINIUM OXIDE (corundum) containing some metal impurities. Now prepared artificially by fusing Al_2O_3. Used as an inert material, e.g. for handling HF.

sarcosine (N-methylglycine, N-methyl-aminoethanoic acid), $C_3H_7NO_2$, $CH_3·NH·CH_2·CO_2H$. Colourless crystals; m.p. 213°C (decomp.). It is obtained on hydrolysing creatine with alkalis. Used in toothpaste.

sarin, $MeP(O)F-O-CH(CH_3)_2$. A nerve gas.

satin white A mixture of $CaSO_4$ and $Al(OH)_3$ ($Al_2(SO_4)_3$ or alum plus $Ca(OH)_2$). Used as a pigment and for surfacing paper.

saturated compound A substance in which the atoms are linked by single bonds, and which has no double or triple bonds, e.g. ethane.

However, the term 'saturated' is often applied to compounds containing double or triple bonds which do not easily undergo addition reactions. Thus ethanoic acid is termed a saturated carboxylic acid and acetonitrile a saturated

nitrile, whereas a SCHIFF BASE is considered to be unsaturated.

sawhorse projections The FISCHER PROJECTION inadequately portrays the spatial relationship between ligands (atoms, groups) attached to adjacent atoms. The sawhorse projection attempts to clarify the relative location of the ligands.

E.g. the Fischer projection of an enantiomer of the simple sugar threose is shown **1** together with a sawhorse projection **2**, or **3** which is obtained by rotation of the rear group through 180°. The Newman projection **4** is an alternative method of representing the structure when viewed along the C_1–C_2 axis in **3**, carbon 2 being represented as the full circle.

Sb Antimony.

SBP spirits Special boiling point spirits.

SBR Styrene–butadiene rubber.

Sc Scandium.

scandium, Sc. Metal, at.no. 21, at.mass 44.956, m.p. 1541°C, b.p. 2836°C, density ρ 2889, crustal abundance 16 p.p.m., electronic configuration [Ar] $3d^14s^2$. Group 3 element. The principal Sc ore is thortveitite ($Sc_2Si_2O_7$) and monazite and other lanthanide ores contain some Sc. Separated from the lanthanides by ion exchange. The metal (ScF_3 + Ca) is fairly reactive and is used in small quantities to strengthen alloys. ScC hardens TiC. Sc_2O_3 is used in electronic components, ScI_3 is added to high-intensity electric light bulbs.

scandium compounds Scandium shows only the +3 oxidation state and its chemistry is much more similar to that of ALUMINIUM than that of the LANTHANIDES

$$Sc^{3+} \rightarrow Sc \ (-1.88 \text{ volts in acid})$$

The colourless Sc^{3+} is much smaller than the lanthanide^{3+} ions and the maximum co-ordination number is 6. Halides, ScX_3, are formed by interaction of the elements, by precipitation from aqueous solution (ScF_3) or by removal of water from, e.g., $ScCl_3.6H_2O$ with, e.g. $SOCl_2$. Compounds such as Sc_5Cl_8, Sc_7Cl_{10} ($ScX_3 + Sc$) contain metal–metal bonds. Complex fluorides of the cryolite type M_3ScF_6 are formed from solution or melts. $ScCl_3$ is hydrolysed by water to a hydrous oxide. $ScO(OH)$ and mixed oxides, e.g. $LiScO_2$ ($Li_2O + Sc_2O_3$), are also known; Sc_2O_3 is very similar to Al_2O_3. Oxyacid salts include the sulphate and the nitrate.

scanning Auger electron microscopy
See AUGER SPECTROSCOPY.

scanning electron microscopy (SEM)
A form of electron microscopy using high-energy electrons. Used particularly to investigate surfaces and detailed structures.

scanning tunnelling microscopy
Used as probe to investigate surfaces down to atomic levels. Electrons tunnel between the probe and the surface.

scavengers
When LEAD TETRAETHYL or TETRAMETHYL is added to gasoline as ANTI-KNOCK ADDITIVES, lead scavengers must be used to ensure that lead oxide is not deposited in combustion chambers. Mixtures of dibromoethane and dichloroethane are added with the lead alkyls in order to produce volatile lead halides which are removed with the exhaust gases. Use being phased out world-wide.

Schäffer's acid (2-hydroxy-7-naphthalene sulphonic acid), $C_{10}H_8O_4S$.
Obtained by sulphonating 2-naphthol with a small amount of sulphuric acid at a higher temperature than is used for the preparation of crocein acid. A valuable dyestuff intermediate.

scheelite
A tungsten mineral $CaWO_4$. An important structure type.

Schiff's bases (anils, *N*-arylimides),
Ar·N=CR$_2$. Prepared by reaction of aromatic amines with aliphatic or aromatic aldehydes and ketones. They are crystalline, weakly basic compounds which give hydrochlorides in non-aqueous solvents. With dilute aqueous acids the parent amine and carbonyl compounds are regenerated. Reduction with sodium and alcohol gives a secondary base. Secondary bases can also be obtained from the reactions with Grignard reagents and with alkyl halides.

Schiff's reagent
A solution of rosaniline in water decolorized with sulphurous acid. Aliphatic aldehydes and aldose sugars give a magenta colour with this reagent; with aromatic aldehydes and aliphatic ketones the colour develops more slowly; aromatic ketones do not react.

Schomaker–Stevenson equation
An empirical equation $r_{A-B} = r_A + r_B - 0.09 (X_A - X_B)$ relating the bond length r_{A-B} to the individual radii r_A and r_B and the electronegativities X_A and X_B of the two atoms concerned in the bond.

schönite, K_2SO_4, $MgSO_4,6H_2O$.
A mineral obtained from the Stassfurt salt deposits and used as a source of potassium sulphate. A series of schönites, analogous to the alums, is formed containing dipositive ions such as Zn^{2+}, Co^{2+}, Ni^{2+}, Fe^{2+}, Cu^{2+}, Mn^{2+}, V^{2+}, in place of magnesium.

Schottky defect
See DEFECT STRUCTURES.

Schrödinger wave equation
The fundamental equation of wave mechanics which relates energy to field.

$$\frac{-\hbar}{2m} \frac{d^2\psi}{dx^2} + V(x)\Psi = E\Psi$$

where $V(x)$ is the potential energy at points $x = h/2\pi$ and E is the total energy.

Schweinfürter green
See EMERALD GREEN.

Schweizer's reagent
The dark blue solution obtained by dissolving $Cu(OH)_2$ in concentrated ammonia solution. Used as a solvent for cellulose; the cellulose is precipitated on acidification. Used in the cuprammonium process for the manufacture of rayon.

scintillation counting
The detection and quantitative estimation of radioactivity by the scintillations produced when the ionizing radiation interacts with a scintillating medium. See DICTIONARY OF PHYSICS.

scleroproteins
See PROTEINS.

(−)scopolamine (hyoscine), $C_{17}H_{21}NO_4$.

O·C(O)·CHPh·CH$_2$OH

Tropane alkaloid obtained from plants of the *Solanaceae* family. Intensely poisonous, its action resembles that of atropine. Sedative in small doses.

screened indicator See COLOUR INDICATORS.

screening Separation of solid material into fractions containing particles whose sizes lie within a certain range, by the use of sieves.

screens Used in SCREENING.

scrubbers Equipment to remove impurity from a gas by washing it with a liquid.

Se Selenium.

seaborgium, Sg. Artificial element at.no. 106 of Group 6. Chemistry appears similar to Mo and W.

sea water An aqueous system of pH 8.0–8.4 (buffered by carbonate, Ca^{2+}, etc.). The major constituents are (p.p.m.) Cl (19 000); Na^+ (10 500); SO_4^{2-} (2650); Mg^{2+} (1270); Ca^{2+} (400); K^+ (380); HCO_3^- (140); Br^- (65); H_3BO_3 (26); Sr^{2+} (8). Used for production of bromine and magnesium. Potential metal sources (Mn, etc.) are available on the seabed.

seaweed colloids The polysaccharide materials agar, algin, CARAGEENIN, prepared by extraction of algae (particularly seaweeds). Used in foodstuffs.

sebacic acid See DECANEDIOIC ACID.

SEC Size exclusion chromatography.

secondary carbon A carbon with two attached carbons.

secondary radiation Radiation produced by the absorption of some other, more energetic, radiation. Thus, when X-RAYS strike a body it emits other X-rays, termed secondary rays, or radiations, characteristic of the atom which acts as the emitter. This phenomenon is also observed with electrons and is then termed secondary emission.

second law of thermodynamics See THERMODYNAMICS, SECOND LAW OF.

second-order spectra See NUCLEAR MAGNETIC RESONANCE.

sedimentation Gravitational settling of solid particles suspended in a liquid so that the original suspension is divided into a reasonably clear effluent and a sludge with an increased solid content. Sedimentation can be made more rapid and more effective by use of a centri-

fuge or ultracentrifuge. See CLARIFICATION, THICKENING, SEGREGATION.

seed crystals Crystals which are added to a supersaturated solution to facilitate crystallization of the whole.

segregation The tendency for solid particles, either dry or in suspension in a liquid, to arrange themselves in different layers according to size.

seignette salt, $C_4H_4O_6KNa.4H_2O$. Potassium sodium tartrate.

selection rules Rules derived from group theory which denote, e.g. whether a particular vibration is i.r. or Raman active, or a particular transition between electronic energy levels is allowed or forbidden.

selective electrodes See ION-SELECTIVE ELECTRODES.

selenates Normally salts containing $[SeO_4]^{2-}$, $[HSeO_4]^-$, i.e. selenates(VI). Prepared from H_2SeO_4 or by electrolytic or HNO_3 oxidation of selenates(IV).

Selenates(VI) resemble sulphates but can be reduced more easily ($E°$ to SeO_3^{2-} +1.15 volts). ALUMS are formed. Pyroselenates, $[O_3SeOSeO_3]^{2-}$ are known.

Selenates(IV) contain the $[SeO_3]^{2-}$ and related species. They are hydrolysed in solution and give complexes.

Selenates are poisonous.

selenic acid, H_2SeO_4. See SELENIUM OXYACIDS.

selenides Binary compounds of selenium with other elements. Generally prepared by reaction of the elements but some result from H_2Se on the element or a derivative. Similar to sulphides but less stable, more easily oxidized and more easily hydrolysed to H_2Se.

selenious acid, H_2SeO_3. See SELENIUM OXYACIDS.

selenite See GYPSUM.

selenites Salts containing $[SeO_3]^{2-}$ ions. See SELENATES(IV).

selenium, Se. Non-metal, at. no. 34, at. mass 78.96, m.p. 221°C, b.p. 685°C, density ρ 4790, crustal abundance 0.05 p.p.m., electronic configuration [Ar] $3d^{10}4s^24p^4$. Occurs as selenide impurities in sulphide ores and as clausthalite (PbSe) and crookesite ((CuTlAg)$_2$Se). Extracted from the flue dusts where sulphide ores are used, dissolved (KCN solution) and precipitated with HCl, and also from the anion sludge from the purification of copper. Stable grey Se is metallic in appearance and contains Se_n chains with some cross-linking. It is a photoconductor (xerography, photo-electric cells). The red form is precipitated from solutions with SO_2 and contains Se_8 units – it is soluble in CS_2. Se reacts with many elements to give, e.g. oxides or halides. Additional uses are as a steel additive, in glass, photography, vulcanization and for dehydrogenation in organic synthesis. Selenium is an essential trace element but is toxic in excess.

selenium, organic derivatives These are similar to the simpler sulphur derivatives. Much of the chemistry involves selenium(II) derivatives although some selenium(IV) organic derivatives are known.

selenium bromides Selenium monobromide, Se_2Br_2. Deep red liquid; b.p. 227°C (decomp.) ($SeBr_4$ or Br_2 on Se). Selenium tetrabromide, $SeBr_4$. Yellow solid readily loses Br_2 to give Se_2Br_2 (Se plus excess Br_2). $[SeBr_6]^{2-}$ formed with alkali bromides. Selenium dibromide, $SeBr_2$, is only stable in the vapour state.

selenium chemistry Selenium is a member of Group 16. It is a non-metal and shows little true cation chemistry (Se_8^{2+}, etc. cations are formed in non-aqueous solvents). The normal oxidation states are +6 (octahedral in SeF_6, tetrahedral in $(SeO_4)^{2-}$), +4 (one lone-pair), +2 (two lone pairs) and –2 (selenides, H_2Se, etc.). Selenium(IV) compounds particularly act as acceptors. Selenium(II) compounds are good donors.

selenium chlorides
Selenium monochloride, Se_2Cl_2. M.p. –85°C, b.p. 130°C (decomp.). Reddish liquid, a good chlorinating agent (Cl_2 plus Se).
Selenium tetrachloride, $SeCl_4$. M.p. 305°C. Yellowish solid (Cl_2 plus Se_2Cl_2). Forms hexachloroselenates containing $(SeCl_6)^{2-}$ in conc. HCl solution.
Selenium dichloride, $SeCl_2$. Only stable in the gas phase.

selenium fluorides
Selenium tetrafluoride, SeF_4. M.p. –14°C, b.p. 106°C. Colourless liquid (F_2 on sublimed Se). Hydrolysed by water.
Selenium hexafluoride, SeF_6. M.p. –39°C, sublimes –47°C. Formed (with Se_2F_{10}) from Se and F_2. Chemically fairly inert.
SeO_2F_2 is also known ($BaSeO_4$ plus HSO_3F).

selenium oxides
Selenium dioxide, SeO_2. Sublimes 315°C. White solid (Se plus O_2) with a polymeric structure.

Forms H_2SeO_3 with water. Used as an oxidizing agent in organic chemistry (α-hydrogens to C=O groups). $SSeO_3$ is formed from Se and molten SO_3.
Selenium trioxide, SeO_3. M.p. 120°C. Formed Se vapour plus O_2 in electric discharge. It is white and gives H_2SeO_4 with water.

selenium oxyacids The system is very simple by comparison with sulphur.
Selenic acid, H_2SeO_4. Formed by oxidation of selenates(IV) (Cl_2, MnO_4^-). A strong acid similar to H_2SO_4 but loses O_2 on heating.
Selenious acid, H_2SeO_3. Formed water on SeO_2. More stable than H_2SO_3, oxidized to H_2SeO_4, reduced to Se by SO_2. Gives selenates(IV) with oxides, hydroxides or carbonates.
Selenosulphuric acid, H_2SeSO_3. An analogue of thiosulphuric acid; the free acid does not exist (Se plus $(SO_3)^{2-}$). Selenosulphates readily decompose.

selenocyanates Derivatives containing the SeCN grouping (KSeCN from KCN plus Se).

selenonium ions Three-co-ordinate selenium-containing cations, e.g. $[Me_3Se]^+$.

self-ionization The process of partial dissociation into cations and anions envisaged for self-ionizing solvents. E.g.

$$2H_2O \rightleftharpoons [H_3O]^+ + OH^-$$
$$2NH_3 \rightleftharpoons [NH_4]^+ + NH_2^-$$

Ionic species containing the characteristic cation – e.g. NH_4Cl – are acids; species containing the characteristic anion are bases, e.g. KNH_2.

SEM Scanning electron microscope.

semicarbazide (aminourea), $NH_2·CO·NH·NH_2$, CH_5N_3O. Colourless crystalline substance; m.p. 96°C. Prepared by the electrolytic reduction of nitrourea. Forms crystalline salts with acids. Reacts with aldehydes and ketones to give semicarbazones. Formerly used for the isolation and identification of aldehydes and ketones.

semicarbazones Organic compounds containing the group >C:N·NH·CONH2. See SEMICARBAZIDE.

semi-conductors Imperfections or impurities in what are normally insulators may give rise to a temperature-dependent conductivity (metallic conductivity decreases with rise in temperature) arising because the highest occupied energy level is very close to an unoccupied level, small bandgap. If foreign atoms are implanted into a pure material the dopant may take electrons from the filled band allowing the remaining electrons to become mobile (*p*-type semi-conductor). If the dopant contains more electrons, the excess are in the otherwise empty bands and *n*-type semi-conductivity results. Semi-conductors, particularly those derived from Si and compounds such as GaAs, are of great importance in electronics and are used in, e.g. transistors, thermistors, solid rectifiers, microprocessors and lasers.

semiochemicals Volatile chemicals released by plants to indicate the presence of herbivores.

semi-permeable membrane See MEMBRANE.

semi-polar bond A semi-obsolete name for a CO-ORDINATE BOND.

semiquinones Stable radicals formed from free radicals and hydroquinones. Basis of food preservatives.

Semtex An explosive, a blend of PETN and RDX.

sensitizing dyes See SPECTRAL SENSITIZATION.

sensors Devices for measurement or identification of specific species, e.g. an oxygen sensor measures the amount of oxygen in food or water.

Sephadex See DEXTRAN.

sequence rules In order to give an organic compound an unambiguous name (see ISOMERISM) or three-dimensional configuration (see OPTICAL ACTIVITY) it becomes necessary to designate the order of groups or atoms attached to a particular atom. Several sequence rules have been introduced but those most widely accepted attach an agreed order of precedence to a ligand thus: **1** ligands are arranged in order of decreasing atomic number. **2** When atoms are of similar atomic number (isotopes) they are arranged in decreasing mass number. **3** When necessary a lone pair of electrons is considered a substituent and given a priority lower than [1]H. **4** When similar atoms are attached to a particular atom, priority order is determined by the nature of the atoms further away. **5** Multiple linkages are treated formalistically as multiple single bonds, e.g. C=O becomes

and C≡N becomes

Thus in order of decreasing priority:
Atoms: I, Br, Cl, F, O, N, C, H, lone pair.
Isotopes: ^{2}H, ^{1}H.
Groups: $-C(CH_3)_3, -CH(CH_3)_2, -CH_2CH_3, CH_3$.
$-C(CH_3)_3, -CH(CH_3)CH_2CH_3, -CH_2CH(CH_3)_2,$
$-(CH_2)_3CH_3$.
$-COOCH_3, -COOH, -CONH_2, -CHO$.
$-C{\equiv}N, -C_6H_5, -C{\equiv}CH, -CH{=}CH_2$.

sequestering agent (complexones) Compounds which are very effective in forming complexes with metal cations (and thus with salts), thus preventing them from acting as simple hydrated cations. Ethylenediaminetetraacetic acid (EDTA) and related derivatives form complexes with most M^{2+} and M^{3+} species. Gluconic acid and other hydroxy acids act similarly. Sequestering agents are used extensively, e.g. for preventing deleterious catalytic effects of traces of Fe or Cu and in analysis.

series, spectroscopic A group of lines the frequencies of which are simply related to each other and may be expressed by a simple mathematical formula (e.g. the BALMER SERIES).

serine (2-amino-3-hydroxypropanoic acid), $C_3H_7NO_3$, $CH_2OH \cdot CHNH_2 \cdot CO_2H$. Colourless crystals; m.p. 228°C (decomp.). Aminoacid.

SERS (surface-enhanced Raman scattering) Technique used for surface studies.

sesame oil Contains oleic, palmitic, myristic, linoleic derivatives. Used in margarine and cosmetics.

sesqui[oxide] A semi-obsolete term to denote a ratio of 3 [oxygen] atoms to 2 of the cationic element (e.g. Fe_2O_3 – now denoted iron(III) oxide).

sesquiterpenes See TERPENES.

sessile dislocation A dislocation in a crystal lattice which cannot glide since the Burger's vector is not in the plane of slip of the dislocation. See DISLOCATION.

sethoxydim Herbicide, oxime derivative.

sewage treatment Waste water from industrial or domestic sources is initially filtered to remove solids and colloidal materials are then removed with micro-organisms which use the waste as food. Methane and other gases are evolved during anaerobic sludge digestion and

SEXAFS (surface extended X-ray absorption fine-structure spectrosco- | 358

can be used as fuels. High-protein animal-feed supplements are also produced by this process.

SEXAFS (surface extended X-ray absorption fine-structure spectroscopy) Surface investigation using synchrotron radiation.

sextet A series of six closely spaced transitions in a spectrum.

Sg Seaborgium. Element 106.

shale oil Oil produced by heating oil shales in continuous retorts at temperatures up to 950 K.

Although estimated world reserves of shale are very high and shales contain other potentially useful materials, oil production has, in the past, been too expensive for large-scale exploitation.

shear structures Crystal structures in which it appears that a regular array of, e.g. corner-linked octahedra, is interrupted along a line by, e.g. edge-sharing octahedra. Found in heavy metal oxides, e.g. tungsten oxides.

shellac A resinous secretion of the insect *Tachardia lacca* used for its content of resinous matter. Shellac yields aleuritic acid, $C_{13}H_{26}O_4$, m.p. 101°C, on extraction with alkali. It is used for the preparation of French polish, as a resin binder for mica-based insulating materials and in sealing waxes, varnishes, buttons and leather treatment.

sherardizing (vapour galvanizing) Coating metal articles (iron) by heating with Zn dust in a closed vessel somewhat below the m.p. of Zn.

shielding N.m.r. effect that has signals at higher field (upfield). Compare DESHIELDING.

shift conversion (shift reaction) An important reaction in which carbon monoxide reacts with steam to produce hydrogen and carbon dioxide

$$CO + H_2O \rightarrow CO_2 + H_2$$

Carbon monoxide and excess steam are normally passed over a cobalt catalyst at about 250–300°C, resulting in greater than 99% conversion of CO to CO_2. This conversion reaction is widely used in oil or solid fuel gasification processes for the production of TOWN GAS or SUBSTITUTE NATURAL GAS.

shift reagents Paramagnetic species which are magnetically anisotropic can cause large n.m.r. shifts in any species with which they form weak complexes. Stable paramagnetic species, particularly lanthanide diketonates, are added to specimens in order to simplify complex n.m.r. spectra.

shikimic acid, $C_7H_{10}O_5$. M.p. 190°C. A 3,4,5-trihydroxy-1-cyclohexene-1-carboxylic acid. Intermediate in pathway for the biogenesis of aromatic compounds in micro-organisms and plants. Important compounds arising by the shikimic acid route include many alkaloids, 4-aminobenzoic acid, phenylalanine, tyrosine, tryptophan and *p*-hydroxybenzoic acid.

Si Silicon.

sialic acid A generic name for *N*- and *O*-acyl derivatives of NEURAMINIC ACID, e.g. acetylneuraminic acid. Amino sugars containing 9 or more carbons, N and O derivatives of neuraminic acid. They are widely distributed in tissues and bacterial cell wall structures. Important in cell development.

sialons Si-Al-O-N systems used in machine tools.

side reactions Few reactions proceed quantitatively to one product. Generally a number of reactions occur. The predominant reaction is the main reaction; other reactions are termed side reactions.

siderophores Microbial ligands principally for Fe(III).

siemens (S) The unit of conductivity, siemens metre^{-1}.

Siemens's process A method of steel manufacture by addition of scrap metal, iron oxides and ferromanganese to molten pig (cast) iron.

sienna Earth containing hydrated Fe_2O_3 used as stain and coloured filler for wood and other materials.

Sievert (Sv) Dose equivalent J kg^{-1} = m^2 s^{-2}.

sigma bond (σ-bonds) Covalent bond

containing two electrons with the major overlap between the orbitals forming the bond being in one region in space directly between the nuclei. E.g. H_2. Single bonds.

sigmatropic reaction A multicentre reaction involving the switching of a single σ-bond to a new position when it is flanked by one or more π-bonds. Thus in

(A)

5-methylcyclopentadiene (A) the original σ-bond between C-5 and the attached hydrogen is replaced by a new σ-bond between the carbon which was C-1 and hydrogen.

silanes See SILICON HYDRIDES.

silazanes Compounds containing Si–N bonds. See SILICON NITRIDES.

silfluofen Insecticide, organosilane derivative.

silica (silicon dioxide), SiO_2. One of the most important constituents of the earth's crust. Polymorphic, quartz (to 573°C), tridymite (to 1470°C), cristobalite (to 1710°C m.p.). Each form has two modifications and all consist of infinitely linked SiO_4 tetrahedra. The melt can solidify to a glass (naturally occurring form obsidian), quartz crystals are non-centro-symmetric. High-pressure forms of SiO_2 are also known; coesite and keatite contain SiO_4 units whilst stishovite contains SiO_6 octahedra in the rutile structure. Naturally occurring silica is frequently coloured, e.g. by Fe; purple quartz is amethyst. Common flint is amorphous silica. Silicaware is used extensively in laboratories. Silica is used in ceramics, water-glass, refractories, abrasives, cement, aluminium, steel and alloys. Diatoms have SiO_2 skeletons, the shape formed by proteins.

silica, vitreous (quartz glass) The amorphous vitreous product formed on cooling fused silica. Transparent to u.v. light down to about 2000 Å, used in optical work. Low coefficient of expansion so that apparatus constructed of vitreous silica may be subjected to irregular heating or to sudden changes of temperature without risk of fracture.

silica gel An amorphous form of hydrated silica produced by the precipitation, flocculation, or coagulation of a silica sol or the decomposition of some silicates. When freshly prepared the silica appears to be in a gelatinous state and, on standing, it will set and form a jelly. After being heated it cannot easily be reconverted into a sol.

Dehydrated silica gel is used commercially as an absorbent in the recovery of solvents, also for drying air, dehydrating gases, refining mineral oils, for filtration, and as a support for various catalysts. Such a gel is in the form of hard granules, chemically and physically almost inert but highly hygroscopic. After use it can be regenerated by heat. Used in the preparation of chromatographic columns.

Cobalt(II) compounds are often incorporated in the gel. When the gel has absorbed a fair amount of moisture it will turn pink but will revert to blue on heating.

silicates A very extensive group of substances, many of which are minerals, derivatives of SiO_2 acting as an acidic oxide. Natural silicates form a major constituent of most rocks. They range in composition from comparatively simple minerals, such as zircon, $ZrSiO_4$, which contain discrete $[SiO_4]^{4-}$ anions, to far more complex structures in which the silicate anion is an extended two- or three-dimensional structure in which oxygen atoms form bridges between the silicon atoms. All silicates are based on the $[SiO_4]$ tetrahedron with partial or complete sharing of oxygen atoms. Al, B, Be, may replace silicon in this anion and the cations are in the holes in the lattice. Typical silicates are feldspar, mica, zeolites, aluminosilicates.

The alkali silicates are soluble in water and are used industrially. See SODIUM SILICATES.

silicic acids Addition of dilute sodium silicate solution to excess of dilute HCl followed by dialysis gives a colloidal solution of hydrated silica which is readily coagulated to a gel. None of the silicic acids can be isolated as such. However, volatile esters, such as $Si(OMe)_4$, b.p. 121°C, are well known.

silicides Binary compounds with silicon, e.g. Mg_2Si. Silicides are related structurally to carbides and are hard and intractable solids. They form a mixture of silicon hydrides and hydrogen with acids.

silicofluorides See FLUOROSILICATES.

silicon, Si. Non-metallic element, at.no. 14, at.mass 28.086, m.p. 1414°C, b.p. 3265°C, density ρ 2330, crustal abundance 277 000 p.p.m., electronic configuration [Ne] $3s^23p^2$. Group 14 element which does not occur native but is the second most abundant element in the earth's crust, occurring as SiO_2 (see SILICA) and in the many silicates. Essential to life. Crude Si is obtained by reduction of SiO_2 with carbon in an electric furnace. For pure Si this is converted to $SiCl_4$ or other halide (Cl_2) which is purified by fractional distillation, reduced to Si on a hot wire and finally purified by zone refining. Si has a diamond lattice, is oxidized on strong heating in air, reacts with F_2 or Cl_2, gives silicates with fused alkalis. The ultra-pure material, doped with, e.g. B or P, is used in transistors and other solid-state devices and silicones. Silicates and SiO_2 are used in glass, refractories, building materials, etc.

silicon, organic derivatives Silicon forms a wide range of organic derivatives containing C–Si bonds (for preparative methods see SILICONES). R_3SiX derivatives, particularly Me_3SiF, are very stable leaving groups (e.g. Me_3SiNMe_2 plus PF_5 gives F_4PNMe_2 plus Me_3SiF).

silicon bromides Silicon forms ranges of compounds containing Si–Br bonds. *Silicon tetrabromide*, $SiBr_4$, m.p. 5°C, b.p. 155°C, is the lowest binary bromide (Si plus Br_2 at red heat).

silicon carbide (carborundum), SiC. A hard refractory material made by heating C and Si or SiO_2 to 2000°C. Used for refractory bricks, as fibres in metals, crucibles, abrasive tools and refractory cement.

silicon chemistry Silicon is a relatively electropositive element of Group 14. It is normally in the +4 oxidation state, 4-co-ordinate (5- and 6-co-ordinate complexes are known). Catenation and bridging through, e.g. C, N, O, is possible so that there is an extensive range of silicon chemistry corresponding to aliphatic-carbon chemistry with cyclization, etc. Terminal Si–X bonds can be exchanged; Si–H are reducing. Cationic species, e.g. $[Ph_3Si–bipyridyl]^+$ are known. Silicon(II) species are important reactive intermediates, see SILICON FLUORIDES. Some unstable silicon(III) species are known. Normally multiply bonded Si – e.g. the C=Si group – is rare, but π-bonding involving the empty d orbitals of Si is common, e.g. linear Si–O–Si units.

silicon chlorides
Silicon tetrachloride, $SiCl_4$. M.p. –70°C, b.p. 57°C. Colourless liquid (Si plus Cl_2 followed by distillation). Hydrolyses in moist air or in water. Used for production of pure Si.

Higher chlorides, Si_2Cl_6 to Si_6Cl_{14} (highly branched – some cyclic), are formed from $SiCl_4$ plus Si or a silicide or by amine-catalysed disproportionations of Si_2Cl_6, etc. Partial hydrolysis gives oxide chlorides, e.g. $Cl_3SiOSiCl_3$. $SiCl_4$ is used for preparation of silicones. Si_2Cl_6 is used for extracting oxygen from an organic derivative.

silicone rubbers, $-(Me_2SiO)_n$ (also other alkyl derivatives). SILICONE derivatives. Retain elastic properties over very wide temperature range, inert to acids, alkalis, oils and oxidizing agents.

silicones Polymeric organosilicon derivatives containing Si–O–Si links. Organosilicon halides are prepared RX plus Cu–Si or Ag–Si alloy; RX plus $SiCl_4$ over Zn or Al; $SiCl_4$ or Si–H compounds plus olefin over catalyst (HYDROSILYLATION). (These derivatives also formed $SiCl_4$ plus, e.g., LiR, RMgX on laboratory scale). The halides are hydrolysed to give polymers with properties depending upon the degree of polymerization (extent of hydrolysis). Silicones are used in greases, hydraulic fluids, lubricants, dielectrics, implants, sealing compounds, enamels, varnishes, resins, SYNTHETIC RUBBERS. Resins are used in electrical insulation and laminates. They are odourless, colourless and inert, immiscible with water and have high flash points.

silicon ester See ETHYL SILICATE.

silicon fluorides *Silicon tetrafluoride*, SiF_4. M.p. –90°C, b.p. –86°C. Colourless gas hydrolysed (fumes) in air gives $F_3SiOSiF_3$ then SiO_2. Prepared $SiCl_4$ plus KF; SiO_2 plus HF or HSO_3F. Forms many complexes including $[SiF_5]^-$ and $[SiF_6]^{2-}$, fluorosilicate ions. Higher fluorides are formed SiF_4 plus Si. Si plus SiF_4 at 1100°C and low pressure gives silicon difluoride, SiF_2. Stable for minutes at low pressure otherwise polymerizes to higher fluorides. Very reactive material, e.g. H_2S gives $HSiF_2(SH)$. Many organosilicon fluorides are known (Si–Cl plus KF), etc.

silicon hydrides (silanes) Monosilane, SiH_4; b.p. –112°C. Formed SiO_2 or $SiCl_4$ plus $LiAlH_4$. Higher hydrides Si_2H_6 to Si_6H_{14} are formed from Mg_2Si and dil. HCl or Mg_2Si plus NH_4Br in liquid NH_3 or by an electric discharge on SiH_4. All are spontaneously inflammable in oxygen or air. Alcohols give SiOR plus H_2. SiH compounds are good reducing agents.

silicon nitrides, Si_3N_4. Formed Si plus N_2 at 1500°C. Grey inert solid, used in wear-resistant ceramics, resistant to acid, gives NH_3 and silicates with alkalis. Many compounds containing Si–N bonds are known, e.g. heptamethyldisilazane, $Me_3SiNMeSiMe_3$ used to introduce NMe groups. Highly polymeric materials are known but they are more reactive than silicones.

silicon oxide chlorides Formed by controlled hydrolysis of SILICON CHLORIDES.

silicon oxides
Silicon monoxide, SiO. Formed SiO_2 plus C in

electric furnace. The impure brown powder is used as a pigment and abrasive (Monex). Stable in vapour phase (Si plus SiO_2).

Silicon dioxide, SiO_2. See SILICA.

silicotungstates Heterotungstates which form heavy solutions used in mineral separations.

silk A protein fibre β-keratin from the cocoons of the silk moth and by spiders.

siloxanes Molecular compounds containing SiOSi groups, e.g. hexachlorosiloxane, $Cl_3SiOSiCl_3$ (controlled hydrolysis of $SiCl_4$; $SiCl_4$ plus air at high temperature). Poly(organosiloxanes) are SILICONES. Dimethylpolysiloxane $[Me_2SiO]_n$ with Me_3Si caps. Antifoaming agent, ointment base.

silver, Ag. Metal, at.no. 47, at.mass 107.87, m.p. 961.78°C, b.p. 2162°C, density ρ 10 500, crustal abundance 0.07 p.p.m., electronic configuration [Kr] $4d^{10}5s^1$. Occurs as the metal and also as sulphide ores (argentite or silver glance, Ag_2S; pyrargyrite or ruby silver, Ag_3SbS_3 and silver–copper glance $(Cu,Ag)_2S$). Silver is frequently recovered after the work-up of Cu and Pb ores and is generally extracted as a complex cyanide or thiosulphate complex and is recovered by reduction with Zn and purified electrolytically. The metal is a pure white ductile material with a ccp structure. It dissolves in HNO_3 and hot conc. H_2SO_4 but is inert to alkalis. The metal is used extensively in jewellery, electrical components particularly conductors; catalysts, contacts and batteries, dental and surgical components. It was formerly used extensively in mirrors, formed by depositing silver on glass by a reduction process, and as a coinage metal. Silver compounds are extensively used in photography.

silver bromide, AgBr. A pale yellow solid, m.p. 420°C, precipitated from aqueous solution. Similar to silver chloride.

silver carbonate, Ag_2CO_3. Prepared by precipitation from a silver nitrate solution. It is a light-sensitive white solid which becomes yellow. It decomposes to Ag_2O above 100°C.

silver chemistry Silver is an element of Group 11. It shows oxidation states +4 and +3, +2, +1. The three former are strongly oxidizing

$$Ag(II) \rightarrow Ag(I) +2.00 \text{ V(in 4M } HClO_4)$$
$$Ag^+ \rightarrow Ag +0.8V$$

but are stable in fluorides, complex fluorides and oxides, and as nitrogen complexes, e.g.

$Ag(bipyridyl)_2(NO_3)_2$. Ag(I) forms a stable hydrated ion and many complexes particularly with N, P and S ligands and olefins. Polymeric $[Ag_7O_8]$,$[Ag_6]^{4+}$ and metal–metal bonded compounds, e.g. Ag_3O and Ag_2F are known.

silver chloride, AgCl. A white solid, m.p. 449°C, precipitated from aqueous solution (the precipitation is used as a test and for analysis of Ag^+ or Cl^- ions; the precipitate is soluble in aqueous ammonia or sodium thiosulphate). Forms ammines with ammonia. Used extensively in photography. Sheets of AgCl are transparent to i.r. radiation and are used as support materials in i.r. spectroscopy. AgCl darkens on exposure to light.

silver fluorides

Silver(I) fluoride, AgF, is prepared by evaporation of a solution of excess Ag_2O in HF after filtration or by heating anhydrous $AgBF_4$. The anhydrous salt is yellow; hydrates are known. It is very soluble in water and in many organic solvents. Used as a mild fluorinating agent. On treatment of a solution with Ag a subfluoride, Ag_2F, is formed.

Black *silver(II) fluoride*, AgF_2 (Ag + F_2), is a strong fluorinating agent.

Complex *silver(III) fluorides* containing $[AgF_4]^-$ or $[AgF_6]^{3-}$ ions and also Cs_2AgF_6 containing Ag(IV) are formed with fluorine. They are also strong fluorinating agents.

silver fluoroborate, $AgBF_4$. Prepared by the action of bromine trifluoride on silver borate or by passing boron trifluoride through a solution or suspension of silver(I) fluoride in acetonitrile, nitromethane, sulphur dioxide or toluene. It is a white solid, very soluble in water and ether, moderately soluble in benzene.

silver halide grains The individual silver halide crystals in a photographic emulsion.

silver nitrate, $AgNO_3$. The most important compound of silver, obtained by dissolving the metal in dilute nitric acid, and crystallizing. On strong heating silver nitrate decomposes into silver, N_2O_4, and oxygen. It has m.p. 212°C.

silver oxides

Silver(I) oxide, Ag_2O, is obtained as a brown amorphous precipitate by the action of alkali on $AgNO_3$ solution. It cannot be obtained pure and is soluble in ammonia solution. Used as catalyst.

Silver(II) oxide, AgO, is a black solid, $Ag^IAg^{III}O_2$, obtained by anodic or persulphate oxidation of an $AgNO_3$ solution. Continued anodic oxidation gives impure Ag_2O_3. Argentates, e.g. KAgO, containing silver(I), and mixed metal oxides containing Ag(I), Ag(II) and Ag(III) are known.

silver perchlorate, AgClO₄. Prepared by the action of HClO₄ on Ag₂CO₃. AgClO₄ is extremely soluble in water and ether and is also soluble in benzene and toluene.

silver phosphate, Ag₃PO₄. Used in photography.

silver salt The commercial name for sodium anthraquinone-2-sulphonate. See ANTHRA-QUINONE SULPHONIC ACIDS.

silver sulphate, Ag₂SO₄. Prepared by dissolving silver in concentrated H₂SO₄ or by addition of a sulphate to AgNO₃. Only sparingly soluble in water. Decomposes on heating to Ag, SO₂ and O₂.

silver sulphide, Ag₂S. Obtained as a black precipitate by the action of H₂S on a solution of any silver salt. It is very insoluble in water.

silylation Formation of R₃Si– derivatives (R = H or organic). See SILICONES.

silyl compounds Derivatives of the H₃Si– group.

SIM Scanning ion microscopy. Used for elemental and isotopic analysis particularly of surface layers.

simazine Herbicide, triazine derivative.

simetryn Herbicide, triazine derivative.

Simmons–Smith reagent Named after the DuPont chemists who discovered that di-iodomethane would react with an active zinc–copper couple in ether to give a reagent with molecular formula ICH₂ZnI. The reagent adds stereospecifically *cis*- to alkenes to give cyclopropanes in high yields.

SIMS secondary ion mass spectroscopy.

simulated moving bed technology (SMB) Automated separation system based on chromatographic columns.

single bond A bond between two atoms with two electrons occupying the bonding orbital and apparently no further electronic interaction between the atoms. Generally a σ-bond, e.g. as in the C–H bonds in CH₄, but can be essentially a π-bond, as in Ni(PF₃)₄.

singlet A single transition in a spectrum with other levels at very different energies.

singlet state See SPIN MULTIPLICITY.

sintering The process of bonding by atomic or molecular diffusion in which powders are heated under pressure but at a temperature below the melting point. Used for metals and high-melting polymers, e.g. polytetrafluoroethene. The process depends upon such mobility of the atoms or molecules in the surface layers as to cause physical diffusion between the lattices of the separate particles and thus physical joining of the lattices.

sitosterol (24-(24-*R*)ethylcholesterol), C₂₉H₅₀O. M.p. 137°C. The main plant sterol but is often accompanied by related sterols, from which it is difficult to separate. The glucoside, m.p. 250°C, is also very common in plants and has been described by a variety of names. Sitosterol is accumulated by many invertebrate marine animals.

size exclusion chromatography See GEL FILTRATION.

size measurement See PARTICLE SIZE MEASUREMENT.

size reduction equipment Plant for particle size reduction which generally falls under the description of *crushers, grinders, mills* or *disintegrators*.

sizing The process of filling the pores of paper and giving it some degree of water repellency. This can be done either in the beaters by the addition of rosin and alum or a colloidal mixture, such as starch and water-glass, when the process is termed engine sizing, or by spraying the surface of the ready-made paper with glue or the like when it is known as surface sizing. See PAPER.

skatole (3-methylindole), C₉H₈N. M.p. 95°C, b.p. 265–266°C. The chief volatile constituent of faeces, formed by the action of the intestinal bacteria on tryptophan. It occurs also in coal tar and beetroot.

slack wax An oily wax obtained from crude oil or oil fractions by chilling or by treatment with methyl ethyl ketone. Slack wax can be cracked to produce higher alkenes (C₈–C₁₈) for use in the production of detergents.

slag A liquid solution of oxides produced in the smelting and refining of metals. The primary function of slag in smelting is to take into solution the gangue of the ore and separate it from the liquid metal. To achieve this fluxes are added which reduce the melting point of the oxide mixture. Common fluxes are alumina, limestone or lime, silica and iron(III) oxide. In addition to removing gangue, the slag can also exert a refining function in smelting processes, e.g. the desulphurization of iron in the blast furnace. In smelting, slag composition depends on the metal being smelted.

Slags are also produced in metal-refining processes where oxidation is used to separate the impurities, e.g. steelmaking, copper and lead refining. The aim in refining is to control the slag composition to effect the removal of impurities and this process invariably means that some of the parent metal is oxidized and passes into the slag. In steelmaking the aim is to remove C, S and P and this requires a slag high in CaO, i.e. a basic slag.

Slags are often sold for use in other industries. Thus blast furnace slag can be used for manufacturing cement, as road metal, as slag wool. The slag from the basic steelmaking process sometimes contains sufficient P_2O_5 to make it suitable as an agricultural fertilizer and it is sold as 'basic slag'; the high lime content of these slags is also useful. Converted to slag-ceram by crystallization of the glass from slag.

slaked lime, $Ca(OH)_2$. See CALCIUM HY-DROXIDE.

slip planes The planes of weakness in crystals corresponding to crystalline boundaries and to planes of atoms with only weak forces between such planes (e.g. the layers in graphite).

slow combustion See COOL FLAMES.

slurry A liquid containing an appreciable quantity of suspended solid.

slush bath See LOW-TEMPERATURE BATHS.

Sm Samarium.

smectic See LIQUID CRYSTALS.

smekal cracks Crystal defects – dislocations.

smelting The melting of an ore with chemical reaction with other materials. A part of the process of winning an element from its ores.

smokeless fuel A fuel which, when burnt, gives rise to minimal smoke. The fuel obtained from the low-temperature carbonization of coal is a smokeless fuel since the constituents which give rise to smoke have already been distilled off, but sufficient volatile matter (about 7–10%) has been left to allow the fuel to be easily ignitable and to burn freely.

Sn Tin.

S_N1 reaction A reaction for which the rate is dependent only on the concentration of one species, from which the leaving group goes before attacking another group (generally a nucleophile).

S_N2 reaction (AnDn reaction) A reaction with the rate dependent on the concentration of one species and also the concentration of the attacking species. For carbon components there is inversion at the central atom with loss of the leaving group. Attack occurs on the opposite side to the leaving group.

SNG Substitute natural gas.

soaps Sodium and potassium salts of fatty acids, particularly stearic, palmitic and oleic acids. Animal and vegetable oils and fats, from which soaps are prepared, consist essentially of the glyceryl esters of these acids. In soap manufacture the oil or fat is heated with dilute NaOH (less frequently KOH) solution in large vats. When hydrolysis is complete the soap is 'salted out', or precipitated from solution by addition of NaCl. The soap is then treated, as required, with perfumes, etc. and made into tablets.

Castile soap is manufactured from olive oil, transparent soap from decolorized fats and liquid green soap from KOH and vegetable oils. Soaps are sometimes 'superfatted' in that they contain some free fatty acid.

Soaps of other metals such as aluminium, calcium, cobalt, lithium, lead or zinc – see METALLIC SOAPS.

soda ash, Na_2CO_3. SODIUM CARBONATE.

soda lime A granular material prepared from quicklime and NaOH solution followed by heating the product until dry. Used as an absorbent for CO_2 and as an alkali in testing for ammonium salts (NH_3 liberated on grinding).

sodamide, $NaNH_2$. See SODIUM AMIDE.

sodium, Na. Alkali metal, at.no. 11, at.mass 22.990, m.p. 97.80°C, b.p. 883°C, density ρ 971, crustal abundance 23 000 p.p.m., electronic configuration [Ne] $3s^1$. Sodium occurs widely, principally as NaCl in sea water and in salt deposits. Sodium metal (bcc) is obtained by electrolysis of fused NaCl or $NaCl-CaCl_2$. The metal is soft and silvery white. It reacts rapidly with water, oxygen, halogens. Dissolves in liquid NH_3 to give a blue reducing solution but ultimately $NaNH_2$. The metal is used in the preparation of some Na compounds. Used as a heat-transfer medium, conductor, in lamps and electric cells, as a reducing agent, particularly as a dispersion, or as Na naphthenide. Sodium compounds are of great importance – see under separate headings. Sodium ions are essential to life, both in the diet and for function.

sodium, organic derivatives Materials which are rarely isolated (Na dispersion plus organic halide) used for olefin polymerization. Sodium cyclopentadienide, NaC_5H_5 and sodium acetylides, e.g. CH_3C_2Na are used in synthesis. Aromatic hydrocarbons, ketones, etc., give radical anions, e.g. $(C_6H_6)^-$ with Na in tetrahydrofuran at low temperatures. Used as reducing agents.

sodium acetate, NaO_2CCH_3. See SODIUM ETHANOATE.

sodium acetylides Formed (e.g. HC_2Na, Na_2C_2) by the action of Na or NaH on alkynes. Used in synthesis.

sodium aluminate, $NaAlO_2$. Used in fabric printing, sizing, soaps.

sodium amide, $NaNH_2$. Formed Na plus NH_3, hydrolysed to ammonia. A strong base, particularly in liquid ammonia, used extensively in organic synthesis.

sodium ammonium hydrogen phosphate (microcosmic salt), $NaNH_4HPO_4$. $4H_2O$. Colourless salt (Na_2HPO_4 plus NH_4Cl or $(NH_4)_2HPO_4$ plus NaCl). The melt solidifies to glasses which give characteristic colours with certain metals. Used for preparation of standard solutions.

sodium antimonyl tartrate, $NaSbOC_4H_4O_6$. Antimony sodium tartrate.

sodium arsenates Sodium arsenates(III), Na_3AsO_3, $NaAsO_2$.; sodium arsenates(V), Na_3AsO_4, $NaAsO_3$, $Na_4As_2O_7$. See ARSENATES. Used in dyeing.

sodium azide, NaN_3. See AZIDES. Used as a propellant in car air bags.

sodium benzoate, $C_6H_5CO_2Na$. White powder, soluble in water, sparingly soluble in ethanol. Used as corrosion inhibitor, preservative and medicinally as an antiseptic and for gout and rheumatism.

sodium bicarbonate, $NaHCO_3$. See SODIUM HYDROGEN CARBONATE. Used in fire extinguishers.

sodium bifluoride (sodium hydrogen fluoride), $NaHF_2$. See SODIUM FLUORIDE.

sodium bismuthate, $NaBiO_3$. Very strong oxidizing agent (Mn^{2+} gives $[MnO_4]^-$) prepared impure by heating Na_2O_2 and Bi_2O_3.

sodium bisulphate See SODIUM SULPHATES.

sodium bisulphite (sodium hydrogen sulphite), $NaHSO_3$. Disinfectant and bleach.

sodium borate, $Na_2B_4O_7\cdot10H_2O$. See BORAX. Used in soldering, glazes, enamels, fireproofing.

sodium borohydride (sodium tetrahydroborate), $NaBH_4$. See BOROHYDRIDES. Prepared NaOMe plus B_2H_6. Used in organic reductions (e.g. ketones, acids to alcohols) preparatively and to eliminate impurities.

sodium bromate(V), $NaBrO_3$. Prepared by electrolysis of aqueous NaBr or Br_2 in hot NaOH solution.

sodium bromide, NaBr. M.p. 757°C. b.p. 1393°C. Formed Na_2CO_3 or NaOH plus HBr or (with $NaBrO_3$) Br_2 plus hot NaOH solution. Has NaCl structure.

sodium carbonate, Na_2CO_3. Manufactured from sodium hydrogen carbonate in the ammonia–soda or Solvay process. Brine (NaCl solution in water) has added NH_3 then CO_2. $NaHCO_3$ is precipitated which decomposes to Na_2CO_3 at 175°C; NH_3 is recovered by treating the NH_4Cl solution with $Ca(OH)_2$ slurry. Forms hydrates. Anhydrous Na_2CO_3 is known as soda ash; $Na_2CO_3.10H_2O$ is washing soda. Na_2CO_3 occurs naturally as trona. Used in glass (55%), soap, chemical detergents and paper industries. U.S. production 1999 11.2 million tonnes.

sodium chemistry Except for the formation of the sodide Na^- ion (e.g. in adducts of Na with cryptates) the chemistry is exclusively that of the +1 state ($E° Na^+ \to Na -2.71V$); predominantly ionic, 6-co-ordinate. Complexes are formed (e.g. $Na_2CO_3,10H_2O$ contains a $[(H_2O)_4Na(\mu-H_2O)_2Na(H_2O)_4]^+$ cation and cryptates are formed by the cyclic polydentate ligands).

sodium chlorate(V), $NaClO_3$. Formed Cl_2 on hot NaOH solution or by electrolysis of NaCl in water and allowing anode and cathode products to come into contact. Used in the pulp industry to generate ClO_2 for bleaching, in the manufacture of chlorates(V) and (VII) (perchlorates) and as a herbicide.

sodium chloride, NaCl. M.p. 801°C, b.p. 1439°C, ρ 2170. Occurs naturally as rock salt

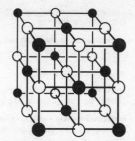

● Na
○ Cl

which is widely distributed and occurs extensively in thick beds and to about 3% in sea water. Obtained commercially by solar evaporation of sea water, by mining or usually as brine by passing water into underground deposits. The brine contains Ca^{2+}, Fe^{2+}, Mg^{2+} as impurities, these are removed by precipitation with Na_2CO_3 and NaOH. Pure NaCl is obtained by passing HCl gas into a saturated aqueous solution of NaCl. Sodium chloride is used for the formation of NaOH and Na_2CO_3 (for the alkali and glass industries) and chlorine (for the chlorocarbon industry) and is one of the chemical industries' most important raw materials. Solid NaCl is used for the removal of ice and snow and as a preservative. The sodium chloride structure is cubic with octahedral co-ordination about Na and Cl (both are ccp). Many other MX compounds crystallize with this structure.

sodium chlorite (sodium chlorate(III)), $NaClO_2$. Used for preparing chlorates(III) and in water purification.

sodium chromate(VI), $Na_2CrO_4.10H_2O$. See CHROMATES.

sodium citrate, $C_6H_5O_7Na_3.2H_2O$(also 5.5 H_2O). Salt of citric acid. Used medicinally as a blood anticoagulant and to reduce blood acidity, to prevent the formation of large curds in the stomach of infants, and in photography and foodstuffs.

sodium cyanate, NaOCN. Used to prepare cyanates.

sodium cyanide, NaCN. M.p. 564°C. Manufactured CH_4 and NH_3 to give HCN followed by neutralization with NaOH. Used to prepare cyanides and as a poison.

sodium cyanoborohydride, $NaBH_3CN$. Prepared by reacting HCN with sodium borohydride. A milder reducing agent than $NaBH_4$, especially valuable for its ability to reduce a given organic functional group in the presence of various other functional groups and for its stability in aqueous media over a broad range of pH, e.g. a $>C=\overset{+}{N}<$ group is reduced more rapidly than $>C=O$.

sodium cyclopentadienide, NaC_5H_5. Formed cyclopentadiene and Na, used in the formation of metal cyclopentadienyl derivatives.

sodium dichromate, $Na_2Cr_2O_7.2H_2O$. See CHROMATES(VI).

sodium dithionite, $Na_2S_2O_4$. Formed by reduction of Na_2SO_3 plus excess SO_2 with Zn. Powerful reducing agent ($E°$ $S_2O_4^{2-} \rightarrow SO_3^{2-}$ in alkaline solution +1.12 volts; used in dyeing), removes O_2 from gases in presence of 2-anthraquinone sulphonate (Fieser's solution).

sodium dodecyl benzene sulphonate, $C_{12}H_{25}NaO_4S$. Anionic detergent, wetting agent.

sodium ethanoate (sodium acetate), $NaO_2CCH_3.3H_2O$. Formed ethanoic acid and Na_2CO_3.

sodium ethoxide (sodium ethylate), $NaOCH_2CH_3$. White solid (Na in EtOH). Decomposed by water, gives ethers with alkyl halides; reacts with esters. Used in organic syntheses, particularly as a base to remove protons adjacent to carbonyl or sulphonyl groups to give resonance-stabilized anions.

sodium ferrocyanide, $Na_4Fe(CN)_6$. Used in formation of Prussian blue, photography, and prevention of caking of NaCl.

sodium fluoride, NaF. M.p. 902°C. Formed Na_2CO_3 or NaOH plus HF; gives HF with H_2SO_4. Forms the hydrogen fluoride, $NaHF_2$, with excess HF. Used as a fluorinating agent.

sodium fluoroacetates (fluoroethanoates) See FLUOROETHANOATES.

sodium (mono)fluorophosphate, Na_2PO_3F. See FLUOROPHOSPHORIC ACIDS.

sodium formaldehyde sulphoxylate dihydrate (rongalite), $HOCH_2SO_2Na.2H_2O$. Reagent for inducing free-radical reactions.

sodium formate, NaO_2CH. Sodium methanoate.

sodium gluconate, $C_6H_{11}NaO_7$. Sequestering agent used in metal plating and tanning.

sodium glutamate (monosodium glutamate, MSG) See L-GLUTAMIC ACID, FOOD ADDITIVES.

sodium hexafluoroaluminate, Na_3AlF_6. Cryolite, used in aluminium production.

sodium hexametaphosphate
See SODIUM PHOSPHATES.

sodium hydride, NaH. Prepared H_2 over Na at 350°C. Has NaCl structure. Used in organic reductions and condensations and in metallurgy as a descaling agent.

sodium hydridotrimethoxyborate, $Na[HB(OMe)_3]$. Reducing and hydroboration agent.

sodium hydrogen carbonate (sodium bicarbonate), $NaHCO_3$. Prepared CO_2 into Na_2CO_3 solution (also in Solvay process for SODIUM CARBONATE); occurs naturally as nahcalite and used in the production of soda ash; sparingly soluble in water. Decomposes to Na_2CO_3 on heating. Used in Na_2CO_3 manufacture, as an ingredient of baking powders, pharmaceuticals, and in fire extinguishers.

sodium hydrogen fluoride See SODIUM FLUORIDE.

sodium hydrogen sulphate, $NaHSO_4$. Formed Na_2SO_4 plus H_2SO_4 other acid salts known. $NaHSO_4$ gives the pyrosulphate, $Na_2S_2O_7$, on heating.

sodium hydrogen sulphite, $NaHSO_3$.

sodium hydrosulphite (sodium hyposulphite), $Na_2S_2O_4$. See SODIUM DITHIONITE. Reducing agent used in bleaching. Also used for $NaHSO_2$.

sodium hydroxide (caustic soda), NaOH. M.p. 318°C, b.p. 1390°C, d 2.13. Manufactured by electrolysis of sodium chloride solution. In the mercury cell process (Castner–Kellner) a stream of Hg is used as cathode and the mercury amalgam formed is reacted with water. In the diaphragm cell process the electrolyte flows from anode to cathode and the diaphragm separates anode and cathode products. In both processes chlorine is the other product. NaOH is a white, translucent, hygroscopic solid; hydrates are known. Gives a strongly alkaline solution in water. Industrial chemical. (World demand 1998 47.9 million tonnes.) Used in the manufacture of other chemicals (50%), rayon, pulp and paper (20%), aluminium (10%), petrochemicals, textiles (5%), soap (5%) and detergents.

sodium hypochlorite (sodium chlorate(I)), NaOCl. Formed (with NaCl) from Cl_2 plus cold NaOH solution. Forms hydrates. Aqueous solutions are used as bleaches, disinfectants and antiseptics.

sodium iodate (sodium iodate(V)), $NaIO_3$. Formed HIO_3 on NaOH or I_2 plus $NaClO_3$. Forms hydrates and acid salts. Used in volumetric analysis.

sodium iodide, NaI. M.p. 660°C, b.p. 1300°C. Obtained aqueous HI plus Na_2CO_3.

sodium lactate, $C_3H_5O_3Na$. Used in calico printing, textile finishing and as plasticizer for casein.

sodium lauryl sulphate, $C_{12}H_{25}NaSO_3$. See TARTRATES. Detergent.

sodium metabisulphite, $Na_2S_2O_5$. See SODIUM PYROSULPHITE.

sodium methanoate, NaO_2CH. Sodium formate.

sodium methoxide, $NaOCH_3$. Similar to SODIUM ETHOXIDE.

sodium molybdate, Na_2MoO_4. Used in metal finishing, pigments and ceramics.

sodium nitrate (Chile saltpetre), $NaNO_3$. Used to prepare sodium nitrite.

sodium nitrite, $NaNO_2$. Prepared $NaNO_3$ plus Pb. M.p. 271°C, decomposes at 320°C. Used in dyestuffs industry and as a corrosion inhibitor, in meat curing and preserving.

sodium oxalate, $Na_2C_2O_4$. Colourless crystals. Sodium hydrogen oxalate, $NaHC_2O_4$.

sodium oxides Sodium normally reacts with O_2 to give Na_2O_2. Sodium monoxide, Na_2O formed Na plus deficiency of O_2 or by reduction of Na_2O_2 (C, Ag, Na) or NaN_3 plus $NaNO_3$. White or yellowish solid, reacts violently with water. Sodium peroxide, Na_2O_2. Formed Na plus O_2. The impure material contains some superoxide, O_2^- ions. Strong oxidizing agent. Forms octahydrate from cold water but generally reacts with water to give NaOH plus H_2O_2.

sodium perborate, $NaBO_3 \cdot 4H_2O$. Used in textile bleaching, dye oxidation, cleaners.

sodium percarbonate See PERCARBONATES.

sodium perchlorate (sodium chlorate(VII)), $NaClO_4$. Formed by heating $NaClO_3$ (see CHLORATES). Used to form perchlorates and in explosives.

sodium periodates (sodium iodates (VII)) Various sodium periodates are known and are used as oxidizing agents and to form periodates. $Na_2H_3IO_6$, $Na_3H_2IO_6$, Na_5IO_6 have octahedrally co-ordinated iodine, $NaIO_4$ contains a tetrahedral $(IO_4)^-$ species.

sodium permanganate, $NaMnO_4$. Impure material used as disinfectant and oxidizing agent.

sodium permonosulphate, Na_2SO_5. Used in bleaching, air purification, disinfectants and as antiviral agent.

sodium peroxide, Na_2O_2. See SODIUM OXIDES.

sodium peroxydisulphate, $Na_2S_2O_8$. Used as oxidizing agent and in preparation of peroxydisulphates and in bleaches.

sodium phosphates Various salts are known, e.g. Na_2HPO_4 (Na_2CO_3 plus H_3PO_4) (used to form $NaH_2PO_4(H_3PO_4$ and $Na_3PO_4(NaOH)$). Heat on Na_2HPO_4 gives the pyrophosphates $Na_2H_2P_2O_7$ and $Na_4P_2O_7$. Heat on NaH_2PO_4 followed by quenching gives a polyphosphate $(NaPO_3)_x$. The 'tripolyphosphate' hexametaphosphate is formed by heating $1NaH_2PO_4:2Na_2HPO_4$. The various sodium phosphates are used principally as detergents but also in water conditioning, foodstuffs and medicine. Sodium pyrophosphates and sodium ammonium phosphates are used in baking powders.

sodium potassium tartrate (rochelle salt, seignette salt), $C_4H_4O_6NaK.4H_2O$. Used as reducing agent (silver mirrors) and saline aperient.

sodium pump Transport mechanism present across membranes of animal cells pumping Na^+ out of cell and K^+ in using energy from ATP. See DICTIONARY OF BIOLOGY.

sodium pyrosulphite (sodium metabisulphite), $Na_2S_2O_5$. Obtained NaOH solution saturated SO_2 at 100°C. Forms 7 and 6 hydrates. Used in photography.

sodium saccharine, $C_7H_4NNaO_3S.2H_2O$. Sweetening agent.

sodium salicylate, $2-C_6H_4(OH)CO_2Na$. White powder, very soluble in water. Antipyretic. Also used as preservative.

sodium selenate, Na_2SeO_4.

sodium silicates Many different sodium silicates are known; some occur as minerals. Silicates prepared from aqueous solution and salts of orthosilicic acid, H_4SiO_4. Fusion of Na_2CO_3 and SiO_2 in a furnace gives a range of products with the Na:Si ratio varying from 4:1 to 1:4. Na_2SiO_3 is sodium metasilicate; those with ratios Na:Si 3.2 to 4 are the water-glasses and are soluble in water. Used in the manufacture of silica gel (treatment with acid) as adhesives, sizes, in preserving eggs, fireproofing, in metal cleaning and in detergent manufacture.

sodium stannate, $Na_2SnO_3.3H_2O$. Prepared by fusing SnO_2 with NaOH. Used as a mordant and in fireproofing.

sodium stearate Used in pharmaceuticals, toothpaste. Usually contains some palmitate.

sodium sulphate (Glauber's salt), Na_2SO_4. M.p. 884°C, b.p. 1429°C. Manufactured from NaCl and H_2SO_4 at high temperature or NaCl, SO_2, air, water vapour; recovered from various chemical processes and natural brines. Forms hydrates and double salts with other sulphates. Forms hydrogen sulphates with excess H_2SO_4. Used in wood pulp production (66%), glass, detergent and chemical manufacture. Used as a mild aperient. U.S. production 1999 665 kilotonnes.

sodium sulphides

Sodium sulphide, Na_2S, formed by reduction Na_2SO_4 with CO or H_2. Aqueous solutions are oxidized to sodium thiosulphate. Used in rubber and dyestuffs manufacture and cotton printing.

Sodium polysulphides, $Na_2S_x (x = 2, 4, 5)$ are formed from Na in liquid NH_3 plus S.

sodium sulphite, Na_2SO_3. Formed SO_2 and NaOH solution, readily oxidized to Na_2SO_4 by air. Forms hydrates. Used to remove chlorine after bleaching and in the paper industry to dissolve lignin, and as a food preservative.

sodium sulphoxylate, $Na_2S_2O_4$. Reducing and bleaching agent. Also applied to $NaHSO_2$, sodium hydrogen sulphite.

sodium tellurate, Na_2TeO_4.

sodium tetraborate, $Na_2B_4O_7$. See BORAX.

sodium tetradecyl sulphate, $C_{14}H_{29}SO_4$. Wetting agent.

sodium tetrahydroborate, $NaBH_4$. See SODIUM BOROHYDRIDE.

sodium tetraphenylborate, $NaB(C_6H_5)_4$. Used in the direct estimation of K^+ which gives insoluble $KB(C_6H_5)_4$.

sodium thiocyanate, NaNCS. Formed fused NaCN plus sulphur. Used to prepare thiocyanates and as a volumetric reagent.

sodium thioglycollate Used in cold waxing of hair and as a depilatory.

sodium thiosulphate, $Na_2S_2O_3$. Manufactured by treating a solution containing Na_2CO_3 and Na_2S with SO_2 or from Na_2SO_3 plus sulphur. Forms many hydrates. Used in photography ('hypo') because it dissolves silver halides. Also used in removing Cl_2, winning silver from ores, tanning, preparation of mordants, as a fermentation preventative in dyeing and in chemical manufacture.

sodium tungstate, $Na_2WO_4.2H_2O$. Used in fireproofing, waterproofing, glass coloration.

soft acids and bases See HARD AND SOFT ACIDS AND BASES.

soft detergents Detergents which are biodegradable. They consist largely of alkylbenzene sulphonates with linear alkyl groups produced by reacting linear 1-alkenes with benzene. See DETERGENTS.

softeners Materials added to the rubber mix to improve the mix. Also materials to enhance or make plant growth more specific.

sol A colloidal solution; term generally used for colloidal dispersions of inorganic solids in liquids.

solder Solders are fusible alloys used for bonding metals. Hard solder or brazing metal is a low-melting brass with m.p. 800°C. Soft solders are normally Pb–Sn eutectic alloys.

soldering Causing a melted metal to adhere to the surface of a solid metal by a process of diffusion brought about by chemical solution. Used to bond metals.

solids, structures The structures of solids depend upon the bonding (ionic, covalent, coordinate, dipole–dipole (hydrogen bonding)) between the atoms and groups. In compounds which are substantially ionic the structures depend on the relative sizes (radius ratios) of the ions. Many structures approximate to close-packed arrangements of the largest species present.

solid foams Solids formed, often in a liquid system, with gas evolution so that a porous network enclosing the gas results. Charcoal is an example and expanded foam plastics are of great importance.

solid-phase synthesis A term applied to the technique of carrying out chemical transformations on a substrate chemically bonded to an insoluble, solid polymer molecule. This technique is used for the synthesis of polypeptides from amino-acids using another amino-acid bonded to a polystyrene matrix. The advantage is that the products at each stage of a multi-step synthesis need not be isolated and purified. Instead the polymer to which the peptide is attached is washed free of excess reagents and impurities and virtually no loss of product occurs. The method has been adapted to automatic control and addition of reagents. At the completion of the desired synthesis the product is liberated from the polymer surface by a suitable reagent.

solid solution A solid solution is formed when two or more elements or compounds share a common lattice. The composition may vary over very wide limits. A substitutional alloy results when atoms of solute replace atoms of solvent in its crystal lattice, e.g. nickel–copper alloys. An interstitial solid solution occurs when a small atom, such as carbon, enters the interstices of the host lattice of larger atoms such as metals.

solid-state electrode See ION-SELECTIVE ELECTRODES.

solid-state reactions Reactions occurring between two or more solid compounds or decompositions occurring from the solid state. Gases may be evolved in such reactions. Most ternary oxides are formed by solid-state reactions although these are now developed to prepare solids by reaction in the presence of steam, etc. under pressure.

solidus curve See LIQUIDUS CURVE.

solketal (2,2-dimethyl-1,3-dioxolane-4-methanol, isopropylidene glycol) Solvent and plasticizer.

solubility The maximum quantity of one phase dissolved by another under specified conditions. The mobility of a saturated solution. In the case of solutions of solids or liquids in liquids. The solubility is often expressed as the weight dissolved in a given weight or volume of the solvent at a specified temperature. See HENRY'S LAW FOR SOLUBILITY OF GASES.

Substances are generally soluble in like solvents. Organic molecules dissolve in molecular solvents such as CCl_4, C_2H_5OH, ether, propanone. Inorganic salts are often soluble in water and less soluble in organic solvents.

solubility constant When a saturated solution of an electrolyte is in contact with some undissolved electrolyte in water the following equilibria may be considered

$$AB_{solid} \rightleftharpoons AB_{dissolved} \rightleftharpoons A^+_{aq} + B^-_{aq}$$

the equilibrium constant for this process is

$$K = \frac{[A^+][B^+]}{[AB]_{dissolved}}$$

However because of the equilibrium between the solid and dissolved undissociated AB, the concentration of the latter may be considered to remain constant and hence

$$[A^+][B^-] = K_{sp}$$

where K_{sp} = solubility constant (product) of the electrolyte. It can be seen that if an excess of A^+ or B^- ions are added to the solution, the K_{sp} is exceeded and the salt will be precipitated to restore the equilibrium.

solubility curve A graphical representation of the variation of the solubility with temperature.

solubility product See SOLUBILITY CONSTANT.

soluble oil Normally a mineral oil containing emulsifiers enabling stable emulsions to be obtained when the oil is added to water. Soluble oils are used in metal cutting and grinding operations for lubrication and cooling purposes. Certain other additives may be used, such as extreme-pressure, anti-freeze or anti-foam agents. See also EMULSION.

solute The substance dissolved in the solvent to yield a solution. See SOLVENT.

solution A liquid or solid may be dispersed through a liquid either by being in suspension, or by being dissolved in it, in the form of a solution. The qualitative characteristics of a solution are homogeneity, and the absence of any tendency for the dissolved substance to settle out again. From these criteria, a suspension of extremely minute particles (colloidal suspension) of a pure chemical compound behaves similarly. A solution is better defined as a homogeneous mixture of substances which is separable into its components by altering the state of one of them (freezing or boiling out one component), and whose properties vary continuously with the proportions of the components between certain limits. Solutions are not confined to liquids and solids; solutions of a gas in liquids, and of gases, liquids and solids in solids

are also formed. The amount of a substance dissolved in a given amount of another substance is described as the concentration of the solution, and may be expressed in grams per litre, or as a molar fraction (see MOLALITY). A solution containing only a small proportion of the dissolved substance is called dilute; one containing a high proportion is said to be concentrated. There is a limit to the solubility of most substances, and a solution containing the limiting amount of the dissolved substance is said to be saturated with respect to the substance, and is termed a saturated solution at a particular temperature. Such solutions are prepared, e.g. by shaking the liquid with excess of the powdered solid until no more is taken up by the solution.

The maximum concentration attainable under such conditions is termed the solubility of the substance at the specific temperature used in the experiment; solubility generally increases with rising temperature. Solubility is often expressed in grams per 100 g of solvent, or grams per 100 g of solution. Sometimes, for practical convenience, it may be expressed in grams per 100 ml of solvent or solution. Molality is more meaningful from a thermodynamic viewpoint.

solvation The process of interaction of a solvent with a solute. The general case corresponding to HYDRATION.

Solvay process See SODIUM CARBONATE.

solvent In a solution, the substance which makes up the bulk of the solution is usually termed the *solvent*. The substance dissolved in the solvent is called the *solute*. These terms are somewhat arbitrary except in special cases when, e.g. a chemical reaction occurs between certain constituents of a solution. In this case the inert medium used to disperse the reactants is termed the inert solvent. Solvents are rarely totally inert and interact electronically with most solutes.

solvent extraction See EXTRACTION.

solvolysis The reaction when a compound reacts with a solvent. In the case of water the special term hydrolysis is used, e.g. the reaction of a salt with water giving the free acid and base.

$$BA + H_2O \rightarrow HA + BOH$$

soman A nerve gas.

somatostatin Growth hormone release factor.

somatropin See GROWTH HORMONE.

sonochemistry The use of ultrasound to modify or promote a chemical reaction.

sorbic acid (2,4-hexadienoic acid), $C_6H_8O_2$, $CH_3 \cdot CH:CH \cdot CH:CH \cdot C(O)OH$. White needles; m.p. 134°C, b.p. 228°C (decomp.). Can be obtained from rowan berries, and prepared by condensing crotonaldehyde and keten. It is a selective growth inhibitor for yeasts, moulds and some bacteria, and is used as a food preservative.

D-sorbitol (D-glucitol), $C_6H_{14}O_6$. M.p. 110°C. The alcohol corresponding to glucose. It is present in many berries.

Manufactured by the reduction of glucose in aqueous solution using hydrogen with a nickel catalyst. It is used in the manufacture of ascorbic acid (vitamin C), various surface-active agents, foodstuffs, pharmaceuticals, cosmetics, dentifrices, confectionery, plasticizers, resins, anti-freezes, adhesives, polyurethane foams, etc.

L-sorbose, $C_6H_{12}O_6$. Colourless crystals, m.p. 160°C. A hexose obtained by the fermentation of sorbitol with certain *Acetobacter*. It is an intermediate in the manufacture of vitamin C.

'sour' products Petroleum fractions, generally gasoline or kerosine, containing mercaptans with an unpleasant odour. 'Sour' fractions are 'sweetened' to convert the mercaptans to disulphides. See SWEETENING.

soxhlet Equipment for the continuous extraction of a solid by a solvent. The material to be extracted is usually placed in a porous paper thimble, and continually condensing solvent allowed to percolate through it, and return to the boiling vessel, either continuously or intermittently. Such extraction procedures are frequently used in the first stages of isolating natural products from raw materials.

space group The whole array of symmetry-elements in a crystal framework or lattice. The total number of possible space groups is 230. The atoms in any crystal are arranged so that they are in one of these groups. The framework composing a space group is termed a SPACE LATTICE.

space lattice A term applied to the array of points in any crystalline structure for which the pattern repeats. By joining these points the crystals can be divided into a series of parallel *unit cells*, each of which contains a complete unit of pattern. The atoms composing the crystal occupy the points in the lattice so that a

space lattice affords a convenient means of representing the atomic structure of a crystalline substance. The position of the points relative to each other is found by means of X-ray analysis of the crystal.

spalling The break-up of a refractory with ultimate mechanical failure. Spalling is likely to occur: **1** when there is a large high temperature gradient through the brick, especially when the brick contains substances of high coefficients of expansion such as magnesite; **2** when bricks are laid badly, giving stresses; **3** when there is reaction between the furnace gas or material in the furnace with the constituents of the refractory.

spandex Polymethine fibres with elastic properties.

spark spectrum See ARC SPECTRUM.

special boiling point spirits (SBP) Refined solvents of specially selected boiling ranges distilled from gasoline fractions. SBP are usually classified by means of their boiling ranges, e.g. SBP 62/82°C. They are used for many industrial purposes such as solvents for rubber and for the extraction of oil from seeds. See WHITE SPIRITS, NAPHTHA.

specific conductance See CONDUCTIVITY.

specific rotatory power See ROTATORY POWER, SPECIFIC.

spectral sensitization The process by which the natural sensitivity of a photographic silver halide emulsion is extended to light of other wavelengths by means of sensitizing dyes (spectral sensitizers). Using appropriate sensitizers, orthochromatic and i.r.-sensitive emulsions are produced. For colour photography, tri-pack materials incorporate three sensitive layers sensitized to red, green and blue light.

spectral sensitizers See SPECTRAL SENSITIZATION. Structurally, the majority of sensitizers are polymethine dyes, e.g. cyanines.

spectrochemical series Ligands arranged in order of their tendency to cause splitting of the d orbitals in complexes (see CRYSTAL FIELD THEORY). The order is obtained from changes in magnetism and the frequencies of d–d transitions in the u.v.–visible–near i.r. spectra. The series is $I^- < Br^- < Cl^- < F^- < OH^- < H_2O <$ pyridine $< NH_3 < en < NO_2^- < CN^-$.

spectrophotometer An instrument that measures (generally automatically) the relation between absorption of electromagnetic radiation and frequency (or wavelength) of that

| **spray ponds**

radiation. Spectrophotometers operate in various regions of the electromagnetic spectrum, e.g. u.v., visible, i.r., microwave.

spectrophotometric methods of analysis Methods which use a spectrophotometer.

speed of light $c = 2.997 \times 10^8$ m s^{-1} in vacuum.

sphalerite (Zn,Fe)S. ZINC BLENDE containing some iron. An important zinc ore.

sphingomyelins Ionic phosphatides generally containing lignoceric acid, though others are found. Sphingomyelins occur abundantly in the brain and to a lesser extent in other animal tissues. They are always found in association with the cerebrosides, which they resemble. On hydrolysis they split up into the bases choline and sphingosine, phosphoric acid and a fatty acid.

sphingosine, $C_{18}H_{37}NO_2, CH_3 \cdot [CH_2]_{12} \cdot CH:CH \cdot CH(OH) \cdot CH \cdot (NH_2) \cdot CH_2OH$. A base, forming part of the molecules of sphingomyelins and the cerebrosides, from which it splits on hydrolysis.

spindle oil Any low-viscosity mineral lubricating oil.

spinel Mineral $MgAl_2O_4$.

spinels A group of mixed metal oxides, generally $M^{2+}M_2^{3+}O_4$ (other combinations of cations are possible) (M^{2+} being usually Mg, Fe, Co, Ni, Zn, Mn; M^{3+} being Al, Fe, Cr, Rh). The crystals are cubic with close-packed oxygen atoms. In a normal spinel each M^{2+} is in tetrahedral co-ordination, in an inverse spinel half of the M^{3+} are in tetrahedral co-ordination, half in octahedral co-ordination. Intermediate arrangements are also found.

spin (s) The intrinsic angular momentum of a particle.

spin density (ρ) The possibility that an unpaired electron is on an atom. Concept used in e.s.r. spectroscopy.

spin label A free radical or other paramagnetic probe (e.g Cu^{2+}, Mn^{2+}) bound to a biological macromolecule in an effort to determine molecular structure or biological function by using e.s.r. techniques. Spin labels are very sensitive to the local environment, allow the measurement of very rapid molecular motion in either optically transparent or opaque solution and do not produce problems of interfering signals from the environment. Substituted organic

nitroxides are the commonest class of free radicals used as spin labels. The relatively unreactive NO group contains the unpaired electron necessary to produce an e.s.r. signal whilst the functional group OH, NH$_2$ or C(O)OH provides the chemical means of attachment to the macromolecule.

spin moment An electron spinning on its own axis produces a magnetic moment, termed the spin moment. For a multielectron system the magnitude of the spin moment, μ_s, is given by

$$\mu_s = g\sqrt{S(S+1)}$$

where g is the gyromagnetic ratio and S is the sum of the spin quantum numbers (m_s) of the individual electrons.

spin multiplicity For an electronic state containing n unpaired electrons, the total spin quantum number, S, is equal to ($n/2$). The spin multiplicity of the state is given by ($2S + 1$). When this is equal to unity, it is called a singlet state; when it is equal to three it is a triplet state, etc.

spin-orbit coupling The interaction of spin and angular momentum.

spin quantum number See ELECTRONIC CONFIGURATION.

spin–spin coupling The process of interaction between nuclear spins giving rise to fine structure in n.m.r. spectra. Generally expressed in Hz. Can be decoupled (electronically) for clarification of spectra.

spirans (spiro-compounds) Bicyclic compounds with one, and only one, atom common to both rings. E.g. spiro[4,5]decane as shown.

$$\begin{array}{c} H_2C — CH_2 \diagdown \quad CH_2 \cdots CH_2 \\ \quad\quad\quad\quad C \\ H_2C — CH_2 \diagup \quad CH_2 — CH_2 \diagup CH_2 \end{array}$$

The atom common to both rings is called the spiro-atom; this may also be nitrogen, phosphorus, etc. Can be chiral.

spirits of salt An old name for hydrochloric acid.

spodumene, $LiAl(SiO_3)_2$. A PYROXENE mineral with SiO_4 tetrahedra linked in chains. An important source of Li also used in glasses and ceramics.

spray ponds A method of cooling water by atmospheric evaporation. The water is sprayed through fine nozzles and collected in open ponds.

spray tower The simplest form of SCRUBBER. The gas to be purified passes up the tower and is met by a stream of liquid introduced by a series of sprays. The space inside the tower is empty, without packing.

squalene E- (*trans*-2,6,10,15,19,23-hexamethyl-2,6,10,14,18-tetracosane-hexaene), $C_{30}H_{50}$. An acrylic triterpene of wide occurrence and significance as the natural precursor of cholesterol and of other sterols and triterpenoids of plants and animals. It is the major hydrocarbon of human skin surface lipid, forming about 10% of total lipid. Natural squalene is a colourless oil, b.p. 261°C/9 mm, m.p. –5°C. Squalene, and the more stable saturated hydrocarbon squalane, are used as vehicles for cosmetics, and for promoting absorption of drugs applied to the skin.

square antiprismatic co-ordination Co-ordination by eight neighbours in a square anti-prismatic arrangement (i.e. the two square faces at 45°). Found in, e.g. $[Mo(CN)_8]^{4-}$. A square antiprism has 2 square and 8 triangular faces.

SQUID (superconducting quantum interference device) Equipment used to determine magnetic susceptibility.

Sr Strontium.

SSF-126 Fungicide, acetamide derivative.

stability constant When a complex is formed in solution between a metal ion and a ligand the equilibrium may be expressed by a constant which is related to the free-energy change for the process:

$$M + A \rightleftharpoons MA; \; K_1 = \frac{[MA]}{[M][A]}$$

where K_1 is the stability constant, the terms in the brackets representing activities. Charges have been dropped for the sake of clarity and

$$-\Delta G = RT \ln K_1$$

Stepwise equilibria may also be expressed in this way, e.g.

$$M + A \rightleftharpoons MA; \; K_1 = \frac{[MA]}{[M][A]}$$

$$MA + A \rightleftharpoons MA_2; \; K_2 = \frac{[MA_2]}{[MA][A]}$$

$$MA_{(n-1)} + A \rightleftharpoons MA_n; \; K_n = \frac{[MA_n]}{[MA_{(n-1)}][A]}$$

$K_1, K_2 \ldots K_n$ being called the stepwise stability constants. Overall stability constants, designated β, may be evaluated by multiplication of the individual stability constants, e.g.

$$M + 2A \rightleftharpoons MA_2; \; \beta_2 = \frac{[MA_2]}{[M][A]^2} = K_1 K_2$$

stained glass Coloured glass produced by introduction of impurities generally by ion-exchange methods. Cu and Ag are widely used but many other elements have been used to produce coloured glasses. Enamels can also be applied to the surface of glasses.

standard cell potential $E^{\ominus}$ The standard Gibbs' energy for the cell reaction.

standard electrode An electrode which produces a standard e.m.f. Electrode potentials are normally quoted relative to the hydrogen electrode, although in practice reference is usually made to the calomel or other ELECTRODE.

standard state Using thermodynamics it is possible to predict whether or not reaction could occur when substances are placed together in a closed system. However, to do this it is necessary to define a standard state of the system as one which contains all the reactants and products: **1** in the pure state if they are crystalline substances or liquids; **2** at one atmosphere pressure if they are gases; **3** for one mol. of reactants; and **4** at a specified temperature, usually, but not necessarily, 298 K.

Thermodynamic quantities which refer to the standard state are denoted by superscript zeros (°), e.g. $\Delta G_T^{\ominus}, \Delta H_T^{\ominus}, \Delta S_T^{\ominus}$, the subscript denoting the temperature T of the system.

standard temperature and pressure (stp) 298.15 K and 1 bar. Reference conditions ambient.

stannane (tin hydride), SnH_4. M.p. –150°C, b.p. –52°C. Colourless gas ($SnCl_4$ plus $LiAlH_4$ in ether at –30°C). Decomposes to tin and H_2 even at 0°C. Acts as a reducing agent. Sn_2H_6 is also known.

stannanes (organotin hydrides), R_nSnH_{4-n}. Useful reducing agents, e.g. $R_3CX \rightarrow R_3CH$.

stannates See TIN OXIDES.

stannic compounds Tin(IV) compounds.

stannite, Cu_2FeSnS_4. A mineral with a structure derived from that of zinc blende. Stannites were formerly mixed oxides containing Sn(II).

stannous compounds Tin(II) compounds.

starch (amylum), $(C_6H_{10}O_5)_x$. The carbohydrate which is being continuously formed and broken down in the living cell and which also serves as a reserve material. It is, like cellulose, made up of a long chain of glucopyranose units joined together through oxygen by α-glucosidic bonds.

It yields glucose alone on complete hydrolysis, maltose when broken down by enzymes and dextrin under other conditions.

Starch consists of amylose, which is water soluble and retrogrades on concentration forming an insoluble precipitate, and amylopectin, a mucilaginous substance with characteristic paste-forming properties. Amylose is composed of long straight chains containing 200–1000 glucose units linked by α-1,4-glycoside links; amylopectin consists of comparatively short chains (about 20 glucose units) cross-linked by α-1,6-glycoside links. Both amylose and amylopectin have been synthesized from glucose-1-phosphate by the action of the enzyme phosphorylase.

Starch is insoluble in cold water, but in hot water the granules gelatinize to form an opalescent dispersion. It is made from corn, wheat, potatoes, rice and other cereals by various physical processes such as steeping, milling and sedimentation. It is used as an adhesive, for sizing paper and cloth, as an inert diluent in foods and drugs, and for many other purposes. World use 1998 15 million tonnes.

Starch can be split into amylose and amylopectin by a commercial process based on selective solubilities. Amylose is used for making edible films, and amylopectin for textile sizing and finishing, and as a thickener in foods.

Stark effect The effect of an external electric field in splitting the energy levels of an atom or molecule.

Stassfurt deposits Deposits of various salts mined at Stassfurt chiefly for potassium (carnallite). Kainite, kieserite and gypsum are also present, Br_2 is obtained as a by-product.

state Physical state is defined by physical properties, e.g. for a gas, volume, amount, pressure, temperature.

state function Description of particular (thermodynamic) properties which depend only on the current state and not on the history.

state symbols Symbols which designate the spin multiplicity and electronic state of a species. E.g. the ground state of a d^2 species is ^3F, the spin multiplicity symbol (3) designates 2 unpaired electrons, the symbol F designates an angular momentum quantum number of 3.

stationary phase See CHROMATOGRAPHY.

stationary state In a reversible chemical reaction, a condition is reached at which the rate of the forward reaction is exactly equal to the rate of the backward reaction. At this point the reaction system has reached a state of equilibrium referred to as a stationary state. In the stationary state the concentration of the reacting species remains constant.

statistical thermodynamics Method of understanding and describing behaviour, generally of gases, on basis of average behaviour of a large number of molecules.

steam Water vapour generally at temperatures above 100°C. Used for power generation, heat distribution, as a chemical reactant (e.g. watergas), in firefighting and for cleaning.

steam distillation Distillation carried out in a current of steam with the object of distilling out a particular component of a mixture, or of avoiding too high a temperature during distillation. As normally carried out, the operation consists of distilling a mixture of two immiscible liquids in a current of steam. The ratio of the weights of the two substances distilling is (m_1p_1/m_2p_2), where m_1 and m_2 are the molecular masses of the two substances and p_1 and p_2 are their respective vapour pressures at the distillation temperature. Substances with a relatively high molecular weight may be steam distilled over in relatively large amounts in spite of the fact that they may have relatively low vapour pressures at the temperature of distillation.

steam reforming The reaction of a hydrocarbon (natural gas) plus steam over a nickel catalyst at 800–1000°C to give CO plus H_2. After the WATER GAS SHIFT REACTION the process is used for HYDROGEN production. See METHANOL.

stearic acid (n-octadecanoic acid), $C_{18}H_{36}O_2$, $CH_3 \cdot [CH_2]_{16} \cdot C(O)OH$. Crystallizes in leaflets; m.p. 70°C, b.p. 376°C (decomp.). Soluble in ether and hot alcohol, insoluble in water. Stearic acid is one of the most common fatty acids, and occurs as glycerides in most animal and vegetable fats, particularly in the harder fats with high melting points. A solid mixture of stearic and palmitic acids, 'stearine', is used for making candles. The soaps are the sodium and potassium salts of stearic and palmitic acids. Used in cosmetics and pharmaceuticals.

stearine See STEARIC ACID, PALMITIC ACID.

stearyl alcohol (1-octadecanol), $C_{18}H_{38}O$. M.p. 58°C, b.p. 210°C. Important fatty alcohol obtained from naturally occurring glycerides. Used in the pharmaceutical and cosmetics industries, textiles, varnishes, and as a gel stabilizer for greases.

steel The name given to alloys of iron and carbon containing 0.05–1.5% C. See IRON.

Alloy steels are steels containing deliberate additions of alloying elements over and above the amounts which are residual from the steelmaking process.

Austenitic steels retain the ccp structure down to room temperature. For this reason these steels cannot be hardened by quenching.

Stainless steel. A group of Cr or Cr–Ni steels showing an unusually high resistance to corrosion by the atmosphere and many chemical reagents. In order to achieve this resistance to corrosion >12% Cr is necessary.

stereochemistry The study of the spatial arrangements of atoms in molecules and complexes.

stereoisomerism See ISOMERISM and CONFORMATION.

stereoselective reaction A reaction which produces a preponderance of one stereoisomer.

stereospecific reaction A reaction which produces mainly or specifically one stereoisomer.

steric hindrance or geometry A term intended to denote the influence exerted on a reacting group by the spatial arrangement of neighbouring atoms. The optical activity of the di-ortho substituted derivatives of diphenyl is probably due entirely to the size of the substituting groups, which obstruct the free rotation of the benzene rings about the single bond joining them and so produce an asymmetric molecule. This is a case of true steric hindrance. Steric hindrance can affect reaction rates and thermodynamic stability.

steroids A generic term for substances having the nuclear carbon skeleton shown, or a very similar structure.

The natural steroids, which arise from the oxidative cyclization of squalene, include sterols, bile acids, sex hormones, adrenocortical hormones, cardiac-active glycosides, sapogenins, alkaloids and insect hormones such as ecdysone. Many synthetic steroids are known, and some of these are important drugs. Semisynthetic steroids are made from, e.g. diosgenin.

sterols Hydroxides of derivatives of steroids. They are found in every animal and plant cell and in some bacteria, often partly esterified with the higher fatty acids (see in particular CHOLESTEROL and SITOSTEROL). Sterols can be isolated by hydrolysing the fat fraction with alcoholic alkali and extracting the unsaponifiable portion with ether. Plant sterols are known as phytosterols.

stibine See ANTIMONY HYDRIDE.

stibnite, Sb_2S_3. The only important antimony ore.

stigmasterol ((24S)-24-ethyl-5,22-cholestadien-3β-ol), $C_{29}H_{48}O$. Isolated from many plant sources where it often co-occurs with SITOSTEROL. It has been used for the synthesis of steroid hormones.

stilbene (trans-1,2-diphenylethene), $C_{14}H_{12}$. Colourless crystals; m.p. 124°C, b.p. 306–307°C. It can be prepared by the action of benzyl magnesium bromide on benzaldehyde, and by many other methods. Derivatives of stilbene are important in the dye industry and some have estrogenic activity (see STILBOESTROL). U.v. light converts stilbene into its *cis* isomer, isostilbene, a yellow oil.

stilbite A zeolite, $Na_2Ca[(Al_2Si_6)O_{16}].6H_2O$, used for softening hard water.

stilboestrol (4,4´-dihydroxy-$\alpha\beta$-diethylstilbene), $C_{18}H_{20}O_2$. White crystals, m.p. 168–171°C. Synthetic estrogen but no longer used because of carcinogenic properties. Formerly used for treatment of cancer of the prostate.

still Any apparatus or plant in which distillation is carried out.

STM Scanning tunnelling microscope.

stoicheiometric (stoichiometric) A stoichiometric compound is one in which the ratio of the number of atoms to each other, as determined from the atomic weights, is a ratio of small whole numbers. In non-stoichiometric compounds there are defects in the crystal lattice, or part replacement of atoms of one element by another.

stoicheiometry (stoichiometry) Generally applied to reacting species. The relative numbers of reacting atoms or molecules and of the products.

stoichiometry Alternative spelling of stoicheiometry.

stopped flow spectrophotometry A technique used to investigate the rates of fast chemical reactions, with half-lives in the range 10 seconds to 5×10^{-3} seconds. The technique involves measuring the optical density or other spectroscopic property of a segment of the resulting solution after the reactants have been mixed.

stp Standard temperature and pressure.

straight-run A term used to denote fractions produced from crude petroleum by distillation but without further processing.

strain hardening WORK HARDENING.

stream double refraction Some colloidal sols such as aged vanadium pentoxide, whose dispersed particles are rod-shaped, are non-polarizing – as can be shown by placing the sol between crossed nicols, there being total extinction. When the liquid is stirred, however, or made to flow, polarization takes place, the stream lines being shown in a brightly illuminated image. This is due to the orientation of the rod-shaped molecules on stirring. See TACTOSOL.

streaming potential The potential set-up when a liquid is forced by pressure through a diaphragm or membrane. The converse of electro-osmotic flow.

Strecker reaction A reaction for synthesis of α-amino-acid derivatives.

strengths of acids and bases The strength of an aqueous acid is its ability to give hydroxonium ions (H_3O^+) and that of an aqueous base its ability to give hydroxyl ions or to accept protons. For Lewis acids and bases, acidic and basic character are measured by the tendency to react with standard bases and acids respectively. Since the degree of dissociation varies with the dilution, it is necessary to compare the relative strengths of acids or bases in solutions of equivalent concentrations. This can be done by measuring the concentrations of hydroxonium ion or of hydroxyl ion. Those acids like hydrochloric and nitric which are largely dissociated into their ions are strong acids; those like ethanoic and tartaric which are only slightly dissociated are weak acids. Potassium, sodium and barium hydroxides which are largely dissociated are strong bases, whilst the slightly dissociated ammonium hydroxide is a weak base.

streptomycin, $C_{21}H_{39}N_7O_{12}$. Antibiotic, bactericide.

stripping The separation of the more volatile component(s) of a liquid mixture from the less volatile component(s) in such a way that the latter is obtained in a pure state, but the former not necessarily so. Stripping may be carried out either by a process of FRACTIONAL DISTILLATION or by bringing the mixture into contact with an inert gas into which the light material will diffuse and be carried away. See RECTIFICATION.

strong electrolytes Materials that are virtually fully ionized in solution – soluble ionic solids and strong acids in water.

strontianite, $SrCO_3$. Mineral used as a source of Sr compounds.

strontium, Sr. Alkaline earth metal, at.no. 38, at.mass 87.62, m.p. 777°C, b.p. 1382°C, density ρ 2540, crustal abundance 370 p.p.m., electronic configuration [Kr] $5s^2$. Occurs as strontianite ($SrCO_3$) and celestine ($SrSO_4$). The metal is obtained by reduction of SrO with Al at 1000°C or electrolysis of fused $SrCl_2$. It is a silvery-white metal, ccp, which reacts vigorously with water and on warming with H_2, O_2, N_2, halogens. Strontium compounds are used in pyrotechnics and flares, in glass and ceramics, for permanent magnets (strontium ferrite).

strontium carbonate, $SrCO_3$. Occurs naturally as strontianite. Prepared CO_2 on SrO or $Sr(OH)_2$ or a Sr^{2+} solution. Decomposes to SrO on strong heating. Used in fireworks, glass and sugar refining.

strontium chemistry Strontium is an alkaline earth of Group 2. It is an electropositive metal ($E°$ Sr^{2+} to Sr in acid -2.89 volts) which readily dissolves in acids. It forms a single range of Sr(II) compounds.

strontium chloride, $SrCl_2$. Anhydrous $SrCl_2$ formed from Sr plus Cl_2 or HCl over SrO or $SrCO_3$. Hydrates are formed.

strontium fluoride, SrF_2. Almost insoluble in water.

strontium hydroxide, $Sr(OH)_2$. Strong base, particularly in aqueous solution. Formed SrO plus water. Hydrates are known. Used in sugar refining.

strontium nitrate, $Sr(NO_3)_2$. Formed from $SrCO_3$ plus HNO_3. Forms 4 and 0 hydrates.

strontium oxide, SrO. M.p. 2430°C. Formed by ignition $SrCO_3$, $Sr(NO_3)_2$ or $Sr(OH)_2$. Forms $Sr(OH)_2$ in water.

Strontium peroxide, SrO_2, is formed from SrO and O_2 at dull red heat under pressure. $SrO_2.8H_2O$ is formed from $Sr(OH)_2$ plus H_2O_2 in cold aqueous solution.

strontium sulphate, $SrSO_4$. Occurs naturally as celestine. Prepared SrO, $Sr(OH)_2$ or $SrCO_3$ plus H_2SO_4. Sparingly soluble in water. Used in ceramics.

strychnine, $C_{21}H_{22}N_2O_2$. An alkaloid of *Nux vomica* and other plants. Colourless prisms, m.p. 270–280°C.

Base, it stimulates all parts of the nervous system and in large doses produces convulsions. Rodenticide.

styphnic acid See 2,4,6-TRINITRO-1,3-BEN-ZENEDIOL. Explosive lead styphrate used as an initiator.

styrene (ethenylbenzene, phenylethene, vinylbenzene), C_8H_8. Colourless aromatic liquid; b.p. 146°C. Telomerizes on heating to a glassy resinous substance, metastyrene, which regenerates styrene on heating. Prepared by treating cinnamic acid with hydriodic acid and then heating the product with aqueous potash. Styrene is manufactured by alkylating benzene with ethene followed by dehydrogenation, or from petroleum reformate, co-production with propylene oxide. Styrene is used almost exclusively for the manufacture of polymers, of which the most important are polystyrene, styrene–butadiene, styrene–acrylonitrite, ABS plastics. Sulphonated polystyrene is used as a cation-exchange resin. U.S. use 2000 4.76 million tonnes.

styrene–butadiene rubber (SBR) The most important synthetic rubber copolymer generally containing about 25% styrene, 75% butadiene.

styrene oxide (1,2-epoxyethylbenzene, phenyloxinane), $PhCHCH_2O$, C_8H_8O. B.p. 194°C. Prepared by epoxidation of styrene with peroxyethanoic acid. Reactions are similar to those of aliphatic epoxides (see, e.g. ethylene oxide). Reacts with alcohols to give monoethers, e.g. $PhCH(OMe)CH_2OH$. Phenols give resins.

styrene polymers Styrene, $PhCH=CH_2$, undergoes polymerization on heating or with free radical initiators. Radiation or Ziegler–Natta catalysts also give useful forms of polystyrene. Polystyrene is used in many copolymers (e.g. acrylontrile–butadiene–styrene, ABS). Polystyrene is a thermoplastic used for moulding (good electrical properties), film, as latexes for paints and paper, in coatings, in laminates, foams, beads.

styroform A branded extruded polystyrene foam used particularly in the construction industry.

suberane The trivial name for cycloheptane, C_7H_{14}, a colourless liquid, b.p. 118°C.

suberic acid (1,6-hexanedicarboxylic acid, octadioic acid), $HO_2C[CH_2]_6CO_2H$, $C_8H_{14}O_4$. Important dicarboxylic acid obtained by oxidizing ricinoleic acid (from castor oil); also obtained by oxidation of cyclo-octene or cyclo-octadiene; formerly obtained from cork. Used in the formation of alkyd resins and polyamides. Esters are used as plasticizers and heavy-duty lubricants and oils.

sublimation The volatilization of a solid substance into the vapour phase without passing through the liquid phase. Also used to describe the process of purification in which the vapour is condensed directly from the vapour phase to a solid (on a 'cold finger' often cooled by refrigerant). In the latter case this substance may melt during the initial vaporization. Used for purification.

sublimation point (temperature) The temperature at which the vapour pressure above a solid is equal to the external pressure.

submicron Particles, in the size range about 1μ, which are invisible in an ordinary microscope, but sufficiently large to appear in an ultramicroscope.

substitute natural gas (SNG) A manufactured fuel produced from coal or oil and used as a complete or peak-demand substitute for natural gas.

substitution reactions Reactions in which there is replacement of one atom or group in a molecule by another atom or group. E.g. the chlorination of benzene in the presence of $AlCl_3$ to give chlorobenzene in which one of the hydrogen atoms is substituted by a chlorine atom. This is an example of electrophilic substitution since the active chlorine species is Cl^+. Nucleophilic reactions are named as S_N1 (substitution, nucleophilic, monomolecular) and S_N2 (bimolecular) reactions. S_N1 reactions proceed via an intermediate carbenium ion and are favoured by t-alkyl halides. If the starting material is chiral a racemic product is formed. S_N2 reactions characterize the reactions of primary alkyl halides (e.g. with cyanide ion) and involve a transition state.

Octahedral substitution reactions (e.g. those involving cobalt(III) complexes) may proceed by both S_N1 and S_N2 mechanisms. In the S_N1 case a slow dissociative mechanism (bond breaking) may take place. Reaction with the substituting ligand then yields the product. In S_N2 reactions direct participation of the incoming group causes the co-ordination number of the transition state intermediate to increase. Here bond making and bond breaking are equally important. Substitution reactions of square planar complex, e.g. those involving platinum(II) complexes, have also been shown to proceed via S_N1 and S_N2 reactions.

substrate The substrate in an enzymic reaction is the substance on which the enzyme acts.

In general terms a substrate is the medium in which the reaction is carried out or the active species held or that provides a surface for adsorption.

succinic acid (butanedioic acid), $C_4H_6O_4$, $HO(O)C(CH_2)_2C(O)OH$. Colourless prisms; m.p. 182°C, b.p. 235°C. Occurs in amber, algae, lichens, sugar cane, beets and other plants, and is formed during the fermentation of sugar, tartrates, malates and other substances by a variety of yeasts, moulds and bacteria. Manufactured by the catalytic reduction of maleic acid or by heating 1,2-dicyanoethane with acids or alkalis. Forms an anhydride when heated at 235°C.

Forms both acid and neutral salts and esters. Used in the manufacture of succinic anhydride and of polyesters with polyols, enzymes, dyestuffs, perfumes, photography. Dimethyl esters, particularly in blends with other dimethyl esters (e.g. adipates) make good solvents.

succinic anhydride (dihydro-2,5-furandione), $C_4H_4O_3$. White crystalline solid; m.p.

$$\begin{array}{c} CH_2CO \\ | \quad\quad O \\ CH_2CO \end{array}$$

120°C, b.p. 261°C. See SUCCINIC ACID. Reduced to butyrolactone by sodium amalgam. Used in the manufacture of dyestuffs and polymers.

succinimide (2,5-pyrrolidine dione), $C_4H_5NO_2$, $\overline{CH_2)_2C(O)\cdot NH\cdot C(O)}$. Colourless plates; m.p. 126°C, b.p. 287°C (decomp.). Prepared by heating ammonium succinate. Acid forms salts with metals; these replace the hydrogen atom of the $>$NH group. Distillation with zinc dust gives pyrrole. Forms a chloroimide used for disinfecting water supplies.

sucrase See INVERTASE.

sucrates See SACCHARATES and SUCROSE.

sucrose (cane or beet sugar, β-D-fructo-furanosyl-α-D-glucose), $C_{12}H_{22}O_{11}$. M.p. 185–186°C. It has no reducing properties. It is

Glucopyranose   Fructofuranose

readily hydrolysed by dilute acids to glucose and fructose and even more rapidly by invertase; the hydrolysed product is termed invert sugar. Sucrose has been synthesized. Readily fermented to, e.g. ethanol; sweetening agent, flavouring and bulking agent for foodstuffs. Forms sucrates with the alkali and alkaline earth metals. Sucrose carboxylates, e.g. octa-acetate (ethanoate), used in adhesives, lacquers, plastics.

sugar of lead, $Pb(O_2CCH_3)_2.3H_2O$. See LEAD ETHANOATES.

sugars The sugars are carbohydrates, the majority of the natural sugars containing six or twelve carbon atoms in the molecule. Natural sugars are right handed. They are crystalline, very soluble in water and generally have a sweet taste. The sugar of commerce, called cane sugar or beet sugar according to its origin, is the disaccharide SUCROSE.

sulcofuron-sodium Insecticide, benzene sulphonate derivative.

sulcotrione Herbicide, hexane derivative.

sulfenatriazone Herbicide, triazole derivative.

sulfluramid Insecticide, sulphonamide derivative.

sulfometuron-methyl Herbicide. Sulphonamide derivative.

sulfosulfuron Herbicide, sulphonyl urea derivative.

sulfotep Insecticide, acaricide, organophosphate derivative.

sulfur North American spelling of SULPHUR.

sulphacetamide (4-aminobenzene sulphonacetamide) Fungicide, acaricide.

sulpha drugs (sulphonamides) Derivatives containing $-S(O)_2NR^1R^2$ grouping. Individual compounds not listed, herbicides, insecticides, antibacterials.

sulphamates Salts of sulphamic acid.

sulphamic acid, $H_3N^+.SO_3^-$. A colourless crystalline solid normally existing as a zwitterion; prepared by the reaction of SO_2Cl_2 or chlorosulphonic acid with ammonia, or, commercially, by the treatment of urea with fuming H_2SO_4. Sulphamic acid is very soluble in water and is a strong acid, forming salts, the sulphamates. It is a good acidimetric primary standard. It is used in herbicides, flameproofing agents, electrolytes in metal deposition processes and sweeteners.

sulphamide, $SO_2(NH_2)_2$, and **sulphimide**, $(SO_2NH)_n$. Colourless crystalline solids formed by the action of ammonia on a solution of sulphuryl chloride in benzene; free sulphimide exists only in the polymerized form. Both compounds are soluble in water and are readily hydrolysed to sulphamic acid and ammonia; the hydrogen atoms are in each case replaceable by metals to form salts. Many derivatives of sulphamide and cyclic derivatives of sulphimide are known.

sulphanes The hydrogen sulphides H_2S_2 to H_2S_8 and beyond (for H_2S see HYDROGEN SULPHIDE). Formed S in aqueous Na_2S and addition to HCl or S_2Cl_2 and H_2S. Decompose into H_2S and S.

sulphanilamides Antibacterials.

sulphanilic acid (4-aminobenzenesulphonic acid), $C_6H_7NO_3S$. Colourless crystals.
It is prepared by heating aniline sulphate for 8 hours at 190°C. It readily diazotizes and is used as first component in a large variety of azo dyes.

sulphatases Enzymes which catalyse the hydrolysis of sulphuric acid esters.

sulphates Formally any compound containing an anionic oxy-sulphur species. Practically derivatives of sulphuric acid; i.e. sulphates(VI) containing $(SO_4)^{2-}$ groups which may be ionic or co-ordinated.

sulphides Compounds of other elements with sulphur. Generally prepared by direct interaction of the elements. Metal sulphides are often very insoluble, occur as minerals and can be used for separation of the elements. Sulphides of the most electropositive elements give H_2S with water. Covalent derivatives formed with the less electropositive elements, e.g. liquid CS_2, S_2Cl_2.

sulphinyl Containing the >SO group.

sulphites Salts of sulphurous acid, i.e. sulphates(IV) containing $[SO_3]^{2-}$ species. Used as preservatives.

sulpholane (tetrahydrothiophen-1,1-dioxide, tetramethylenesulphone), $C_4H_8O_2S$. A viscous liquid, b.p. 285°C, which

readily solidifies to a colourless solid, m.p. 28°C when freed from water, with which it is miscible. Used as a selective solvent for liquid-vapour extractions especially by gas-liquid chromatography. Prepared by reduction of 2,5-dihydrothiophene-1,1,-dioxide, which is obtained from butadiene and sulphur dioxide.

sulphonamides Compounds containing the grouping $-SO_2N<$. They are obtained by the action of ammonia and primary and secondary amines on sulphonyl halides. Sulphanilamide is the active portion of the prontosil molecule. More effective derivatives of sulphanilamide have been introduced. They are prepared in general by condensing 4-acetamidobenzenesulphonyl chloride with the appropriate amine and removing the acetyl group by hydrolysis.

Sulphonamides are bacteriostatic (inhibiting growth) and not bactericidal, acting by allowing the natural body mechanisms to destroy the bacteria.

The sulphonamides are antagonized by 4-aminobenzoic acid and they act by inhibiting the uptake and utilization of 1-aminobenzoic acid by bacterial cells, which require this as a precursor of folic acid.

sulphonation Introduction of sulphonic groups. See SULPHONIC ACIDS.

sulphones Organic compounds containing an $\ge SO_2$ group linked directly to two carbon atoms, e.g. sulpholane. They are produced by oxidation of organic sulphides with nitric acid or potassium permanganate. They are colourless solids and are very stable. Many are used as solvents.

sulphonic acids (aliphatic) Prepared alkane with SO_3, or by oxidation of disulphides, e.g. MeSSMe gives CH_3SO_3H. Methane sulphonic acid, m.p. 20°C, b.p. 122°C, used as solvent, polymerization and alkylation catalyst.

sulphonic acids (aromatic) Organic compounds containing the –SO_3H group.

The sulphonic acids are usually prepared by the action of sulphuric acid, chlorosulphonic acid or oleum upon a compound. Sulphur trioxide complexed to pyridine or dioxan can be used with reactive substrates. Aminosulphonic acids such as sulphanilic and naphthionic acids are most conveniently prepared by heating the sulphate of the amine at 180°C.

Prolonged sulphonation of benzene gives 1,3-benzenedisulphonic acid. In the case of naphthalene, sulphonation gives 1-substitution at low temperatures and 2-substitution at higher temperatures.

The sulphonic acids are strongly acidic compounds, very soluble in water and readily give water-soluble metallic salts.

Converted into phenols by fusion with caustic alkalis. When they are fused with potassium cyanide, nitriles are obtained, e.g. benzonitrile from benzenesulphonic acid.

Sulphonation is widely used for the solubilization of insoluble azo dyestuffs.

sulphonphthaleins A group of dyes, many of which are used as indicators, made by heating 2-sulphobenzoic anhydride with phenols or substituted phenols. See, e.g. PHENOL RED.

sulphonyl halides Organic compounds $RS(O)_2X$, where R is an alkyl or aryl group and X is a halogen atom. A particularly important example is toluene-4-sulphonyl chloride (R = 4-$CH_3 \cdot C_6H_4$, X = Cl). This is prepared by the action of chlorosulphonic acid on toluene. U.S. production 1999 14.8 million tonnes. The properties of sulphonyl halides resemble those of carboxylic acid chlorides. See ETHANOYL CHLORIDE, TOLUENE-4-SULPHONYL CHLORIDE.

sulphoxides Organic compounds containing an $\ge SO$ group linked to two C atoms. Obtained by the oxidation of organic sulphides with hydrogen peroxide. They are usually hygroscopic liquids and form addition compounds with acids. Oxidized to sulphones. dimethyl sulphoxide is a useful solvent and chemical intermediate.

sulphoxylic acid, H_2SO_2. Two forms, symmetric, $S(OH)_2$, by hydrolysis of $S(NR_2)_2$ or $S(OR)_2$ (thiosulphite esters plus NaOEt) and asymmetric HS(O)OH (reaction intermediate and some esters). Cobalt(II) sulphoxylate is formed NaHSO_3 plus cobalt ethanoate plus NH_3. Used as reagents.

sulphur, S. Non-metal, at.no. 16, at.mass 32.066, m.p. 119.6°C, b.p. 444.6°C, density ρ 2070, crustal abundance 260 p.p.m., electronic configuration [Ne] $3s^23p^4$. Yellow non-metallic element of Group 16. Occurs native (extracted from subterranean sources with superheated steam – Frasch process). Also occurs as sulphates, e.g. gypsum ($CaSO_4.2H_2O$), anhydrite ($CaSO_4$) and sulphides, e.g. iron and copper pyrites (used as SO_2). Occurs as H_2S in some natural gas and oil (recovered by reaction with SO_2 over a catalyst). Sulphur exists in various allotropic forms, α-S, rhombic sulphur, contains puckered S_8 cyclic units. β-S, monoclinic sulphur also contains S_8 rings (the conversion temperature α $\rightleftharpoons$ β is 95.5°C). γ-S (also monoclinic) is deposited from some solvents and also contains S_8 rings; ρ-sulphur, rhombohedral, is formed from $Na_2S_2O_3$ and conc. HCl at 0°C followed by extraction with toluene, it contains S_6 chair rings. Melting of sulphur gives S_λ still containing S_8 rings but above 160°C spiral chains of μ-S are formed. π-S (probably S_6) is also present. Plastic sulphur (molten S into water) contains μ-S chains. Sulphur vapour contains S_2 (predominates, paramagnetic), S_4, S_6, S_8 species. Sulphur is a reactive element, combining with most other elements. The principal use is in H_2SO_4 production (90%) but also used in the wood pulp industry, photography, leather, rubber, dyestuffs industries, in detergents, fertilizers, insecticides and fungicides.

sulphur bromide (sulphur monobromide), S_2Br_2. The only known sulphur bromide.

sulphur chemistry Sulphur is an element of Group 16. It shows oxidation states +6, +4, +2, 0, –2. It behaves principally as a non-metal, forming covalent species (S_5^+, S_4^{2-}, etc. are also formed). It shows a great tendency to catenation. 'Onium cations are readily formed (e.g. Ph_3S^+).

sulphur chlorides Sulphur and chlorine react in all proportions.

Sulphur monochloride, S_2Cl_2. M.p. –80°C, b.p. 138°C. Yellow liquid hydrolysed by water to SO_2, HCl and S. Lower halides S_xCl_2 (x up to 5) are formed S_2Cl_2 plus H_2 at a hot surface with freezing of products. S_2Cl_2 is used in the rubber industry as a solvent for S. Polymerization agent.

Sulphur dichloride, SCl_2. B.p. 59°C. Red liquid.

Sulphur tetrachloride, SCl_4. M.p. –30°C. Stable only in the solid. The oxide chloride of sulphur(IV) is $SOCl_2$, THIONYL CHLORIDE.

Sulphur(VI) forms the oxide chloride SO_2Cl_2, sulphuryl chloride (see SULPHURYL HALIDES).

sulphur dyes Dyestuffs which are applied in a reduced state from solutions containing S^{2-}, SH^- or polysulphide. The dyestuff is reoxidized on the fibre. The actual dyestuff may or may not contain sulphur.

sulphuretted hydrogen, H_2S. See HYDROGEN SULPHIDE.

sulphur fluorides *Sulphur hexafluoride*, SF_6. M.p. –51°C. Formed S plus F_2. Very inert material used as an inert dielectric, for tracing gas movements, lasers. S_2F_{10} (toxic) is also formed from S plus F_2 and there is an extensive chemistry of SF_5 derivatives (e.g. SF_5Cl, ClF plus SF_4).

The oxide fluorides SO_2F_2 and SOF_4 and $N\equiv SF_3$ are known.

Sulphur tetrafluoride, SF_4. M.p. –121°C, b.p. –40°C. Reactive gas (SCl_2 plus NaF in $MeCN$). Used as fluorinating agent (e.g. $>C=O$ gives $>CF_2$). Derivatives, e.g. $F_3SN(CH_2CH_2OCH_3)_2$ are often more convenient to handle.

The oxide fluoride SOF_2 ($SOCl_2$ plus NaF) and $(NSF)_4$ are known.

Disulphur difluoride, S_2F_2. Two isomers, FSSF and $S=SF_2$, formed AgF plus S. Reactive material.

sulphuric acid, H_2SO_4, $S(O)_2(OH)_2$. M.p. 10°C, b.p. 340°C, d 1.83. Very important industrial acid, top tonnage chemical. Manufactured from SO_2 plus air over a Pt or V_2O_5 catalyst (500–600°C then 400–450°C). Emergent gases, SO_3, (100°C), are absorbed in 97% acid (contact process). In the virtually obsolete lead chamber process SO_2, NO_2 air and steam are sprayed into lead chambers to produce 62–68% acid. SO_2 is obtained from combustion of sulphide ores (approximately 10% from copper ores) or H_2S or by heating $CaSO_4$ with C and clay. H_2SO_4 is miscible in all proportions with water and forms an azeotrope b.p. 339°C containing 98.3% H_2SO_4. A solution of SO_3 in H_2SO_4 is fuming sulphuric acid (see SULPHURIC ACID, FUMING). It decomposes to SO_3 and H_2O above the b.p. It is a very powerful dehydrating agent (the reaction with H_2O is very violent and exothermic). The hot concentrated acid is a strong oxidizing agent (e.g. C gives CO_2). It is a strong acid although dissociation of $[HSO_4]^-$ is not always complete. Low volatility, and therefore displaces most other acids (or their anhydrides) from their salts. Can be handled in glass, cast iron or steel. Used in the production of phosphate (70% used for fertilizers), ammonium sulphate, aluminium sulphate, explosives, pigments, petrochemicals, pickling iron and steel, dyestuffs. U.S. production 2000 43.6 million tonnes. The salts of sulphuric acid are the SULPHATES, used as herbicides.

sulphuric acid, fuming (Nordhausen sulphuric acid, oleum) A solution of SO_3 in H_2SO_4 containing some $H_2S_2O_7$, etc. SO_3 distils from the acid on heating. Used in organic synthesis.

sulphur nitrogen derivatives There is an extensive chemistry of sulphur nitrogen derivatives, many of them being cyclic derivatives. S_4N_4 (S_2Cl_2 plus NH_4Cl, S plus NH_3 in CCl_4) gives a series of derivatives, e.g. $SOCl_2$ gives $(S_4N_3)Cl$. $(SN)_x$ is a polymer which is a good electrical conductor. S_7NH, $S_5(NH)_3$ and $S_4(NH)_4$ (S_2Cl_2 plus NH_3) all have structures related to S_8. See SULPHONAMIDES.

sulphurous acid, $OS(OH)_2$. Sulphur dioxide in water gives $SO_2.7H_2O$, a gas hydrate, and free H_2SO_3 does not exist. The solutions react with bases to give sulphites, sulphates(IV), containing the $[SO_3]^{2-}$ or $[HSO_3]^-$ ions. Sulphites react with acids to give SO_2. Sulphites are reducing agents.

sulphur oxide halides See SULPHUR HALIDES, SULPHURYL and THIONYL HALIDES. SO_2F_2, SOF_4, SOF_2, SO_2Cl_2, $SOCl_2$ are known.

sulphur oxides *Disulphur monoxide*, S_2O, SSO. Unstable, glow discharge on SO_2. *Disul-*

phur trioxide, S_2O_3. Blue–green solid S plus SO_3.

Sulphur dioxide, SO_2. M.p. $-72.7°C$, b.p. $-10°C$. Colourless gas with characteristic smell. Formed by burning S, metal sulphides, H_2S in air or acid on a sulphite or hydrogen sulphite. Powerful reducing agent, particularly in water. Dissolves in water to give a gas hydrate; the solution behaves as an acid – see SULPHUROUS ACID. Used in the production of SO_3 for sulphuric acid, food preservative, disinfectant and bleach. Air pollutant, formed by combustion of sulphur-containing fuels.

Sulphur trioxide, SO_3. M.p. $17°C$, b.p. $49°C$. Formed SO_2 plus O_2 over a catalyst (contact process – see SULPHURIC ACID). The solid exists in several forms. The gaseous form is monomeric planar but less reactive trimers and polymers are formed on standing. Readily absorbed by water (to give sulphuric acid) or H_2SO_4 (to give fuming sulphuric acid).

sulphur oxyacids Sulphur forms an extensive range of oxyacids which are dealt with under separate headings. In many cases the free acid is not known. The acids are:

$S(OH)_2, HS(O)OH$	sulphoxylic acid[†]
$OS(OH)_2$	sulphurous acid[†]
$O_2S(OH)_2$	sulphuric acid
$S(O)S(OH)_2$	thiosulphuric acid
$O(HO)SS(OH)O$	dithionous acid
$O(HO)SS_nS(OH)O$	polythionic acids
$O_2(HO)SS(OH)O$	pyrosulphurous acid,[†] disulphurous acid[†]
$O_2(HO)SS(OH)O_2$	dithionic acid
$O_2(HO)SOS(OH)O_2$	pyrosulphuric acid, disulphuric acid
$H_2S_3O_{10}$	trisulphuric acid
$H_2S_4O_{13}$	tetrasulphuric acid
$O_2S(OH)(OOH)$	peroxymonosulphuric acid
$O_2(HO)SOOS(OH)O_2$	peroxydisulphuric acid

[†] Free acid not known. Halosulphur acids, e.g. HSO_3F, are known.

sulphuryl Containing the $>SO_2$ group.

sulphuryl halides, SO_2X_2. Sulphuryl chloride, SO_2Cl_2, b.p. $69°C$ (SO_2 plus Cl_2 plus light) is used as a chlorinating agent. SO_2F_2 and SO_2Br_2 are also known. Sulphuryl fluoride is used as an insecticide.

sulphydryl The –SH group, also called thiol.

sulprofos Insecticide, organophosphate derivative.

sunscreens Materials, e.g. TiO_2 or organic materials, which protect the skin against u.v. light.

super acids Mixtures, e.g. $HF-SbF_5$, HSO_3F-SbF_5, which act as very strong acids and, e.g. induce carbenium ion formation, protonation, hydride abstraction.

superconductivity Below a transition temperature Tc certain metals and intermetallic compounds and ceramic oxides possess superconductivity, have zero electrical resistivity and undetectable magnetic permeability. They are of great engineering interest. See DICTIONARY OF PHYSICS.

supercooling The phenomenon of cooling a liquid to a temperature below that at which crystallization would normally occur, without the separation of solid. E.g. pure water may be cooled to several degrees below its freezing point provided crystallization nuclei, such as suspended dust particles, are absent. Glass is an extreme example of supercooling in which solidification has occurred without crystallization. Superheating is the comparable phenomenon on boiling. Solutions which have been cooled below the saturation temperature for the quantity of solute present, without crystallization occurring, are said to be supersaturated.

supercritical fluid extraction The use of supercritical fluids for processing or reaction. Carbon dioxide widely used.

superheating See SUPERCOOLING.

superheavy elements See POST-ACTINIDE ELEMENTS.

superoxide dismutase (SOD) Enzymes catalyse the reaction

$$2O_2^- + 2H^+ \rightarrow H_2O_2 + O_2$$

Have been isolated from many plants, animal and bacteria cells. Organisms that utilize oxygen for respiration produce some superoxide radicals and superoxide dismutase protects cells against the deleterious action of these radicals. See also XANTHINE OXIDASE.

superoxides Compounds containing the paramagnetic O_2^- ion, e.g. KO_2. Can be harmful to life.

superphosphates Insoluble calcium phosphate, $Ca_3(PO_4)_2$, treated with concentrated sulphuric acid gives calcium hydrogen phosphate, $Ca(H_2PO_4)_2$, which is soluble in water and is a very effective fertilizer. The dry product consists of the soluble calcium salt together with calcium sulphate and sulphates of other metals present in the raw material, and is known as 'superphosphate'. The grades of superphosphate most widely used have about 13.7% of phosphoric acid (30% tricalcium phosphate) and 16–17% of phosphoric acid. See CALCIUM PHOSPHATES.

supersaturation See SUPERCOOLING.

supported catalyst See CATALYST.

supramolecular chemistry Self-assembly of ordered groups of molecules or ions. Often present in surface-active agents.

surface-active agents Water, possessing powerful intermolecular attractive forces, has a high surface tension (72.8 dynes/cm at 20°C). Many soluble substances (mainly organic) when dissolved in water reduce the surface tension even in very low concentrations.

These surface-active agents have weaker intermolecular attractive forces than the solvent, and therefore tend to concentrate in the surface at the expense of the water molecules. The accumulation of adsorbed surface-active agent is related to the change in surface tension through the Gibbs adsorption energy. Surface-active agents, which include certain dyestuffs, wetting agents, detergents, antibiotics, bactericides, etc., may be electrically neutral, nondissociating substances (non-ionic), or they may ionize to give surface active anions or cations with small oppositely charged counterions (usually metallic ions and halogen ions respectively). Used in mineral treatment.

surface activity A phenomenon which arises from the non-symmetrical distribution of attractive forces between molecules (and ions) residing within the surface layer. This results in their being attracted away from the surface towards the interior, and with liquids leads to the contraction of the surface to the smallest possible area. Work has therefore to be performed to cause the surface to expand, the amount being numerically equal to the surface tension. The surface activity of solids, particularly when finely divided, often leads to gas adsorption, with which catalysis is frequently associated.

surface combustion The burning of gaseous fuel and air in such a way that catalytic oxidation occurs at the surface of a porous refractory. Combustion is rapid and complete within a very thin flame section and very high rates of heat release are obtained. Many industrial furnaces and kilns employ this method.

surface compounds The surfaces of many solids become coated with a unimolecular layer of chemisorbed gas in which the gas molecules form chemical bonds with the surface atoms leading to the formation of surface compounds. Such layers often give the surface its characteristic properties. Thus many metals become coated with an oxide layer which render them stable to further attack.

surface energy The surface energy of a system is defined as the work necessary to increase the surface by unit area against the force of surface tension.

surface orientation When a film of an organic material is formed on water, the molecules in the surface layers usually orientate themselves according to their structure. Groups such as carboxylate or sulphonate will tend to confer solubility upon the molecule and interact with water. Such groups are said to be hydrophilic, i.e. attracted to the water, and the molecule will be anchored to the surface by those groups which will tend to penetrate below the surface of the water. Conversely, hydrocarbon groups are hydrophobic and will tend to remove themselves from the water. Thus, e.g. long-chain fatty acids, will tend to orientate themselves so that the hydrocarbon groups are perpendicular to the water surface into which the acid groups are dipping. Compare MICELLES.

surface plasma resonance (SPR) Thin gold films functionalized with receptor molecules. Bonding molecules change refractive index and shift SPR wavelength.

surface potential The change in work function, relative to a clean surface, when an adsorbed layer is present on the surface. The difference between potentials at various points close to an electrode.

surface pressure Thin films of spreading oils, e.g. oleic acid, lower the surface tension of the liquid (generally water) on which they are spread. Consequently, they exert a force of repulsion on floating particles. The force exerted by a spreading film per cm of boundary is numerically equal to the lowering of surface tension; it is called the surface pressure of the film by analogy (in two dimensions) with the pressure of a gas and can be measured directly. Study of surface pressure can give information on the type of film (e.g. unimolecular or larger). The surface pressure exerted by a solid or liquid present in excess is termed its spreading pressure.

surface tension The molecules in the body of a liquid are equally attracted in all directions by the other molecules. Those at the surface, on the other hand, will have a residual inward attraction. This is associated with a *surface tension* which acts in a direction parallel to the boundary surface and tends to reduce that surface to a minimum. Work has to be done to increase the area of the surface against this force. The unit of surface tension is measured in energy metre^{-2} or Newtons metre^{-1}. See SURFACE ENERGY, SURFACE ACTIVE AGENTS.

surface viscosity Unimolecular films on liquid surfaces may be either readily mobile (e.g. a 'gaseous' film) or slow to flow under the action of a two-dimensional stress (see SURFACE PRESSURE). They therefore possess surface viscosity, the two-dimensional analogue of ordinary viscosity. Some surface films show surface plasticity, i.e. they behave as solids until a critical shearing stress is applied.

surfactants Surface-active agents which must be soluble in the solvent (generally water), contain groups of opposite polarity and solubilizing tendencies, form oriented monolayers at phase interfaces, form micelles, have detergency, foaming, wetting, emulsifying and dispersing properties. Long-chain sulphonic acids are typical surfactants; fluorocarbon derivatives are even more active. Sugar-based surfactants and alcohol ethoxylates have useful surfactant properties.

surgical spirit Industrial METHYLATED SPIRITS further denatured by the addition of castor oil (2.5%), diethyl phthalate (2.0%) and methyl salicylate (0.5%).

suspended transformation A system in a METASTABLE STATE.

suspending agents The suspension of a small amount of material in a large volume of liquid is required for many purposes. To prepare this the material to be suspended should be ground as finely as possible and given an electrical double layer by adsorption of an ionic detergent. A further method is to obstruct sedimentation by the use of a bulky hydrated precipitate which will hold up the solid to be suspended. Bentonite and related clays may be used for this purpose.

suxamethonium chloride (succinyl choline chloride, succinyl dicholine), $C_{14}H_{30}Cl_2N_2O_4$. White powder prepared from dimethylaminoethanol and succinyl chloride, followed by methylation. Neuromuscular blocking agent.

Suzuki reaction See BORONATES. Coupling reaction to form diacyls, etc.

Sv Sievert.

sweating The process of removing entrapped oil and low-melting waxes from the filter cake obtained in dewaxing processes.

sweetening The process of treating 'SOUR' gasoline and kerosine fractions for the purpose of converting or extracting mercaptans to remove objectionable odours. All involve oxidation of mercaptans to dialkyl disulphides.

sweetening agents The sugars, particularly sucrose, fructose, lactose and maltose are the classic sweetening agents. Sorbitol and mannitol are sweet-tasting polyols. Synthetic sweeteners have no food value. Cyclamates are possibly carcinogenic. Saccharin and aspartame are widely used.

swelling of colloids When a piece of dry gelatin is exposed to moist air or placed in water, it becomes hydrated and swells considerably. This is an effect common to all hydrophilic colloids – most natural organic matter, e.g. cellulose, wood, carbohydrates. The swelling produces considerable pressure in the colloid and, in consequence, still occurs even when opposed by an external pressure of many atmospheres.

The swelling of gels is markedly affected by the presence of electrolytes.

sydnones Heterocyclic compounds for which satisfactory formulae using normal covalent bonds cannot be written. Termed 'mesoionic' and are pseudoaromatic. Undergo aromatic substitution. Sydnone itself can be represented as:

Sydnones are neutral, highly crystalline, stable compounds, soluble in most organic solvents. *N*-Arylsydnones typically are obtained by treating *N*-nitroso-*N*-arylglycines with ethanoic anhydride. The parent glycine is regenerated when the sydnone is heated with dilute alkali.

sylvestrene, $C_{10}H_{16}$. Monocyclic terpene, formed by the isomerization of Δ^3- and Δ^4-carene. Two double bonds.

sylvine (sylvite), KCl. Occurs in salt deposits – an important source of potassium salts.

sylvinite, KCl–NaCl. A source of potassium salts.

symmetry elements The operators which give repetition and pattern in molecules and lattices. Symmetry elements comprise:

1 Axes of symmetry. An axis about which rotation of the body through an angle of $2\pi/n$ (where n is an integer) gives an identical pattern; 2-fold, 3-fold, 4-fold and 6-fold axes are known in crystals; 5-fold axes are known in molecules. In a lattice the rotation may be accompanied by a lateral movement parallel to the axis (screw axis).

2 Planes of symmetry. Planes through which there is reflection to an identical point in the pattern. In a lattice there may be a lateral movement parallel to one or more axes (glide plane).

3 Identity operator. Present in all lattices and molecules.

4 Centre of symmetry. A point through which there is reflection to an identical point in the pattern.

symproportionation (conproportionation) The opposite of DISPROPORTIONATION.

syn Having particular substituents on the same side of a molecule. See ISOMERISM.

synchrotron radiation Intense radiation from a synchrotron used for structure determination, induction of a reaction.

syndiotactic See SYNTACTIC POLYMERS.

syneresis The separation of liquid from a gel on standing. All gels show syneresis to some extent and the phenomenon may be regarded as a continuation of the process of gelation, the gel material adhering at more and more places and, in consequence, shrinking and squeezing out a portion of the dispersion medium. It is often the opposite of swelling of colloids. Viscose gels synerize to a large extent. The phenomenon is commonly met in the cooking of foods and is thought to play an important part in gland secretion.

synergism The phenomenon by which the additive properties of two components are much greater than the sum of the effects of the two separately. The opposite of antagonism.

synergist A compound which, whilst formally inactive or weakly active, in a particular context can give enhanced activity to another compound present in the system.

syntactic polymers Polymers with a stereoregular structure in which the substituent groups lie alternately above and below the plane of the carbon chain. Such structures are also referred to as syndiotactic configurations.

synthesis In a chemical sense the building up of complex molecules from smaller ones.

synthesis gas A mixture of hydrogen and carbon monoxide, normally produced from the reaction of methane and steam under pressure or, less usually, by the partial oxidation of methane. Synthesis gas is used in the manufacture of methanol, oxo alcohols and hydrocarbons. See FISCHER–TROPSCH REACTION, OXO REACTIONS.

synthetic fibres See FIBRES.

synthetic rubbers See RUBBER, ELASTOMERS, etc.

synthon Molecule or molecule type suitable for synthesis.

systemic insecticides Compounds which when applied to the leaves, stems and sometimes the roots of plants, are absorbed and translocated in the plant in the course of normal nutrition in concentrations safe for the plants but lethal to insects feeding on the plants.

SZ 1-121 Acaricide tetrazine derivative.

Szilard–Chalmers effect See RECOIL ATOM.

T Tritium, hydrogen isotope $_1^3$H.

Ta Tantalum.

tactosols Colloidal sols which contain non-spherical particles, which are capable of orientating themselves in a streaming potential. E.g. aged vanadium pentoxide sol separates into rod-shaped particles in a concentrated tactosol and a dilute isotropic atactosol. In a magnetic field the particles of such a sol arrange themselves along the lines of force.

talc, $Mg_3(OH)_2Si_4O_{10}$. A silicate with layers of linked SiO_4 tetrahedra; the layers are electrically neutral so that there is only weak interaction between layers. The crystals are soft and used as a lubricant (French chalk) and dusting powder. Also used as an extender and filler in paint, rubber, insecticides, and tar and asphalt roofing materials, as a constituent of toilet powders and in some types of porcelain. Steatite is a pure compact variety, soapstone is a massive impure form. See CLAYS.

tall oil A major by-product of the kraft or sulphite pulping process of wood. Mainly fatty acids but contains some unsaponifiable material. Used in protective coatings (30%), soaps and detergents (15%), esters, plasticizers (30%), adhesives, paper.

tallow Fatty tissue from animals. Mainly fatty acid glycerides.

talose See HEXOSE.

tamazepam The most commonly prescribed benzodiazepine.

tamoxifen Anti-estrogen. Used in cancer therapy.

Tanabe–Sugano diagram Graphs showing the relation between the splitting of electronic energy levels (E/B) and crystal-field stabilization energy (Δ/B) where B is a Racah parameter. Used in assigning electronic transitions in complexes and obtaining values of Δ and B.

tanacetyl alcohol See THUJYL DERIVATIVES.

tannase A number of moulds, e.g. *Aspergillus niger*, contain tannase, an enzyme which splits tannins of the ester type, of which methylgallate is the simplest equivalent.

tannic acid (gallotannic acids, tannin) (a) Derivatives of flavonols, (b) esters of sugars, usually glucose. Contain trihydroxybenzene carboxylic acid residues. Obtained from oak galls. Used as mordant in inks, sizing, fabric finishing, clarifying.

tanning The process of changing skins and hides into useful leather. Skins are cleaned and soaked in water, treated with lime, and hair and flesh removed. Neutralized with buffering salts [$(NH_4)_2SO_4$, NH_4Cl] and proteolytic enzyme and finally tanned with tannin, or with a basic chromium(III) sulphate.

tanning development The use of the products of photographic development, dihydroxybenzenes, to react with the photographic gelatin to produce cross-linking and hardness in the developed areas which remain stable whilst the rest of the gelatin may be washed away. Used for the preparation of printing plates.

tannins A large class of amorphous substances present in plants. They have an astringent taste, give a blue or green colour with iron salts and are precipitated from water solution by proteins and by alkaloids. They are condensation products of various phenols, of which the most important are pyrogallol and catechol. The tannins have similar structures to TANNIC ACID. Condensed tannins are flavonol derivatives, hydrolysable tannins are sugar (mainly glucose) esters with one or more trihydroxybenzene carboxylic acids. The tannins can be obtained by extracting the raw materials with water or other solvents and precipitating with lead ethanoate. It is their property of precipitating gelatin which is responsible for the use of tannins in the treatment of hides to make leather. They are also used as mordants in the textile industry.

tantalates See TANTALUM OXIDES.

tantalite, $(Fe,Mn)(Nb,Ta)_2O_6$ with excess Ta. The principal ore of TANTALUM.

tantalum, Ta. Metal, at.no. 73, at.mass 180.95, m.p. 3017°C, b.p. 5458°C, density ρ 17 754, crustal abundance 2 p.p.m., electronic configuration [Xe] $4f^{14}3d^36s^2$. An element of Group 5. Occurs with Nb in tantalite $(Fe,Mn)(Nb,Ta)_2O_6$. The ores are treated by NaOH fusion followed by acid washing to give mixed pentoxides. Ta is separated by liquid extraction from HF with methyl isobutyl ketone. The metal is prepared from Ta_2O_5 by reduction with alkali metals or by electrolysis of fused fluorides; it is bcc, remaining lustrous in air but reacts with oxygen or steam at high temperatures and dissolves in HNO_3–HF. Used in resistant alloys, particularly construction for the chemical industry, capacitors and in surgical appliances. World demand 2001, 2600 tonnes.

tantalum chemistry Largely that of the +5 oxidation state and very similar to NIOBIUM CHEMISTRY. The major difference is the absence of stoichiometric lower oxides for Ta and minor differences in complex species, e.g. $[TaCl_6]^-$ and $[NbOCl_5]^{2-}$ in HCl solutions. Shows oxidation states –3 and –1 to +5. Ta–Ta bonds common in oxidation states +2 and +3.

tantalum halides Known for all TaX_5 and TaX_4 (except TaF_4). TaX_5 ($Ta + X_2$; $Ta_2O_5 + C + Cl_2$ or CCl_4 give $TaCl_5$) contain bridged polymeric species ($TaCl_5$ dimer). TaF_5, m.p. 95°C, b.p. 229°C; $TaCl_5$, m.p. 211°C, b.p. 241°C. Hydrolysis gives oxide halides, e.g. $TaOF_3$, TaO_2F. Complexes are formed with many donors and complex fluorides, e.g. $[TaF_6]^-$, $[TaF_7]^{2-}$, $[TaF_8]^{3-}$, $[TaCl_6]^-$ are formed in acid solution. TaX_4 ($TaX_5 + Ta$) are dark-coloured solids with some Ta–Ta interaction. Adducts are formed with various ligands. Lower halides, e.g. $[Ta_6Cl_{12}]Cl_{4/2}$, $[Ta_6Cl_{12}]Cl_{6/2}$ (not fluorides) are formed by further reduction – they contain metal-cluster species and there is an extensive chemistry based on, e.g. $[Ta_6Cl_{12}]^{n+}$ ($n = 2, 3$ or 4) species. TaF_3 (Ta + HF at 225°C) has an essentially ionic structure.

tantalum oxides Ta_2O_5 is the most important oxide and is prepared by dehydration of the hydrated oxides formed by hydrolysis of halides. It is a dense white inert material, reduced to $Ta_{2.5}O_5$. The pentoxide forms tantalates, e.g. $LiTaO_3.K_3TaO_4$, as mixed metal oxides by fusion with oxides or carbonates, reduction gives bronzes. Isopolytantalates, e.g. $[H_xTa_6O_{19}]^{(8-x)-}$ ($x = 0, 1$ or 2) are formed by fusing Ta_2O_5 with alkali and dissolving the melt in water.

tapiolite, $FeTa_2O_6$. A tantalum mineral.

TAPS (TAPSO) Propanesulphonic acid derivatives, used as biological buffers.

tar The non-aqueous liquid condensate resulting from the destructive distillation of carbonaceous materials. Types are: **1** High-temperature tar from coke ovens. **2** Low-temperature tar from the low-temperature carbonization of coal. **3** High-temperature gasworks tar. **4** Oil tar from water-gas processes. **5** Wood tar or Stockholm tar from wood carbonization.

tar acids These consist largely of phenols, and are obtained from tar by reaction with sodium carbonate to obtain the sodium phenolate and treatment of this compound with CO_2 to regain the phenol. See CRESYLIC ACIDS.

tar oils Heavy tar oils, mainly aromatic hydrocarbons. Insecticides, herbicides, fungicides.

tar sands See ATHABASCA TAR SANDS.

tartar emetic See ANTIMONY POTASSIUM TARTRATE.

tartaric acid (2,3-dihydroxybutanedioic acid), $C_4H_6O_6$, $HO_2CCH(OH)CH(OH)CO_2H$. Occurs in two optically active and two inactive forms. (+)-Tartaric acid, m.p. 170°C, very soluble in water. It occurs in the grape and a few other fruits. Manufactured from tartar argol and wine lees (potassium salts) by precipitation as the calcium salt and decomposition of this with sulphuric acid. Chief use is as foodstuff, acidulant in the manufacture of effervescent drinks but also used in textiles, dyeing, metal cleaning. (–)-Tartaric acid is obtained from racemic acid by fractional crystallization of the cinchonine salt.

Racemic acid, (±)-tartaric acid, is a compound of the two active forms. M.p. 273°C (with $1H_2O$), m.p. 205°C (anhydrous). Formed, together with mesotartaric acid, by boiling (+)-tartaric acid with 30% NaOH solution, or by oxidation of fumaric acid.

Mesotartaric acid is achiral, m.p. 140°C (anhydrous). Very soluble in water. Obtained from the mother-liquors in the preparation of racemic acid; or by oxidation of maleic acid.

tartrazine A pyrazolone dyestuff.

tartronic acid (hydroxymalonic acid, hydroxypropanedioic acid), $C_3H_4O_5$, $HO_2CCH(OH)CO_2H$. Colourless crystals with $1H_2O$ lost at 60°C. M.p. 160°C (decomp.). Prepared by heating dinitrotartaric acid in aqueous alcohol.

taurine (β-aminoethylsulphonic acid), $C_2H_7NO_3S$, $NH_2 \cdot CH_2 \cdot CH_2 \cdot SO_3H$. Crystallizes in columns, decomposing at 317°C. In combination with cholic acid it forms one of the bile acids. It is formed in the liver from cysteine.

taurocholic acid (cholyltaurine), $C_{26}H_{45}NO_7S$. The ester of taurine and cholic acid. It occurs in the bile as the sodium salt. See BILE SALTS.

tautomerism (dynamic isomerism) See ISOMERISM.

tau value (τ) A measure of shift in NUCLEAR MAGNETIC RESONANCE.

taxol Important anti-cancer drug originally isolated from yew trees.

Tb Terbium.

236-TBA Herbicide, benzoic acid derivative.

TBA$_E$ (*t*-butyl acetate (ethanoate)).

Tc Technetium.

TCA-sodium Herbicide, trichloroacetate derivative.

TCDD See DIOXIN.

TCNE Tetracyanoethylene.

TDI Tolylene-2,4-diisocyanate.

Te Tellurium.

tear gas Materials that cause pain and burning to the eyes and skin, used in riot control.

tebucnazole Fungicide, azole derivative.

tebufenozide Insecticide, hydrazine derivative.

tebufenpyrad Acaricide, pyrazole derivative.

tebupirimfor Insecticide, organophosphate derivative.

tebutam Herbicide, amide derivative.

tebuthuron Herbicide, diazole urea derivative.

technetium, Tc. Artificial metallic element, at.no. 43, at.mass ^{99}Tc 98.906, m.p. 2157°C, b.p. 4265°C, density ρ 11 500, (calc.) electronic configuration [Kr] $4d^55s^2$. ^{99}Tc (half-life 2×10^5 years) is obtained from fission products. All Tc isotopes are radioactive. The metal is obtained by hydrogen reduction of Tc compounds. It is silvery-grey in colour and has an hcp structure. It dissolves in H_2O_2, nitric and sulphuric acids and is attacked by hot Cl_2, S and O_2. Pertechnetates, TcO_4^-, are effective corrosion inhibitors. $^{99}Tc^{III}$ is used in clinical medicine.

technetium chemistry Technetium is an element of Group 7 of the transition series. Its chemistry is very similar to that of RHENIUM although the higher oxidation states are slightly more strongly oxidizing.

The halides known are yellow TcF_6 and TcF_5 (both Tc and F_2). Red–brown $TcCl_4$ (Tc + Cl_2) is the only known chloride. Oxides Tc_2O_7, TcO_3 and TcO_2 are known. The complexes are similar to those of Re; K_2TcH_9 is known. Some metal–metal bonded species known.

tecloftalam Bactericide, phthalamic acid derivative.

tecnazene Fungicide, plant growth regulator, nitrobenzene derivative.

teeth Mainly calcium phosphate in the form of carbonated hydroxyapatite.

Teflon See POLYTETRAFLUOROETHENE and FLUORINE-CONTAINING POLYMERS. See also KEL-F.

teflubenzuron Insecticide, urea derivative.

tefluthrin Insecticide, pyrethroid.

teichoic acids Polymers of sugar alcohols linked by phosphate groups and found in bacteria.

TEL, Et_4Pb. LEAD TETRAETHYL.

tellurates Formally salts containing oxyanions of tellurium. Tellurates(VI) contain octahedral $[TeO_6]^{6-}$ species and derivatives of these. Tellurates(IV), tellurites, contain $[TeO_3]^{2-}$ species.

telluric acid, $Te(OH)_6$. See TELLURIUM OXYACIDS.

tellurides Binary derivatives of Te with other elements. Similar structurally to sulphides.

tellurites TELLURATES(IV).

tellurium, Te. Non-metal, at.no. 52, at.mass 127.60, m.p. 450°C, b.p. 988°C, density ρ 6240, crustal abundance 0.005 p.p.m., electronic configuration [Kr] $4d^{10}5s^25p^4$. Group 16 element occurring naturally as tellurides in sulphide ores. Recovered from the flue dust after combustion of sulphide ores (particularly Au, Ag, Cu). It is taken into solution with H_2SO_4 and reduced to Te with zinc. The grey 'metallic' form has a chain structure with low electrical conductivity. The amorphous form is grey-black in colour. Te combines readily with O_2, halogens and metals, and dissolves in oxidizing acids. Used in alloys (particularly with Pb and also in Cu and stainless steel), semiconductors, glass. Te compounds are poisonous.

tellurium, organic derivatives Tellurium forms organic derivatives in the +2 and +4 states. The +2 compounds are similar to divalent sulphur derivatives although less stable. Tellurium(IV) derivatives are comparatively unstable.

tellurium bromides
Tellurium dibromide, $TeBr_2$. M.p. 210°C, b.p. 339°C. Unstable compound formed Te plus $TeBr_4$ in Et_2O.
Tellurium tetrabromide, $TeBr_4$. M.p. 380°C, b.p. 420°C. Red solid, Te plus excess Br_2. The vapour is dissociated to $TeBr_2$ and Br_2. Forms many complexes, including $[TeBr_6]^{2-}$ ions.

tellurium chemistry Tellurium is a Group 16 element. It is a non-metal showing oxidation states −2, +2, +4 and +6 but is much more electropositive than sulphur giving Te_4^{2+} (and other cluster) compounds and almost cationic character in TeO_2. Te(IV) derivatives form complexes, e.g. $[TeCl_6]^{2-}$. Forms polymeric cations, e.g. $[Te_2S_8]^{2+}$.

tellurium chlorides
Tellurium dichloride, $TeCl_2$. Unstable solid $TeCl_4$ plus Te.
Tellurium tetrachloride, $TeCl_4$. Colourless solid, m.p. 225°C, b.p. 390°, obtained Te plus excess Cl_2. Hydrolysed by water, forms complexes including $[TeCl_6]^{2-}$ ions.

tellurium fluorides
Tellurium tetrafluoride, TeF_4. M.p. 130°C. Colourless deliquescent solid prepared SeF_4 plus TeO_2. Hydrolysed by water.
Tellurium hexafluoride, TeF_6. M.p. −35.5°C, sublimes −39°C. Stable gas prepared Te plus F_2. Reacts to give many derivatives, e.g. TeF_5OR, TeF_5OH (forms $[TeF_5O]^-$ derivatives).
Te_2F_{10}, and oxide fluorides, e.g. TeF_5OTeF_5, are also formed during the fluorination of tellurium oxides.

tellurium oxides
Tellurium monoxide, TeO. Black solid resulting from thermal decomposition of $TeSO_3$ (Te plus SO_3).

Tellurium dioxide, TeO_2. Obtained by burning Te in air or heating tellurates(IV). Has rutile or brookite structures. Almost insoluble in water but gives tellurates(IV) with bases and, e.g., $TeCl_4$ with acids.
Tellurium trioxide, TeO_3. Yellow or grey solid formed by heating telluric acid, H_6TeO_6. Oxidizing agent being reduced to TeO_2. Gives tellurates(VI) with hot concentrated alkalis.

tellurium oxyacids
Tellurous acid, H_2TeO_3. Does not exist as TeO_2 is insoluble in water. *Tellurates(IV)*, tellurites containing $[TeO_3]^{2-}$ species can be formed from TeO_2 and bases.
Telluric acid, $Te(OH)_6$. Is formed from Te in aqua regia and a chlorate(V). Forms tellurates, e.g. $Na[TeO(OH)_5]$, $Na_2[TeO_2(OH)_4]$.
Fluorotellurates containing $\{OTeF_5\}^-$ groups are also known from $HOTeF_5$ – see TELLURIUM FLUORIDES.

tellurous acid See TELLURIUM OXYACIDS.

telomerase Enzyme which acts on telomers. See DICTIONARY OF BIOLOGY.

telomerization Polymerization to a very low mol·wt. polymer, or condensation involving a few molecules of monomer. E.g. $[C_2F_4]_4$ is a telomer; $[C_2F_4]_n$ (n large) is a polymer.

temazapam (dibenzodiazapam) Sedative, controlled substance.

TEMED N,N,N',N'-tetramethyldiaminoethane.

temephos Insecticide, organophosphate derivative.

temperature (T) Relative property that determines the direction of heat flow between objects. Thermodynamic temperature is measured in degrees Kelvin, practical temperatures in degrees Celsius, centigrade or Kelvin. See DICTIONARY OF PHYSICS.

tempering The process of reheating a quenched steel to relieve internal stresses, reduce hardness and tensile strength, and restore ductility and resistance to impact loading. Tempering is carried out within the temperature range 200–550°C for plain carbon steels. For alloy steels the temperatures may be somewhat higher. The higher the tempering temperature the greater the loss in hardness.

template reaction A reaction, often a condensation reaction or a polymerization reaction, carried out in the presence of a metal species which, by co-ordination, induces a particular configuration and a particular product in the reaction.

teratogen A chemical which produces malfunction, generally in the form of mutations or tumours.

terbacil Herbicide, uracil derivative.

terbium, Tb. Lanthanide metal, at.no. 65, at. mass 158.93, m.p. 1356°C, b.p. 3230°C, density ρ 8230, crustal abundance 1.1 p.p.m., electronic configuration [Xe] $4f^9 6s^2$. The metal is hcp. The oxide used as a phosphor for television and in lasers. Organoterbium compounds used in polymerization.

terbium compounds Shows +3 and +4 oxidation states. Tb(III) compounds are typical LANTHANIDE COMPOUNDS.

Tb^{3+} (f^8 pale pink) $\rightarrow$ Tb (–2.39 volts in acid)

TbF_4 ($ThF_3 + F_2$ at 300°C) and $Cs_3 TbF_7$ are known and Tb oxides intermediate between Tb_2O_3 and TbO_2 are formed by heating Tb compounds in air. These higher oxides liberate oxygen from water. Tb_2Cl_3 and TbCl ($TbCl_3$ plus Tb) have metal–metal bonding.

terbufos Insecticide, nematocide, organophosphate derivative.

terbumeton Herbicide, triazine derivative.

terbuthylazine Herbicide, triazine derivative.

terebene Dipentene and other terpene hydrocarbons obtained from tall oil or turpentine. B.p. 160–190°C. It has a pleasant characteristic odour and is used medicinally and for treatment of cellulosic materials.

terephthalic acid (1,4-benzenedicarboxylic acid), $C_8H_6O_4$. Crystallizes in colourless needles; m.p. 300°C (sublimes). Manufactured by the oxidation of 4-xylene and used in the production of Terylene (see also POLYESTERS). Forms polyesters with glycols. Polyethylene terephthalate important plastic used for bottles for carbonated drinks, fibres, film, etc.

term symbols The spectroscopic notation used to denote an electronic state.

ternary compound A compound containing three elements, e.g. $KNbO_3$.

terpenes In the strict sense terpenes are volatile hydrocarbons of the empirical formula $C_{10}H_{16}$. In a wider sense the term includes sesquiterpenes, $C_{15}H_{24}$, diterpenes, $C_{20}H_{32}$, and

higher-mass derivatives. In a still looser sense the term includes various oxygen-containing compounds derived from the terpene hydrocarbons, such as alcohols, ketones and camphors. The terpenes are of great scientific and industrial importance, being characteristic products of many varieties of vegetable life and important constituents of most odorants, natural and synthetic, employed in perfumery. Many of them are of pharmaceutical importance. Practically without exception the terpene hydrocarbons may be considered as polymers of isoprene, C_5H_8, and may be either open-chain compounds or may contain one or more C_6 and other rings. They are chemically unsaturated, very reactive and in many cases form characteristic derivatives.

1,8-terpin (terpine, *p*-menthane-1,8-diol, dipentene glycol), $C_{10}H_{20}O_2$.

Monocyclic terpene dialcohol obtained with α-terpineol by the action of dilute alkali on *trans*-dipentene dihydrochloride, by the cyclization of linalool, geraniol and nerol.

terpinenes, $C_{10}H_{16}$. Related to 1,8 terpin and to terpineols. Some used in perfumes, and in processing fats. Monocyclic unsaturated terpenes, $C_{10}H_{16}$, which all yield the same dihydrochloride by addition of 2HCl.

tertiary carbon A carbon atom with three other carbon substituents.

testosterone (17β-hydroxy-4-androsten-3-one), $C_{19}H_{28}O_2$. M.p. 154°C. Testosterone is the androgenic hormone formed in the testes: it controls the development and maintenance of the male sex organs and secondary sex characteristics. Used medicinally and as performance-enhancing drug. Controlled substance.

tetraborane (10), B_4H_{10}. See BORON HYDRIDES.

tetrabromomethane, CBr_4. High-density liquid used with other bromoalkanes in microscopy and mineral separation. Could be carcinogenic.

tetrabutylammonium fluoride, Bu_4NF.
Fluorinating agent and catalyst.

tetracene (1-(5-tetrazolyl)-4-guanyl tetrazene hydrate), $C_2H_8N_{10}O$. Pale yellow solid formed from nitrous acid and amino-guanidine. Used as a detonator.

2,3,7,8-tetrachlorodibenzo-*p*-dioxin (dioxin) See DIOXIN.

sym-tetrachloroethane (acetylene tetrachloride), $HCCl_2CHCl_2$. Colourless toxic liquid with a chloroform-like odour, b.p. 146°C. Manufactured from chlorine and ethyne. Reacts with dilute alkalis to give trichloroethene. Important solvent, used in synthesis.

tetrachloroethene (tetrachloroethylene, polychloroethylene, perchloroethylene), $CCl_2:CCl_2$. M.p. −22°C, b.p. 121°C. Manufactured by reacting pentachloroethane, (Cl_2 and $CHCl:CCl_2$), with $Ca(OH)_2$ by the direct chlorination of light hydrocarbons at 500–650°C; by the direct chlorination of ethene and by oxidation of *sym*-tetrachloroethane. Dry-cleaning solvent, metal degreasing, and textile processing.

tetrachloromethane See CARBON TETRA-CHLORIDE.

tetrachlorophthalic anhydride, $C_8O_3Cl_4$. M.p. 255–257°C. Prepared by the direct chlorination of phthalic anhydride in 50–60% oleum in the presence of iodine.

It is used instead of phthalic anhydride in the preparation of certain eosin dyes (phloxines, etc.).

tetrachlorvin phos Insecticide, acaricide, organophosphate derivative.

tetraconazole Fungicide, azole derivative.

tetracosanoic acid See LIGNOCERIC ACID.

tetracyanoethylene (TCNE), $(NC)_2C:C(CN)_2$. A sublimable white crystalline solid, m.p. 200°C, which has high thermal and oxidative stability. Reacts with most compounds containing an active hydrogen and with dienes in typical Diels–Alder additions.

It is probably the strongest π-acid known; it forms a series of coloured complexes with aromatic hydrocarbons and gives salts of the radical anion with many metals, e.g. K^+TCNE^-.

7,7,8,8-tetracyanoquinodimethane (TCNQ)

M.p. 296°C. π-acid forms charge transfer complexes with some donors. Used for colorimetric determination of free radical precursors, replacement of MnO_2 in aluminium solid electrolytic capacitors, construction of heat-sensitive resistors and ion-specific electrodes and for inducing radical polymerizations.

tetracyclines An important group of broad spectrum ANTIBIOTICS isolated from *Streptomyces* spp., having structures based on a naphthacene skeleton.

tetracyclone See TETRAPHENYLCYCLOPENTA-DIENONE.

tetradecanoic acid See MYRISTIC ACID.

tetradecanol See MERISTYL ALCOHOL.

tetradec-11-en-1-yl acetate Insecticide, pheromone.

tetradifon Acaricide, sulphone derivative.

tetraethylene glycol, $O(CH_2CH_2OCH_2CH_2OH)_2$. Used as solvent, particularly as methyl and ethyl esters.

tetraethyl lead (TEL) See LEAD TETRA-ETHYL.

tetraethyl silicate See ETHYL SILICATE.

tetrafluoroberyllates Containing the $[BeF_4]^{2-}$ ion.

tetrafluoroborates Containing the $[BF_4]^-$ ion.

tetrafluorohydrazine, N_2F_4. See NITROGEN FLUORIDES.

tetraglym (pentaoxapentadecane), $CH_3O(CH_2CH_2O)_4CH_3$. Solvent.

tetraglyme (tetraethylene glycol dimethyl ether), $CH_3(OCH_2CH_2)_4OCH_3$. B.p. 275°C.

tetragonal system The CRYSTAL SYSTEM with a 4-fold axis as principal axis of symmetry. The unit cell has its z-axis parallel to the 4-fold axis and the other two equal axes mutually perpendicular at right angles to this axis. E.g. rutile, TiO_2.

tetrahedral co-ordination Regular co-ordination by four ligands situated at the corners of a tetrahedron.

Tetrahedron

Methane, CH_4, has tetrahedral co-ordination about the carbon. A tetrahedron has four triangular faces.

tetrahedron See TETRAHEDRAL CO-ORDINATION.

tetrahydrocannabinol, $C_{21}H_{29}O_2$. Pyran derivative, active component of hashish (marihuana). Used medicinally. Controlled substance.

tetrahydrofuran (diethylene oxide, THF), C_4H_8O. Colourless liquid, b.p. 66°C, m.p. –108.5°C. It is widely used in industry as a solvent for resins, many plastics (especially of the PVC type) and elastomers; as an ether type solvent for chemical reactions it often produces increased reaction rates and yields. Also used in tetrahydrofuran polymers. Prepared by acid hydrolysis of the polysaccharides in oat husks or by catalytic hydrogenation of maleic anhydride.

tetrahydrofuran polymers (poly(oxytetramethylene) glycols, poly(tetramethylene oxide), polytetrahydrofuran) Tetrahydrofuran undergoes ring opening and polymerization with carbenium, diazonium and trialkyloxonium ions, and also with Lewis acid halides, generally to give low mol.wt. diol polymers (1–3000) used as plasticizers. These polymers are used with polyurethanes to give thermoplastics and elastomers.

tetrahydrofurfuryl alcohol (tetrahydro-2-furanmethanol), $C_5H_{10}O_2$. Solvent for fats, waxes, resins and used in synthesis.

tetrahydrothiophen (THT, tetramethylene sulphide) Colourless mobile liquid, with a penetrating odour; b.p. 120°C. Obtained by the catalytic reduction of thiophen or tetrahydrothiophen-1,1-dioxide (sulpholane). Used to give an odour to domestic gas supplies.

N,N′,N′′,N′′′-tetrakis (2-hydroxypropyl) ethylenediamine (EDTP) B.p. 175–180°C. Prepared propylene oxide on ethylene diamine. Cross-linking agent for epoxy resins; used as humectant, plasticizer, emulsifier.

tetrakis(triphenylphosphine) metallates, $(Ph_3P)_4M$ (M = Ni, Pd, Pt). Important catalysts for, e.g. aryl coupling reactions.

tetralin (1,2,3,4-tetrahydronaphthalene), $C_{10}H_{12}$. Colourless liquid; b.p. 207°C. Obtained by the catalytic hydrogenation of naphthalene. Owing to the presence of one aromatic ring it can be nitrated and sulphonated. It is non-toxic and is used as a solvent for fats, oils and resins and for degreasing.

tetralite See TETRYL.

tetramethrin (and derivatives) Insecticides, pyrethroids.

1,2,3,4-tetramethylbenzene Prehnitene.

1,2,3,5-tetramethylbenzene Isodurene.

1,2,4,5-tetramethylbenzene Durene.

3,3′,5,5′-tetramethylbenzidine, $C_{16}H_{20}N_2$. M.p. 168–169°C. Used for blood detection, and medicinally. Controlled substance.

N,N,N′,N′-tetramethyl-1,2-diaminoethane (N,N,N′,N′-tetramethylethylene diamine, TMEDA, TEMED), $(CH_3)_2NCH_2CH_2N(CH_3)_2$. B.p. 122°C. Hygroscopic base which forms a hydrocarbon-soluble stable chelate with lithium ions and promotes enhanced reactivity of compounds of lithium, e.g. $LiAlH_4$, LiC_4H_9, due to enhanced kinetic basicity of the chelate. Used in polymerization catalysts.

tetramethyl lead (TML) See LEAD TETRAMETHYL.

tetramethylsilane (TMS), $(CH_3)_4Si$. Used as an internal reference standard for ^1H and ^{13}C nuclear magnetic resonance spectroscopy. On the ∂-scale TMS protons absorb at 0.00 ∂ and most protons bonded to carbon then appear downfield (positive ∂-value) from TMS.

tetramethylthiuram disulphide See THIURAM DISULPHIDES.

tetranitromethane, $C(NO_2)_4$. Colourless liquid; b.p. 126°C. It is made by nitration of ethanoic anhydride with anhydrous nitric acid. Decomposed by alcoholic solutions of KOH to nitroform $HC(NO_2)_3$. Reacts with alcohols in the presence of alkali to give the nitrate of the alcohol. Used in rocket propellants and explosives.

tetraphenylboron sodium, $NaBPh_4$. Precipitating agent for heavier Group 1 cations, source of tetraphenylborates.

tetraphenylcyclopentadienone (tetracyclone) An intensely purple crystalline solid; m.p. 219–220°C. One of the few monomeric cyclopentadienone derivatives, most of which spontaneously undergo Diels–Alder type self-dimerization. Used as a diene in many studies of various aspects of the DIELS–ALDER REACTION.

tetrathionic acid, $HO(O)_2SS_2S(O)_2OH$. See POLYTHIONIC ACIDS.

tetrazoles Contain CN_4 ring. Used as precursors for heterocycles in pharmaceuticals, etc.

tetrazolium salts Quaternary salts derived from tetrazoles. Colourless or yellow substances, in the presence of a trace of alkali oxidize aldoses, ketoses and other α-ketols and are themselves reduced to insoluble formazan pigments, e.g. 'blue tetrazolium' gives an intense blue pigment. This sensitive test distinguishes between reducing sugars and simple aldehydes. Tetrazolium salts are used in colour photography.

tetritol See ERYTHRITOL.

tetrole See FURAN.

tetrolic acid The trivial name for 1-carboxyprop-1-yne $CH_3–C\equiv C–CO_2H$.

tetrose A carbohydrate with four carbon atoms. There are four aldo-tetroses, the two stereoisomers of erythrose and threose, and two possible keto-tetroses.

tetryl (tetralite, nitramine, N-methyl-N-2,4,6-trinitronitramine, N-methyl tetralite), $C_7H_5N_5O_8$. Pale yellow, m.p. about 129°C. Obtained by nitrating dimethylaniline in concentrated sulphuric acid. Very stable explosive much used as a primer or booster. Used as an indicator.

textiles Apart from the natural fibres (wool, cotton) or synthetics (acrylics, polyesters, Nylons, etc.), many other chemicals are used in textiles. Dyebath additives are principally dyestuffs but include solvents (tetrachloroethene, aromatics), dispersing agents (e.g. sulphuric acid, quaternary ammonium salts), anti-foams (e.g. silicones), acids (e.g. ethanoic), wetting agents (ethoxylated alcohols). Finishing agents include non-fibrous polymers, stain repellants (e.g. fluorochemicals), durable press agents (glyoxal), softeners and lubricants (silicones), flame retardants. Printing chemicals include thickeners (e.g. starch, alginates), binders (acrylates), acids, wetting agents, bactericides, solvents (hydrocarbons, urea), softeners and dispensing agents.

Th Thallium.

thallates Mixed oxides of thallium(III).

thallation See THALLIUM(III) TRIFLUORO-ACETATE.

thallic Thallium(III) compounds.

thallium, Tl. Metal, at.no. 81, at.mass 204.38, m.p. 304°C, b.p. 1473°C, density ρ 11 850, crustal abundance 0.6 p.p.m., electronic configuration [Xe] $4f^{14}5d^{10}6s^26p^1$. Element of Group 13, occurs in some sulphide and selenide ores – e.g. crookesite (Ag,Cu,Tl)–Se, recovered from the flue dusts after oxidation of sulphide ores and reduced to the metal electrolytically. The soft metal is similar to Pb. Fairly reactive metal – combines halogens, S, only slowly soluble in dil. acids as Tl(I) salts insoluble, dissolves rapidly in HNO_3. The metal itself is not used; Tl compounds are used in poisons, glass and some electronic applications. Thallium compounds are very poisonous.

thallium, organic derivatives Formed TlX_3 plus Grignard reagents; $ArTlX_2$ derivatives can be formed by direct thallation (see THALLIUM TRIFLUOROACETATE). Simple derivatives, e.g. Me_3Tl, are spontaneously inflammable in air. The $[R_2Tl]$ groups (linear) are stable. TlC_5H_5 (polymeric) is formed from TlOH and C_5H_6; it is used in the synthesis of metal cyclopentadienides. TlOR (e.g. TlOMe) are polymeric.

thallium bromides
Thallium(I) bromide, TlBr. Insoluble in water.
Thallium(III) bromide, $TlBr_3$. Formed Tl plus Br_2. Unstable, decomposes to $Tl^I[TlBr_4]^-$.

thallium(I) carbonate, Tl_2CO_3. Formed by CO_2 on solution of TlOH.

thallium chemistry Thallium is a Group 13 element. The group oxidation state is strongly oxidizing ($E°$ Tl(III) to Tl +1.25 volts). Thallium(I) is the stable state ($E°$ Tl(I) to Tl –0.33 volts in acid solution) and is very similar in its chemistry to Ag(I). Compounds which have stoichiometries corresponding to other than Tl(III) or Tl(I) contain mixed oxidation states, e.g. $TlBr_2$ is $Tl^I[Tl^{III}Br_4]$. Thallium(I) forms few complexes; Tl(III) is readily complexed.

thallium chlorides
Thallium(I) chloride, TlCl. M.p. 430°C, b.p. 806°C. Precipitated from aqueous solution of a Tl(I) salt by HCl. White solid.
Thallium(III) chloride, $TlCl_3,4H_2O$. Formed by

passing Cl_2 through a suspension of TlCl in water. Hygroscopic, loses Cl_2 at 100°C. The $[TlCl_2]^+$ ion is stable; chloro complexes up to $[TlCl_6]^{3-}$ are formed.

thallium chromate, Tl_2CrO_4. Yellow powder precipitated from aqueous solution of Tl(I) salt by Na_2CrO_4. Used to estimate Tl(I).

thallium fluorides
Thallium(I) fluoride, TlF. Colourless solid soluble in water (TlOH in HF). Forms $TlHF_2$.

Thallium(III) fluoride, TlF_3. White solid (Tl_2O_3 plus F_2), immediately hydrolysed by water.

thallium hydroxide, $TlOH.H_2O$. Formed Tl_2SO_4 plus $Ba(OH)_2$. Crystallizes from aqueous solution as yellow needles. Soluble in water to give strongly basic solution (used to prepare other Tl(I) salts). The solution absorbs CO_2 to give Tl_2CO_3.

thallium iodides
Thallium(I) iodide, TlI. Yellow or red solid precipitated from aqueous solution.

Thallium(I) triiodide, $Tl[I_3]$. Formed from Tl(III) solution plus KI, readily loses I_2.

thallium nitrates
Thallium(I) nitrate, $TlNO_3$. Formed by dissolving Tl, Tl_2CO_3 or TlOH in HNO_3; soluble in water. Decomposes at 300°C.

Thallium(III) nitrate, $Tl(NO_3)_3.3H_2O$. Formed Tl plus conc. HNO_3. Immediately hydrolysed by water but is soluble in dilute mineral acids, alcohols and diglyme. Used in OXYTHALLATION.

thallium oxides
Thallium(I) oxide, Tl_2O. Black powder formed by heating TlOH, gives Tl(I) salts with acids. Used in glass.

Thallium(III) oxide, Tl_2O_3. Dark solid obtained by Tl plus O_2 or Tl(III) salt plus base. Insoluble in water and alkalis but dissolves in acid (often gives Tl(I) salts).

thallium sulphate
Thallium(I) sulphate, Tl_2SO_4. Formed Tl plus hot conc. H_2SO_4 or TlOH plus H_2SO_4. Moderately soluble in water; forms alums and double sulphates.

Thallium(III) sulphate, $Tl_2(SO_4)_3.7H_2O$. Probably formed Tl_2O_3 in conc. H_2SO_4. Very readily hydrolysed by water.

thallium trifluoroacetate, $Tl(O_2CCF_3)_3$. Solid formed Tl_2O_3 plus CF_3CO_2H. Used in thallation of aromatics; Ar–H plus $Tl(O_2CCF_3)_3$ gives $ArTl(O_2CCF_3)_2$. The reaction proceeds specifically depending upon the substituents present and the thallium can then be replaced specifically, e.g. KI gives ArI. See OXYTHALLATION.

thenyl See THIENYL RING.

thenylchlor Herbicide, acetanilide derivative.

theobromine (3,7-dimethylxanthine), $C_7H_8N_4O_2$. M.p. 337°C. Xanthine alkaloid.

theophylline (1,3-dimethylxanthine), $C_7H_8N_4O_2$. Bronchodilator.

theoretical plate A theoretical plate in a distillation or ABSORPTION COLUMN is one on which perfect liquid–vapour contacting occurs, so that the two streams leaving it are in equilibrium with each other.

In the case of a PLATE COLUMN the performance of a real plate is related to the performance of a theoretical one by the PLATE EFFICIENCY. In the case of a PACKED COLUMN the *height equivalent to a theoretical plate* (*HETP*) gives a measure of the contacting efficiency of the packing.

thermal analysis Analytical techniques used for identification and also for investigation of the products of the action of heat on a compound or complex. In differential thermal analysis (DTA) the sample is heated, frequently in an inert atmosphere, and weight is measured as a function of temperature. In differential scanning calorimetry (DSC) electrical heat is added or removed from the sample as the temperature is raised; enthalpy changes due to thermal decomposition are thus followed.

thermal cracking Formerly, the thermal cracking of gas oil and similar petroleum fractions was used to increase the gasoline yield from a crude oil. This involved subjecting the petroleum fraction to temperatures of about 500°C at pressures up to 25 bar to yield gases, hydrocarbons in the gasoline and gas oil range and heavier fractions. While catalytic cracking has replaced thermal cracking for the production of gasoline the latter process is still used for certain purposes such as coke production, or the production of feedstocks for special purposes, e.g. the cracking of wax to produce long-chain alkenes. See CATALYTIC CRACKING.

thermal reforming See REFORMING.

thermite reaction A reduction process involving Al powder and a metal oxide (generally Fe_2O_3) used to produce heat for welding or incendiary devices. Generally ignited with Mg ribbon; the reaction gives, e.g. Al_2O_3 and Fe.

thermobalance A balance in which the sample can be heated to allow THERMAL ANALYSIS.

thermochemistry Most chemical reactions occur with the evolution or absorption of heat. Thermochemistry is that branch of chemistry concerned with the heat changes accompanying chemical reactions.

thermochroism Change in colour (and spectrum) with temperature.

thermodynamic control The products of a reaction at equilibrium are determined by the relative stability (free energies) of the products and reactants. Kinetic contol can override thermodynamic control, e.g. the hydrolysis of SF_6 which does not proceed for kinetic reasons.

thermodynamic data Standard enthalpies of formation (ΔH_f°), standard free energies of formation (ΔG_f°) and absolute entropies of selected substances at 298 K.
 By definition all pure elements have $\Delta H_f^\circ = 0$ and $\Delta G_f^\circ = 0$. For tables of thermodynamics see KAYE AND LABY.

thermodynamics The study of the transformation and translation of energy.

thermodynamics, first law of This law is a consequence of the law of conservation of energy, which states that mechanical energy and work are quantitatively interconvertible. Alternatively, 'In any process energy can be changed from one form to another, but it is neither created nor destroyed.' Considering the internal energy U, $\Delta U = q + w$ where w is the work done on the system and q the energy transferred as heat.

thermodynamics, second law of This states that 'Heat cannot of itself pass from a colder to a warmer body.' Thus, if heat is to be transferred from the cold to the hot body, work must be provided by an external agency. An alternative and useful expression of the second law is; 'Any system of its own accord will always undergo change in such a way as to increase the entropy.'

thermodynamics, third law of This states that 'for a perfect crystal at absolute zero on the Kelvin scale the entropy is zero'. This follows the predictions of Einstein (1907) that the specific heats of all substances would approach zero at 0 K, and the conclusions of Planck (1912) that the entropy of all pure solids and liquids approaches zero at this temperature.

thermogram A plot of weight against temperature obtained in THERMAL ANALYSIS.

thermolysis Reactions initiated by heat.

thermonuclear reaction See ATOMIC ENERGY.

thermoplastic resins Resins (plastics) which can be repeatedly softened by heating and harden again on cooling. The most important types are the various forms of polyethene, polyvinyl chloride, polypropene and polystyrene, polyamides and polyesters. U.S. production 2000 29.8 million tonnes. Major uses packaging 29%, construction 15%.

thermosetting resins Resins (plastics) which once formed in their final moulded state are infusible and insoluble. They are said to be thermocured. The most important types are phenolic resins, polyesters, urea resins, epoxide resins and melamine resins.

thexyl Trivial name for the 1,2,3-trimethylpropyl group.

THF Tetrahydrofuran.

thiabondazole Fungicide, thiazole derivative.

thiamine (aneurine, vitamin B_1) The anti-neuritic factor, the absence of which from the diet of man leads to the disease beri-beri, and in mammals and birds to polyneuritis, the most fundamental symptom of which is a general nervous atrophy. Component of a number of enzymes that catalyse decarboxylation reactions. The richest natural sources of the vitamin are yeast, eggs and the germ of cereals. It is not present in polished rice and other highly purified cereal products.

thiamine diphosphate (thiamine pyrophosphate) The biochemically active derivative of thiamine. It is a coenzyme which is concerned in a number of important metabolic processes including the decarboxylation of α-keto acids

thiazole dyes Dyestuffs containing the thiazole ring. Often show good dyeing properties to cellulosic substrates.

thiazole ring The ring numbered as shown.

thiazopyr Herbicide, pyridine derviative.

thiazyl ions Species, e.g. $[[S_2N]^+]^+$; cationic derivatives of sulphur nitrogen compounds.

thickener Additive for, e.g. defoamers, foodstuffs, mineral processes, shampoos, to give particular physical properties. Thickeners include starches and carbohydrate derivatives.

thickening Treatment of a slurry or suspension to recover the solids in the form of a more concentrated slurry. A thickener is equipment for such an operation.

thidiazuron Plant growth regulator, phenyl urea derivative.

thienyl ring The ring system numbered as shown.

thifensulfuron-methyl Herbicide, sulphonylurea derivative.

thifluzamide Fungicide, thiazole derivative.

thiirane 3-Membered sulphur heterocycle.

The sulphur analogue of an epoxide (oxirane).

thin-layer chromatography See CHROMATOGRAPHY.

thio- Sulphur-containing, e.g. thioantimonates (antimony sulphides), thioarsenates (arsenic sulphides).

thiobendazole Fungicide, benzimidazole derivative.

thiobenicarb Herbicide, thiocarbamate derivative.

thiocarbanilide (sym-diphenylthiourea), $C_{13}H_{12}N_2S$, PhNHC(S)NHPh. Colourless flakes; m.p. 151°C. Prepared by boiling aniline with carbon disulphide. It is used in vulcanization and formation of sulphur dyes.

thiocarbonates Salts of THIOCARBONIC ACID. For ROC(S)SH see XANTHATES.

thiocarbonic acid, H_2CS_3. Formed, mixed with H_2CS_4, by acidifying thiocarbonates. Na_2CS_3 formed from solution of Na_2S plus CS_2. $(NH_4)_2CS_3$ formed CS_2 plus concentrated aqueous NH_3. $BaCS_3$ (relatively stable) formed $Ba(HS)_2$ plus CS_2. Mixed oxide thiocarbonates containing, e.g.

$[COS_2]^{2-}$ ions are also known. Thionocarbonates containing the $[SC(OR)_2]^{2-}$ group can be cleaved to alkenes with phosphorus(III) compounds. Thiocarbonates are slowly hydrolysed and react with CO_2 to CS_2 and sulphides. Used in destroying the fungus phylloxera which occurs on vines.

thiocarbonyl complexes Complexes containing CS ligands analogous to carbonyls. As CS does not exist formed from CS_2, e.g. $(C_5H_5)Mn(CO)_2(CS)$.

thiocyanates Salts of thiocyanic acid, HSCN. Complexes and esters can be N or S bonded. Formed cyanides plus S. Used in estimation of Ag^+ (precipitate of AgSCN) and Fe^{3+} (red coloration). Photographic sensitizers. With oxidizing agents (MnO_2, Br_2) give thiocyanogen, $(SCN)_2$, which polymerizes above 0°C. Thiocyanates have antibacterial and insecticidal properties.

thiocyanic acid, HSCN. Prepared $KHSO_4$ plus KSCN. Gas at room temperature but rapidly polymerized. Forms thiocyanic esters RSCN and isothiocyanic esters or mustard oils, RNCS.

thiocyanogen, $(SCN)_2$. See THIOCYANATES.

thiocyclam Insecticide, trithiol derivative.

thiodicarb Insecticide, molluscide, oxime carbonate.

thiofanox Insecticide, acaricide, oxime carbonate.

thioglycollic acid (mercaptoethanoic acid), $HS \cdot CH_2 \cdot COOH$, $C_2H_4O_2S$. B.p. 123°C/29 mm. Obtained by electrolytic reduction of dithioglycollic acid formed by the action of sodium hydrosulphide on chloroethanoic acid. It oxidizes readily in the air and is very reactive. Adds across double bonds of unsaturated compounds. The hydrogen of the –SH group is replaceable by metals. It is an important constituent of the cold process for waving hair. The calcium salt is a depilatory.

thiokols Polysulphide rubbers.

thiols, RSH. See MERCAPTANS.

thiometon Insecticide, acaricide, organophosphate derivative.

thiomolybdates Species such as $[MoS_4]^{2-}$, $[MoO_2S_2]^{2-}$. Can act as bidentate ligands. Thiotungstates are similar.

thionic acids, $H_2S_nO_6$. See POLYTHIONIC ACIDS.

thionocarbonates See THIOCARBONIC ACID.

thionyl Containing $\geq$SO groups.

thionyl chloride, $SOCl_2$. M.p. $-105°C$, b.p. $79°C$. Formed SO_2 plus PCl_5 or S_2Cl_2 plus SO_3. Hydrolysed by water to HCl, SO_2 and H_2O. Used to convert COH groups to CCl and for the dehydration of metal chloride hydrates.

thionyl halides, SOF_2. ($SOCl_2$ plus SbF_3 or anhydrous HF); SOClF (as for SOF_2); $SOBr_2$ ($SOCl_2$ plus HBr); $SOCl_2$ (see THIONYL CHLORIDE).

thiopentone sodium, $C_{11}H_{17}N_2O_2SNa$. Monosodium salt of 5-ethyl-5-(1-methylbutyl)-2-thiobarbituric acid. Anaesthetic, controlled substance.

thiophanate-methyl Fungicide, carbamate derivative.

thiophene, C_4H_4S. A colourless liquid with a faint odour resembling that of benzene, m.p. $-38°C$, b.p. $84°C$. It occurs to the extent of about 0.5% in commercial benzene and can be prepared by heating sodium succinate with P_2S_5. It is manufactured from butane and sulphur. It can be nitrated, sulphonated and brominated, and gives rise to two series of monosubstituted derivatives (2 and 3). In derivatives the ring system shown is called the thienyl ring.

Solvent for resins, dyestuff intermediate.

thiophenol (phenylthiol, phenylmercaptan, mercaptobenzene), C_6H_5SH, PhSH. The sulphur analogue of phenol. A colourless, foul-smelling liquid; b.p. $168°C$. Manufactured by the reduction of benzene sulphonyl chloride and forms a series of complexes (mercaptides) with metal salts.

thiosinamine See PROPENYLTHIOUREA.

thiosulphates Salts of THIOSULPHURIC ACID.

thiosulphuric acid, $H_2S_2O_3$. The free acid is only known (SO_3 plus H_2S) at $-78°C$ as an etherate. Thiosulphates, containing the $[S_2O_3]^{2-}$ ion are formed from sulphites (sulphates(IV)) plus S. Form complexes – particularly with Ag^+. Thiosulphates are used in photography. React quantitatively with I_2; reaction used in volumetric analysis. Acids on $[S_2O_3]^{2-}$ give SO_2 and S.

thiotungstates See THIOMOLYBDATES.

thiourea, CH_4N_2S, $S:C(NH_2)_2$. Colourless crystals, m.p. $172°C$. Decomposed on heating with water to give ammonium thiocyanate, but in many of its properties it resembles urea. It

forms complexes with many metal salts. It is manufactured by the action of hydrogen sulphide on cyanamide. It is used as a photographic sensitizer and vulcanization accelerator.

thiram See THIURAM DISULPHIDES. Fungicide, dithiocarbamate.

third law of thermodynamics See THERMODYNAMICS, THIRD LAW OF.

thiuram disulphides, $R_2NC(S)SSC(S)NR_2$. Obtained by oxidation of dithiocarbamates (by, e.g. H_2O_2, Cl_2) and used as polymerization initiators and vulcanization accelerators. The tetramethyl derivative TMTD is Antabuse, the agent for treatment of alcoholism, and in agricultural application to prevent the 'damping off' of seedlings. TMTDS is tetramethyithiuram disulphide, m.p. $155°C$.

thixotropy The isothermal gel-sol transformation biought about by shaking or other mechanical means. It may be regarded as a packing phenomenon. E.g. $Fe(OH)_3$ gels on shaking form sols which rapidly set again when allowed to stand. Thixotropy is commonly found with suspensions of clay or oil paints and emulsions and in some alloys where it can be used for component manufacture. Compare DILATANCY.

thorium, Th. Radioactive metal, at.no. 90, at. mass 232.04, m.p. $1750°C$, b.p. $4788°C$, density ρ 11 720, crustal abundance 12 p.p.m., electronic configuration [Rn] $6d^27s^2$. The most important ores are thorite and thorogummite (silicates), thorianite (ThO_2) and monazite (a mixed phosphate, the principal lanthanide ore). The ores are decomposed with alkali or acid and Th is extracted from acid solution using solvents such as tributyl phosphate. The metal has been prepared by Ca reduction of ThF_4; it is tarnished in air and slowly attacked by hot water and dilute acids; low temperature ccp, higher temperature bcc. ThO_2 was used in the production of incandescent gas mantles; the fissile isotope ^{233}U is obtained from ^{232}Th and thermal neutrons; ThO_2 is an important catalyst (Fischer–Tropsch) and is used in strengthening nickel. The metal is used as an oxygen remover (getter) in the electronics industry.

thorium compounds The first actinide, thorium chemistry is almost entirely that of the +4 state

$$Th^{4+} \xrightarrow{-1.9 \text{ V}} Th$$

Compounds which are formally in lower oxidation states ThI_3 and ThI_2 (prepared from ThI_4

and Th in a sealed container) contain Th^{4+} ions, I^- ions and electrons; Th(IV) compounds are partially hydrolysed in water. Salts crystallized from water are heavily hydrated; O- and N-donors also form complexes. High co-ordination numbers (e.g. 11 in $[Th(NO_3)_4.5H_2O]$ and 10 in $[Th(NO_3)_4(OPPh_3)_2])$ are common. Thorium diketonates (e.g. acetylacetonates) are particularly stable. Some organometallic derivatives of thorium are known e.g. $Th(C_5H_5)_4$ is formed from $ThCl_4$ and KC_5H_5.

thorium halides All ThX_4 are known

$$ThO_2 + C_2Cl_2F_4 \text{ or HF} \xrightarrow{400°C} ThF_4$$
$$ThCl_4aq. \xrightarrow{HF} ThF_4,$$
$$Th \xrightarrow{Br_2 \ 700°C} ThBr_4,$$
$$Th \xrightarrow{I_2 \ 400°C} ThI_4$$

ThF_4 is insoluble in water (but $ThF_4.4H_2O$ can be formed); the other halides are soluble in water. Halide oxides, $ThOX_2$, are formed from ThO_2 and ThX_4. Complex halides, e.g. K_5ThF_9, Na_4ThF_8, Rb_2ThCl_6 are formed from acid solution and have high co-ordination numbers about the metal. For lower halides see THORIUM COMPOUNDS.

thorium oxide, ThO_2. Thorium salts are hydrolysed in aqueous solution and the hydrated oxide is precipitated by alkalis. Ignition of this hydrated oxide, not other oxyacid salts, gives white ThO_2.

thorium oxyacid salts Commercially available thorium salts include the nitrate ($4H_2O$ and H_2O), the oxalate ($6H_2O$) and sulphate (0, 4, 6, 8 and $9H_2O$). The anion is generally co-ordinated to the metal; the salts are eventually hydrolysed in aqueous solution.

thortveitite, $Sc_2Si_2O_7$. An important source of Sc. Contains $Si_2O_7^{6-}$ ions.

three-centre bond A bonding orbital formed by two electrons between three atoms. See MULTICENTRE BONDS.

threitols See ERYTHRITOL.

threo- See ERYTHRO-.

threonine (2-amino-3-hydroxybutanoic acid), $C_4H_9NO_3$, $CH_3CHOHCHNH_2C(O)OH$. Amino-acid; m.p. 251–252°C. Occurs widely in proteins. Essential component of foods.

threose   **(2,3,4-trihydroxybutanol)**, $C_4H_8O_4$. A tetrose sugar. The D-form has been

obtained crystalline, m.p. 126–132°C. It is very hygroscopic and soluble in water and alcohol.

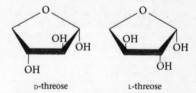

d-threose   l-threose

threshold limit values (TLV) A list of values for chemical substances in the form of time-weighted average concentrations, representing conditions which can be tolerated in working environments. For some materials the designation 'C' is added to indicate that the value shown is a ceiling value which should not be exceeded. E.g. benzene –25 C (i.e. 25 p.p.m.); xylene –100; pentane –500.

thrombin Enzyme which catalyses the conversion of fibrinogen to fibrin in clotting blood.

thromboxanes Thromboxanes are cellular regulatory compounds similar to PROSTAGLANDINS, both of them being derived biosynthetically from prostaglandin endoperoxides.

thujyl derivatives Terpene derivatives based on the structure shown.

Thujone is used in flavouring. Thujyl alcohol is also known as tanacetyl alcohol.

thulium, Tm. Lanthanide metal, at.no. 69, at. mass 168.93, m.p. 1545°C, b.p. 1950°C, density ρ 9321, crustal abundance 0.48 p.p.m., electronic configuration [Xe] $4f^{13}6s^2$. The metal is hcp. The metal or its oxides are widely used as a portable source of X-radiation after neutron irradiation.

thulium compounds Thulium forms a series of +3 compounds which are typical LANTHANIDE COMPOUNDS. Cp_2Tm in THF can reduce N_2.

Tm^{3+} (f^{12} pale green) $\rightarrow$ Tm (–2.28 volts in acid)

thymidine The nucleoside of thymine and deoxyribose, i.e. the *β-N*-DEOXYRIBOFURANO-SIDE OF THYMINE. See NUCLEOSIDES.

thymine (5-methyluracil, 5-methyl-2,4 (1H,3H)pyrimidine-dione), $C_5H_6N_2O_2$. M.p. 321–325°C. Thymine is a constituent of deoxyribose nucleic acid.

thymol (5-methyl,-2-(1-methylethyl) phenol), $C_{10}H_{14}O$. Colourless plates; m.p. 51.5°C, b.p. 233.5°C. It has the pungent odour of thyme. It is a constituent of oil of thyme and other oils. It is manufactured from piperitone and used to a limited extent in antiseptic mouthwashes and gargles.

thyroid hormones Various hormones produced in vertebrate thyroids influencing metabolic rate, tissue growth and development. See DICTIONARY OF BIOLOGY.

thyroxine, $C_{13}H_{11}I_4NO_4$. Iodine-containing thyroid hormone influencing metabolic rate.

Ti Titanium.

tiglic acid (E-2-methyl-2-butenoic acid, *cis*-1,2-dimethylacrylic acid, *trans*-(2-methyl-2-butenoic acid)), $C_5H_8O_2$.

M.p. 64°C, b.p. 198.5°C. Occurs in Roman oil of cumin and croton oil. Used in perfumes and flavouring. See ANGELIC ACID.

tin, Sn. Metallic element, at.no. 50, at.mass 118.71, m.p. 231.93°C, b.p. 2602°C, density ρ 5750, crustal abundance 2.2 p.p.m., electronic configuration [Kr] $4d^{10}5s^25p^2$. Group 14 metal easily obtained from its ores and hence known since prehistoric times. Occurs as cassiterite or tinstone, SnO_2, which is reduced with C (also recovered electrolytically). The element has three allotropic forms; metallic tin is silvery in colour, soft and malleable, tetragonal (distorted cp lattice). Below 13.5°C the stable form is grey tin (α) diamond lattice, brittle powder (the phase change is slow in the absence of grey tin). γ-Sn, another brittle modification is stable above 161°C. The metal is attacked by halogens (Cl_2

used in Sn recovery) but because of high over-voltage is not attacked by cold dilute acids. Concentrated acids will dissolve Sn. Used in coating steel (tin plating) and in ALLOYS (SOLDER, BRONZE) (Sn–Nb alloys are superconductors). Compounds are used in glass coatings and as fungicides.

tin, organic derivatives Tin(IV) derivatives are formed by use of Grignard reagents on $SnCl_4$ or in some cases directly; catenated derivatives, e.g. $Ph_3SnSnPh_3$ are known (Ph_3SnCl plus Na). Used as fungicides, wood preservatives, as stabilizers in plastics (e.g. PVC) and as catalysts (particularly for the formation of polyurethanes) but are poisonous and are facing bans. There is an extensive organotin chemistry, e.g. hydrides R_3SnH are good reducing agents.

tin alloys Tin is a component of many alloys including SOLDERS, Britannia metal, fusible metal, anti-friction metal, Babbitt metal, bell metal, bronze, gunmetal and pewter.

tin bromides
Tin(II) bromide, $SnBr_2$. M.p. 215°C, b.p. 619°C. Very similar to tin(II) chloride.

Tin(IV) bromide, $SnBr_4$. M.p. 33°C, b.p. 203°C, prepared from the elements. Forms many complexes, including $[SnBr_6]^{2-}$ ions.

tincal Natural BORAX, $Na_2B_4O_7.10H_2O$.

tin chemistry Tin is a Group 14 element, electronic configuration $5s^25p^2$. It shows the group oxidation state of +4 (largely covalent except, e.g., SnO_2) and also the +2 state which has some reducing properties; SnX_2 compounds can coordinate through the lone pair. Both oxidation states form complexes (Sn(IV) compounds 5- or 6-co-ordinate). Sn–Sn bonds are formed but are much less stable than for C and Si. Has no pure cationic chemistry, polymeric species, e.g. $[Sn_3(OH)_4]^{2+}$ are formed by hydrolysis of Sn(II) solutions.

tin chlorides
Tin(II) chloride, $SnCl_2$. Stannous chloride. M.p. 247°C. White solid (Sn plus gaseous HCl), forms hydrates ($SnCl_2.2H_2O$ is tin salt) from Sn and aqueous HCl. Acts as acceptor in complexes and forms complexes with transition metals. Used as a mordant.

Tin(IV) chloride, $SnCl_4$. Stannic chloride. M.p. –33°C, b.p. 114°C. Colourless fuming liquid (Sn plus Cl_2) hydrolysed in water but forms $SnCl_4.5H_2O$ and $[SnCl_6]^{2-}$ from acid solutions, soluble in organic solvents. Used as a mordant.

tincture of iodine Alcoholic solution of iodine containing about 2½% iodine and 2½% potassium iodide, used as an antiseptic.

tin fluorides

Tin(II) fluoride, SnF$_2$. Stannous fluoride. M.p. 213°C. Formed from Sn and aqueous HF. Forms ternary derivatives, e.g. MSnF$_3$. Used in toothpastes and in water fluoridation.

Tin(IV) fluoride, SnF$_4$. Stannic fluoride. Polymeric solid formed Sn plus F$_2$ or SnCl$_4$ plus HF. Very hygroscopic, forms fluorostannates(IV) containing [SnF$_6$]$^{2-}$ ions.

tin hydrides, SnH$_4$. B.p. –52°C, Sn$_2$H$_6$ – see STANNANE. Compounds containing SnH bonds, e.g. Me$_3$SnH, are good reducing agents.

tin iodides *Tin(II) iodide*, SnI$_2$. *Tin(IV) iodide*, SnI$_4$. Very similar to tin chlorides.

tinning The coating of iron with tin, chiefly used for the manufacture of tin cans. The sheet iron is pickled and passed into a bath of molten tin; the surface of the bath is protected from oxidation by a ZnCl$_2$–NH$_4$Cl melt. Sheet iron was formerly dipped in a tallow bath before tinning. A cheap form of tinplate, 'terne plate' is coated from a Pb–Sn bath.

tin oxides

Tin(II) oxides. Lower tin oxides: SnO (white, NH$_4$OH to SnCl$_2$ solution; black, heat on white SnO; red), form a complex system.

Tin(IV) oxide, SnO$_2$. Occurs naturally as cassiterite or tinstone. Precipitated as hydrate from Sn(IV) solution; can be dehydrated to SnO$_2$. Forms mixed metal oxides and species containing [Sn(OH)$_6$]$^{2-}$ ions (stannates). Used, under the name putty powder, for polishing glass and metal.

tin salt, SnCl$_2$.2H$_2$O. See TIN(II) CHLORIDE.

tinstone, SnO$_2$. Cassiterite. See TIN OXIDES.

tin sulphate, SnSO$_4$. Tin(II) sulphate.

tin sulphides

Tin(II) sulphide, SnS. Grey solid, Sn plus S at 900°C.

Tin(IV) sulphide, SnS$_2$. Precipitated from Sn(IV) solution with H$_2$S or Sn plus S under pressure. NH$_4$Cl, Sn, S heated gives a yellow solid (mosaic gold). Used as a pigment.

tiocarbazil Herbicide, thiocarbamate derivative.

titanates See TITANIUM OXIDES. Organic titanates, Ti(OR)$_4$, are used in catalysis, for cross-linking resins etc.

titanium, Ti. Metal, at.no. 22, at.mass 47.867, m.p. 1668°C, b.p. 3287°C, density ρ 4500, crustal abundance 5600 p.p.m., electronic configuration [Ar] 3d^{2}4s^2. The most important minerals are ilmenite, FeTiO$_3$, and rutile, TiO$_2$. The metal is obtained electrolytically (some by use of TiO$_2$ pellets in molten CaCl$_2$) by reducing TiCl$_4$ with Mg. It is used as a lightweight construction material as electrodes, frequently electroplated or surface coated with platinum alloys; TiO$_2$ is used extensively as a white pigment and in catalysts, electronic components. The massive metal has a protective layer of oxide in air. Titanium alkoxides are used as polymerization catalysts. Titanium nitride, TiN, is used in hard coatings.

titanium, organic compounds Many TiR$_4$ containing bulky groups, e.g. Ti(CH$_2$SiMe$_3$)$_4$ are known although TiMe$_4$ is only stable to –40°C. Cyclopentadienyls, e.g. Cp$_2$TiCl$_2$, are stable. Organotitanium compounds are used or implicated in Ziegler–Natta catalysis.

titanium alloys Used in the aircraft industry, engines, chemical plant. Additives include Al, Mn, Cr, Fe. Bronzes.

titanium carbide, TiC. Steel-grey solid of high m.p., formed TiO$_2$ plus charcoal in an electric furnace. Very resistant to chemical attack. Used in tool tips and as a deoxidizer in steel manufacture. Ti$_{14}$C$_{13}$ has been detected in space.

titanium chemistry Titanium is a Group 4 transition element, electronic configuration 3d^{2}4s^2. The most stable oxidation state is +4 (e.g. TiO$_2$) largely covalent. Co-ordination numbers 4 in halides, 6 (in [TiF$_6$]$^{2-}$) and higher ($E°$ [TiO]$^{2+}$ to Ti(III) in acid solution –0.1 volt). Titanium(III) is more ionic, the [Ti(H$_2$O)$_6$]$^{3+}$ ion is violet; titanium(III) compounds are reducing agents. Titanium(II) is strongly reducing and has no aqueous chemistry. Ti(0) and Ti(–1), ([Tibipy$_3$]$^-$) are generally octahedral.

titanium chlorides

Titanium(II) chloride, TiCl$_2$. Black powder (TiCl$_4$ plus Ti or heat on TiCl$_3$). Strong reducing agent, immediately reduces water. Forms some complexes.

Titanium(III) chloride, TiCl$_3$. Violet or brown solid (TiCl$_4$ plus H$_2$ at 700°C; TiCl$_4$ plus AlR$_3$ (brown form)). Forms violet 6 hydrate. Used as a reducing agent. The fibrous brown form is an active agent in the Ziegler–Natta stereoregular polymerization of olefins.

Titanium(IV) chloride, TiCl$_4$. M.p. –25°C, b.p.

136°C, ρ 1760. Colourless liquid, important intermediate in Ti production (TiO$_2$ plus C plus Cl$_2$). Hydrolysed via oxide chlorides to TiO$_2$. Forms many complexes including [TiCl$_6$]$^{2-}$.

titanium dioxide, TiO$_2$. Occurs naturally in three modifications: rutile, brookite and anatase. A most important white pigment formed by hydrolysis of purified TiOSO$_4$ or TiCl$_4$, or TiCl$_4$ plus O$_2$ through flame. A thin film of TiO$_2$ is transparent, of use in coating of glass, ceramics, fabrics, paper, etc. Used as a white pigment (66%) and in paper, rubber, fabrics, plastics, leather, printing inks, sunscreens. See TITANIUM OXIDES. U.S. production 1999 1.5 million tonnes.

titanium fluorides
Titanium(IV) fluoride, TiF$_4$. White solid (TiCl$_4$ plus anhydrous HF). Forms [TiF$_6$]$^{2-}$ ion.
   Titanium(III) fluoride, TiF$_3$. Formed Ti or TiH$_{1.7}$ plus anhydrous HF at 700°C. Blue stable solid. Complexes containing [TiF$_6$]$^{3-}$ ions are known.

titanium halides In addition to the CHLORIDES and FLUORIDES, TiBr$_4$, TiBr$_3$, TiI$_4$, TiI$_3$, TiI$_2$ are known.

titanium hydride, TiH$_2$. Formed from TiO$_2$ and CaH$_2$. Wetting agent and source of hydrogen.

titanium oxides
Titanium(IV) oxide, TiO$_2$. See TITANIUM DIOXIDE. Dissolves in concentrated alkali hydroxides to give titanates. Mixed metal oxides, many of commercial importance, are formed by TiO$_2$. CaTiO$_3$ is perovskite. BaTiO$_3$, perovskite-related structure, is piezoelectric and is used in transducers in ultrasonic apparatus and also as a polishing agent. Other mixed oxides have the ilmenite structure (e.g. FeTiO$_3$) and the spinel structure (e.g. Mg$_2$TiO$_4$).
   Titanium(III) oxide, Ti$_2$O$_3$. Violet, formed by reducing TiO$_2$ with H$_2$ at high temperatures.
   Titanium(II) oxide, TiO. Has the NaCl structure but is non-stoichiometric (TiO$_2$ plus Ti).

titanium sulphates
Titanium(IV) sulphate, Ti(SO$_4$)$_2$. Also TiO(SO$_4$). Readily hydrolysed salts formed TiCl$_4$ plus conc. H$_2$SO$_4$.
   Titanium(III) sulphate, Ti$_2$(SO$_4$)$_3$.8H$_2$O. Formed as violet salt TiCl$_3$ plus dil. H$_2$SO$_4$ with exclusion of O$_2$.

titration A process for determining the volume of one solution required to react quantitatively with a given volume of another in which one solution is added to the other, a small amount at a time until just sufficient has been added to complete the reaction (EQUIVALENCE POINT). The equivalence point is determined as an END POINT by, e.g. cessation of precipitation (Ag$^+$ plus Cl$^-$), colour change (oxalate plus MnO$_4^-$), use of an INDICATOR, or by use of electrical methods (ELECTROMETRIC, CONDUCTOMETRIC). Titrations may be carried out by hand from a burette or automatically.

Tl Thallium.

TLC Thin-layer CHROMATOGRAPHY.

TLV THRESHOLD LIMIT VALUES.

Tm Thulium.

TMA See TRIMELLITIC ACID.

TMEDA Tetramethyldiaminoethane.

TML Lead tetramethyl.

TMS Tetramethylsilane.

TMTD, TMTDS See THIURAM DISULPHIDES.

TNT Trinitrotoluene.

toad venoms, batrochotoxins The poisonous substances of the toad secreted by the skin glands. Alkaloids. Similar toxins found in the feathers of some tropical birds. Bufotoxin is an example.

tobacco The raw material for cigarettes, pipe tobacco and cigars. Implicated in medical problems. The principal active materials are polycyclic aromatic hydrocarbons (PAH), tobacco specific nitroamines (TSNA), CO, nitrogen oxides.

tocopherols Various methylated derivatives of tocol. Have vitamin E and anti-oxidant activity.
   They are widely distributed in vegetable lipids, and in the body fat of animals, though animals cannot synthesize them. Have some vitamin E activity.

tolan (diphenylethyne), C$_5$H$_5$.C≡C·C$_6$H$_5$. Obtained as volatile colourless crystals, m.p. 61°C, by HgO oxidation of benzil bis-hydrazone.

tolclofos-methyl Fungicide, organophosphate derivative.

tolidines, C$_{14}$H$_{16}$N$_2$. Amino derivative of ditoluenes formed by alkaline reduction of nitrotoluene with zinc followed by rearrangement with HCl. Most used is *o*-tolidine, 3,3'-dimethylbenzidine, 3,3'-dimethyl-4,4'-diaminobiphenyl. A light-sensitive white solid, m.p.

130–131°C. Used as a colour-testing reagent for blood.

Tollens reagent An ammoniacal solution of 10% $AgNO_3$ and 10% NaOH used as a test for aldehydes, which, unlike ketones, cause the deposition of a silver mirror.

toluene, methylbenzene, C_7H_8, $C_6H_5CH_3$. B.p. 111°C, m.p. –95°C. Colourless refractive liquid with characteristic smell. Burns with a smoky flame. Very volatile in steam.

It is produced from petroleum fractions rich in naphthenes by catalytic reforming in the presence of hydrogen (hydroforming); in this process dehydrogenation and dealkylation take place simultaneously and a mixture of aromatics is produced. Also obtained from the carbonization of coal.

Chromic or nitric acids cause oxidation to benzoic acid; milder oxidation gives benzaldehyde and benzoic acid. The action of chlorine on boiling toluene gives mainly side-chain substitution to benzyl chloride, benzal alcohol, benzal chloride and benzotrichloride, all used for the preparation of benzyl alcohol, benzaldehyde and benzoic acid. Cold chlorination in the presence of iron causes nuclear substitution to a mixture of 2-and 4-chlorotoluenes. Easily nitrated in the cold to mixture of 2- and 4-nitrotoluenes; more vigorous conditions produce ultimately 2,4,6-trinitrotoluene, which is widely used as an explosive (TNT). Sulphonates to a mixture of 2- and 4-isomers, the latter preponderating; the sulphonic acids are used for the preparation of saccharine and chloramine-T respectively. Toluene is used as a constituent of high-octane aviation and motor gasolines, as a solvent and as a raw material in the manufacture of benzene, caprolactam, phenol, many dyestuffs and various other chemicals.

toluene-2,4-diisocyanate (TDI), $C_9H_6N_2O_2$, $CH_3C_6H_3(NCO)_2$. B.p. 120°C. Manufactured by nitration of toluene, reduction to the diamine and condensation with phosgene. Used in condensation reactions with glycols to give polyurethanes.

4-toluene sulphonic acid (4-methylbenzene sulphonic acid), $CH_3C_6H_4SO_3H$. M.p. 106–107°C. Prepared from toluene and sulphuric acid. Used in synthesis and dye chemistry.

toluene-4-sulphonyl chloride (tosyl chloride), $C_7H_7ClO_2S$, $4\text{-}CH_3C_6H_4SO_2Cl$. Colourless crystals, m.p. 71°C, formed by the action of chlorosulphonic acid on toluene. Es-

ters of toluenesulphonic acid are frequently called tosylates and their formation tosylation.

2-toluidine (2-aminotoluene), C_7H_9N. Colourless liquid; b.p. 198°C. Prepared by the reduction of 2-nitrotoluene. It is basic, forms a stable hydrochloride and sulphate. The sulphate is converted at 200°C into toluidine sulphonic acid. Readily diazotizes and in this form used as first component in many dyestuffs.

4-toluidine (*p*-aminotoluene), C_7H_9N. Colourless leaflets; m.p. 45°C, b.p. 200°C. It is basic and gives well-defined salts with mineral acids. Prepared by reduction of 4-nitrotoluene with iron and hydrochloric acid. Easily sulphonated. It is used as a first component in azo dyes.

toluyl The group $CH_3C_6H_4CO-$.

tolyl The group $CH_3C_6H_4CH_2-$.

tolylene-2,4-diisocyanide (4-methyl-1,3-phenylene diisocyanate, TDI) M.p. 12.5°C, b.p. 115°/10 mm. Lacrymator, moisture sensitive. Intermediate for synthesis of polyurethanes.

tolylfluamid Fungicide, sulphamide derivative.

***N-m*-tolylphthalamic acid** Plant growth regulator.

TON See TURNOVER NUMBER.

tonne One million gram, 1000 kg.

topaz, $Al_2SiO_4(F,OH)_2$. Occurs in igneous rocks and pegmatites, contains discrete SiO_4 tetrahedra. Used in glass and glazes and in slags in the steel industry and after heating as a refractory. Coloured transparent crystals are used as gemstones.

topotactic A reaction in which the product has an ordered geometric relation to the reactant.

torr A pressure unit. One torr is 1 mm Hg pressure and 1.33×10^{-4} MPa.

torsional barrier The barrier to rotation in a molecule caused by atomic and electronic repulsions.

tosyl See TOLUENE-4-SULPHONYL CHLORIDE.

tosylation A chemical transformation which substitutes the TOLUENE-4-SULPHONYL (tosyl) group into a molecule.

total pressure In a mixture of gases and vapours the total pressure exerted by the mixture is equal to the sum of the partial pressures of the constituents. See PARTIAL PRESSURE.

total reflux A distillation column is said to be operating under total reflux when all the vapour leaving the column is condensed and returned to it. In this case no products are withdrawn from the system and the reflux ratio is infinity. Total reflux is frequently employed when starting up a column. See RECTIFICATION.

town gas Manufactured gas of CALORIFIC VALUE about 18MJ m^{-3}. Various processes have been used, e.g. coal carbonization at about 1250°C; the complete gasification of coal, as in the LURGI PROCESS; and oil-based processes, such as the catalytic steam reforming of naphtha.

All these fuel gases contain more than 50% hydrogen and 10–30% methane, the other main components being CO, higher hydrocarbons, CO_2 and N_2. NATURAL GAS of calorific value of approximately 38 MJ m^{-3} has become the most widely used gaseous fuel.

toxicity Generally measured as the concentration for 50% mortality, LC_{50} or lethal dose, LD_{50} but other definitions, e.g. MAK and TLV (threshold limit values) are used.

toxic wastes Released either accidentally or by dumping. The most common in water are methanol, ammonia, toluene, nitrate compounds and heavy metals.

toxins Poisonous substances of animal, microbial and plant origin.

tracer A substance added to a system to enable the experimenter to follow the course of a process, e.g. reaction or self-diffusion, without seriously altering the conditions. More particularly, the term is chiefly used for isotopic species added as tracers, especially radioactive isotopes or rare isotopes of the lighter elements (e.g. ^{18}O). The former are detected and determined with the aid of a counter, the latter with a mass spectrometer. As these detectors are very sensitive, only small proportions of tracer are needed.

tractor vapourizing oil (TVO) Kerosine used for spark-ignition engines of low compression ratio.

tragacanth See GUMS. Used as emulsifier, thickener, adhesive.

tralkoxydim Herbicide, oxine derivative.

tralomethrin Insecticide, pyrethroid.

trans Term used in designating isomers to indicate isomer with non-adjacent (opposite) like ligands. Compare *cis*.

trans-planar   *trans*-octahedral

trans-actinides See POST-ACTINIDE ELEMENTS.

transaminases Enzymes catalysing the transfer of amino groups.

transcription The production of complementary single-stranded messenger RNA by RNA polymerase enzymes. Proteins that bond DNA in a highly specific manner.

trans effect In substitution in complexes the

proportions of *cis* and *trans* isomers formed depend markedly upon the other ligands present. Thus in square-planar platinum complexes the proportion of the two products A and B depends upon the relative *trans*-directing effects of L and L'. Ligands can be placed in a series of ability to direct *trans* substitution; some common ligands in order are $H_2O < NH_3 < Cl^- < CO$, both rate effect and thermodynamic effect.

transferrin A protein which is the main mode of transport of iron.

transfluthrin Insecticide, pyrethroid.

transformer oils Highly refined oil of low viscosity, high resistance to oxidation, used for cooling and insulation in electrical installations, e.g. transformers and switch gear. Often replaced by synthetic materials.

transition elements The elements of the periodic table in which there is a partially filled (including empty and full) d shell. The elements from Sc to Zn, Y to Cd, La to Hg and Rf–112 are classed as transition elements, the first-named element in each series having one d electron and the last ten d electrons. Transition elements are characterized by variable oxidation state, coloured ions and a tendency to form complexes. Many of the elements and their salts are paramagnetic.

transition interval The normally observable visible colour change of an indicator is limited to a definite range of hydrogen-ion concentrations or of potential. The region between the limiting values, expressed in terms of pH or rH, is the transition interval, or region of change of the indicator, e.g. litmus changes from red to blue in the interval pH 5.0–8.0. See HYDROGEN-ION CONCENTRATION; INDICATORS, OXIDATION–REDUCTION.

transition point Transition temperature.

transition state theory Since most rate processes (e.g. diffusion or electrical conduction through solid and liquids, viscous flow in gases and liquids and chemical reactions) obey the ARRHENIUS EQUATION, it is implied that they proceed by thermal activation via an activated complex. The transition state (or activated state) is the high-energy configuration through which the reactants must pass before becoming products.

A quantitative theory of rate processes has been developed on the assumption that the activated state has a characteristic enthalpy, entropy and free energy; the concentration of activated molecules may thus be calculated using statistical mechanical methods. Whilst the theory gives a very plausible treatment of very many rate processes, it suffers from the difficulty of calculating the thermodynamic properties of the transition state.

transition temperature The temperature at which one (allotropic) form of a substance is converted into another (allotropic) form. E.g. rhombic sulphur is converted into monoclinic sulphur at the transition temperature of 95.6°C. Below this temperature the rhombic form is the stable allotrope; the monoclinic allotrope is the stable form above this temperature. Transition temperatures vary with the pressure exerted on the system.

transmethylases Enzymes which catalyse the transfer of methyl groups in biological systems.

transmission electron microscopy ((S)TEM) See SCANNING ELECTRON MICROSCOPY.

transmutation The process of transforming one element into another. It has occupied the attention of scientists since the earliest times. The search for the Philosopher's Stone, a sub-

stance by means of which base metals like iron could be converted into gold, was one of the chief preoccupations of the alchemists, but it is only since 1919 that successful transmutation of any kind has been achieved using particle accelerators and reactors. A proton fired with a high velocity at an atom of lithium, may yield two atoms of helium.

$$\mathrm{^7_3Li + \,^1_1p \rightarrow 2\,^4_2He}$$

Numerous nuclear transformations have been induced by processes in which atoms have been bombarded with neutrons, protons, deuterium, carbon and heavier atoms and ions.

Neptunium and plutonium are produced in quantity by neutron bombardment of uranium. Subsequently many isotopes have been obtained by transmutation and synthetic isotopes of elements such as Ac and Pa are in practice more easily obtained than the naturally occurring species. Synthetic species of lighter elements, e.g. Tc and Pm are also prepared.

The transformations of the radioactive elements, whereby, e.g. uranium ultimately becomes lead, are not usually regarded as instances of transmutation because the processes are spontaneous, and cannot be controlled by the experimenter.

transport number The fraction of the total current carried by a particular ion during electrolysis.

traumatic acid (trans-2-dodecenedioic acid), $C_{12}H_{20}O_4$. M.p. 165°C. A plant growth hormone, which is produced in damaged plant tissue, and on diffusing into adjacent intact tissue cells stimulates them to divide.

tray See PLATE.

tray column See PLATE COLUMN.

trehalose ((α-D-glucosido)-α-D-glucoside), $C_{12}H_{22}O_{11},2H_2O$. M.p. 97°C. A nonreducing disaccharide, which forms the principal carbohydrate of insect haemolymph and in fungi, e.g. *Amanita muscaria*. Generally replacing sucrose in plants lacking chlorophyll and starch.

tremolite, $(OH)_2Ca_2Mg_5(Si_4O_{11})_2$. An amphibole, an asbestos mineral. Used in acid-resistant filters.

triacetin See ACETINS.

triacetoneamine (2,2,6,6-tetramethyl-γ-piperidone), $C_9H_{17}NO$. Colourless needles, m.p. 35°C, b.p. 205°C. Prepared from diacetoneamine and propanone or by passing NH_3 gas into propanone containing fused $CaCl_2$.

triacontanol See MELISSYL ALCOHOL.

triademefon Fungicide, triazole derivative.

triadimenol Fungicide, triazole derivative.

triallate Herbicide, thiocarbamate derivative.

triangular diagram A graphical representation of the phase-rule data for a system of three components.

triarylmethane dyes See TRIPHENYLMETHANE DYES.

triaziflam Herbicide, triazine derivative.

triazine herbicides Derivatives of s-triazine (from cyanuric chloride). Various derivatives are used as herbicides, etc.

triazines Compounds containing $sym\text{-}C_3N_3$ rings. See e.g. MELAMINE, DIALLYLMELAMINE. Used in dyestuffs, herbicides, etc.

triazole (pyrrodiazole)

triazophos Insecticide, nematocide, triazole derivative.

triazoxide Fungicide, triazine derivative.

tribasic acid An acid with three replaceable hydrogen atoms which may yield three series of salts, e.g. H_3PO_4 gives NaH_2PO_4, Na_2HPO_4 and Na_3PO_4.

tribenuronmethyl Herbicide, sulphonylurea derivative.

triboluminescence The emission of light by solids on crushing or grinding.

tribromoethanal (tribromoacetaldehyde, bromal), CBr_3CHO. An oily liquid; b.p. 174°C. Prepared by action of bromine on ethanol and decomposed by alkali to bromoform $CHBr_3$.

tribromomethane (bromoform), $CHBr_3$. Colourless liquid; m.p. 8°C; b.p. 151°C (slight decomp.). Sparingly soluble in water; miscible with alcohol and ether. Prepared by the action

of bromine and sodium hydroxide on ethanol or propanone or by warming bromal with alkalis. Decomposed by light and air more readily than chloroform; stabilized by 4% alcohol. Converted to carbon monoxide and potassium bromide by potassium hydroxide.

tribufos Plant growth regulator, organophosphate derivative.

tributyl citrate M.p. 157°C. Plasticizer and solvent, anti-foaming agent, used in polishes.

tributyl phosphate See PHOSPHATE ESTERS and ORGANOPHOSPHATES.

tricarboxylic acid cycle See CITRIC ACID CYCLE.

trichloroacetic acid See CHLOROETHANOIC ACIDS.

1,1,1-trichloroethane (methyl chloroform), $CCl_3.CH_3$. Prepared by chlorination of ethane or HCl on vinyl chloride. Used in metal cleaning and as a solvent.

trichloroethanoic acid See CHLOROETHANOIC ACIDS.

trichloroethene (trichloroethylene), $CHCl\text{:}CCl_2$. Colourless liquid, b.p. 87°C. It is non-inflammable but probably carcinogenic.

Manufactured by the dehydrochlorination of tetrachloroethane derived from the chlorination of ethyne with lime or by vapour-phase cracking.

Trichloroethylene is stable to dilute acids or alkalis, but when heated with sodium hydroxide under pressure it yields sodium glycollate. In the presence of light and oxygen dichloroethanoyl chloride is formed, which can react with any moisture present to give small amounts of highly corrosive HCl.

Most of the trichloroethylene produced is used for metal degreasing. Other important uses are in the scouring of wool and as an extractive solvent for natural oils. Minor uses are as a heat transfer medium, anaesthetic, insecticide and fumigant, paint remover and fire extinguisher.

trichlorofluoromethane, Cl_3CF. A colourless liquid; b.p. 24°C. Used as an internal standard with $\partial = 0$ for ^{19}F n.m.r. spectroscopy. Prepared by fluorination of CCl_4 with $SbCl_5$ and HF. Has been used in refrigeration.

trichlorofon Insecticide, organophosphate derivative.

trichloroisocyanuric acid, [(CO)N(Cl)]₃. Chlorinating agent, disinfectant.

trichloromethane (chloroform), CHCl₃. Colourless liquid with a sweet, pleasant odour; b.p. 60–61°C. Very slightly soluble in water. Manufactured by the chlorination of methane. Oxidized by air and sunlight to phosgene. An excellent solvent for oils, fats, waxes, rubber and many other organic substances. Forms phenyl isocyanide when it is warmed with aniline and potassium hydroxide. Bromoform, iodoform and chloral also give this CARBYLAMINE REACTION.

Chloroform was used as a potent volatile anaesthetic, but is now little used due to its hepatotoxicity. It was used principally for the manufacture of chlorofluorocarbon refrigerants ('Arctons' and 'Freons') and certain polymers.

trichloronitromethane, CCl₃NO₂. See CHLOROPICRIN.

2,4,6-trichlorophenol, C₆H₃Cl₃O. Colourless crystalline substance, m.p. 68°C. Used medicinally as an antiseptic and disinfectant. Probably carcinogenic.

tricine (N(tris-hydroxymethyl] methyl glycine) Biological buffer.

triclinic system The CRYSTAL SYSTEM without symmetry planes or axes. A unit cell has three unequal axes at mutually oblique angles. Example CuSO₄.5H₂O.

triclopyr Herbicide, pyridyl carboxylic acid derivative.

tricresyl phosphate See PHOSPHATE ESTERS.

tricylazole Fungicide. Thiazole derivative.

tridec-4-en-1-yl acetate Insecticide, pheromone.

tridecyl benzene Detergent, surfactant, can be sulphonated to enhance solubility in water.

tridemorph Fungicide, morpholine derivative.

tridesan, (2,4-Cl₂C₆H₃)₂O. Biocide used in many household cleaning agents.

tridymite, SiO₂. The form of SILICA stable between 573°C and 1470°C. Desirable constituent of silica bricks because of its low coefficient of expansion.

trietazine Herbicide, triazine derivative.

triethanolamine (2,2′,2″-nitrilo triethanol) Surfactant, emulsifier. Quaternizing agent.

triethoxymethane See TRIETHYL ORTHOFORMATE.

triethylaluminium
See ALUMINIUM ORGANIC DERIVATIVES.

triethylamine See ETHYLAMINES.

triethylene glycol (triglycol), HOCH₂CH₂OCH₂CH₂OCH₂CH₂OH. Similar to dihydroxyethane. Used as desiccant and in humidity control; esters are used as plasticizers.

triethylene melamine (2,4,6-tris-(1-aziridenyl)-s-triazine) Used in resins and textiles.

triethylene tetramine, H₂NCH₂CH₂NHCH₂CH₂NHCH₂CH₂NH₂. Used in resins.

triethyl orthoformate (orthoformic ester, triethoxymethane), HC(OC₂H₃)₃. B.p. 146°C; prepared by reaction between chloroform and sodium ethoxide in ethanol. A valuable reagent for the introduction of the formyl group, –CHO (as its diethylacetal – CH(OC₂H₅)₂) into aliphatic and aromatic substrates. Other uses include acetal formation, ethyl ester formation and heterocyclic synthesis.

triethyl phosphate, (C₂HSO)₃PO. Colourless liquid, b.p. 215–216°C, soluble in water and used as a polar reaction medium. Prepared from POCl₃ and ethanol. Ethylating agent.

triethyl phosphite, (C₂H₅O)₃P. Colourless liquid, b.p. 156°C, insoluble in water. Prepared from PCl₃ and ethanol. Reacts with alkyl halides to form quaternary salts [(EtO)₃P–R]⁺X⁻ which readily lose C₂H₅Cl to form phosphonates (EtO)₂P(O)R which may be used in modified Wittig reactions. Used as a deoxygenating desulphurizing agent.

triflate Containing the CF₃SO₃⁻ group, a strongly acidic group.

triflic acid, CF₃SO₃H. Trifluoromethylsulphonic acid.

triflumizole Fungicide, azole derivative.

triflumuron Insecticide, urea derivative.

trifluoroethanoic acid (trifluoroacetic acid), CF₃CO₂H. Colourless liquid, b.p. 72.5°C, fumes in air. Very strong acid (pK_a = 0.3) and used extensively for acid-catalysed reactions, especially ester cleavage in peptide synthesis.

It forms an anhydride $(CF_3CO)_2O$, b.p. 39.5°C, with P_2O_5. Used for trifluoroacetylation of –OH and –NH_2 groups. Used as catalyst. Found in seawater.

trifluoroiodomethane, CF_3I. The parent compound for many trifluoromethyl derivatives. Prepared from silver trifluoroethanoate and iodine.

trifluoromethyl derivatives Compounds containing the electronegative CF_3– group. This confers important pharmaceutical properties on compounds and may radically alter properties, e.g. CF_3SO_3H is a very strong acid.

trifluralin Herbicide, dinitroaniline derivative.

triforine Fungicide, azole derivative.

triglyme, $C_8H_{18}O_4$,
$CH_3OCH_2CH_2OCH_2CH_2OCH_2CH_2OCH_3$. The dimethyl ether of triethylene glycol.

trigol The trivial name for triethylene glycol, $(HO(CH_2)_2O(CH_2)_2O(CH_2)_2OH)$, a colourless hygroscopic liquid, b.p. 285°C. Used as a solvent.

trigonal bipyramidal co-ordination

Co-ordination of five ligands at the corners of a trigonal bipyramid. Occurs in PF_5. A trigonal bipyramid has six triangular faces. The three equatorial ligands are not equivalent to the two apical but many trigonal bipyramidal species are fluxional and the apical and equatorial ligands cannot be identified separately by n.m.r. spectroscopy.

trigonal prismatic co-ordination Co-ordination of six ligands at the corners of a trigonal prism. Occurs for Mo in MoS_2. A trigonal prism has two triangular and three rectangular faces.

trigonal prism

trigonal system The CRYSTAL SYSTEM with a 3-fold axis as principal axis. The unit cell is generally taken with the z-axis parallel to the 3-fold axis and the other two axes mutually inclined at 60° and perpendicular to the 3-fold axis. Rhombohedral unit cells (three equal axes equally at an angle [not 90°]) can be expressed in terms of trigonal cells. Example trigonal $CaCO_3$.

trihydroxypropane See GLYCEROL.

triiodomethane (iodoform), CHI_3. Yellow crystalline solid having a powerful and characteristic odour, m.p. 119°C. Volatile in steam. Manufactured by the electrolysis of a solution of an iodide in dilute alcohol or propanone. Reacts with KOH to give methylene iodide. Decomposed slowly by light and air to CO_2, CO, I_2 and H_2O. With NaOH can co-ordinate hydrocarbons.

3,5,3´-tri-iodothyronine, $C_{15}H_{12}I_3NO_4$. Thyroid hormone.

Trilene TRICHLOROETHYLENE.

trimellitic acid (benzene-1,2,4-tricarboxylic acid) M.p. 218°C. Obtained by oxidizing pseudocumene. Converted to trimellitic-1,2-anhydride, TMA, m.p. 168°C, used in plasticizers, alkyd resins, polyesters.

trimeric acid Benzene 1,3,5-tricarboxylic acid.

trimethacarb Insecticide, molluscide, carbamate derivative.

trimethylamine See METHYLAMINES.

trimethylamine oxide, C_3H_9NO, Me_3NO. Crystallizes from water in colourless needles containing $2H_2O$. Occurs in fish and animal tissues. Prepared by treating an aqueous solution of trimethylamine with hydrogen peroxide.

1,2,3-trimethylbenzene Hemimelletene.

1,2,4-trimethylbenzene
See PSEUDOCUMENE.

1,3,5-trimethylbenzene See MESITYLENE.

trimethylborate, $C_3H_9O_3B$, $(MeO)_3B$. Solvent for waxes, resins. Catalyst for formation of ketones.

trimethylene See CYCLOPROPANE.

trimethylolethane (2-hydroxymethyl-2-methyl-1,3-propenediol), $C_5H_{12}O_3$, $CH_3C(CH_2OH)_3$. Colourless solid, m.p. 202°C, b.p. 283°C, prepared by condensation of methanal with CH_3CH_2CHO in the presence of NaOH. Used for alkyd resins particularly for coatings.

trimethylolpropane (2-ethyl-2-hydroxymethyl-1,3-propanediol), $C_6H_{14}O_3$, $EtC(CH_2OH)_3$. Hygroscopic solid; m.p. 57–59°C, b.p. 160°C/15 mm. Prepared by condensation of methanal with butyraldehyde. Used (as esters) in the production of polyurethane foams.

trimethyleneoxide (oxetane) Cyclic, $(CH_2)_3O$.

N,N-2,3-trimethyl-2-isopropylbutamide Used in foodstuffs, m.p. 63°C.

2,2,4-trimethylpentane See OCTANE.

trimethyl phosphine, Me_3P. Readily forms metal complexes. AgI complex good source of Me_3P. Used in catalysts.

2,4,6-trimethylpyridine (s-collidine), $C_8H_{11}N$. B.p. 171°C; a base, generally superior to pyridine for elimination reactions.

trimethylsilyl derivatives, Me_3SiX. Used for introduction of X groups. $(Me_3Si)_3Si$ – useful sterically hindered group.

trimyristin (trimeristryl glyceride) Constituent of many fats.

trinexapac-ethyl Plant growth regulator, cyclohexane carboxylate.

1,3,5-trinitrobenzene, $C_6H_3N_3O_6$. A yellow crystalline solid; m.p. 122°C. Best prepared by oxidizing trinitrotoluene to 2,4,6-trinitrobenzoic acid with sodium dichromate, and then boiling a suspension of the acid in water.

Forms characteristic charge-transfer complexes with aromatic hydrocarbons.

2,4,6-trinitro-1,3-benzenediol (styphnic acid), $C_6H_3N_3O_8$. Yellow crystals, m.p. 178°C, obtained by nitrating 1,3-dihydroxy-

benzene; sparingly soluble in water, soluble in alcohol and ether. Similar to picric acid. Explosive; lead styphnate is an initiator.

2,4,6-trinitrotoluene (TNT), $C_7H_5N_3O_6$. Yellow crystals; m.p. 81°C. It forms charge-transfer complexes with aromatic hydrocarbons and their derivatives.

Prepared by the direct nitration of toluene with a mixture of nitric and sulphuric acids. TNT is a very unstable, violent and powerful high explosive.

triolein (9-octadecanoic acid-1,2,3-propantriyl ester, glyceryl trioleate) Chief constituent of some vegetable oils and fats. Used to prepare octadecen-1-ol.

triose A carbohydrate with three carbon atoms. Three trioses are known, dihydroxyacetone and two stereoisomers of glyceric aldehyde.

trioxane, $(CH_2O)_3$. See PARAFORMALDEHYDE.

trioxymethylene See PARAFORMALDEHYDE.

tripentaerythritol,
$(HOCH_2)_3CCH_2OCH_2C(CH_2OH)_2CH_2OCH_2C-(CH_2OH)_3$. White compound; m.p. 250°C. Prepared during the preparation of pentaerythritol. Used in the preparation of surface-coating materials.

triphenylmethane dyes A major class of dyestuffs, formally part of the triarylmethane series, the most useful dyestuffs are derivatives of triphenylmethane, Ph_3CH, and diphenyl-naphthylmethane. The dyestuffs have brilliant hue but poor fastness. The dyestuffs are carbenium ions $[Ar_3C]^+X^-$ and are reduced to the 'leuco-dyes', the triarylmethanes, Ar_3CH. Widely used in copying papers and as phospho-tungstates or molybdates. Many of the dyestuffs are also used as antiseptics.

triphenylmethyl, Ph_3C. The anion, cation and radical. The radical was announced by Gomberg in 1900. It can be isolated only in the bimolecular form, hexaphenylethane, a colourless solid, m.p. 145–147°C. In solution an equilibrium is established between colourless hexaphenylethane and the yellow triphenylmethyl radical. The proportion of the paramagnetic radical increases with dilution or rise in temperature. The equilibrium mixture is obtained by heating triphenylmethyl chloride with mercury, silver or zinc in the complete absence of oxygen. The free radical reacts with oxygen to give a peroxide and with iodine to give triphenylmethyl iodide.

If triphenylmethyl chloride in ether is treated with sodium, a yellow colour is produced due to the presence of the anionic species Ph$_3$C$^-$. Alternatively, if triphenylmethyl chloride is treated with silver perchlorate in a solvent such as THF, the triphenylmethyl cation is obtained. More conveniently, triphenylmethyl salts, Ph$_3$C$^+$X$^-$, can be obtained as orange–red crystalline solids from the action of the appropriate strong acid on triphenylcarbinol in ethanoic or propanoic anhydride solution. The perchlorate, fluoroborate and hexafluorophosphate salts are most commonly used for hydride ion abstraction from organic compounds (e.g. cycloheptatriene gives tropylium salts). The salts are rather easily hydrolysed to triphenylcarbinol.

The radical and ions have stability due to resonance.

triphenylphosphate, (PhO)$_3$P. Formed from P$_2$O$_5$ and phenol. Used in plastics.

triphenylphosphine, (C$_6$H$_5$)$_3$P. Prepared from PhMgBr and PCl$_3$. Colourless crystalline solid; m.p. 82°C, sublimes easily. It is used widely as a stabilizing ligand for low oxidation states of many transition metal derivatives (many of the complexes are used as catalysts), and, *via* the appropriate phosphonium salt, for conversion to a phosphorane used in the Wittig alkene synthesis.

triphosgene See PHOSGENE.

triple bond A bond using three bonding pairs of electrons between two atoms.

triple point In a one-component system three phases can only co-exist in equilibrium at one temperature and pressure. Thus, e.g. in the system ice, water and water vapour, the three phases are in equilibrium at a pressure of 611 Pa and 273.16 K. This point is termed the triple point for water.

triplet Three closely spaced transitions in a spectrum.

triplet state See SPIN MULTIPLICITY.

tripolite A variety of DIATOMITE.

triptycene, C$_{20}$H$_{14}$. Tribenzobicyclo-[2,2,2]-octatriene. Colourless solid, m.p. 255–256.5°C,

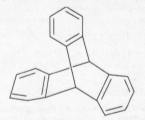

which is best obtained by a Diels–Alder reaction between benzyne and anthracene.

tris(hydroxymethyl)aminomethane (trimethamine, TRIS), C(CH$_2$OH)$_3$.NH$_2$. Surfactant, emulsifier. Used in leather, textile production and as biological buffer.

tristerin (glyceryl tristearate) Used in sizing and textiles.

tris(triphenylphosphine)rhodium(I) chloride See WILKINSON'S CATALYST.

trisulfuron Herbicide, sulphonylurea derivative.

triterpenes Triterpenoid compounds are widely distributed in nature. They contain 30 C atoms and are derived from the hydrocarbons C$_{30}$H$_{50}$. They include squalene and tetracyclic compounds such as lanosterol.

trithiocarbonates, M$_2$CS$_3$. Carbonothioic acid derivatives. Prepared MSH plus CS$_2$. Contain planar [CS$_3$]$^{2-}$ ions.

trithionic acid, H$_2$S$_3$O$_6$. See POLYTHIONIC ACIDS.

triticonazole Plant growth regulator, triazole derivative.

tritium, $_1^3$H. The heaviest isotope of hydrogen, occurring in a proportion of less than 1 in 10^{17} in natural hydrogen. It is β-active (half-life 12.26 years) and is used as a radioactive tracer and in thermonuclear reactions. It may be prepared by the deuterium bombardment of various light atoms in a light-water reactor or by the thermal neutron reaction ^{6}Li$(n,\alpha)^3$H in a linear accelerator. Widely used tracer, used in fusion weapons.

tritolyl phosphate Tricresyl phosphate. See PHOSPHATE ESTERS.

Triton B Trade name for benzyltrimethylammonium hydroxide, usually as a 40% solution in methanol. A strong base, soluble in many solvents; used as a catalyst. See PHASE TRANSFER CHEMISTRY.

trityl The TRIPHENYLMETHYL GROUP. Triphenylmethyl chloride reacts with pyridine solutions of certain hydroxy compounds and sugar derivatives to form trityl ethers.

trona A naturally occurring sodium carbonate, $Na_2CO_3.NaHCO_3.2H_2O$, formed by evaporation of soda lakes. Known also as urao. An important natural source of sodium carbonate.

tropilidine See 1,3,5-CYCLOHEPTATRIENE.

tropine (3-tropanol), $C_8H_{15}NO$. M.p. 63°C, b.p. 229°C. It is formed by hydrolysis of its esters, which are called tropeines, and are important medically. See ATROPINE, SCOPALAMINE.

tropolones 2-Hydroxy derivatives of cycloheptatrienone (tropone). The parent compound,

tropolone, forms colourless needles, m.p. 49–50°C. Forms metal derivatives. Prepared from cyclohepta-1,2-dione by bromination followed by dehydrobromination. A number of tropolones occur naturally. Not an aromatic species.

tropylium The trivial name given to the symmetrical cation $(C_7H_7)^+$ of 1,3,5-cycloheptatriene, obtained when a hydride ion is abstracted from 1,3,5-cycloheptatriene by, e.g. trityl salts. It is an aromatic system having 6 π electrons distributed equally over seven carbon atoms, and is an unusually stable carbonium ion. Alkaline hydrolysis produces tropone. Complexes of tropylium with many transition metal carbonyls, etc. are known.

Trouton's rule The standard entropy of vaporization of normal unassociated liquids is approximately 85 J K^{-1}mol^{-1}.

tryparsamide (tryparsone), $C_8H_{10}AsN_2NaO_4$. Sodium salt of N-phenylglycineamide-4-arsenate. It is used in the treatment of trypanosomiasis.

trypsin Important digestive enzyme, formed in the pancreas as its inactive precursor, trypsinogen, which is activated by enterokinase or, autocatalytically, by trypsin itself. Converts proteins to peptides. See DICTIONARY OF BIOLOGY.

tryptamine (1H-indole-3-ethanamine) Present in plants.

tryptophan (1-amino-2-indolylpropionic acid), $C_{11}H_{12}N_2O_2$. M.p. 289°C. L-Tryptophan is an essential amino-acid; its presence in the food of animals is necessary for proper growth.

tung oil (china-wood oil) A rapidly DRYING OIL obtained from the nuts of trees of the euphorbia order. Tung oil is used with natural resins in the preparation of oil varnishes; when it is heated with certain metallic oxides, DRIERS are formed; these are known as tungates.

tungstates Formed by dissolving WO_3 in NaOH solution and subsequently by cation exchange. The normal tungstates, e.g. $Na_2WO_4.2H_2O$, contain tetrahedral WO_4^{2-} ions. Most metal salts other than alkali metal salts are insoluble. Acidification of a solution of WO_4^{2-} ions gives polymeric anions – isopolytungstates – containing WO_6 octahedra joined corner to corner or edge to edge; $[HW_6O_{21}]^{5-}$ and $[H_2W_{12}O_{42}]^{10-}$ are well established. If the acidification is carried out in the presence of another oxyanion, heteropolytungstates in which MO_6 or MO_4 groups (M = heteroatom) are incorporated into the polyanion, are formed, e.g. $[FeW_{12}O_4]^{5-}$ and $[PW_{12}O_{40}]^{3-}$. It seems unlikely that free acids of any of these forms are stable. Further acidification of isopolytungstates gives $WO_3.2H_2O$. Heteropolytungstates are used to form W-containing catalysts by heating.

tungsten, W. At.no. 74, at.mass 183.84, m.p. 3422°C, b.p. 5555°C, density ρ 19 300, crustal abundance 1 p.p.m., electronic configuration [Xe] $4f^{14}5d^46s^2$. The main ore of tungsten is wolframite, $(Fe,Mn)WO_4$; scheelite $(CaWO_4)$ and stolzite $(PbWO_4)$ are also important. The concentrated ores are fused with NaOH and, after water extraction, WO_3 is precipitated with acid and reduced to the metal with H_2. The metal is bcc. The fused metal is lustrous silver–white and is attacked only slowly even by HNO$_3$–HF. Dissolves in KNO$_3$–NaOH or Na$_2O_2$, attacked by O_2 at red-heat. The metal is used extensively in steel alloys, in electric lamps and heating filaments and in electrical contacts. Tungsten carbides are very hard and are used in cutting tools.

tungsten alloys The most important alloys are the tungsten steels which contain up to 18% of tungsten. Stellite, an extremely hard alloy of W with Cr and Cu, is used for high-speed cutting tools. Electric contacts for switchgear are made from Cu and Ag–W alloys.

tungsten blue   Partially reduced WO_3 or tungstates.

tungsten bronzes   Coloured compounds M_nWO_3 (M = unipositive metal $0 < n < 1$) prepared by reducing tungstates with, e.g., Na which contain W^{VI} and W^{V}. See TUNGSTEN OXIDES.

tungsten carbides W_2C and WC are obtained by heating W powder with C. Both compounds are extremely hard and are used for making cutting tools and dies. Ternary carbides are also used in cutting tools and other metallic carbides, borides, silicides or nitrides are added to improve the mechanical properties of the mixture. Of importance as catalysts.

tungsten carbonyl, $W(CO)_6$.   Very similar to MOLYBDENUM CARBONYL but less readily substituted.

tungsten chemistry   Tungsten is a typical element of transition Group 6 and shows oxidation states from +6 to –2 and, particularly in its oxides, forms many non-stoichiometric compounds. There is little aqueous chemistry except that of complex oxyanions and some complex halides. The hexahalides are molecular but lower halides are polymeric and the lowest halides show extensive W–W bonding (more than Mo). Carbonyl and phosphine derivatives are typical low oxidation state compounds. Complexes are formed, particularly by O- and S-ligands in higher oxidation states and by P-ligands in low oxidation states. Complex cyanides are well established.

tungsten halides   The known fluorides are colourless WF_6 (W + F_2), b.p. 17°C, mixed tungsten(VI) chloride fluorides (e.g. WF_5Cl) and derivatives of WF_6 (e.g. WF_5OMe), tetrameric WF_5 (W + WF_6) and WF_4 (heat on WF_5). Complex fluorides of W(VI) and W(V) contain ions such as $[WF_7]^-$ and $[WF_6]^-$. The known chlorides are WCl_6 (W + Cl_2), an olefin metathesis reagent, green WCl_5 (W + Cl_2), involatile WCl_4 (WCl_6 + Al), WCl_3 and WCl_2 (both WCl_6 + H_2). WCl_3 and WCl_2 both contain metal clusters $[W_6Cl_{12}]Cl_6$ and $[W_6Cl_8]Cl_4$ respectively. Complex chlorides containing $[WCl_6]^-$, $[WCl_6]^{2-}$, $[W_2Cl_9]^{3-}$ are known. ($[W_2Cl_9]^{3-}$ contains three bridging Cl and a W–W bond). All of the halides form extensive ranges of complexes with various ligands. Hydrolysis of the higher oxidation state fluorides and chlorides gives oxide halides and these compounds may also be prepared directly

from oxides and halogenating agents. WOF_4, $[WOF_5]^-$, $[WOF_4]^{2-}$, $[WO_2F_3]^{3-}$, $WOCl_4$, WO_2Cl_2, $WOCl_3$ and $[WOCl_5]^{2-}$ are known and their chemistry is similar to that of the halides. Tungsten bromides, WBr_6, WBr_4, WBr_3, WBr_2 and iodides WI_4, WI_3 and WI_2 are similar to the chlorides.

tungsten, organic derivatives   Generally similar to the molybdenum derivatives but higher oxidation state alkyls, e.g. WMe_6 (WCl_6 + LiMe) are known.

tungsten oxides   Yellow WO_3 is the final product from heating tungsten or other tungsten oxides in excess oxygen. It is used in yellow glazes for ceramics, X-ray screens and fireproofing. Brown WO_2 (WO_3 + H_2) and many intermediate phases containing SHEAR STRUCTURES (e.g. $W_{18}O_{49}$) are formed by reduction of WO_3. Mixed metal oxides are formed by fusing WO_3 with other metal oxides.

tungstic acid   See TUNGSTATES.

tunnelling microscopy   Current controls distance between sample and very sharp point. Gives relief map of surface. Atomic force spectroscopy. Force measured mechanically. Can be used to write patterns almost at atomic levels.

turbidimetry   A method of quantitative analysis which involves the spectrophotometric estimation of absorption by a colloidal dispersion of a precipitate. See NEPHELOMETRY.

turbidity indicators   The use of, e.g. a weak organic acid which flocculates in the presence of a slight excess of hydrogen ions as an indicator for pH titrations. The applications are restricted but are used in titrations of, e.g. glycine where the change in pH near the equivalence-point is slow.

turbidity point   The point in a titration at which a TURBIDITY INDICATOR shows flocculation.

turkey-red oil   A sulphonated castor oil, consisting chiefly of ricinoleosulphuric acid, ricinoleic acid and anhydrides of ricinoleic acid. It is used in the preparation of the cotton fibre for dyeing with turkey red and as an anti-foaming agent.

Turnbull's blue   Various blue CYANO-FERRATES.

turnover number (TON)   A measure of efficiency of a catalyst; the number of molecular conversions per molecule of catalyst.

turpentine Various types of turpentine are produced; all are light volatile essential oils obtained as exudates or by distillation from coniferous trees. Turpentine is a mixture of cyclic terpene hydrocarbons, chiefly α-pinene; it begins to boil at 150°C, has d about 0.85–0.875. It is used as a thinner for paints and varnishes, as a solvent and in polishes.

turquoise A blue basic hydrated aluminium phosphate, $Al_2(OH)_3PO_4.H_2O$ containing some Cu. Usually microcrystalline but amorphous in appearance with a waxy lustre; used as a gemstone.

Tutton salts The isomorphous salts $M_2^ISO_4M^{II}SO_4.6H_2O$ where M^I is an alkali metal and M^{II} is a dipositive transition metal.

twinning A term used in crystallography to denote a set of crystals that have one or more planes or faces in common but whose main developments away from that plane are not those of a single crystal.

two-component system Most commonly met in problems involving the distillation of two liquids.

Tyndall effect When a beam of light is passed through a disperse system, the path of light is visible because of the scattering (Rayleigh scattering) caused by the particles of the disperse phase. This is the Tyndall effect, which is shown both by colloidal sols as well as coarse suspensions. A common example is the illumination of dust particles by sunlight. The light produced by Tyndall scattering is polarized. The effect is greatest for light of short wavelengths and the Tyndall cone is thus bluish in colour.

tyramine (4(2-aminoethyl)phenol), $C_8H_{11}NO$. Colourless crystals, m.p. 164°C. A base found in ergot, and in putrefying animal and vegetable material and certain cheeses where it is the decarboxylation product of tyrosine.

tyrian purple, $C_{16}H_8Br_2N_2O_2$. A purple vat dye of great antiquity. Occurs in the shell fish *Murex brandaris* from which it was once extracted for making royal purple.

tyrosinase Copper-containing oxidase responsible for the darkening of the cut surface of potatoes and other plant tissues. Tyrosine is first converted to a red compound and then to black melanin.

tyrosine (4-hydroxyphenylalanine, 2-amino-3(4-hydroxyphenyl)-propanoic acid), $C_9H_{11}NO_3$. Colourless crystals; m.p. 314–318°C. The naturally occurring substance is laevorotatory. Non-essential for humans.

U

u UNGERADE.

U Uranium.

ubiquinones (coenzymes Q) Play an important rôle in the ELECTRON TRANSPORT CHAIN. Electron-transporting coenzymes. See DICTIONARY OF BIOLOGY.

ulexite, $NaCaB_5O_9.8H_2O$. An important source of BORATES.

Ullman reaction The synthesis of diaryls by coupling aromatic halides with themselves or other aromatic halides, with the concomitant removal of halogens by a metal, e.g. copper powder: thus bromobeazene gives diphenyl. Also used to prepare diaryl ethers and diaryl thioethers by coupling a metal phenolate with an aryl halide.

ulose Suffix denoting a keto-sugar.

ultimate analysis A determination of the proportion of every element in a compound without regard to its molecular structure.

ultracentrifuge A device used to bring about the sedimentation of colloidal sols by subjecting them to large centrifugal forces. An ultracentrifuge may be used to determine the molecular weights and size distribution of sol particles.

ultrafiltration A method of pressure-driven filtration in which particles of colloidal dimensions are separated from molecular and ionic substances by drawing the sol liquid through fine membranes.

The mechanism of ultrafiltration is not simply a sieve effect, but depends also upon the electrical conditions of both the membrane and the colloid.

ultramarine An aluminosilicate with an $(Al,Si)O_2$ framework containing sulphur, some of which is present as S_2 groups. It has a blue colour and is manufactured as a pigment from china clay, sulphur, Na_2CO_3, SiO_2 and some minor constituents. Used in laundry blue and some printing processes.

ultrasonics Ultrasonic vibrations have frequencies of the order of 2×10^5 Hz; they may be set up by applying an alternating current to quartz, tourmaline or Rochelle salt, or by the effect of oscillating magnetism on a rod of magnetic material immersed in a liquid. They can be used to bring about the liquefaction of gels, the depolymerization of macromolecules and to bleach solutions of coloured dyestuffs. With their help it is possible to make an emulsion of mercury and water, or to sterilize milk. They are also used in descaling metals, during soldering for crack detection and creating nanoparticles of metals.

Ultrasonic absorption is used in the investigation of fast reactions in solution. If a system is at equilibrium and the equilibrium is disturbed in a very short time (of the order of 10^{-7} seconds) then it takes a finite time for the system to recover its equilibrium condition. This is called a relaxation process. When a system in solution is caused to relax using ultrasonics, the relaxation time of the equilibrium can be related to the attenuation of the sound wave. Relaxation times of 10^{-4} to 10^{-9} seconds have been measured using this method and the rates of formation of many mono-, di- and tripositive metal complexes with a range of anions have been determined. Ultrasonics can be used to initiate or modify chemical reactions. Sonochemistry.

ultrasound See SONOCHEMISTRY.

ultraviolet absorbers (light stabilizers) Compounds which absorb u.v. light and dissipate energy harmlessly. Used to protect polymers against photolytic degradation and also in suntan lotions. Most light stabilizers are aromatic compounds with the aromatic ring conjugated with a C=O or N system.

ultraviolet light (u.v.) Radiation beyond the visible violet light in the spectrum at wavelengths less than about 420nm is termed ultraviolet light. Radiation of yet shorter wavelength are the vacuum u.v. and X-rays. Ultraviolet light possesses much greater energy than visible radiation, and is generally much more effective in inducing photochemical reactions, but possesses much less penetrating power. Ordinary glass is opaque to light of wavelength less than about 360nm. Quartz is transparent down to about 180nm, and is therefore used for making prisms and lenses in ultraviolet optical apparatus.

ultraviolet photoelectron spectroscopy (UPS) Photoelectron spectroscopy using u.v. radiation of less than 200 nm.

umbellic acid (2,4-dihydroxycinnamic acid), $C_9H_8O_4$. A yellow powder, darkening at 240°C and decomposing at 260°C. It occurs in asafoetida as an ester with its anhydride, umbelliferone.

umbelliferone (7-hydroxycoumarin), $C_9H_6O_3$. Crystals; m.p. 223–224°C. Aglucone; occurs in many plants. Used as sunscreen and fluorescent indicator.

umber Natural $Fe(OH)_3$ often containing MnO_2. A brown powder used as a pigment.

uncertainty principle See HEISENBERG UNCERTAINTY PRINCIPLE.

undecane (hendecane), $C_{11}H_{24}$. B.p. 195°C.

2-undecanone (methyl nonyl ketone) Used in natural oils and fragrances.

undecenoic acid (hendecenoic acid, Δ^{10}-undecylenic acid), $CH_2{:}CH[CH_2]_8C(O)OH$. Pale yellow liquid, m.p. 20–24°C. Prepared by the vacuum distillation of castor oil and ricinoleic acid. Fungicide.

ungerade (u) The subscript indicating uneven parity of an orbital (behaviour under inversion). There is no centre of symmetry for inversion for the orbital. Gerade–g–(even parity) orbitals have a centre of inversion.

uniaxial Crystals which have one principal axis, the other two axes being equivalent. Hexagonal, tetragonal and rhombohedral crystals are uniaxial.

uniconazole Plant growth regulator, triazole derivative.

unimolecular films When certain insoluble oils and fats (e.g. stearates) are dispersed on the surface of water it is possible to prepare these as unimolecular monolayers. In these unimolecular films the hydrophobic group is orientated away from the water; the hydrophilic group is orientated towards it. In some cases the molecules form close-packed arrays, whilst in other cases the molecules are well separated, thus forming different types of two-dimensional solids, liquids or gases. Monolayer fibres can be transferred to solid surfaces for study. See LANGMAN–BLODGETT FILMS.

unimolecular reaction Chemical reactions in which the mechanism involves only single species of the reactant. Although many reactions obey first-order kinetics (see ORDER OF REACTION), relatively few true unimolecular reactions are known. The radioactive decay process is an example, as is the thermal decomposition of certain alkyl halides.

unit cell The characteristic portion of the crystal which, if repeated indefinitely in the directions of the crystal axes, reproduces the crystal. The pattern of the crystal is contained within a parallelopiped called the unit cell, whose sides are parallel to the crystal axes.

universal indicator See COLOUR INDICATORS.

unsaturated compound See SATURATED COMPOUND.

unsaturated polyesters (polyester alkyds) Low mol.wt. polymers containing carboxylic ester groupings and double bonds derived from unsaturated diols and saturated acids. Used as copolymers with styrene and other vinyl monomers in free-radical polymerization to give very strong cross-linked structures.

uracil (2,6-dioxytetrahydropyrimidine), $C_4H_4N_2O_2$. Colourless crystalline powder,

m.p. 338°C (decomp.). Constituent of ribose nucleic acid. Derivatives have pharmaceutical importance. 5-Fluorouracil is used in cancer treatment.

uranates See URANIUM OXIDES.

uranium, U. At.no. 92, at.mass 238.03, m.p. 1135°C, b.p. 4131°C, density ρ 18 950, crustal abundance 2.4 p.p.m., electronic configuration [Rn] $4f^3 6d^1 7s^2$. Uranium is widely spread in nature (all isotopes are radioactive) and important ores are pitchblende (U_3O_8), uraninite, carnotite ($KUO_2VO_4.1.5H_2O$), coffinite (a silicate) and autunite and torbernite (hydrated double phosphates). Extraction is by flotation, acid extraction under oxidizing conditions, ion exchange, precipitation as a hydrated oxide and purification by solvent extraction. The metal is best prepared by reduction of UF_4 with Mg at 700°C. The metal is rapidly tarnished in air and attacked by hot water or acids. Both the fissile isotope ^{235}U and the more common isotope ^{238}U (half-life 4.51×10^9 years) are of importance in nuclear reactions and in the formation of the actinide elements. Fissile ^{235}U is used in nuclear weapons. The isotopes are separated commercially by gaseous diffusion or centrifugation of UF_6.

Uranium is used as an additive in steels, armaments, radiation shields; uranium carbide is a catalyst in the formation of NH_3.

uranium chemistry Uranium compounds occur in the oxidation states +3 to +6

$$UO_2^{2+} \text{ (yellow)} \xrightarrow{-0.063} UO_2^+ \text{ (blue)} \xrightarrow{+0.58}$$
$$U^{4+} \text{ (green)} \xrightarrow{-0.63} U^{3+}$$
$$\text{(red-brown)} \xrightarrow{-1.80} U$$
$$\text{(volts in 1M } HClO_4\text{)}$$

The ultimate product of oxidation in moist air is the linear UO_2^{2+} species which is present in many uranium(VI) compounds. The UO_2^+ ion disproportionates in aqueous solution although U(V) can be stabilized by, e.g., fluoride and alkoxide groups (UF_5 and $U(OR)_5$). U(III) is rapidly oxidized by air or water. Many uranium compounds show high co-ordination numbers, e.g. $UX_4(8)$, $UO_2Cl_2(7)$. Organometallic derivatives include $U(C_5H_5)_4$ and $U(C_5H_5)_3X$ (X = halides, or alkyls and aryls) ($UCl_4 + KC_5H_5$) and uranocene, $U(C_8H_8)_2$,($K_2C_8H_8 + UCl_4$) which has uranium sandwiched between two planar, parallel, cyclo-octatetraenyl rings.

uranium halides The +6 halides are UF_6 (U + F_2, UF_4 + O_2, m.p. 64°C, sublimes 57°C) used in separation of ^{235}U and UCl_6 (UCl_4 plus Cl_2) both of which are immediately hydrolysed by water to, e.g. UOF_4 and UO_2F_2. Other uranyl halides, UO_2X_2, are formed directly (e.g. UCl_4 + $O_2/350°C$ gives UO_2Cl_2) and O-bonded complexes, including hydrates are readily formed

and complex fluorides are also known. The +5 halides are UF_5, UCl_5 and UBr_5 (e.g. UF_6 + HBr gives UF_5) and the $[UF_6]^-$ ion is stable in aqueous HF. All of the tetrahalides, UX_4, are known (UO_2 + HF/500°C gives UF_4; U_3O_8 + C_3Cl_6) and all form O- and N-bonded complexes. A wide range of complex fluorides is known. The +3 halides UF_3, UCl_3 and UBr_3 are formed by H_2 reduction of UX_4 hydrates. Ammines and other N-bonded complexes are known.

uranium hydride, UH_3. Made from U plus H_2 at 300°C. Pyrophoric black powder. Useful in synthesis of uranium compounds.

uranium oxides Uranium forms an extensive series of oxides including UO (semi-metallic), UO_2 (used in nuclear fuels), U_3O_8 (the stable oxide in air), U_4O_9, $U_{2.25}$–$U_{2.40}$ and UO_3. The hydrated peroxide $UO_4.2H_2O$ is precipitated from an aqueous solution of a U(VI) salt by H_2O and on heating gives UO_3. Complex insoluble uranates(VI), generally polymeric mixed oxides, are precipitated from solutions of uranium(VI) salts or are formed by solid state reactions; many uranates(VI) contain linear UO_2 groupings. U(V) and U(IV), mixed metal oxides are formed by solid state reactions (e.g. $BaUO_3$ + $UO_3/550°C$ gives $Ba(UO_3)_2$). So-called sodium di- or pyro-uranate, uranium yellow, $Na_2U_2O_7.6H_2O$, is an insoluble sodium salt used in staining glass.

uranyl derivatives Uranium(VI) species containing the (linear) UO_2 grouping. Amongst commercially available compounds are the ethanoate, hydrogen phosphate, magnesium ethanoate, nitrate, perchlorate, sulphate and zinc ethanoate. Many compounds are hydrated, show high co-ordination numbers with co-ordinated anions, and water may be replaced with other N- and O-donors. Zinc uranyl ethanoate forms a precipitate from water with Na^+ ions. Most uranyl derivatives have a strong yellow colour and show a bright green fluorescence when irradiated. Used in glazes.

urao See TRONA.

urea, carbamide, CH_4N_2O, $(H_2N)_2C=O$. Colourless crystals, m.p. 132°C. It is a weak base, and forms salts with strong acids. Urea was first obtained synthetically by Wöhler in 1828 by evaporating a solution of ammonium cyanate. It can be made by the addition of water to cyanamide, by passing water into fused phenylcyanamide, and by numerous other methods. Commercially it is made by reacting carbon dioxide with ammonia at 200°C at pressures up to

400 atm., the ammonium carbamate formed initially largely dehydrating to urea.

Urea occurs in the urine of all mammals and in small quantities in the blood of mammals and fish (see UREA CYCLE).

Urea is largely used as a fertilizer (75%), and as a non-protein feed supplement for sheep and cattle. The most important chemical use is in the manufacture of urea–formaldehyde resins. It is also used in the manufacture of adhesives, treatment of paper and cellulose, pharmaceuticals, dyes and disinfectants (with H_2O_2). U.S. production 1999 8.4 megatonnes.

urea adduction The basis of processes for the separation of normal alkanes, especially those from petroleum distillates in the C_{25}–C_{30} range. In the presence of n-alkanes urea crystallizes in a form which allows only the normal alkanes to be clathrated. The clathrate formed is decomposed into two phases by heating with water at 80°C.

urea cycle The cyclic sequence of reactions occurring in the liver, which results in the excretion of excess nitrogen as urea via carbamyl phosphate. See DICTIONARY OF BIOLOGY.

urea–formaldehyde resins Typical AMINO-RESINS used in adhesives foams and moulding. See UREA.

urease Nickel-dependent enzyme which catalyses the hydrolysis of urea to ammonia and CO_2. It is present in many plants but not in vertebrates.

ureides Compounds formed between urea and organic acids by either or both nitrogens. Dicarboxylic acids or α-hydroxy acids give cyclic ureides, e.g. barbituric acid, hydantoin. Colourless crystalline solids, sparingly soluble in water, soluble in ethanol, decomposed by alkalis, form salts.

urethane (ethyl carbamate), $NH_2.COOC_2H_5$. Colourless prisms; m.p. 49–50°C, b.p. 184°C. Prepared by the action of am-

monia on ethyl chloroformate. Good solvent, used in synthesis.

urethanes Esters of carbamic acid, $NH_2.COOH$. Colourless crystalline solids which distil without decomposition. They are prepared by the action of ammonia on carbonic or chlorocarbonic esters, or by treating alcohols with urea nitrate or isocyanates. When heated with ammonia they form urea and an alcohol. Phenyl- and α-naphthylurethanes were used for characterizing alcohols and phenols. The term is also used for polyurethanes, major plastics used in foams (85%) and elastomers. World market 2000 8.2 megatonnes.

uric acid (2,6,8-trihydroxypurine), $C_5H_4N_4O_3$. A colourless microcrystalline powder, odourless and tasteless, which decomposes on heating above 250°C. It is dibasic and forms two series of salts. It occurs in small quantities in human urine as the end product of nucleic acid metabolism and in the excrement of birds and reptiles. It is prepared commercially from guano by extracting with alkali and reprecipitating with acid. It can be synthesized by fusing glycine with urea, and by other processes.

uricase Enzyme which cleaves the purine skeleton.

uridine (1-β-D-ribofuranosyluracil) See NUCLEOSIDES.

uronic acids Formed by oxidation of the primary alcohol groups of sugars, and named after the sugar from which they are formed, thus glucuronic acid from glucose. Only D-glucuronic, D-mannuronic and D-galacturonic acids are important. Polyuronides are constituents of the gums and mucilages; alginic acid is a polymer of mannuronic acid and pectic acid of galacturonic acid.

ursolic acid Polycyclic acid from cranberries. Emulsifying agent.

u.v. Ultraviolet.

V

V Vanadium.

vacancy Missing atom in a crystal lattice. May be compensated by alteration of oxidation state, ion elsewhere in lattice or by presence of an electron.

vaccenic acid (trans-11-octadecenoic acid), $CH_3 \cdot [CH_2]_5 \cdot CH{:}CH \cdot [CH_2]_9 \cdot CO_2H$.

vancomycin Glycopeptide antibiotic.

vacuum crystallizer See CRYSTALLIZERS.

vacuum pump A pump designed to extract gas from a space where the pressure is considerably below atmospheric and discharge it to atmosphere. Mechanical pumps used for this purpose are normally of the rotary, positive displacement type and can provide pressures down to about $1.3\,N\,m^{-2}$. In the chemical industry STEAM JET EJECTORS are widely used, especially with distillation columns and evaporators, pressures of 0.13 to $6.5 \times 10^{-9}\,N\,m^{-2}$ being attained, depending on the number of stages. For very high vacua, as in molecular distillation, DIFFUSION PUMPS are employed.

valence isomerization The isomerization of molecules which involve structural changes resulting only from relocation of single and double bonds. If a dynamic equilibrium is established between the two isomers it is also

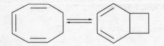

referred to as valence tautomerism, e.g. for the cyclo-octa-1,3,5-triene system.

valency shell electron pair repulsion theory (VSEPR theory) Theory for the prediction of molecular shape – see VALENCY, THEORY OF.

valency A term formerly used to denote OXIDATION STATE. Frequently used to designate the number of bonded neighbour atoms, see CO-ORDINATION NUMBER.

valency electrons The outer electrons of an atom which determine the chemical reactivity of an element, the compounds formed by that element and the geometry of those compounds. Generally the electrons which are additional to those of the preceding noble gas.

valency, theory of The oxidation states, co-ordination number and stereochemistry of any atom are determined by the outermost or valency electrons of the atom. The elements in which the outer s or p shells are full are relatively inert chemically and simple valency theory is based on the concept that elements tend to gain, lose or share electrons in order to complete their outer electron shells. The elements having these shells filled are the noble gases and other elements tend to approach the noble gas structures.

For the transition metals it is often impossible to reach a noble gas structure except in covalent compounds (see effective ATOMIC NUMBER RULE, WADES' RULES) and it is found that relative stability is given by having the sub-shells (d or f) filled, half-filled or empty.

The *electrovalent bond* is formed by electrostatic attraction between oppositely charged ions. Thus Na, with one outer electron, loses this electron to achieve the noble gas Ne structure, while Cl with seven outer electrons, gains one electron to achieve the Ar structure.

$$Na(1s^2 2s^2 2p^6 3s^1) \rightarrow Na^+(1s^2 2s^2 2p^6)$$
$$Cl(1s^2 2s^2 2p^6 3s^2 3p^5) + e \rightarrow$$
$$Cl^-(1s^2 2s^2 2p^6 3s^2 3p^6)$$

Crystalline sodium chloride consists of Na^+ and Cl^- ions.

Covalent bonds are formed by the sharing of electrons. Thus the carbon atom, with four equivalent electrons, shares with the electrons from four hydrogen atoms:

$$4H_x + \cdot \overset{..}{\underset{..}{C}} \cdot \longrightarrow H \overset{H}{\underset{\overset{..}{H}}{\overset{..}{C}}} H$$

The carbon atom has a share in eight electrons (Ne structure) whilst each hydrogen atom has a share in two electrons (He structure). This is a gross simplification of covalent bonding, since the actual electrons are present in molecular orbitals which occupy the whole space around the five atoms of the molecule. Multicentre bonds involve more than two atoms. See BORON HYDRIDES.

Co-ordinate bonds are formed by the sharing of electrons, both electrons being donated by the same atom. Thus the hydrogen ion, H+, has no outer electrons whilst ammonia has eight, six shared with hydrogen atoms and one lone pair. This lone pair is donated to the hydrogen ion and the ammonium ion is formed:

$$H \overset{H}{\underset{\overset{..}{H}}{\overset{..}{N}}} \cdot \cdot + H^+ \longrightarrow \left[H \overset{H}{\underset{\overset{..}{H}}{\overset{..}{N}}} H \right]$$

The N retains its share in eight electrons (Ne structure) and each hydrogen atom has a share in two electrons (He structure). Each hydrogen atom is equivalent.

The shapes of covalent compounds are determined by the tendency for bonding pairs to be as far apart as possible whilst lone pairs have a greater effect than bonding pairs (VSEPR theory).

The co-ordination number in ionic compounds is determined by the radius ratio – a measure of the necessity to minimize cationic contacts. More subtle effects are the JAHN–TELLER EFFECT (distortions due to incomplete occupancy of degenerate orbitals) and metal–metal bonding.

valeric acids See PENTANOIC ACIDS.

validamycin Fungicide, streptomycin derivative.

valine (2-aminoisovaleric acid, L-2-amino-3-methylbutanoic acid),
$C_5H_{11}NO_2$. Crystallizes in white shiny leaves; m.p. 315°C. The naturally occurring substance is dextrorotatory. It is one of the amino-acids obtained in the hydrolysis of proteins.

valium See DIAZEPAM.

valve tray See PLATE (TRAY).

vamidothion Insecticide, acaricide, organophosphate derivative.

vanadates Anionic derivatives of vanadium. Practically vanadates(V) although vanadates (IV), mixed metal oxides, are also known. Free vanadic acids are not known. *Vanadates(V)* are oxy-salts of vanadium(V). Vanadium pentoxide, V_2O_5, dissolves in strong bases to give hydroxy species – e.g. $[VO_3(OH)]^{2-}$, $[VO_2(OH)_2]^-$ – which are colourless. On acidification these give orange and then yellow species, e.g. $[HV_{10}O_{28}]^{5-}$. Orthovanadates, e.g. $Na_3VO_4.12H_2O$, $K_3VO_4.6H_2O$ can be crystallized from alkaline solution and contain tetrahedral $[VO_4]$ units. The polymeric species of KVO_3 contain chains of tetrahedra. The $[V_{10}O_{28}]^{6-}$ unit from acid solution (decavanadate) is an isopolyanion with VO_6 units sharing edges. Solid-state mixed metal oxides derived from V_2O_5 are also known.

vanadic acids See VANADATES.

vanadium, V. At.no. 23, at.mass 50.942, m.p. 1910°C, b.p. 3407°C, density ρ 6100, crustal abundance 160 p.p.m., electronic configuration [Ar] $3d^34s^2$. Group 5 transition element occurring naturally as patronite (V_2S_5 plus S), carnotite ($KUO_2VO_4.1.5H_2O$ – also a uranium ore), and vanadinite ($Pb_5(VO_4)_3Cl$ – apatite structure). Also occurs in some crude oils. Obtained from the above ores or from ores worked for Fe or Ti production. The ore is converted to vanadate(V) by roasting with NaCl, leached and acidified to V_2O_5 which is purified via NH_4VO_3. V_2O_5 is reduced with Ca in a $CaCl_2$ flux and refined by electrolysis of fused $NaCl–LiCl–VCl_2$ or by a van Arkel process. Often used as ferrovanadium (Al reduction in the presence of Fe). The metal is bcc; soft grey or silver. It burns on heating in O_2 or Cl_2 and dissolves in HNO_3 and slowly in hot conc. H_2SO_4 and fused alkalis. Used mainly as a steel additive (80%) and in V–Al alloys (10%). Used in catalysts. Essential element for many organisms. ^{50}V is radioactive.

vanadium bromides
Vanadium(II) bromide, VBr_2. Reddish-brown crystals formed from VBr_3 and H_2.

Vanadium(III) bromide, VBr_3. Dark green or black (V plus Br_2) gives green solution in water and green crystalline $VBr_3,6H_2O$. Forms many complexes.

Vanadium oxide tribromide, $VOBr_3$. Dark red deliquescent liquid formed by heating V_2O_3 plus Br_2.

Vanadium oxide dibromide, $VOBr_2$. Yellow powder obtained by heating $VOBr_3$ or Br_2 plus S_2Br_2 over heated V_2O_5.

vanadium carbides VC (NaCl structure) and V_4C_3 are formed from the elements and are important in the constitution of vanadium steels. High-melting species.

vanadium chemistry Vanadium shows oxidation states –1 to +5. The +5 state is largely covalent or in complexes; it is colourless or yellow–red in oxides, etc. Vanadium(IV) is blue in aqueous solution. Vanadium(IV) is blue in aqueous solution; its chemistry is part ionic and there is a very stable $[VO]^{2+}$ group particularly in square pyramidal co-ordination. Vanadium(III) is green in aqueous solution, predominantly ionic, octahedral or tetrahedral. Vanadium(II) is violet in aqueous solution, ionic but strongly reducing. Vanadium (1), (0) and (–1), e.g. [V-dipy₃]³⁺, $[V(CO)_6]$, $[V(CO)_6]^-$, are largely covalent. All states readily form complexes. Redox potentials are $[VO_3]^-$ to $[VO]^{2+}$ +1.0 volts; $[VO]^{2+}$ to $[V(H_2O)_6]^{3+}$ +0.3 volts; $[V(H_2O)_6]^{3+}$ to $[V(H_2O)_6]^{2+}$ –0.25 volts. V^{2+} to V –1.13 volts.

vanadium chlorides
Vanadium(IV) chloride, VCl_4. B.p. 154°C. Reddish brown liquid formed V plus Cl_2. Decomposes slowly to VCl_3 and Cl_2, hydrolysed by water.

Vanadium(III) chloride, VCl_3. Violet solid formed V plus HCl gas. $VCl_3.6H_2O$ formed from solution. Forms many complexes.

Vanadium(II) chloride, VCl_2. Green solid, formed VCl_4 plus H_2.

Vanadium oxide trichloride, $VOCl_3$. Vanadyl chloride. Readily prepared yellow liquid, b.p. 127°C, formed Cl_2 plus heated V_2O_5 plus C. Readily hydrolysed by water.

Vanadium oxide dichloride, $VOCl_2$. Green crystals $VOCl_3$ plus H_2.

vanadium fluorides
Vanadium pentafluoride, VF_5. M.p. 19.5°C, b.p. 48°C. White solid immediately hydrolysed by water via VOF_3 and VO_2F. VF_5 prepared V plus F_2. Forms hexafluorovanadates(V), MVF_6, most easily in BrF_3.

Vanadium tetrafluoride, VF_4. Green solid prepared HF on VCl_4. Forms hexafluorovanadate(IV) ion $[VF_6]^{2-}$. VOF_2 formed by hydrolysis.

Vanadium trifluoride, VF_3. Yellow green (VCl_3 plus HF at 600°C). Forms $[VF_6]^{3-}$ ion and $VF_3.6H_2O$ from aqueous HF.

Vanadium difluoride, VF_2. Blue solid formed H_2 plus HF on VF_3 or HF on VCl_2 at 600°C.

vanadium oxides
Vanadium pentoxide, V_2O_5. Orange–yellow oxide formed by heating NH_4VO_3. Only sparingly soluble in water; dissolves alkalis to give VANADATES(V). V_2O_5 dissolves in acids to give vanadyl salts. V_2O_5 is used in catalyst systems

for the contact process (sulphuric acid) and some other reactions. Reduction of V_2O_5 gives lower oxides V_nO_{2n+1} then V_nO_{2n-1}.

Vanadium dioxide, VO_2. Is dark blue (V_2O_5 plus SO_2) but is readily reduced further. VO_2 gives the $[VO]^{2+}$ ion with acids and vanadates(IV) with alkalis and as mixed metal oxides.

Vanadium trioxide, V_2O_3. Black powder (V_2O_5 plus H_2 under heat). Readily reoxidized to V_2O_5. Stable down to $VO_{1.35}$.

Vanadium monoxide, VO. Formed V plus V_2O_3, stable over composition range $VO_{0.85}$ to $VO_{1.15}$.

vanadium sulphates The only important oxyacid salts.
Vanadium(IV) forms blue $VOSO_4$, (0, 3 and $5H_2O$), vanadyl sulphate, which forms ranges of double salts. Prepared by SO_2 reduction of V_2O_5 in H_2SO_4. Used as mordant and in coloured glass.

Vanadium(III) forms green $V_2(SO_4)_3$ (electrolytic or Mg reduction of V_2O_5 in H_2SO_4). Alums and other double salts are formed.

Vanadium(II) forms violet $VSO_4.7H_2O$ (electrolytic or Na/Hg reduction). Forms double salts.

vanadyl species Compounds containing the blue VO^{2+} group, e.g. $VOSO_4$, $VOCl_2$.

van Arkel–de Boer process Process for purification of metals, e.g. Ti, where TiI_4 is vaporized and decomposed on a hot wire.

van der Waals' adsorption See ADSORPTION, PHYSICAL.

van der Waals' equation Because the molecules in a gas have a definite volume and exert forces of attraction on each other, real gases do not exactly obey the ideal gas laws. The effects of real gases were taken into account by van der Waals in his equation of state:

$$P = \frac{nRT}{V - nb} - a\left(\frac{N}{V}\right)^2$$

where a and b are van der Waals' coefficients characteristic of the gas under consideration. The term (a/V^2), called internal pressure, allows for the reduction of pressure from the ideal value due to intermolecular attractive forces; b is the 'co-volume' and allows for the volume occupied by the molecules, it has a value of about four times the actual volume of the molecules.

Other equations of state also take deviations from ideality into account, e.g. Berthelot, Dieterici, Beattie–Bridgeman, virial equations.

van der Waals' forces The weak forces between molecules arising from weak electronic coupling. Act only over a very short distance.

vanillin (4-hydroxy-3-methoxybenz-aldehyde), $C_8H_8O_3$. White needles, m.p. 82°C, b.p. 285°C, strong vanilla odour. It occurs extensively in nature, and is the odoriferous principle of the vanilla pod; it can be obtained from the glucoside coniferin. Vanillin is made commercially from the lignosulphonic acid obtained as a by-product in the manufacture of wood pulp or from guaiacol derived from petroleum. It is one of the most important flavouring and perfuming materials, and large quantities are used in the manufacture of foodstuffs, toilet goods and pharmaceuticals.

vanilloids Irritant and spicy materials. Used in pain-killing creams, defence sprays, etc. Capsaicin (red pepper), zingerone (ginger) are examples.

van Slyke determination Estimation of amino groups using nitrous acid.

van't Hoff equations In a reversible chemical reaction change of the equilibrium constant, K, with temperature is given by:

$$\frac{d \ln K}{dT} = \frac{\Delta_r H^\ominus}{RT^2} \text{ or } \frac{d \ln K}{dT^{-1}} = \frac{\Delta_f H^\ominus}{R}$$

Also: $\pi = \frac{n_B}{V} \cdot RT$

the equation for the osmotic pressure of a dilute solution. See VAN'T HOFF ISOCHORE.

van't Hoff isochore The name given to the equation:

$$\frac{d \ln K}{dT} = \frac{\Delta H}{RT^2}$$

where K is the equilibrium constant of a reversible reaction, ΔH the enthalpy of reaction, T the absolute temperature and R the gas constant. This equation permits the calculation of enthalpies of reaction by measuring the variation of K with temperature.

vapour density The ratio of the weight of a given volume of its vapour to that of the same volume of hydrogen, measured under identical conditions of temperature and pressure. Now little used.

vapour galvanising See SHERARDIZING.

vapour lock The restriction of supply of volatile fluids being pumped through pipe systems due to the formation of bubbles of vapour at critical points in the system. This can be caused by high-temperature or low-pressure conditions.

vapour phase osmometry Comparison of vapour pressure between a drop of solution and

solvent followed by temperature rise from vaporization. Used to measure molar mass.

vapour pressure Over every liquid or solid there is a certain pressure of its vapour. In a closed vessel after sufficient time, equilibrium is established, so that as many molecules leave the liquid surface to form vapour as return from the vapour to the liquid. At equilibrium at the specified temperature the pressure exerted by the liquid or solid is termed the vapour pressure at that particular temperature.

Vaseline A trade name for soft paraffin, petrolotum.

Vaska's compound, *trans*-[Ir(CO)(Ph₃P)₂Cl]. Undergoes ready oxidative addition to give Ir(III) complexes. Catalyst.

vasopressin A cyclic peptide hormone secreted by the posterior lobe of the pituitary gland. Used medicinally.

vat dyes Water-insoluble dyestuffs that become water-soluble, and in a reactive state towards fibres, on reduction in alkaline solution. On reoxidation, generally with air, the insoluble dyestuff is precipitated within the fibre. Sodium dithionite, $Na_2S_2O_4$, is generally used as reducing agent. Vat dyes include anthraquinone and indigoid types.

Végard's law In phases in which there is a range of composition of solid solution, the cell dimensions vary linearly with the molar proportions of the constituents. The law is rarely followed exactly.

velocity of reaction The velocity of a chemical reaction is the amount of reactants transformed, or of product produced, per unit time. Thus velocity is usually expressed in moles, or atoms, s^{-1}. See RATE OF REACTION.

veratraldehyde (3,4-dimethoxybenz-aldehyde), $C_9H_{10}O_3$. Methylation of vanillin gives veratraldehyde which may be oxidized to veratric acid.

veratric acid (3,4-dimethoxybenzoic acid) Identified as a degradation product of the alkaloid papaverine. See VERATRALDEHYDE.

veratrole 1,2-Dimethoxybenzene.

verbenone, $C_{10}H_{14}O$. Unsaturated terpenoid ketone.

verdigris The green compound to which copper or bronze is converted on exposure to the atmosphere. It is usually a basic copper carbonate, but near the sea will be a basic chloride. The term is also used for basic copper ethanoate which is used as a pigment.

vermiculite A hydrated Mg–Al–Fe SILICATE CLAY. Used as a catalyst, thermal insulator and in agriculture as a rooting medium and soil additive.

vermilion red, HgS. Cinnabar. The red pigment in Chinese lacquer painting; red ink and temple paint are due to vermilion. Eventually Pb_3O_4 replaced HgS as red pigment.

vermolate Herbicide, thiocarbonate derivative.

vesicles Molecular aggregates. Stacked bilayers.

vetivone, $C_{15}H_{22}O$. Stereoisomeric dimethylmethylene ketones from oil of vetiver. Give vetivazulene, $C_{15}H_{18}$.

Vetrocoke process See CATACARB PROCESS.

vibrational–rotational spectrum See VIBRATIONAL SPECTRUM.

vibrational spectrum When a molecule absorbs energy the vibrational energy of the constituent atoms relative to each other may be increased. Conversely, when a transition occurs from a higher vibrational level to a lower one, it corresponds to the emission of energy, generally as radiation of frequency v, where the energy loss E is related to v by the equation $E = hv$. Vibrational transitions, usually associated with simultaneous rotational transitions, occur in the i.r. and near i.r., and give rise to bands characteristic of the vibrational and rotational changes. The vibrational–rotational spectrum may consist of a number of branches characteristic of the molecule. OVERTONES $2v$, $3v$, etc. are also observed.

vic- The prefix *vic-* (short for vicinal) in the name of an organic compound indicates that substituent atoms or groups are bonded to adjacent carbon atoms. Thus 1,2,3-trichloropropane is *vic*-trichloropropane.

Victor Meyer method for vapour densities Method using the volume occupied by a known weight of vapour (generally measured as expelled air).

Vilsmeier reagent, $ClCH=NMe_2Cl$. The reagent obtained from $POCl_3$ mixed with either N,N-dimethylformamide or N-methylformanilide. Used for introducing the methanoyl (formyl) (–CHO) group into activated aromatic substrates.

vinclozalin Fungicide, carboximide derivative.

vinegar A dilute (4–10%) solution of ethanoic acid prepared by the oxidation of alcoholic liquors by various species of *Acetobacter*. In the U.K. vinegar is largely obtained from malt, the sugars of which are converted to alcohol by yeast which is then treated with *Acetobacter*. Synthetic ethanoic acid is also used for flavouring. In Europe much vinegar is made from poorquality wine and is known as wine vinegar.

vinyl acetate (ethenyl ethanoate, vinyl ethanoate), $CH_2:CHO(O)C\cdot CH_3$. Colourless liquid, b.p. 73°C, with an ethereal odour. Manufactured by the vapour-phase reaction of ethyne and ethanoic acid in the presence of a zinc ethanoate catalyst at 200°C or by a solution phase reaction of ethene and ethanoic acid in the presence of palladium(II) chloride catalyst. Purified vinyl acetate absorbs free oxygen, reacting to give ethanal and the free acid. In the absence of catalysts, vinyl acetate shows little tendency to polymerize, but conversion to polyvinyl acetate occurs with free-radical catalysts. Used in the manufacture of adhesives (30%), paints (20%), coatings (10%), paper and textile finishes (15%), moulding compounds, etc., polyvinyl acetate is of great importance as an intermediate in the manufacture of polyvinyl alcohol and polyvinyl acetals. Copolymers of vinyl acetate, particularly those with vinyl chloride, are also of commercial importance. See VINYL ESTER POLYMERS.

vinyl acetylene See MONOVINYLACETYLENE.

vinyl alcohol (ethenyl alcohol), $CH_2:CHOH$. The enol form of ethanal. It cannot be isolated in the pure state but vinyl esters and ethers are known and hydrolysis of these gives either ethanal or a polymer of vinyl alcohol. See POLYVINYL ALCOHOL.

vinylamines, $CH_2:CHNH_2$. Generally unstable.

vinylation A term used to describe the reaction between ethyne and compounds such as alcohols, carboxylic acids and amines which contain active hydrogen atoms. With catalysts, addition occurs across the triple bond of ethyne to give vinyl (ethenyl) compounds, e.g.

$$H-C\equiv C-H + HX \rightarrow CH_2=CHX$$

vinyl (ethenyl) The $CH_2=CH-$ group.

vinyl chloride (monochloroethylene, chloroethene), $CH_2:CHCl$. Colourless gas, b.p. −14°C, carcinogenic. It is manufactured by reacting ethyne with HCl over a catalyst, by the pyrolysis of ethene dichloride or by the hydrolysis of dichloroethene using dilute NaOH.

Vinyl chloride is used almost exclusively for the manufacture of polymers and copolymers. World production 1999 25 million tonnes (pipes and fittings 34%). See VINYL CHLORIDE POLYMERS.

vinyl chloride polymers Polymers containing $(CH_2-CHCl)_n$ units including polyvinyl chloride (PVC) and many copolymers. Polymerized in aqueous systems containing emulsifying agents or suspension stabilizer or by solution and bulk processes. It is used in rigid materials (e.g. for construction), flooring, gaskets, shoes, electrical coverings, clothing, furnishing, packaging, toys, luggage. Chlorination gives a material with greater solubility in organic solvents, used in lacquers and fibres.

vinyl ester polymers The most important such ester is polyvinyl acetate (ethanoate). VI-NYL ACETATE is polymerized by free-radical processes. Used in adhesives, thickeners, solvents, plasticizers, in textiles, in concrete additives and particularly for the production of polyvinyl alcohol by hydrolysis. Other vinyl esters are produced by transvinylation with the appropriate acid. Polymers derived from vinyl propanate (b.p. 95°C), vinyl caproate (b.p. 166°C), vinyl laurate (b.p. 142°C/10mm), vinyl stearate (b.p. 187°C/4.3 mm), or vinyl benzoate (b.p. 203°C) are important.

vinyl esters See VINYL ESTER POLYMERS, VINYL ACETATE.

vinyl ethanoate See VINYL ACETATE.

vinyl ether (divinyl ether, diethenyl ether), C_4H_6O, CH_2:CH·O·CH:CH_2. Prepared by the action of KOH on 2,2'-dichlorodiethyl ether. It is unstable, breaking down to methanal and methanoic acid.

vinyl ether polymers See VINYL ETHERS.

vinyl ethers, ROCH=CH_2. Prepared commercially by reaction between alkynes (generally C_2H_2) and alcohols at about 150°C under pressure and in the presence of base; also prepared by thermal cracking of acetals (formed aldehydes plus alcohols plus acid catalysts or alkynes plus alcohols). Hydrolysed in presence of acid to alcohol plus aldehyde. Polymerized to homopolymers by acidic catalysts (e.g. BF_3, $AlCl_3$). Vinyl ether polymers are used as adhesives, in processing, lubricants, fibres, film and in moulding components. Important copolymers are formed with maleic anhydride, vinyl chloride, methyl acrylate. Vinyl ethers of importance include methyl vinyl ether,

MeOCH=CH_2, m.p. –222°C, b.p. 5.5°C; ethylvinyl ether EtOCH=CH_2, m.p. –115°C, b.p. 36°C; isobutyl vinyl ether C_4H_9OCH=CH_2, m.p. –132°C, b.p. 83°C; and octadecylvinyl ether $C_{18}H_{37}$OCH=CH_2, m.p. 30°C, b.p. 182°C.

vinyl fluoride, CH_2=CHF. M.p. –160°C, b.p. –72°C. Prepared HF plus ethyne or, in the laboratory, CF_2H·$CBrH_2$ plus zinc or RMgX or pyrolysis of CF_2H·CH_3. It is relatively stable to polymerization but Ziegler–Natta catalysts give homopolymers. Used in coatings and laminates particularly where there is outdoor exposure. See FLUORINE-CONTAINING POLYMERS.

vinylidene The CH_2:CH< group.

vinylidene chloride (1,1-dichloroethene), CH_2:CCl_2. Colourless liquid, b.p. 32°C, manufactured by the dehydrochlorination of trichloroethane. In the presence of light and air, decomposes with the evolution of HCl, phosgene, and methanal and deposition of some polyvinylidene chloride. Stored away from light and in the presence of dissolved inhibitors (such as phenols and amines). Under the influence of free radical and ionic polymerization catalysts, the monomer can be readily converted to give the industrially important VINYLIDENE CHLORIDE POLYMERS.

vinylidene chloride polymers (Saran polymers) Produced by free-radical polymerization of CH_2=CCl_2. Homopolymers and copolymers are used. A thermoplastic used in moulding, coatings and fibres. The polymers have high thermal stability and low permeability to gases, and are self-extinguishing.

vinylidene fluoride, CH_2=CF_2. M.p. –144°C, b.p. –84°C. Prepared by dehydrohalogenation or dehalogenation of, e.g. CH_3CClF$_2$ (prepared C_2H_2 plus HF then Cl_2 or CH_2=CCl_2 or CH_3CCl$_3$ plus HF). Vinylidene fluoride is normally stable but can be polymerized with free-radical initiators. The polymer is very stable and is used in electrical applications, as exterior coatings and as elastomeric copolymers with hexafluoropropene, CCIF=CF_2. etc.

vinylogs Compounds related by the introduction or removal of –CH=CH– units in a chain. Vinylogous functional groups have two parts of a functional group separated by a –CH=CH– unit.

vinyl polymers and resins Polymers derived by polymerization of CH_2=CHX. See various entries under vinyl and polyvinyl.

vinyl pyridines The important vinyl pyridines are 2-vinylpyridine, b.p. 158°C; 4-vinylpyridine, b.p. 65°C at 15mm; and 2-methyl-5-vinylpyridine, b.p. 75°C at 15mm. Prepared by catalytic dehydrogenation or dehydration of ethyl or hydroxyethyl derivatives. Polymerized similarly to styrene by free-radical and anionic mechanisms. Used as copolymers for synthetic rubbers and fibres.

N-vinylpyrrolidone (1-ethenyl-2-pyrrolidone), C_6H_9NO. Prepared from pyrrolidone and ethyne in the presence of a base. B.p. 96°C at 14mm, m.p. 13.5°C. Polymerizes in the

presence of free-radical initiators so that the monomer is generally stored with 0.1% NaOH to prevent hydrolysis and polymerization. Poly (vinylpyrrolidone) and other N-vinyl amides are water soluble. Poly(vinylpyrrolidone) (1-ethenyl-2-pyrrolidone polymer) is used in adhesives, surfactants, detergents, pharmaceuticals, toiletries and in the textile industry for fibre treatment. It has been used in blood plasma.

violaxanthin, $C_{40}H_{56}O_4$. A carotenoid pigment, esters of which are present in viola and other blossoms. Reddish-brown spears, m.p. 20°C.

violuric acid (5-isonitrosobarbituric acid, alloxan-5-oxine), $C_4H_3N_3O_4$. Pale yellow crystalline substance; m.p. 203–204°C (decomp.). Prepared by treating barbituric acid with a solution of sodium nitrite or by the action of hydroxylamine on alloxan. A chelating agent, it dissolves in water to give a violet solution; forms intensely coloured salts with metals. Reduced to uramil which has NH_2 for NOH.

virial equations See VAN DER WAALS' EQUATION. Equations of state for gases which relate pressure, volume and temperature, and involve terms of V^{-1}, V^{-2} etc.

viridian, Cr_2O_3. See CHROMIUM(III) OXIDE.

viscoelasticity Plastics, rubbers and other long-chain macromolecular compounds which show rheological properties intermediate between those of an elastic solid and a Newtonian liquid. In rubber-like plastics, the application of a stress immediately produces a large strain ('high elasticity') which is recoverable only if its duration is very short. If the stress is maintained, the material acquires a permanent strain. It is as though the chain molecules, at first simply extended, slowly flow past one another as in a very viscous liquid. Flow increases with temperature.

At the other extreme are materials which flow like liquids but show a tendency to recoil when the stress is removed. This has been called 'flow-elasticity' or 'elastico-viscosity'. Such liquids (e.g. molten Nylon) can be spun to form threads.

viscose Sodium cellulose xanthogenate solution. Intermediate in rayon production.

viscosity All fluids show a resistance to flow, called viscosity. Mobile liquids, e.g. water, have a low viscosity whereas liquids such as oil or treacle have a high viscosity and flow only with difficulty. Viscosity measurements are often used to determine the molecular weights of polymeric substances.

vitamins Vitamins are specific substances other than proteins, carbohydrates, fats and mineral salts, that are essential constituents of the food of animals but needed only in small amounts. In their absence the animal develops certain deficiency diseases or other abnormal conditions. Vitamins might be defined as substances that play an essential part in animal metabolic processes, but which the animal cannot normally synthesize, although certain animals can synthesize certain vitamins and all animals needing vitamin D can manufacture it from ergosterol in the presence of u.v. light. Most vitamins are groups of related compounds. See DICTIONARY OF BIOLOGY.

vitamin A (retinol), $C_{20}H_{30}O$. The original fat-soluble vitamin. Its absence from the diet leads to a loss in weight and failure of growth, and to a general susceptibility to infections. Vital as a component of rhodopsin, the visual pigment of the eye, and absence can lead to eye damage and damage to the mucous membrane. Present in vegetables and fats, often in the form of carotenoids, which are converted to retinol in the gut. The vitamin is present in the liver of fish and animals.

vitamin B The B vitamins include vitamin B_1 (thiamine), vitamin B_{12} (cyanocobalamine), vitamin B_c folic acid (pteroylglutamic acid), vitamin B_6 (pyridoxine), pantothenic acid and biotin.

vitamin B_1 (thiamine, aneurine), $C_{12}H_{17}ClN_4OS$. Found in foods including wheat-germ and pork. Deficiency leads to beri-beri. Thiamine pyrophosphate is vital to oxidative decarboxylation reactions in the Krebs cycle. See THIAMINE.

vitamin B₂ (riboflavin), $C_{17}H_{20}N_4O_6$.
Found in flavoproteins, respiratory enzymes and as coenzymes. Deficiency leads to cracking of the lips and to eye problems.

vitamin B₃ (niacin), $C_6H_5NO_2$. Found in wheat germ, yeast, liver as a component of NAD, nicotinamide adenine phosphate.

vitamin B₅ (pantothenic acid), $C_9H_{17}NO_5$. Constituent of coenzyme involved in many processes including fatty acid metabolism.

vitamin B₆ (5-hydroxy-6-methylpyridine dimethanol hydrochloride) Substances related to pyridoxine and its phosphates. Cofactors to enzymes involved in metabolism of proteins and amino-acids.

vitamin B₁₂ (cyanocobalamine), $C_{63}H_{88}CoN_{14}O_{14}P$. Essential growth factor involved in biosynthesis. Cobalt cyanide complex.

vitamin Bc (folic acid, pteroylglutamic acid), $C_{19}H_{19}N_7O_6$. Involved in amino-acid metabolism and with vitamin B_{12} in the formation of red blood cells.

vitamin C (L-ascorbic acid), $C_6H_8O_6$.
Probably involved in the function of collagen and the absorption of iron. Deficiency leads to scurvy. Found in fruits.

vitamin D Group of related components involved in calcium uptake in the body. Deficiency can cause rickets; humans exposed to sufficient sunlight can synthesize sufficient vitamin D.

vitamin E (tocopherols and tocotrienols) Essential for fertility and reproduction. Occur in seed embryos and their oils and green leaves. Have antioxidant properties.

vitamin H (biotin), $C_{10}H_{16}N_2O_3S$. Functions as the prosthetic group that catalyses carbonylation.

vitamin K Group of naphthaquinone derivatives derived from menadione. Of importance in blood clotting. Found in leafy vegetables.

vitamin L Factors necessary for lactation.

vitamin T (tegotin) Growth-promoting substance.

vitamin V Anti-ulcer vitamin.

vitellin The chief protein of egg yolk. It is a phosphoprotein and can be obtained as a yellow powder, insoluble in water, neutral salts and dilute acids.

vitreous state The glassy state. See GLASS.

VOC (volatile organic compound) Term used particularly in respect of paints, polymers.

Volhard method Titration of Ag^+ with NCS^- in the presence of Fe^{3+}. A deep red colour is formed at the end-point.

volt Potential in electrochemistry. $1V = 1$ joule $coulomb^{-1}$.

voltammetry Study of electrode processes through the current whilst potential is varied.

VSEPR theory See VALENCY, THEORY OF.

vulcanite See EBONITE.

vulcanization The process of conferring more cross-linking upon a plastic (generally rubber) and so altering its structure that it becomes less plastic and sticky, more resistant to swelling by organic liquids and has enhanced elasticity. The rubber is heated with a vulcanizing agent, generally sulphur although other compounds, e.g. S_2Cl_2, tetramethyl thiuram disulphide, Se, Te, organic peroxides are used with additives to affect the rate of vulcanization.

VX A potent nerve gas (o-ethyl-S-2-diisopropyl-aminoethyl methyl-phosphonodithiolate).

W

W Tungsten.

Wacker process The oxidation of ethene to ethanal by air and a $PdCl_2$ catalyst in aqueous solution. Using ethene in ethanoic acid at 70–80°C it is possible to prepare vinyl acetate (ethenyl ethanoate) and vinyl alcohol can be formed in alcohols.

Wade's Rules See POLYHEDRAL SKELETAL ELECTRON PAIR THEORY. Used to predict and understand structures of cluster compounds.

Wagner–Meerwein rearrangement A rearrangement of the carbon skeleton of a compound occurring in the course of an addition reaction to an alkene, an alkene-forming elimination or a substitution reaction. The rearrangement may involve a migration of an alkyl group or a change in ring structure, and it is frequently encountered in the chemistry of the bicyclic terpenes. The mechanism is essentially the same as that of the PINACOL–PINACOLONE REARRANGEMENT.

Walden inversion When one of the atoms or groups attached to the asymmetric carbon atom in an optically active compound is replaced by a different atom, the product is sometimes a derivative of the chiral isomer of the original compound and it is possible to pass from one isomer to the other without the formation and separation of a racemic compound. A Walden inversion occurs at a tetrahedral carbon atom when the entry of the reagent and the departure of the leaving group are synchronous – a bimolecular nucleophilic substitution mechanism. Since the reagent must approach from the side of the molecule opposite to that of the leaving group an *inversion* of optical configuration results.

warfarin Rodenticide, coumarin derivative.

washing soda Sodium carbonate decahydrate, $Na_2CO_3.10H_2O$.

waste heat boiler A boiler which produces steam by utilizing the heat in the gases or liquid from a chemical process, e.g. a calcining operation.

water, H_2O. M.p. 0°C, b.p. 100°C, density at 0°C 999.87. The simplest oxygen hydride, a colourless or very pale blue–green liquid. Maximum density at 4°C. Solid water (ice) has an essentially tetrahedral arrangement of oxygens about each other with hydrogen bridges (there are different forms depending upon the detailed arrangement). Hydrogen bonding and residual order is carried through into solution. Forms hydrates with many salts (e.g. $CuSO_4.5H_2O$); bonding to the cation is through the oxygen lone pairs; bonding to anions and H_2O is by hydrogen bonding. In many solids the essential structure is determined by the hydrogen bonding. Behaves as a neutral oxide; dissociation to $[H_3O]^+$ and $[OH]^-$ is small. Reacts with electropositive metals (e.g. Na, Ca) to give hydrated oxides or hydroxides and hydrogen; reacts with non-metal oxides (e.g. SO_3, P_2O_5) to form acids and with oxides of the most electropositive metals to form hydroxides (e.g. NaOH). Reacts with many halides, particularly non-metal halides and metal halides in high oxidation states with hydrolysis – e.g. $SOCl_2$ gives SO_2 and HCl. Of major industrial importance as a reagent for cooling and as a solvent. Domestically important for drinking, cleaning and sanitation. Sea water can give fresh water (desalination) by evaporation, freezing, extraction, reverse osmosis and ionic processes but these are all expensive. Industrial waste water is purified for re-use and domestic water is also purified, see SEWAGE TREATMENT.

water gas (blue gas) A mixture of CO (42%) and H_2 (51%), produced by passing steam through a bed of incandescent coke followed by an air blast to counteract the temperature drop due to the endothermic water gas reaction, N_2

and CO_2 along with a trace of methane are present. Formerly used as a fuel and to enrich coal gas. Used in ammonia production. The water gas shift reaction is a source of industrial hydrogen by, e.g. reaction of carbon with H_2O in the presence of a catalyst.

water-glass See SODIUM SILICATES.

waterproofing (water repellency) Waterproofing renders a fabric impermeable to water. Impregnation with oils, varnishes, rubber, etc., can confer waterproof properties. Water repellent coatings repel liquid water but allow passage of air and water vapour, e.g. Goretex – a fluorocarbon polymer. Silicones and fluorocarbon materials are widely used to confer water repellency.

watt Unit of electrical power, 1 watt = 1 joule sec^{-1}.

wavelength (λ) The distance between corresponding points in the profile of wave motion. The frequency used in spectroscopy is λ^{-1}.

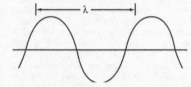

wave function (ψ) The wave that is described in wave (quantum) mechanics. The square modulus of ψ gives the probability.

wave mechanics The electronic theory of matter is confronted by certain difficulties which have only been overcome by assumptions of a very arbitrary nature. To overcome these difficulties by a general theorem, de Broglie and Schrödinger independently devised systems of wave mechanics which are essentially mathematical, and may only be very approximately described in terms of physical pictures. Wave mechanics gives allowed energy levels (orbitals). Transitions between these states give rise to various forms of electromagnetic radiation. Wave mechanics has led to a considerable clarification of many difficulties inherent in the older quantum theory and in the electronic theory of matter, but it is still necessary to use these concepts because of their more concrete nature.

wave number The reciprocal of wavelength, i.e. the number of waves in a given length (gen-erally 1 cm). Often referred to as frequency. $1000 \, cm^{-1} = 12.5 \, kJ \, mol^{-1}$.

waxes Formally fatty acid esters with mono-hydric fatty alcohols having wax-like properties but now arbitrarily extended to any organic material having similar physical properties. Waxes are water repellent and have plasticity. They are used in paper coating, polishes, electrical insulation, textiles, leather, pharmaceuticals. Typical examples are cetyl palmitate in spermaceti and melissyl palmitate in beeswax. Synthetic waxes include some forms of poly-ethers.

WDXRF See X-RAY FLUORESCENCE.

weak electrolytes Species which are not, or behave as if not, fully ionized in solution.

weedkillers Herbicides used to control select-ively unwanted vegetation on agricultural land.

Weston cell (cadmium cell) A widely used standard cell Cd(Hg)/CdSO$_4$.8/3H$_2$O(s)// Hg$_2$SO$_4$(s)/Hg. The electrolyte is a saturated solution of CdSO$_4$. At 20°C E = 1.01485 volts and the cell has a low temperature coefficient of e.m.f.

wetting agents Water, because of its power-ful intermolecular attractive forces, does not readily spread over many surfaces. The addition of surface-active agents, which decrease the contact angle between the water and the other surface, thereby lowering the surface tension, permits the wetting of the surface. Such surface-active agents, e.g. detergents, surfactants, usu-ally consist of molecules which have an oil-attracting hydrophobic group (e.g. an alkyl chain) and a water-attracting (hydrophilic) group (e.g. a negatively charged carboxylic or sulphonic acid group).

white arsenic Crude ARSENIC(III) OXIDE, As$_2$O$_3$.

white lead, Pb$_3$(OH)$_2$(CO$_3$)$_2$. See LEAD CARBONATE.

white oils Oils used in medicinal and toilet preparations obtained by drastic refining of lubricating oil stocks. See LIQUID PARAFFIN.

white spirits, mineral solvents A mixture of hydrocarbons, usually from a petroleum source, b.p. 130–220°C. White spirits are used as paint thinners and as solvents for a variety of purposes.

whiting A form of CaCO$_3$ made by grinding chalk and collecting the finer sediments from water. It is extensively used in the chemical and many other industries.

Wilkinson's catalyst, [Rh(Ph$_3$P)$_3$Cl]. An important catalyst for homogeneous hydrogenation which also catalyses oxidation by O_2 and CO abstraction from organic derivatives.

Williamson ether synthesis Alkyl halides react with sodium or potassium alkoxides or phenoxides to give ethers.

$$RX + NaOR' \rightarrow R\text{-}O\text{-}R' + NaX$$

Although the yields are often poor, especially for halides other than primary alkyl halides, it remains a valuable method for synthesizing unsymmetrical ethers.

Wiswesser line notation A method for expressing the more usual graphical structure of a chemical compound as a linear string of symbols. The resulting alternative notation is unambiguous, short and suitable for computer processing and retrieval.

witherite, BaCO$_3$. The white mineral form of barium carbonate. Used as a source of Ba compounds and in the brick and ceramic industries.

Wittig reaction The reaction between an alkylidene phosphorane, RR'C:PR'$_3$, and an aldehyde or ketone, R''R'''C(O), to produce an alkene, RR'C=CR''R'''. Although the position of the double bond in the alkene is known, the reaction is not completely stereospecific and can produce *cis–trans* mixtures. See PHOSPHONIUM SALTS.

WLN Wiswesser line notation.

wolfram An alternative name for TUNGSTEN.

wolframite, (Fe,Mn)WO$_4$. The principal ore of TUNGSTEN.

wood Wood is a complex substance with cell walls containing carbohydrates (cellulose, hemicellulose, etc.) and lignin together with some inorganic and many organic materials (hydrocarbons, phenols, aldehydes, ketones, terpenes, etc.). Wood is used in construction (may be treated to lower flammability), in the pulp and paper industries, for charcoal, rayon and for various chemicals – see TALL OIL, ROSIN, TERPENES, TANNIC ACID.

wood flour Finely ground timber used as filler in plastics, linoleum and also in sheet metal work.

wood naphtha (wood spirit) See METHANOL.

Woodward–Hoffmann rules Orbital symmetry rules (see FRONTIER ORBITAL SYMMETRY) relating to ring formation and ring opening.

Woodward's reagent (2-ethyl-5-(3-sulphophenyl) isoxazolium iron salt) Peptide bond former.

wool A natural protein fibre (keratin) mainly from the fleece of sheep.

wool fat Lanolin.

work Work is done when an object is moved against a force or, e.g. a chemical reaction drives a current through a resistance or when a surface is formed.

work function The energy required to remove an electron from the highest occupied level inside a solid to a point in vacuo outside the surface. See DICTIONARY OF PHYSICS.

work hardening Increase in strength and hardness, decrease in ductility and toughness of a metal alloy resulting from plastic deformation at a temperature below its recrystallization temperature (normally room temperature). Beneficial in production of high tensile wire by cold drawing and spring strip by cold rolling. Results from mutual interference of dislocations moving on intersecting slip planes and also of grain boundary obstruction of dislocations.

Wurster's salts Stable radical cations formed by one-electron oxidation of, e.g. *N,N,N',N'*-tetramethyl-4-phenylendiamine and isolated as salts.

wurtzite A form of ZnS which gives its name

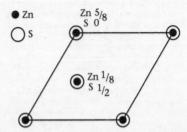

to an important structural type. Hexagonal lattice with both Zn and S tetrahedrally coordinated. Has hcp S with Zn in tetrahedral holes. Other compounds with this structure include ZnO, BeO, AlN, GaN.

Wurtz synthesis Alkyl halides react with sodium in dry ethereal solution to give coupled hydrocarbons. If equimolecular amounts of two different halides are used, then a mixture of three hydrocarbons of the types R–R, R–R' and R'–R', where R and R' represent the original radicals, will be formed. The yields are often poor owing to subsidiary reactions taking place.

X

XAFS X-ray absorption fine structure.

xanthates Salts or esters of unstable acids, ROC(S)SH, where R may be either an alkyl or aryl group. The sodium salts are formed by treating an alcohol or some hydroxy compounds with carbon disulphide and sodium hydroxide or by treating a sodium alkoxide with carbon disulphide. The esters are formed by treating the sodium salts with alkyl halides. The free acids are very unstable; other metallic xanthates are formed by double decomposition with the sodium salts; the copper(I) salts are yellow and gave rise to the name 'xanthate'. Cellulose xanthate is formed in the viscose process for making artificial silk. Other xanthates are used in flotation processes and for the detection of certain metals. Used in curing and vulcanizing rubber.

xanthen gum Polysaccharide produced by bacteria. Used in foodstuffs.

xanthin Hydrocolloid used as a thickener.

xanthine (2,6-oxypurine, 3,7-dihydro-1H-purine-2,6-dione), $C_5H_4N_4O_2$. Colourless crystalline powder with $1H_2O$. A breakdown product of nucleic acid metabolism. It is itself oxidized in the body to uric acid.

xanthine oxidase Oxidoreductase enzyme class. Xanthine oxidase produces superoxide radicals (O_2^-) during the course of its oxidation of substrates and is thus inhibited by SUPER-OXIDE DISMUTASE.

xanthone (dibenzo-4-pyrone), $C_{13}H_8O_2$.

Colourless crystals; m.p. 174°C. It is obtained by the action of heat on phenyl salicylate. It may be reduced to xanthene (reduction of C=O). It is the parent substance of the xanthone group of dyestuffs.

xanthophyll, $C_{40}H_{56}O_2$. Carotenoid pigments. Oxidation products of carotenes.

xanthosine, xanthine riboside See NUCLEOSIDE. Formed by deamination of guanosine.

xanthdrol (9-hydroxyanthrene), $C_{13}H_{10}O_2$. Colourless crystals; m.p. 122°C. It is prepared by reducing an alcoholic solution of xanthone with sodium amalgam.

XDE-105 Insecticide, natural product.

Xe Xenon.

xenon, Xe. At.no. 54, at.mass 131.29, m.p. –111.79°C (4 Pa), b.p. –108.12°C, density liquid 352 g l^{-1}, crustal abundance 2×10^{-6} p.p.m., electronic configuration [Kr] $4d^{10}5s^25p^6$. One of the noble gases (% abundance 8.7×10^{-6} in air) separated by fractional distillation of liquid air. Used in lamps and discharge tubes and in bubble chambers. Moderately soluble in water.

xenon chemistry Xenon, noble gas of Group 18. Shows oxidation states +2 [XeF$_2$ (Xe plus F$_2$ plus light), XeCl$_2$ (Xe plus Cl$_2$ plus electric discharge) both linear], +4 [XeF$_4$ (Xe plus F$_2$), square planar and its adducts with acid fluorides], +6 [XeF$_6$ (Xe plus F$_2$), explosive XeO$_3$ (XeF$_6$ plus H$_2$O)] and +8 [(XeO$_6$)$^{4-}$ ((HXeO$_4$)$^-$ plus base)]. Other derivatives including oxyfluoro anions, e.g. [XeO$_3$F]$^-$, and oxide fluorides, e.g. XeOF$_2$ and XeOF$_4$. Organoxenon(II) and (IV) compounds known. The [AuXe$_4$]$^{2+}$ cation is known.

xenylamine (4-aminobiphenyl), $C_{12}H_{11}N$. Colourless crystals, m.p. 53–54°C. Obtained by the iron, or tin and aqueous acid reduction of 4-nitrobiphenyl.

Substituted derivatives are called xenylamines. All xenylamines are suspected carcinogens.

xerogels A classification of gels. Xerogels are relatively free of the dispersion medium; lyogels are rich in the dispersant.

xerography A reprographic process in which a photoconductor (generally Se) on a surface is charged, exposed to the image which discharges all except the image, removal of Se from the non-image areas, developing the image with an oppositely charged pigment, transferring the image to paper electrostatically and fixing the image on paper by baking.

XMC Insecticide, carbamate derivative.

XPS X-ray photoelectron spectroscopy.

X-ray diffraction The technique for the study of crystal structure in which the atomic nuclei in a crystal lattice act as diffraction sites; the planes of atoms have spacings of a few nm, which are comparable with the wavelength of X-rays. Strong scattering of the rays by the crystal therefore occurs in certain directions, according to BRAGG'S EQUATION. Various techniques are available for applying X-ray diffraction to the study of proteins, single crystals, powders, fibres, etc., and by computation it is possible to work out 3-dimensional electron-density maps of the solid lattice from the recorded X-ray patterns.

X-ray fluorescence, XRF A method of analysis used to identify and measure heavy elements in the presence of each other in any matrix. The sample is irradiated with a beam of primary X-rays of greater energy than the characteristic X-radiation of the elements in the sample. This results in the excitation of the heavy elements present and the emission of characteristic X-ray energies, which can be separated into individual wavelengths and measured. The technique is not suitable for use with elements of lower atomic number than calcium. WDXRF, wavelength dispersive XRF, is used for quantative measurements. EDXRF, energy diffusive technique, provides qualitative and quantitative information.

X-ray spectrometer An apparatus used in the X-ray study of crystals in which a fine beam of monochromatic X-rays impinges at a measured angle on a single crystal mounted in its path, and in which the intensity of the X-rays diffracted in various directions is measured.

X-ray spectroscopy Analytical method by which a sample is irradiated with X-rays, characteristic radiation being emitted after scattering from the specimen.

X-ray tube The apparatus employed for producing X-rays.

X-rays Electromagnetic radiation of wave length 1nm to 1pm. Generated in various ways, including the bombarding of solids with electrons, emitted as a result of electron transitions in the inner orbits of the atoms bombarded. Each element has a characteristic X-ray spectrum.

X-rays may be detected either photographically or with an ionization counter. They have great penetrating power which increases with their frequency, and are used to photograph the interior of solid objects, including the human body, and in monitoring for faults in construction.

X-rays find wide applications in X-ray photography (radiography) and in crystallography. Prolonged exposure of the human body to the rays induces a dangerous form of dermatitis, and even sterility, but controlled exposures are applied to alleviate cancer.

xylans Occur in association with cellulose in lignified cell walls, in the wood of deciduous trees, in the bran and straw of cereals and in similar plant substances. Polysaccharides, linear and branched D-xylopyranosyl units. Polysulphates, fentosan polysulphates, used therapeutically.

xylene (dimethylbenzene), C_8H_{10}. As usually obtained, xylene is a colourless refractive liquid of characteristic smell, burns with a smoky flame and is a mixture of the three possible isomers, 1,2-xylene, b.p. 144°C, 1,3-xylene, b.p. 139°C and 1,4-xylene, b.p. 138°C. Oxidation with chromic acid or permanganate gives the corresponding dicarboxylic acids.

Commercially xylene is obtained by the catalytic reforming of naphthenes in the presence of hydrogen (see TOLUENE) and was formerly obtained from coal tar. The material so produced is suitable for use as a solvent or gasoline ingredient, these uses accounting for a large part of xylene consumption. If xylene is required as a chemical, separation into the isomers is usually necessary, and, although the 1,2-compound can be readily removed by fractional distillation, the separation of the 1,3- and 1,4-isomers requires fractional crystallization, solvent extraction or clathration. 1,2-Xylene is used in the manufacture of phthalic anhydride, and the 1,3- and 1,4-isomers in the manufacture of isophthalic and terephthalic acid respectively. U.S. production 1999, 1,2-xylene 5.2 million tonnes, 1,4-xylene 3.6 million tonnes.

xylenols, $C_8H_{10}O$, $C_6H_3(CH_3)_2OH$. Hydroxydimethylbenzenes of which six isomers are possible. The pure substances are low-melting solids having the general properties of phenols. Xylenol is the name given to a mixture of the isomers separated from the phenolic fraction of coal tar and used as a solvent. 2,6-xylenol is used in resins and chlorinated derivatives as disinfectants.

xylic acid The trivial name given to 2,4-dimethylbenzoic acid.

xylidenes, $C_8H_{11}N$, $C_6H_3(CH_3)_2NH_2$. A mixture of isomeric xylenes is usually nitrated without separation of the isomers. Reduction of the nitroxylenes with iron and hydrochloric acid gives a mixture of five aminoxylenes or xylidines.

Used as a first component of azo-dyes.

D-xylose (wood sugar), $C_5H_{10}O_5$. M.p. 144°C. The pentose sugar of straw, cotton-seed hulls and various hemicelluloses, and of some glycosides, including the primeverosides. It is not fermentable and behaves chemically as other sugars. Used in tanning and dyeing.

xylylcarb Insecticide, carbamate derivative.

Y Yttrium.

Yb Ytterbium.

yellow ammonium sulphide See AMMO-NIUM SULPHIDES.

ylides Internal salts, e.g.

$$R_2C = PPh_3 \leftrightarrow R_2C^- - P^+Ph_3$$

with the heteroatom formally in a cationic state. (N, P, S are amongst elements forming ylides.) Intermediates in the Wittig reaction.

ytterbium, Yb.   At.no. 70, at.mass 173.94, m.p. 819°C, b.p. 1196°C, density ρ 6903, crustal abundance 3.3 p.p.m., electronic configuration [Xe] $4f^{14}6s^2$. A typical LANTHANIDE. The metal is ccp (to 798°C) and bcc (to m.p.). Some Yb-based garnets are used as synthetic gemstones. Used in n.m.r. shift reagents.

ytterbium compounds Ytterbium shows +3 and +2 oxidation states. Yb(III) species are typical LANTHANIDE COMPOUNDS.

Yb^{3+} (f^{13} colourless) $\rightarrow$ Yb (–2.27 volts in acid)

The dihalides, YbX_2, are yellow (YbX_3 + Yb) and are genuine Yb^{2+} salts.

$$Yb^{3+} \rightarrow Yb^{2+} \text{ (–1.15 volts)}$$

yttrium, Y.   Metal, at.no. 39, at.mass 88.906, m.p. 1522°C, b.p. 3345°C, density ρ 4469, crustal abundance 30 p.p.m., electronic configuration [Kr] $4d^15s^2$. An element of Group 3. The principal ore is gadolinite (a silicate also containing lanthanides). Y_2O_3 containing Eu is used as a red phosphor in colour television. Yttrium iron garnets are used as microwave filters and mixed oxides as phosphors and artificial gemstones. Y_2O_3 is used as an undercoat on steel.

yttrium compounds Yttrium only forms cornpounds in the +3 oxidation state

Y^{3+} (colourless) $\rightarrow$ Y (–2.37 volts in acid)

The chemistry is similar to that of the LANTHANIDES.

Z

Z Description of stereochemistry about a double bond. Compare E.

zeaxanthin, zeaxanthol, $C_{40}H_{56}O_2$. Yellow crystals; m.p. 215°C.

Carotenoid pigment present in various leaves, seeds and fruits, and in yolk of egg.

Zeeman effect The splitting of the absorption and emission lines in an atomic spectrum in a magnetic field. The polarization of the lines is also affected. Effect used in flame atomic absorption spectroscopy.

zein A protein of the prolamine class present in maize. It contains no lysine or tryptophan. Zein has been used in plastics, paints, paper, foodstuffs.

Zeise's salt, $K[PtCl_3(C_2H_4)]$. A π-ethene complex, the first such complex isolated.

zeolites Aluminosilicates containing $(Si,Al)_nO_{2n}$ frameworks with a negative charge which is balanced by cations present in the cavities. The cations are easily exchanged and water and gases can be selectively absorbed into the cavities. Various types of zeolites are known including analcite, chabazite, faujasite, the synthetic zeolite A, natrolite and thomsite. Sodium-containing zeolites are used to soften water, replacing Ca^{2+} with Na^+; the zeolite (permutit) is regenerated with concentrated NaCl solution. Zeolites are used to remove molecules of specific sizes by absorption into the pores of the zeolite (molecular sieves); for drying solvents, absorbing gases, as catalysts and in washing powders.

zero-point energy The lowest possible energy of a particle. Not zero.

zeta (ζ) potential An alternative name for the ELECTROKINETIC POTENTIAL.

Ziegler catalysts Complex catalysts prepared by interaction between an organometallic derivative and a transition metal derivative. A typical catalyst is the product of the interaction of $TiCl_4$ and $AlBu_3$, but chromium-based catalysts are also important. These catalysts polymerize olefins, particularly ethylene, to polyolefins, the polymerization generally being in a stereoregular manner.

Ziegler–Natta polymerization Stereospecific polymerization of olefins using a ZIEGLER CATALYST. See TITANIUM(III) CHLORIDE.

zinc, Zn. Metal, at.no. 30, at.mass 65.39, m.p. 419.5°C, b.p. 907°C, density ρ 7133, crustal abundance 75 p.p.m., electronic configuration [Ar] $3d^{10}4s^2$. Transition element occurring as zinc blende, sphalerite $(Zn,Fe)S$, calamine or smithsonite $(ZnCO_3)$, willemite (Zn_2SiO_4) and franklinite $(ZnFe_2O_4)$. Extracted by roasting to ZnO and reduction with carbon. The metal is bluish-white (deformed hcp), fairly hard and brittle. Burns in air and combines with halogens and sulphur. Reacts with steam at red heat. Dissolves in dilute acids and hot solutions of alkalis. Used in alloys (e.g. brass, solder), for die-casting and in protecting (galvanizing) steel by coating. Used in anodes for protection against corrosion. ZnO is used in batteries, paints, rubber, plastics, textiles, etc., and as an inert filler and in electrical components. ZnS is used in luminous paint, X-ray and TV screens, fluorescent lights. World production 2000 8.8 million tonnes. The Zn^{2+} ion is the active site in many enzymes including carbonic anhydrases and dehydrogenases and in transcription factors.

zinc, organic derivatives, RZnX, R_2Zn. Zinc dialkyls and diaryls result from Zn or a Zn–Cu couple with RI in an autoclave at 150°C, also prepared using LiR, RMgX and HgR_2. ZnR_2 are immediately oxidized by air. Used, with $TiCl_4$, as polymerization catalysts for olefins and as intermediates in synthesis.

zinc amalgam A solid Zn–Hg solution, frequently used as a reducing agent.

zincates Ions such as $[Zn(OH)_4]^{2-}$, $[Zn(OH)_3H_2O]^-$ formed by the action of excess base on Zn^{2+} solutions.

zinc blende This cubic form of zinc sulphide gives its name to an important crystal structure type. The Zn and S are tetrahedrally co-ordinated. Has ccp S with Zn in tetrahedral holes:

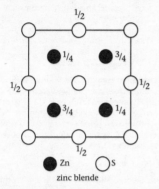

zinc blende

Other compounds with this structure include BN, SiC, AlP, CuCl.

zinc borates Compounds of rather indefinite composition prepared by treating a solution of borax with zinc sulphate. Widely used for fireproofing textiles and in the ceramic and pharmaceutical industries.

zinc carbonate A basic carbonate is precipitated by addition of an alkali carbonate to a solution containing Zn^{2+}. $ZnCO_3$ may result from the use of, e.g., $NaHCO_3$. Hydrolysed to ZnO on boiling with Na_2CO_3 solution. The basic carbonate is used medicinally for external application. See CALAMINE.

zinc chemistry Zinc is an electropositive Group 12 transition element. ($E°$ $Zn^{2+} \rightarrow$ Zn in acid solution –0.76 volts). Apart from possible Zn^+ and Zn_2^{2+} in some melts, all zinc compounds are in the +2 state, generally in octahedral or tetrahedral co-ordination. Readily forms complexes, particularly with O and N ligands.

zinc chloride, $ZnCl_2$. White mass (Zn plus HCl gas, or in solution, Zn or ZnO plus aqueous HCl). Forms hydrates. Used as wood preservative, in batteries, as a dehydrating agent, in dental stoppings.

zinc chromate Yellow pigment precipitated from $ZnSO_4$ solution by K_2CrO_4. Mixed with Prussian blue gives zinc greens which are particularly stable and light fast. Zinc chromates are used in rust-inhibiting paints.

zinc dithionite (zinc hydrosulphite), ZnS_2O_4. Made SO_2 on aqueous suspension of Zn dust. Used in bleaches and the vat dyeing process.

zinc ethanoate (zinc acetate), $Zn(O_2CCH_3)_2.2H_2O$. Also forms basic salt $Zn_4O(O_2CCH_3)_6$.

zinc fluoride, $ZnF_2.4H_2O$. Forms ZnF_2 at 100°C, (ZnO plus aqueous HF or Zn plus HF or F_2). Used as catalyst, wood preservative, in phosphors, glazes.

zinc fluorosilicate, $ZnSiF_6.6H_2O$. Water-soluble salt used as wood preservative and fungicide and in plastics industry.

zinc hydrosulphite, ZnS_2O_4. See ZINC DITHIONITE.

zinc hydroxide, $Zn(OH)_2$. Precipitated from Zn^{2+} solutions by, e.g. NaOH. Occurs hydrated and loses water on heating. Dissolves in acids and in excess alkali (to give zincates). Used as rubber filler and as an absorbent in surgical dressings.

zinc nitrate, $Zn(NO_3)_2.xH_2O$.

zinc oxide, ZnO. Soft powder. White cold, yellow hot. Obtained by burning Zn or from ZnS ores. Dissolves in acids and bases. Used as filler in rubber and in plastics for assisting curing, in ceramics, and medicinally for external application and as a white pigment (Chinese white). *Zinc peroxide*, ZnO_2. Formed Zn^{2+} solution plus Na_2O_2 at low temperatures. Used as a disinfectant in pharmacy.

zinc perchlorate, $Zn(ClO_4)_2.6H_2O$.

zinc sulphate (white vitriol), $ZnSO_4$. Forms 7, 6, 1 and 0 hydrates (ZnO plus H_2SO_4). Used in the textile industry and in arsenical sprays in agriculture.

zinc sulphide, ZnS. Occurs naturally as ZINC BLENDE (sphalerite and wurtzite). Formed from Zn plus S or precipitated from a Zn^{2+} solution by ammonium sulphide. Impure ZnS is phosphorescent. Used as a pigment and in luminous paints.

zineb Fungicide, dithiocarbamate derivative.

zingiberene, $C_{15}H_{24}$. Major constituent of ginger oil; b.p. 134°C/14mm. Used in cosmetics, pharmaceuticals, lacquers.

zintl phases Species e.g. $Ca_{31}Sn_{20}$ formed by Group 1 or 2 elements with moderately electronegative elements.

ziram Fungicide, bird and rodent repellant, dithiocarbamate derivative.

zircon, $ZrSiO_4$. A major mineral source of Zr but contains some Hf which must be removed if Zr is to be used in nuclear technology.

zirconium, Zr. Metal, at.no. 40, at.mass 91.224, m.p. 1855°C, b.p. 4405°C, density ρ 6506, crustal abundance 190 p.p.m., electronic configuration [Kr] $4d^2 6s^2$. A transition element of Group 4. Mineral sources are baddeleyite (ZrO_2) and zircon ($ZrSiO_4$). Generally occurs with Hf which is separated by ion exchange. The metal ($ZrCl_4$ + Mg at 1150°C) is hcp (to 862°C) and bcc (to m.p.). The metal is steel-like in appearance, very resistant to corrosion but is dissolved by aqua regia or HF and burns in oxygen at high temperatures. The element has low neutron absorption and alloys are used in reactor construction, the metal is also used for laboratory ware. Nb–Zr alloys are used in superconducting magnets.

zirconium compounds Most zirconium compounds are in the +4 oxidation state and the aqueous chemistry is confined to this state. The Zr^{4+} ion is extensively hydrolysed in aqueous solution and polymeric species, e.g. $[Zr_3(OH)_4]^{8+}$, are present. Zr(IV) compounds have high co-ordination numbers and show a wide variety of stereochemistries. Lower oxidation state halides $ZrX_3(Cl,Br,I)$, $ZrX_2(Cl,Br,I)$ and $ZrCl(ZrCl_4 + Zr)$ are only stable in the solid state. Zirconium(II) compounds often contain Zr_6X_{12} species. Complexes formed by the tetrahalides, e.g. $ZrCl_4(OPCl_3)_2$, and O-bonded complexes, e.g. zirconium diketonates, are stable. Organometallics, e.g. $Zr(CH_2C_6H_5)_4$, and relatively stable cyclopentadienyls, e.g. $(C_5H_5)_2ZrCl_2$ and $(C_5H_5)_2Zr(CO)_2$, are known. Cp_2ZrHCl reacts with alkenes and alkynes to give Zr alkyls, which can be converted to alcohols and other products. Alkoxides and alkylamides are easily formed as polymers.

zirconium halides Formed by interaction of the elements ZrF_4 (gaseous HF on ZrO_2) $ZrCl_4$ (sublimes 331°C, ZrO_2 + CCl_4 or C + Cl_2). ZrF_4 has m.p. 932°C and forms 1 and 3 hydrates. The halides other than ZrF_4 are readily hydrolysed in water and, e.g. $ZrOCl_2$ and polymeric hydroxo cations are formed. Complex halides are known for F^-, Cl^- and Br^-. ZrF_8^{2-} is 8-co-ordinate square antiprism, ZrF_7^{3-} has pentagonal bipy-

ramidal ions and polymeric species are known, e.g. K_2ZrF_6 (8-co-ordinate) $(NH_4)_2ZrF_6$ (7-co-ordinate). Lower halides (Cl, Br, I) are formed in the solid state (ZrX_4 + Zr); there is extensive metal–metal bonding in ZrCl.

zirconium oxides The dioxide is the stable species and occurs naturally as baddeleyite. It is very inert and has been used as a refractory and insulator. ZrO_2 exists, in one form with 7-co-ordinate Zr and also one form with the FLUORITE structure – the latter form is stabilized by addition of CaO and this is used in refractories. Hydrated species are precipitated from aqueous Zr(IV) salts. Mixed metal oxides are formed by solid-state reactions. ZrO_2 is used in ceramics, glazes and coatings. A lower oxide, ZrO, is known.

zirconium oxyacid salts The nitrate ($5H_2O$) is obtained from conc. HNO_3 solution, the anhydrous compound ($ZrCl_4$ + N_2O_5) is volatile and an oxide nitrate is available commercially. Oxide sulphates, perchlorates and ethanoates are available as hydrates – all have high coordination numbers for Zr.

Zn Zinc.

zone electrophoresis See CHROMATOGRAPHY.

zone refining A method of refining metals and some inorganic and organic compounds depending on the difference in solubility of impurities in the liquid and solid states. A narrow molten zone is caused to move along a bar or column of the substance; those impurities which lower the melting point will tend to remain in solution and will be moved in the direction of zone travel. Solutes which raise the melting point will preferentially freeze out and will move in a direction opposite to that of zone travel. By successive traverses the substance can be obtained in a very pure state. Solvent can be added. The method is applied to the production of pure germanium and other metals required very pure but in small quantities for, e.g. semi-conductors.

Zr Zirconium.

zwitterion A zwitterion is an electrically neutral ion with both a positive and a negative charge. Thus the amino-acid glycine exists in solution at the isoelectric point as the zwitterion $^+H_3N\cdot CH_2\cdot COO^-$.

ZXI 8901 Insecticide, pyrethroid.

zymase The enzyme present in yeast which converts glucose to ethanol but can also induce other reactions.